BALLISTICS

Theory and Design of Guns and Ammunition

Third Edition

BALLISTICS

Theory and Design of Guns and Ammunition

Third Edition

Donald E. Carlucci
Sidney S. Jacobson

CRC Press
Taylor & Francis Group
Boca Raton London New York

CRC Press is an imprint of the
Taylor & Francis Group, an **informa** business

CRC Press
Taylor & Francis Group
6000 Broken Sound Parkway NW, Suite 300
Boca Raton, FL 33487-2742

Printed on acid-free paper

International Standard Book Number-13: 978-1-138-05531-5 (Hardback)

Library of Congress Cataloging-in-Publication Data

Names: Carlucci, Donald E., author. | Jacobson, Sidney S., author.
Title: Ballistics : theory and design and ammunition / Donald E. Carlucci & Sidney S. Jacobson.
Description: Third edition. | Boca Raton : Taylor & Francis, 2018. | Includes bibliographical references and index.
Identifiers: LCCN 2017044302 | ISBN 9781138055315 (hardback) | ISBN 9781315165967 (ebook)
Subjects: LCSH: Ballistics.
Classification: LCC UF820 .C28 2018 | DDC 623/.51–dc23
LC record available at https://lccn.loc.gov/2017044302

Visit the Taylor & Francis Web site at
http://www.taylorandfrancis.com

and the CRC Press Web site at
http://www.crcpress.com

Contents

Section II Exterior Ballistics

Preface to the Third Edition

The third edition of *Ballistics: Theory and Design of Guns and Ammunition* is significantly different from the second edition. The huge number of equations that are present in the book have made it difficult to eliminate all of the errors and the second edition still contained quite a few errors. We have done the best we could in continuing the process of making corrections. For that, we are still indebted to the students and readers who have taken the time and effort to point these out. It continues to be a humbling experience.

Unlike the second edition, very few problems were added. This was a conscious decision so that we would be able to put in more material. This material came from a variety of sources. We added a section in Chapter 3 on cubical and spherical grains since this was a question constantly asked of us. We included a section in Chapter 5 on the design of gas-operated weapons that was developed by the authors at the request of Brian Vuksanovich and has helped him in some of his design work. It is very impressive to see that he has taken this and run with it in his consulting. He is a pleasure to work with and we thank him for inspiring us.

Some new assistance for us arrived on the scene as well. Dr. Eric Kathe from Benet Laboratories and the United States Army Armament Graduate School assisted us by revamping the gun dynamics section of Chapter 5 and added a complete chapter on recoilless weapons, which has become Chapter 6. He is certainly an expert in this area, and we are indebted to him for his enthusiasm and hard work in putting this new material together.

Nathan Okun, a former Navy civilian, added a chapter on the penetration of steel armors by ogival projectiles. Included in this new chapter is a great deal of historical background, which the authors find extremely interesting. This interest is attested to by the large number of terminal ballistics problems found in both of the previous editions that involved naval gunfire and penetration. This new chapter delves into the historical metallurgy of naval armors as well as presents the formulas for penetration of ogival projectiles into these armors. Additionally, Mr. Okun's codes are provided on the CRC website for use. We have decided to hold off on including the equations for face-hardened armor, capped projectiles, etc., for a future edition. Even though we did not include problems using these formulas and codes (this time around), the interested reader might explore the differences in results obtained if the formulas and/or codes are used in some of the problems that are currently in the book. Mr. Okun's knowledge on naval armors and projectiles of all nations is encyclopedic, and we sincerely appreciate him taking the time to write the chapter. The plots provided in this new chapter were created by Mr. Mark Sproul by using the codes. There was significant code work on his part to generate these plots and we are grateful for it.

Many other people assisted us in this edition. This assistance ranged from providing peer reviewing to consulting to critical commentary on the book. They include (but are not limited to): LTC (Dr.) Josh Keena (US Military Academy); Dr. Ryan Decker, Dr. Stephen Recchia, David Geissler, Dr. Gordon Cooke, Dr. Thomas Recchia, Yin Chen, Adam Foltz, Carlton Adam, Sandy Einstein, Don Chiu, Jonathan Jablonski, Steven Doremus, Jeff Ranu, Elizabeth Reisman, Ken Klingaman, Pasquale Carlucci, Dr. Robert Dillon, Dr. Sam Sopok, Dr. Robert Carson, Dr. Brian Fuchs, and Ed Rudnicki (all of the US Army Armament, Research, Development and Engineering Center); Dr. Jon Yagla, James Poyner, and Dr. Carl Dyka (all of the US Naval Surface Warfare Center, Dahlgren Division); Don Levin (Aberdeen Test Center); Dr. Tom Mason (Los Alamos National Laboratories); Dr. Clive Woodley

(QinetiQ); Dr. Michael Nusca (US Army Research Laboratory); Dr. Kirk Vanden (Air Force Research Laboratory); Byron Angel, Chris Carlson, Jack Joyner, and Dr. Kent Crawford (US Gunnery Fire Control Group); and Joseph Backofen. One of the dangers with listing people by name is that we will invariably forget someone who played a part. If we did omit someone—you know how to contact us. Let us know and we will make things right.

We are again indebted to Dr. Jonathan Plant from our publisher, CRC Press, of the Taylor & Francis Group, for once again encouraging us to undertake the revision and facilitating its publication. We also thank Hector Mojena III, our project editor, for helping us through the publication process.

Finally, the authors would like to thank Raymond Sicignano, an outstanding professional artist, who created the cover art for this edition. Not only did he create the artwork, but he also actually *painted* it on wood! This talented man's work can be seen at his website: http://www.raizart.com.

Disclaimer

The design, fabrication, and use of guns, ammunition, and explosives are, by their very nature, dangerous. The techniques, theories, and procedures developed in this book should not be utilized by anyone without the proper training and certifications. In the checking and editing of these techniques, theories, and procedures, every effort has been made to identify potential hazardous steps, and safety precautions have been inserted where appropriate. However, these techniques, theories, and procedures must be exercised at one's own risk. The authors and the publisher, its subsidiaries and distributors, assume no liability and make no guarantees or warranties, express or implied, for the accuracy of the contents of this book or the use of information, methods, or products described within. In no event shall the authors, the publisher, its subsidiaries or distributors be liable for any damages and expense resulting from the use of information, methods, or products described in this book.

Preface to the Second Edition

When we were asked to create a second edition of *Ballistics: Theory and Design of Guns and Ammunition*, we were a bit concerned about what could or should be added or changed. One thing that was certain was that the numerous errors that we became aware of through teaching and student observations had to be corrected. For that, we are indebted to the students and readers who have taken the time and effort to point these out. It was certainly a humbling experience. Hopefully we got them all this time. With the errors corrected, the task of deciding what we could put in to add more value to the book while maintaining an introductory treatment rose to the forefront.

A great deal of work went into this edition that might not be apparent on the surface. We added a large number of new problems and updated the solutions manual accordingly. In fact, the solutions manual increased in size from 247 pages to 543 pages! One might notice that the new problems do not have answers in the text. This was intentionally done to allow instructors to assign these problems as tests or graded homework. We also added some Mathcad® codes to the CRC website to assist our readers with solution of the problems. These are available at http://www.crcpress.com/product/isbn/9781466564374.

The more obvious changes include an increase in topical areas. We decided to add a section to Chapter 9 that would, in a general sense, discuss the topic of probability of first round hit for direct fire weapons. This treatment is general and certainly does not even attempt to reflect the state of the art since that data are not appropriate for a general audience. Another area of major improvement is the additional sections added to Chapter 18, which discuss explosive equations of state. Dr. Ernie Baker, our professional colleague and recognized expert in the field of explosives effects and insensitive munition design, added this section. We are extremely grateful to him for this. His schedule was incredibly busy, yet he found the time to assist us. Finally, there was a significant update to Chapter 21 on wound ballistics. During any course on terminal ballistics, it seems that this topic area is the one people seem most interested in. Granted, the treatment is still general, and, as is common in our experience, any statement made about the wounding process will be hotly debated, we decided to put more flesh on the bones of this section. Hopefully it is of help, keeping in mind the introductory nature of the treatment.

Michael S. L. Hollis is the creator of the cover art, which depicts a typical kinetic energy penetrator in the midst of discarding its sabot. The sabot consists of the three "petals," each with bow shocks forming ahead of them, peeling away from the subprojectile. The flames emanating from the rear of the subprojectile represent the tracer that allows the firer to visualize its trajectory downrange. These projectiles are generally fired at high Mach numbers and are used in weapons with bore diameters that range from 20–120 mm.

We are indebted to Jonathan Plant of CRC Press, Taylor & Francis Group, for encouraging (i.e., gently prodding) us to undertake the work of revision and facilitating its publication.

Preface to the First Edition

This book is an outgrowth of a graduate course taught by the authors for the Stevens Institute of Technology at the Picatinny Arsenal in New Jersey. Engineers and scientists at the arsenal have long felt the need for an armature of the basic physics, chemistry, electronics, and practice on which to flesh out their design tasks as they go about fulfilling the needs and requirements of the military services for armaments. The Stevens Institute has had a close association with the arsenal for several decades, providing graduate programs and advanced degrees to many of the engineers and scientists employed there. It is intended that this book be used as a text for future courses and as a reference work in the day-to-day business of weapons development.

Ballistics as a human endeavor has a very long history. From the earliest developments of gunpowder in China more than a millennium ago, there has been an intense need felt by weapon developers to know how and why a gun works, how to predict its output in terms of the velocity and range of the projectiles it launched, how best to design these projectiles to survive the launch, fly to the target and perform the functions of lethality, and the destructions intended.

The discipline over the centuries has divided itself into three natural regimes: Interior ballistics or what happens when the propellant is ignited behind the projectile until the surprisingly short time later when the projectile emerges from the gun; exterior ballistics or what happens to the projectile after it emerges and flies to the target and how to get it to fly there reproducibly shot after shot; and terminal ballistics or once it is in the vicinity of the target, how to extract the performance from the projectile for which the entire process was intended, usually lethality or destruction.

Ballisticians, those deeply involved in the science of ballistics, tend to specialize in only one of the regimes. Gun and projectile designers, however, must become proficient in all the regimes if they are to successfully field weapons that satisfy the military needs and requirements. The plan of this book is bilateral: first, an unfolding of the theory of each regime in a graduated ascent of complexity, so that a novice engineer gets an early feeling for the subject and its nomenclature and is then brought into a deeper understanding of the material; second, an explanation of the design practice in each regime. Most knowledge of weapon design has been transmitted by a type of apprenticeship with experienced designers sharing their learning with newer engineers. It is for these engineers that this work is intended, with the hope that it will make their jobs easier and their designs superior.

Authors

Donald E. Carlucci has been an engineer at the US Army Armament, Research, Development and Engineering Center, Picatinny Arsenal, Morris County, New Jersey, since May 1989. He is currently the US Army senior scientist for computational structural modeling based at Picatinny. He holds a doctor of philosophy degree in mechanical engineering (2002) and a master of engineering (mechanical) (1995) degree from the Stevens Institute of Technology, Hoboken, New Jersey. In 1987, he received his bachelor of science degree in mechanical engineering from the New Jersey Institute of Technology, Newark, New Jersey. Dr. Carlucci is an adjunct professor of mechanical engineering at the Stevens Institute of Technology, founding and senior member of the International Ballistics Society, and chancellor of the US Army's Armament Graduate School.

Sidney S. Jacobson was a researcher, designer, and developer of ammunition and weapons at the US Army's Picatinny Arsenal in Morris County, New Jersey, for 35 years. He rose from junior engineer through eight professional levels in research and development laboratories to become associate director for research and development at the arsenal. His specialty for most of his career was in the development of large-caliber tank munitions and cannons. Many of these weapons, such as the long rod, kinetic energy penetrators (armor-piercing fin-stabilized discarding sabot rounds), and shaped charge, cannon-fired munitions (high-explosive antitank rounds), have become standard equipment in the US Army. For these efforts and successes, he earned several awards from the army, including, in 1983, the Department of the Army Meritorious Civilian Service Medal. In 1972, he was awarded an Arsenal Educational Fellowship to study continuum mechanics at Princeton University, where he received his second master of science degree (1974). He earned a master of science degree in applied mechanics from the Stevens Institute of Technology (1958) and a bachelor of arts degree in mathematics from Brooklyn College (1951).

He retired in 1986 but maintains his interest in the field through teaching, consulting, and lecturing. He holds two patents and was a licensed professional engineer in New Jersey.

Section I

Interior Ballistics

1

Introductory Concepts

The subject of ballistics has been studied for centuries by people at every level of academic achievement. Some of the world's greatest mathematicians and physicists, such as Newton, Lagrange, and Bernoulli, solved problems in mathematics and mechanics that were either directly or indirectly applied to the various ballistic disciplines. At the other end of the academic scale, there are individuals such as James Paris Lee (inventor of the Lee–Enfield rifle), who developed his first weapon (not the famous Lee–Enfield) at age 12 with no formal education.

The dominant characteristic of any of the ballistic disciplines is the "push–pull" relationship of experiment and analysis. It is a rare event, even as of this writing, when an individual can design a ballistic component or device, either digitally or on paper, and have it function "as designed" in the field. Some form of testing and consequent tweaking of the design are always required. This inseparable linkage between design and test is due to three things: the stochastic nature of ballistic events, the infinite number of conditions into which a gun–projectile–charge combination can be introduced, and the lack of understanding of the phenomena.

The stochastic behavior that dominates all the ballistic disciplines stems from the tremendous number of parameters that affect muzzle velocity, initial yaw, flight behavior, etc. These parameters can be as basic as how or when the propellant was produced to what was the actual diameter of the projectile measured to 0.0001 in. Even though, individually, we believe that we understand the effect of each parameter, when all parameters are brought together, the problem becomes intractable. Because of this parameter overload condition, the behavior is assumed to be stochastic.

The number of battlefield and test conditions that a gun–projectile–charge combination can be subjected to is truly infinite. For safety and performance estimates, the US Army is often criticized for demanding test conditions, which could not possibly occur. While this may be true, it is simply a means of overtesting a design to assure that the weapon system is safe and reliable when the time comes to use it. This philosophy stems from the fact that you cannot test every condition and because soldiers are an ingenious bunch and will invent new ways to employ a system beyond its design envelope.

The lack of understanding of the phenomena may seem rather strong wording even though there are instances where this is literally true. In most cases, we know that parameters are present that affect the design. We also know how they should affect the design. Some of these parameters cannot be tested because there is some other, more fundamental variable that affects the test setup to a far greater degree.

The overall effect of ballistic uncertainty, as described earlier, is that it will be very unusual for you to see the words *always* or *never* when describing ballistic phenomena in this work.

1.1 Ballistic Disciplines

The field of ballistics can be broadly classified into three major disciplines: interior ballistics, exterior ballistics, and terminal ballistics. In some instances, a fourth category named intermediate ballistics has been used.

Interior ballistics deals with the interaction of the gun, projectile, and propelling charge before the emergence of the projectile from the muzzle of the gun. This category would include the ignition process of the propellant, the burning of propellant in the chamber, pressurization of the chamber, first-motion event of the projectile, engraving of any rotating band and obturation of the chamber, in-bore dynamics of the projectile, and tube dynamics during the firing cycle.

Intermediate ballistics is sometimes lumped together with interior ballistics, but has come into its own category of late. Intermediate ballistics deals with the initial motion of the projectile as it is exiting the muzzle of the tube. This generally includes initial tip-off, tube and projectile jump, muzzle device effects (such as flash suppression and muzzle brake venting), and sabot discard.

Exterior ballistics encompasses the period from when the projectile has left the muzzle until impact with the target. One can see the overlap here with intermediate ballistics. In general, all that the exterior ballistician is required to know is the muzzle velocity and tip-off and spin rates from the interior ballistician and the physical properties (shape and mass distribution) from the projectile designer. In exterior ballistics, one is generally concerned with projectile dynamics and stability, the predicted flight path and time of flight, and angle, velocity and location of impact. More often, now than in previous years, the exterior ballistician (usually called an aeroballistician) is also responsible for designing or analyzing guidance algorithms carried onboard the projectiles.

Terminal ballistics covers all aspects of events that occur when the projectile reaches the target. This means penetration mechanics, behind armor effects, fragment spray patterns and associated lethality, blast overpressure, nonlethal effects, and effects on living tissue. This last topic is becoming more and more important because of the great interest in less-than-lethal armaments, and indeed, it has been categorized into its own discipline known as wound ballistics.

1.2 Terminology

Throughout this work, we will be using the word *gun* in its generic sense. A gun can be loosely defined as a one-stroke internal combustion engine. In this case, the projectile is the piston, and the propellant is the air–fuel mixture. Guns themselves can be classified in four broad categories: a "true" gun, a howitzer, a mortar, and a recoilless rifle.

A true gun is a direct-fire weapon that predominantly fires a projectile along a relatively flat trajectory. Later on, we will decide what is truly flat and what is not. Notice the word *predominantly* crept in here. A gun, say, on a battleship, can fire at a high trajectory sometimes. It is just usually used in the direct-fire mode. A gun can be further classified as rifled or smooth bore, depending upon its primary ammunition. Guns exhibit a relatively high muzzle velocity commensurate with their direct-fire mission. Examples of guns include tank cannon, machine guns, and rifles.

A howitzer is an indirect-fire weapon that predominantly fires projectiles along a curved trajectory in an attempt to obtain improved lethal effects at well-emplaced targets. Again, howitzers can and have been used in a direct-fire role; it is simply not one at which they normally excel.

A mortar is a tube that is usually man portable used to fire at extremely high trajectories to provide direct and indirect support to the infantry. Mortars generally have much shorter ranges than howitzers and cannot fire a flat trajectory at all.

A recoilless rifle is a gun designed with very little weight. They are usually mounted on light vehicles or man emplaced. They are used where there is insufficient mass to counteract the recoil forces of a projectile firing. This is accomplished by venting the high-pressure gas out of a rear nozzle in the breech of the weapon in such a way as to counter the normal recoil force.

A large listing of terminology unique to the field of ballistics is included in the glossary in Appendix A.

1.3 Units and Symbols

The equations included in the book may be used with any system of units. That being said, one must be careful of the units chosen. The literature that encompasses the ballistic field uses every possible system and is very confusing for the initiate engineer. The US practice of mixing the International System of Units, US Customary System, and centimeter–gram–second units is extremely challenging for even the most seasoned veteran of these calculations. Because of this, an emphasis has been placed on the units in the worked-out examples and cautions are liberally placed in the book.

Intensive and extensive properties (where applicable) are denoted by lowercase and uppercase symbols, respectively. In some instances, it is required to use the intensive properties on a molar basis. These will be denoted by an overscore tilde. In all cases, the reader is advised to always be sure of the units.

2

Physical Foundation of Interior Ballistics

2.1 Ideal Gas Law

The fundamental means of exchanging the stored chemical energy of a propellant into the kinetic energy of the projectile is through the generation of gas and the accompanying pressure rise. We shall proceed in a disciplined approach, whereby we introduce concepts at their simplest level and then add the complications associated with the real world.

Every material exists in some physical state of either solid, liquid, or gas. There are several variables that we can directly measure and some that we cannot but that are related to one another through some functional relationship. This functional relationship varies from substance to substance and is known as an equation of state.

Thermodynamically, the number of independent properties required to define the state of a substance is given by the so-called state postulate, which is described by Wark [1]. For all the substances examined in this chapter, we shall assume that they behave in a simple manner. This essentially means that the equilibrium state of all our substances can be defined by the specification of two independent, intrinsic properties. In this sense, an intrinsic property is a property that is characteristic of (in other words, governed by) molecular behavior.

The ideal gas law is essentially a combination of three relationships [2]. Charles's law states that the volume of a gas is directly proportional to its temperature. Avogadro's principle states that the volume of a gas is directly proportional to the number of moles of gas present. Boyle's law states that volume is inversely proportional to pressure. If we combine these three relationships, we arrive at the famous ideal gas law, which states in extensive form

$$p\tilde{v} = N\mathfrak{R}T \tag{2.1}$$

where p is the pressure of the gas; $\tilde{v}$ is the molar specific volume; N is the number of moles of the gas; $\mathfrak{R}$ is the universal gas constant; and T is the absolute temperature.

The units of Equation 2.1 are not always convenient to work with. For this reason, the form of the ideal gas law that we shall use most often in this text is

$$pv = RT \tag{2.2}$$

In this case, p is the pressure of the gas; v is the specific volume (in mass units as we are used to); R is the specific gas constant, unique to each gas; and T is again the absolute temperature

The specific gas constant can be determined from the universal gas constant by dividing the latter by the molar mass:

$$R = \frac{\mathfrak{R}}{M} \tag{2.3}$$

where M is the molar mass of the gas (e.g., 15.994 lbm/lb mol for oxygen). There are many other variants of the ideal gas law, which differ only in units. The other two versions that we occasionally utilize are

$$p\mathrm{V} = m_g RT \tag{2.4}$$

and

$$p = \rho RT \tag{2.5}$$

In these equations, V (nonitalicized) is the volume the gas occupies; m_g is the mass of the gas; and ρ is the gas density.

One should always check the units when using these equations.

The pressure in a vessel filled with gas is caused by innumerable collisions of the gas molecules on the walls of the vessel [2]. The more tightly packed the molecules are, the more collisions occur—the higher the pressure is. Similarly, temperature excites the gas molecules so that they move faster, collide more—thus also increasing pressure. It is these collisions, among other things, that must be handled somehow by our equation of state.

The ideal gas law relies upon the fact that the gas molecules are very far apart relative to one another [3]. If the molecules linger in the neighborhood of one another, they will be influenced by strong intermolecular forces, which can either attract or repel them from one another. Thus, the ideal gas law ignores this effect. The ideal gas law further assumes that intermolecular collisions occur completely elastically (i.e., like billiard balls). These assumptions must be kept in mind when using the ideal gas law. We shall soon see that under the pressures and temperatures in a gun, these assumptions are invalid. Nevertheless, they provide us with a point of departure and a useful stepping-stone for our studies.

To use the ideal gas law to determine the state of the gas in a gun, we need to invoke classic thermodynamic relationships. The second law of thermodynamics can be stated as follows:

$$Q = \Delta U + W + \text{losses} \tag{2.6}$$

where Q is the energy added to the system; ΔU is the change in internal energy; W is the work done on the system; and the losses term contains all the energy that cannot be recovered if, say, we pushed the projectile back to its starting position in the gun tube.

Our sign convention shall be that a Q will be positive when energy is added to the system, ΔU will be positive if the internal energy of the system is increased, and W will be positive if work is done on the system. Losses always remove energy from the system.

If we tailor Equation 2.6 to a gun launch situation, then Q would be the energy released by burning our propellant; ΔU would be the change in internal energy of the propellant; and W would be the work done on the projectile.

Let us further define the work term in the classical sense. It is typical of a first-year engineering curriculum to define the work as follows:

$$W = \int \mathbf{F} \cdot \mathrm{d}\mathbf{x} \tag{2.7}$$

In Equation 2.7, work is defined as a scalar that results from the vector dot product of force **F** with the distance over which the force acts, also a vector, **dx** (note that all vectors are characterized by bold type in this book). If we restrict our analysis to a gun system, we can see that given pressure acting on the base of a projectile, it only has one direction to travel due to the constraints of the gun tube. If we imagine that this gun tube is perfectly straight (it never is) and we align a coordinate system with the axis of the tube, then the displacement vector **dx** must be aligned with force vector **F** (i.e., the angle between **F** and **dx** is zero; therefore, the cosine of the angle between them is unity), and our relation for a dot, or more formally, the scalar product of these two vectors, gives us

$$\mathbf{F} \cdot \mathbf{dx} = |\mathbf{F}| \cdot |\mathbf{dx}| \cdot \cos(0) = F\mathrm{d}x \tag{2.8}$$

Our work definition for this case is then

$$W = \int F\mathrm{d}x \tag{2.9}$$

This relationship for work has to be somewhat refined to fulfill our needs. We will need to put the force acting on the projectile in terms of the pressure and would sometimes like the volume to be included in the equation. If we look at the ideal gas equation of state in the form of Equation 2.4, we do not see a force in there, but we do see a pressure term and a volume term.

We know from the mechanics of materials [4] that

$$F = pA \tag{2.10}$$

This has not been written in vector form so as to keep things simple (we will write it differently later). Equation 2.10 states that the resultant force F on a body is equal to the average pressure p on that body times the area A over which the pressure acts. So we can rewrite Equation 2.9 using this result as

$$W = \int pA\,\mathrm{d}x \tag{2.11}$$

We now need to get volume in there somehow. We shall use the fact that except for the chamber of a gun (and a few notable exceptions with the bore), the area over which the pressure acts is constant and equal to the bore cross-sectional area that we have defined as A earlier. The area of the rifling grooves does contribute here if the tube is rifled, but let us assume a nice smooth cylindrical bore for now. If A is the cross-sectional area and dx is a differential element of length, then the differential element of volume, dV, can be defined as

$$\mathrm{dV} = A\mathrm{d}x \tag{2.12}$$

We can now write Equation 2.11 in terms of pressure and volume as

$$W = \int p\mathrm{dV} \tag{2.13}$$

You may recall this form of the definition of work from thermodynamics [5].

We now have two equations and a definition at our disposal as a pedagogical device that can help illustrate the energy exchange mechanism in a gun. The equations are an ideal gas equation of state, Equation 2.4; and the second law of thermodynamics, Equation 2.6; and the definition of how we defined work in Equation 2.13.

Let us imagine that we have a simple gun as depicted in Figure 2.1.

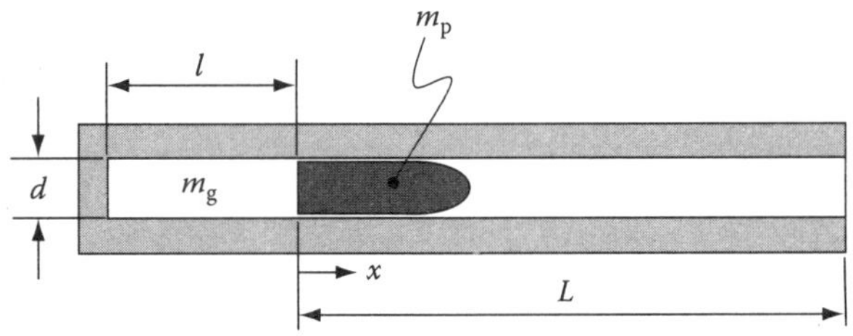

FIGURE 2.1
Simple gun system.

We shall assume that we have somehow placed a mass m_g of a gas that behaves according to the ideal gas equation of state in the tube and compressed it, adiabatically, using the projectile, and no leakage has occurred. We shall further assume that there is no friction between the projectile and the tube wall. Thus, in the situation depicted by Figure 2.1, we have an ideal gas trapped between the projectile and the breech, compressed to some pressure p at some absolute temperature T. We shall further assume that the projectile of mass m_p is somehow held at position $x = 0$ and no gas or energy can escape. In this situation, the volume the gas occupies, which we shall call the chamber volume V_c is given by

$$V_c = \frac{\pi d^2}{4} l \tag{2.14}$$

What we have essentially done is compressed the projectile against an imaginary spring (the gas), which now has a potential energy associated with it. From a thermodynamic standpoint, we can reduce Equation 2.6 to

$$0 = \Delta U + W \tag{2.15}$$

Recapping, we note that $Q = 0$ because there was no heat lost through the tube wall (adiabatic compression), and there is no propellant per se that will burn to generate heat. The losses were zero because we have no friction.

Now that everything is set, we need to release our projectile and see what happens. If we substitute Equation 2.4 into Equation 2.13, we can write

$$W = \int m_g RT \frac{dV}{V} \tag{2.16}$$

This equation now shows how much work is being done on the projectile as a function of the volume. It is noteworthy here that we are assuming that the gas that is actually pushing on the projectile is massless. By this, we mean that no energy is being applied to accelerate the mass of the gas. We will remove this assumption later in our studies. What we do not like about Equation 2.16 is that temperature still appears as a variable.

By our earlier assumptions, we stated that the process was frictionless and adiabatic. Recall, again from thermodynamics, that this actually defines an isentropic process [1]. For a closed system (one with constant mass), it can be shown [6] that the absolute temperature T of our system is related to the initial temperature of the gas T_i through

$$T = T_i \left(\frac{V_c}{V} \right)^{(\gamma - 1)} \tag{2.17}$$

where V is the volume at a given time t; V_c is the initial chamber volume; and γ is the specific heat ratio of the gas (defined later).

If we substitute Equation 2.17 into Equation 2.16, we can write

$$W = m_g R T_i V_c^{(\gamma-1)} \int_{V_c}^{V} V^{-\gamma}\, dV \tag{2.18}$$

This equation is easy to work with because we know most of the terms on the right-hand side (RHS) when we set up our pedagogical gun. We know the mass m_g of the gas. We know R and γ because we picked which gas it was. We know the initial temperature of the gas and we know the chamber volume.

Now that we did all this work with volumes, we want to convert these back to distances. A typical output desired by ballisticians is the pressure vs. travel (i.e., distance) curve. This plot helps the gun designer determine where to make his/her tube thick and where he/she can get away with thinning the wall. If we again recognize that our gun has a constant inner diameter, we can use Equation 2.14 to write Equation 2.18 as

$$W = m_g R T_i l^{(\gamma-1)} \int_{0}^{L} (l + x)^{-\gamma} dx \tag{2.19}$$

If we perform this integration, we obtain

$$W = \frac{m_g R T_i l^{(\gamma-1)}}{(1-\gamma)} \left[(l + L)^{(1-\gamma)} - l^{(1-\gamma)} \right] \tag{2.20}$$

We need to recall from dynamics that the kinetic energy of the projectile can be written as

$$\mathrm{KE}_{\mathrm{projectile}} = \frac{1}{2} m_p V_m^2 \tag{2.21}$$

where V_m is the muzzle velocity. If we assume that all the energy of the gas is converted with no losses into kinetic energy of the projectile, then we can use Equation 2.15 to state that

$$\mathrm{KE}_{\mathrm{projectile}} = W \tag{2.22}$$

We can make use of Equations 2.20 and 2.21 to write this as

$$\frac{1}{2} m_p V_m^2 = \frac{m_g R T_i l^{(\gamma-1)}}{(1-\gamma)} \left[(l + L)^{(1-\gamma)} - l^{(1-\gamma)} \right] \tag{2.23}$$

This is an important result as it relates muzzle velocity to the properties, amount of the gas used, and the mass of the projectile and includes the effect of tube length. We can use this equation to estimate muzzle velocity. So a convenient form of this equation is

$$V_m = \sqrt{2 \frac{m_g}{m_p} \frac{R T_i l^{(\gamma-1)}}{(1-\gamma)} \left[(l + L)^{(1-\gamma)} - l^{(1-\gamma)} \right]} \tag{2.24}$$

In some instances, we would like to use these relationships to determine the state of the gas or velocity of the projectile at some point in the tube other than the muzzle. If this is the case, the procedure would be as follows:

1. Solve for the work term up to the position of interest x_{proj} using

$$W\left(x_{\text{proj}}\right) = m_g R T_i l^{(\gamma-1)} \int_0^{x_{\text{proj}}} (l + x)^{-\gamma} \mathrm{d}x \tag{2.25}$$

2. Determine the volume at the position of interest by using

$$V\left(x_{\text{proj}}\right) = \frac{\pi d^2}{4}\left(l + x_{\text{proj}}\right) \tag{2.26}$$

3. Determine the gas temperature at this position from Equation 2.17.
4. Determine pressure from the ideal gas Equation 2.4.

This procedure is relatively straightforward.

If, as an example, we look at an idealized 155 mm compressed air gun and assume the following parameters:

- Projectile weight = 100 lbm
- Initial pressure = 45 MPa (approximately 6500 psi)
- Tube length = 6 m

From Figures 2.2 through 2.4, we can depict the results of a calculation for temperature, pressure, and velocity vs. distance for this idealized situation.

Problem 1

Assume that we have a quantity of 10 g of 11.1% nitrated nitrocellulose ($C_6H_8N_2O_9$), and it is heated to a temperature of 1000 K assuming it changes from solid to gas somehow without changing chemical composition. If the process takes place in an expulsion cup with a volume of 10 in.^{3}, assuming ideal gas behavior, what will the final pressure be in pounds per square inch?

Answer: $p = 292\left[\frac{\text{lbf}}{\text{in.}^2}\right]$

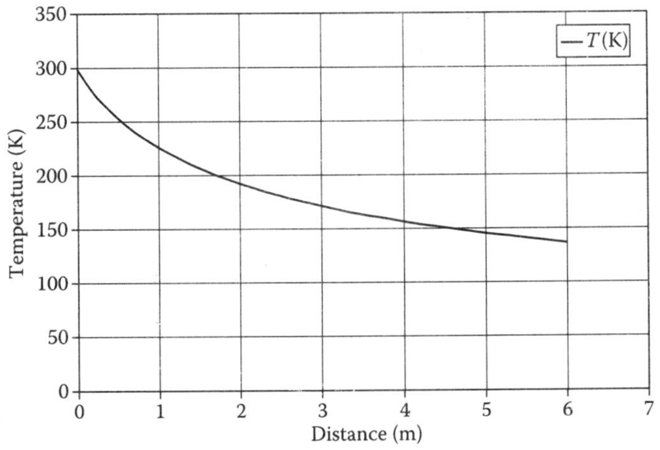

FIGURE 2.2
Temperature vs. distance in an ideal gas gun.

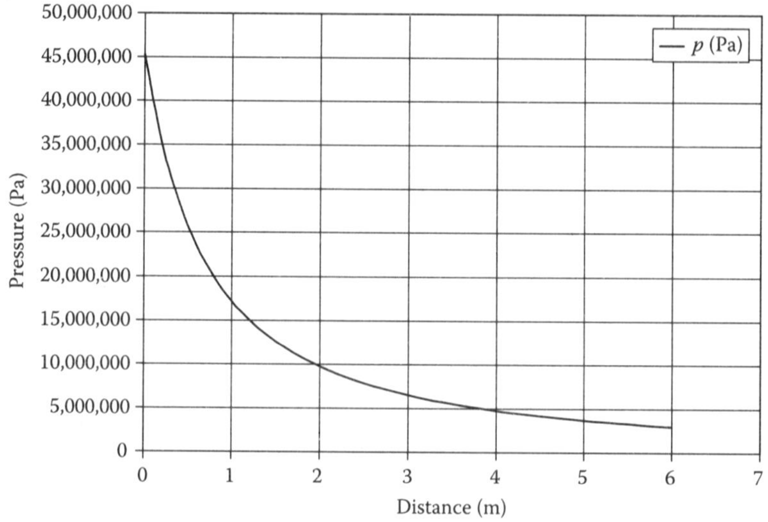

FIGURE 2.3
Pressure vs. distance in an ideal gas gun.

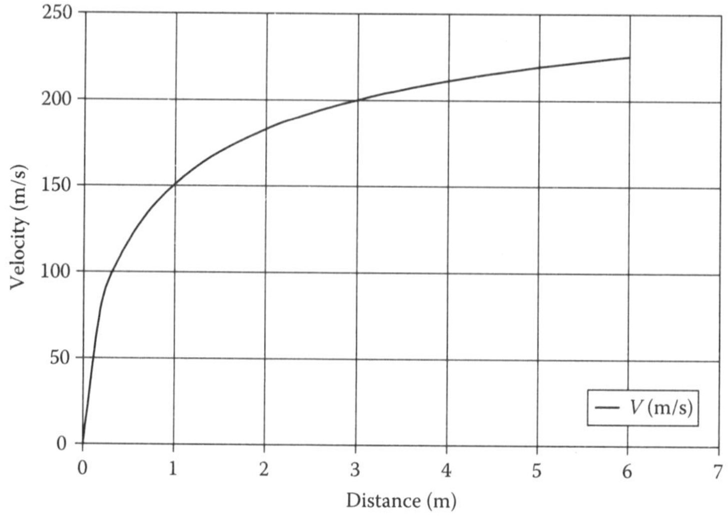

FIGURE 2.4
Velocity vs. distance in an ideal gas gun.

2.2 Other Gas Laws

There are many times when ideal gas behavior is insufficient to model real gases. This is certainly true under the pressures and temperatures of gun launch. Although there are many models that attempt to account for the deviation of real gases from ideal or perfect behavior [2,3], we shall examine only two, the simplest of which we shall use.

Ideal gas behavior is approached when the distance between molecules (known as the mean free path) is large. Thus, molecules do not collide or interact with one another very often. Temperature is a measure of the internal energy of the gas. Thus, when the temperature is high, the molecules are moving around faster and have more of an opportunity to interact with one another. Pressure is a result of how closely the molecules are packed together; thus, a higher pressure tends to put the molecules in close proximity. It is for these reasons that we cannot normally use the ideal gas law in gun launch applications.

The Noble–Abel equation of state is given by

$$p\left(V - m_g b\right) = m_g RT \tag{2.27}$$

where p is the pressure of the gas; V is the volume the gas occupies; m_g is the mass of the gas; R is the specific gas constant; T is the absolute temperature; and b is the covolume of the gas.

The covolume of the gas has been described as a parameter that takes into account the physical size of the molecules and any intermolecular forces created by their proximity to one another. Think of it as not having physical meaning but as simply a number that allows for a better fit to observed experimental data. The units of the covolume are cubic length per mass unit. Usually, the gas covolume is provided in the literature, but an estimation tool has been provided by Corner [7], which will not be repeated here since actual data exist.

Occasionally, the Noble–Abel equation of state is insufficient to suit our needs. At these times, it is typical to use a van der Waals equation of state given by

$$p = \frac{\tilde{R}T}{\tilde{v} - b'} - \frac{a'}{\tilde{v}^2} \tag{2.28}$$

In this case, p is again the pressure of the gas; $\tilde{v}$ is the molar specific volume; $\tilde{R}$ is the molar specific gas constant, unique to each gas; T is again the absolute temperature; and a' and b' are constants particular to the gas.

The Noble–Abel equation of state is the basis for nearly all our work in this chapter; therefore, Equation 2.27 is very important. At times, we may write it a little differently, but you will always be reminded of where it originated.

Problem 2

Perform the same calculation as in Problem 1, but use the Noble–Abel equation of state and assume the covolume to be 32.0 in.^{3}/lbm.

Answer: $p = 314.2 \left[\frac{\text{lbf}}{\text{in.}^2}\right]$

Problem 3

A hypothetical "air mortar" is to be made out of a tennis ball can using a tennis ball as the projectile. The can has a 2–1/2 in. inside diameter and is 8 in. long. If a tennis ball of the same diameter weighs 2 oz and initially rests against the rear of the can, to what air pressure must one pressurize the can in order to achieve a 30 ft/s launch of the tennis ball? Assume that the tennis ball can be held against this pressure until released and that it perfectly obturates and assume an isentropic process and ideal gas behavior with $\gamma = 1.4$ for air.

2.3 Thermophysics and Thermochemistry

The main energy exchange process of conventional interior ballistics is through combustion. Once ignited, the chemical energy of the propellant is released through an oxidation reaction. This energy release will be in the form of heat, which, in turn, increases the pressure in the volume behind the projectile (i.e., in a combustion chamber). The pressure exerts a force on the projectile, which accelerates it to the desired velocity.

In general, combustion requires three main ingredients to commence: a fuel, an oxygen source, and heat. In a common combustion reaction, such as an internal combustion engine like the one in your car, oxygen is supplied to the reaction independently of the fuel. The heat in this case is generated by a spark ignition and the burning of the air–fuel combination that ensues.

A gun chamber has very little room for oxygen once it is stuffed with propellant. It is important to note that for other reasons, there is always free volume in the chamber (called ullage)—we will explain this later. For now, we should understand that although there is some oxygen in the chamber, the amount is insufficient to completely combust the propellant. It is for this reason that propellants are formulated to contain both the fuel and the oxidizer. In general, the propellant burning is an underoxidized reaction. This has some implications as the propellant gases leave the muzzle—again, we shall discuss this in more detail later.

This brief introduction should make clear the reason to examine thermochemistry, thermophysics, and combustion phenomena. To proceed, we shall first define each field of study. The definitions by Cooper [8] shall be used here to describe the first two topics as they are extremely straightforward and clear. Thermophysics is defined as the quantification of changes in the energy state of a substance caused by changes in the physical state of the material. An example of this would be the determination of the amount of energy required to vaporize water in your teapot. Thermochemistry is then the quantification of changes in a the energy state of a substance caused by changes in the chemical composition of the molecules of the material. An example of this would be the energy required to dissociate (break up) water molecules into hydrogen and oxygen. Combustion is defined by Wark [1] as the quantification of the energy associated with oxidizer–fuel reactions. Thus, combustion is a natural outgrowth of thermophysics and thermochemistry.

Now that we have categorized these three fields of study, we shall attack them in a somewhat jumbled order. The reason for this is that from our perspective, we really need not distinguish between any of them, and all them appear in our gun launch physics. It is also important to realize that whether the energy change comes from a chemical reaction or a phase change from solid to gas, as long as we can calculate the extent of the energy change, we can perform a valuable analysis.

Energy to all intents and purposes consists of two types: potential and kinetic. Potential energy can be considered as stored energy. There are many ways to store energy. We can store energy by compressing a steel bar or spring, by lifting a mass to a higher elevation in the gravitational field of the earth and by chemically preparing a compound that, whether by combustion or chemical reaction, will release energy. Each of these forms of potential energy, elastic strain, gravitational potential, and chemical potential energy, has a different method of storing and releasing the energy, but they are all potential energies. There are other forms of potential energy, but we need not deal with them in this context.

Kinetic energy is the energy of a mass in motion. It can be observed in objects that are in translational motion or in rotational motion. To extract some or all this energy, it is necessary to slow or stop the moving mass that has the kinetic energy. The energy in a spinning flywheel is an example of rotational kinetic energy.

The field of thermodynamics is the study of energy transformations. It quantifies the balance of energy between kinetic and potential. In thermodynamics, it is common to see two energy transformation mechanisms: heat and work.

Heat transfer is essentially an exchange of energy through molecular motion. As we shall soon see that molecules of a substance are always in motion. The faster they are in motion, the hotter the substance is. These molecules can influence other molecules when they are placed in contact with them, thus giving up some of their energy and increasing the energy of the contacted substance. Temperature is a sensible measure of the internal energy of an object.

Work is a means of increasing the energy an object by the application of a force through a distance. This method of energy transfer can create either potential energy, as in compressing a spring, or kinetic energy as applied to a free, rigid mass. While the equations for heat transfer can be the subject of entire texts (e.g., the book by Sucec [9]), work can be defined through the vector equation

$$W = \mathbf{F} \cdot d\mathbf{x} \tag{2.29}$$

where W is the work done on or by the system; $\mathbf{F}$ is the force vector; and $d\mathbf{x}$ is the vector distance through which the force acts, known as the displacement vector.

We must note that this is a *vector* equation. The work term is a scalar because the dot product of two vectors results in a scalar. Because of the dot product term, the sign of W is dependent upon the cosine of the angle between $\mathbf{F}$ and $d\mathbf{x}$. Recall the definition of a dot product as

$$\mathbf{A} \cdot \mathbf{B} = AB\cos\theta \tag{2.30}$$

where A and B are the scalar magnitudes of the vectors $\mathbf{A}$ and $\mathbf{B}$ (Figure 2.5). If we use Equation 2.30 with the variables of Equation 2.29, this tells us that if the angle between the force vector and the displacement vector is between 0° and 90° or 270° and 0°, the work is positive, i.e., it is work performed on the system. If, however the angle is between 90° and 270°, the work is negative and therefore work is performed by the system.

The internal energy U of a substance can be considered a form of potential energy. Some authors [5] categorize the internal energy separately from potential and kinetic energies. This can clearly be done in general, but for the application of gun launch, it seems proper to group it as a potential energy. The internal energy of a substance is manifested in the molecular motions within that substance. These motions are generally translational or

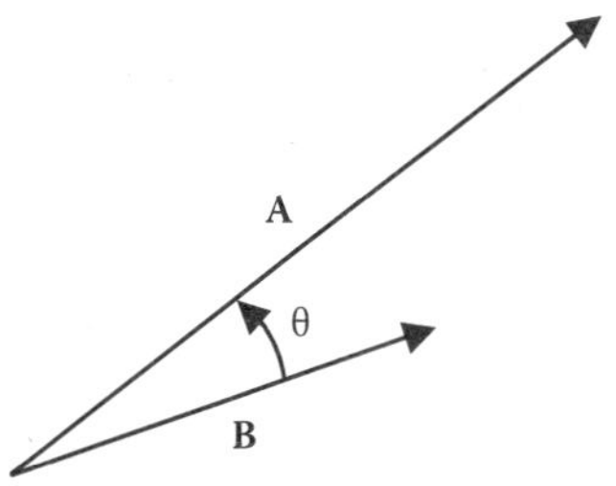

FIGURE 2.5
Depiction of two vectors for scalar product definition.

vibrational in nature. The molecules of a substance are attracted to and repelled by one another and are in some degree of translational motion. Additionally, the attractive or repulsive forces within a molecule itself allow us to use an analogy of springs holding the atoms together. Imagine a structure of a water molecule, for instance, as depicted in Figure 2.6. If the oxygen and hydrogen atoms are assumed to be steel balls, and the molecular, bond springs, we could pick this molecule up, hold the oxygen atom, and shake it. If the springs were really stiff in bending and much less so in tension or compression, we would see the hydrogen atoms oscillating in and out at some frequency. The greater the frequency, the more energy we would need to put into the system. Even though the springs are stiff in bending, it does not mean that they cannot bend. This just takes more energy. Like springs, we can store energy in the molecules this way.

This simple model of a molecule is a crude but useful approximation. Imagine now that we put our model on a frictionless surface, such as an ice hockey rink. If we hit the molecule in a random way, we will excite these vibrational modes as well as create translational and rotational motion. Now, if we fill the ice hockey rink with models, well, you get the idea. As stated previously, the level of this interaction (collisions) must be represented somehow. The metric used is internal energy with the level of activity defined as zero at the temperature known as absolute zero (0° in the Kelvin or Rankine scales).

The internal energy also includes the energy required to maintain a particular phase of the material such as solid, liquid, or gas. Additionally, certain phases associated with molecular structure, such as face-centered cubic (FCC) and body-centered cubic (BCC), are accounted for in the internal energy.

Quite often we shall see internal energy and what is commonly known as "$p\mathrm{d}V$" work terms together in our energy balance equations. The term is called $p\mathrm{d}V$ work because it is special and separate from work generated by, say, a paddle wheel moving fluid around. This work term arises from pressure pushing on a given volume. If the volume changes by an infinitesimal amount $\mathrm{d}V$ we essentially have force acting through a distance. To prove this to yourself, look at the units. Because we see these terms together so often, it is convenient for us to group them into one term, which we will call enthalpy H. Mathematically, the enthalpy is defined as

$$H = U + pV. \tag{2.31}$$

Notice here that we have removed the differential from the work term. The reason for this is that considering both enthalpy and internal energy, we are concerned with changes in

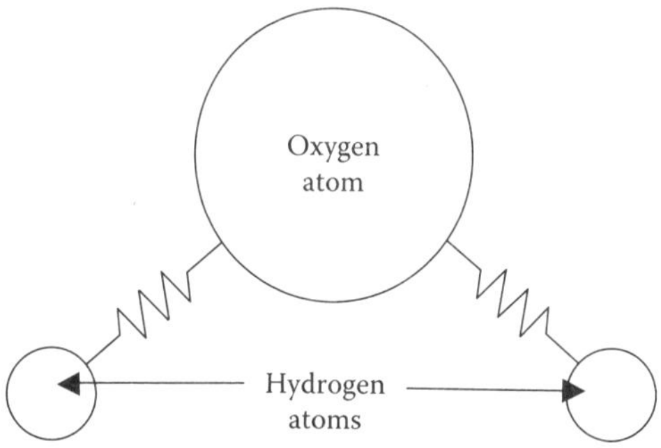

FIGURE 2.6
Model of a water molecule.

H and U. Therefore, the differential appears when we write the entire equation in differential form as

$$dH = dU + p dV. \tag{2.32}$$

For proof of this result, refer to any thermodynamics text (e.g., the books by Wark [1] and Van Wylen and Sonntag [5]). An example of the difference between internal energy and enthalpy is the rigid container or piston container. Consider a rigid container that has some amount of gas in it. Assume the container is sealed so that matter cannot enter or leave. Let us also assume that the container will allow energy to be transferred to and from the gas. If we transfer heat (energy) to the gas, the temperature will rise as will the pressure. Since the volume of the container is fixed, no work can be done; thus, all the energy added to the gas is internal energy. From Equation 2.32, we see that in this case, the change in enthalpy would be exactly equal to the change of internal energy.

Now we assume that instead of our container being rigid, the roof of the container is a sealed yet moveable piston. In this case, once again, matter cannot escape; however, the volume is able to change. Now the only thing holding up the roof is the pressure of the gas acting to just counteract the weight of the roof itself. Let us add the same amount of heat that we added to the original, rigid container. In this case, the temperature of the gas will increase (but less than before) and the volume will increase because the piston is moveable, and the pressure must remain constant and just sufficient to counteract the weight of the roof. In this instance, the enthalpy would be greater than the internal energy because it includes the work done in lifting the piston.

When a substance changes form, chemically or physically, energy is either absorbed or released. The method that we use to quantify this energy change is through heats of formation and the like. Although called a "heat," what is really implied is an enthalpy change. We shall proceed through these different enthalpy changes, attempting to list some of the more common ones. For greater detail, the reader is encouraged to consult thermodynamics texts in addition to the descriptions provided by Cooper [8]. Specific values for text problems will be given as needed. It is not the intent of the authors to tabulate the different energy parameters of different materials.

When a substance is formed, atomic bonds in the constituent molecules are destroyed and then recreated (at least, this is a clean way to think of it from a bookkeeping perspective). The energy absorbed or generated by this process is commonly called the heat of reaction ΔH_r^0. The Δ reminds us that we are always concerned with changes in enthalpy from a particular reference state (usually standardized as 25°C and 1 atm). The superscript 0 is a convenient reminder that this is from a reference state of 1 atm. As the subscript, sometimes we see "298" meaning 298 K. Although 298 K and 25°C are the same value, one must always be wary of the reference state chosen by a particular author.

The heat of formation ΔH_f^0 is the energy required to form a particular substance from its individual component atoms. The heats of formation are the building blocks that determine the heat of reaction. Any elemental substance in its stable configuration at standard conditions has a heat of formation equal to zero at that state. For instance, diatomic nitrogen N_2 has $\Delta H_f^0 = 0$ at 25°C and 1 atm. We will provide an example of the heat of formation calculation in a later section.

Now that with the aforementioned quantities defined, we can write an equation for the heat of reaction

$$\Delta H_r^0 = \sum_{\text{products}} \Delta H_f^0 - \sum_{\text{reactants}} \Delta H_f^0 \tag{2.33}$$

Equation 2.33 states that the heat of reaction for a given substance is equal to the sum of the heats of formation of the final products of the reaction that created the substance minus the sum of the heats of formation of the materials that had to be reacted together to create the new substance. This is further reinforcement of the definition of the heat of reaction. Recall that we stated that the atomic bonds of the molecules were destroyed and then remade. This is essentially what Equation 2.33 is saying. The energy it took to create each of the reactants has to be accounted for, and then the energy it takes to create the new substances from the constituents is calculated—energy is conserved. If the heat of reaction is a negative number, heat is liberated by the reaction; otherwise, it is absorbed.

When a compound is specifically combusted with sufficient oxygen to attain its most oxidized state, the heat of reaction has a special name: the heat of combustion. The heat of combustion is identified by the symbol ΔH_c^0. The heat of combustion is typically what is obtained when the propellant is burned in a closed bomb. The equation for the heat of combustion mirrors that of the heat of reaction, the only difference being as noted earlier:

$$\Delta H_c^0 = \sum_{\substack{\text{fully oxidized} \\ \text{products}}} \Delta H_f^0 - \sum_{\text{reactants}} \Delta H_f^0 \tag{2.34}$$

The heats of detonation and explosion have meanings that seem to be reversed. The heat of detonation is the heat of reaction taken when detonation products are formed from an explosive compound during a detonation event. The formula for the heat of detonation is given by

$$\Delta H_d^0 = \sum_{\substack{\text{detonation} \\ \text{products}}} \Delta H_f^0 - \sum_{\substack{\text{original} \\ \text{explosive}}} \Delta H_f^0 \tag{2.35}$$

What is termed the heat of explosion is the amount of energy released when a propellant or explosive is burned (not detonated) and is given by

$$\Delta H_{exp}^0 = \sum_{\substack{\text{burning} \\ \text{products}}} \Delta H_f^0 - \sum_{\substack{\text{original} \\ \text{propellants}}} \Delta H_f^0 \tag{2.36}$$

The heat of afterburn is another type of heat of reaction that occurs often in propellants and explosives. Because the composition of propellants and explosives usually force an underoxidized reaction, the reaction products will often combine with the oxygen present in the air outside the gun or explosive device, given sufficient temperature and pressure. This secondary reaction results in a second pressure wave or blast and a fireball. The heat of afterburn can be mathematically described as

$$\Delta H_{AB}^0 = \sum_{\substack{\text{fully oxidized} \\ \text{products}}} \Delta H_c^0 - \sum_{\substack{\text{remaining} \\ \text{detonation} \\ \text{products}}} \Delta H_d^0 \tag{2.37}$$

Not all energy changes involve chemical reactions. We mentioned earlier that changes in physical state and structure require energy. When a solid melts to form a liquid or a liquid solidifies, we call the energy required, the latent heat of fusion λ_f. These values are tabulated in any chemistry book or thermodynamics text. Some authors use different symbols so one must, as always, be careful.

In a similar vein, the energy required to vaporize a liquid to a gas or condense a gas to a liquid is known as the latent heat of vaporization and given by the symbol λ_{fg}.

If a material changes the structure of its atoms, say, from BCC to FCC, the energy is known as the heat of transition λ_t.

There are many other types of material transitions that require energy. The types described earlier cover the needs of this work.

2.4 Thermodynamics

The combustion process that occurs in a gun is a thermodynamic process. The term *thermodynamics* is a bit misleading because it implies that the dynamics of the combustion process is examined. This is not quite true. Classical thermodynamics is based on the examination of the various processes through equilibrium states. This is somewhat akin to frames of a motion picture. We examine the state of the system before some event, and we usually examine it at some point, later in time, we are interested in.

Some of the concepts of thermodynamics were introduced in earlier sections, work and energy being the major ones. Here we shall look in detail at two ways of describing thermodynamic systems to proceed with our study.

We shall define energy for an arbitrary system as

$$E = U + \frac{1}{2}mV^2 + mgz \tag{2.38}$$

Equation 2.38 is our extensive form of the definition of the system energy E. In this equation, U is the internal energy, m is the system mass, V is the system velocity, g is a gravitational constant, and z is some height above a reference datum. The second and third terms on the RHS of the equation are the kinetic and potential energies, respectively. If we examine this equation, it is easy to see why some authors group the internal energy as a separate energy type. However, in the case of a gun launch, the potential energy term is insignificant. This focuses us on the transfer of energy between internal and kinetic.

We sometimes write Equation 2.38 in its intensive form as

$$e = u + \frac{1}{2}V^2 + gz \tag{2.39}$$

Recall from our earlier discussions that an intensive property is the associated extensive property divided by mass.

We shall now examine the first law of thermodynamics as it is applied to two different types of systems: a fixed mass of material and a fixed volume of space through which material flows. The first type of analysis, where the material is a fixed mass, is known as a Lagrangian approach, while the fixed or control volume (CV) approach is known as Eulerian. Both are important from a ballistic analysis standpoint and are prevalent in interior, exterior, and terminal ballistic studies.

For a fixed mass of material, undergoing some thermodynamic process, the first law of thermodynamics can be written as

$$Q_{1-2} + W_{1-2} = \Delta E_{1-2} \tag{2.40}$$

where Q is the heat or energy added to the system; W is the work performed on or by the system; and ΔE is the change in the energy state of the material.

The subscript 1–2 simply lets us know that the process began at some state 1 and ends at some state 2. The signs on the terms are very important. We assume a positive change in energy comes about by adding heat to the system and doing work on the system. Thus, work performed on the system is positive and heat added is positive. Different thermodynamics texts write the first law slightly differently, but if you understand that the net result of work on the system or heat transfer to the system is to increase its energy, then few mistakes will be made.

An interesting observation of Equation 2.40 is that the energy state change has an infinite number of paths that lead to the same result. For instance, if we wanted to add 24 kJ of energy to some arbitrary system, we could do it by adding 12 kJ of heat and performing 12 kJ of work on the system. We could obtain the same result by adding 36 kJ of heat and extracting 12 kJ of work from the system. The possibilities are limitless. This reinforces our assertion that thermodynamics is really only concerned with end states.

Caution is warranted at this point. Equation 2.40 does not say how the energy, once added to the system, is partitioned between potential (internal) energy and kinetic energy. This reveals something. Heat and work are added to or removed from the system at the system boundaries while the distribution of energy between internal or kinetic energy is done within the system.

We shall now explicitly write out Equation 2.40 for a Lagrangian system

$$Q_{1-2} + W_{1-2} = m\left[\left(u_2 + \frac{1}{2}V_2^2\right) - \left(u_1 + \frac{1}{2}V_1^2\right)\right] \tag{2.41}$$

Here we have neglected the gravitational potential energy terms and used the intensive form of the energy, multiplied by the system mass. As previously stated many times, we would like to use enthalpies instead of internal energies. If this is the case, we can rewrite Equation 2.41 using our relationship between the two from Equation 2.40. We shall use the intensive form of Equation 2.40 to yield

$$Q_{1-2} + W_{1-2} = m\left[\left(h_2 - pv_2 + \frac{1}{2}V_2^2\right) - \left(h_1 - pv_1 + \frac{1}{2}V_1^2\right)\right] \tag{2.42}$$

Here we note that h is the specific enthalpy and v is the specific volume.

We shall now examine the first law of thermodynamics in the Eulerian frame of reference. Recall that in the Eulerian frame, we chose a CV (real or imaginary) and observed how the energy within the volume changes based upon the energy carried into or out of it by any entering or exiting substance as well as any heat or work done at the system boundaries. It is convenient for us to write the first law in terms of the time rate of change of energy, heat, and work. We start by writing Equation 2.40 as a rate equation

$$\frac{dQ}{dt} + \frac{dW}{dt} = \frac{dE}{dt} \tag{2.43}$$

or

$$\dot{Q} + \dot{W} = \dot{E} \tag{2.44}$$

Here the dots over the heat and work terms indicate the time rate of change of the variable. Proper thermodynamics terminology would require us to use the δ'' instead of d in Equation 2.43 because of path dependency considerations, but for our purposes, we shall ignore this fact. The reader is advised to consult any thermodynamics text for a better understanding of the difference.

The substitutions that were performed to arrive at Equation 2.41 are not as straightforward in this case. Because we have material entering and leaving the CV, we can imagine that this material can enter or leave with a different pressure and density as it interacts with our fixed CV. Because of this, we must account for the energy used to make these changes. Alternatively, one can envision the material coming in at a higher pressure or density and wanting to push our imaginary CV outward, but since we fixed our CV, it cannot. The energy from this must go somewhere so it works on the fluid in and around our CV. Mathematically, this results in the energy term in Equation 2.44 having to include a pv term. This is sometimes known as flow work [10]. With this in mind, Equation 2.44 can be written as

$$\dot{Q} + \dot{W} = \dot{m}_{out}(e_{out} + p_{out}v_{out}) - \dot{m}_{in}(e_{in} + p_{in}v_{in}) \tag{2.45}$$

Here, by multiplying the intensive properties by the mass flow rate $\dot{m}$, we have the rate of change of the energy terms. We have also arbitrarily assumed one inlet and one outlet. If more inlets or outlets in our CV were present and they had different mass flow rates or pressures, we would have to consider each with a term identical to our outlet or inlet terms given earlier. We now can make the substitution for our energy terms to yield

$$\dot{Q} + \dot{W} = \dot{m}_{out}\left(u_{out} + \frac{1}{2}V_{out}^2 + p_{out}v_{out}\right) - \dot{m}_{in}\left(u_{in} + \frac{1}{2}V_{in}^2 + p_{in}v_{in}\right) \tag{2.46}$$

In this case, we have also assumed a uniform velocity over the inlets and outlets. With one inlet and outlet, the mass flow in must equal the mass flow out so we can write Equation 2.46 as

$$\dot{Q} + \dot{W} = \dot{m}\left[\left(u_{out} + \frac{1}{2}V_{out}^2 + p_{out}v_{out}\right) - \left(u_{in} + \frac{1}{2}V_{in}^2 + p_{in}v_{in}\right)\right] \tag{2.47}$$

Substitution of enthalpy into the aforementioned equation puts it into a compact form:

$$\dot{Q} + \dot{W} = \dot{m}\left[\left(h_{out} + \frac{1}{2}V_{out}^2\right) - \left(h_{in} + \frac{1}{2}V_{in}^2\right)\right] \tag{2.48}$$

In many fluid dynamics texts, there are wonderful examples of how these equations are used with multiple inlets and outlets [11]. You may be asking yourself how useful are these equations if we only use one inlet or outlet? The answer is that they are very useful. Except for flows through muzzle devices or through internal ports such as bore evacuators and ports for automatic weapons, a gun is a right circular tube that contains the propellant gas. Any flow field analysis we perform on the moving gases will have just one inlet (toward the breech) and one outlet (toward the projectile). Thus, as we develop our equations later for in-bore motion, we can use these simple equations in the aforementioned form.

As a review, we have two equations that state the first law of thermodynamics. For a fixed mass of material (Lagrangian frame), we have

$$Q_{1-2} + W_{1-2} = m\left[\left(h_2 - pv_2 + \frac{1}{2}V_2^2\right) - \left(h_1 - pv_1 + \frac{1}{2}V_1^2\right)\right] \tag{2.42}$$

and for a fixed volume that material can flow in and out of (Eulerian frame), we have

$$\dot{Q} + \dot{W} = \dot{m}\left[\left(h_{\text{out}} + \frac{1}{2}V_{\text{out}}^2\right) - \left(h_{\text{in}} + \frac{1}{2}V_{\text{in}}^2\right)\right] \tag{2.48}$$

These equations have been repeated here because of their critical importance to our work.

In many instances, we will find that we require a relationship between internal energy or enthalpy and temperature. If we have a gas that is not reacting and intermolecular forces are small enough to ignore, we can consider the gas to be thermally perfect [12]. The implications of this are that internal energy and enthalpy are functions of the temperature alone. With this model, we can write expressions for internal energy and enthalpy as follows:

$$du = c_v \, dT \tag{2.49}$$

$$dh = c_p \, dT \tag{2.50}$$

where c_v is the specific heat at constant volume and c_p is the specific heat at constant pressure.

Normally, c_p and c_v vary with temperature. In many practical cases, this variation is small, and we can further assume that the gas is calorically perfect, which results in the aforementioned equations being written as

$$u = c_v T \tag{2.51}$$

$$h = c_p T \tag{2.52}$$

For a thermally or calorically perfect gas (not a reacting gas), there is a relationship between c_p, c_v, and R. If we define γ as the ratio of specific heats where

$$\gamma = \frac{c_p}{c_v} \tag{2.53}$$

then we can write the aforementioned relationships as

$$c_p - c_v = R \tag{2.54}$$

$$c_p = \frac{\gamma R}{\gamma - 1} \tag{2.55}$$

$$c_v = \frac{R}{\gamma - 1} \tag{2.56}$$

The second law of thermodynamics defines the concept of entropy for us [1]. We know from the second law of thermodynamics that

$$Tds = du + pdv \tag{2.57}$$

or if we insert the definition of enthalpy,

$$Tds = dh - vdp \tag{2.58}$$

If we evaluate Equations 2.57 and 2.58 under the assumptions of a calorically perfect gas, we obtain

$$s_2 - s_1 = c_p \ln\left(\frac{T_2}{T_1}\right) - R\ln\left(\frac{p_2}{p_1}\right) \tag{2.59}$$

$$s_2 - s_1 = c_v \ln\left(\frac{T_2}{T_1}\right) + R\ln\left(\frac{v_2}{v_1}\right) \tag{2.60}$$

In these expressions, the subscripts 1 and 2 indicate the initial and final states of the substance, respectively. An isentropic process is a process in which there is no entropy change. This is also known as a reversible process. In a real system, entropy must always increase or, at best, stay constant. Many processes have slight enough entropy increases as to be considered isentropic. Isentropic processes are also excellent for examining theoretical limits on real processes. If we examine Equations 2.59 and 2.60 under an isentropic assumption, we see that the left-hand side (LHS) is zero in both. This has implications that allow us to write (for an isentropic process)

$$\frac{p_2}{p_1} = \left(\frac{\rho_2}{\rho_1}\right)^{\gamma} = \left(\frac{\upsilon_2}{\upsilon_1}\right)^{-\gamma} = \left(\frac{T_2}{T_1}\right)^{\frac{\gamma}{\gamma-1}} \tag{2.61}$$

Problem 4

The M898 Sense and Destroy Armor (SADARM) projectile weighs 102.5 lb. The projectile was fired from a 56-caliber, 155 mm weapon, and a pressure–time trace was obtained. The area under the pressure–time curve was (after converting the time to distance) calculated to be 231,482 psi m. Calculate the muzzle energy of the projectile in megajoules. Assume the bore area to be 29.83 in.^{2}

Answer: $E = 30.7$ [MJ].

Problem 5

An 8 in. Mk. 14 Mod. 2 Navy cannon is used at Naval Surface Warfare Center Dahlgren, Virginia, for "canister" firings. These firings are used to gun harden electronics that are carried in an 8 in. projectile. The projectile used weighs 260 lb. The measured muzzle velocity is around 2800 ft/s. Calculate the muzzle energy of the projectile in megajoules. Assume the bore area to be 51.53 in.^{2} The rifled length of the tube (distance of projectile travel) is 373.65 in.

Answer: $E \approx 43$ [MJ].

2.5 Combustion

As stated in the previous two sections, combustion is the process through which the energy of the solid propellant is converted to useful work. The purpose of this section is to quantify the oxidation reaction. The tactic we shall employ is to examine the more common, everyday combustion processes that combine (relatively) simple fuels with air to produce work. In this way, we shall hopefully bring to mind the combustion thermodynamics that has been taught at an undergraduate level and perhaps has been forgotten or not exercised since it was first learned.

If we utilize the concept of a fixed CV, we can imagine a combustion chamber as depicted in Figure 2.7. In this CV, we can envision a mass of fuel entering as well as some mass of air. The two are then combusted with one another and the gaseous products leave as a mixture. We can write the first law of thermodynamics for this system then as in Equation 2.46, which we shall repeat here with subscripts that reflect Figure 2.7:

$$\dot{Q}+\dot{W}=\dot{m}_{\text{products}}\left(h_{\text{products}}+\frac{1}{2}V_{\text{products}}^2\right)-\dot{m}_{\text{air}}\left(h_{\text{air}}+\frac{1}{2}V_{\text{air}}^2\right)-\dot{m}_{\text{fuel}}\left(h_{\text{fuel}}+\frac{1}{2}V_{\text{fuel}}^2\right) \tag{2.62}$$

In Equation 2.62, we can see how the heat and energy generated are affected by the amount of mass flow, the enthalpies, and the velocities of the fuel, the oxidizer (air in this case), and the product gases. We require some means of determining the energy converted through the chemical reaction. We achieve this through the balancing of the chemical reaction. We shall return to Equation 2.62 once we have discussed chemical reactions.

One of the most important compounds in the study of combustion is air. We shall adopt a convention that is standard in many thermodynamics texts [5,13,14] that model air as 21% diatomic oxygen (O_2) and 79% diatomic nitrogen (N_2). This means that every mole of oxygen carries with it 3.76 mol of nitrogen. This relationship comes about because

$$\frac{0.79\left[\dfrac{\text{moles N}_2}{\text{mole air}}\right]}{0.21\left[\dfrac{\text{moles O}_2}{\text{mole air}}\right]}=3.76\left[\frac{\text{moles N}_2}{\text{mole O}_2}\right] \tag{2.63}$$

As can be seen in Table B.1, the molecular weight for our simple model of air is 28.97 kg/kg mol.

The balancing of a chemical reaction determines what the molecular composition of the combustion products will be and furthermore helps us quantify the amount of energy absorbed or released. If energy is absorbed in a chemical reaction, in other words, if we had to add energy to force the reaction to completion, the reaction is said to be endothermic. If heat is liberated, the reaction is said to be exothermic [15].

A reaction can be said to be theoretically or stoichiometrically balanced if the reaction goes to completion and there is no excess oxygen in the products [1]. We shall define a

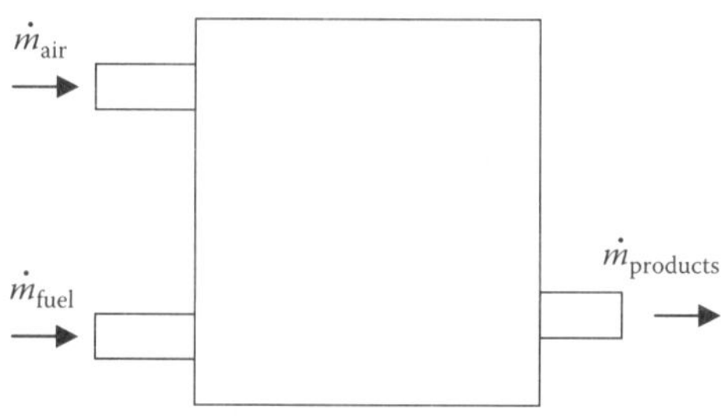

FIGURE 2.7
Fixed CV combustion chamber.

complete reaction as one in which all the oxygen combines first with all the hydrogen to form steam and then with all the carbon to form carbon dioxide. Oxygen has a greater affinity for combining with hydrogen than with carbon [1]. The only time that carbon monoxide (CO) will be formed is if there is insufficient oxygen. We must keep in mind that in any real reaction, there will usually be some amounts of carbon monoxide and other compounds such as nitric oxide (NO) in the combustion products. We shall return to this issue later. For the time being, we shall assume that the only reaction products in the stoichiometric reaction are CO_2 and H_2O. The balancing of these chemical reactions is an important part of our study of the combustion process, which we shall now examine.

We shall use two convenient forms of chemical equations: a molar-based equation and a mass-based equation. In the molar-based equation, we shall usually combust 1 mol of fuel with some amount of air. The result may be multiplied by the number of moles of fuel actually burned to obtain a final answer. When the mass-based equation is employed, we generally use one mass unit of fuel (pound-mass or kilograms) and some amount of air, again multiplying the solution by whatever the actual mass of fuel happens to be. The techniques just described are applicable to a system where the mass is fixed. The same equations can be used with mass or molar flow rates if the system happens to be a steady flow or open system.

It is informative to balance the chemical reactions in the context of everyday systems that combust a fuel with air. Usually, this fuel is a hydrocarbon composition. The stoichiometric amount of air required would be enough so that all the carbon combusts with sufficient oxygen to form CO_2 and all the hydrogen combusts to form water or steam.

If we had a hydrocarbon fuel of chemical composition C_xH_y, we would like to find the number of moles a of air required to completely combust the fuel, and we would write the balanced chemical reaction as

$$C_xH_y + a(O_2 + 3.76N_2) \rightarrow xCO_2 + \frac{y}{2}H_2O + 3.76aN_2 \tag{2.64}$$

We could solve for a to yield

$$a = x + \frac{y}{4} \tag{2.65}$$

As an example, let us say we have 1 mol of benzene (C_6H_6) that we would like to burn in air. The balanced, stoichiometric equation would be found by first determining a from Equation 2.65

$$a = 6 + \frac{6}{4} = 7.5 \tag{2.66}$$

Now the balanced equation is found using Equation 2.64:

$$C_6H_6 + 7.5(O_2 + 3.76N_2) \rightarrow 6CO_2 + 3H_2O + 28.2N_2 \tag{2.67}$$

This is an example of a stoichiometrically balanced equation using a molar basis. There are times when a particular fuel is burned with too much air (overoxidized) or too little air (underoxidized). The latter is usually the case with propellants in the chamber of a gun. When a fuel is overoxidized, we usually categorize it by stating how much excess air is included in the reaction. For instance, 50% excess air used in the reaction of Equation 2.67 would alter the balanced equation to be written as

$$C_6H_6 + (1.5)(7.5)(O_2 + 3.76N_2) \rightarrow 6CO_2 + 3H_2O + 3.75O_2 + 42.3N_2 \tag{2.68}$$

If the fuel were burned with 50% deficient air, we would have

$$C_6H_6 + (0.5)(7.5)(O_2 + 3.76N_2) \rightarrow 4.5CO + 3H_2O + 1.5C + 14.1N_2 \tag{2.69}$$

In this case, we have used the rules set forth earlier where steam is formed first then carbon monoxide. At this point, all the oxygen has been used up, so solid carbon is formed. From this simple example, you can see that the amount of air used in the combustion is critical to determination of the products.

We can now define an air–fuel ratio as the ratio mass of air combusted to the mass of fuel combusted. This is given mathematically by

$$A-F = \frac{m_{\text{air}}}{m_{\text{fuel}}} = \frac{\dot{m}_{\text{air}}}{\dot{m}_{\text{fuel}}} \tag{2.70}$$

If we continue using our three examples, we could find the mass fuel ratio for each of the reactions defined in Equations 2.67 through 2.69. If we note here that the molar mass of benzene is 78.11 lbm/lb mol and the molar mass of air is 28.97 lbm/lb mol, we have for the stoichiometric reaction

$$A-F_{\text{Stoich}} = \frac{(7.5)\left[\text{mol}_{\text{air}}\right](4.76)(28.97)\left[\frac{\text{lbm}}{\text{lb mol}}\right]}{(1)\left[\text{mol}_{C_6H_6}\right](78.11)\left[\frac{\text{lbm}}{\text{lb mol}}\right]} = 13.24\left[\frac{\text{lbm}_{\text{air}}}{\text{lbm}_{C_6H_6}}\right] = 13.24 \tag{2.71}$$

For the reaction with 50% excess air, we have

$$A-F_{50\%\text{excess}} = \frac{(1.5)(7.5)\left[\text{mol}_{\text{air}}\right](4.76)(28.97)\left[\frac{\text{lbm}}{\text{lb mol}}\right]}{(1)\left[\text{mol}_{C_6H_6}\right](78.11)\left[\frac{\text{lbm}}{\text{lb mol}}\right]} = 19.85\left[\frac{\text{lbm}_{\text{air}}}{\text{lbm}_{C_6H_6}}\right]$$

$$= 19.85\left[\text{mol}_{\text{air}}\right] \tag{2.72}$$

For the reaction with 50% deficient air, we have

$$A-F_{50\%\text{deficient}} = \frac{(0.5)(7.5)\left[\text{mol}_{\text{air}}\right](4.76)(28.97)\left[\frac{\text{lbm}}{\text{lb mol}}\right]}{(1)\left[\text{mol}_{C_6H_6}\right](78.11)\left[\frac{\text{lbm}}{\text{lb mol}}\right]} = 6.61\left[\frac{\text{lbm}_{\text{air}}}{\text{lbm}_{C_6H_6}}\right] = 6.61 \tag{2.73}$$

Now that we have introduced the process of chemical equation balancing and some of the mathematics required, we must quantify the energy released (or absorbed) by the chemical reaction. We have already introduced the concept of enthalpy as well as defined the enthalpy of formation. We shall pause here to examine how a heat of formation is obtained.

We shall consider carbon dioxide for our example. If we have a combustion chamber in which we react pure oxygen with solid carbon, we can put the two substances into the container at 25°C and start the reaction somehow. The balanced equation on a molar basis would be

$$C(s) + O_2 \rightarrow CO_2 \tag{2.74}$$

The first law of thermodynamics states that

$$Q + W = N_{\text{products}}\bar{h}_{\text{products}} - N_{\text{reactants}}\bar{h}_{\text{reactants}} \tag{2.75}$$

Here we have used specific values so that everything is on a molar basis. Since the container is rigid, there is no work performed on or by the system; thus, Equation 2.75 reduces to

$$Q = N_{\text{products}}\bar{h}_{\text{products}} - N_{\text{reactants}}\bar{h}_{\text{reactants}} \tag{2.76}$$

If we were to perform this experiment, we would find that the container would get hot. Theoretically, we could extract this heat from the container until the temperature returned to 25°C; if we were to do this, we would find that 393,546 kJ/kg mol of energy would have been produced. An examination of Table B.1 reveals that this is exactly the value of the heat of formation of carbon dioxide recalling that a negative value denotes heat given off by the reaction.

The enthalpy of a substance allows us to quantify the energy state of a material. The enthalpy of formation was defined as the energy required to form a particular composition from its basic elements resulting in the compound as a product at some reference temperature and pressure (we shall use 25°C or 298 K and 1 atm as this reference condition). If we were to take this compound and arbitrarily increase its temperature or pressure by some amount and if there were no phase change or change in composition, we will have increased its enthalpy. If we restrict our analysis to an ideal gas, it can be shown [1] that the enthalpy is a function of temperature only. With this, we can write for a composition

$$\bar{h}_T = \bar{h}_f^0 + \Delta\bar{h}_{298\to T} \tag{2.77}$$

where $\bar{h}_T$ is the enthalpy of the material at temperature T; $\bar{h}_f^0$ is the enthalpy of formation; and $\Delta\bar{h}_{298\to T}$ is the change in enthalpy from the reference state to the temperature T.

We define $\Delta\bar{h}_{298\to T}$ as

$$\Delta\bar{h}_{298\to T} = \bar{h}(T) - \left(\bar{h}_{298}^0\right) \tag{2.78}$$

Tables of enthalpies are located in Appendix B at the end of the book. As an example, consider carbon monoxide at 2000 K. The enthalpy of this compound using Tables B.1 and B.2 would be

$$h_{CO_{2000K}} = -110{,}541\left[\frac{\text{kJ}}{\text{kg mol}}\right] + 56{,}737\left[\frac{\text{kJ}}{\text{kg mol}}\right] = -53{,}804\left[\frac{\text{kJ}}{\text{kg mol}}\right] \tag{2.79}$$

Now that we have worked with enthalpies a bit, we can begin to apply what we have learned. We shall look at an example of these principles applied first to a closed bomb where there is no work performed and then to a gun where there is.

For a closed bomb, we shall tailor Equation 2.42 to our needs. If we consider a closed vessel, we realize that there is no velocity into or out of the CV, and there is no work performed on or by the system. This allows us to write Equation 2.42 as

$$Q_{1-2} = m[(h_2 - pv_2) - (h_1 - pv_1)] = m(u_2 - u_1) \tag{2.80}$$

If we write this equation on a molar basis as limit to ideal gas behavior, we can state that

$$Q = \sum_i N_i\left(\bar{h}_{\text{prod}} - R_u T_{\text{prod}}\right) - \sum_i N_i\left(\bar{h}_{\text{reac}} - R_u T_{\text{reac}}\right) \tag{2.81}$$

This relationship is important because it tells us that the heat given off by the closed bomb is affected by the enthalpy change of the chemical reaction and the temperature of the products.

We shall examine a pressure vessel containing 0.001 kg of methane (CH_4) and 0.002 kg of air. The enthalpy of formation for methane is –74,850 kJ/kg mol and its molecular weight is 16.04 kg/kg mol. The reaction will begin at 298 K, and we shall remove enough heat from the vessel that the final temperature becomes 1500 K. We would like to determine how much heat is given off.

We need to balance the chemical reaction on a molar basis, so we shall determine how many moles of methane and air we have in the container. For methane, we have

$$N_{CH_4} = \frac{(0.001)\left[kg_{CH_4}\right]}{(16.04)\left[\frac{kg}{kg\,mol}\right]} = 6.23 \times 10^{-5}\left[kg\,mol_{CH_4}\right] \tag{2.82}$$

For the air, we have

$$N_{air} = \frac{(0.002)\left[kg_{air}\right]}{(28.97)\left[\frac{kg}{kg\,mol}\right]} = 6.90 \times 10^{-5}\left[kg\,mol_{air}\right] \tag{2.83}$$

Our balanced reaction is then

$$\begin{aligned}&(6.23 \times 10^{-5})CH_4 + (6.90 \times 10^{-5})((0.21)O_2 + (0.79)N_2)\\ &\rightarrow (2.9 \times 10^{-5})H_2O + (9.56 \times 10^{-5})H_2 + (6.23 \times 10^{-5})C(s) + (5.45 \times 10^{-5})N_2\end{aligned} \tag{2.84}$$

We shall examine the reactants first. For the methane, we have

$$\begin{aligned}&N_{CH_4}\left(\bar{h}_f^0 + \Delta\bar{h}_{298\to T} - R_u T_{CH_4}\right)\\ &= \left(6.23\times10^{-5}\right)\left[kg\ mol\right]\left(-74{,}850\left[\frac{kJ}{kg\ mol}\right] + 0 - (8.314)\left[\frac{kJ}{kg\ mol\ K}\right](298)[K]\right)\\ &N_{CH_4}\left(\bar{h}_f^0 + \Delta\bar{h}_{298\to T} - R_u T_{CH_4}\right) = -4.82[kJ]\end{aligned}$$

For the oxygen and nitrogen, we have

$$\begin{aligned}&N_{N_2}\left(\bar{h}_f^0 + \Delta\bar{h}_{298\to T} - R_u T_{N_2}\right) = (0.79)\left(6.90\times10^{-5}\right)\left[kg\ mol\right]\\ &\left(0 + 0 - (8.314)\left[\frac{kJ}{kg\ mol\cdot K}\right](298)[K]\right)\end{aligned}$$

$$N_{N_2}\left(\bar{h}_f^0 + \Delta\bar{h}_{298\to T} - R_u T_{N_2}\right) = -0.135[kJ]$$

The enthalpies of the reactants are therefore

$$\sum_i N_i\left(\bar{h}_{reac} - R_u T_{reac}\right) = -4.82\,[kJ] - 0.036\,[kJ] - 0.135\,[kJ] = -4.99\,[kJ]$$

For the products, we have (using the tables in Appendix B)

$$N_{H_2O}\left(\bar{h}_f^0 + \Delta\bar{h}_{298\to T} - R_u T_{H_2O}\right)$$
$$= \left(2.9\times10^{-5}\right)[\text{kg mol}]\left(-241{,}845 + 48{,}181 - (8.314)\left[\frac{\text{kJ}}{\text{kg mol K}}\right](1{,}500)[\text{K}]\right)$$

$$N_{H_2O}\left(\bar{h}_f^0 + \Delta\bar{h}_{298\to T} - R_u T_{H_2O}\right) = -5.98[\text{kJ}]$$
$$= \left(9.56\times10^{-5}\right)[\text{kg-mol}]\left(0 + 36{,}307 - (8.314)\left[\frac{\text{kJ}}{\text{kg-mol}\cdot\text{K}}\right](1{,}500)[\text{K}]\right)$$
$$N_{H_2}\left(\bar{h}_f^0 + \Delta\bar{h}_{298\to T} - R_u T_{H_2}\right) = +2.28[\text{kJ}]$$

$$N_{C}\left(\bar{h}_f^0 + \Delta\bar{h}_{298\to T} - R_u T_{C}\right)$$
$$= \left(6.23\times10^{-5}\right)[\text{kg mol}]\left(0 + 23{,}253\left[\frac{\text{kJ}}{\text{kg mol}}\right] - (8.314)\left[\frac{\text{kJ}}{\text{kg mol K}}\right](1{,}500)[\text{K}]\right)$$

$$N_{N_2}\left(\bar{h}_f^0 + \Delta\bar{h}_{298\to T} - R_u T_{N_2}\right)$$
$$= \left(5.45\times10^{-5}\right)[\text{kg mol}]\left(0 + 38{,}404\left[\frac{\text{kJ}}{\text{kg mol}}\right] - (8.314)\frac{\text{kJ}}{\text{kg mol K}}(1{,}500)[\text{K}]\right)$$

The enthalpies of the products are then given by

$$\sum_i N_i\left(\bar{h}_{\text{prod}} - R_u T_{\text{prod}}\right) = -5.98\,[\text{kJ}] + 2.28\,[\text{kJ}] + 0.67\,[\text{kJ}] + 1.41\,[\text{kJ}] = -1.62\,[\text{kJ}]$$

The heat given off by the reaction is then calculated through Equation 2.81 as

$$Q = (-1.62)\,[\text{kJ}] - (-4.99)\,[\text{kJ}] = +3.37\,[\text{kJ}] \tag{2.85}$$

This illustrates the process of calculating the amount of energy given off by a closed bomb reaction as well as the effect of temperature on the reaction products. In this case, energy has to be added to get the reaction to go to completion. It must be noted that had we decided to lower the temperature of the products, energy would eventually have to have been removed. This will be examined as a problem at the end of the chapter.

If we apply the same principles to a gun launch, we can determine the amount of energy imparted to the projectile and, in so doing, obtain a feeling for the process of energy conversion between propellant chemical energy and projectile kinetic energy.

Unlike the fixed boundary examined in the closed-bomb problem, a gun launch involves a boundary that is moving (the base of the projectile). This problem is similar to a piston of an internal combustion engine that undergoes one stroke. We have defined work earlier as a form of energy, and if we assume that all the energy of the propellant goes into heating of the gaseous products, kinetic energy of the projectile, and a loss term (including friction and swelling of the gun tube), we can write the first law of thermodynamics as given in Equation 2.75. Rewriting this by assuming the velocity of the seated projectile is zero, we obtain our thermodynamic equation for a gun launch as

$$Q + \frac{1}{2}mV^2 = \sum_i N_i\left(\bar{h}_{\text{prod}}\right) - \sum_i N_i\left(\bar{h}_{\text{reac}}\right) + \text{losses} \tag{2.86}$$

We have neglected potential energy changes here because they are usually quite small relative to the other terms. We shall examine an example in the form of a potato gun to illustrate the use of Equation 2.86 and the other methods of this chapter.

A potato gun is a device that people use to project potatoes at targets. These devices can be very dangerous to the operator as well as the target. We would like to calculate the muzzle velocity of a half-pound potato projectile used in a particular gun. This gun is made of 2 in. diameter polyvinyl chloride pipe (a very good insulator). The projectile rests on a stop when loaded through the muzzle so that there is a 6 in. long chamber. The device in question was injected with 0.005 oz. (mass) of lighter fluid as a gas (*n*-butane—C_4H_{10} (g) $\bar{h}_f^0 = -124,733$ kJ/kg mol, $n = 58.123$ kg/kg mol) to fire the potato. We shall assume the potato perfectly obturates and that there is no bore friction. The travel of the potato in the gun tube is 24 in. The weapon is fired under standard conditions of 77°F and 14.7 psi. Assume that the reactants and the products both exist at these conditions. We would like to determine the velocity of the potato at the completion of combustion in feet per second assuming no losses. Note that the combustion here is not modeled and likely is idealized like an air standard Otto instantaneous combustion resulting in the mean value of the pressure as the potato exits the tube. Refer to Problem 8 as well.

The chamber was 6 in. long and 2 in. in diameter, so our chamber volume is

$$\text{V}_i = Al = \pi\frac{(2)^2}{4}\left[\text{in.}^2\right](6)\left[\text{in .}\right] = 18.85\left[\text{in.}^3\right] \tag{2.87}$$

The air weighs 28.97 lbm/lb mol, and if we assume ideal gas behavior, the density of air is calculated from

$$pv = RT \rightarrow \rho = \frac{p}{RT} \tag{2.88}$$

$$\rho = \frac{(14.7)\left[\frac{\text{lbf}}{\text{in.}^2}\right](28.97)\left[\frac{\text{lbm}}{\text{lb mol}}\right]}{(1545)\left[\frac{\text{ft lbf}}{\text{lb mol } R}\right](12)\left[\frac{\text{in .}}{\text{ft}}\right](537)\left[\text{R}\right]} = 0.0000428\left[\frac{\text{lbm}}{\text{in.}^3}\right]$$

So the amount of air we actually have is

$$m_{\text{air}} = \rho\text{V}_i = (0.0000428)\left[\frac{\text{lbm}}{\text{in.}^3}\right](18.85)\left[\text{in.}^3\right] = 0.0008068\left[\text{lbm}\right] \tag{2.89}$$

The amount of fuel was given in ounces:

$$m_{\text{fuel}} = (0.005)\left[\text{oz}\right](0.0625)\left[\frac{\text{lbm}}{\text{oz}}\right] = 0.0003125\left[\text{lbm}\right]$$

For the actual combustion, we need to use our mass information and convert it to molar values, recognizing that the molar mass is the same whether it is in kilograms per kilogram-mole or pound-mass/pound-mole. For the fuel and air, we have

$$N_{\text{fuel}} = \frac{m_{\text{fuel}}}{n_{\text{fuel}}} = (0.0003125)\,[\text{lbm}]\,\frac{1}{(58.123)\left[\dfrac{\text{lbm}}{\text{lb mol}}\right]} = 0.0000054\,[\text{lb mol}] \tag{2.90}$$

$$N_{\text{air}} = \frac{m_{\text{air}}}{n_{\text{air}}} = (0.0008068)\,[\text{lbm}]\,\frac{1}{(28.97)\left[\dfrac{\text{lbm}}{\text{lb mol}}\right]} = 0.0000278\,[\text{lb mol}] \tag{2.91}$$

For each pound-mole of air, we know that 1/4.76 lb mol of it is oxygen, so we have

$$N_{O_2} = \frac{1}{4.76}(0.0000278)\,[\text{lb mol}] = 0.0000058\,[\text{lb mol}]$$

$$N_{N_2} = \frac{3.76}{4.76}(0.0000278)\,[\text{lb mol}] = 0.0000220\,[\text{lb mol}]$$

Now we can write our combustion equation as

$$(0.0000054)C_4H_{10}(g) + (0.0000058)O_2 + (0.0000220)N_2 \rightarrow$$

$$(0.0000116)H_2O + (0.0000216)C + (0.0000154)H_2 + (0.0000220)N_2$$

To determine the muzzle velocity, we start with our first law of thermodynamics equation, simplified by the fact that there is no heat transfer and no shaft work. Then the energy of the fuel–air mixture equals the work done on the projectile plus the energy of the products of combustion.

$$H_R = H_p + W_p \tag{2.92}$$

Let us look at the internal energies for each of the reactants:

Reactant	Enthalpy of Formation (kJ/kg mol)	Enthalpy of Formation (in. lbf/lb mol)
C_4H_{10} (*g*)	–124,733	–500,728,155
O_2	0	0
N_2	0	0

The conversion used here is as follows:

$$(x)\left[\frac{\text{kJ}}{\text{kg mol}}\right](0.4299)\left[\frac{\dfrac{\text{BTU}}{\text{lb mol}}}{\dfrac{\text{kJ}}{\text{kg mol}}}\right](778.16)\left[\frac{\text{ft lbf}}{\text{BTU}}\right](12)\left[\frac{\text{in.}}{\text{ft}}\right] \rightarrow 4014.4x\left[\frac{\text{in. lbf}}{\text{lb mol}}\right] \tag{2.93}$$

For the products, we have

Product	Enthalpy of Formation (kJ/kg mol)	Enthalpy of Formation (in. lbf/lb mol)
H_2O (*g*)	−241,845	−970,862,568
N_2	0	0
C_2	0	0
H_2	0	0

We will rearrange our first law equation as follows:

$$W_p = H_R - H_p$$

We calculate H_R first:

$$H_R = N_{C_4H_{10}}\left(\bar{h}_f^0 + \Delta\bar{h}_{298\to T}\right) + N_{O_2}\left(\bar{h}_f^0 + \Delta\bar{h}_{298\to T}\right) + N_{N_2}\left(\bar{h}_f^0 + \Delta\bar{h}_{298\to T}\right)$$

Plugging in the numbers we have, we get

$$H_R = (0.0000054)\,[\text{lb mol}]\,(-500,728,155 + 0)\left[\frac{\text{in. lbf}}{\text{lb mol}}\right]$$

$$+ (0.0000058)\,[\text{lb mol}]\,(0 + 0)\left[\frac{\text{in. lbf}}{\text{lb mol}}\right]$$

$$+ (0.0000220)\,[\text{lb mol}]\,(0 + 0)\left[\frac{\text{in. lbf}}{\text{lb mol}}\right]$$

$$H_R = -2704\,[\text{in. lbf}]$$

We calculate H_p in a similar manner:

$$H_p = N_{H_2O}\left(\bar{h}_f^0 + \Delta\bar{h}_{298\to T}\right) + N_{H_2}\left(\bar{h}_f^0 + \Delta\bar{h}_{298\to T}\right) + N_{N_2}\left(\bar{h}_f^0 + \Delta\bar{h}_{298\to T}\right) + N_C\left(\bar{h}_f^0 + \Delta\bar{h}_{298\to T}\right)$$

$$H_p = (0.0000116)\,[\text{lb mol}]\,(-970,862,568 + 0)\left[\frac{\text{in. lbf}}{\text{lb mol}}\right]$$

$$+ (0.0000154)\,[\text{lb mol}]\,(0 + 0)\left[\frac{\text{in. lbf}}{\text{lb mol}}\right]$$

$$+ (0.0000220)\,[\text{lb mol}]\,(0 + 0)\left[\frac{\text{in. lbf}}{\text{lb mol}}\right]$$

$$+ (0.0000216)\,[\text{lb mol}]\,(0 + 0)\left[\frac{\text{in. lbf}}{\text{lb mol}}\right]$$

$$H_p = -11,262\,[\text{in. lbf}]$$

Then the work done on the projectile is

$$W_p = -2704\,[\text{in. lbf}] - (-11,262)\,[\text{in. lbf}] = 8558\,[\text{in. lbf}]$$

Since this work equals the muzzle energy of the projectile,

$$W_p = \frac{1}{2} m V^2 = 8,558 \text{ [in. lbf]}$$

Therefore,

$$V = \sqrt{\frac{2W_p}{m}} = \sqrt{\frac{(2)(8558)\text{ [in. lbf] }(32.2)\left[\frac{\text{lbm ft}}{\text{lbf s}^2}\right]}{(0.5)\text{ [lbm]}(12)\left[\frac{\text{in.}}{\text{ft}}\right]}} = 303\left[\frac{\text{ft}}{\text{s}}\right]$$

Wow! That is pretty fast, but we used a lot of butane, assumed that the products return to ambient conditions quickly, and neglected things. Also, note that the length of the tube did not come into play. We would definitely have to account for this as we shall later see.

One important parameter in determining the amount of energy transferred to the projectile is the temperature of the product gases. As you can see from our example, an increase in the temperature of the product gases will result in a decrease in the projectile velocity because H_p goes up. Typically, we can assume the product gases exit at a temperature between $0.6T_0$ and $0.7T_0$, where T_0 is the adiabatic flame temperature of the product gases [7]. The adiabatic flame temperature of a gas is the temperature that is achieved if the gases burn to completion in the absence of any heat transfer or work being performed [1]. The calculation of the adiabatic flame temperature is relatively straightforward but requires iteration. This is beyond the scope of this chapter, but the reader is referred to the references at the end of this chapter for a complete description of the procedure. In addition, there are several commercially available codes (including some that come with the purchase of textbooks now, for instance, the book by Cengel and Boles [13]). To achieve our objectives, the temperature of the reaction products will always be given.

Problem 6

Calculate the A–F ratio for the combustion of the following fuels. Calculate the ratio with both theoretical air and 10% excess air.

1. Benzene (C_6H_6)
   Answer: 13.24 and 14.56
2. *n*-Butane (C_4H_{10})
   Answer: 15.42 and 16.96x
3. Ethyl alcohol (C_2H_5OH)
   Answer: 8.98 and 9.88

Problem 7

Let us examine a pressure vessel identical to the example problem in the text containing 0.001 kg of methane (CH_4) and 0.002 kg of air. The enthalpy of formation for methane is –74,850 kJ/kg mol and its molecular weight is 16.04 kg/kg mol. The reaction will begin at 298 K, and we shall remove enough heat from the vessel so that the final temperature becomes 1000 K.

1. Determine the maximum heat given off.
   Answer: $Q = +0.606$ [kJ].
2. Compare the result in question 1 with the example problem in this chapter.
   Answer: This situation removes 2.764 kJ more energy than the example.

Problem 8

A really interesting person takes the tennis ball mortar we built in Problem 3 and modifies it—squirting in and igniting 0.003 oz. of acetylene gas (C_2H_2 (g)). If we assume that the combustion kinetics is fast enough such that the energy release occurs before the ball can move, we want to determine the muzzle velocity of the tennis ball. Proceed along the following steps:

1. Balance the stoichiometric reaction equation for acetylene.
2. Balance the actual equation neglecting the volume the acetylene occupies in the chamber. Assume the air initially in the chamber is at 14.7 psia and 77°F.
3. Determine the increase in internal energy of the gas.
4. Assuming the gas is calorically perfect ($\Delta U = m_g c_v \Delta T$) and that $c_v = 0.33$ BTU/lbm R for the mixture, determine the increase in temperature of the gas.

Note: you will have to do 3 and 4 by iteration, first assuming a final reaction temperature, carrying out the calculation for ΔU, and seeing if the ΔT you get matches—if not, iterate again once you get an answer within say 10% that is good enough.

5. Based on the result of 4, determine the initial pressure on the tennis ball assuming the specific gas constant of the products is $R = 80$ ft lbf/lbm R.
6. Use the result of 5 and possibly your results from Problem 2 to determine the muzzle velocity of the tennis ball. Assume $\gamma = 1.4$.
7. Determine the temperature of the gases at shot exit.

For acetylene, $n = 26.038$ lbm/lb mol and $\Delta h_f^0 = +97,477$ BTU/lb mol.

Problem 9

A potato is stuffed into the 3 in. diameter exhaust pipe of a car that is not running too well. Incompletely combusted combustion products (assume gaseous heptane) (1 g) mixes with a stoichiometric amount of air behind the potato and ignites. If the potato is wedged 4 in. into the exhaust (i.e., it has 4 in. of travel) and weighs 0.25 lbm, and assuming the combustion takes place before the potato moves, determine the theoretical maximum "muzzle" velocity of the potato. Also, calculate the muzzle velocity assuming an isentropic expansion. For gas expansion purposes, you can assume that the volume available initially behind the potato is equal to the volume between the point of obturation and the end of the exhaust and assume a "smeared" specific heat ratio of 1.3 for the product gases. Also assume that the combustion begins at 500 K and completes at 1500 K. Assume that the total enthalpy at 500 K for n-heptane (C_7H_{16}) is −120,000 kJ/kg mol.

Problem 10

Assume we have a quantity of 29 lb of 11.1% nitrated nitrocellulose ($C_6H_8N_2O_9$), and it is placed in an empty chamber of a gun at 77°F and 14.7 psia. The chamber is 1160 in.^{3} in volume. The propellant density is 0.060 lbm/in.^{3} If the air in the chamber is *not* neglected and assuming the volume is fixed,

1. Write the balanced equation for this combustion (assume that the oxygen goes preferentially into CO_2 instead of CO this time—you will find the difference later).
2. Using the tables in Appendix B, estimate the adiabatic flame temperature of the resultant gas (*Hint*: recall the definition of adiabatic flame temperature).

2.6 Solid Propellant Combustion

Now that we have examined the background of the thermochemistry and thermodynamics of combustion, we shall see how this applies to the behavior of a burning solid propellant. We shall endeavor, in this section, to come up with definitions and relationships that will allow us to define the state of the propellant behind a projectile at any given time. The process we will use is somewhat simplified because the real situation behind a moving projectile is generally a two-phase, reacting flow field. Some of our assumptions, although not necessarily valid in the purest sense, are good enough to predict bulk behavior of the propelling gas.

In the previous sections, we have discussed how energy is evolved by the propellant. We saw that thermodynamic properties were not dynamic at all, but merely means of accounting for energy knowing the initial and end states and making assumptions on the process between them. This section will allow us to add in some time dependency to the equations to somewhat understand the rates at which combustion is occurring.

Solid propellants are generally nitrocellulose compounds that are manufactured by nitrating through immersion in acid. The details of this process for various materials can be examined in detail in the books by Corner [7], Cooper [8], Cooper and Kurowski [16], Hayes [17], and Eringen et al. [18]. This material is then chopped and worked into a doughy substance and pushed through dies to form various shapes. The material then has solvents removed, and it is dried. When this process is complete, the propellant has the consistency of uncooked (i.e., hard and somewhat brittle) pasta. Although this statement is general, there are, as always, exceptions.

The burning of solid propellant is a surface phenomenon. The rate of gas evolution is dependent upon the amount of surface area of the propellant. Because of this, the shape that the propellant takes is extremely important. Burning is the mechanism of transforming the solid propellant to a gas. The burn rate of a propellant is highly dependent upon the pressure at which the burning reaction takes place. Essentially, the greater the pressure, the faster the propellant burns. These two behavioral observations tell us that if we can control the geometry and confinement of a given propellant, we can, to a large degree, control the rate of gas evolution.

We shall examine a single propellant grain to gain an understanding of how the geometry affects the rate of evolution of gas. Consider a long cylinder of solid propellant that is commonly referred to as a grain. If the cylinder were long enough, we could see that most of the surface area would be located along the circumference and length. In other words, we can neglect the two small surface areas that comprise the ends. This is illustrated in Figure 2.8. If we neglect the burning of the end surfaces, it allows us to examine the geometry through simple mathematical relationships.

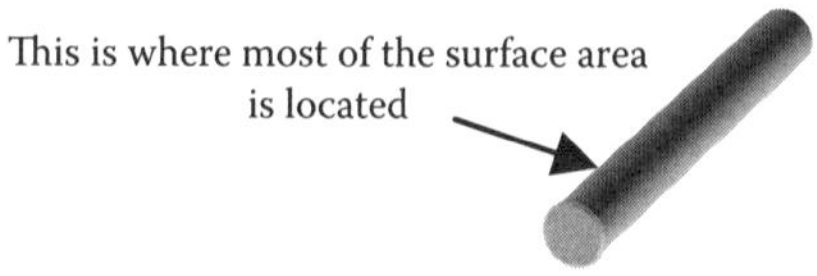

FIGURE 2.8
Long cylindrical propellant grain.

As our grain begins to burn, solid material will be evolved into gas. Thus, we can imagine the solid surfaces shrinking toward the centerline of the grain. If we examine our grain from the end looking down its axis, we would see a circular section as depicted in Figure 2.9. We could then write an expression for the surface area of our grain as a function of its diameter and length:

$$A(t) = \pi d(t) l \tag{2.94}$$

In this expression, $A(t)$ is the surface area of the grain, $d(t)$ is the diameter, and l is the length. We have denoted the surface area and diameter as functions of time to remind us of our assumption of no burning at the ends of the grain. After some time t, the grain surface will have regressed such that our diameter has decreased. This is depicted in Figure 2.10. This graphically shows us that at time t_1, the grain clearly has more surface

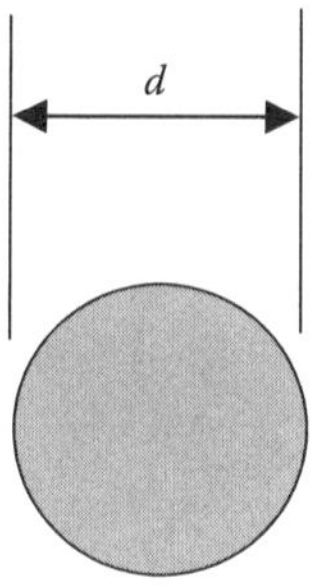

FIGURE 2.9
Propellant grain cross section.

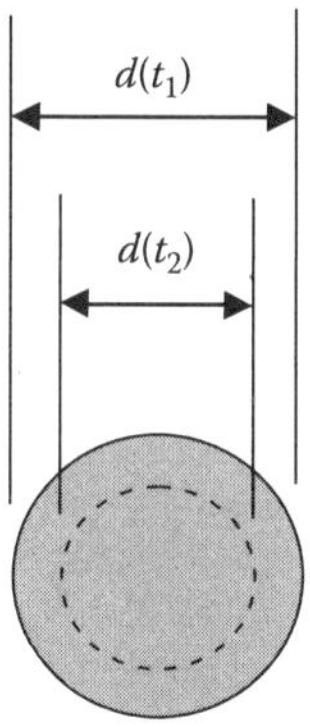

FIGURE 2.10
Propellant grain cross section at two times.

area than at time t_2; therefore, as burning progresses, the rate of evolution of gas slows down. This is commonly called regressive burning.

Propellant geometry is characterized by a quantity known as the web thickness or simply the web. The symbol used for the web is D. The web is the smallest thickness of the initial propellant grain. In the case of our cylindrical grain, it would be the initial diameter.

In the interior ballistics analysis of a gun system, we need to track how much gas is evolved and how much solid is remaining. This is important because we have seen that all our equations of state are dependent upon volume as well as pressure and temperature, and these, in turn, affect the burning rate. The amount of solid propellant remaining is tracked through use of the web fraction f. The web fraction is the fraction of web remaining at a given time t. Through use of this web fraction, we can write an expression for the amount of propellant remaining at any time as a function of the web:

$$d(t) = fD \tag{2.95}$$

This is illustrated for a grain with a single perforation (colloquially known as a perf) in Figure 2.11. It is important to note here that for a single perf grain, the web is defined as the outside radius minus the inside radius. This is sometimes confusing for new ballisticians since we use D as the web thickness. Also one can see from the figure that an advantage of a single perf grain is that it burns from both the inside out and the outside in, thus decreasing the surface on the outside while increasing the surface on the inside—known as neutral burning behavior.

The use of the web fraction is convenient because, mathematically, it is a function that varies from unity to zero. The manner in which it varies may be somewhat complex, but at least, the end states are well defined. An example plot of web fraction vs. time is shown in Figure 2.12. In this figure, t_B is the time at which all the propellants have evolved into gas—the burnout time.

Many times, we are more interested in the volume of the propellant that has evolved into gas rather than the fraction of the web remaining. It should be clear that the two quantities are related since the gas had to come from the solid material and conservation of mass states that we can neither destroy nor create mass. This is handled through use of the fraction of propellant burnt ϕ. Since ϕ is a function of f and f is a function of time, we see that ϕ must also be a function of time. Since propellant geometries can be fairly complicated, ϕ can be a rather complicated function of f. For simple shapes, this relationship is straightforward. For instance, a single perforated grain has the functional relationship that

$$\phi(t) = 1 - f(t) \tag{2.96}$$

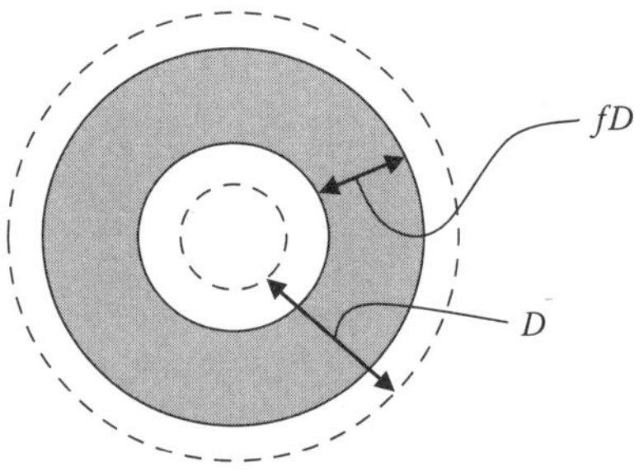

FIGURE 2.11
Burning of a single perforated propellant grain.

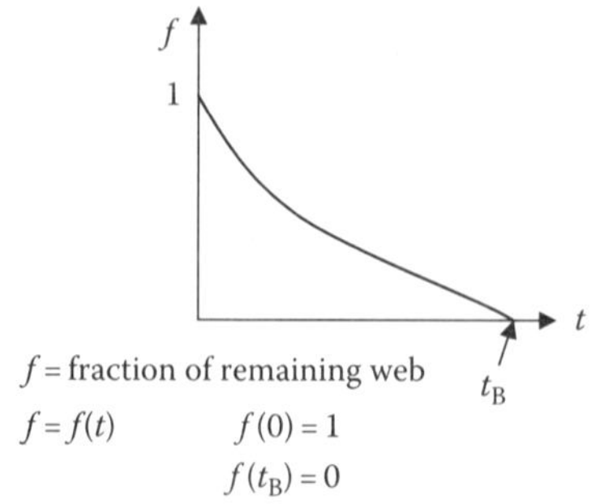

FIGURE 2.12
Fraction of remaining web vs. time.

Most shapes can be simplified to express ϕ as a quadratic function of f through use of a shape function θ:

$$\phi(t) = [1 - f(t)][1 + \theta f(t)] \tag{2.97}$$

This expression allows us to cover almost any simple geometry, the most notable exception being a sphere. Figure 2.13 depicts how variation in the shape function affects the relationship between f and ϕ.

With the formulations earlier, we have been able to mathematically define the effect of propellant geometry on the rate of gas evolution. The second important parameter in this generation of gas was stated to be the effect of pressure on burning. Whenever a propellant burns, say in a fixed volume, two competing processes are happening: the volume into which the gaseous propellant is moving is increasing because there is less solid material—this decreases the pressure—and the more and more propellant gas is being pushed into a confined space—this increases the pressure. The rate at which the surface area decreases

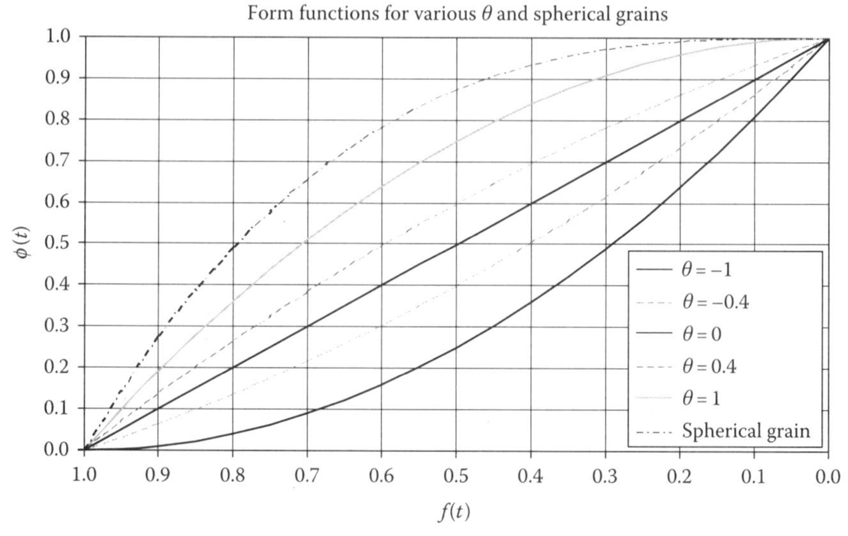

FIGURE 2.13
Effect of different values of θ on ϕ and f.

affects this relationship. The simplest model for the relationship between burn rate and pressure is given by

$$D\frac{df}{dt} = -\beta p_B(t) \tag{2.98}$$

where $D(df/dt)$ is the time rate of change of the web (i.e., the burning rate); β is a burn rate coefficient; and p_B is the pressure (we will discuss the subscript later).

The negative sign comes about because the amount of propellant would be increasing if $D(df/dt)$ were to result in a positive number. This simple relationship facilitates our analysis of propellant behavior in a gun. Other relationships can more accurately describe propellant behavior, but their complexity is such that computer codes must be used to obtain answers with them. Two very common burn relationships are

$$D\frac{df}{dt} = -\beta[p_B(t)]^{\alpha} \tag{2.99}$$

$$D\frac{df}{dt} = -\beta[p_B(t) - P_1] \tag{2.100}$$

Equation 2.99 is by far the most commonly used in computer codes. Caution must be exercised when using burn rate data from the literature as the units will be an indicator of the proper burn rate form of the governing equation. If we examine the units of $D\,df/dt$, we see that they are in terms of [length]/[time]. This type of data is usually obtained from a strand burner. A strand burner is a device that can accurately measure the rate of linear burning in a propellant. The book by Kubota [19] contains an excellent diagram of a strand burner.

If we consider a pressure vessel so thick as to be rigid and the amount of propellant so small such that we can neglect its contribution to the volume, we can describe the burning of the propellant as a constant-volume process. This is the essence of closed-bomb testing. We can further assume that this pressure vessel can be thermally isolated and the gas behavior is ideal. In this case, our closed bomb, with internal volume V, would resemble Figure 2.14. Since we assumed ideal gas behavior, we can write an expression for pressure as a function of volume and temperature:

$$p_B V = m_g R T \tag{2.101}$$

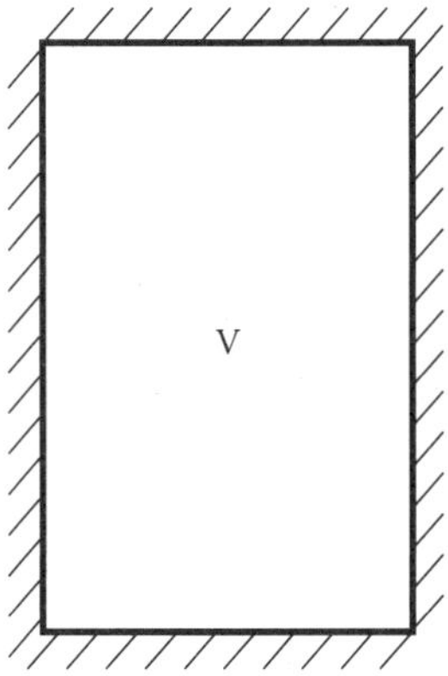

FIGURE 2.14
Diagram of a closed bomb.

where p_B is the pressure; V is the volume; m_g is the mass of the gas; R is the specific gas constant; and T is the temperature.

When we place our solid propellant into the closed bomb, it has an initial weight that we would call c. So initially, we can write

$$c = \rho V_{solid} \tag{2.102}$$

where ρ is the density of the solid propellant and V_{solid} is its volume.

If we now assume that the propellant is cylindrical, we can write its volume as the product of its cross-sectional area and its length. The initial diameter is the web for a cylindrical grain, so

$$V_{cyl.grain} = \pi \frac{D^2}{4} l \tag{2.103}$$

This volume at any time t can be expressed as

$$V(t)_{cyl.grain} = \pi \frac{[d(t)]^2}{4} l \tag{2.104}$$

Because mass is conserved, the amount of solid propellant burned is equal to the amount of gas generated. This is an important concept. If we started with 1 lbm of propellant and completely burned it, we would be left with 1 lbm of gas. Based on this, we can write for the mass of the gas as

$$m_g(t) = \rho\left[V_{cyl.grain} - V(t)_{cyl.grain}\right] = \rho \frac{\pi}{4} l \{D^2 - [d(t)]^2\} \tag{2.105}$$

We discussed the fraction of propellant burnt ϕ earlier. We are now in a position to formally define it as

$$\phi(t) \equiv \frac{m_g(t)}{c} \tag{2.106}$$

If we substitute Equations 2.102, 2.103, and 2.105 into Equation 2.106, we obtain

$$\phi(t) = \frac{\rho \frac{\pi}{4} l \{D^2 - [d(t)]^2\}}{\rho \frac{\pi}{4} l D^2} = \left\{1 - \frac{[d(t)]^2}{D^2}\right\} \tag{2.107}$$

Now we insert Equation 2.95 into Equation 2.107 to yield

$$\phi(t) = 1 - \left(\frac{fD}{D}\right)^2 = 1 - f^2 = (1 + f)(1 - \theta f) \tag{2.108}$$

Comparing this expression (derived for a cylindrical grain) to Equation 2.97 shows that the shape factor $\theta = 1$ for a cylindrical grain. Also, by comparison to Equation 2.96, we see that the shape factor $\theta = 0$ for a single perforated grain. Essentially, any shape factor can be derived using this same procedure. So up to this point, we have determined that the shape factor

- $\theta = 0$ for single perforated grains
- $\theta = 1$ for cylindrical grains

An interesting thing has happened. We started this section attempting to find a relationship for the mass of gas evolved from the solid propellant, and we have come around to finding the relationship between ϕ and f again. The key procedure here is now to rearrange Equation 2.106:

$$m_g(t) = c\phi(t) \tag{2.109}$$

This is the relationship that governs the amount of gas evolved from the burning propellant. It looks rather simple, but consider that ϕ is a function of f and t, and f is a function of p_B and t. We shall return to this later.

The burning propellant in our closed bomb must generate pressure. To take this further, we need to rearrange Equation 2.98 into

$$p_B(t) = -\frac{D}{\beta}\frac{df}{dt} \tag{2.110}$$

In this expression, we know that D is the initial web and, therefore, a constant, and we shall assume that β is a constant (β actually increases somewhat with pressure).

Because we want to work with masses of substances, f is not a convenient variable. We shall use a relationship to express it in terms of ϕ. At this point, caution must be exercised. Recall that the relationship between ϕ and f varies with propellant geometry. We shall proceed using our cylindrical grain relationship (Equation 2.108). Rearranging Equation 2.108, we obtain

$$f(t) = \sqrt{1 - \phi(t)} \tag{2.111}$$

If we differentiate this relationship with respect to time, we obtain

$$\frac{df}{dt} = -\frac{1}{2\sqrt{1 - \phi(t)}}\frac{d\phi}{dt} \tag{2.112}$$

This form allows us to rewrite Equation 2.110 as

$$p_B(t) = -\frac{D}{2\beta\sqrt{1 - \phi(t)}}\frac{d\phi}{dt} \tag{2.113}$$

We now have all the expressions we need to bring this together. We have an equation of state:

$$p_B(t)V = m_g(t)RT(t) \tag{2.114}$$

We have an expression for conservation of mass (relationship between m_g and ϕ):

$$m_g(t) = c\phi(t) \tag{2.115}$$

and we have an expression that relates the amount of pressure generated to the amount of propellant burnt (burn rate equation):

$$p_B(t) = -\frac{D}{2\beta\sqrt{1 - \phi(t)}}\frac{d\phi}{dt} \tag{2.116}$$

All these expressions are in terms of constants we know beforehand or f, ϕ, and T.

To describe the temperature of gas, we need to define a parameter often used in interior ballistics, the propellant force λ. Propellant force is a constant that is defined as the amount of energy released from a propellant under adiabatic conditions. In other words, it is the most energy one can obtain by burning a propellant. Mathematically, we express it as

$$\lambda \equiv RT_0 \tag{2.117}$$

where R is the specific gas constant and T_0 is the adiabatic flame temperature of the gas.

This constant has units of energy per unit mass. Sometimes, T_0 is referred to as the uncooled explosion temperature. In our development, we shall assume that all gases are evolved at the adiabatic flame temperature. There are many theories that describe combustion. Introductory treatments are provided by Turns [20] and Borman and Ragland [21], but all the references in the end of this section cover the topic to some degree. Yang et al. [22] and Kuhl et al. [23,24] treat the topic in great detail. If we utilize this reactive assumption, we can rewrite Equation 2.114 using Equation 2.117 to give us

$$p_B(t)V = \lambda m_g(t) \tag{2.118}$$

Now we can combine Equations 2.118 and 2.116 to yield (for a cylindrical grain)

$$\frac{\lambda m_g(t)}{V} = -\frac{D}{2\beta\sqrt{1-\phi(t)}}\frac{d\phi}{dt} \tag{2.119}$$

We then substitute Equation 2.115 into the aforementioned expression, resulting in

$$\frac{\lambda c\phi(t)}{V} = -\frac{D}{2\beta\sqrt{1-\phi(t)}}\frac{d\phi}{dt} \tag{2.120}$$

This can be rearranged to yield

$$\frac{1}{\phi(t)\sqrt{1-\phi(t)}}\frac{d\phi}{dt} = -\frac{2\beta\lambda c}{DV} \tag{2.121}$$

This is a separable, first-order, nonlinear, differential equation, which can be written in integral form as

$$\int_0^1 \frac{d\phi}{\phi(t)\sqrt{1-\phi(t)}} = -\frac{2\beta\lambda c}{DV}\int_0^{t_B} dt \tag{2.122}$$

the solution of which is

$$\ln\left(\frac{\sqrt{1-\phi}-1}{\sqrt{1-\phi}+1}\right)\Bigg|_0^1 = -\frac{2\beta\lambda c}{DV}t\Big|_0^{t_B} \tag{2.123}$$

This expression is somewhat problematic because of its singular behavior at $\phi = 1$ and $\phi = 0$. The equation was numerically approximated to yield

$$t_B \approx 350\frac{DV}{\beta\lambda c} \tag{2.124}$$

In this case, the solution to this expression was problematic; however, in many cases, it can be more readily evaluated. The techniques that will follow are much simpler from a hand calculation standpoint.

Even though the closed bomb may seem academic, it is actually quite a useful device for determining propellant parameters. If we consider Equations 2.110 and 2.113, we see that since we know the initial web D, and we can measure the pressure, the only thing missing is β and ϕ or f. Equation 2.114 tells us that if we measure p_B and T and know V, we can get ϕ or f. Thus, the closed bomb is useful for determining the burn rate coefficient β.

Problem 11

M1 propellant is measured in a closed bomb. Its adiabatic flame temperature is 3906°F. Its molar mass is 22.065 lbm/lb mol. What is the effective mean force constant in feet pound-force per pound-mass?

Answer: $\lambda = 305,709 \left[\frac{\text{ft-lbf}}{\text{lbm}}\right]$

Problem 12

M15 propellant was tested in a strand burner to determine the linear burning rate. The average pressure evolved was 10,000 psi. If the burning exponent α was known to be 0.693 and the pressure coefficient β was known to be 0.00330 in./s/psi$^{0.693}$, determine the average linear burning rate B in inches per second.

Answer: $B(p) = 1.952 \left[\frac{\text{in.}}{\text{s}}\right]$

Problem 13

Derive the functional form of ϕ in terms of f for a flake propellant. Assume cylindrical geometry.

Hint: Flake propellant consists of grains that have thicknesses much smaller than any other characteristic dimension.

Answer: $\phi(t) = 1 - f$

Problem 14

An M60 projectile is to be fired from a 105 mm M204 Howitzer. The propellant used in this semifixed piece of ammunition is 5.5 lbm of M1 propellant. M1 propellant consists of single perforated grains ($\theta = 0$) with a web thickness of 0.0165 in., if the average pressure (over the launch of this projectile) developed in the weapon is 20,455 psi. Calculate the average burning rate coefficient in cubic inches per pound-force-second if the burn rate is (we use a negative sign in the burn rate to make the form come out right later)

$$\frac{df}{dt} = -185.9 \left[\text{s}^{-1}\right]$$

Answer: $\beta = 1.50 \times 10^{-4} \left[\frac{\text{in.}^3}{\text{lbf-s}}\right]$

Problem 15

β is actually a function of pressure and temperature (it is really given in tables at 25°F at this value). For simplification (and illustration), we will assume that it is constant. Given this assumption, calculate the functional form of the web fraction f from Problem 14.

Answer: $f = 1 - \frac{\beta p_{avg}}{D} t$

Problem 16

Given the data provided in Problems 14 and 15, determine the proper form of the fraction of charge burnt.

Answer: $\phi(t) = 185.9t$

Problem 17

You are asked to characterize a commercial propellant. In order to do this, you take one grain of the propellant and place it in a closed bomb of 0.5 in.^{3} volume, initially evacuated. You have a temperature and pressure sensor in the device. After 0.063 s, you decide that the propellant has fully combusted. You read the data—pressure was measured to be 3.706 psi (this is not a big value, but it was only one small grain of propellant), but it looks as though the temperature sensor is broken. The initial propellant grain weighed 0.003189 grains, and it was 0.1 in. long by 0.01 in. diameter. Based on these data only—

1. Estimate the propellant force λ in feet pound-force per pound-mass.
2. Estimate the linear burn rate coefficient β in inches per second per pounds per square inch.
3. List all assumptions and explain why you believe these estimates are too high or too low (certain assumptions may make the estimates high while others make them low)—there are at least four assumptions used here.

2.7 Fluid Mechanics

The entire field of ballistics is steeped in the principles of fluid mechanics. The flow of propellant gases in the gun tube, the flow of the propellant gases through a muzzle brake upon shot exit, the flow of the air around the projectile in flight, and, even, as we shall see, the flow of target material during a penetration event can many times be modeled as a fluid. This section is devoted to a basic treatment of fluid mechanics principles. We will use some of these very soon; others will be used at a later time. All them are important in the study of ballistics.

A fluid differs from a solid in its behavior when placed in shear. In general, fluids can support little or no shear loads or tensile stress. Fluids are generally characterized by their behavior under shear stress. Because a fluid will, in general, flow readily under a shear stress, this behavior is normally plotted in a graph of rate of deformation vs. shear stress as depicted in Figure 2.15.

A fluid is considered to exhibit Newtonian behavior if there is a linear relationship between shear stress and rate of deformation. A fluid is non-Newtonian otherwise. Some fluids such as an ideal plastic or a thixotropic material actually do exhibit a yield stress. In the case of an ideal plastic, after a certain yield stress is achieved, the material exhibits a linear relationship between stress and deformation rate. A thixotropic material exhibits a nonlinear relationship after yield stress is reached. An ideal fluid is one where the material will flow and continue to accelerate regardless of the amount of shear stress applied.

Many of the fluids we will deal with are Newtonian. Mathematically, the relationship between the applied shear stress and deformation rate is given by

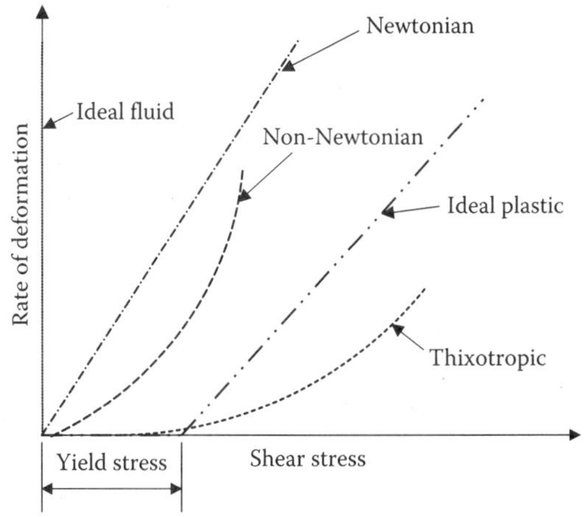

FIGURE 2.15
Rate of deformation vs. shear stress.

$$\tau = \mu \frac{\partial u}{\partial y} \tag{2.125}$$

where τ is the applied shear stress; μ is the dynamic viscosity of the fluid; and $\partial u/\partial y$ is the deformation gradient (change in velocity with respect to a spatial coordinate).

The ratio of the dynamic viscosity to the fluid density occurs so often that it is customary to define a kinematic viscosity as

$$\nu = \frac{\mu}{\rho} \tag{2.126}$$

where ν is the kinematic viscosity and ρ is the density of the fluid.

In the section on thermodynamics, we introduced the concept of a Lagrangian or control mass approach and a Eulerian or CV approach to solving transport problems. In examination of the behavior of a fluid, we need to develop both of these techniques. Our plan of attack will be to develop these equations in a CV and provide equations to change the reference frame afterward. For a more complete treatment, the reader is referred to the books by Fox and McDonald [11], Anderson [12], White [25], Panton [26], Currie [27], and White [28].

The basis for our development of the following equations are the tenets that (1) mass must be conserved and (2) Newton's second law must hold true. Newton's second law can be written as

$$\sum \mathbf{F} = \frac{\mathrm{d}}{\mathrm{d}t}(m\mathbf{V}) \tag{2.127}$$

where $\Sigma\mathbf{F}$ is the vector sum of all the forces acting on a body (or blob of fluid or CV); m is the mass of the body; and $\mathbf{V}$ is the vector velocity of the body.

It is important to note that throughout this work, V is volume (a scalar), V is velocity (as a scalar quantity), and $\mathbf{V}$ is velocity (as a vector quantity).

Since CVs can be oriented in an arbitrary manner, it is important to understand that only that component of velocity normal to the control surface (CS) (i.e., the boundary of the CV) transports material or energy into the CV. If we examine Figure 2.16 where we have broken the velocity vectors into normal and tangential components (denoted as V_n and V_t, respectively), we can clearly see why this is so.

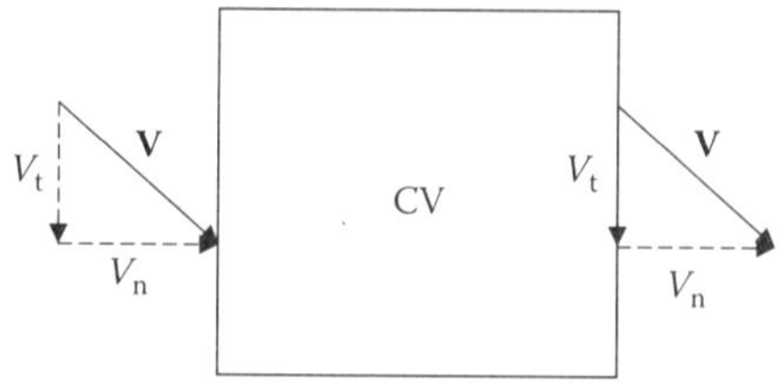

FIGURE 2.16
Depiction of normal and tangential velocity components with respect to an arbitrary CV.

Consider an arbitrary property N of a substance. We would like to see how this property is transported into and out of a CV. If we define an intensive property η such that

$$\eta = \frac{N}{m} \quad \text{or} \quad N = \eta m \tag{2.128}$$

then we can write

$$\frac{dN}{dt} = \frac{\partial}{\partial t}\int_{CV} \eta \rho dV + \int_{\text{outflow area}} \eta \rho \mathbf{V} \cdot d\mathbf{A} + \int_{\text{inflow area}} \eta \rho \mathbf{V} \cdot d\mathbf{A} \tag{2.129}$$

This equation defines how a property of interest is transported into and out of the CV. If we look at each of the terms, we see that this is an intuitively satisfying equation. The term on the LHS is the time rate of change (decrease) of any property of the CV over a time of interest. The first term on the RHS tells us how much of that property is stored in the CV over this time. The second term on the RHS tells us how much material has left the CV, while the third term tells us how much material has entered.

Now wait a minute! If we look at the signs on the second and third terms, they seem to be incorrect—should not the stuff leaving have a negative sign and the stuff entering have a positive sign? The answer to this is yes, but Equation 2.129 is written correctly. The key to this seemingly inconsistent sign convention lies in the fact that the dot product in the second term is positive when we define the area as a vector that points outward and is normal to the surface. Similarly, the inflow term will always lead to a negative number since the velocity vector points inward and the area vector points outward.

We shall now examine the flow of propellant gases in a suitable CV located somewhere behind a projectile at an instant in time. This will serve to foster the understanding of the CV approach.

Consider a CV in a gun tube located somewhere behind a moving projectile as depicted in Figure 2.17. There will be a velocity associated with the propelling gases (we will see this later) such that the gases are flowing in one side and out the other, but no gases flow through the walls.

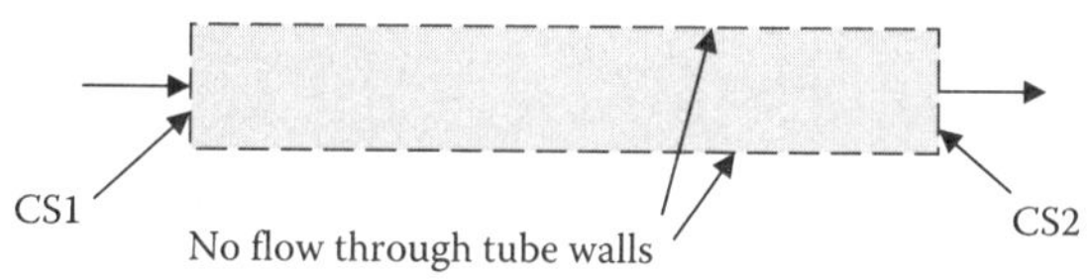

FIGURE 2.17
Typical gun tube CV (two-dimensional representation).

The ends of this cylindrical CV are designated as CSs. The inlet side is CS1 and the outlet side is CS2. If we would like to write an equation for how mass is transferred into or out of this CV, we set N, the flow variable in Equation 2.129, equal to m, the property of interest. When this is done, Equation 2.128 tells us that

$$\eta = \frac{N}{m} = \frac{m}{m} = 1 \tag{2.130}$$

So for this case, we can write

$$\frac{\mathrm{d}m}{\mathrm{d}t} = \frac{\partial}{\partial t}\int_{\text{CV}} \rho \mathrm{dV} + \int_{\text{outflow area}} \rho \mathbf{V}\cdot \mathrm{d}\mathbf{A} + \int_{\text{inflow area}} \rho \mathbf{V}\cdot \mathrm{d}\mathbf{A} \tag{2.131}$$

We know mass can neither be created nor destroyed, so $\mathrm{d}m/\mathrm{d}t = 0$, then we arrive at what is commonly called the equation of conservation of mass, or the continuity equation. In a general form, it is given as

$$\frac{\partial}{\partial t}\int_{\text{CV}} \rho \mathrm{dV} + \int_{\text{outflow area}} \rho \mathbf{V}\cdot \mathrm{d}\mathbf{A} + \int_{\text{inflow area}} \rho \mathbf{V}\cdot \mathrm{d}\mathbf{A} = 0 \tag{2.132}$$

The first term on the LHS states how the mass in the CV is changing with time. The second term is the amount of mass exiting the CV, and the third term is the amount of mass entering the CV.

The flow inside a gun tube is never steady or uniform. Nevertheless, it is informative to look at this expression using these two assumptions to gain some physical insight into the nature of the terms. The steady flow assumption means that there is no increase or decrease in material flow into or out of our CV. This implies that the first term is zero. So for the special case of steady flow, we have

$$\int_{\text{outflow area}} \rho \mathbf{V}\cdot \mathrm{d}\mathbf{A} + \int_{\text{inflow area}} \rho \mathbf{V}\cdot \mathrm{d}\mathbf{A} = 0 \tag{2.133}$$

Simply put, this equation states that what comes into the CV equals what goes out of the CV.

Uniform flow is a special case where fluid viscosity effects are neglected. This results in a constant velocity across the CS's. In essence, the velocity at the wall of the gun tube is the same as the velocity on the centerline of the tube. We will discuss this and its implications in more detail later.

When we apply this assumption to Equation 2.133 and note that $\mathbf{V}\cdot \mathrm{d}\mathbf{A}$ is negative at CS1 (because the vectors have opposite directions) and positive at CS2, we obtain the following simple relationship:

$$\rho_1 V_1 A_1 = \rho_2 V_2 A_2 = \dot{m} \tag{2.134}$$

Thus, under the steady flow assumption, the mass flow rate $m.$ is constant.

We shall now examine the use of momentum $m\mathbf{V}$ as our flow variable. The use of Equation 2.128 with this flow variable yields

$$\eta = \frac{N}{m} = \frac{m\mathbf{V}}{m} = \mathbf{V} \tag{2.135}$$

Now we can include this into Equation 2.129 to obtain

$$\frac{\mathrm{d}(m\mathbf{V})}{\mathrm{d}t} = \frac{\partial}{\partial t}\int_{\mathrm{CV}} \mathbf{V}\rho \mathrm{d}V + \int_{\text{outflow area}} \mathbf{V}\rho\mathbf{V}\cdot \mathrm{d}\mathbf{A} + \int_{\text{inflow area}} \mathbf{V}\rho\mathbf{V}\cdot \mathrm{d}\mathbf{A} \tag{2.136}$$

Through Newton's second law, we know that the term on the LHS (time rate of change of momentum) equals the forces on the system. The first term on the RHS is the change in the momentum of the system through storage in the CV. The second and third terms are the momentum leaving and momentum entering the CV, respectively. It is again informative to examine the steady flow case, which reduces our equation to

$$\mathbf{F} = \int_{\text{outflow area}} \mathbf{V}\rho\mathbf{V}\cdot \mathrm{d}\mathbf{A} + \int_{\text{inflow area}} \mathbf{V}\rho\mathbf{V}\cdot \mathrm{d}\mathbf{A} \tag{2.137}$$

Here we have replaced the time rate of change of momentum term with the force. Once again, we shall use the uniform flow assumption to facilitate our understanding of this equation. Consider the same gun tube CV as earlier, drawn slightly differently in Figure 2.18.

As discussed earlier, the velocity and area scalar products result in a negative sign on the inflow and a positive sign on the outflow side. With this uniform flow assumption (recall that we also included a steady flow assumption to reduce the equation to the form of Equation 2.137), our Equation 2.137 would become

$$\mathbf{F} = \rho_2 V_2 \mathbf{V}_2 A_2 - \rho_1 V_1 \mathbf{V}_1 A_1 \tag{2.138}$$

Note that this is still a vector equation with the vectors $\mathbf{V}_1$ and $\mathbf{V}_2$ determining the direction of $\mathbf{F}$. If we had already worked out or it was obvious what direction the resultant force would be in, then we could write

$$F = \rho_2 V_2^2 A_2 - \rho_1 V_1^2 A_1 \tag{2.139}$$

Equation 2.138 only tells us part of the story. It tells us the inertial reaction of the CV to the forces arising from a fluid passing through it. There are two types of forces that occur on the LHS in response to or independent of this, body forces and surface tractions.

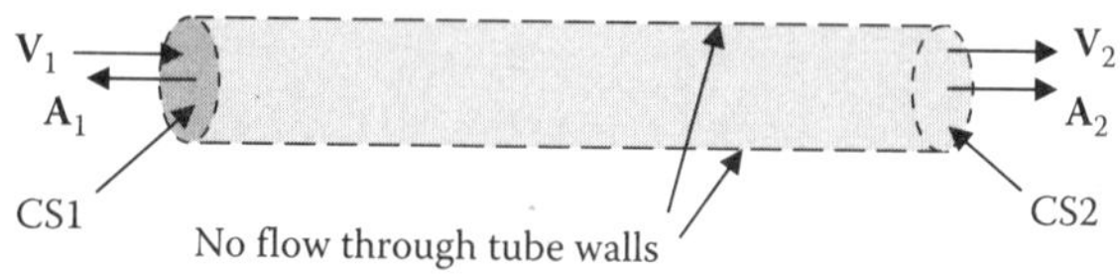

FIGURE 2.18
Typical gun tube CV (three-dimensional representation).

Body forces are those that act through the bulk of the material (i.e., directly affecting every molecule). Examples of this are gravitational loads and electromagnetic loads. It is customary to write these loads on a unit mass basis to be consistent with the rest of the equation. In many cases, these are small and are neglected.

Surface tractions are forces that act on the CS. These forces tend to be large and can be categorized into normal forces and shear forces. As the name implies, normal forces act normal to the CS. Pressure is the most common normal force. Because pressure cannot be negative, it always acts opposite to the surface area vector.

Shear stresses are a result of the propensity of fluid to stick to a solid (or other fluid) surface. The fluid viscosity, as defined earlier, is a measure of the intensity of these stresses. Shear stresses always act opposite to the direction of flow and along the CS. If a fluid is modeled as inviscid, there can be no shear stresses.

Picking up from Equation 2.138, if we model a flow as steady with no viscosity, there will still be pressure forces present. This is depicted in Figure 2.19.

Since pressure forces always act opposite to the area vector, it is customary to define the pressure forces as

$$\mathbf{F}_{\mathrm{p}} = -\int_{\text{outflow area}} p d\mathbf{A} - \int_{\text{inflow area}} p d\mathbf{A} \tag{2.140}$$

In Equation 2.140, the signs of the area vectors would define the direction of the force.

Before we establish a CV with viscous forces acting, it is instructive to describe these viscous forces and their effect on the flow field. As previously established, viscosity is a property of a fluid. The greater the viscosity of a fluid is, the more difficult it is to shear the material. If the viscosity is high enough or the flow velocity is low enough, a fluid will exhibit what is known as laminar flow. Laminar flow is a very orderly shearing of the fluid from a solid surface where the fluid sticks to the boundary. In a tube or pipe, after some entrance length required for the flow to establish itself, the fluid will achieve a parabolic velocity distribution as depicted in Figure 2.20.

FIGURE 2.19
CV with no viscous forces acting.

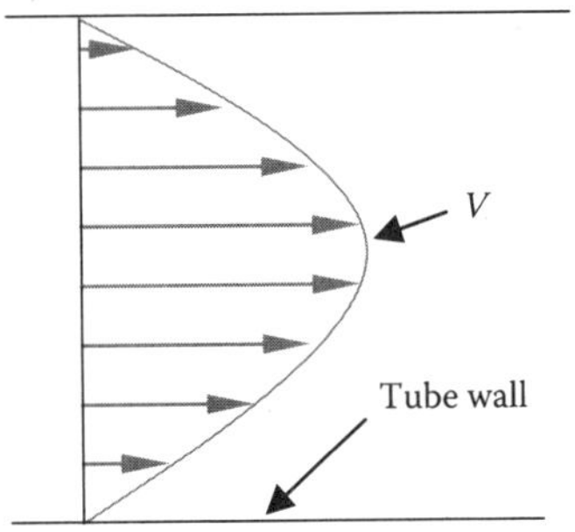

FIGURE 2.20
Laminar velocity profile in a tube.

The laminar profile in Figure 2.20 is in stark contrast to the uniform profile that we had assumed in our previous discussions depicted in Figure 2.21. If the flow velocity is high enough or the viscosity low enough, the flow will transition from laminar flow to what is known as turbulent flow. Turbulent flow is characterized by a large number of eddies that swirl around in the flow. These eddies are important in that they tend to distribute momentum, energy, and matter throughout the fluid resulting in better mixing and very different transport properties. Many more flows are turbulent than laminar. The dimensionless parameter that governs this behavior is known as the Reynolds number and is given by

$$\mathrm{Re}_d = \frac{\rho V d}{\mu} = \frac{V d}{\nu} \tag{2.141}$$

where Re_d is the Reynolds number and is dimensionless; ρ is the fluid density; V is the fluid velocity; d is a relevant characteristic length of the system (an internal diameter of a pipe, a length of a projectile, etc.); and μ and ν are the dynamic and kinematic viscosities of the fluid, respectively.

If the Reynolds number is high enough, the flow will be turbulent. This demarcation is, in general, a range of values that also depends on whether the flow is an internal one (such as the gas flow in a gun tube) or an external one (such as the flow about a projectile). The velocity profile of a turbulent flow is depicted in Figure 2.22. Here we can see that the effect of the eddies is to distribute the momentum, resulting in a profile that is flatter and more akin to our inviscid flow model of Figure 2.21.

If we now return to our discussion on the surface tractions, we can discern that the effect of fluid viscosity is to create a shear stress at the boundary between the fluid inside a gun tube and the solid tube itself (i.e., on our CS). If we consider the diagram in Figure 2.19, we can redraw this figure to include the effect of shear stresses as depicted in Figure 2.23.

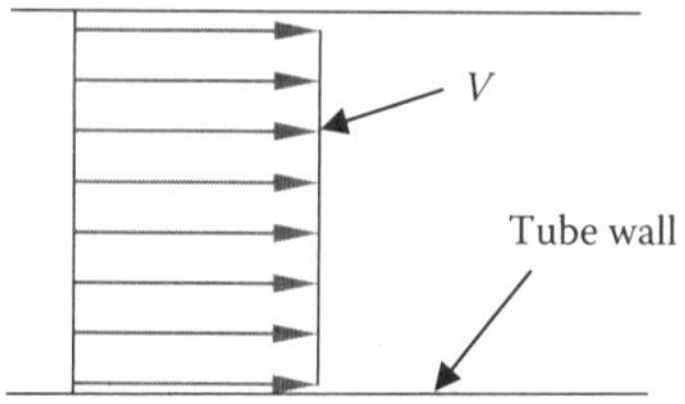

FIGURE 2.21
Uniform velocity profile in a tube.

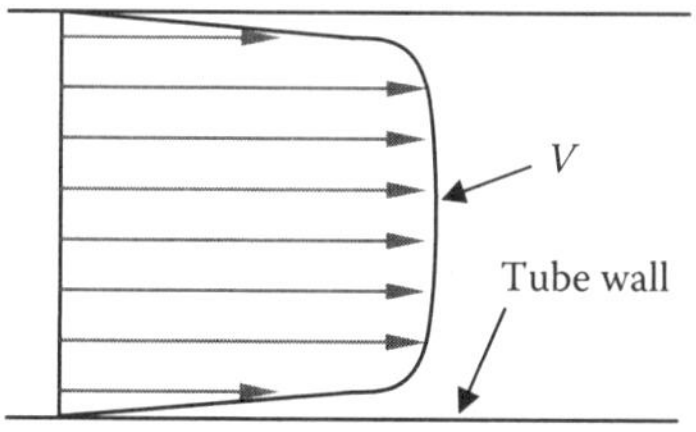

FIGURE 2.22
Turbulent velocity profile in a tube.

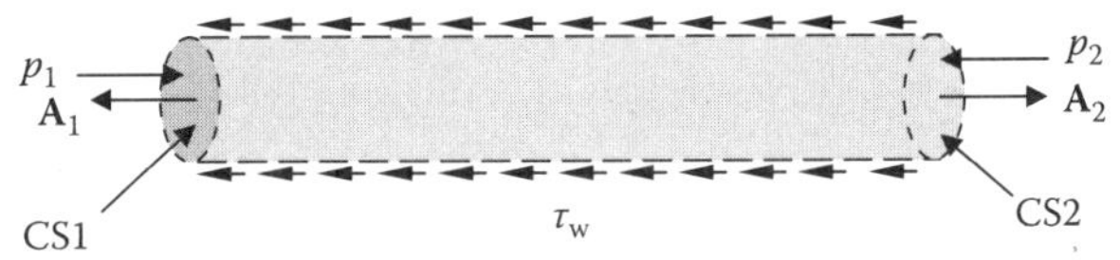

FIGURE 2.23
Surface tractions on a gun tube CV.

Since the shear stress τ_w acts all over the area of our CV, we can add a term for this into Equation 2.140 to obtain an expression for all the surface forces as follows:

$$\mathbf{F}_{\text{surface}} = -\int_{\text{outflow area}} p\mathrm{d}\mathbf{A} - \int_{\text{inflow area}} p\mathrm{d}\mathbf{A} - \int_{\text{surface area}} \tau_w \mathrm{d}\mathbf{A} \tag{2.142}$$

We can insert this into our expression for the conservation of momentum Equation 2.137 to obtain, for steady flow,

$$-\int_{\text{outflow area}} p\mathrm{d}\mathbf{A} - \int_{\text{inflow area}} p\mathrm{d}\mathbf{A} - \int_{\text{surface area}} \tau_w \mathrm{d}\mathbf{A} = \int_{\text{outflow area}} \mathbf{V}\rho\mathbf{V}\cdot\mathrm{d}\mathbf{A} + \int_{\text{inflow area}} \mathbf{V}\rho\mathbf{V}\cdot\mathrm{d}\mathbf{A} \tag{2.143}$$

or, in a more general sense,

$$-\int_{\text{outflow area}} p\mathrm{d}\mathbf{A} - \int_{\text{inflow area}} p\mathrm{d}\mathbf{A} - \int_{\text{surface area}} \tau_w \mathrm{d}\mathbf{A} = \frac{\partial}{\partial t}\int_{\text{CV}} \mathbf{V}\rho\,\mathrm{dV} + \int_{\text{outflow area}} \mathbf{V}\rho\mathbf{V}\cdot\mathrm{d}\mathbf{A} + \int_{\text{inflow area}} \mathbf{V}\rho\mathbf{V}\cdot\mathrm{d}\mathbf{A} \tag{2.144}$$

The next transport property we shall examine is that of energy. In Sections 2.4 and 2.5, this was discussed to a degree. The objective of this section is to demonstrate that we can use the same transport Equation 2.129 to come up with the energy equations we have used earlier. We start by recognizing that our transport variable is energy E. With this in mind, Equation 2.128 can be rewritten as

$$\eta = \frac{E}{m} = e \tag{2.145}$$

Recall that lower case letters are intensive properties. Then we can write

$$\frac{\mathrm{d}E}{\mathrm{d}t} = \frac{\partial}{\partial t}\int_{\text{CV}} e\rho\,\mathrm{dV} + \int_{\text{outflow area}} e\rho\mathbf{V}\cdot\mathrm{d}\mathbf{A} + \int_{\text{inflow area}} e\rho\mathbf{V}\cdot\mathrm{d}\mathbf{A} \tag{2.146}$$

This states that the change in energy of a system is equal to the change in energy stored in the system minus that which is advected away plus that which is advected into the system. Recall from Equation 5.6 that

$$\frac{dQ}{dt} + \frac{dW}{dt} = \frac{dE}{dt} \tag{2.147}$$

From our definition of work, we know that

$$W = \int p\mathrm{dV} \tag{2.148}$$

But volume is nothing more than a length times an area. This allows us to write

$$W = \int p\mathbf{x} \cdot d\mathbf{A} \tag{2.149}$$

If we take the derivative of this expression with respect to time assuming pressure is an average value over the time increment, we can write

$$\frac{dW}{dt} = \int p\frac{d\mathbf{x}}{dt} \cdot d\mathbf{A} = \int p\mathbf{V} \cdot d\mathbf{A} \tag{2.150}$$

There are many types of work terms. The aforementioned term happens to be called $p\mathrm{dV}$ work or pressure work. The other types of work, such as shaft work, are usually not present in a gun launch so we shall neglect them. The insertion of Equation 2.150 into Equation 2.146 and rearranging yields

$$\frac{dQ}{dt} = \frac{\partial}{\partial t}\int_{CV} e\rho\mathrm{dV} + \int_{\text{outflow area}} \left(e + \frac{p}{\rho}\right)\rho\mathbf{V} \cdot d\mathbf{A} + \int_{\text{inflow area}} \left(e + \frac{p}{\rho}\right)\rho\mathbf{V} \cdot d\mathbf{A} \tag{2.151}$$

In the section on thermodynamics, we defined the specific energy through Equation 2.39. If we insert this definition into the aforementioned expression, we obtain

$$\frac{dQ}{dt} = \frac{\partial}{\partial t}\int_{CV} e\rho\mathrm{dV} + \int_{CS} \left(gz + \frac{V^2}{2} + u + \frac{p}{\rho}\right)\rho\mathbf{V} \cdot d\mathbf{A} \tag{2.152}$$

Here we have combined the last two terms on the RHS of Equation 2.151 with the understanding that the integral of the last term in Equation 2.152, being an integral over the entire CS, accounts for the difference between inflow and outflow. It is informative to look at this equation with respect to a gun launch. The term on the LHS represents the transfer of heat to or from the system. The first term on the RHS represents the change in stored energy of the system (such as energy released by propellant combustion). The last term on the RHS is the change in energy of the system. Since gravitational potential energy, the product gz, is small relative to the other energy terms, it is usually neglected allowing us to rewrite the expression as

$$\frac{dQ}{dt} = \frac{\partial}{\partial t}\int_{CV} e\rho\mathrm{dV} + \int_{CS} \left(\frac{V^2}{2} + u + \frac{p}{\rho}\right)\rho\mathbf{V} \cdot d\mathbf{A} \tag{2.153}$$

Earlier in this section, we introduced the common practice of characterizing a fluid based on its behavior under shear stress. This allowed us to come up with a relationship between applied shear stress and deformation rate. Another distinction has to be made between fluids with respect to the density. If the density is considered constant in a fluid or solid that we model, we call this material incompressible. If the density varies, we must analyze the problem with the assumption of compressible material. This has many ramifications. The most significant ramification is that if the material is incompressible, then the energy equation is decoupled from the momentum equation and we can solve them independently [25]. This makes problem solving much simpler. We do not have this luxury when the density significantly varies.

In fluid flows, such as those that we shall study later, a dimensionless parameter known as the Mach number Ma is used as a measure to determine the effect of compressibility, among its other uses. The Mach number is given by

$$\mathrm{Ma} = \frac{V}{a} \tag{2.154}$$

where V is some characteristic velocity in the material and a is the speed of sound in the material.

In general, if the Mach number is below 0.3, the deviation from incompressible flow is small so the assumption of incompressibility leads to an acceptably small error [16]. In an ideal gas, the speed of sound is given by the relation

$$a = \sqrt{\gamma RT} \tag{2.155}$$

where γ is the specific heat ratio; R is the specific gas constant; and T is the absolute temperature (i.e., in Rankine degree or Kelvin).

The speed of sound in any material is formally defined as

$$a = \sqrt{\left.\frac{\partial p}{\partial \rho}\right|_s} \tag{2.156}$$

That is to say that the speed of sound in a material is equal to the square root of the partial derivative of pressure with respect to density evaluated with constant entropy. The interested reader is referred to any of the following books for the detailed proof of this equation: Masterson et al. [15], Cooper and Kurowski [16], and White [25].

The speed of sound is essentially the fastest speed at which a disturbance can be propagated by molecular interaction. If a disturbance is created that is strong enough, a shock will form. This shock must always move faster than the speed of sound in the material. We will discuss this in detail later.

In the study of compressible flows, it is common practice to utilize stagnation values in many of our calculations. Stagnation values are the values of the enthalpy, pressure, temperature, and density that are achieved by adiabatically slowing a flow down to zero velocity. The assumption of adiabatic behavior is warranted in many of the situations we will examine, particularly in exterior ballistics. The stagnation enthalpy is given by

$$h_0 = h + \frac{1}{2}V^2 \tag{2.157}$$

In this and the following equations, the subscript 0 indicates the stagnation value; V is the velocity of the flowing fluid, and the values without the subscript are the static value; in the case of Equation 2.157, h is the static enthalpy. Equation 2.157 holds for any material. If the material is an ideal gas, we can define the stagnation temperature, pressure, and density as

$$T_0 = h + \frac{1}{2}\frac{V^2}{c_\mathrm{p}} \quad \text{or} \quad \frac{T_0}{T} = 1 + \frac{\gamma - 1}{2}\mathrm{Ma}^2 \tag{2.158}$$

$$\frac{p_0}{p} = \left(1 + \frac{\gamma - 1}{2}\mathrm{Ma}^2\right)^{\frac{\gamma}{\gamma - 1}} \tag{2.159}$$

$$\frac{\rho_0}{\rho} = \left(1 + \frac{\gamma - 1}{2}\mathrm{Ma}^2\right)^{\frac{1}{\gamma - 1}} \tag{2.160}$$

In each of these cases, thermodynamic relations have been used for an ideal gas (Equation 2.61).

Shock waves are formed in materials when disturbances of sufficient strength propagate through the medium. *Sufficient strength* is a term that we throw about rather loosely to describe conditions where shocks are formed—it can be cast in terms of flow velocities or pressures (the two are linked as we shall see). Shocks can be classified as normal or oblique, depending upon the direction of material flow into them. They can also be analyzed as steady or transient. In general, shocks can take curved and rather complex shapes, but the simple analytical tools we have allow us to look at them only under simplified geometries. More complex geometries require the assistance of a computer.

We shall only examine normal shocks in this brief review and direct the reader to Cooper and Kurowski [16] for the handling of oblique shocks. The best way to examine the behavior of a shock is to look at a shock tube. This simple device will allow us to introduce all the materials necessary for the introductory study of ballistics and set the stage for later work when we discuss stress waves in solids.

Before we look at a shock tube, we need to discuss the principle of superposition as applied to shock waves. Consider two shocks as depicted in Figure 2.24. One of these cases is a stationary shock where we could consider ourselves "riding on the wave," while in the other case, we can consider ourselves to be sitting on the ground watching the shock pass by. If, in both cases, the shock were passing into a stagnant medium, we would see some important correlations. The passage of a shock wave always induces motion that follows the wave. Consider the situation where we are sitting on the ground; the air about us is stagnant, and all of a sudden, a shock passed by us just as is shown in Figure 2.24b. If the shock were moving at velocity U, we would feel an induced motion, a wind, immediately afterward moving at velocity u_p in the same direction that the shock was moving. If we experienced this same situation, but instead, were riding on the shock, we would feel a wind of velocity U coming toward our face. This would be analogous to the situation in Figure 2.24a. In this situation, velocity V_1 would be equal to U. Note the direction of the velocity vectors in the figure. The velocity vector of magnitude V_2 is moving away from the wave. The figure is drawn correctly, but in the case that was just described, based on superposition, since U is larger than u_p (and it always is), if we were riding on the wave, we would see

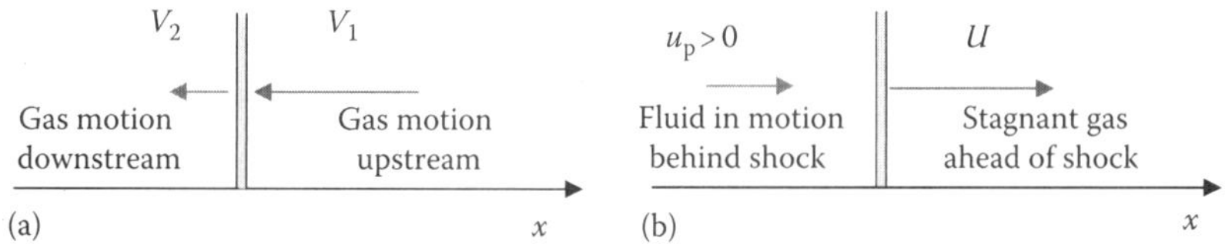

FIGURE 2.24
(a) Stationary and (b) moving shock waves.

material leaving us at velocity ($U - u_p$). When we examine a shock wave in the frame of Figure 2.24b, we are said to be using a Eulerian frame of reference, if we analyze the very same situation as shown in Figure 2.24a, we are using a Lagrangian reference frame.

The difference between Lagrangian and Eulerian reference frames is important because we sometimes prefer to solve a problem in one frame or the other because the mathematics are simpler. As long as the reference frame motion is accounted for, solving in one frame or the other leads to the same answer.

We shall now use the Lagrangian approach to examine the governing equations for a stationary normal shock wave. Consider the situation in Figure 2.25, where a shock wave is moving to the left at velocity U. Since we would like to examine the behavior of this shock, we will put ourselves in a reference frame attached to the shock itself. We form a CV enclosing the shock only. We observe, while riding on this shock, that fluid enters the CV at velocity U and leaves at velocity u_2. We can write the conservation of mass, momentum, and energy equations for this system as follows:

- Conservation of mass (continuity equation):

$$\rho_1 U = \rho_2 u_2 \tag{2.161}$$

- Conservation of momentum:

$$p_1 + \rho_1 U^2 = p_2 + \rho_2 u_2^2 \tag{2.162}$$

- Conservation of energy:

$$h_1 + \frac{1}{2}U^2 = h_2 + \frac{1}{2}u_2^2 = h_{01} = h_{02} = h_0 = \text{constant} \tag{2.163}$$

We see from the last equation that across a shock wave, the stagnation enthalpy must remain constant. This falls out directly from the fact that we assumed the shock wave was adiabatic. These equations are coupled through a material model such as the ideal gas equation of state (relates p, V, and T) and the calorically perfect assumption (relates h to T). If we consider the special case where the shock under examination is moving into a stagnant fluid as depicted in Figure 2.26, we can write the aforementioned three equations as

$$\rho_1 U = \rho_2 \left(U - u_p\right) \tag{2.164}$$

$$p_1 + \rho_1 U^2 = p_2 + \rho_2 \left(U - u_p\right)^2 \tag{2.165}$$

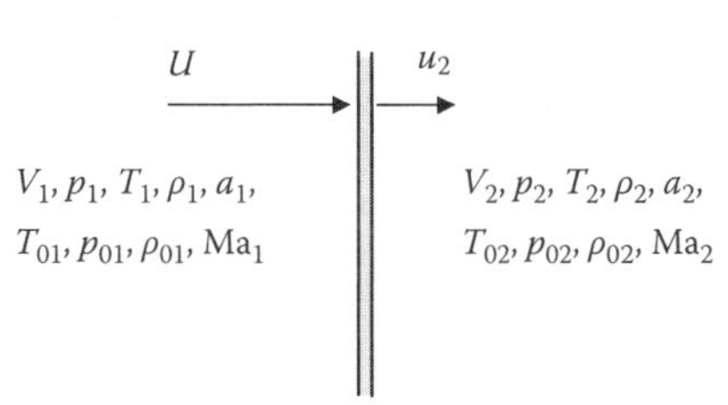

FIGURE 2.25
Stationary shock wave.

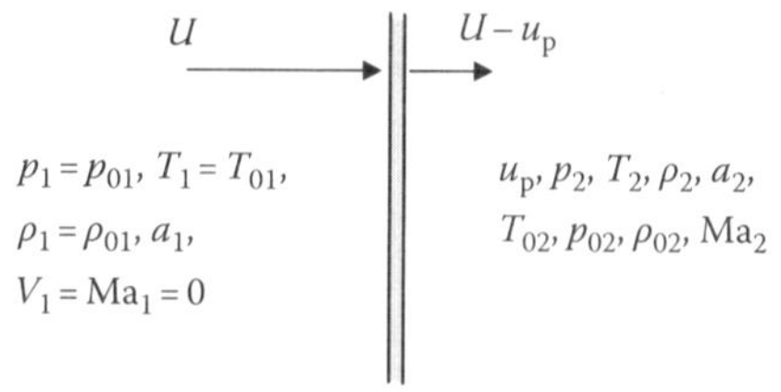

FIGURE 2.26
Stationary shock wave moving into a stagnant fluid.

$$h_1 + \frac{1}{2}U^2 = h_2 + \frac{1}{2}\left(U - u_p\right)^2 = h_0 = \text{constant} \tag{2.166}$$

The conservation of mass, momentum, and energy equations can be combined as detailed by Kays and Crawford [10] and Cooper and Kurowski [16] to yield the Rankine–Hugoniot relationship. This relationship determines how the energy changes across a normal shock wave. It is very important and will appear again when the terminal ballistics material is discussed. It can be written in terms of total specific energy e, or if some of the energy components are negligible, it can be written in terms of enthalpy h. At this stage, we will use the latter expression, but we shall switch when we discuss shock in the terminal ballistics section. Writing the Rankine–Hugoniot relationship in terms of enthalpy, we have

$$h_2 - h_1 = \frac{1}{2}(p_2 - p_1)\left(\frac{1}{\rho_2} - \frac{1}{\rho_1}\right) \tag{2.167}$$

The strength of a shock is normally assessed by the change in pressure across it. In other words, its strength is given by the ratio p_2/p_1. If we assume the material through which this shock is propagating is an ideal gas, Equations 2.164 through 2.166 can be combined with the relationships provided in Equation 2.61 to yield expressions that relate all the values ahead of the shock to values after the passage of the shock. The details of this are available in the book by Cooper and Kurowski [16]. These expressions are as follows:

$$\frac{T_2}{T_1} = \frac{p_2}{p_1}\left[\frac{\frac{\gamma+1}{\gamma-1} + \frac{p_2}{p_1}}{1 + \frac{\gamma+1}{\gamma-1}\left(\frac{p_2}{p_1}\right)}\right] \tag{2.168}$$

$$\frac{\rho_1}{\rho_2} = \frac{1 + \frac{\gamma+1}{\gamma-1}\left(\frac{p_2}{p_1}\right)}{\frac{\gamma+1}{\gamma-1} + \frac{p_2}{p_1}} \tag{2.169}$$

The real power of these equations lies in the fact that with just the strength of the shock known, we can determine all the other items of interest. In the aforementioned equations, we have seen that given the pressure ratio (i.e., the strength) of the shock, we know the temperature behind the wave and the increase in density across the wave. We can also determine the wave speed U and induced velocity u_p through

$$U = a_1\sqrt{\frac{\gamma+1}{2\gamma}\left(\frac{p_2}{p_1}-1\right)+1} \tag{2.170}$$

$$u_{\mathrm{p}} = U\left(1-\frac{\rho_1}{\rho_2}\right) = \frac{a_1}{\gamma}\left(\frac{p_2}{p_1}-1\right)\sqrt{\frac{\dfrac{2\gamma}{\gamma+1}}{\dfrac{\gamma-1}{\gamma+1}+\dfrac{p_2}{p_1}}} \tag{2.171}$$

If we change reference frames to one in which we are stationary and the shock is moving, then the assumption of constant stagnation enthalpy h_0 is no longer valid. The reason is best illustrated by an example. Consider the gas ahead of the shock wave. It was initially motionless, so $h_1 = h_{01}$. After the wave passes, we know that the temperature must increase, so $h_2 > h_1$. Additionally, the gas is now moving at velocity u_{p} so that we can see

$$h_1 = h_{01} < h_{02} = h_2 + \frac{1}{2}u_{\mathrm{p}}^2 \tag{2.172}$$

It is by this very same logic that the stagnation pressure, temperature, and density must also increase.

We have discussed some governing equations, but let us break for a moment to discuss why a gas shocks up. If we closely examine Equation 2.170, we see that a higher pressure causes a faster motion of the wave. If we imagine a shock wave as depicted in Figure 2.27 moving to the right, we can pick out three points that we shall follow for some time. Point A is essentially the beginning of the pressure increase and at the unshocked initial pressure. Point B is at some pressure in between the peak pressure of the shock and the initial pressure of the material into which the shock is propagating. Point C is at the peak shock pressure. From Equation 2.170, we see that the local velocity of point B must be greater than point A and that the local velocity of point C is greater still. This means that at some time t, these points must converge, thereby forming a step discontinuity in pressure. This step discontinuity is the way we model the shock—there is actually a very small distance over which a shock will develop so that the pressure increase is rapid, but continuous. With this information, we see that compression shocks are the only admissible shocks. Later we will introduce rarefactions that are the converse of shocks. Since the pressure decreases in a rarefaction wave, the wave will tend to spread out over time and distance.

Now that we have the governing equations, we shall examine the behavior of a shock wave in a shock tube. A shock tube is a device as depicted in Figure 2.28 that contains two

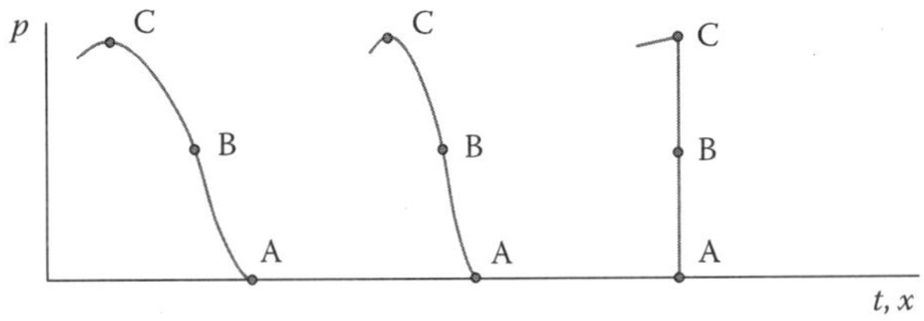

FIGURE 2.27
Formation of a shock wave.

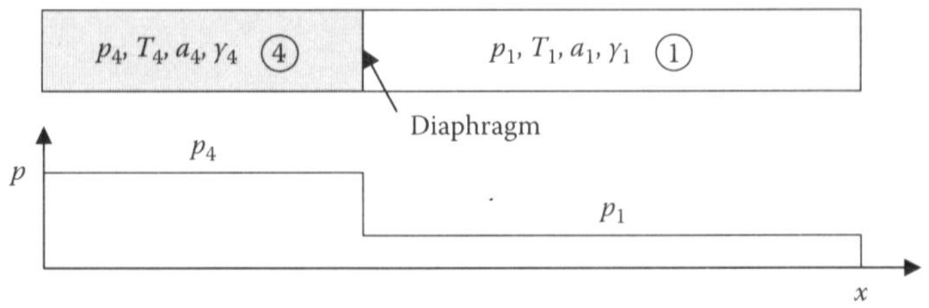

FIGURE 2.28
Shock tube in its initial state. (From Anderson, J. D., *Modern Compressible Flow with Historical Perspective*, McGraw-Hill, New York, 2003. With permission.)

regions of gas. These regions are separated by a diaphragm that can burst very quickly and contain one gas at high pressure and another at lower pressure. The gases could be different (thus, all their properties as well) as can their temperatures. Below the graphic of the shock tube is a pressure vs. distance plot showing that the pressure in region 4 (the high-pressure region) is greater than that in region 1, and the diaphragm divides the two regions. If the diaphragm bursts, then a shock will propagate into the lower pressure region, increasing the pressure, and a rarefaction wave (to be discussed later) will propagate into the high-pressure region, decreasing the pressure. If we examine the shock tube after some very short time t, the situation will appear as shown in Figure 2.29 with the corresponding pressure–distance profile. One of the most interesting aspects of compressible fluid flow is that if we know what the initial states of the ideal gases in the shock tube are, we can predict the pressures and temperatures of the unsteady motion afterward by Equations 2.161 through 2.171. In fact, we can predict the pressure behind the initial shock from

$$\frac{p_4}{p_1} = \frac{p_2}{p_1}\left\{1 - \frac{(\gamma_4 - 1)\left(\dfrac{a_1}{a_4}\right)\left(\dfrac{p_2}{p_1} - 1\right)}{\sqrt{2\gamma_1\left[2\gamma_1 + (\gamma_1 + 1)\left(\dfrac{p_2}{p_1} - 1\right)\right]}}\right\}^{\frac{-2\gamma_4}{(\gamma_4 - 1)}} \tag{2.173}$$

Equation 2.173 needs to be solved for the initial shock strength p_2/p_1, but afterward, Equations 2.161 through 2.171 can be directly used to calculate the parameters of interest. The details of this derivation can be found in the book by Cooper and Kurowski [16].

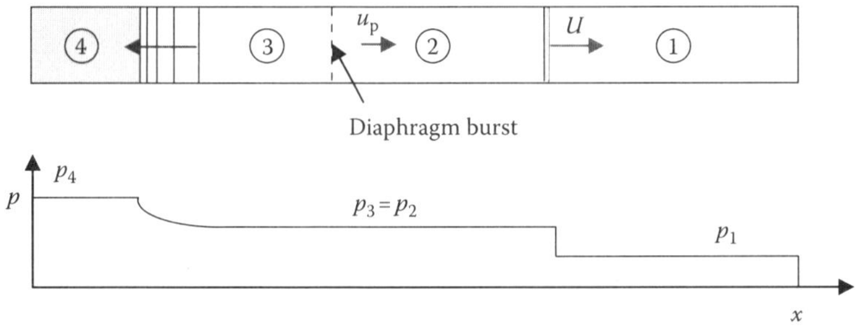

FIGURE 2.29
Shock tube after some short time t. (From Anderson, J. D., *Modern Compressible Flow with Historical Perspective*, McGraw-Hill, New York, 2003. With permission.)

You can see from Figures 2.28 and 2.29 that the shock tube is not infinite in extent. At some point, the shock produced by the bursting of the diaphragm will reach the right end of the tube. When this occurs, the condition at the wall is such that no flow through it is possible. Consider that all the fluid behind the shock wave is moving with induced velocity u_p toward the wall. Clearly, this situation is at odds with the wall-imposed boundary condition of zero velocity. Nature handles this issue by creating a shock wave of strength U_R that propagates back into the fluid that is heading toward the wall at velocity u_p. Notice that we have used a velocity here to define the strength of the shock—it should be clear by now that if we know either the velocity of the shock or the pressure ratio, we can find the other. The net effect of this reflected shock is that it stagnates the fluid between it and the fixed end of the shock tube as depicted in Figure 2.30. In this figure, we shall assume that the tube is extremely long on the rarefaction side so we do not have to discuss rarefaction reflections, yet. If we look at our conservation Equations 2.161 through 2.163 and consider that the shock wave sees material coming into it at velocity $U_R + u_p$, we can write equations for the reflected shock that are analogous to Equations 2.164 through 2.166 for the incident wave. These are

$$\rho_2\left(U_R + u_p\right) = \rho_5 U_R \tag{2.174}$$

$$p_2 + \rho_2\left(U_R + u_p\right)^2 = p_5 + \rho_5 U_R^2 \tag{2.175}$$

$$h_2 + \frac{1}{2}\left(U_R + u_p\right)^2 = h_5 + \frac{1}{2}U_R^2 \tag{2.176}$$

A simple method for the determination of the speed of the reflected shock is to first determine the Mach number of the incident pulse Ma_s through

$$\text{Ma}_s = \frac{U}{a_1} \tag{2.177}$$

The relationship between the incident shock velocity and the reflected velocity is derived Cooper and Kurowski [16] and given by

$$\frac{\text{Ma}_R}{\text{Ma}_R^2 - 1} = \frac{\text{Ma}_s}{\text{Ma}_s^2 - 1}\sqrt{1 + \frac{2(\gamma - 1)}{(\gamma + 1)^2}\left(\text{Ma}_s^2 - 1\right)\left(\gamma + \frac{1}{\text{Ma}_s^2}\right)} \tag{2.178}$$

Here Ma_R is the Mach number of the reflected shock that can be converted to a velocity through use of

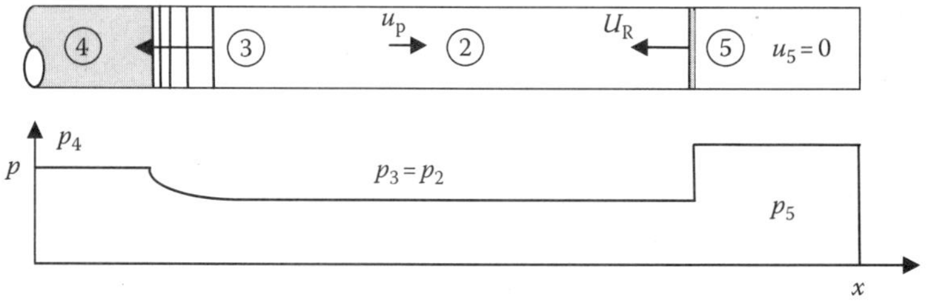

FIGURE 2.30
Shock tube after a reflection of the incident wave.

$$\mathrm{Ma}_R = \frac{U_R + u_p}{a_2} \tag{2.179}$$

In our discussions on shock waves throughout the terminal ballistics sections, we will make use of time–distance diagrams, so-called *t–x* plots. It is prudent to introduce them here as reinforcement of the shock wave discussion. A *t–x* plot places distance on the abscissa and time on the ordinate. Because of this placement, which is opposite to the normal function vs. time plots, we need to adjust some of our logic that we have used to. For instance, slopes of straight lines on these diagrams are reciprocal velocities. If we consider the situation in Figure 2.30 and draw a *t–x* plot for it with the origin starting from the initial diaphragm location, we would have a plot as depicted in Figure 2.31. We shall examine the shocks in this diagram first. If we assume that the incident shock immediately forms (this is not really true, as we learned earlier, but close enough for our purposes), it propagates toward the wall that is located at point x_2 in our figure. If we wanted to determine what the velocity distribution was in this device at any time t, we would examine a horizontal line in the figure. For instance, if we examined the situation at time t_1, we would see that the material in the unshaded region up to point x_1 would have a velocity u_p, and everything between x_1 and x_2 (the wall) would have zero velocity. Once the incident shock reflects off the wall, a new shock of velocity U_R propagates back into the fluid. This is depicted by the upper line in the diagram. Note that the slope is greater on this reflected shock, indicative of a lower velocity than the incident wave. The material in the shaded region behind this wave has been stagnated to zero velocity. We can use a *t–x* diagram to determine how a particle moves over time. Consider a particle initially located at location x_1. It remains stationary until the shock wave passes by at time t_1, as indicated by a vertical line. At time t_1, the incident shock passes it and induces a velocity u_p to the particle. When the particle moves at velocity u_p, it will trace out a line on the diagram that has a slope of $1/u_p$. While this particle is moving at velocity u_p, the shock interacts with the wall and reflects at time t_2. While the reflected shock is approaching, the observed particle has no idea that anything is about to happen and continues to move at velocity u_p until the reflected shock passes by at time t_3.

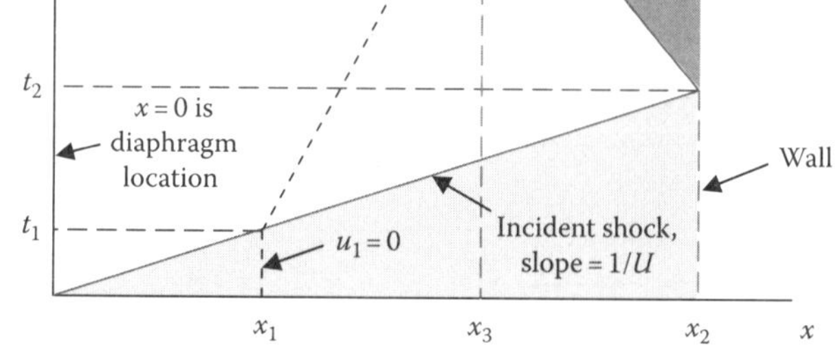

FIGURE 2.31
t–x plot for the reflection of a shock wave. (From Anderson, J. D., *Modern Compressible Flow with Historical Perspective*, McGraw-Hill, New York, 2003. With permission.)

This passage of the reflected shock stagnates the particle to zero velocity, and its motion (or lack thereof) traces out a vertical line. A final point of interest regarding t–x plots is that we can actually see the compression of the material. If we consider all the materials initially between points x_1 and x_2, we see that after the passage of the shock and its reflection, it has all been compressed to the region between x_3 and x_2. With this information, the basis for our future discussions using t–x plots is established.

A rarefaction wave, sometimes known as an expansion or relief wave, is the means by which nature handles a sudden drop in pressure. As we stated earlier, compression waves (also known as condensations) eventually coalesce into shocks that are analyzed as step discontinuities in pressure. This coalescence was brought about by the fact that the local velocity increases with increasing pressure. In a rarefaction, the opposite is true. A rarefaction increases over time because the pressure at the head of the wave is greater than that at the tail of the wave. In the case of our shock tube, the head of the rarefaction will propagate at the local speed of sound in the material (a_4 in Figure 2.29), while the tail will propagate at a velocity ($u_3 - a_3$) that is equal to ($u_p - a_2$). This is schematically depicted in Figure 2.32. Throughout the rarefaction wave, the velocity continuously decreases between these two values. Because of this continuous decrease in velocity, it is common to model the decrease as a series of wavelets. The more wavelets we include, the smoother the curve. If we use Figure 2.32 to trace a particle path after the bursting of the diaphragm, we see that the particle would not move until the head of the rarefaction wave passed by it. After the passage of the head of the wave, the velocity would continuously increase until the passage of the tail of the wave, after which it would be moving at velocity u_p. The length of the rarefaction can be determined at any time by scribing a horizontal line through the diagram. If we do this at two points in time on the diagram, we can see how the length of the wave increases.

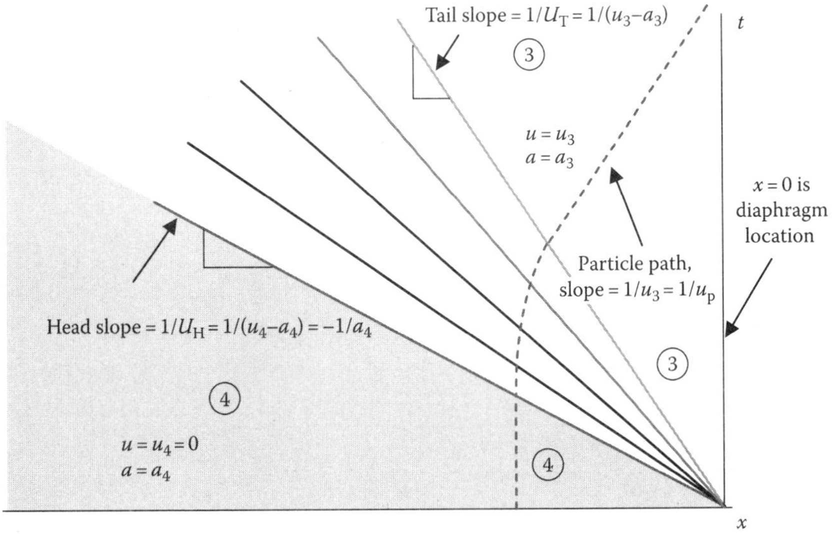

FIGURE 2.32
(See color insert.) t–x plot for a rarefaction wave.

What is depicted in Figure 2.32 is a simple, centered rarefaction wave. A wave is considered simple if all the characteristics (the rays emanating from the origin) are straight. Reflections of a rarefaction are somewhat more complicated than those of a shock. The reflection of the head of the rarefaction wave must pass through the characteristics of the rest of the wave being both affected by as well as affecting them. The result is that the characteristics tend to bend, making the calculations somewhat more complex. We will handle this in a simplified fashion later, but the interested reader is directed to Cooper and Kurowski [16] for an outstanding treatment for handling these situations.

We now have sufficient information to handle the fluid mechanics of interior and exterior ballistics. We shall treat the formation of shocks and rarefactions as necessary in the terminal ballistics section.

Problem 18

The principle behind a muzzle brake on a gun is to utilize some of the forward momentum of the propelling gases to reduce the recoil on the carriage. In the simple model that follows (Figure 2.33), the brake is assumed to be a flat plate with the jet of gases impinging upon it. If the jet diameter is 105 mm and the velocity and density of the gas (assume air) are 750 m/s and 0.457 kg/m^3, respectively, find the force on the weapon in Newtons assuming the gases are directed 90° to the tube and the flow is steady.

Answer: −2225.9 [N]

Problem 19

Some engineer gets the idea that if deflecting the muzzle gases to the side is a good idea, then deflecting it rearward would be better (until, of course, an angry gun crew gets hold of him/her). If the jet diameter is again 105 mm and the velocity and density of the gas (again assume air) are 750 m/s and 0.457 kg/m^3, respectively, find the force on the weapon in Newtons assuming the gases are directed 150° to the tube and the flow is steady (Figure 2.34).

Answer: −4153.5 [N]

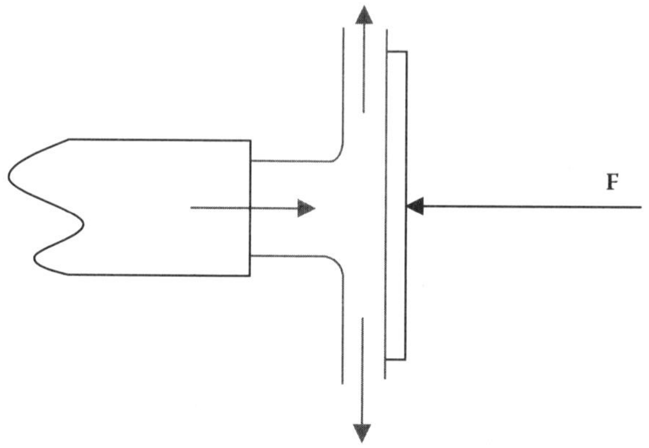

FIGURE 2.33
Normal deflection of flow through a muzzle brake.

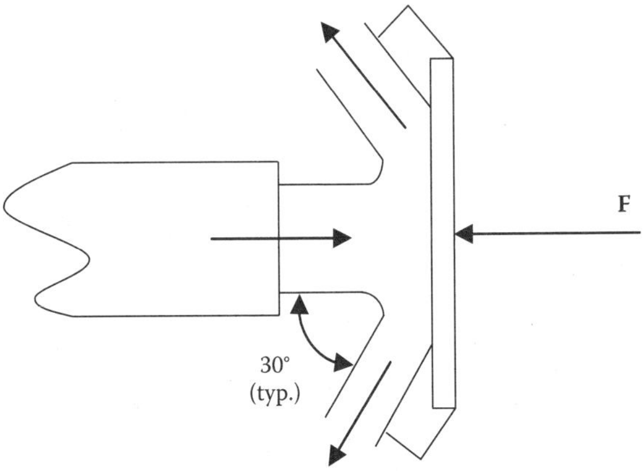

FIGURE 2.34
Rearward deflection of flow through a muzzle brake.

Problem 20
Consider a shock tube that is 6 ft long with a diaphragm at the center. Air is contained in both sections ($\gamma = 1.4$). The pressure in the high-pressure region is 2000 psi. The pressure in the low-pressure region is 14.7 psi. The temperature in both sections is initially 68°F.

1. When the diaphragm bursts, determine the following:
   a. The velocity that the shock wave propagates into the low–pressure region
      Answer: 2798 [ft/s]
   b. The induced velocity behind the wave
      Answer: 1946 [ft/s]
   c. The velocity of a wave normally reflected off the wall (relative to the laboratory)
      Answer: 1232 [ft/s]
   d. The temperature behind the incident wave
      Answer: 657 [°F]
2. Draw a *t*–*x* diagram of the event. Include the path of a particle located 2 ft from the diaphragm.

Problem 21
An explosion generates a shock wave in still air. Assume we are far enough from the initial explosion that we can model the wave as a one-dimensional shock. Assume that the pressure generated by the explosion was 10,000 psi and the ambient atmospheric pressure, density, and temperature are 14.7 psi, 0.06 lbm/ft^3, and 68°F, respectively. Determine the following:

1. The static pressure behind the wave (assume $\gamma = 1.4$, and since we are far away from the effects of the explosion, assume $a_1/a_4 \approx 0.5$)
   Answer: $p_2 = 376.6$ [psi]

2. The velocity that the wave propagates in still air
   Answer: $U = 5294$ [ft/s]
3. The induced velocity that a building would see after the wave passes
   Answer: $u_p = 4212$ [ft/s]
4. The velocity of a wave normally reflected off a building
   Answer: $U_R = 1921$ [ft/s]

References

1. Wark, K., *Thermodynamics*, 5th ed., McGraw-Hill, New York, 1988.
2. Jones, L., and Atkins, P., *Chemistry, Molecules, Matter, and Change*, 4th ed., W. H. Freeman, New York, 2003.
3. Masterson, W., Slowinski, E., and Stanitski, C., *Chemical Principles*, 6th ed., Saunders College Publishing, Philadelphia, PA, 1985.
4. Beer, F. P., and Johnson, R., *Mechanics of Materials*, 2nd edn., McGraw-Hill, New York, 1992.
5. Van Wylen, G. J., and Sonntag, R. E., *Fundamentals of Classical Thermodynamics*, 3rd ed., John Wiley & Sons, New York, 1986.
6. Fermi, E., *Thermodynamics*, Dover Publications, New York, 1956.
7. Corner, J., *Theory of the Interior Ballistics of Guns*, John Wiley & Sons, New York, 1950.
8. Cooper, P. W., *Explosives Engineering*, Wiley-VCH, New York, 1996.
9. Sucec, J., *Heat Transfer*, William C. Brown, Dubuque, IA, 1985.
10. Kays, W. M., and Crawford, M. E., *Convective Heat and Mass Transfer*, 3rd ed., McGraw-Hill, New York, 1993.
11. Fox, R. W., and McDonald, A. T., *Introduction to Fluid Mechanics*, 4th ed., John Wiley & Sons, New York, 1992.
12. Anderson, J. D., *Modern Compressible Flow with Historical Perspective*, 3rd ed., McGraw-Hill, New York, 2003.
13. Cengel, Y. A., and Boles, M. A., *Thermodynamics and Engineering Approach*, 4th ed., McGraw-Hill, New York, 2002.
14. Moran, M. J., and Shapiro, H. N., *Fundamentals of Engineering Thermodynamics*, 5th ed., John Wiley & Sons, New York, 2004.
15. Masterson, W. L., Slowinski, E. J., and Stanitski, C. L., *Chemical Principles*, 5th ed., Saunders College Publishing, Philadelphia, PA, 1981.
16. Cooper, P. W., and Kurowski, S. R., *Introduction to the Technology of Explosives*, Wiley-VCH, New York, 1996.
17. Hayes, T. J., *Elements of Ordinance*, John Wiley & Sons, New York, 1938.
18. Eringen, A. C., Liebowitz, H., Koh, S. L., and Crowley, J. M. (Eds), *Mechanics and Chemistry of Solid Propellants, Proceedings of the Fourth Symposium on Naval Structural Mechanics*, Pergamon Press, London, 1965.
19. Kubota, N., *Propellants and Explosives*, Wiley-VCH, New York, 2002.
20. Turns, S. R., *An Introduction to Combustion*, 2nd ed., McGraw-Hill, New York, 2000.
21. Borman, G. L., and Ragland, K. W., *Combustion Engineering*, WCB-McGraw-Hill, New York, 1998.
22. Yang, V., Brill, T. B., and Ren, W.-Z., Solid propellant chemistry, combustion, and motor interior ballistics, *Progress in Astronautics and Aeronautics*, Vol. 185, American Institute of Astronautics and Aeronautics, Reston, VA, 2000.

23. Kuhl, A. L., Leyer, J. C., Borisov, A. A., and Sirignano, W. A., Dynamics of deflagrations and reactive systems flames, *Progress in Astronautics and Aeronautics*, Vol. 131, American Institute of Astronautics and Aeronautics, Washington, DC, 1989.
24. Kuhl, A. L., Leyer, J. C., Borisov, A. A., and Sirignano, W. A., Dynamics of gaseous combustion, *Progress in Astronautics and Aeronautics*, Vol. 151, American Institute of Astronautics and Aeronautics, Washington, DC, 1993.
25. White, F. M., *Fluid Mechanics*, 5th ed., McGraw-Hill, New York, 2003.
26. Panton, R. L., *Incompressible Flow*, 2nd ed., John Wiley & Sons, New York, 1995.
27. Currie, I. G., *Fundamental Mechanics of Fluids*, 2nd ed., McGraw-Hill, New York, 1993.
28. White, F. M., *Viscous Fluid Flow*, 3rd ed., McGraw-Hill, New York, 2006.

3

Analytic and Computational Ballistics

Chapter 2 has provided us with the necessary background to discuss procedures that calculate the behavior of projectiles and propellant in the gun tube. The chapter had to be brief because detailed treatment of any one of the subjects could be (and is) collected into complete texts in its own right. The reader is directed to the references at the end of the chapter if a more complete background in the individual subject is felt to be necessary.

Much like other introductory texts on difficult subjects, this chapter shall begin with fundamental treatments that will allow the reader to perform meaningful calculations of interior ballistic problems. This simplified treatment will, by its very nature, not provide exact answers but answers which are reasonable from an engineering viewpoint. As will be discussed, more exact methods require a varying degree of computer assets.

3.1 Computational Goal

The interior ballistician is charged with devising a propellant charge that will deliver the projectile of interest to the gun muzzle intact, with the desired muzzle velocity, with no damage to the weapon from excess pressure and with high probability that successive charges propelling the same projectiles will produce the same results. To do this, the ballistician must be able to predict a priori what the charge will do, i.e., what pressures will both the gun and the projectile experience during travel down the bore and what the velocity and acceleration profile would be during the travel to the muzzle. Over the centuries, ballisticians, including some quite eminent mathematicians and physicists, have devised computational schemes that can be used to make such predictions. We intend to explore a few of these analytic tools in sufficient depth so that the physics and mathematics become clear to the user, who would then also be able to discern reasonable answers from patently erroneous ones.

It is important to understand how predictions of pressure and velocity are experimentally verified in real guns. Such understanding has led to the development of pressure ratios that allow the gun and projectile designers to know what pressures are acting on the gun and on the projectile at locations that practical instrumentation has some difficulty capturing. Pressure is most readily measured at the base of the gun chamber, where the gas flow is minimal or nonexistent. When pressure taps are introduced along the bore to take measurements while the projectile is traveling and the gases are flowing, it has been found that turbulent flow and shock waves make such measurements difficult to interpret. Copper crusher gauges are used in which small copper cylinders are crushed to a barrel shape in the gauge by the applied pressure and the distortion of the cylinders measured. These gauges are placed in the base of the charge and recovered after firing. Distortion is checked against a calibration chart and the pressure is quickly read. Of course, pressure measured in this way is representative only of the maximum pressure sensed by the gauge, which gives no indication of its profile in time or in travel.

Even such a primitive measurement was and still is of use; because the designer would know the maximum pressure, the projectile and gun would have to contend with an indication that piezo-type pressure gauges are properly functioning. These gauges are still widely used to check the pressure consistency of already developed charges. Knowledge of how that copper pressure was related to pressures at other locations during the travel was a great advance. When pressure ratios were devised that related chamber pressure to the pressure at the base of the projectile during its travel down the bore, these were greatly appreciated by the designers. Even better was the introduction of electronic piezo gauges installed through the breech that allowed the measurement of pressure over time so that a pressure–time profile could be available. The study of a few of the computational theories that develop these ratios follows in succeeding sections.

3.2 Lagrange Gradient

To determine the time-dependent motion of the projectile, we need to make some assumptions about the behavior of the gas pushing it out of the gun. These assumptions will involve the pressure, mass, and density distribution of the gas. We shall refer to the sketch in Figure 3.1 in the text that follows. We shall continue to use x as the distance from the projectile base position at the seating location to its position at all later times with the time derivative defined as

$$\frac{dx}{dt} = \dot{x} = V \tag{3.1}$$

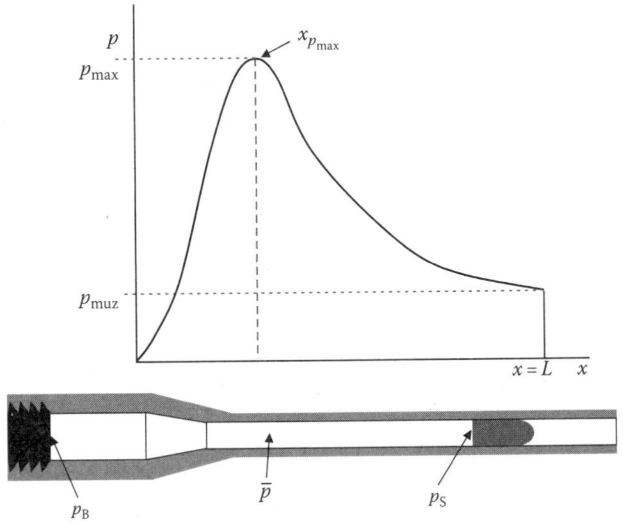

FIGURE 3.1
Pressure–distance relationship in a typical gun firing.

We will first assume that the gas density is uniform in the volume behind the projectile at time t. We can then write, for any time t, that

$$\rho = \rho\left(x_g, t\right) \tag{3.2}$$

In this equation, x_g is the x-location of the gas mass center behind the projectile. We shall also assume that there is no spatial gradient in density at any time; thus,

$$\left.\frac{\partial \rho}{\partial x_g}\right|_t = 0 \tag{3.3}$$

We can also write the continuity equation for a compressible fluid as

$$\frac{\partial \rho}{\partial t} + \frac{\partial}{\partial x_g}\left(\rho V_{x_g}\right) = 0 \tag{3.4}$$

We can expand the continuity in Equation 3.4 as

$$\frac{\partial \rho}{\partial t} + \frac{\partial \rho}{\partial x_g} V_{x_g} + \rho \frac{\partial V_{x_g}}{\partial x_g} = 0 \tag{3.5}$$

Inserting our assumption of the absence of a spatial density gradient allows us to simplify this expression to

$$\frac{\partial \rho}{\partial t} + \rho \frac{\partial V_{x_g}}{\partial x_g} = 0 \tag{3.6}$$

Now because we stated that the density was not a function of x, we can remove the partial derivative notation from the temporal term and rearrange to yield

$$\frac{1}{\rho}\frac{\mathrm{d}\rho}{\mathrm{d}t} = \frac{\partial V_{x_g}}{\partial x_g} \tag{3.7}$$

Assume at this point that the solid propellant in the charge has all turned to gas; then, what was initially a solid propellant of charge weight c is now a gas of identical weight c. So the gas density is this weight divided by the volume the gas occupies or

$$\rho(t)|_{\mathrm{c}} \frac{c}{\mathrm{V}(t)} \tag{3.8}$$

Here, the subscript c refers to conditions after the charge has burned out, i.e., all the solid has evolved into gas. If the base of the projectile has moved a distance x and the bore area is A, then the volume behind the projectile containing gas is

$$\mathrm{V}(t) = Ax(t) \tag{3.9}$$

If we insert Equation 3.9 into Equation 3.8 and then take the derivative with respect to time, the result can be simplified to

$$\frac{1}{x}\frac{\mathrm{d}x}{\mathrm{d}t} = \frac{\partial V_{x_g}}{\partial x_g} \tag{3.10}$$

Note that there is a difference here between x and x_g:

- x is the location of the base of the projectile.
- x_g is the location of the mass center of the gas.

If we integrate Equation 3.10 with respect to x_g and use the boundary conditions of $V_{x_g} = 0$ when $x_g = 0$, then we get

$$\frac{x_g}{x}\frac{dx}{dt} = V_{x_g}\left(x_g\right) \tag{3.11}$$

Now, since x is the position of the base of the projectile at time t, we see that dx/dt is the velocity of the projectile at time t, so we can write

$$\frac{V}{x} = \frac{V_{x_g}}{x_g} \tag{3.12}$$

This implies that the gas particle velocity linearly varies from the breech face to the projectile base and is a fundamental tenet of the Lagrange* approximation. We can describe the kinetic energy of the gas stream as

$$KE_g = \frac{1}{2} m_g V_{x_g}^2 \tag{3.13}$$

But, as described earlier, the mass of the gas is its density times the volume it occupies at time t; therefore,

$$KE_g = \int_0^x \frac{1}{2} \rho A V_{x_g}^2 \, dx_g \tag{3.14}$$

Moving the spatially constant terms $\rho A/2$ outside the integral and performing the integration gives us

$$KE_g = \frac{\rho A}{2} \frac{V^2}{x^2} \frac{x_g^3}{3} \Bigg|_0^x = \frac{1}{6} \rho A x V^2 \tag{3.15}$$

But we know from our earlier work that

$$\rho A x = c \tag{3.16}$$

So we can write

$$KE_g = \frac{1}{6} c V^2 \tag{3.17}$$

The total kinetic energy of the system (neglecting recoil) is

$$KE_g = \frac{\rho A}{2} \frac{V^2}{x^2} \frac{x_g^3}{3} \Bigg|_0^x = \frac{1}{6} c V^2 \tag{3.18}$$

But the kinetic energy of the projectile is

$$KE_{shot} = \frac{1}{2} w_p V^2 \tag{3.19}$$

---

* Joseph-Louis Lagrange, 1736–1813, Italian French mathematician.

where w_p is the projectile mass.

So the Lagrange approximation for kinetic energy is

$$\text{KE}_{\text{tot}} = \frac{1}{2} w_p V^2 + \frac{1}{6} c V^2 = \frac{1}{2}\left(w_p + \frac{c}{3} \right) V^2 \tag{3.20}$$

In this development, the volume of gas is assumed to be a cylinder of cross-sectional area A. In reality, it is not; while the bore is cylindrical, the chamber is not. Chamber diameters can be much greater than bore diameters. To account for this, an effort to modify the Lagrange gradient approximations has been performed [1]. This will be subsequently explored. The changes from the Lagrange gradient will be found to be small but not insignificant and the so-called chambrage gradient will be explained in Section 3.3 and incorporated in the discussion of numerical methods in Section 3.4.

We can describe the linear momentum of the gas stream as

$$\text{Mom}_g = m_g V_{x_g} \tag{3.21}$$

But, again, the mass of the gas is its density times the volume it occupies at time t; therefore,

$$\text{Mom}_g = \int_0^x \rho A V_{x_g} \, dx_g \tag{3.22}$$

We can use our continuity relationship in Equation 3.11 to write

$$\text{Mom}_g = \rho A \int_0^x \left(\frac{x_g}{x} \frac{dx}{dt} \right) dx_g = \rho A \int_0^x \left(\frac{x_g}{x} V \right) dx_g \tag{3.23}$$

Performing the integration gives us

$$\text{Mom}_g = \rho A \frac{V}{x} \frac{x_g^2}{2} \Bigg|_0^x = \frac{1}{2} \rho A x V \tag{3.24}$$

If we recall Equation 3.16, we can write

$$\text{Mom}_g = \frac{1}{2} c V \tag{3.25}$$

The total linear momentum of the system (neglecting the weapon) is

$$\text{Mom}_{\text{tot}} = \text{Mom}_{\text{shot}} + \text{Mom}_g \tag{3.26}$$

The linear momentum of the projectile is

$$\text{Mom}_{\text{shot}} = w_p V \tag{3.27}$$

So the Lagrange approximation for linear momentum is

$$\text{Mom}_{\text{tot}} = w_p V + \frac{1}{2} c V = \left(w_p + \frac{c}{2} \right) V \tag{3.28}$$

Because we are looking for the parameters, we can readily measure breech pressure and muzzle velocity, and we must develop predictive equations for them, i.e., equations for pressure in terms of charge parameters and equations of motion of the projectile. To do this, we adopt a Lagrangian approach to track the motion of a particle of gas. What follows is a derivation for the equation of motion for an element of gas. For a rigorous, complete treatment, see any text on fluid mechanics, for example, that of Panton [2].

For differentiation that tracks a fluid element (the Lagrangian approach), the following differential operator (called the substantial derivative or material derivative) is used:

$$\frac{\mathrm{D}}{\mathrm{D}t} = \frac{\partial}{\partial t} + u\frac{\partial}{\partial x} + v\frac{\partial}{\partial y} + w\frac{\partial}{\partial z} \tag{3.29}$$

where u, v, and w are the velocity components in the x, y, and z directions, respectively.

If we consider a one-dimensional flow operating on the velocity $V_{x_g}(x)$ (here V_{x_g} is the axial velocity and replaces u in Equation 3.29),

$$\frac{\mathrm{D}V_{x_g}}{\mathrm{D}t} = \frac{\partial V_{x_g}}{\partial t} + V_{x_g}\frac{\partial V_{x_g}}{\partial x} \tag{3.30}$$

In vector notation, the gradient of a function is

$$\nabla = \mathbf{i}\frac{\partial}{\partial x} + \mathbf{j}\frac{\partial}{\partial y} + \mathbf{k}\frac{\partial}{\partial z} \tag{3.31}$$

Force is the time rate of change of momentum:

$$\mathbf{F} = \frac{\partial}{\partial t}(m\mathbf{v}) \tag{3.32}$$

It can be shown using Gauss's theorem [3] that the rate of change of linear momentum of the fluid inside a surface S in changing to surface S' in time dt is

$$\int_{\mathrm{V}} \rho\frac{d\mathbf{v}}{dt}\mathrm{dV} \tag{3.33}$$

From the equations of motion for an inviscid fluid, we know that the total force equals the pressure on the boundary element integrated over the boundary plus the body force **F** integrated over the mass in S or

$$\int_{S} p\mathbf{n}\mathrm{d}S + \int_{\mathrm{V}} \mathbf{F}\rho\mathrm{dV} = -\int_{\mathrm{V}} \nabla p\mathrm{dV} + \int_{\mathrm{V}} \mathbf{F}\rho\mathrm{dV} \tag{3.34}$$

Because by Gauss's theorem,

$$\int_{S} p\mathbf{n}\mathrm{d}S = -\int_{\mathrm{V}} \nabla p\mathrm{dV} \tag{3.35}$$

By setting the right-hand side (RHS) of Equation 3.35 equal to Equation 3.33, we get

$$\int_{\mathrm{V}} \left[\mathbf{F}\rho - \nabla p - \rho\frac{\mathrm{d}\mathbf{v}}{\mathrm{d}t}\right]\mathrm{dV} = 0 \tag{3.36}$$

Since V is arbitrarily chosen, the sum in brackets must equal zero:

$$\mathbf{F}\rho - \nabla p - \rho\frac{\mathrm{d}\mathbf{v}}{\mathrm{d}t} = 0 \tag{3.37}$$

In the absence of a body force **F**, we can rewrite this as

$$\frac{d\mathbf{v}}{dt} = \frac{1}{\rho}\nabla p \tag{3.38}$$

We can write Equation 3.38 as follows for one-dimensional flow and negligible body forces:

$$\frac{dp}{dx_g} = -\rho\left[\frac{\partial V_{x_g}}{\partial t} + V_{x_g}\frac{\partial V_{x_g}}{\partial x_g}\right] \tag{3.39}$$

Note here that we have used the substantial derivative for the velocity of the gas stream.

If we insert the relationship for the gas stream velocity we obtained through the continuity Equation 3.11 into Equation 3.39, we can write

$$\frac{dp}{dx_g} = -\rho\left[\frac{\partial}{\partial t}\left(\frac{x_g}{x}\frac{dx}{dt}\right) + V_{x_g}\frac{\partial V_{x_g}}{\partial x_g}\right] \tag{3.40}$$

or

$$\frac{dp}{dx_g} = -\rho\left[\frac{\partial}{\partial t}\left(\frac{x_g}{x}\frac{dx}{dt}\right) + \left(\frac{x_g}{x}\frac{dx}{dt}\right)\frac{\partial}{\partial x_g}\left(\frac{x_g}{x}\frac{dx}{dt}\right)\right] \tag{3.41}$$

We can combine terms in Equation 3.41 as follows:

$$\frac{dp}{dx_g} = -\rho\left[-\frac{x_g}{x^2}\left(\frac{dx}{dt}\right)^2 + \frac{x_g}{x}\frac{d^2x}{dt^2} + \frac{x_g}{x^2}\left(\frac{dx}{dt}\right)^2\right] \tag{3.42}$$

Simplifying the expression gives us

$$\frac{dp}{dx_g} = -\rho\frac{x_g}{x}\frac{d^2x}{dt^2} \quad \text{or} \quad \frac{dp}{dx_g} = -\rho\frac{x_g}{x}\ddot{x} \tag{3.43}$$

If we use our relationship between density and charge weight in Equation 3.8, we can write

$$\frac{dp}{dx_g} = -\frac{cx_g}{Ax^2}\ddot{x} \tag{3.44}$$

We can integrate this expression with respect to the gas mass center as

$$\int_0^{x_g}\frac{dp}{dx_g}dx_g = -\frac{c}{Ax^2}\ddot{x}\int_0^{x_g}x_g dx_g \tag{3.45}$$

Performing the integration yields

$$p = -\frac{cx_g^2}{2Ax^2}\ddot{x} + \text{constant} \tag{3.46}$$

Let us now define p_S as the pressure at the projectile base; p_B as the pressure at the breech; $\bar{p}$ as the mean pressure in volume behind projectile; and p_R as the pressure resisting projectile motion (force/bore area).

We will develop the equations of motion both with a resistive force in the bore (such as friction and the air being compressed in front of the projectile) and neglecting the resistance. If we write Newton's second law for a projectile being acted upon by propellant gases, we have

$$w\ddot{x} = Ap_S \tag{3.47}$$

Writing this in terms of the acceleration, we get

$$\ddot{x} = \frac{A}{w}p_S \tag{3.48}$$

where w is the projectile mass. Since the base of our projectile is at location x and the local pressure on the base is p_S, we can substitute these values into Equation 3.46 for x_g and p to obtain

$$p_S = -\frac{c}{2A}\ddot{x} + \text{constant} \tag{3.49}$$

Keep in mind that this is a local condition that we applied to the gas in the vicinity of the base (that the mass center of the gas is approximately at x). We can rearrange Equation 3.49 to yield our constant of integration:

$$\text{constant} = p_S + \frac{c}{2A}\ddot{x} \tag{3.50}$$

If we use Equation 3.48, we obtain

$$\text{constant} = p_S + \frac{c}{2A}\frac{A}{w}p_S = \left(1 + \frac{c}{2w}\right)p \tag{3.51}$$

Inserting this constant back into Equation 3.46 gives us

$$p = -\frac{cx_g^2}{2Ax^2}\ddot{x} + \left(1 + \frac{c}{2w}\right)p_S = -\frac{cx_g^2}{2Ax^2}\left(\frac{A}{w}\right)p_S + \left(1 + \frac{c}{2w}\right)p_S \tag{3.52}$$

or

$$p = p_S + p_S\left(1 - \frac{x_g^2}{x^2}\right)\frac{c}{2w} \tag{3.53}$$

This equation relates the pressure at the base of the projectile to that at the location of the gas mass center. By similar logic, at the breech, $x_g = 0$ and the pressure $p = p_B$, so we can substitute the values into Equation 3.53 to obtain a relationship between the breech pressure and the pressure at the projectile base:

$$p_B = p_S + p_S\frac{c}{2w} = p_S\left(1 + \frac{c}{2w}\right) \tag{3.54}$$

The space-mean pressure is formally defined as

$$\bar{p} = \frac{1}{x}\int_0^x p\,dx_g \tag{3.55}$$

If we insert Equation 3.53 into this equation, we get

$$\bar{p} = \frac{1}{x}\int_0^x \left[p_S + p_S\left(1 - \frac{x_g^2}{x^2}\right)\frac{c}{2w}\right]dx_g \tag{3.56}$$

Solving this integral, inserting the limits of integration, and simplifying yields

$$\bar{p} = \frac{1}{x}\left[p_S x_g + p_S \frac{c}{2w} x_g - p_S \frac{c}{2w}\frac{x_g^3}{3x^2}\right]_0^x \tag{3.57}$$

Inserting the limits of integration gives us

$$\bar{p} = p_S + p_S \frac{c}{2w} - \frac{1}{3} p_S \frac{c}{2w} \tag{3.58}$$

By simplifying, we get

$$\bar{p} = p_S\left(1 + \frac{c}{3w}\right) \tag{3.59}$$

This equation relates the space-mean pressure to the base pressure acting on the projectile. We now have equations that relate breech pressure to base pressure (Equation 3.54) and space-mean pressure to base pressure (Equation 3.59). What is missing is a relationship between breech pressure and space-mean pressure. We can arrive at the desired result by dividing Equation 3.59 by Equation 3.54, simplifying to yield

$$\frac{\bar{p}}{p_B} = \frac{p_S\left(1 + \frac{c}{3w}\right)}{p_S\left(1 + \frac{c}{2w}\right)} \tag{3.60}$$

For easier manipulation, it is sometimes desirable to expand Equation 3.60 in a Taylor series, which, neglecting higher-order terms, would be

$$\frac{\bar{p}}{p_B} = 1 - \frac{c}{6w} + \cdots \tag{3.61}$$

To account for the effects of bore resistance, we again write Newton's second law for a projectile being acted upon by propellant gases and bore friction as

$$w_1 \ddot{x} = A(p_S - p_R) \tag{3.62}$$

Here we have used w_1 to represent the mass of the projectile (you will see why later) and have included a resistive pressure p_R that fights the gas pressure. Note that the resistive pressure is simply the resistive force divided by the bore cross-sectional area so that the terms in the preceding equation can be conveniently grouped—it is not actually a pressure at all. Writing this in terms of the acceleration, we get

$$\ddot{x} = \frac{A}{w_1}(p_S - p_R) \tag{3.63}$$

Again, since the base of our projectile is at location x and the local pressure on the base is p_S, we can substitute these values into Equation 3.46 for x_g and p to obtain

$$p_S = -\frac{c}{2A}\ddot{x} + \text{constant} \tag{3.64}$$

Remember that this is a local condition that we applied to the gas in the vicinity of the base where the mass center of the gas is approximately at x.

Following the same procedure that we used to arrive at a general expression for pressure, but now with bore resistance, we rearrange Equation 3.64 to find the constant of integration and, with simplification, arrive at

$$\text{constant} = p_S + \frac{c}{2A}\ddot{x} \tag{3.65}$$

If we use Equation 3.63, we obtain

$$\text{constant} = p_S + \frac{c}{2A}\left[\frac{A}{w_1}(p_S - p_R)\right] = \left(1 + \frac{c}{2w_1}\right)p_S - \frac{c}{2w_1}p_R \tag{3.66}$$

Inserting this constant back into Equation 3.64 gives us

$$p = -\frac{cx_g^2}{2Ax^2}\ddot{x} + \left(1 + \frac{c}{2w_1}\right)p_S - \frac{c}{2w_1}p_R \tag{3.67}$$

or

$$p = -\frac{cx_g^2}{2Ax^2}\left(\frac{A}{w_1}\right)(p_S - p_R) + \left(1 + \frac{c}{2w}\right)p_S - \frac{c}{2w_1}p_R \tag{3.68}$$

or

$$p = p_S + (p_S - p_R)\frac{c}{2w_1}\left(1 - \frac{x_g^2}{x^2}\right) \tag{3.69}$$

which relates the pressure at the base of the projectile to the pressure at the gas mass center, but with the effect of bore friction included. Having this general equation, we can again proceed as we did earlier to find equations that relate breech to base pressure, space-mean to base pressure, and space-mean to breech pressure for the bore friction case. These are

$$p_B = p_S + \frac{c}{2w_1}p_S - \frac{c}{2w_1}p_R \tag{3.70}$$

$$\bar{p} = p_S\left(1 + \frac{c}{3w_1}\right) - p_R\frac{c}{3w_1} \tag{3.71}$$

$$\frac{\bar{p}}{p_B} = \frac{1 + \left(1 - \frac{p_R}{p_S}\right)\frac{c}{3w_1}}{1 + \left(1 - \frac{p_R}{p_S}\right)\frac{c}{2w_1}} \tag{3.72}$$

If we plot breech, space-mean, and base pressure vs. x, the position of the projectile base, we shall see that a gradient of pressure exists in which the breech pressure is always the greatest and the base pressure is always the smallest. This is the so-called Lagrange gradient and is fundamental to our modeling of the propellant gas. There are instances where this gradient is reversed, and this usually means that we have a problem—a so-called negative delta-p. This is indicative of a fragmented propellant charge caused by poor ignition. A charge designed to move with the accelerating projectile, the traveling charge, is a notable exception.

We are essentially prepared now to treat the **F** in the equation $\mathbf{F} = m\mathbf{a}$, which is, in its simplest form, the base pressure times the base area. We now need to determine what generates the pressure, what the acceleration of the projectile will be, and how the acceleration and the ever-increasing volume behind the projectile affect the pressure. To do this, we shall review the equations from our initial discussions of propellant burning as well as revisit our notation before moving on to combining everything into the equations of motion of the projectile.

We have previously defined the following quantities and shall simply list them here for ease of reference. The first quantity is the acceleration $\ddot{x}$ of the projectile. The pressure acting on the base of the projectile is the stimulus that causes the acceleration: $p_S(t)$ is the pressure at the base of the projectile at time t.

We usually measure pressure at the breech of the weapon, and it is this pressure that we are determining when we examine the burning of the propellant. We need to constantly refer this breech pressure to the base pressure. We do this by invoking the Lagrange gradient assumption, keeping in mind that we begin by neglecting bore resistance:

$$p_B = p_S\left(1+\frac{c}{2w}\right) \tag{3.73}$$

We can write Newton's second law for the force on the projectile base as

$$w\ddot{x} = p_S A \tag{3.74}$$

If we substitute our Lagrange gradient into this equation to put it in terms of the breech pressure and the projectile velocity, we can write

$$w\frac{dV}{dt} = \frac{p_B}{\left(1+\dfrac{c}{2w}\right)}A \tag{3.75}$$

If we want to include losses, w can be replaced by w_1, an effective projectile mass that can be thought of as an added mass due to the combination of resistance of bore friction, engraving by the rifling, resistance due to compression of the air ahead of the shot, etc. Then, we have

$$\left(w_1+\frac{c}{2}\right)\frac{dV}{dt} = p_B A \tag{3.76}$$

The burning of the propellant generates the pressure that pushes on the projectile. Let us now recall the equation that relates the amount of propellant turned to gas

$$\phi = (1-f)(1+\theta f) \tag{3.77}$$

Also recall that the rate of gas evolution (burning) is a function of the pressure

$$D\frac{df}{dt} = -\beta\bar{p} \approx \beta p_B \tag{3.78}$$

In our earlier study of solid propellant combustion, we developed an equation of state for the gas that related ϕ to the pressure and the distance the projectile traveled

$$p_B(x+l) = \frac{c\lambda\phi}{A}\left[\frac{1+\dfrac{c}{2w_1}}{1+\dfrac{c}{3w_1}}\right] \tag{3.79}$$

Finally, we have our equation of motion for the projectile:

$$\left(w_1 + \frac{c}{2}\right)\frac{\mathrm{d}V}{\mathrm{d}t} = p_{\mathrm{B}}A \tag{3.80}$$

whose initial conditions are $x = 0$, $V = 0$, and $f = 1$ at $t = 0$.

These equations may be manipulated to determine the parameters of interest as functions of the fraction of the remaining web $f = f(t)$: x is the projectile travel; V is the projectile velocity; and p_{B} is the breech pressure

If we combine Equations 3.79 and 3.80, eliminating the breech pressure between them, we can write

$$-\frac{D}{\beta}\frac{\mathrm{d}f}{\mathrm{d}t} = \frac{w_1}{A}\left(1+\frac{c}{2w_1}\right)\frac{\mathrm{d}V}{\mathrm{d}t} \tag{3.81}$$

We can rearrange this to get the equation in terms of the projectile acceleration:

$$-\frac{DA}{\beta w_1}\left[\frac{1}{1+\dfrac{c}{2w_1}}\right]\frac{\mathrm{d}f}{\mathrm{d}t} = \frac{\mathrm{d}V}{\mathrm{d}t} \tag{3.82}$$

This can be integrated, resulting in

$$V = -\frac{AD}{\beta w_1\left(1+\dfrac{c}{2w_1}\right)}f + \text{constant} \tag{3.83}$$

If we insert the initial conditions that $V = 0$ when $f = 1$, Equation 3.83 yields

$$\text{constant} = \frac{AD}{\beta w_1\left(1+\dfrac{c}{2w_1}\right)} \tag{3.84}$$

This gives us

$$V(t) = \frac{AD}{\beta w_1\left(1+\dfrac{c}{2w_1}\right)}(1 - f(t)) \tag{3.85}$$

From the preceding equation, we can rearrange Equation 3.82 as follows:

$$\frac{\mathrm{d}V}{\mathrm{d}t} = -\frac{AD}{\beta w_1\left(1+\dfrac{c}{2w_1}\right)}\frac{\mathrm{d}f}{\mathrm{d}t} \tag{3.86}$$

We can now substitute our relationship between velocity and fraction of web remaining (Equation 3.86) into our projectile equation of motion (Equation 3.80), algebraically simplifying it and inserting the relationship for base pressure (Equation 3.79) to yield

$$-\frac{D}{\beta}\frac{\mathrm{d}f}{\mathrm{d}t} = \frac{c\lambda\phi}{A(x+l)}\left[\frac{1+\dfrac{c}{2w_1}}{1+\dfrac{c}{3w_1}}\right] \tag{3.87}$$

This may be rearranged to obtain

$$\frac{df}{dt} = -\frac{c\lambda\phi\beta}{AD(x+l)}\left[\frac{1+\dfrac{c}{2w_1}}{1+\dfrac{c}{3w_1}}\right] \tag{3.88}$$

Using the chain rule transformation between distance and time,

$$\frac{df}{dt} = \frac{df}{dx}\frac{dx}{dt} = V\frac{df}{dx} \tag{3.89}$$

This can be written as

$$\frac{df}{dx} = \frac{1}{V}\frac{df}{dt} \tag{3.90}$$

Now let us substitute Equations 3.85 and 3.88 into Equation 3.90 and simplify the result to yield

$$\frac{df}{dx} = -\frac{w_1 c\lambda\phi\beta^2}{A^2D^2(x+l)(1-f)}\left[\frac{\left(1+\dfrac{c}{2w_1}\right)^2}{1+\dfrac{c}{3w_1}}\right] \tag{3.91}$$

To examine the rate of change of f, the fraction of web remaining, with the travel distance x, we take the reciprocal of Equation 3.91:

$$\frac{dx}{df} = -\frac{A^2D^2(1-f)}{w_1 c\lambda\phi\beta^2}\left[\frac{1+\dfrac{c}{3w_1}}{\left(1+\dfrac{c}{2w_1}\right)^2}\right](x+l) \tag{3.92}$$

Here l is an initial chamber length, to be described subsequently.
By inserting the relationship between ϕ and f, from Equation 3.77, we get

$$\frac{dx}{df} = -\frac{A^2D^2}{w_1 c\lambda\beta^2}\left[\frac{1+\dfrac{c}{3w_1}}{\left(1+\dfrac{c}{2w_1}\right)^2}\right]\frac{(x+l)}{(1+\theta f)} \tag{3.93}$$

Equation 3.93 is cumbersome, and following Corner [4], we find that we can define a dimensionless central ballistic parameter M that is a function of the gun, the charge, and the projectile, i.e., the system

$$M = \frac{A^2D^2}{w_1 c\lambda\beta^2}\left[\frac{1+\dfrac{c}{3w_1}}{\left(1+\dfrac{c}{2w_1}\right)^2}\right] \tag{3.94}$$

This simplifies our distance–web fraction relationship to

$$\frac{dx}{df} = -M\frac{(x+l)}{(1+\theta f)} \tag{3.95}$$

The dimensionless nature of M can be shown if we note that c and w_1 are mass units. We can also write the units of the burning rate coefficient as

$$[\beta] = \left[\frac{D}{p_B}\frac{df}{dt}\right] \Rightarrow [\beta] = \left[\frac{L^2T}{M}\right] \tag{3.96}$$

The units of the propellant force λ are

$$[\lambda] = \left[\frac{\text{energy}}{\text{mass}}\right] = \left[\frac{ML}{T^2}\times\frac{L}{M}\right] = \left[\frac{L}{T}\right]^2 = [\text{velocity}]^2 \tag{3.97}$$

Using these in our definition of the central ballistic parameter, we can show

$$[M] = \left[\frac{L^6}{\left(\frac{L^2T}{M}\right)^2 M\times M\left(\frac{L}{T}\right)^2}\right] = [1] \tag{3.98}$$

thereby demonstrating that M is dimensionless. Equation 3.95, repeated here,

$$\frac{dx}{df} = -M\frac{(x+l)}{(1+\theta f)} \tag{3.95}$$

shows how M relates the burning of the propellant f with the expansion of the volume represented by x, the travel. A similar concept appears in all interior ballistic theories.

We are now in a position to compute the parameters that the interior ballistician really seeks, the velocity of the projectile and the concomitant instantaneous breech pressure for each point along its travel down the tube. If we wish to know the pressure on the base of the projectile or the space-mean pressure in the volume behind the projectile, we only need to apply the appropriate Lagrange approximation to the breech pressure. This is an extraordinary result. By simply understanding the amount of propellant burnt and some gun or propellant or projectile data, we have determined everything we need to know about the interior ballistics.

We can now take the distance–web fraction relationship and directly integrate it. But we must examine two distinct cases for θ, the form factor of the grain. One where $\theta \neq 0$ and one where $\theta = 0$. Let us separate the variables in Equation 3.95 to obtain

$$\frac{dx}{(x+l)} = -M\frac{df}{(1+\theta f)} \tag{3.99}$$

Then, we can write for $\theta \neq 0$

$$\int_0^x \frac{dx}{(x+l)} = -M\int_0^f \frac{df}{(1+\theta f)} \tag{3.100}$$

or for $\theta = 0$

$$\int_0^x \frac{dx}{(x+l)} = -M\int_0^f df \tag{3.101}$$

The evaluation of the integral Equation 3.100 for $\theta \neq 0$ gives us

$$\ln(x + l) = -\frac{M}{\theta}\ln(1 + \theta f) + \ln(K) = \ln\left[K(1 + \theta f)^{-\frac{M}{\theta}}\right] \tag{3.102}$$

Solving for K with the initial conditions $f = 1$ at $x = 0$, we get

$$K = l(1 + \theta)^{\frac{M}{\theta}} \tag{3.103}$$

This constant, when inserted into the original Equation 3.102, gives us

$$x + l = l\left(\frac{1+\theta}{1+\theta f}\right)^{\frac{M}{\theta}} \tag{3.104}$$

In a similar fashion, we can evaluate Equation 3.101 to give us the distance–remaining web fraction relation for $\theta = 0$:

$$x + l = le^{M(1-f)} \tag{3.105}$$

We now know how the web fraction f varies with distance and have, incidentally, shown the algebraic simplification inherent in the central ballistic parameter M. We can now pursue a relationship between pressure and web fraction. If we look at Equation 3.88, we see the quotient on the RHS and note that this frequently occurs. We define it as our Lagrange ratio R_L, another simplification:

$$R_L = \frac{1 + \dfrac{c}{2w_1}}{1 + \dfrac{c}{3w_1}} \tag{3.106}$$

This will allow us to rewrite Equation 3.79 in a simpler form as

$$p_B(x + l) = \frac{c\lambda\phi}{A}R_L \tag{3.107}$$

We will make an assumption that the chamber and bore diameters are the same and relate the volume behind the projectile to a fictitious chamber length l. (We will correct this subsequently when we examine the chambrage gradient.)

$$V_i = Al = U - \frac{c}{\delta} \tag{3.108}$$

In this expression, U is the empty chamber volume and c/δ is the volume occupied by the solid propellant charge.

We continue by substituting Equations 3.108 and 3.77 into Equation 3.107 and rearranging to give our relationship between the breech pressure and the fraction of remaining web for $\theta \neq 0$.

$$p_B = \frac{\lambda c}{V_i}\frac{\left(1+\dfrac{c}{2w_1}\right)}{\left(1+\dfrac{c}{3w_1}\right)}(1-f)(1+\theta f)\left(\frac{1+\theta f}{1+\theta}\right)^{\frac{M}{\theta}} \tag{3.109}$$

We can also proceed in similar fashion for $\theta = 0$ by substituting Equations 3.105 and 3.77 into Equation 3.107 to find the relationship between the breech pressure and the fraction of remaining web.

$$p_B = \frac{\lambda c R_L}{V_i}(1-f)(1+\theta f)\exp[-M(1-f)] \quad \text{for } \theta = 0 \tag{3.110}$$

In summary, we now have the definition of the central ballistic parameter (Equation 3.94) and equations that relate velocity as a function of remaining web (Equation 3.85) and travel as a function of remaining web for different form functions (Equations 3.104 and 3.105) as well as breech pressure as a function of remaining web for different form functions (Equations 3.109 and 3.110). With these, we can now integrate the governing equations and find solutions for velocity at peak pressure, at all-burnt point of travel, and at muzzle exit.

Equations 3.109 and 3.110 are somewhat cumbersome to work with, so we shall define a parameter Q as follows:

$$Q = \frac{\lambda c}{V_i}\frac{\left(1+\dfrac{c}{2w_1}\right)}{\left(1+\dfrac{c}{3w_1}\right)} = \frac{\lambda c}{V_i}R_L \tag{3.111}$$

Then we can rewrite Equation 3.109 in a more compact way:

$$p_B = Q(1-f)\,(1+\theta f)\left(\frac{1+\theta f}{1+\theta}\right)^{\frac{M}{\theta}} \tag{3.112}$$

The maximum or peak pressure attained is then found by taking the first derivative of p_B with respect to f and setting it equal to zero:

$$\frac{dp_B}{df} = Q\left[(1-f)\left(\frac{M}{\theta}+1\right)\theta(1+\theta f)^{\frac{M}{\theta}} - (1+\theta f)^{\frac{M}{\theta}+1}\right] = 0 \tag{3.113}$$

Let us solve Equation 3.113 for f. By introducing the subscript m to denote maximum, we obtain the product of two terms:

$$(1+\theta f_m)[(M+\theta)(1-f_m) - (1+\theta f_m)] = 0 \tag{3.114}$$

In solving this, we have two choices here, either

$$(1+\theta f_m) = 0 \quad \text{or} \quad [(M+\theta)(1-f_m) - (1+\theta f_m)] = 0 \tag{3.115}$$

The first would only be admitted for the special case of $\theta = -M$; thus, our criterion for determination of f_m is

$$(M+\theta)(1-f_m) - (1+\theta f_m) = 0 \tag{3.116}$$

and

$$f_m = \frac{M+\theta-1}{M+2\theta} \tag{3.117}$$

Equation 3.117 works for all values of θ. If we want to determine ϕ_m, the fraction of propellant burnt at peak pressure, we call on our relationship between f and ϕ (Equation 3.77). Here we have denoted peak values with the subscript m:

$$\phi = (1-f)(1+\theta f) \tag{3.118}$$

Substitution of Equation 3.117 into the preceding equation yields

$$\phi_{\mathrm{m}}=1-\left(\frac{M+\theta-1}{M+2\theta}\right)\left[1+\theta\left(\frac{M+\theta-1}{M+2\theta}\right)\right] \tag{3.119}$$

This, when simplified, gives

$$\phi_{\mathrm{m}}=\frac{(1+\theta)[M+\theta+\theta(M+\theta)]}{[M+2\theta]^2}=\frac{(1+\theta)[(M+\theta)(1+\theta)]}{[M+2\theta]^2} \tag{3.120}$$

or the following (valid for all θ):

$$\phi_{\mathrm{m}}=\frac{(M+\theta)(1+\theta)^2}{[M+2\theta]^2} \tag{3.121}$$

In designing a gun (and for other reasons), it is desirable to know where a projectile is in its travel down the bore when the pressure is at a maximum. This involves substitution of Equation 3.117 into Equation 3.104, and for the case where $\theta \neq 0$, this yields

$$x_{\mathrm{m}}+l=l\left[\frac{1+\theta}{1+\theta\left(\frac{M+\theta-1}{M+2\theta}\right)}\right]^{\frac{M}{\theta}} \tag{3.122}$$

Simplifying, we finally get

$$x_{\mathrm{m}}+l=l\left[\frac{(M+2\theta)}{(M+\theta)}\right]^{\frac{M}{\theta}} \quad \text{for} \quad \theta \neq 0 \tag{3.123}$$

For the case where $\theta = 0$, we substitute Equation 3.117 into Equation 3.105, which we rewrite as follows:

$$x_{\mathrm{m}}+l=l\exp[M(1-f_{\mathrm{m}})] \tag{3.124}$$

On substitution, we get

$$x_{\mathrm{m}}+l=l\exp\left[M\left(1-\frac{M+\theta-1}{M+2\theta}\right)\right] \tag{3.125}$$

Simplifying this result and substituting $\theta = 0$ into it gives us

$$x_{\mathrm{m}}+l=l\mathrm{e} \quad \text{for} \quad \theta=0 \tag{3.126}$$

for a zero form factor.

Knowing now the position of the peak pressure in the bore, we can then ask what the breech pressure would be at this point. We can insert the value we have for the fraction of remaining web at peak pressure f_{m} back into the breech pressure equation for $\theta \neq 0$ (Equation 3.109):

$$p_{\mathrm{Bm}}=Q(1-f_{\mathrm{m}})(1+\theta f_{\mathrm{m}})\left(\frac{1+\theta f_{\mathrm{m}}}{1+\theta}\right)^{\frac{M}{\theta}} \tag{3.127}$$

With considerable algebraic simplification including substituting the values for Q, and the Lagrange ratio R_L for the case $\theta \neq 0$, we finally arrive at

$$p_{B_m} = \frac{\lambda c}{V_i}\left(\frac{1+\dfrac{c}{2w_1}}{1+\dfrac{c}{3w_1}}\right)\left(\frac{(1+\theta)^2(M+\theta)^{\frac{M}{\theta}+1}}{(M+2\theta)^{\frac{M}{\theta}+2}}\right) \tag{3.128}$$

Following a similar procedure, we now insert the value we have for the fraction of remaining web at peak pressure f_m into the breech pressure equation for $\theta = 0$ (Equation 3.110):

$$p_{B_m} = Q(1-f_m)\exp[-M(1-f_m)] \tag{3.129}$$

Then, substituting for Q and R_L and simplifying, we see that we have characterized the breech pressure at the instant that peak pressure is achieved down the bore:

$$p_{B_m} = \frac{\lambda c}{V_i}\left(\frac{1+\dfrac{c}{2w_1}}{1+\dfrac{c}{3w_1}}\right)\left(\frac{1}{M\exp[1]}\right) \tag{3.130}$$

Determining the breech pressure and travel when the solid grains have been completely consumed is also of considerable interest. We shall use the subscript c to represent charge burnout. If the charge is designed properly, it will burn out somewhere in the bore which allows us to extract most energy from the propellant and reduce the muzzle blast. Recall from our previous discussions that at $t = 0$, $x = 0$, $f = 1$, and $\phi = 0$, but at the all-burnt point (subscript c), $t = t_c$, $x = x_c$, $f = 0$, and $\phi = 1$. If we substitute $f = 0$ in Equations 3.109 and 3.110, we obtain the breech pressure at the instant of charge burnout:

$$p_{B_c} = \frac{\lambda c}{V_i}\frac{\left(1+\dfrac{c}{2w_1}\right)}{\left(1+\dfrac{c}{3w_1}\right)}\left(\frac{1}{1+\theta}\right)^{\frac{M}{\theta}} \quad \text{for } \theta \neq 0 \tag{3.131}$$

and

$$p_{B_c} = \frac{\lambda c}{V_i}\frac{\left(1+\dfrac{c}{2w_1}\right)}{\left(1+\dfrac{c}{3w_1}\right)}\exp[-M] \quad \text{for } \theta = 0 \tag{3.132}$$

The travel of the projectile at burnout is a data point we usually want to know because if this distance turns out to be longer than the barrel length, then the charge is not completely burnt when the projectile exits. If we substitute $f = 0$ in Equations 3.104 and 3.105, we obtain the position of the projectile at the instant of charge burnout

$$x_c + l = l(1+\theta)^{\frac{M}{\theta}} \quad \text{for} \quad \theta \neq 0 \tag{3.133}$$

and

$$x_c + l = le^M \quad \text{for} \quad \theta = 0 \tag{3.134}$$

It is a good idea to use these equations first to see whether the propellant burns out in the tube with the parameters we have designed into the grain. Still-burning grains leaving the tube signify a poorly designed charge. For completeness, however, if charge burnout happens outside the bore, the pressure at the breech location when the projectile leaves the muzzle may be calculated by evaluating f at the muzzle through Equation 3.104 or 3.105 and using this value to calculate p_B from Equation 3.109 or 3.110. The muzzle velocity could then be obtained from Equation 3.85.

If charge burnout is, as desired, in the bore, recall that there is still a net force (pressure) pushing on the projectile. A simple means of calculating this pressure is to assume that the process occurs so quickly that it is essentially adiabatic and that the gas behaves as an ideal gas. With these assumptions and the initial conditions that the pressure is p_{Bc} and the distance is x_c, we have a closed form solution to the problem. It is vitally important to note that the expansion of the gas after charge burnout is neither adiabatic nor isentropic; however, the result is usually within about 5% with respect to pressure. The isentropic relationships for an ideal gas are

$$\frac{p}{p_0} = \left(\frac{\rho}{\rho_0}\right)^{\gamma} = \left(\frac{\frac{1}{v}}{\frac{1}{v_0}}\right)^{\gamma} = \left(\frac{v_0}{v}\right)^{\gamma} = \left(\frac{v}{v_0}\right)^{-\gamma} \tag{3.135}$$

This equation relates pressure to specific volume in a general way, but we need to involve the projectile travel as well. We can express the volume behind the projectile as a function of distance as

$$\mathrm{V}(x) = (x + l)A \tag{3.136}$$

Then, the specific volume of the gas is this value divided by the mass of the gas, which we know is still c after burnout. Thus, we can write Equation 3.136 in its intensive form as

$$v(x) = (x + l)\frac{A}{c} \tag{3.137}$$

Furthermore, we can specialize this to the point at which the charge burns out and write

$$v(x_c) = (x_c + l)\frac{A}{c} \tag{3.138}$$

We can now tailor Equation 3.135 to our needs by substituting the conditions at burnout as our reference conditions

$$\frac{\bar{p}(x)}{\bar{p}_c} = \left(\frac{v(x)}{v(x_c)}\right)^{-\gamma} = \left(\frac{x + l}{x_c + l}\right)^{-\gamma} \tag{3.139}$$

This condition occurs often, so we define

$$r(x) = \frac{x + l}{x_c + l} \tag{3.140}$$

which can be written in a more compact form as

$$\frac{\bar{p}(x)}{\bar{p}_c} = r(x)^{-\gamma} \tag{3.141}$$

A sketch of this situation is depicted in Figure 3.2.

These extensive preparations have finally brought us to the goal of interior ballistics and the design of a gun system—imparting a desired velocity to a projectile and being able to repeat that process at will. We have developed the means for predicting how the propellant burns over time, how the breech, space-mean, and base pressures vary with time, and where the projectile moves to in relation to these pressures. Now we will focus on the velocity of the projectile during this ballistic cycle. Recall that the kinetic energy of the projectile plus the gas losses was written as

$$\text{KE}_{\text{tot}} = \frac{1}{2} w_p V^2 + \frac{1}{6} c V^2 = \frac{1}{2}\left(w_p + \frac{c}{3}\right) V^2 \tag{3.20}$$

The work done on the projectile and the gas from charge burnout to the point of interest (usually muzzle exit) is

$$W = A \int_{x_c}^{x} \bar{p}\, dx \tag{3.142}$$

Combining Equations 3.20 and 3.142 and inserting Equation 3.139 yields

$$\frac{1}{2}\left(w_1 + \frac{c}{3}\right)\left[V^2(x) - V^2(x_c)\right] = A\bar{p}_c \int_{x_c}^{x} \left(\frac{x+l}{x_c+l}\right)^{-\gamma} dx \tag{3.143}$$

We must keep in mind that we are using space-mean pressure here because the work is being done on both the projectile and the gas. We can use any of breech, space-mean, and base pressures (with the appropriate relationship) because we know each in terms of the others. Integrating and rearranging, we get

$$\frac{1}{2}\left(w_1 + \frac{c}{3}\right)\left[V^2(x) - V^2(x_c)\right] = A\bar{p}_c \frac{1}{(1-\gamma)(x_c+l)^{-\gamma}} (x+l)^{-\gamma+1}\Bigg|_{x_c}^{x} \tag{3.144}$$

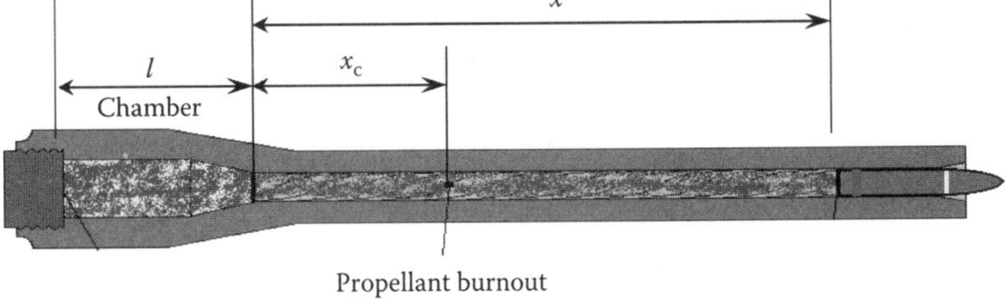

FIGURE 3.2
Position of projectile at charge burnout.

Evaluation of the limits of integration yields

$$\frac{1}{2}\left(w_1+\frac{c}{3}\right)\left[V^2(x)-V^2(x_c)\right]=A\bar{p}_c\frac{1}{(1-\gamma)(x_c+l)^{-\gamma}}\left[(x+l)^{1-\gamma}-(x_c+l)^{1-\gamma}\right] \quad (3.145)$$

By rearranging and inserting Equation 3.132 into the preceding equation, we get a velocity relationship after burnout for $\theta = 0$:

$$V^2(x)-V^2(x_c)=\frac{2A\lambda ce^{-M}(x_c+l)^{1-\gamma}}{V_i(1-\gamma)(x_c+l)^{-\gamma}(w_1+\frac{c}{3})}\left[\frac{(x+l)^{1-\gamma}}{(x_c+l)^{1-\gamma}}-1\right] \quad (3.146)$$

But recall that the volume $V_i = Al$, and if we define

$$\Phi=\frac{2}{(1-\gamma)}\left[\left(\frac{x+l}{x_c+l}\right)^{1-\gamma}-1\right] \quad (3.147)$$

we can write

$$V^2(x)-V^2(x_c)=\frac{\lambda c(x_c+l)\exp[-M]}{l\left(w_1+\dfrac{c}{3}\right)}\Phi \quad (3.148)$$

Here again we revert to the cases of the form factor being zero or not zero and examine the former first. Recall Equation 3.105. For conditions after charge burnout, there is no remaining web ($f = 0$), so we can write

$$x_c+l=le^{M} \quad \text{for} \quad \theta=0 \quad \text{and} \quad f=0 \quad (3.149)$$

Rearranging this and substituting it into Equation 3.149 yields

$$V^2(x)-V^2(x_c)=\frac{\lambda c}{\left(w_1+\dfrac{c}{3}\right)}\Phi \quad (3.150)$$

This allows us to calculate the velocity of a projectile after burnout of the web for $\theta = 0$. The case for nonzero θ requires further examination and manipulation. In Equation 3.85, we had a general expression for velocity as a function of remaining web. After burnout, this becomes

$$V(x_c)=\frac{AD}{\beta\left(w_1+\dfrac{c}{2}\right)} \quad (3.151)$$

Since we are working with kinetic energy, squaring this gives

$$V^2(x_c)=\frac{A^2D^2}{\beta^2\left(w_1+\dfrac{c}{2}\right)^2} \quad (3.152)$$

We defined the central ballistic parameter M in Equation 3.94 and can rearrange it for our purposes into the form

$$\frac{c\lambda M}{\left(w_1+\frac{c}{3}\right)}=\frac{A^2D^2}{\beta^2\left(w_1+\frac{c}{2}\right)^2} \tag{3.153}$$

When this is compared with Equation 3.152, we conclude that

$$V^2(x_c)=\frac{\lambda cM}{\left(w_1+\frac{c}{3}\right)} \tag{3.154}$$

This is an important result—it says that just by knowing the physical parameters of the weapon, projectile, and charge, one can predict the projectile velocity at charge burnout. With this result and continuing the examination for nonzero θ, we can then say that Equation 3.150 is valid for any θ. If we solve Equation 3.150 for the velocity at any point $V(x)$, insert Equation 3.154, and rearrange the terms, we get

$$V^2(x)=\frac{c\lambda}{\left(w_1+\frac{c}{3}\right)}(M+\Phi) \quad \text{for } \theta\neq 0 \tag{3.155}$$

This result along with our earlier work now allows us to determine projectile velocity at all points in the gun for charge grains of all form factors both before and after burnout.

We have been through many derivations that have led us to the essentials of interior ballistics—breech pressure and velocity in terms of projectile travel. These results, furthermore, are in closed form, accessible to computation by hand calculator. Specialized pressures, space-mean pressure, and projectile base pressure may be computed from the breech pressure data by using the Lagrange approximations. Projectile design and gun design proceed from these equations. In the following sections, we shall discuss refinements to the Lagrange formulation with an emphasis on the use of modern computer programs that take the drudgery out of hand calculation and provide the ability to iterate solutions for small changes in the parameters.

Problem 1

You are asked to analyze the pressure of a charge zero (igniter) firing in an M31 boom for a 120 mm mortar projectile. You decide to examine it as a closed bomb first. Assume we have 59 g of M48 propellant (properties given in the following). The volume of the closed bomb is 5.822 in.^{3}. The propellant grains are balls (roughly spherical) with a diameter (web) of 0.049 in.

M48 propellant properties:

- Density $\rho = 0.056 \left[\frac{\text{lbm}}{\text{in.}^3}\right]$
- Ratio of specific heats $\gamma = 1.21$
- Covolume $b = 26.72 \left[\frac{\text{in.}^3}{\text{lbm}}\right]$

- Isochoric flame temperature $T_0 = 3720°F$
- Burn rate exponent $\alpha = 0.9145$
- Average burn rate coefficient $\beta = 0.0095\left[\frac{\text{in.}}{\text{s psi}^{0.9145}}\right]$
- Burn rate $D\frac{df}{dt} = 40.341\left[\frac{\text{in.}}{\text{s}}\right]$
- Force constant $\lambda = 391{,}000\left[\frac{\text{ft lbf}}{\text{lbm}}\right]$

1. Come up with the equation for the web fraction f as a function of time.

   Answer: $f = (1 - 823.29t)$ [%]
2. For a sphere, the fraction of propellant burnt has the functional form $\phi = 1 - f^3$; write this in terms of time and ϕ.

   Answer: $\phi = 2470t - 2{,}033{,}419t^2 + 558{,}031{,}251t^3$
3. Determine how long it will take for the propellant to burn halfway through and all the way through.

   Answer: The time to burn through halfway is 0.6 ms
4. Using the Noble–Abel equation of state, determine the pressure in the vessel when half of the propellant is burnt and when all the propellant is burnt. Note that this cannot occur usually as the propellant is a charge zero firing that is vented into the main ullage volume behind the mortar bomb (significantly greater volume).

   Answer: $p = 258{,}746\left[\frac{\text{lbf}}{\text{in.}^2}\right]$

Problem 2

If we use the Lagrange approximation in the examination of a 155 mm projectile launch, what is the average pressure in the volume behind a 102 lbm projectile if the breech pressure is 55,000 psi? The propelling charge weighs 28 lbm.

Answer: $\bar{p} = 52{,}787\left[\frac{\text{lbf}}{\text{in.}^2}\right]$

Problem 3

A 120 mm projectile is to be examined while in the bore of a tank cannon at a time of 4 ms from shot start. Over this time period, the projectile has acquired an average velocity of 1000 ft/s. The propellant grain (M15) is single perf ($\theta = 0$) with a 0.034 in. initial web. The covolume of the propellant is 31.17 in.^{3}/lbm. The density of the propellant is 0.06 lbm/in.^{3}. If the projectile weighs 50.4 lbm, the propellant weighs 12.25 lbm and the chamber volume is 330 in.^{3}. At this time, 0.02 in. of the web remains. The propellant force is 337,000 ft·lbf/lbm. Determine the breech pressure in the weapon. Be careful with the units!

Answer: $p_B = 21{,}784\left[\frac{\text{lbf}}{\text{in.}^2}\right]$

Problem 4

The Paris gun was a monstrous 210 mm weapon designed by Germany during the First World War to bombard Paris from some 70 mi away. It was unique in that it fired the first exoatmospheric projectile ever designed. The weapon had a chamber volume of 15,866 in.^{3}. Very little of the projectile protrudes into the chamber after it seats (so ignore the volume the base occupies). The length of travel for the projectile from shot start to shot exit is 1182 in. The projectile weighs 234 lb. The propelling charge weighs 430.2 lb. The propellant used was specially designed and was similar to US M26 propellant. It consisted of 64–68% nitrocellulose (NC), 25–29% nitroglycerin (NG) with 7% Centralite (symmetrical diethyl diphenylurea [$C_{17}H_{20}N_2O$]), and some other additives. The propellant was single perforated with a web thickness described in the following. Assume the propellant has the following properties. (Note that these are the authors' guesses—a better estimate of the properties can be found in the book by Bull and Murphy [5]).

- Adiabatic flame temperature $T_0 = 2881$ K
- Specific heat ratio $\gamma = 1.237$
- Covolume $b = 1.06\ \text{cm}^3/\text{g}$
- Density of solid propellant $\rho = 1.62\ \text{g/cm}^3$
- Propellant burn rate coefficient $\beta = 0.0707$ (cm/s)/MPa
- Web thickness $D = 0.217$ in.
- Propellant force $\lambda = 1019$ J/g

1. Using the preceding data, determine (a) the projectile base pressure in pounds per square inch, (b) velocity in feet per second, and (c) distance down the bore of the weapon in inches for peak pressure.

   Answer:

   a. $p_{S_{max}} = 31{,}548 \left[\frac{\text{lbf}}{\text{in.}^2}\right]$

   b. $V_{p_{max}} = 2880 \left[\frac{\text{ft}}{\text{s}}\right]$

   c. $x_{p_{max}} = 270.1$ [in.]

2. Determine the pressure in pounds per square inch at a point 3 in. behind the projectile base when the charge burns out

   Answer: $p_{x-3} = 28{,}637 \left[\frac{\text{lbf}}{\text{in.}^2}\right]$

3. Assuming that the gas behaves according to the Noble–Abel equation of state, determine the muzzle velocity of the projectile in feet per second.

   Answer: $V = 5791 \left[\frac{\text{ft}}{\text{s}}\right]$

Problem 5

A British 14 in. Mark VII gun has a chamber volume of 22,000 in.^{3}. A 5 in. of the projectile protrude into the chamber after it seats. The length of travel for the projectile from shot start to shot exit is 515.68 in. The weapon has a uniform twist of 1 in 30. The projectile weighs

1590 lb. The propelling charge weighs 338.25 lb. The propellant used is called SC and consists of 49.5% NC (12.2% nitrated) and 41.5% NG with 9% Centralite. Assume that SC propellant has the following properties:

- Adiabatic flame temperature $T_0 = 3090$ K
- Specific heat ratio $\gamma = 1.248$
- Covolume $b = 26.5$ in.^{3}/lbm
- Density of solid propellant $\rho = 0.0567$ lbm/in.^{3}
- Propellant burn rate $\beta = 0.000331$ (in./s)/(psi)
- Web thickness $D = 0.25$ in.
- Specific molecular weight $n = 0.04262$ lb-mol/lbm

1. Determine the force constant λ in feet-pound-force per pound-mass (ft.lbf/lbm).
2. Determine the central ballistic parameter for this gun–projectile combination.
3. Using the preceding data, determine the projectile base pressure, velocity, and distance down the bore of the weapon for both peak pressure and charge burnout assuming the grain is a cylindrical propellant ($\theta = 1$).

Answers:

$$\lambda = 366,246\left[\frac{\text{ft lbf}}{\text{lbm}}\right]$$

$$M = 1.933$$

$$p_{B_m} = 43,281\left[\frac{\text{lbf}}{\text{in.}^2}\right]$$

$$p_{B_c} = 26,350\left[\frac{\text{lbf}}{\text{in.}^2}\right]$$

$$p_{S_m} = 39,120\left[\frac{\text{lbf}}{\text{in.}^2}\right]$$

$$p_{S_c} = 23,817\left[\frac{\text{lbf}}{\text{in.}^2}\right]$$

$$V_{p_m} = 1082\left[\frac{\text{ft}}{\text{s}}\right]$$

$$V_c = 2128\left[\frac{\text{ft}}{\text{s}}\right]$$

$$x_{p_m} = 75.6[\text{in.}]$$

$$x_c = 293.7[\text{in.}]$$

Problem 6
Verify that Equation 3.150 is valid for any θ.

Problem 7
You are asked to design a gun to propel a fragment for an explosive initiation test at 3000 ft/s. The diameter of the chamber and bore is to be 0.50 in. You have on hand some M10 flake propellant with the ballistic properties given in the following. Because of space limitations, the device (chamber and bore) cannot exceed 5 ft in length. The fragment plus sabot weighs 0.1 lbm. Assume that the propellant has the following properties:

- Adiabatic flame temperature $T_0 = 3000$ K
- Specific heat ratio $\gamma = 1.2342$
- Covolume $b = 27.76$ in.^{3}/lbm
- Density of solid propellant $\delta = 0.0602$ lbm/in.^{3}
- Propellant burn rate coeff. $\beta = 0.00002468$ (in./s)/(psi)
- Web thickness $D = 0.011$ in.

Propellant force $\lambda = 339{,}000$ ft·lbf/lbm

1. Using the preceding data, determine the bore and chamber length for the weapon as well as the amount of propellant required—be careful to leave some air space in the chamber.
2. Once the system is established in 1, determine the central ballistic parameter, value of peak breech pressure, and location of the projectile when peak pressure occurs.

Problem 8
A test of an experimental propellant grain yields a parametric relation for the fraction of web remaining of

$$f(t) = C_1 t^2 + C_2 t + 1$$

In preceding equation, C_1 and C_2 are constants. The burn rate model for this propellant was determined to be

$$D\frac{\mathrm{d}f}{\mathrm{d}t} = -\beta p_{\mathrm{B}}^{\alpha}$$

Determine the expression for the velocity of the projectile as a function of time for a given gun-projectile system. Describe the assumptions used and why they are relevant.

Problem 9
A black powder charge is to be designed to throw a 3 in. diameter firework canister that weighs 2 lbs at 100 ± 2 ft/s. The tube is 3 in. in diameter and the chamber where the propellant charge sits is 6 in. long. The projectile travel distance can be up to 18 in. long, but you may size the tube length. The propellant grains are spherical and obey the relation

$$\phi(t) = 1 - f^3$$

The average diameter of the propellant grains (they vary quite a bit and they are not exactly spherical) is 0.02 in. The tube can handle a maximum pressure of 2000 psi.

Assume that the linear burn rate for the black powder is 2.22 in./s measured at 3329 psi and the force constant is 105,000 ft·lbf/lbm. Assume a specific heat ratio of 1.21 and a solid density of 0.06 lbm/in.^{3}. Listing all assumptions, do the following:

1. Develop the proper equations for the motion of the projectile.
2. Size the charge so that the system will function as required—make sure that it fits too!
3. Determine the peak pressure in pounds per square inch and the location of peak pressure in inches.
4. Determine the location of charge burnout in inches and the velocity at burnout in feet per second. Please note that if your design allows unburnt propellant to exit the tube, please state that and calculate muzzle exit pressure and velocity in this case.
5. Determine the muzzle velocity in feet per second.

Problem 10

A 7 perf grain is a very common geometry in weapons. The geometry is such that the web between the outer diameter of the grain and the inside diameter of the outer perforations and the web between the perforations themselves is equal.

1. Determine, up to the point of slivering, the equation for $\phi(t)$ for any perforation size and any web.
2. If possible, obtain an estimate for the value of θ, assuming that the perforations are ¼ of the web thickness; if not, show why not.

End effects may be neglected.

Problem 11

A Japanese 18.1 in. type 94 gun was the largest weapon ever mounted on a warship. The gun had a chamber volume of 41,496 in.^{3}. An estimated 8 in. of the projectile protrude into the chamber after it seats. The length of travel for the projectile from shot start to shot exit is 806.3 in. The weapon has a uniform twist of 1 in 28. The type 91 armor-piercing projectile weighs 2998 lbs. The propelling charge weighs 794 lbs. The propellant used is called DC_1 and consists of 51.8% NC (11.85% nitrated), 41.0% NG with 4.5% Centralite (symmetrical diethyl diphenylurea [$C_{17}H_{20}N_2O$]), 2.0% orthotolyl urethane (added as an improvement to the Centralite), and 0.7% mineral matter (salts for wear and flash reduction) [6]. Assume that the DC_1 propellant has the following properties:

- Adiabatic flame temperature T_0 = 3000 K
- Specific heat ratio γ = 1.23
- Covolume b = 27.0 in.^{3}/lbm
- Density of solid propellant δ = 0.059 lbm/in.^{3}
- Propellant burn rate β = 0.000300 (in./s)/(psi)
- Web thickness D = 0.184 in.

- Propellant force λ = 284,000 ft·lbf/lbm

1. Determine the central ballistic parameter for this gun/projectile combination.
2. Using the preceding data, determine the projectile base pressure, velocity, and distance down the bore of the weapon for both peak pressure and charge burnout assuming the grain is single perforated propellant.
3. Determine the muzzle velocity of the weapon and the pressure acting on the projectile at muzzle exit.

Problem 12

A British "2 pounder" (so-called because the projectile weighed about 2 lbs) was their main antitank weapon for the first two and a half years of the Second World War. It fired a 40 mm projectile that weighed 1.94 lbm. The gun had a chamber volume of 23.0 in.^{3}. Since the shot had a flat base, when crimped to the cartridge case, none of the projectile protrudes into the chamber. The length of travel for the projectile from shot start to shot exit is 70 in. The weapon has a uniform right-hand twist of 1 in 30. The propelling charge weighs 0.583 lbs. The propellant used was single perforated cordite. Assume that the cordite propellant has the following properties:

- Adiabatic flame temperature T_0 = 2442 K
- Specific heat ratio γ = 1.21
- Covolume b = 31.32 in.^{3}/lbm
- Density of solid propellant δ = 0.059 lbm/in.^{3}
- Propellant burn rate β = 0.00024 (in./s)/(psi)
- Web thickness D = 0.0197 in.
- Propellant force λ = 318,000 ft·lbf/lbm

1. Determine the central ballistic parameter for this gun/projectile combination.
2. Using the preceding data, determine the projectile base pressure, velocity, and distance down the bore of the weapon for both peak pressure and charge burnout.
3. Determine the muzzle velocity of the weapon and the pressure acting on the projectile at muzzle exit.
4. Create pressure–travel and velocity–travel curves for this system. Annotate the location of charge burnout on the pressure–travel curve.

Problem 13

Develop the equations required to model a propellant/charge/gun system where the propellant behaves according to a βp^{α} burn rate. Do this for both $\theta = 0$ and $\theta \neq 0$. Describe, in bulletized form, how you would solve these equations numerically.
Hint: Start with Equations 3.80 and 3.77 through 3.79, making sure that you alter Equation 3.78 to the new burn rate.

Problem 14

Write a code using any software you want to solve Problem 13. Only write the code so it solves the interior ballistics problem up to charge burnout. Check the code by setting $\alpha = 1$ and show that you get answers close to that obtained in Problem 12.

Problem 15

Use your code developed in Problem 14 to obtain a solution to the same problem assuming black powder is the propellant. Only use your code to take the problem to the all-burnt point. Assume the properties of black powder are as follows:

- Adiabatic flame temperature T_0 = ~2300 K
- Specific heat ratio γ = ~1.22
- Covolume b = ~31 in.^{3}/lbm
- Density of solid propellant $\delta = 0.060$ lbm/in.^{3}
- Propellant burn rate $\beta = 0.04044$ (in./s)/(psi$^{0.511}$)
- Web thickness $D = 0.0197$ in.
- Propellant force $\lambda = 105{,}000$ ft·lbf/lbm

Problem 16

A Japanese (designed by the British firm of Vickers) 14 in./45 cannon is to be examined. It fired a 14 in. (36 cm) projectile that weighed 1485 lbm. The gun had a chamber volume of 17,996 in.^{3} [6]. Assume 4 in. of the projectile protrudes into the chamber. The length of travel for the projectile from shot start to shot exit is 540.8 in the study by Campbell [6]. The weapon has a uniform right-hand twist of 1 in 28. The propelling charge has four increments, where each weighs 78.45 lbs. The propellant used was DC, which consisted of 64.8% NC, 30% NG, 4.5% Centralite, and 0.7% mineral matter [6]. Assume that the propellant geometry is such that $\theta = 0.1$. Assume that the DC propellant has the following properties:

- Adiabatic flame temperature T_0 = 3200 K
- Specific heat ratio γ = 1.23
- Covolume b = 27.0 in.^{3}/lbm
- Density of solid propellant $\delta = 0.059$ lbm/in.^{3}
- Propellant burn rate $\beta = 0.000298$ (in./s)/psi
- Web thickness $D = 0.165$ in.
- Propellant force $\lambda = 365{,}000$ ft lbf/lbm

The weapon was "zoned" to fire by using two, three, and four bags which were called, "weak," "reduced," and "full" [7]. For each of these charge configurations, do the following:

1. Determine the central ballistic parameter for this gun–projectile–propellant combination.
2. Using the preceding data, determine the projectile breech pressure for both peak pressure and charge burnout.
3. Using the preceding data, determine the projectile base pressure, velocity, and distance down the bore of the weapon for both peak pressure and charge burnout.
4. Determine the muzzle velocity of the weapon and the pressure acting on the projectile at muzzle exit.
5. Plot the pressure vs. distance, based on the preceding results at the instant of peak pressure and muzzle exit.

3.3 Lagrange Gradient for Spherical and Cubic Grains

In our derivation of the Lagrange gradient approximations, we assumed a relation for the fraction of the propellant burnt according to Equation 3.77. Over the years, formulations for the handling of spherical and cubic grains have been developed (e.g., Venkatesan [8]) that allow us to handle the problem in an analytic manner, although iteration is usually involved. We shall examine a method that is fairly straightforward and follows from our previous discussions.

We begin by expressing the fraction of propellant burnt for either a cube or a sphere as

$$\phi = \left(1 - f^3\right) \tag{3.156}$$

If we now pick up our derivation from Equation 3.92, which we repeat here for convenience,

$$\frac{dx}{df} = -\frac{A^2D^2(1-f)}{w_1 c\lambda\phi\beta^2}\left[\frac{1+\dfrac{c}{3w_1}}{\left(1+\dfrac{c}{2w_1}\right)^2}\right](x+l) \tag{3.92}$$

inserting Equation 3.156 into Equation 3.157 yields

$$\frac{dx}{df} = -\frac{A^2D^2(1-f)}{w_1 c\lambda\left(1-f^3\right)\beta^2}\left[\frac{1+\dfrac{c}{3w_1}}{\left(1+\dfrac{c}{2w_1}\right)^2}\right](x+l) \tag{3.157}$$

which we can expand to

$$\frac{dx}{df} = -\frac{A^2D^2(1-f)}{w_1 c\lambda(1-f)\left(1+f+f^2\right)\beta^2}\left[\frac{1+\dfrac{c}{3w_1}}{\left(1+\dfrac{c}{2w_1}\right)^2}\right](x+l) \tag{3.158}$$

Using the definition of the central ballistic parameter M and simplifying allows us to write

$$\frac{dx}{df} = -\frac{M}{(1+f+f^2)}(x+l) \tag{3.159}$$

We now separate the variables in Equation 3.159 and integrate to obtain

$$\int_0^x \frac{dx}{(x+l)} = -M\int_1^f \frac{df}{(1+f+f^2)} \tag{3.160}$$

By performing the integration, we obtain

$$\ln(x+l) = -M\left[\frac{2\tan^{-1}\left(\dfrac{1+2f}{\sqrt{3}}\right)}{\sqrt{3}}\right]+K \tag{3.161}$$

By inserting the limits of integration and noting that at $x = 0$ and $f = 1$, we arrive at

$$K = \ln l + M\frac{2}{\sqrt{3}}\tan^{-1}\left(\frac{3}{\sqrt{3}}\right) = \ln l + M\frac{2}{\sqrt{3}}\left(\frac{\pi}{3}\right) \tag{3.162}$$

Then,

$$\ln(x+l) = -\frac{2M}{\sqrt{3}}\left[\tan^{-1}\left(\frac{1+2f}{\sqrt{3}}\right) - \left(\frac{\pi}{3}\right)\right] + \ln l \tag{3.163}$$

Taking the antilog of both sides yields

$$x + l = l\exp\left[-\frac{2M}{\sqrt{3}}\left[\tan^{-1}\left(\frac{1+2f}{\sqrt{3}}\right) - \left(\frac{\pi}{3}\right)\right]\right] \tag{3.164}$$

If we now insert this and Equation 3.156 into Equation 3.107, we obtain a relation for pressure of the form

$$p_{\mathrm{B}}\left\{l\exp\left[-\frac{2M}{\sqrt{3}}\left(\tan^{-1}\left(\frac{1+2f}{\sqrt{3}}\right) - \left(\frac{\pi}{3}\right)\right)\right]\right\} = \frac{\lambda c}{A}\left(1-f^3\right)R_{\mathrm{L}} \tag{3.165}$$

By rearranging, we obtain our relationship between pressure and the fraction of the web remaining.

$$p_{\mathrm{B}} = \frac{\lambda c}{V_i}\left(1-f^3\right)R_{\mathrm{L}}\left\{\exp\left[\frac{2M}{\sqrt{3}}\left(\tan^{-1}\left(\frac{1+2f}{\sqrt{3}}\right) - \left(\frac{\pi}{3}\right)\right)\right]\right\} \tag{3.166}$$

We now proceed along the path taken earlier to determine the relation for the values of the various parameters at peak pressure. The equivalent form of Equation 3.112 for a cubic or spherical grain is obtained by substituting Equation 3.111 into Equation 3.166 to yield

$$p_{\mathrm{B}} = Q\left(1-f^3\right)\left\{\exp\left[\frac{2M}{\sqrt{3}}\left(\tan^{-1}\left(\frac{1+2f}{\sqrt{3}}\right) - \left(\frac{\pi}{3}\right)\right)\right]\right\} \tag{3.167}$$

Differentiating this and setting the result equal to zero results in a maximum or minimum value:

$$\frac{\mathrm{d}p_{\mathrm{B}}}{\mathrm{d}f} = Q\left\{\exp\left[\frac{2M}{\sqrt{3}}\left(\tan^{-1}\left(\frac{1+2f}{\sqrt{3}}\right) - \left(\frac{\pi}{3}\right)\right)\right]\right\}\left[\frac{4M}{3}\frac{\left(1-f^3\right)}{1+\left\{\tan^{-1}\left(\frac{1+2f}{\sqrt{3}}\right)\right\}^2} - 3f^2\right] = 0 \tag{3.168}$$

Because this equation is cubic in f, we need to find the value of f that lies between 0 and 1. This is best accomplished through either iteration or numerical means. It involves solving the equation

$$\frac{4M}{3}\frac{\left(1-f_m^3\right)}{1+\left\{\tan^{-1}\left(\frac{1+2f_m}{\sqrt{3}}\right)\right\}^2}-3f_m^2=0 \quad (3.169)$$

The result is the value of f at peak pressure, again demoted as f_m. Once this is obtained, we can determine the fraction of propellant burnt, peak pressure, position, and velocity through the following:

$$\phi_m = \left(1-f_m^3\right) \quad (3.170)$$

$$p_{B_m}=Q\left(1-f_m^3\right)\left\{\exp\left[\frac{2M}{\sqrt{3}}\left(\tan^{-1}\left(\frac{1+2f_m}{\sqrt{3}}\right)-\left(\frac{\pi}{3}\right)\right)\right]\right\} \quad (3.171)$$

$$x_m+l=l\exp\left[-\frac{2M}{\sqrt{3}}\left[\tan^{-1}\left(\frac{1+2f_m}{\sqrt{3}}\right)-\left(\frac{\pi}{3}\right)\right]\right] \quad (3.172)$$

$$V(x_m)=\frac{AD}{\beta w_1\left(1+\frac{c}{2w_1}\right)}\left(1-f_m\right) \quad (3.173)$$

At charge burnout we simply substitute $f = 0$ in the preceding set of equations to obtain

$$\phi_c = 1 \quad (3.174)$$

$$p_{B_c}=Q\left\{\exp\left[\frac{2M}{\sqrt{3}}\left(\tan^{-1}\left(\frac{1}{\sqrt{3}}\right)-\left(\frac{\pi}{3}\right)\right)\right]\right\}=Q\left\{\exp\left[-\frac{\pi M}{3\sqrt{3}}\right]\right\} \quad (3.175)$$

$$x_c+l=l\exp\left[-\frac{2M}{\sqrt{3}}\left[\tan^{-1}\left(\frac{1}{\sqrt{3}}\right)-\left(\frac{\pi}{3}\right)\right]\right]=l\exp\left[\frac{\pi M}{3\sqrt{3}}\right] \quad (3.176)$$

$$V(x_c)=\frac{AD}{\beta w_1\left(1+\frac{c}{2w_1}\right)} \quad (3.177)$$

All the other calculations for the determination of the projectile pressure, position, and velocity after charge burnout are identical to those introduced earlier. This includes the methodology for determining whether the charge burns out in the tube or not

Problem 17

The following are the data for the three different powders used in the 8 in. rifled muzzle loader. Initial data were provided by Kent Crawford. Some data were provided in the literature brought to the attention of the authors by Byron Angel [9].

Assume the following:

- Projectile weight: 180 lbm
- Propellant weight: see the table "Data Provided"
- Projectile and bore diameter: 8 in. (in actuality the projectile diameter was 7.92 in.)
- Adiabatic flame temperature: 2600 K
- Covolume: 5.02–5.68 in.^{3}/lbm
- Specific heat ratio: 1.24
- Burn rate coefficient: see the table "Model Inputs and Outputs"
- Propellant force: see the table "Model Inputs and Outputs"
- Chamber length: see the table "Model Inputs and Outputs"
- Projectile travel: see the table "Model Inputs and Outputs"

Data Provided

Powder	Form	Propellant Weight (lbm)	Propellant Density (g/cm^3)	Expected Peak Pressure (tons/in.^{2})	Expected Travel to Peak Pressure (ft)	Expected Muzzle Velocity (ft/s)
Pellet	1 in. diameter sphere	30	1.77	17.3	0.3	n/a
Pebble	0.625 in. cube [9]	35	1.8	15.3	0.5	1391
RLG	0.14 in. diameter sphere [9]	30	1.77	n/a	n/a	1330

Model Inputs and Outputs

Powder	Burn rate Coefficient[a] β (in./s)/(psi)	Propellant Force λ (ft lbfl/bm)	Chamber Length (in.)	Projectile Travel Length (in.)	Calculated Peak Pressure (tons/in.^{2})	Calculated Travel to Peak Pressure (in.)	Calculated Muzzle Velocity (ft/s)
Pellet	0.0068	101,100	19.5	96	17.3	4.435	1435
Pebble	0.0036	80,900	19.5	96	15.3	4.344	1391
RLG	0.0014	84,800	18.5	97	20.8	3.081	1330

Note: RLG, Rifle Large Grain.

[a] The burn rate coefficient is assumed to have a linear relationship with pressure, the value for the exponent of black powder is usually assumed to be 0.511. This introduces an error of about 5% in the pressure.

Given the preceding information, verify the calculated peak pressure, calculated travel to peak pressure, and calculated muzzle velocity.

3.4 Chambrage Gradient

In our derivation of the Lagrange gradient approximations, we assumed that the chamber of the gun was simply an extension of the bore. The volume of the chamber was converted to a cylinder of bore diameter, and the tube was appropriately lengthened behind the projectile. In doing this, we neglected the effects of short, larger-diameter chambers (the definition of chambrage is the ratio of the diameter of the chamber to the bore inner diameter), and all calculations that are functions of distance from the breech $x - x_s$ are inaccurate in the distance term. If we account for these differences by deriving a chambrage gradient, we find that the two methods yield similar but close answers. Nevertheless, one should understand how the answers relate to each other and to the real problem. Fredrick W. Robbins of the Army Research Laboratory, who has allowed us to base this section on his excellent work, derived the chambrage gradient formulation as follows.

The formulation of the chambrage gradient follows much the same pattern that was used in the development of the Lagrange gradient. It leads, however, to an algorithm that is best applied with the aid of a computer. Small increments of time (hence, distance) are chosen, and computations of pressure (breech, mean, and base), velocity, acceleration, and distance traveled are made for the end point of the interval. The calculation is then repeated for the next increment of time. This is done until the projectile exits the bore. A representation of the situation is shown in Figure 3.3 for a chosen time step.

The definitions of the terms used in Figure 3.3 are shown in Figure 3.4.

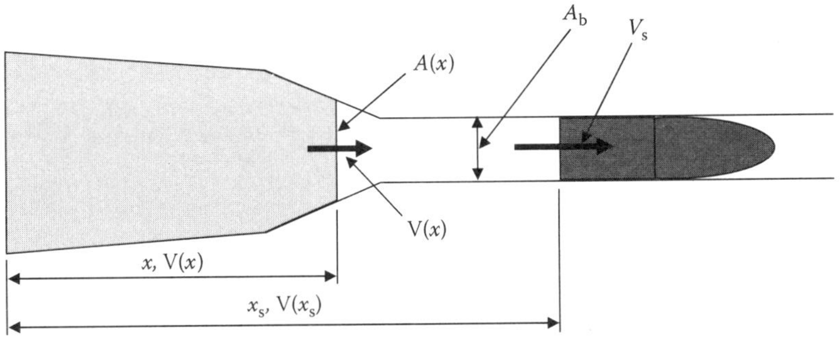

FIGURE 3.3
Chamber with large chambrage.

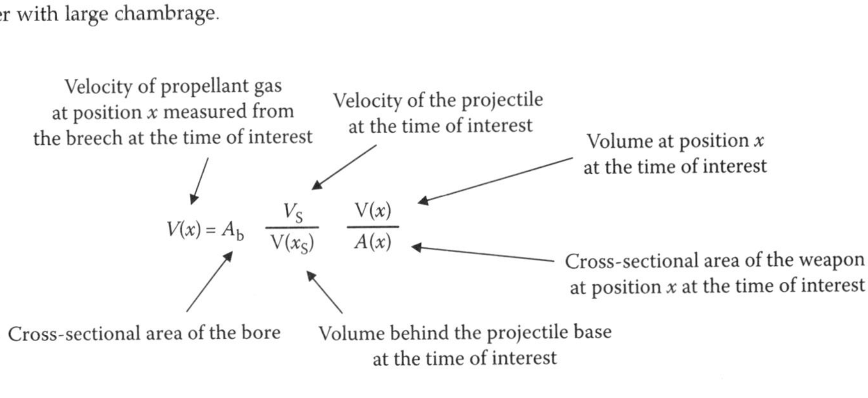

FIGURE 3.4
Definitions of terms used in chambrage gradient development.

In Robbins' derivation, certain integrals called J integral factors are developed and must be computed. They are as follows:

$$J_1(x_0) = \int_0^{x_0} \frac{V(x)}{A(x)} dx \tag{3.178}$$

$$J_1(x_s) = J_1(x_0) + \frac{1}{A_b}\left[V(x_0)(x_s - x_0) + \frac{A_b}{2}(x_s - x_0)^2\right] \tag{3.179}$$

$$J_2(x_s) = \left[\frac{V(x_0) + A_b(x_s - x_0)^2}{A_b^2}\right] \tag{3.180}$$

$$J_3(x_s) = J_3(x_0) + A_b J_1(x_0)(x_s - x_0) + \frac{V(x_0)}{2}(x_s - x_0)^2 \frac{A_b}{6}(x_s - x_0)^3 \tag{3.181}$$

$$J_4(x_s) = J_4(x_0) + \frac{\left[V(x_0) + A_b(x_s - x_0)^3\right] - [V(x_0)]^3}{3A_b^2} \tag{3.182}$$

The acceleration at any point a appears in one of our algorithm factors explicitly:

$$a(t) = a_1(t) + a_2(t)p_s \tag{3.183}$$

where

$$a_1(t) = \frac{cA_b}{[V(x_s)]^2}\left[\frac{AbV_s^2}{V(x_s)} + \frac{cA_b p_{resist}}{m_p}\right] \tag{3.184}$$

$$a_2(t) = \frac{cA_b^2}{m_p[V(x_s)]^2} \tag{3.185}$$

Another factor required in the algorithm is $b(t)$ derived as

$$b(t) = -\frac{cA_b^2 V_s^2}{2[V(x_s)]^3} \tag{3.186}$$

The way the algorithm is used is (roughly) as follows:
At each time step

- The breech pressure is calculated from the burning rate equations.
- J_1 through J_4 are calculated.
- $a(t)$ and $b(t)$ are calculated.
- The projectile acceleration, velocity, and distance down the bore are calculated.
- The volume behind the projectile is updated.
- The process moves to the next time step.

This gradient, while only slightly more accurate than the Lagrange gradient in the computed distance from the breech, is used in some modern interior ballistic computer codes.

3.5 Numerical Methods in Interior Ballistics

In this section, we shall briefly discuss methods for solving the interior ballistics problem through the use of computational tools. In recent decades, computational capabilities have increased at an astronomical rate. One of the most famous early uses of the computer to solve the exterior ballistics problem (firing tables) was the use of the electronic numerical integrator and computer machine during and immediately after the Second World War. In this case, the computer was used to solve tedious exterior ballistics problems in rapid order.

In the field of interior ballistics, the computer revolution has given the individual ballistician the tools (although some commercial packages can be expensive) to solve extremely complicated interior ballistics problems and quickly optimize a system. The complexity of these tools is driven by the physics that are incorporated in the particular code. We shall discuss some general categories of software, their uses, and their limitations.

Many interior ballistics codes are of the zero-dimensional variety. In these types of codes, the density of the propellant gas (as stipulated by the Lagrange approximation) is considered constant in the volume between the breech and the projectile. The Lagrange pressure gradient is assumed to be in effect and results in a nice, always well-behaved launch. These codes are extremely useful for predictive applications because they run fast. One of the features of these codes that make them so useful is that we can easily include and track burn characteristics of multiple propellant types (both geometry and chemical composition). This allows us to tailor the burn characteristics so that a particular pressure–distance distribution is achieved while maintaining a particular muzzle velocity.

Another excellent feature of this type of code is that heat transfer to the weapon can be accounted for in the energy balance. This provides a more realistic muzzle velocity than if it is neglected and can be of great value to the gun designer. Friction and blowby effects can be fudged in, and burn rate parameters varied to replicate actual tests. Additionally, the effects of the regression of all surfaces (recall that we neglected end effects in our hand calculation methods) can be simulated and accounted for. Zero-dimensional codes can also include the effects of inhibitors on the propellant grains as well as highly nonlinear pressure–burn rate relationships. Since zero-dimensional codes track the pressure, it is simple enough to use them to develop recoil models as well. All in all, zero-dimensional codes are probably the most effective tools at the disposal of the interior ballistician for basic ballistics design work. Once a set of experiments have been conducted to validate these codes, their accuracy is excellent.

A quasi-one-dimensional code is one in which the density of the propellant gas behind the projectile is a known function of some other variable. An example of this would be a zero-dimensional code that incorporated the chambrage gradient. Essentially, beyond the ability to track the effect of variable chamber or bore area on the density, the limitations and benefits of this type of code are the same as discussed in the zero-dimensional section.

A one-dimensional interior ballistics code allows density to vary based on the physical equations and conservation laws in the axial direction only. Thus, at a given cross section, the density is considered constant throughout the radial direction. These codes are very good at predicting pressure waves and can therefore estimate the pressure differential along the volume behind the projectile. The benefit of this is that since propellant generally burns faster under higher pressure, the local burn rate and therefore the amount of gas evolved can be tracked. This allows the user to see pressure waves develop and propagate. The disadvantage is that since the code can only track pressure waves in the axial direction, unless the charge fully fills the volume behind the projectile, it is difficult to completely match the physics of the firing. This occurs because the presence of solids and gases in the chamber is generally not uniform—the solids are usually at the bottom of the chamber. This affects the gas dynamics. Solids will also be entrained by the gas flow down the bore, and some modeling of their motion has to be accomplished (or ignored). In most cases, the propellant bed is assumed to be a monolithic mass that regresses and stretches as the propellant is burned. Although these codes are usually very good, the user should completely understand the assumptions on how the propellant is allowed to move before using them.

A two-dimensional model is one where the density can vary in the radial direction as well. These models are better at predicting pressure waves but take somewhat longer time to run than one-dimensional models. Pressure can be tracked in the radial direction and the propellant motion included. The same issues with propellant motion are present as they were in the one-dimensional models, although it is possible to track propellant motion.

A three-dimensional model has it all. Because of this, they usually take an excruciatingly long time to set up and run. This time constraint makes them generally reserved for failure investigations rather than predictive simulations. Individual propellant grains can regress and be tracked, and one can imagine the difficulty with this in the sense of model validation. With suitable stress and failure models, grain fracture can also be examined. If erosion models are incorporated, the effect of gas wash on propellant burn rate can even be included. One has to ask oneself if all this is really necessary. In some cases, these models are crucial; in other cases, they are certainly overkill. The usefulness of this type of model is still somewhat limited by computer speed, but as computers become faster, the limitation will change to a lack of accurate physical models for motion, surface regression, propellant and gun tube erosion, grain fracture, etc. These issues are certainly solvable, but finding a proponent who will fund the research is difficult.

Now that we have described the general types of models, it is important to explain their use further. In general, all of them are used in a similar manner. We shall use the zero-dimensional model as an example and leave the rest to the readers' imagination (and budget restrictions). Typically, a propellant formulation and geometry is chosen as a point of departure given that we have a preliminary gun design and a projectile to work with. This propellant is then further developed in terms of geometry or chemical composition. Some zero-dimensional codes are provided with optimization subroutines so that particular characteristics of the ballistic cycle can be achieved. The pressure–time, acceleration–time, and pressure–distance curves are examined and, if suitable, some experimental charges are made up. The configuration is then fired and the results checked against the code. These results can then be used to adjust burn rates and resistive characteristics, and the model can be used to predict all future firings and design iterations.

A particular example of the power of these codes is their usefulness in assessing the interior ballistics of systems that widely vary in matters of scale, for example, in mass of projectile, diameter of bore, and muzzle velocity. In the 1960s, ballisticians J. Frankle and M. Baer at the Ballistics Research Laboratories at Aberdeen, Maryland [10], and others elsewhere devised codes largely based on Corner's zero-dimensional analysis that we described in detail in Section 3.2. Among these, the Frankle–Baer simulation, still in use today, which examined and expanded on the basic energy equation,

Energy released by burning propellant

= internal energy of gases + work done on the projectile + secondary losses

or

$$Q = U + W + \text{losses} \tag{3.187}$$

developed equations of state of the propellant gases based on more recent thermodynamic theories and refined the losses term from new experimental data. This led to more refined ratios for breech, mean, and shot base pressures and more accurate equations of motion for the projectile.

To examine the effects of scale, we computed the relevant pressure ratios for three widely different gun–projectile combinations. We show these combinations, and the resultant ratio values are shown in Tables 3.1 through 3.5. What is noteworthy is the applicability of the theory over the range of size, projectile mass, and propellant type and volume. Notice also

TABLE 3.1

Inputs for Comparison of Corner and Frankle–Baer

Parameter	**Expression or Value (J. Corner)**	$\begin{bmatrix}\textbf{M735}\\\textbf{M1}\\\textbf{M193}\end{bmatrix}$	**Expression or Value (Frankle–Baer)**	$\begin{bmatrix}\textbf{M735}\\\textbf{M1}\\\textbf{M193}\end{bmatrix}$
Charge weight	c	$\begin{bmatrix}13.125\\9.000\\4.020\times10^{-3}\end{bmatrix}$	c	$\begin{bmatrix}13.125\\9.000\\1.020\times10^{-3}\end{bmatrix}$
Projectile weight	W	$\begin{bmatrix}12.78\\31.97\\7.86\times10^{-3}\end{bmatrix}$	w_P	$\begin{bmatrix}12.78\\31.97\\7.86\times10^{-3}\end{bmatrix}$
Propellant type	–	$\begin{bmatrix}\text{M30}\\\text{M1}\\\text{Ball}\end{bmatrix}$	–	$\begin{bmatrix}\text{M30}\\\text{M1}\\\text{Ball}\end{bmatrix}$

TABLE 3.2

Burn Characteristic Inputs for Numerical Comparison of Corner and Frankle–Baer

Parameter	**Expression or Value (J. Corner)**	$\begin{bmatrix}\textbf{M735}\\\textbf{M1}\\\textbf{M193}\end{bmatrix}$	**Expression or Value (Frankle–Baer)**	$\begin{bmatrix}\textbf{M735}\\\textbf{M1}\\\textbf{M193}\end{bmatrix}$
Propellant impetus (force)	λ	$\begin{bmatrix}3.64\times10^{5}\\3.05\times10^{5}\\3.32\times10^{5}\end{bmatrix}$	λ	$\begin{bmatrix}3.64\times10^{5}\\3.05\times10^{5}\\3.32\times10^{5}\end{bmatrix}$
Specific heat ratio	γ	$\begin{bmatrix}1.2385\\1.2592\\1.26\end{bmatrix}$	γ	$\begin{bmatrix}1.2385\\1.2592\\1.26\end{bmatrix}$
Polytropic index	$\dfrac{1}{\gamma-1}$		n	

TABLE 3.3

Pressure Gradient Calculations for Numerical Comparison of Corner and Frankle–Baer

Parameter	Expression or Value (J. Corner)	Expression or Value (Frankle–Baer)
$\frac{\bar{p}}{p_S}$	$1+\frac{c}{3w}$	$1+\frac{1}{\delta}\frac{c}{w_p}$
$\frac{1}{\delta}$	n/a	$\frac{1}{2n+3}\left[1+\alpha n\left(\frac{1+c_1\beta n}{1+c_1 n}\right)\right]$
$\frac{1}{a_b}$	n/a	$\left[\frac{2n+3}{\delta}+\frac{2(n+1)}{c/w_p}\right]$
$\frac{\bar{p}}{p_B}$	$\left[1-\frac{1}{6}\frac{c}{w}\right]$	$\left[1-\frac{1}{\delta}\left(\frac{c}{w_p}\right)\right](1-a_b)^{n+1}$
$\frac{p_B}{p_S}$	$\left[1+\frac{1}{2}\frac{c}{w}\right]$	$[(1-a_b)^{-(n+1)}]$

TABLE 3.4

Specific Frankle–Baer Computations for the M735, M1, and M193 Projectiles

Projectile	α	β	c_1	$\frac{1}{\delta}$	$\varepsilon=\frac{c}{w_p}$	$\frac{1}{a_b}$	$(1-a_b)^{n+1}$
M735 (105 mm KE)	0.56	1.07	1.05	0.333	1.027	11.002	0.683
M1 (105 mm HE)	0.63	1.01	1.02	0.322	0.282	37.811	0.876
M193 (5.56 mm ball)	0.60	1.03	1.04	0.315	0.511	22.325	0.800

TABLE 3.5

Specific Gradient Comparison of Corner and Frankle–Baer for the M735, M1, and M193 Projectiles

	$\frac{\bar{p}}{p_S}$		$\frac{\bar{p}}{p_B}$		$1+\frac{1}{2}\frac{c}{w}$	
Projectile	Corner	Frankle–Baer	Corner	Frankle–Baer	Corner	Frankle–Baer
M735 (105 mm KE)	1.342	1.342	0.829	0.917	1.514 (1.619)	1.464 (1.463)
M1 (105 mm HE)	1.094	1.091	0.953	0.956	1.141 (1.148)	1.142 (1.141)
M193 (5.56 mm ball)	1.170	1.161	0.915	0.929	1.256	1.250 (1.250)

the closeness of the pressure ratios for each projectile between the Corner and the Frankle–Baer simulations.

3.6 Sensitivities and Efficiencies

Having explored the detailed development of theories of interior ballistic events, we will now probe the outcome of varying some of the parameters that are under the control of the charge designer. To do this, we will be referring back to definitions and equations developed under Section 3.2.

A useful quantity for our analysis is the dimensionless central ballistic parameter M

$$M = \frac{A^2 D^2}{w_1 c \lambda \beta^2} \left[\frac{1 + \dfrac{c}{3w_1}}{\left(1 + \dfrac{c}{2w_1}\right)^2} \right] \tag{3.188}$$

Of particular importance in this are the variables D and β, the original web dimension and the burning rate coefficient, respectively. If we examine

$$p_{B_m} = \frac{\lambda c}{V_i} \left(\frac{1 + \dfrac{c}{2w_1}}{1 + \dfrac{c}{3w_1}} \right) \left(\frac{1}{Me} \right) \tag{3.189}$$

we can see that, at least for the case of $\theta = 0$, M is in the denominator, and as the ratio D/β decreases, M decreases, and from Equation 3.190, the peak pressure p_B increases. That is, if the original web size is decreased, the peak pressure will increase. This is a parameter much under the control of the designer.

Referring again to Equation 3.188, we see that if the charge mass (weight) c is increased, then M decreases (the c's in the gradient term largely cancel out and c in the first term denominator governs). In Equation 3.189, c appears in the numerator and M in the denominator causing p_B, the peak pressure, to again rise.

Let us now examine the shift in location of x_m, the point in travel where the peak pressure exists.

$$x_m + l = l \left[\frac{(M + 2\theta)}{(M + \theta)} \right]^{\frac{M}{\theta}} \tag{3.190}$$

Equation 3.190 relates x_m to M. If the ratio D/β or the charge mass c decreases, then M decreases and, consequently, x_m is reduced (it moves toward the breech). This kind of shift is important in gun design since wall thickness and center of mass are important considerations for weapon mounting.

The sensitivity of muzzle velocity V to changes in web size or charge weight can be seen in

$$V^2(x) - V^2(x_c) = \frac{\lambda c(x_c + l)e^{-M}}{l\left(w_1 + \frac{c}{3}\right)f} \tag{3.191}$$

Here M is the governing term and is decreased as we showed earlier, if charge mass is increased or web size reduced. Because M has a negative exponent in the equation, its reduction drives an increase in V.

Finally, the influence of travel on muzzle velocity can be shown to be quite weak. The computation is complex and will not be shown here. But, for example, by doubling the travel, velocity increases only by a factor of about a tenth, hardly worth the effort in the real world.

There are two measures of efficiency that are of interest to the interior ballistician: piezometric or pressure efficiency and ballistic or energy efficiency.

Piezometric efficiency ε_p is the ratio of the average pressure during the entire ballistic cycle to the peak pressure during the cycle:

$$\varepsilon_p = \frac{\bar{p}}{p_{B_m}} \tag{3.192}$$

An illustration of the space-mean pressure and maximum breech pressure is provided in Figure 3.5.

Increasing ε_p implies that the muzzle pressure will be high (usually an undesirable trait) and that the charge burnout point will move toward the muzzle (hopefully never outside the muzzle). High piezometric efficiency usually means poor regularity, i.e., round-to-round muzzle velocity repeatability is poor (an undesirable trait). For powerful, high-velocity cannons, this efficiency is usually in the 50–60% range. Other cannons are lower. High piezometric efficiency also implies that the expansion ratio, the ratio of total gun volume to chamber volume, will be low: powerful guns have large chambers and consume lots of propellant.

Ballistic efficiency ε_b is defined as the ratio of the kinetic energy of the projectile as it exits the muzzle to the total potential energy of the propellant charge.

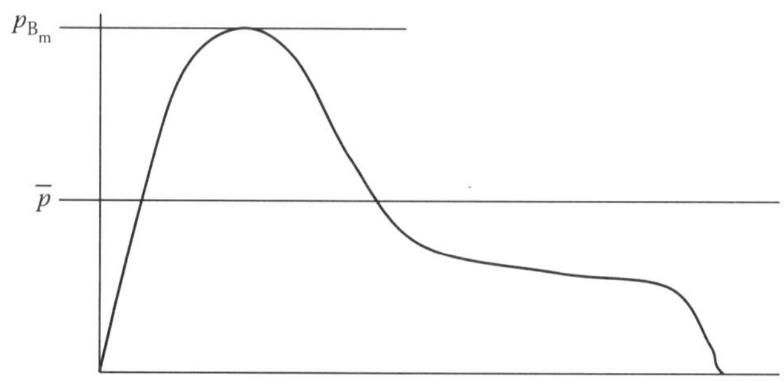

FIGURE 3.5
Average and maximum breech pressure for a typical gun firing.

$$\varepsilon_b = \frac{\text{muzzle KE}}{\text{propellant PE}} = \frac{\frac{1}{2}wV^2}{\frac{\lambda c}{\gamma - 1}} = \frac{(\gamma - 1)wV^2}{2\lambda c} \quad (3.193)$$

because the potential energy is defined as

$$\text{propellant PE} = \frac{RT_0}{(\gamma - 1)} \quad \text{and} \quad \lambda = RT_0 \quad (3.194)$$

Increasing ε_b tends to shift the all-burnt position toward the breech and increases the expansion ratio. Reducing the central ballistic parameter M by going to a smaller web will also increase ε_b. The ballistic efficiency of most guns is approximately 0.33.

References

1. Robbins, F., *Interior Ballistics Course Notes*, Self-published, Aberdeen, MD, 2002.
2. Panton, R. L., *Incompressible Flow*, 2nd ed., John Wiley & Sons, New York, 1995.
3. Currie, I. G., *Fundamental Mechanics of Fluids*, 2nd ed., McGraw-Hill, New York, 1993.
4. Corner, J., *Theory of the Interior Ballistics of Guns*, John Wiley & Sons, New York, 1950.
5. Bull, G. V., and Murphy, C. H., *Paris Kanonen—The Paris Guns (Wilhelmgeschutze) and Project HARP*, Verlag, E.S. Mittler & Sohn GmbH, Herford und Bonn, 1988.
6. Campbell, J., *Naval Weapons of World War Two*, Naval Institute Press, Annapolis, MD, 1985.
7. Jordan, John (Ed.), *Warship 2012*, Conway, Publishing, London, 2012.
8. Venkatesan, N. S., *On the Solution of the Equations of Internal Ballistics with a Cubic Form Function*, Proceedings of the India National Science Academy, Vol. 22, No. 3, pp. 129–136, 1956.
9. Very, Edward W., *Navies of the World*, John Wiley & Sons, New York, 1880, p. 207.
10. Frankle, J. M., *Interior Ballistics of High Velocity Guns Experimental Program, Phase I, BRL Memorandum Report 1879*, US Army Ballistic Research Laboratory, Aberdeen Proving Ground, MD, November 1967.

4

Ammunition Design Practice

4.1 Stress and Strain

Before proceeding with our examination of design practices, a discussion of the fundamentals of the general state of stress in materials is in order. Consider an arbitrary cube of material under load as depicted in Figure 4.1. The state of stress can be completely defined by six stress components: σ_x, σ_y, σ_z, τ_{xy}, τ_{yz}, and τ_{zx}. Here, we have used a Cartesian coordinate system where the normal stresses are denoted by σ and the shear stresses are denoted by τ.

The first subscript represents the plane in which the stress acts (defined by its normal vector), while the second subscript indicates the direction of action. These components form the stress tensor, which is actually a 3 × 3 matrix of nine elements, except that we have assumed that $\tau_{xy} = \tau_{yx}$, $\tau_{zy} = \tau_{yz}$, and $\tau_{xz} = \tau_{zx}$. When written as a tensor, the state of stress in a material is defined as

$$\sigma = \begin{bmatrix} \sigma_x & \tau_{xy} & \tau_{zx} \\ \tau_{xy} & \sigma_y & \tau_{yz} \\ \tau_{zx} & \tau_{yz} & \sigma_z \end{bmatrix} \tag{4.1}$$

It can be shown that the coordinate system in which we measure the stresses can be rotated so that the shear stresses vanish. The three remaining stresses are normal stresses, known as the principal stresses, and are denoted as σ_1, σ_2, and σ_3.

These stresses are important because, regardless of what coordinate system we view the component in, the stress state is uniquely determined. Also, in some materials, these stresses are associated with failure and fracture.

These points are sometimes graphically shown through use of Mohr's circle. The determination of the principal stresses will be discussed later in this section.

It is also very important to understand this when we try to experimentally examine the stress levels in a part with a strain gage. Stress is a point function defined by force per unit area expressed as

$$\sigma = \frac{F}{A} \tag{4.2}$$

where σ is the stress; F is a force; and A is the cross-sectional area of the component.

The same equation also holds if we use the symbol τ signifying a shear stress.

When we examine a structure, we are normally given the loads that are imposed on it. We then either choose a material or evaluate a given material to see how it will behave under

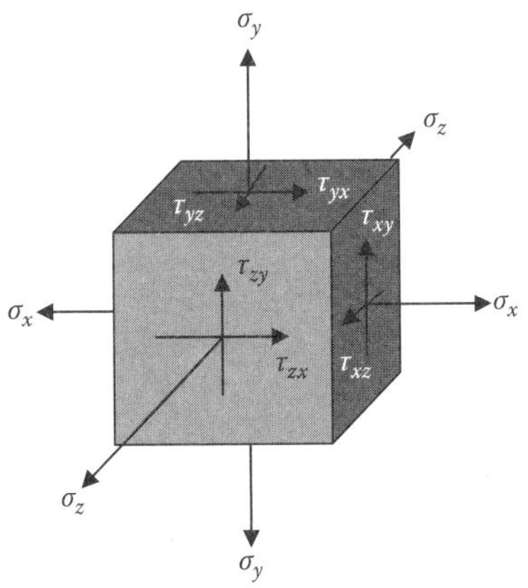

FIGURE 4.1
Cartesian stress components.

the applied loads. This process requires us to convert the external loads to stress. These stresses will cause the movement of the material in the form of either stretching (tension) or compression. This movement is the actual displacement of the material. There is an intermediate analytical step between these two where we need to define the strain of the material. The strain in the material is defined as the change in length of a part over its initial, unstressed length. Mathematically, this is expressed as

$$\varepsilon = \frac{\Delta l}{l} \tag{4.3}$$

We require a relationship between stress and strain to evaluate material behavior under a load. The link between stress and strain is called a stress–strain relationship. The most common and simplest stress–strain relationship is that for a linear-elastic material. This is known as Hooke's law and is given for small deformations and uniaxial loading by

$$\varepsilon = \frac{\sigma}{E} \tag{4.4}$$

where E is the modulus of elasticity, sometimes known as Young's modulus. In a linear-elastic material, any loading and unloading of the structure occurs along a curve in stress–strain space that has a slope equal to the modulus of elasticity. Under the assumption of general loading, the material will be "pulled in" in the transverse directions as it is longitudinally stretched. The ratio of lateral strain to axial strain is denoted as ν and called Poisson's ratio and is given for an isotropic material as

$$\nu = -\frac{\varepsilon_y}{\varepsilon_x} = -\frac{\varepsilon_z}{\varepsilon_x} \tag{4.5}$$

This assumption of general loading changes our Hooke's law relation as follows:

$$\varepsilon_x = \frac{\sigma_x}{E} - \frac{\nu\sigma_y}{E} - \frac{\nu\sigma_z}{E} \tag{4.6}$$

$$\varepsilon_y = -\frac{\nu\sigma_x}{E} + \frac{\sigma_y}{E} - \frac{\nu\sigma_z}{E} \tag{4.7}$$

$$\varepsilon_z = -\frac{\nu\sigma_x}{E} - \frac{\nu\sigma_y}{E} + \frac{\sigma_z}{E} \tag{4.8}$$

While we have defined ε to represent longitudinal strain in a material, a different type of strain can be examined—shear strain. Shear strain γ is defined as the angular deviation of a material from its original, undeformed shape. Shear strain is given by its own version of Hooke's law as

$$\gamma = \frac{\tau}{G} \tag{4.9}$$

where G is known as the shear modulus of the material.

In an isotropic material, E, ν, and G are not independent. The relationship that links them is

$$G = \frac{E}{2(1 + \nu)} \tag{4.10}$$

When we perform hand calculations, it is customary to convert the loads to stresses, then the stresses to strains, and finally strains to deformations. The process is somewhat different (i.e., reversed) in a finite element analysis (FEA).

The determination of the principal stresses is important in several failure criteria. When a part is being experimentally examined during a gun launch, it is customary to utilize a strain gage. A strain gage measures the change in the length of a part using the fact that resistance increases in a conductor as it is stretched. Strain gages are not always placed along the directions in which it is desired to compute stress however. Since strain gages only measure in-plane stress, it is common to transform this two-dimensional (2D) measurement into a desired in-plane direction. To transform stress from the strain gage coordinate system to the desired coordinate system, we use the following equations:

$$\sigma_{x'} = \sigma_x \cos^2\theta + \sigma_y \sin^2\theta + 2\tau_{xy} \sin\theta\cos\theta \tag{4.11}$$

$$\sigma_{y'} = \sigma_x \sin^2\theta + \sigma_y \cos^2\theta - 2\tau_{xy} \sin\theta\cos\theta \tag{4.12}$$

$$\tau_{x'y'} = \tau_{xy}\left(\cos^2\theta - \sin^2\theta\right) + \left(\sigma_y - \sigma_x\right)\sin\theta\cos\theta \tag{4.13}$$

In each of these equations, the primed variables are those in the desired direction and the unprimed variables are those measured by the strain gages. This is depicted in Figure 4.2.

The rotation of coordinate systems in three dimensions is covered in excellent detail by Budynas [1]. It was stated earlier that a rotation can be made such that the shear stresses

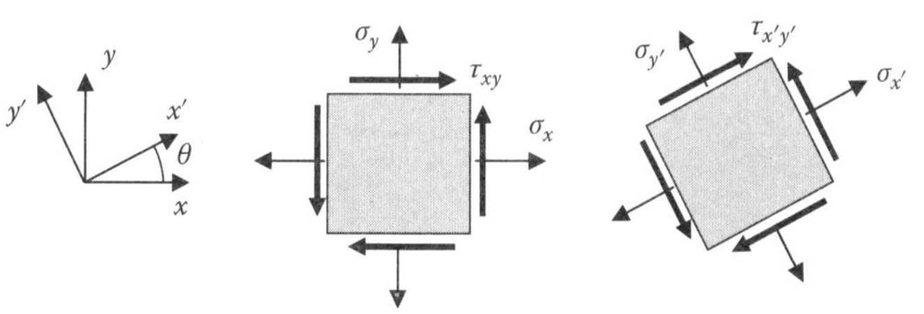

FIGURE 4.2
Transformation of stress components.

vanish, and this results in what are known as principal stresses [2]. To determine the values of the principal stresses, we determine the stress invariants through the solution of the eigenvalue problem. The three stress invariants are given by

$$I_1 = \sigma_x + \sigma_y + \sigma_z \tag{4.14}$$

$$I_2 = \sigma_x\sigma_y + \sigma_y\sigma_z + \sigma_z\sigma_x - \tau_{xy}^2 - \tau_{yz}^2 - \tau_{zx}^2 \tag{4.15}$$

$$I_3 = \sigma_x\sigma_y\sigma_z - \sigma_x\tau_{yz}^2 - \sigma_y\tau_{zx}^2 - \sigma_z\tau_{xy}^2 + 2\tau_{xy}\tau_{yz}\tau_{zx} \tag{4.16}$$

Once these invariants are obtained, the principal stresses are obtained through

$$\sigma_1 = \frac{I_1}{3} + \frac{2}{3}\sqrt{I_1^2 - 3I_2}\cos\phi \tag{4.17}$$

$$\sigma_2 = \frac{I_1}{3} + \frac{2}{3}\sqrt{I_1^2 - 3I_2}\cos\left(\phi + \frac{2\pi}{3}\right) \tag{4.18}$$

$$\sigma_3 = \frac{I_1}{3} + \frac{2}{3}\sqrt{I_1^2 - 3I_2}\cos\left(\phi + \frac{4\pi}{3}\right) \tag{4.19}$$

In Equations 4.17 through 4.19, the quantity ϕ is calculated through

$$\phi = \frac{1}{3}\cos^{-1}\left[\frac{2I_1^3 - 9I_1I_2 + 27I_3}{2\left(I_1^2 - 3I_2\right)^{\frac{3}{2}}}\right] \tag{4.20}$$

Now we have all the basic information necessary to discuss failure criteria. Limits of space prevent a more in-depth treatment of this topic. The reader is referred to the references at the end of this chapter for a more detailed treatment.

Problem 1

For the state of stress in the following, find the principal stresses and the maximum shear stress:

$$[\sigma] = \begin{bmatrix} 20 & 15 & 0 \\ 15 & 4 & 0 \\ 0 & 0 & -9 \end{bmatrix} [\text{MPa}]$$

Answer: $\tau_{max} = 19$ [MPa]

4.2 Failure Criteria

When embarking on the design of a particular projectile component, we must initially determine certain characteristics of the material contemplated for the design: Will we use a metal or a plastic? Does it have a distinct yield point? Is it brittle or very ductile?

Such determinations will govern which criteria we use when we calculate the stresses that will cause the failure of the component. There are three commonly used criteria for yield or failure: von Mises, which is also known as the maximum distortion energy criterion; Tresca, which is known as the maximum shear stress criterion; and Coulomb, which uses a maximum normal stress criterion. Other materials may require unique failure criteria, for example, composites or nonisotropic metals may require Tsai–Wu or Tsai–Hill criterion.

The von Mises or maximum distortion energy criterion is typically used when the component is to be made of metal. It assumes that the energy required to change the shape of the material is what causes yielding and that a hydrostatic state of stress will not result in failure. The materials for which it is used should have a distinct yield point. Our convention shall follow that of structural engineers in which we shall assume tensile stress to be positive. By this criterion, we assume that the distortion of the material will precipitate the failure. We shall order the stresses with 1 as largest to 3 being smallest and state the following:

$$(\sigma_1 - \sigma_2)^2 + (\sigma_2 - \sigma_3)^2 + (\sigma_3 - \sigma_1)^2 = \text{constant} \tag{4.21}$$

We set this constant equal to $2\sigma_Y^2$ or $6K^2$. Here σ_Y is the yield stress in simple tension and K is the yield stress in pure shear. This implies that

$$\frac{1}{3}\sigma_Y^2 = K^2 \tag{4.22}$$

or

$$K = \frac{2}{\sqrt{3}}\left(\frac{\sigma_Y}{2}\right) = 1.155\left(\frac{\sigma_Y}{2}\right) \tag{4.23}$$

σ_Y is also known as the equivalent stress, and either σ_Y or K can be experimentally found. In σ_1–σ_2–σ_3 space, the criterion is represented by an ellipsoidal surface whose inner region symbolizes stress states that are safe (nondistorting). This is two-dimensionally shown in Figure 4.3.

The Tresca or maximum shear stress criterion is used when the material is known to have great ductility. It assumes that the failure mechanism is by slippage along shear planes

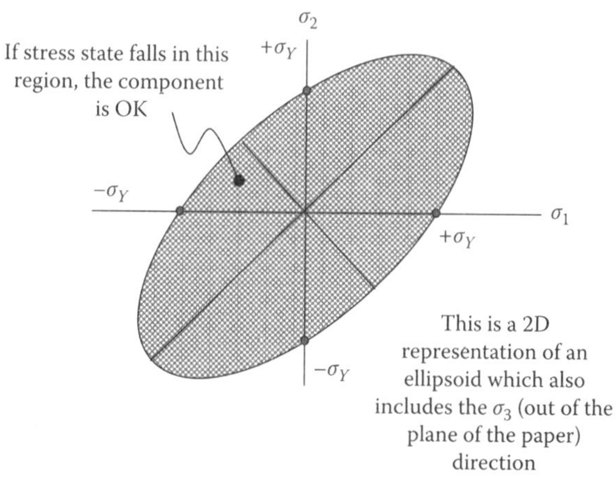

FIGURE 4.3
von Mises failure surface.

generated by the shear stress in the material. This assumption says that the material will not fail unless the shear stress it is experiencing is greater than that exhibited by a tensile test specimen of the same material at its failure point. Again, we assume that tensile stress is positive and orders the stresses with 1 being the largest to 3 being the smallest and state the following:

$$\frac{(\sigma_1 - \sigma_3)}{2} = \text{constant} \tag{4.24}$$

We set this constant equal to σ_Y^2 or K. Here σ_Y is the yield stress in simple tension and K is the yield stress in pure shear. This implies that for a component not to exhibit failure,

$$(\sigma_1 - \sigma_3) < \sigma_Y \tag{4.25}$$

and

$$\sigma_Y = 2K \tag{4.26}$$

as well as

$$(\sigma_1 - \sigma_2) < \sigma_Y, \quad (\sigma_1 - \sigma_3) < \sigma_Y, \quad \text{and} \quad (\sigma_2 - \sigma_3) < \sigma_Y \tag{4.27}$$

Once again, either σ_Y or K can be experimentally found. In σ_1–σ_2–σ_3 space, this is represented by a polyhedral surface whose inner region includes all stress states that are safe (nonfailing). This is shown as a 2D sketch in Figure 4.4.

The Tresca criterion is slightly more conservative if used for metals than von Mises. The Tresca polyhedron is contained within (circumscribed by) the ellipsoid of von Mises.

The third failure criterion we will examine is the Coulomb or maximum normal stress criterion. Here, we assume that the normal stress in the material will precipitate the failure. Tensile stress is again assumed to be in the positive direction, and stresses from 1 to 3 are again in order of decreasing magnitude. In this criterion, we require that for a material that does not exhibit failure,

$$\sigma_1, \sigma_2, \sigma_3 < \sigma_U \tag{4.28}$$

That is, all the principal stresses must be less than the ultimate stress σ_U in the material in that particular direction. Recall that we use this for brittle materials where there is no yield

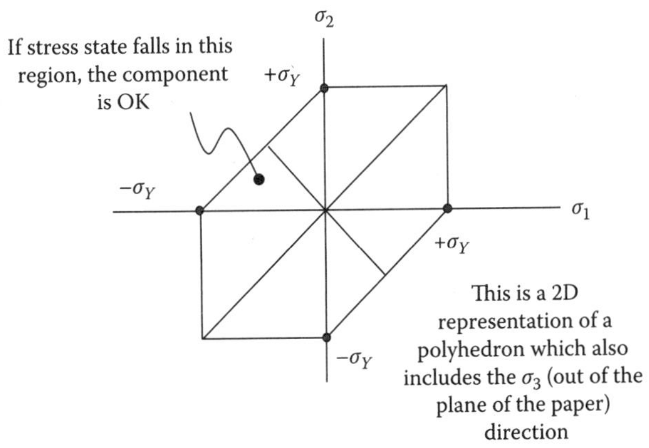

FIGURE 4.4
Tresca failure surface.

point or yielding behavior. The failure surface is a rectangular polyhedron whose edges are the ultimate stresses in each principal direction. Stress levels within the polyhedron will not cause failure. A 2D representation is depicted in Figure 4.5. Even though this figure is shown as a square, in many materials, the compressive strength is much greater than the tensile strength, resulting in different limits and thereby changing the appearance (and sometimes resulting in a name change as well to a Mohr–Coulomb criterion) of the failure surface. In this instance, the failure surface would look like Figure 4.6. In the Mohr–Coulomb failure criterion, a greater compressive normal stress allows the material to carry more load. This is caused by the locking of slip planes akin to the sliding friction of a block causing greater resistance when the block gets heavier (i.e., an increase in normal stress on the slip plane). When this is applicable, our criteria result in an equation for the failure surface as follows:

$$\max[|\tau| - \lambda \cdot (\sigma)] = \sigma_E \tag{4.29}$$

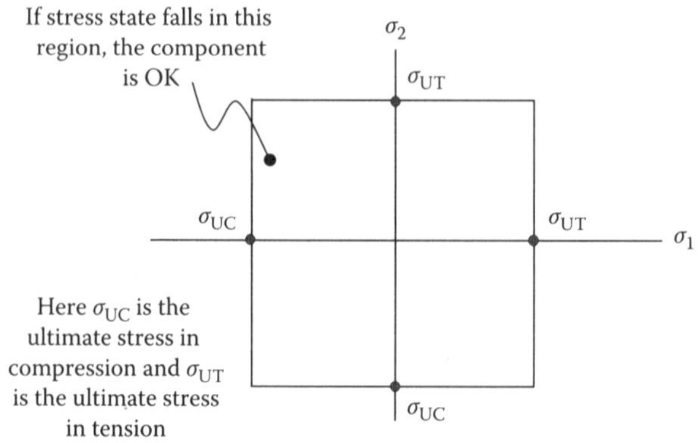

FIGURE 4.5
Coulomb failure surface.

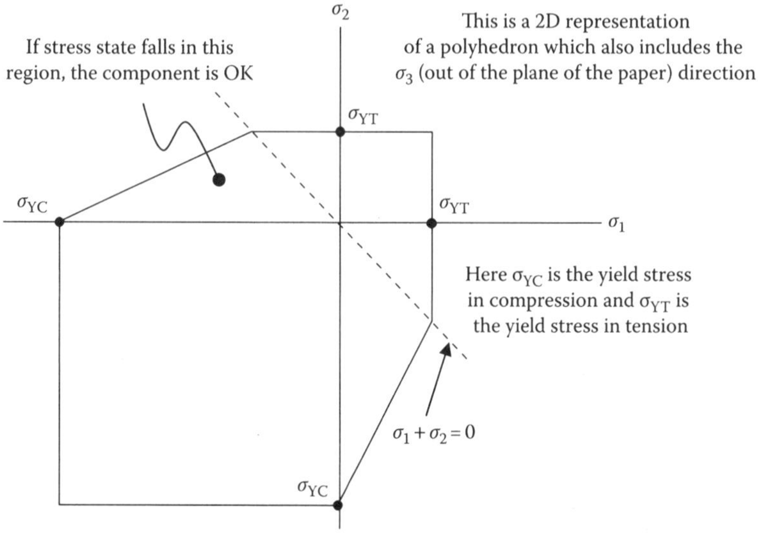

FIGURE 4.6
Mohr–Coulomb failure surface.

This equation results in a greater stress to failure due to the internal friction coefficient λ. Since compressive strength is negative and λ is a positive quantity, the equivalent failure stress σ_E is greater with greater normal stress σ.

Occasionally, it will be essential that we combine two or more of these criteria due to a change in material behavior. We shall describe this in due course.

Problem 2

A component has principal stress values of 20,000, 56,000, and −220,000 psi (note that negative means compressive stress); if the yield strength in a simple tension test of the material was found to be 180,000 psi, will the part survive based on the von Mises failure criterion?

Answer: No, the part will fail.

4.3 Ammunition Types

Just as weapons are categorized by their usage as guns (low angle, line of sight, direct fire), howitzers (high angle, beyond line of sight, indirect fire), or mortars (very high angle, short range, indirect fire), the munitions for them are also categorized, not by use, but by their construction or assembly methods. Ammunition can be fixed, separable, or separate loaded.

Fixed ammunition, usually called a cartridge, consists of a container for the propellant charge, called the cartridge case, which is firmly attached to the projectile by crimping or cement and that remains in the weapon after firing and is ejected near it or is consumed during firing, and the projectile that flies downrange to the target. The charge, priming, and ignition system are assembled inside the case and are not alterable. This type of ammunition is characteristically used in tank, antiaircraft, aircraft weapons, and in most small arms (rifles and pistols).

Separable ammunition (also called semifixed ammunition) also consists of the cartridge case and projectile, but the case is not firmly attached to the projectile and can be removed in the field to adjust the charge, which can be incrementally changed. This type of ammunition was used in older howitzers and is still used in shotguns.

Separate-loaded ammunition (sometimes called separated ammunition) consists of the projectile, which is loaded first into the weapon; the propellant charge loaded next; and finally, the primer and igniter loaded last. The charge, which is supplied to the weapon site, is in bagged increments and is altered, along with the quadrant elevation of the weapon, to vary the range. The primer is usually loaded into the breechblock of the weapon. The block is self-sealing and assumes this function, which in fixed ammunition is done by the cartridge case. Ammunition of this type is used in howitzers and large naval guns.

Mortar ammunition is essentially of the separated type. The charge is incremental to help vary the range by altering the muzzle velocity. The charge increments are held in place on the projectile body by clips or holders. Increments may be added or deleted in the field by the gunner. Priming is done through an integral attachment to the projectile (a boom). Primer initiation is by a firing pin in the weapon that strikes an initiator in the boom at the termination of the fall of the projectile as it is dropped down the tube from the muzzle end. Trigger firing is also possible in some weapon designs.

The practical design of fixed ammunition cartridges, which is what we will mainly dwell on, encompasses the design of the propellant charge ignition system, the construction of the main body of the propellant charge, the design of the projectile body itself, including its shape and mass distribution, and its obturation and stabilization components. The design must also incorporate into the projectile the ancillary systems necessary for its intended functioning, for example, fuzes, expulsion charges, explosive trains, and, in modern projectiles, guidance, and control.

4.4 Propellant Ignition

Energetic devices that are combined in a specific manner into the ignition train accomplish the initiation of combustion of the propellant charge. The first of these elements is the highly sensitive, detonator cup filled with material such as lead azide, which itself receives an energy pulse from a trigger mechanism that delivers the pulse in the form of a mechanical (spring actuated) or electrical (hot wire or laser) impulse. The sensitive mix is detonated by the impulse and flashes into the main, less sensitive ignition charge. This material is contained in the primer head.

Secondary ignition takes place in the main primer body, where the ignition material known as the primer charge is stored. This material has traditionally been fine-grained black powder, which is known to have certain undesirable properties such as hygroscopicity. Attempts have been made to replace black powder, but it still remains the chief secondary ignition material. Two basic forms of primer charge are used in large caliber munitions: flat base-pad igniter charges are used with separate-loaded bagged propellant charges and in fixed, stick propellant charges; central core or bayonet-type primer bodies are used in most fixed, loose, granular propellant charges.

The design goal of all ignition systems is to provide rapid but smooth ignition of the main propellant charge avoiding at all cost pressure surges or spikes. Such surges can crush individual grains or sticks causing large, uncontrolled increases in burning surfaces, and uncontrolled burning of the main charge. Symptoms of such burning are negative delta pressure ($-\Delta p$) waves, that is, negative gradients of pressure along the length of the chamber. One cause of pressure surges are the so-called blind primers, where vent holes are missing along the length of the primer body tube. The pressure buildup in the tube can rupture it causing asymmetric ignition and a $-\Delta p$.

Other caveats are to avoid overly sensitive detonator mixes and to provide gas flow space in the main propellant charge. Ignition and burning are surface phenomena, and too tightly packed charges do not provide the necessary surfaces.

4.5 Gun Chamber

To the rear of the long cylindrical portion of the gun (the bore) is the chamber, shown in Figure 4.7. The tapers shown facilitate the removal rearward of the spent cartridge case that hugs the chamber wall. During the firing cycle, the case swells because of the internal

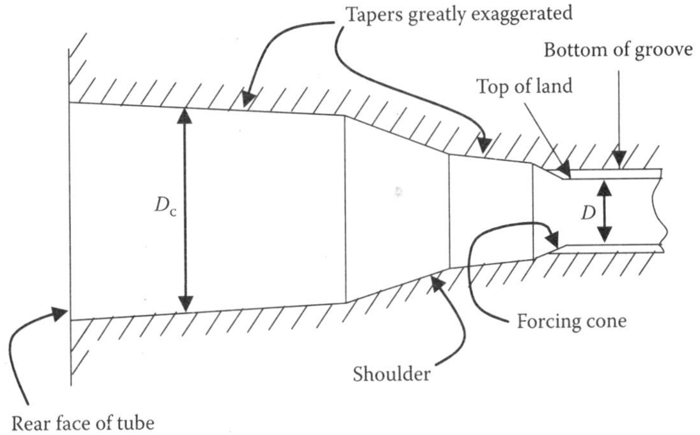

FIGURE 4.7
Chamber geometry.

pressure and firmly contacts the chamber wall sealing the gases from exiting rearward. When the pressure decays, a properly designed case comes away from the wall and the tapers ensure that it does not stick in the chamber. When a fixed round of ammunition is loaded into the chamber, the rear face of the tube provides the stop and seat for the rim of the case. During the expansion of the pressurized case, the forcing cone of the chamber forms the seal for the hot gases by the extrusion and engraving of the rotating band in a rifled bore or the extrusion of the obturating band in a smooth bore. The ratio D_c/D is known as the chambrage, an important characteristic of the design. Large values of the chambrage tend to cause turbulent flow of the gases as they enter the bore. Such turbulence contributes to the erosion of the bore surfaces.

The gun designer is caught in a curious bind: for a desired volume of propellant, a large chambrage provides a shorter cartridge length, frequently a highly desirable parameter in the tight confines of a turret, for example; on the other hand, large chambrage values subject the bore to more erosion. Some of this difficulty has been overcome by the use of erosion-reducing coolants. It has been found that much of the erosive wear in high-performance guns and howitzers can be ameliorated by the introduction of a cool liquid, gaseous, or particulate layer between the hot propellant gases and the bore. Materials such as titanium dioxide, wax, talc, or silicone oil have proven efficacious. If these materials are assembled in the body of the propellant charge so that the gas flow keeps the coolant at the bore wall, a substantial decrease of erosion results. This is called laminar flow and is observed in low-chambrage guns. Thus, a compromise may have to be made in the chambrage to reduce the turbulent flow.

4.6 Propellant Charge Construction

In fixed cartridges, the most common practice is to fill a metallic cartridge case with perforated granular propellant grains around a bayonet-type primer that has already been inserted in the case. The grains commonly have seven perforations for progressive burning. In high-performance rounds, vibrating the case to help settle the grains maximizes the loading density of the charge. Tank munitions are often loaded with perforated stick propellant. The sticks are bundled and carefully laid up around the boom and fin components that intrude into the depth of the case. Supplementary granular propellant is occasionally added to the stick bundles to further boost the charge mass and increase the progressivity of burning. Rocket grain configurations with complex star and slit perforations have been tried as well as 19-perf grains to raise the burning rate, but these are difficult to make and are not standard.

Howitzer (separate-loaded) charges are made up of bagged increments that are ignited by the last increment loaded, the base-pad igniter. A primer in the breechblock sets off the igniter. In these and in the fixed ammunition charges, coolants are strategically emplaced to promote erosion resistance. With bagged propelling charges, since there is no cartridge case present, it is extremely important that all the materials be combusted. Great care is taken in selecting materials—silk was used for many years in the navy—to assure that there are no burning embers left in the weapon after it was fired. It is typical for a howitzer crew member to look down the bore and shout "bore clear" during firing operations. If burning materials are present and a fresh charge is inserted into the bore, the propellant may ignite and cause serious injury to the gun crew. This has been termed *cook-off*.

There have been extensive efforts to take advantage of the convenience of stowage, low cost, and inherent safety of liquid bipropellants. However, severe operational and performance problems have prevented their adoption. These problems have centered on combustion instability that manifests itself in destructive, unpredictable pressure peaks, particularly in bulk-loaded systems. Attempts to get around these so-called Taylor instabilities have had some success with regenerative pressurized systems that atomize the pumped-in liquids, ignite this cloud, and avoid the pressure wave unpredictability of an ignited bulk of liquid. This concept, even though it has shown promise, may still not be able to overcome the poor low-temperature properties of the liquid propellants. They show marked increases in viscosity at low temperatures causing severe flow and pumping problems.

Two other concepts of gun propulsion should be mentioned. These are the use of electromagnetically generated force to propel a projectile down a gun and the idea of using a low-molecular weight gas to propel the projectile—the light gas gun. At the time of this writing, neither concept has shown the ability to progress beyond the laboratory stage to a fieldable weapon, although light gas guns are in common use in laboratories to reach velocities with small projectiles approaching meteorite entry speeds.

4.7 Propellant Geometry

The geometry of the propellant grain is one of the parameters available to the interior ballistician to tailor the pressure curve in the gun. The production of gas from a grain depends on the evolution of the total surface of the grain as the burning proceeds. If the surface area increases with time, the grain is considered progressive. If the total surface remains constant over time, the grain is neutral, and if the surface decreases with time, the grain is considered regressive. The perforations in the grain affect the surface area and, therefore, the burning characteristics. In cylindrical grains, the number of perforations is usually one of the numbers in the sequence: 1, 7, 19, and 37. The largest number in use in the United States is 19, and this is rarely found because of the difficulty of manufacture. The various types of grains

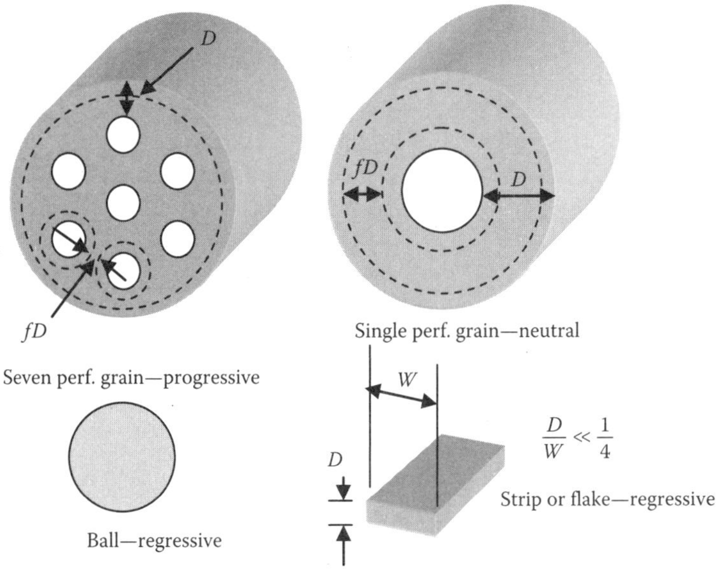

FIGURE 4.8
Typical propellant grain geometries.

are shown in Figure 4.8. The web D that is the smallest thickness of propellant between any two surfaces is one of the major parameters in interior ballistic computations.

4.8 Cartridge Case Design

The design of a metallic cartridge case must fulfill four basic roles: the case must seal or obturate the gun breech so that gases do not stream backward out of the gun; it must serve as a protective container for the propellant charge; it must act as a structural member of the cartridge assembly to allow for vigorous handling during shipping, stowage, and loading into the chamber; and it must be easily extractable from the chamber after the round is fired. Metallic cases have been used for much more than a hundred years, and the design practices are well established to fulfill these roles. Yet difficulties still arise in the extraction of the case after firing—it can stick in the chamber, rendering the weapon useless until it is removed. The case by itself cannot sustain the gun pressure and is intended to be supported by the chamber walls. Yet, the case must be designed with sufficient clearance to permit loading and ramming. The analysis of sticking that follows must be part of the design engineer's overall task before a new weapon can be fielded.

It is possible, through the use of some relatively simple equations, to determine if a cartridge case will expand enough to stick in the chamber of the weapon after firing. Graphically, we can depict this as shown in Figure 4.9. In this figure, we see the effect when a case with a low yield strength is loaded to the same levels as a good case. The expansion and contraction of the gun tube itself must be taken into account when the cartridge case is designed. This condition can be approximated using a bilinear, kinematic hardening model where the stress–strain curves of the case material are modeled, as depicted in Figure 4.10.

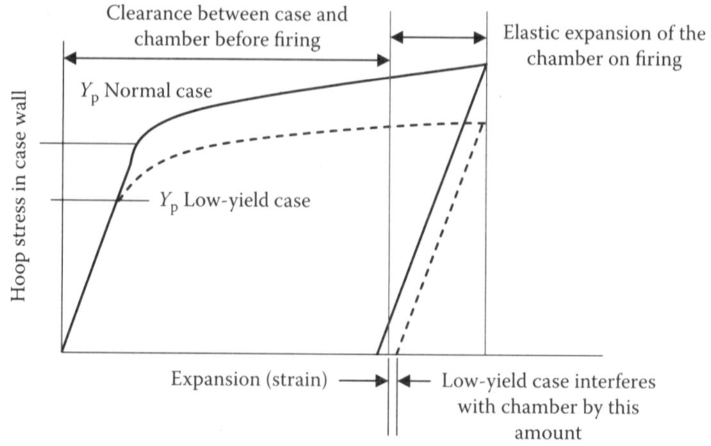

FIGURE 4.9
Stress–strain diagram of a normal case and one with low yield strength.

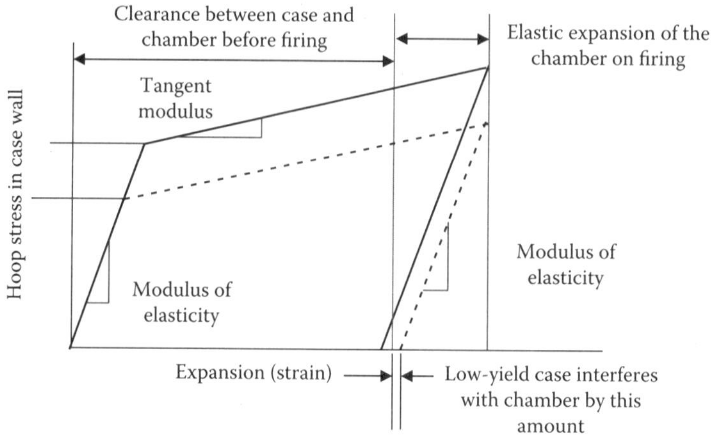

FIGURE 4.10
Stress–strain diagram of a normal case and one with low yield strength modeled as bilinear kinematic hardening materials.

The first step in this procedure is to model the gun tube. In this case, we assume that the material is perfectly elastic—which will be the case for any properly designed tube—and we can determine the radial expansion through [2]

$$u_{\text{tube}} = \frac{a'}{E_{\text{tube}}\left(b^2 - a'^2\right)}\left[(1-\nu)\left(p_1 a'^2 - p_2 b^2\right) + (1+\nu)b^2(p_1 - p_2)\right] \tag{4.30}$$

In this equation (which has been tailored from a previous formula for a thick-walled cylinder because the point we are interested in is on the inside radius of the tube wall), a' is the inner radius of the chamber, b is the outer radius of the gun tube, p_1 is the internal pressure, p_2 is the external pressure (usually conservatively taken as 0), ν is Poisson's ratio for the tube material, and E_{tube} is the modulus of elasticity.

We now calculate the stress, strain, and displacement of the case through use of the thin-walled cylinder equations [3]

$$u_{\text{case}} = \frac{a^2 p_1}{E_{\text{case}} h} \tag{4.31}$$

$$\sigma_{\theta\theta} = \frac{a p_1}{h} \tag{4.32}$$

$$\varepsilon_{\theta\theta} = \frac{\sigma_{\theta\theta}}{E_{\text{case}}} \tag{4.33}$$

In these equations, u_{case} is the radial expansion of the case, $\sigma_{\theta\theta}$ is the hoop stress in the case, $\varepsilon_{\theta\theta}$ is the hoop strain, a is the outside radius of the case, and h is the case wall thickness. Now, the gun tube will stop the case from expanding further once contact is made so the maximum expansion of the case will be as follows:

$$r_{\text{case}} + u_{\text{case}_{\max}} = r_{\text{tube}} + u_{\text{tube}} = r_{\text{case}} + a\varepsilon_{\theta\theta_{\max}} \tag{4.34}$$

Because we know the pressure and the tube dimensions and, therefore, the value of u_{tube}, we can calculate $\varepsilon_{\theta\theta_{\max}}$. We can then use this value to calculate the stress in the case at the maximum expansion:

$$\varepsilon_{\theta\theta_{\max}} - \varepsilon_Y = \frac{\sigma_{\theta\theta_{\max}} - \sigma_Y}{E_{\text{case-tangent}}} \tag{4.35}$$

In this expression, the subscript Y indicates yield values and $E_{\text{case-tangent}}$ is the tangent modulus of the cartridge case material. Once we determine the stress at the maximum expansion, we need to recall that a material which has yielded will retract along its original elastic modulus. Thus, we can write

$$\varepsilon_{\text{return}} = \frac{\sigma_{\theta\theta_{\max}}}{E_{\text{case}}} \tag{4.36}$$

Now the residual strain in the case is given by

$$\varepsilon_{\text{residual}} = \varepsilon_{\theta\theta_{\max}} - \varepsilon_{\text{return}} \tag{4.37}$$

We can then find the permanent radial displacement through

$$u_{\text{residual}} = a\varepsilon_{\text{residual}} \tag{4.38}$$

If we now add u_{residual} to the original radius of the case a, we can see that if

$$u_{\text{residual}} + a \geq a', \quad \text{the case will stick} \tag{4.39}$$

or if

$$u_{\text{residual}} + a < a', \quad \text{the case will not stick} \tag{4.40}$$

Over the last 20 years, the metallic case of drawn brass, extruded steel, or spirally wrapped steel has been replaced in certain systems by a fully combustible or consumable case. These cases are manufactured of felted nitrocellulose and usually consist of a base and sidewall that are assembled with cement, filled with granular or stick propellant, and attached to the projectile with clamps and cement. Since the cases are completely consumed, they do not seal the breech. With these munitions, a self-sealing breech must be designed for the weapon. For guns with nonsealing breeches that are already fielded for use with conventional metallic cartridge cases, a case that has a metallic stub and a combustible sidewall has been devised to take advantage of the small volume of the ejected stub in the confines of a tank turret, for example, and the overall reduction in the cost and weight of the round. While systems with the combustible case have been fielded, the success of this development has not been complete. Occasional problems with the incomplete combustion of a case that leaves smoldering residue capable of igniting the next loaded round (cook-off) have required scavenging systems for the chamber to be installed. The inherent structural weakness of nitrocellulose has also posed problems of case attachment and handling. Yet the obvious advantages of the combustible case have kept the concept in the weapon designer's toolbox for possible use.

Problem 3

A design for a 105 mm weapon is being considered. The chamber is stated to withstand the desired 35,000 psi and is essentially a steel cylinder of 4.5 in. inside diameter (ID) and 7 in. OD ($E_{tube} = 30 \times 10^6$ psi, $\nu = 0.3$). We have decided to use brass with an outside diameter (OD) of 4.490 in. If we use a bilinear, kinematic hardening model where the brass has a modulus of elasticity of 15×10^6 psi, a local tangent modulus of 12.5×10^6 psi, a yield stress of 15,000 psi (yield occurs in this material at $\varepsilon = 0.001$), and an ultimate tensile strength of 45,000 psi, with the information given, what is the radial clearance between the case and the chamber after firing neglecting thermal effects?

Answer: Approximately 0.004 in. radial clearance

4.9 Projectile Design

While propulsion systems are fairly straightforward in design because their intended use is simple, projectiles widely vary in use, and as a consequence, their designs are complex and demanding. The propulsion system must get the projectile through the launch environment with consistent muzzle velocities, but without undue stress to the gun or the projectile. The projectile, on the other hand, must withstand the forces of launch; be efficient, consistent, and precise in its flight environment; and deliver its intended utility at the target. We will explore only projectile design for launch in this section, reserving design for flight and terminal effects until later.

Projectiles may be classified into two general types: cargo carriers and pure kinetic energy deliverers. The cargo carriers include shells that deliver high explosives (HE), submunitions and mines, pyrotechnics, smart munitions, and other specialized lethal systems, e.g., shaped charges (high-explosive anti-tank [HEAT]) and explosively formed penetrator (EFP) shells. The kinetic energy delivery systems, used chiefly for the attack of armor, are monobloc steel shot (armor piercing), saboted, long-rod, heavy metal penetrators (armor-piercing, fin-stabilized, discarding sabot [APFSDS]), and older types of spin-stabilized, saboted (armor-penetrating, discarding sabot [APDS]) projectiles.

The stresses induced into a projectile during launch are chiefly due to the acceleration that the gases impart to it. The cargo carriers are shells whose stresses are due to relatively low accelerations and which, except for the tank cannon-fired HEAT shell, achieve only moderate muzzle velocities. We will therefore explore the kinds of stresses and failures inherent in shell-like structures under load in Section 4.10. Kinetic energy munitions, on the other hand, are subject to extremely high accelerations and have high muzzle velocities. For these types, we will explore the driving mechanism stresses and other aspects of these designs.

The gamut of topics in projectile design is almost unlimited. However, several suggest themselves because of their general applicability or timely interest. Shell design is a ubiquitous problem and will be explored in depth in Section 4.10. The use of buttress threads is so common in projectile and gun design that it warrants its own in Section 4.11. Sabot design is more specialized as are the problems of kinetic energy rods and their buttress driving grooves. These will be explored in Section 4.12.

Modern projectiles employ a variety of electronic and electromechanical devices for fuzing, target detection, and guidance and control. This relatively new engineering discipline called "gun hardening" deals with designing these devices to survive the harsh environment of gun launch.

4.10 Shell Structural Analysis

Most cargo-carrying projectiles, whether fin- or spin-stabilized, are designed with cargo bodies in the shape of an axisymmetric cylindrical shell. Because the loads on these cylinders are the result of spin and acceleration of the shells and their contents, the stresses encountered are highly variable along and through their walls. These stresses will be examined as will the consequences of failure criteria.

The symbols and definitions of the constants and variables of shell loading are listed next:

A—Bore area of the gun

a—Linear acceleration

d—Diameter of bore (across lands), diameter of assumed shear circle in base of shell

d_i—ID of projectile

d_o—OD of projectile

F_b—Maximum force on base of projectile and rotating band

F_T—Maximum tangential force on projectile wall

F_{TR}—Hoop tension (force) in wall of projectile resulting from rotation of the shell

F'_T—Tangential force at section of shell

f'—Setback force

g—Acceleration due to gravity

h—Total depth of filler from nose

h'—Total depth of filler from nose end of cavity to section under consideration

I_{zz}—Polar moment of inertia

I'_{zz}—Polar moment of inertia of metal parts forward of section when section is ahead of rotating band and aft of it when section is aft of the rotating band

n—Twist of the rifling

p_b—Maximum propellant pressure

p_h—Filler pressure due to setback
p_{rot}—Filler equivalent pressure due to rotation, includes wall inertia
r_i—Inside radius of projectile
r_o—Outside radius of projectile
S—Compressive strength of the rotating band
S_1—Longitudinal stress
S_2—Tangential stress
S_3—Radial stress
τ—Shear stress
σ_Y—Static yield stress in tension
T—Torque applied to the projectile
t—Base thickness, wall thickness
V—Muzzle velocity
w—Total projectile weight
w'—Weight of metal parts forward of section under consideration
w'_f—Weight of filler forward of section under consideration
α—Angular acceleration
ρ_m—Density of projectile material
ρ_f—Density of filler material
ω—Angular velocity
r_b—Radius of band seat
p_{band}—Band pressure

We distinguish between thin-walled and thick-walled cylinders in this analysis so that the designer may run quick, ballpark estimates of the stress levels encountered. In practice, FEA is usually conducted on the components, but as emphasized earlier, the designer should have a good idea of the bounds of the answer before beginning the FEA.

We begin with a review of basic mechanics of materials as applied to cylinders. If a cylinder is subjected to an axial load and does not buckle, the axial stress can be determined from

$$S_1 = -\frac{F_{Axial}}{A} = -\frac{F_{Axial}}{\pi\left(r_o^2 - r_i^2\right)} \tag{4.41}$$

The stress–strain relationships for a cylinder are as follows:

$$\varepsilon_{rr} = \frac{1}{E}\left[\sigma_{rr} - \nu(\sigma_{\theta\theta} + \sigma_{zz})\right] \tag{4.42}$$

$$\varepsilon_{\theta\theta} = \frac{1}{E}\left[\sigma_{\theta\theta} - \nu(\sigma_{rr} + \sigma_{zz})\right] \tag{4.43}$$

$$\varepsilon_{zz} = \frac{1}{E}\left[\sigma_{zz} - \nu(\sigma_{\theta\theta} + \sigma_{rr})\right] \tag{4.44}$$

where ν is Poisson's ratio; σ_{rr} is the radial stress; $\sigma_{\theta\theta}$ is the transverse (hoop) stress; σ_{zz} is the axial stress; and E is Young's modulus.

If a cylinder is subjected to a torsional load, it will twist. We typically assume that this deformation is small and plane sections remain plane. Thus, when we apply a torque T to a cylinder of length L with shear modulus G, and polar moment of inertia J, the structure will rotate through an angle ϕ (in radians):

$$\phi = \frac{TL}{JG} \tag{4.45}$$

For a hollow cylinder,

$$J = \frac{1}{2}\pi\left(r_o^4 - r_i^4\right) \tag{4.46}$$

For a material that behaves according to Hooke's law,

$$G = \frac{E}{2(1+\nu)} \tag{4.47}$$

Such a material under pure torsion will only exhibit shear stress according to

$$\tau_{\theta z} = \frac{Tr}{J} \tag{4.48}$$

While the thick-walled cylinder analysis, which we describe next, is an exact solution, a quick way to assess the major stresses if the wall thickness is less than 10% of the cylinder radius is to assume that the stresses in the radial direction S_3 are negligible.

Thus, we examine only the meridional or longitudinal and the circumferential or hoop stresses. We define S_1 as the longitudinal stress, S_2 as the hoop stress, and p as the pressure depicted in Figure 4.11.

If the cylinder has closed ends, then internal pressure can cause a longitudinal stress

$$S_1 = \sigma_{zz} = \frac{pr}{2t} \tag{4.49}$$

Otherwise, $S_1 = 0$. Internal (or external) pressure always causes hoop stress

$$S_2 = \sigma_{\theta\theta} = \frac{pr}{t} \tag{4.50}$$

In practical shell design, we always perform a thick-walled cylinder analysis assuming that the stresses in the radial direction are significant enough to be considered. Thus, we must examine longitudinal, hoop, and radial stresses. We again define $S_1 = \sigma_{zz}$ = longitudinal

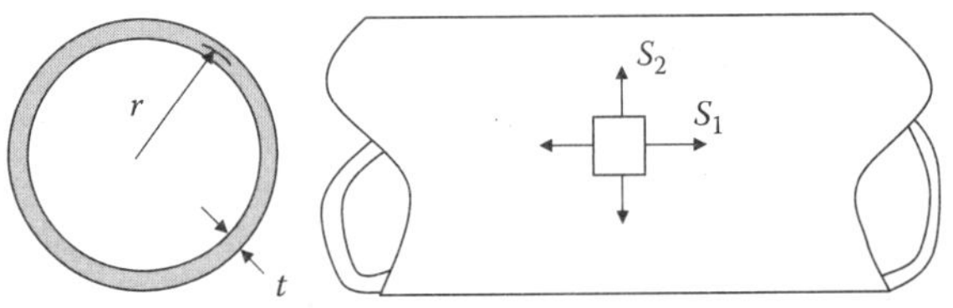

FIGURE 4.11
Thin-walled cylinder geometry.

stress, $S_2 = \sigma_{\theta\theta}$ = hoop stress, $S_3 = \sigma_{rr}$ = radial stress, and p = pressure. This is depicted in Figure 4.12.

The following solutions are known as the Lamé formulas and assume open ends, which implies $S_1 = 0$ if no axial loads are present. If axial loads are present, they must be accounted for. Internal (or external) pressure always causes hoop stress. (Note that the subscripts o and i refer to the outer and inner surfaces, respectively.)

$$S_2 = \sigma_{\theta\theta} = \frac{1}{\left(r_o^2 - r_i^2\right)}\left[p_i r_i^2 - p_o r_o^2 - \frac{r_i^2 r_o^2 (p_o - p_i)}{r^2}\right] \tag{4.51}$$

With a maximum at $r = r_i$. The radial stress can be calculated from

$$S_3 = \sigma_{rr} = \frac{1}{\left(r_o^2 - r_i^2\right)}\left[p_i r_i^2 - p_o r_o^2 + \frac{r_i^2 r_o^2 (p_o - p_i)}{r^2}\right] \tag{4.52}$$

with a maximum again at the inner surface $r = r_i$ and equal to $S_3 = -p_i$.

Initially, we will analyze the state of stress caused by the centrifugal loading induced by the rotation of a projectile in a rifled gun tube. In a spin-stabilized projectile, besides the longitudinal loads induced by the acceleration through the tube, the rotation of the projectile, which is dependent upon the axial velocity and the twist of the rifling in the tube, induces stresses in the walls. The twist of the rifling is usually measured in revolutions per caliber of travel (i.e., a twist of 1 in 20 means the projectile makes one revolution in 20 calibers of travel [$n = 20$]). The units of n are calibers per revolution. If we multiply n by the diameter, d, we get units of length per revolution:

$$n\left[\frac{\text{caliber}}{\text{revolution}}\right] \times d\left[\frac{\text{length}}{\text{caliber}}\right] = nd\left[\frac{\text{length}}{\text{revolution}}\right] \tag{4.53}$$

Since there are 2π radians per revolution, the angular velocity a projectile has attained is defined as

$$\omega = \frac{2\pi V}{nd} = \frac{\left[\frac{\text{radians}}{\text{revolution}}\right]\left[\frac{\text{length}}{\text{time}}\right]}{\left[\frac{\text{length}}{\text{revolution}}\right]} = \left[\frac{\text{radians}}{\text{time}}\right] = \left[t^{-1}\right] \tag{4.54}$$

The centrifugal force directed radially outward on an element of material at radius r is

$$F_c = ma_r = \frac{w}{g} r\omega^2 \tag{4.55}$$

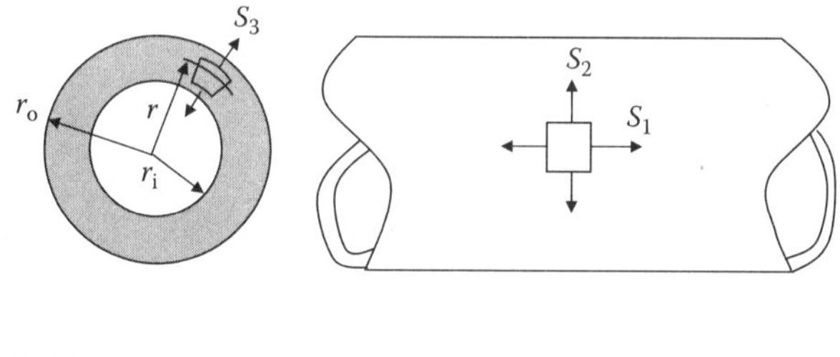

FIGURE 4.12
Thick-walled cylinder geometry.

In the tangential direction, the inertial forces on an element of material can be determined from

$$F_t = ma_t = \frac{w}{g} r\alpha \tag{4.56}$$

We can determine the centrifugal force on the cylinder wall caused by spinning the cylinder in the absence of other loads by integrating Equation 4.55 from the inner diameter to the outer diameter. To do this, we consider the differential element as depicted in Figure 4.13. From this diagram, we see that the mass of an infinitesimal annular ring of material is

$$\mathrm{d}m = \frac{\mathrm{d}w}{g} = \rho\, \mathrm{d}V = \rho l 2\pi r\, \mathrm{d}r \tag{4.57}$$

Inserting Equation 4.57 into Equation 4.55 yields

$$\mathrm{d}F_c = a_r\, \mathrm{d}m = \rho l 2\pi r^2 \omega^2\, \mathrm{d}r \tag{4.58}$$

which, when integrated from the inner to the outer radius, gives

$$F_{c\mathrm{WALL}} = 2\pi\rho l\omega^2 \int_{r_i}^{r_o} r^2 \mathrm{d}r = \frac{2\pi\rho l\omega^2}{3}\left(r_o^3 - r_i^3\right) \tag{4.59}$$

This is the radial force on the wall due to the inertia of the wall material only. If the projectile is filled with material, we need to account for this filler as well. Thus, if we integrate from the centerline to the inner radius of the projectile wall, we obtain

$$F_{c\mathrm{FILL}} = 2\pi\rho_{\mathrm{FILL}} l\omega^2 \int_{0}^{r_i} r^2 \mathrm{d}r = \frac{2\pi\rho_{\mathrm{FILL}} l\omega^2}{3}\left(r_i^3\right) \tag{4.60}$$

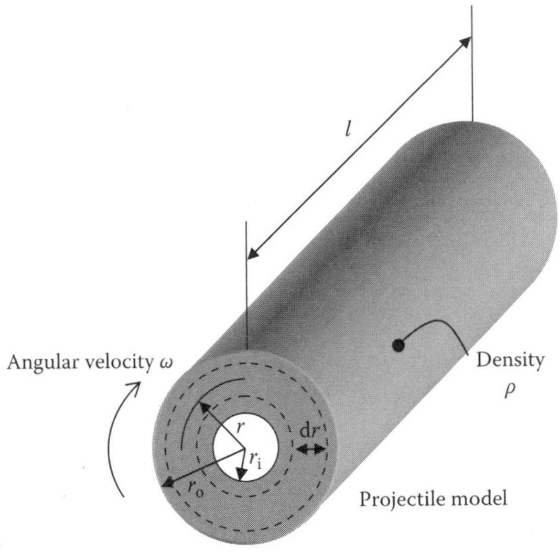

FIGURE 4.13
Differential thickness element geometry.

The total force acting on the projectile wall due to spin is then

$$F_c = F_{cWALL} + F_{cFILL} = \frac{2\pi l \omega^2}{3}\left[\rho\left(r_o^3 - r_i^3\right) + \rho_{FILL}\left(r_i^3\right)\right] \tag{4.61}$$

For stress computations, we require an internal pressure; thus, we need to convert the centrifugal forces to an equivalent internal pressure. If we assume that our centrifugal forces are acting on the interior of the shell, pushing radially outward, the area for our equivalent pressure is

$$A_{rad} = 2\pi r_i l \tag{4.62}$$

Thus, our equivalent pressure can be written as

$$p_{rot} = \frac{F_c}{A_{rad}} = \frac{\omega^2}{3r_i}\left[\rho\left(r_o^3 - r_i^3\right) + \rho_{FILL}\left(r_i^3\right)\right] \tag{4.63}$$

In Equation 4.56, we determined the tangential force arising from the angular acceleration. If we perform a similar analysis to that which developed Equation 4.61, we will obtain an expression for the torque as follows:

$$T = M_{WALL} + M_{FILL} = \frac{1}{2}\pi\alpha l\left[\rho\left(r_o^4 - r_i^4\right) + \rho_{FILL}\left(r_i^4\right)\right] \tag{4.64}$$

The derivation of this is left as an exercise for the interested reader and is included as a problem at the end of the chapter.

The formulas for calculating the tangential and radial stresses at radial location r in a rotating cylinder where $r_o > 10(r_o - r_i)$ can be given as

$$\sigma_{\theta\theta} = rw^2\left(\frac{3+\nu}{8}\right)\left(r_i^2 + r_o^2 + \frac{r_i^2 r_o^2}{r^2} - \frac{1+3\nu}{3+\nu}r^2\right) \tag{4.65}$$

$$\sigma_{rr} = \rho\omega^2\left(\frac{3+\nu}{8}\right)\left(r_i^2 + r_o^2 + \frac{r_i^2 r_o^2}{r^2} - r^2\right) \tag{4.66}$$

We are reminded that the longitudinal stress (assuming the structure does not buckle) is simply the axial acceleration multiplied by the weight of all the materials forward of the location of interest divided by the shell cross-sectional area—we will discuss this presently. These formulas were developed for the centrifugal loading of a spinning projectile by forces that act during both in-gun setback and flight. The axial load on a projectile, however, is for the most part only present during acceleration in the tube, is a function of time, and occurs whether the projectile is spinning or not. Beyond this, there is also an applied torque due to the angular acceleration, which is applied through the rotating band or slip obturator. The setback load and (if spinning) the centrifugal and torsional loads must all be superimposed on the projectile to determine its state of stress.

The axial force on the projectile during firing is given by

$$F = p_s A, \tag{4.67}$$

where p_s is the pressure acting on the base of the projectile defined by the Lagrange approximation

$$p_s = p_B\left(\frac{1}{1+\dfrac{c}{2w}}\right) \tag{4.68}$$

The d'Alembert force is the force due to acceleration that exactly equals this pressure force

$$a = \frac{p_s A g}{w} \tag{4.69}$$

At any axial position, the force on the cross-sectional area can be shown to be proportional to the weight of material forward of the section:

$$f' = \frac{w'}{w} p_s A \tag{4.70}$$

To calculate the force (or really the pressure) in the filler material, we usually resort to a hydrostatic model

$$p_h = \rho h a = \rho h \frac{p_s A g}{w} \tag{4.71}$$

where ρ is the density of the filler; h is the filler head height; and p_h is the hydrostatic pressure that is developed.

In a spin-stabilized projectile, the angular acceleration α is proportional to the linear acceleration a, where

$$\alpha = Ka \tag{4.72}$$

Then,

$$\alpha = K \frac{p_s A g}{w} \tag{4.73}$$

where K has units of length−1 and is dependent upon the twist n (in calibers of travel per turn), and the bore diameter d; thus,

$$K = \frac{\tan\theta}{\pi d} = \frac{2\pi}{nd} \tag{4.74}$$

where θ is the angle between the circumferential twist distance and the axial distance traveled. From Equations 4.73 and 4.74, we get

$$\alpha = \frac{2\pi}{nd} \frac{p_s A g}{w} \tag{4.75}$$

If we define a tangential force applied to the rotating band of the projectile as F_T, then the torque on the projectile is

$$T = F_T \frac{d}{2} \tag{4.76}$$

We know that the torque is equal to the product of the polar moment of inertia of the projectile and its angular acceleration

$$T = I_{zz}\alpha \tag{4.77}$$

Solving for the angular acceleration in terms of the tangential force, we get

$$\alpha = \frac{F_T}{I_{zz}} \frac{d}{2} \tag{4.78}$$

Inserting this into Equation 4.56 and solving for F_T yields

$$F_T = \frac{\pi^2 I_{zz} p_s}{nw} \tag{4.79}$$

Since

$$A = \pi \frac{d^2}{4} \tag{4.80}$$

The force that is applied by the rifling to the rotating band is transmitted through the structure to regions both forward and aft of the rotating band. These forces are proportional to the moment of inertia of the sections ahead of or behind the application of the torque load I'_{zz}. We assume that this force acts over a mean diameter of the outer and inner wall surfaces of the shell, and then we get

$$F'_T = \frac{16\pi I'_{zz}}{n(d_o + d_i)^2} \frac{p_s A}{w} \tag{4.81}$$

Because the rotating band is intended to act as a gas seal (obturator) as well as the rotational driver, designs typically exhibit a diameter over the band that is slightly larger than the groove diameter of the weapon. The engraving action of the gun lands and the interference fit in the grooves causes a plastic flow of the band, resulting in a pressure on the band seat as well as a developed reaction in the gun wall. This pressure can be greater than the gas base pressure on the projectile. Measurements of this pressure have been obtained by strain gaging of the gun tube and computing the stress at the inner diameter of the weapon. The pressure required to cause this stress is called the interface pressure. It has been shown that cannelures or circumferential grooves cut into the band surface reduce this substantially pressure by allowing room for band material to flow rather than being loaded in a quasi-hydrostatic condition. This is depicted in Figure 4.14. The composition/material of the rotating band can have a dramatic effect upon the behavior of the projectile in the tube as well as tube wear. An excellent example of this relationship is contained in the report by Montgomery [4].

We have the forces on the projectile structure but must now translate these into stresses that allow us to determine how much design margin is present. Once determined, these stresses are then linked to well-established failure criteria to determine the failure point of the material. Since projectiles may be made of a variety of materials, specialized criteria may

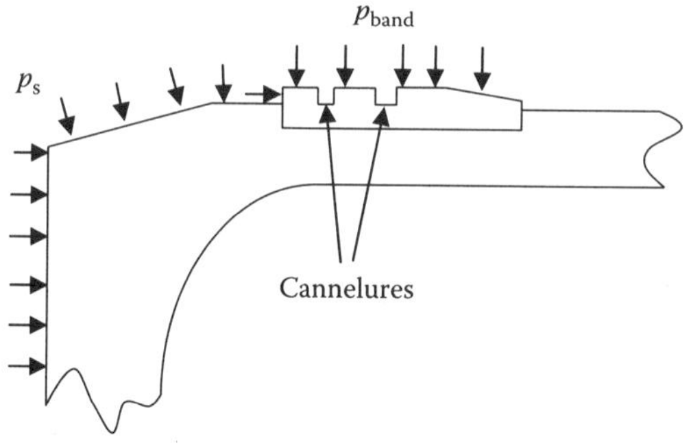

FIGURE 4.14
Rotating band pressure.

have to be used on each material. This full procedure is somewhat complicated and beyond the scope of this book, but we will attempt to describe the basics through an examination of a simple M1, highly explosive projectile structure depicted in Figure 4.15.

Assume a thick-walled cylinder as shown for stress calculations where S_{1j} is the longitudinal stress at the *j*th location; S_{2j} is the hoop stress at the *j*th location; S_{3j} is the radial stress at the *j*th location; τ_{11} is the longitudinal shear at the base; and τ_{2j} is the torsional (shear) stress at the *j*th location

It is helpful to recap here all the loads on an element of projectile wall material at a generalized location (such as point A) in the diagram. This element of material is

- Compressed in the axial direction due to the axial acceleration
- Loaded in tension in the hoop direction because of the wall mass being pulled radially outward due to the spin
- Loaded in tension in the hoop direction because of the filler material moving outward due to the setback load and the spin
- Loaded in shear due to the rotating band accelerating the projectile in an angular direction
- Loaded in shear due to the greater stress in the outer wall than on the inside wall

Note that when including mass forward of a particular section, we must include all mass transmitting loads to the section, e.g., fuze, bushings, and cups. The pressures applied to our model of the M1 projectile are shown in Figure 4.16.

Now let us examine specific locations of interest along the shell where experience tells us that failures might occur. For convenience, these have been tabulated in Table 4.1 and tailored to each individual location with the symbol, source load, and type of stress noted.

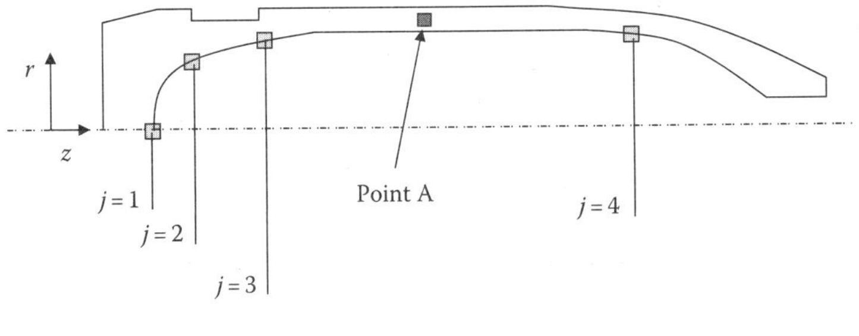

FIGURE 4.15
Stress locations in an M1 highly explosive projectile.

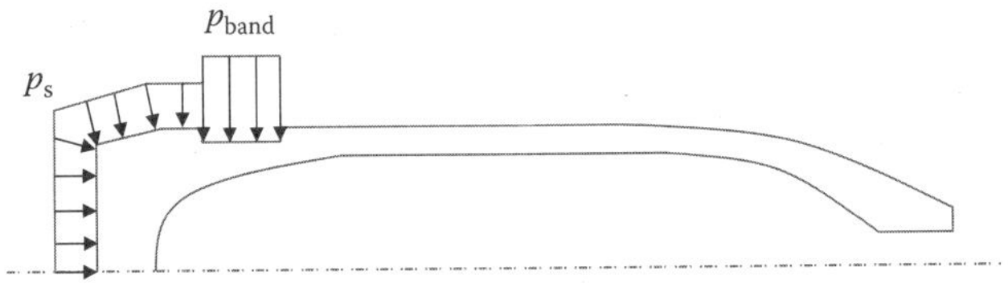

FIGURE 4.16
Load conditions for an M1 highly explosive projectile.

TABLE 4.1

Typical Stresses in a Highly Explosive Projectile and Their Sources

Type of Stress	Symbol	Source of Load
Compressive load on base	S_{11}	Setback of filler
Radial stress on base at centerline	S_{31}	Moments of filler setback and base pressures (flat base)
Radial stress on base at centerline	S'_{31}	Moments of filler setback and base pressures (round base)
Hoop stress at rear of the band	S_{22}	Setback of filler, rotation, and external pressure (band and gas)
Radial stress at ends of band and maximum ID	S_{32}, S_{33}, S_{34}	Rotation of projectile, filler setback, and filler rotation
Longitudinal stress at ends of band and maximum ID	S_{12}, S_{13}, S_{14}	Setback of metal parts in wall (filler contribution usually neglected)
Hoop stress at forward end of band and maximum ID	S_{23}, S_{24}	Filler pressure and rotation of wall
Shear stress through thickness t	τ_{11}	Moments of filler setback and base pressures (round base)
Torsional shear in projectile wall	τ_{22}, τ_{23}, τ_{24}	Setback of filler, rotation, and external pressure (band and gas)

At location 1, these are the formulas used to calculate stresses due to the setback of filler on base, the moments caused thereby, and by gas pressure on base:

$$S_{11} = -p_\text{h} \tag{4.82}$$

$$S_{21} = 0 \tag{4.83}$$

$$S_{31} = \frac{r_\text{o}^2}{t^2}(p_\text{s} - p_\text{h}) \tag{4.84}$$

$$S'_{31} = -p_\text{s}\left[\frac{3r_\text{o}^3}{2\left(r_\text{o}^3 - r_\text{i}^3\right)}\right] + p_\text{h}\left[\frac{r_\text{o}^3 + 2r_\text{i}^3}{2\left(r_\text{o}^3 - r_\text{i}^3\right)}\right] \tag{4.85}$$

Equation 4.82 is the axial component stress. We can see that it is just driven by the reaction of the fill and shell to the axial acceleration. Since this is a centerline location, by definition, there is no hoop stress, which is defined by Equation 4.83. Equation 4.84 specifies the radial stress assuming the base is flat faced. This comes about from the difference in the base pressure reacting against the internal forces and attempting to push the center of the base into the fill. Equation 4.85 is the radial stress equation assuming the base is a rounded bottom (i.e., with the concave portion enclosing the fill). We can see from this equation that the stresses are much lower as it carries the load more efficiently than a flat bottom shell. The drawback is that a base of this type requires a skirted boat tail which is more expensive to manufacture but saves considerable weight.

Moving to location 2, these are the stresses due to setback of filler, filler rotation, wall rotation, and band pressure:

$$S_{12} = -\frac{w' + w'_\text{f}}{w}\left[\frac{p_\text{s}A}{\pi\left(r_\text{o}^2 - r_\text{i}^2\right)}\right] + \left[\frac{(p_\text{h} + p_\text{rot})r_\text{o}^2}{\left(r_\text{o}^2 - r_\text{i}^2\right)}\right] \tag{4.86}$$

$$S_{22} = (p_h + p_{rot})\left(\frac{r_o^2 + r_i^2}{r_o^2 - r_i^2}\right) - p_{band}\left(\frac{r_b^2}{r_b^2 - r_i^2}\right) \tag{4.87}$$

$$S_{32} = -(p_h + p_{rot}) \tag{4.88}$$

At this location, we see that the axial stress defined by Equation 4.86 has two parts. The first term on the right-hand side (RHS) is the inertia of all the fill and shell material ahead of this location. The second term is the axial stress caused by the internal pressure of the fill expanding. In Equation 4.87, the first term on the RHS is the contribution of spin to the hoop stress and the second term is the restoring force caused by the gun tube pushing in on the rotating band. Equation 4.88 is simply the radial stress caused by the rotation and compression of the fill and wall.

Further forward on the shell at location 3, the stresses due to setback of filler, filler rotation, wall rotation, and band pressure have identical formulas to location 2 but with, of course, different values of the variables due to the lower hydrostatic pressure component:

$$S_{13} = -\frac{w' + w_f'}{w}\left[\frac{p_s A}{\pi(r_o^2 - r_i^2)}\right] + \left[\frac{(p_h + p_{rot})r_o^2}{(r_o^2 - r_i^2)}\right] \tag{4.89}$$

$$S_{23} = (p_h + p_{rot})\left(\frac{r_o^2 + r_i^2}{r_o^2 - r_i^2}\right) - p_{band}\left(\frac{r_b^2}{r_b^2 - r_i^2}\right) \tag{4.90}$$

$$S_{33} = -(p_h + p_{rot}) \tag{4.91}$$

Finally, at location 4, near the nose of the shell, the stresses due to setback of filler, filler rotation, and wall rotation are as follows:

$$S_{14} = -\frac{w' + w_f'}{w}\left[\frac{p_s A}{\pi(r_o^2 - r_i^2)}\right] + \left[\frac{(p_h + p_{rot})r_o^2}{(r_o^2 - r_i^2)}\right] \tag{4.92}$$

$$S_{24} = (p_h + p_{rot})\left(\frac{r_o^2 + r_i^2}{r_o^2 - r_i^2}\right) \tag{4.93}$$

$$S_{34} = -(p_h + p_{rot}) \tag{4.94}$$

At each location, one must be certain to use the proper head height of filler and the proper inner and outer radii of the shell.

We must also account for the shear stresses which are most severe at location 1. For simplicity, we will assume a flat base and calculate the shear stress due to wall torsion.

Wherever these calculations are done on the shell, the proper I_{zz} and the proper inner and outer diameters must be used:

$$\tau_{11} = \frac{(p_s - p_h)\pi r_i^2}{2\pi r_i t} = \frac{(p_s - p_h) r_i}{2t} \tag{4.95}$$

$$\tau_{22}, \tau_{23}, \tau_{24} = \frac{F_T'}{\frac{\pi}{4}(d_o^2 - d_i^2)} = \frac{64 I_{zz}'}{n(d_o + d_i)^3 (d_o - d_i)} \frac{p_s A}{w} \tag{4.96}$$

A typical loading of the shell using known weights, pressures, and acceleration is shown in Figure 4.17 and Table 4.2.

The common practice currently used in projectile design is to dispense with the hand calculations and go right to an FEA. While this is usually very accurate and saves a good deal of time, there are instances when one would like to check the answers through a hand calculation. Let us examine one location on this 105 mm M_1 High Explosive (HE) projectile fired from an M2A2 cannon at 145°F.

Projectile data:

- Shell material: HF-1 steel
  - Density—0.283 lbm/in.^{3}
  - Projectile OD—4.10 in.
  - Projectile ID (average)—2.95 in.
  - Projectile effective (including friction) mass (fuzed)—42 lb
  - Projectile base intrusion into cartridge case—85.43 in.^{3}
  - I_{zz}—80.24 lbm in.^{2}
- Fill material: TNT
  - Density—0.036 lbm/in.^{3}
  - Total length of explosive column—13.44 in.
  - I_{zz}—5.17 lbm in.^{2}

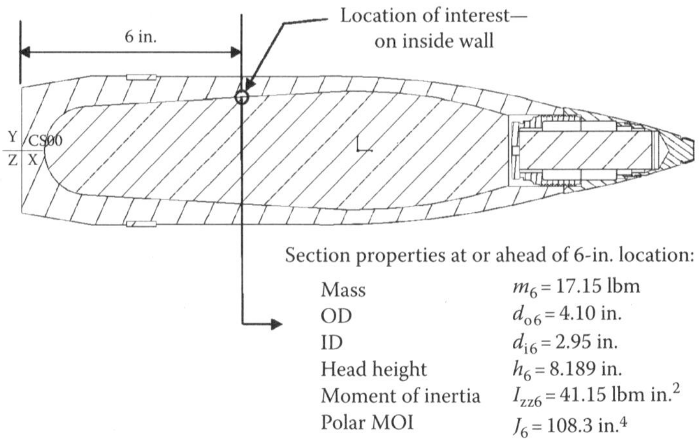

FIGURE 4.17
Location of interest on a 105 mm M1 projectile.

TABLE 4.2
Typical Values for Use in a Highly Explosive Projectile Design

Component	Weight (lbm)	Loads	
Fuze	2.1	Breech pressure (psi)	38,400
Body	34.0	Spin rate—maximum p (H_z)	82.4
Rotating band	0.4	Base pressure (psi)	37,150
Filler (TNT)	5.5	Acceleration (g)	11,873
Total	42.0	Angular acceleration (rad/s_2)	348,600

- Average fill cross-sectional area—6.49 in.^{2}
- Fill surface area—124 in.^{2}

The M1 projectile fired from our cannon is depicted in Figure 4.17. The properties of the section ahead of the location of interest are provided in Figure 4.17. We shall determine the stress tensor at the location shown. We shall assume that the projectile perfectly obturates and that there is no friction between the projectile and the tube.

To begin, we should always draw a free-body diagram of an infinitesimal element at the point of interest.

Let us look at the hoop direction first. We shall use Equation 4.51:

$$\sigma_{\theta\theta} = \frac{1}{\left(r_o^2 - r_i^2\right)}\left[p_i r_i^2 - p_o r_o^2 - \frac{r_i^2 r_o^2 (p_o - p_i)}{r^2}\right] \tag{4.97}$$

In this case, $r = r_i$ and $p_o = 0$, so we can write

$$\sigma_{\theta\theta} = \frac{1}{\left(r_o^2 - r_i^2\right)}\left[p_i\left(r_i^2 + r_o^2\right)\right] \tag{4.98}$$

The internal pressure is found through our equivalent pressure technique mentioned earlier:

$$p_{rot} = \frac{\omega_{p_{max}}^2}{3r_i}\left[\rho\left(r_o^3 - r_i^3\right) + \rho_{fill} r_i^3\right] \tag{4.99}$$

$$p_{rot} = \frac{(82.4)^2\left[\frac{rev}{s}\right]^2 (2\pi)^2 \left[\frac{rad}{rev}\right]^2}{(3)(1.475)\,[in.]\,(12)\left[\frac{in.}{ft}\right](32.2)\left[\frac{lbm\text{-}ft}{lbf\text{-}s^2}\right]}$$

$$\times\left\{(0.283)\left[\frac{lbm}{in.^3}\right]\left[(2.05)^3 - (1.475)^3\right]\left[in.^3\right] + (0.036)\left[\frac{lbm}{in.^3}\right](1.475)^3\left[in.^3\right]\right\}$$

$$p_{rot} = 258\left[\frac{lbf}{in.^2}\right]$$

For the hydrostatic component of the equivalent pressure, we know that

$$p_h = \rho_{fill} a_{p_{max}} h_6, \tag{4.100}$$

$$p_h = \frac{(0.036)\left[\frac{lbm}{in.^3}\right](382,300)\left[\frac{ft}{s^2}\right](8.189)\,[in.]}{(32.2)\left[\frac{lbm\text{-}ft}{lbf\text{-}s^2}\right]}$$

$$p_{\mathrm{h}} = 3500\left[\frac{\mathrm{lbf}}{\mathrm{in.}^2}\right]$$

The equivalent internal pressure is then

$$p_{\mathrm{eq}} = p_{\mathrm{rot}} + p_{\mathrm{h}}, \tag{4.101}$$

$$p_{\mathrm{eq}} = p_{\mathrm{i}} = 3758\left[\frac{\mathrm{lbf}}{\mathrm{in.}^2}\right]$$

The hoop stress is then

$$\sigma_{\theta\theta} = \frac{1}{\left[\left(\frac{4.10}{2}\right)^2 - \left(\frac{2.95}{2}\right)^2\right][\mathrm{in.}^2]}\left\{(3758)\left[\frac{\mathrm{lbf}}{\mathrm{in.}^2}\right]\left[\left[\left(\frac{4.10}{2}\right)^2 - \left(\frac{2.95}{2}\right)^2\right][\mathrm{in.}^2]\right]\right\}$$

$$\sigma_{\theta\theta} = 11,830\left[\frac{\mathrm{lbf}}{\mathrm{in.}^2}\right]$$

Now let us look at the axial stress. This is the stress at the point due to two things: the axial inertia of all the materials ahead of the cut setting back and the effective internal pressure caused by the rotation of the projectile and the hydrostatic compression of the fill material:

$$\sigma_{zz} = \frac{\left(p_{\mathrm{i}}r_{\mathrm{i}}^2 - p_{\mathrm{o}}r_{\mathrm{o}}^2\right)}{\left(r_{\mathrm{o}}^2 - r_{\mathrm{i}}^2\right)} - \frac{F_{\mathrm{Axial}}}{\pi\left(r_{\mathrm{o}}^2 - r_{\mathrm{i}}^2\right)} \tag{4.102}$$

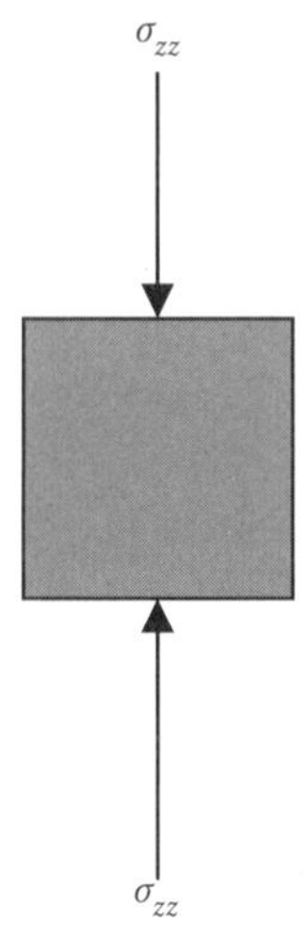

We shall use the radii given in the problem statement. We put negative sign in the aforementioned equation to denote compressive stress because only the axial component loads the inner wall in compression. The force acting on the section of interest due to setback is given by

$$F_{\mathrm{Axial}} = m_6 a_{\mathrm{p_{max}}} \tag{4.103}$$

$$F_{\text{Axial}} = \frac{(382{,}300)\left[\frac{\text{ft}}{\text{s}^2}\right](17.15)[\text{lbm}]}{(32.2)\left[\frac{\text{lbm-ft}}{\text{lbf-s}^2}\right]} = 203{,}600\ [\text{lbf}]$$

Using this result, we have

$$\sigma_{zz} = \frac{(3758)\left[\frac{\text{lbf}}{\text{in.}^2}\right]\left(\frac{2.95}{2}\right)^2[\text{in.}^2]}{\left[\left(\frac{4.10}{2}\right)^2-\left(\frac{2.95}{2}\right)^2\right][\text{in.}^2]} - \frac{(203{,}600)[\text{lbf}]}{\frac{\pi}{4}\left[(4.10)^2-(2.95)^2\right][\text{in.}^2]}$$

$$\sigma_{zz} = -27{,}940\left[\frac{\text{lbf}}{\text{in.}^2}\right]$$

Many times, we neglect the first term in the aforementioned equation for conservatism. In the radial direction, we only have our equivalent pressure pushing radially outward, and our location of interest is on the ID, so

$$\sigma_{rr} = -p_{\text{eq}} \tag{4.104}$$

$$\sigma_{rr} = -3758\left[\frac{\text{lbf}}{\text{in.}^2}\right]$$

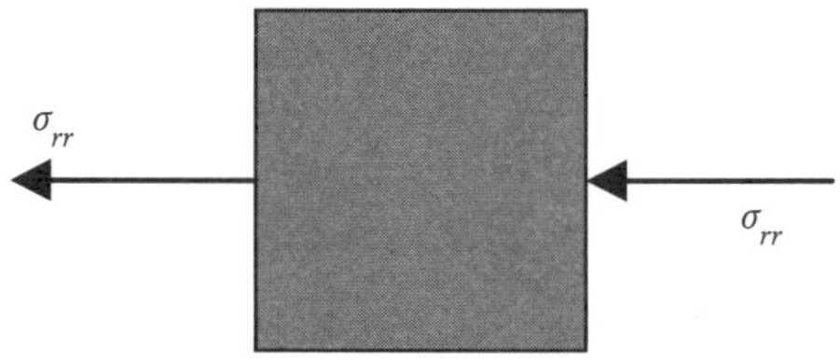

The angular acceleration will generate a torque through the rotating band that results in a shear stress in the plane normal to the axis of the projectile.

The torque on the projectile is also the opposite of the torque on the gun tube and comes directly from Equation 4.77:

$$T_6 = I_{zz_6}\alpha_{p_{\max}} \tag{4.105}$$

The moments of inertia were provided, and we must use the angular acceleration calculated at peak pressure provided earlier. Now the torque comes about through

$$T_6 = (41.15)\left[\text{lbm-in.}^2\right](348{,}600)\left[\frac{\text{rad}}{\text{s}^2}\right]\left(\frac{1}{32.2}\right)\left[\frac{\text{lbf-s}^2}{\text{lbm-ft}}\right]\left(\frac{1}{12}\right)\left[\frac{\text{ft}}{\text{in.}}\right]$$

$$T_6 = 37{,}130\,[\text{lbf-in.}]$$

The in-plane shear stress is given by

$$\tau = \frac{Tr}{J} \tag{4.106}$$

Then, we have

$$\tau_{z\theta} = \frac{(37{,}130)\,[\text{lbf-in.}]\left(\frac{2.95}{2}\right)[\text{in.}]}{(108.3)\,[\text{in.}^4]} = 506\left[\frac{\text{lbf}}{\text{in.}^2}\right]$$

The shear stress caused by the rotation is generated by the shell trying to spin up the explosive fill. The torque on the explosive fill is determined through

$$T_{\text{fill}} = I_{zz_{\text{fill}}}\alpha_{p_{\max}} \tag{4.107}$$

$$T_{\text{fill}} = (5.17)\,[\text{lbm-in.}^2]\,(348{,}600)\left[\frac{\text{rad}}{\text{s}^2}\right]\left(\frac{1}{32.2}\right)\left[\frac{\text{lbf-s}^2}{\text{lbm-ft}}\right]\left(\frac{1}{12}\right)\left[\frac{\text{ft}}{\text{in.}}\right]$$

$$T_{\text{fill}} = 4644\,[\text{lbf in.}]$$

This generates a force at the internal radius of

$$F_{\text{fill}} = \frac{T_{\text{fill}}}{r_{\text{i}}} \tag{4.108}$$

$$F_{\text{fill}} = \frac{(4644)\,[\text{lbf-in.}]}{\left(\frac{2.95}{2}\right)[\text{in.}]} = 3162\,[\text{lbf}]$$

Smearing this over the entire internal surface area gives us

$$\tau_{r\theta} = \frac{(3162)\,[\text{lbf}]}{(124)\,[\text{in.}^2]} = 25.5\left[\frac{\text{lbf}}{\text{in.}^2}\right]$$

The axial shear is approximated as a worst case by calculating the hydrostatic pressure at the bottom of the explosive column, transforming it into a force, and smearing that force over the entire internal cavity area. We know that the entire explosive column height is

$$h = 13.44\,[\text{in.}]$$

Then the peak hydrostatic pressure of the fill is

$$p_{\text{h}} = \rho_{\text{fill}}a_{p_{\max}}h \tag{4.109}$$

$$p_{\text{h}} = \frac{(0.036)\left[\frac{\text{lbm}}{\text{in.}^3}\right](382{,}300)\left[\frac{\text{ft}}{\text{s}^2}\right](13.44)\,[\text{in.}]}{(32.2)\left[\frac{\text{lbm-ft}}{\text{lbf-s}^2}\right]}$$

$$p_{\text{h}} = 5744\left[\frac{\text{lbf}}{\text{in.}^2}\right]$$

By calculating this pressure over the average cross-sectional area of the projectile, we obtain

$$F_{\text{base}} = p_{\text{h}} A_{\text{avg}_{\text{fill}}}, \tag{4.110}$$

$$F_{\text{Axial}} = (5744)\left[\frac{\text{lbf}}{\text{in.}^2}\right](6.49)\left[\text{in.}^2\right] = 37,280\,[\text{lbf}]$$

Now this force smeared over the interior surface area will yield the stress

$$\tau_{rz} = \frac{-F_{\text{axial}}}{A_{\text{fill}}} = -\frac{(37,280)\,[\text{lbf}]}{(124)\,[\text{in.}^2]} = -300\left[\frac{\text{lbf}}{\text{in.}^2}\right] \tag{4.111}$$

We can now write our stress tensor

$$\sigma = \begin{bmatrix} \sigma_{rr} & \tau_{r\theta} & \tau_{rz} \\ \tau_{r\theta} & \sigma_{\theta\theta} & \tau_{\theta z} \\ \tau_{rz} & \tau_{\theta z} & \sigma_{zz} \end{bmatrix} = \begin{bmatrix} -3,758 & 25.5 & -300 \\ 25.5 & 11,830 & 506 \\ -300 & 506 & -27,940 \end{bmatrix}\left[\frac{\text{lbf}}{\text{in.}^2}\right]$$

It must be noted that these equations assumed that there were no other forces acting on the projectile. For instance, in some projectiles with poorly designed rotating bands, leaking of the propellant gases (known as blowby) causes the exterior of the projectile to be pressurized. This load must be considered because it has been known to collapse projectiles in development. Another point is that while it is common to check a projectile at peak acceleration, the spin rate at this location is not a maximum. Maximum spin occurs at the exit of the muzzle of the weapon where the velocity is the highest. It is always good practice to check a projectile for maximum spin with no axial acceleration to simulate this.

Problem 4

A highly explosive projectile is to be designed for a 155 mm cannon using a $\frac{1}{2}$ in. thick steel wall with trinitrotoluene (TNT) as the filler material. Assume the shell and filler are a cylinder 0.75 m in length. It is to be capable of surviving a worn-tube torsional impulse (angular acceleration) of 440,000 rad/s^2.

1. Derive the expression to calculate the torque on the projectile that achieves this acceleration if the torque is applied at the OD of the shell.
2. Calculate the value of the torque assuming that the density of steel is 0.283 lbm/in.^{3} and TNT is 0.060 lbm/in.^{3}

Hint: Start from $F_{\text{T}} = ma_{\text{T}}$.

Answer:

1. $T = M_{\text{WALL}} + M_{\text{FILL}} = \frac{1}{2}\pi\alpha l[\rho(r_{\text{o}}^4 - r_{\text{i}}^4) + \rho_{\text{FILL}} r_{\text{i}}^4]$
2. $T = 796,600[\text{lbf-in.}]$

Problem 5

To participate in a failure investigation of an explosive, someone asks you to look at their design of a cylinder that was supposed to hold the explosive during a 155 mm howitzer launch. Assume the explosive sticks completely to the interior wall. The firing conditions at the time of the failure were as follows:

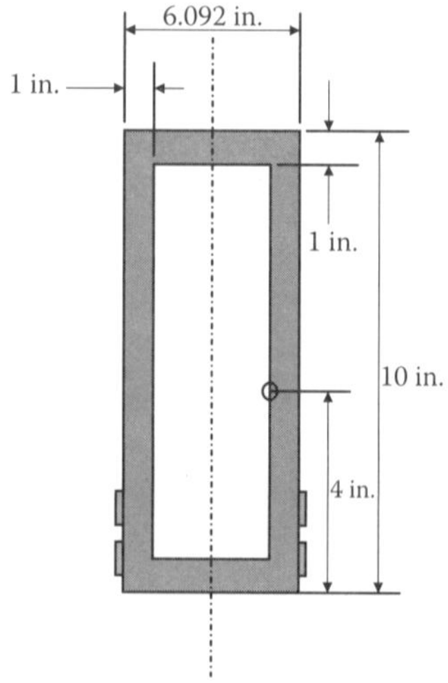

- Axial acceleration = 10,000 g
- Angular acceleration = 300,000 rad/s^2
- Angular velocity = 100 Hz

The projectile was as shown earlier:
The wall is AISI 4140:

$$E = 30 \times 10^6 \left[\frac{\text{lbf}}{\text{in.}^2}\right]$$

$$v = 029,$$

$$\rho = 0.283 \left[\frac{\text{lbf}}{\text{in.}^3}\right]$$

The explosive is composition B:

$$\rho_{\text{fill}} = 0.71 \left[\frac{\text{g}}{\text{cm}^3}\right]$$

Write the stress tensor for a point on the ID, 4 in. from the base.

Answer:

$$\sigma = \begin{bmatrix} \sigma_{rr} & \tau_{r\theta} & \tau_{rz} \\ \tau_{r\theta} & \sigma_{\theta\theta} & \tau_{\theta z} \\ \tau_{rz} & \tau_{\theta z} & \sigma_{zz} \end{bmatrix} = \begin{bmatrix} -2265 & 21 & -266 \\ 21 & 5564 & -2864 \\ -266 & -2864 & -19{,}307 \end{bmatrix} \left[\frac{\text{lbf}}{\text{in.}^2}\right]$$

Problem 6

A 155 mm projectile is fired from a tube with a 1 in 20 twist. Its muzzle velocity is 1000 m/s. What is the spin rate at the muzzle in hertz?

Answer: 322.6 [Hz]

Problem 7

It is requested that a brass slip ring be constructed for a spin test fixture to allow electrical signals to be passed (although really noisy) to some instrumentation. The design requirements are for the ring to have an ID of 4 in. and a length of 2 in. and be capable of supporting itself during a 150 Hz spin test. How thick does the ring have to be? The properties of brass are as follows: yield strength of 15,000 psi and density of approximately 0.32 lbm/in.^3

Answer: 1/4 in. thickness will work, but it can be thinner.

Problem 8

A 155 mm projectile is to be designed with a rotating band designed to discard as the projectile leaves the muzzle of the weapon. The rotating band is fixed to the projectile (so it transmits the proper torque to the projectile) with splines that prevent rotational motion relative to the projectile while allowing the band to expand in the radial direction for proper discard. This function must occur at the highest as well as the lowest spin rates. The two extreme muzzle velocities are 250 and 800 m/s, respectively, with corresponding peak axial accelerations of 2,000 and 15,500 g, respectively. The projectile mass is 98 lbm and the axial moment of inertia is 41 lbm–in.^2 Geometry constraints require the band to have an engraved OD of 6.2 in., and the outer diameter of the band seat (band ID) is 5.5 in. The band is 2.5 in. long. The 48 rifling lands are 0.05 in. high and are 0.2 in. wide. Consider both a copper and a soft iron band with the properties provided next and determine whether the bands will.

1. Withstand the shear at peak angular acceleration
2. Break up upon muzzle exit
3. If one (or both) designs fail to work properly, what can be done with the analysis and/or design to make it work? Is there anything we must be careful of?

Assume the following: The weapon has a 1 in 20 twist. At peak acceleration, only shear on the rotating band need be considered. Ignore the increase in shear area caused by the rifling helix. Ignore the stress concentration developed by the engraving for discard calculations. Ignore the effect of the splines. For conservatism, on muzzle exit, ignore the mass of material above the rifling marks (i.e., use an OD of 6.1 in. for the band).

The properties for the copper and steel are as follows:

	Copper	Soft Iron
Ultimate tensile strength (psi)	35,000	40,000
Shear strength (psi)	19,250	22,000
Density (lbm/in.^3)	0.316	0.263
Poisson's ratio	0.28	0.34

Problem 9

An experimental 40 mm gun has an average chamber inner diameter of 60 mm. The weapon is expected to develop a maximum breech pressure of 35,000 psi. If we would like the weapon to withstand 10,000 cycles at this pressure and given the properties of the steel given later, determine the OD of the chamber. Without proper design experience, an interference fit can sometimes be catastrophic. If we were instead to design this chamber out of two tubes, each at half of this thickness but with the outer tube compressing the inner tube by 0.002 in. diametrally, what is the maximum pressure the design will accommodate and still function for the 10,000 cycles?

Assume AISI 4340 steel with a yield strength (S_Y) of 100,000 psi. The endurance stress (S'_n) for 4340 is 0.875S_Y for the amount of cycles desired. Assume the following factors from our cyclic loading discussion: C_R = 0.93, C_G = 0.95, and C_S = 0.99. Assume the chamber is open ended. The modulus of elasticity and Poisson's ratio are 30 × 106 psi and 0.3, respectively.

Problem 10

You are to design a fragment-throwing gun system for another organization to be used in fragment impact testing. The gun is to throw a 22 g, 0.500 in. diameter, cylindrical fragment at 2300 m/s. Your design must use a brass cartridge case to assist with obturation of the breech. Other assumptions and information are

1. You do not need to design the breech—assume it will hold the cartridge case in properly (in reality, we can always add more threads to the design).
2. Even though there will be a slight taper on the chamber (which must be larger than the bore diameter for seating purposes), assume, for calculation purposes, that the chamber is cylindrical at its maximum diameter.
3. The tube is to be steel and assume that the yield strength is 60,000 psi (this accounts for the effect of cyclic loading). The modulus of elasticity is 29 × 10^6 psi. Poisson's ratio is 0.29.
4. Assume the propellant is either cylindrical or single perforated (and state your assumption).
5. Choose from the following propellants:

Propellant	Linearized Burn Rate β (in./s/psi)	Solid Density δ (lbm/in.^{3})	Adiabatic Flame Temperature T_0 (°R)	Propellant Force λ (ft-lbf/lbm)	Specific Heat Ratio γ
IMR	0.000132	0.0602	5103	327,000	1.2413
M12	0.000137	0.0600	5393	362,000	1.2326
Bullseye	0.000316	0.0590	6804	425,000	1.2523
Red dot	0.000153	0.0593	5774	375,000	1.2400
Navy pyro	0.000135	0.0566	4477	321,000	1.2454

6. Assume the cartridge case is brass and use a bilinear kinematic hardening model where the brass has a modulus of elasticity of 15 × 10^6 psi, a local tangent modulus of 13 × 10^6 psi, and a yield stress of 16,000 psi (yield occurs in this material at ε = 0.002).

7. Weight is not a major concern; however, you should make the design light enough to be moved using reasonable test range equipment.

The design is to proceed as follows (not necessarily in the order given):

1. Interior ballistics design

   a. Size the chamber length and diameter.
   b. Determine the amount of propellant needed based on your choice of the aforementioned propellants and propellant geometry (make sure it fits in the chamber).
   c. Determine a web thickness for the propellant.
   d. Determine the length of the gun.
   e. Determine V, p_B, and x for the projectile at peak pressure.
   f. Determine Vc, p_{Bc}, and x_c for the projectile at charge burnout.
   g. Determine the muzzle velocity of the projectile.

2. Gun tube design

   a. Based on the calculations of part 1 develop a pressure–distance curve to use as criteria for your gun design.
   b. Determine the OD of the gun tube. To keep the design light as possible, use the design rules provided in this chapter and taper the tube toward the muzzle. If needed, over the chamber, you may shrink fit cylinders to build up a composite tube.
   c. Determine the weight of your gun and comment on if it is reasonable.

3. Cartridge case design

   a. Determine a thickness and tolerance for your cartridge case.
   b. Determine the OD and tolerance for the cartridge case.
   c. Decide on a tolerance for your chamber ID.

Note that for these calculations show that the case may be easily extracted at the limits of the tolerance.

It is important that you write down all your assumptions. It is also highly likely that as you proceed further along with your design, you may come upon a situation that requires you to revisit an assumption you made earlier—this is to be expected, and it is part of the design process.

4.11 Buttress Thread Design

There is a variety of instances where a buttress thread form is the desired means of transmitting loads between mating components. In some instances, the thread form is not the usual continuous spiral associated with a normal thread but a series of discontinuous grooves that exhibit the cross-sectional form of the buttress. In this section, we will discuss a true thread with lead-ins and partial thread shapes, but we will assume that the basic analysis will apply to buttress grooves as well.

Buttress threads are designed to maximize the load-carrying capability in one direction of a threaded joint. There are many variations on such threads, but on ammunition components, we predominantly use threads with a pressure flank angle (described later) of 7° as shown in Figure 4.18. Thread callouts on drawings usually appear, e.g.,

2. 750-4UNC-2A LH Buttress

The meanings of these callouts are as follows:

- First number is the major diameter of the thread (here it is in inches).
- Second number is how many threads per inch.
- The letters are the thread form callout (UNC = Unified National Coarse).
- The last number is the class of fit of the thread related to clearances in the engagement (3 is the tightest fit, 1 the loosest).
- The last letter determines whether the thread is male (A) or female (B) (mnemonic – A = Adam = male).
- LH means left handed (there will be no callout if the threads are right-hand twist or if the thread is a groove and not a continuous spiral).

Thread nomenclature of relevance is as follows:

- The major diameter is the largest diameter of the thread form.
- The minor diameter is the smallest diameter of the thread form.
- The pitch diameter is the diameter where there is 1/2 metal and 1/2 air.

We use buttress threads for several reasons: most important is to improve the directional loading characteristics of the thread; also to allow for a more repeatable, controllable shear during an expulsion event, i.e., if we want the thread to intentionally and controllably fail allowing separation of the components; and to prevent thread slip in joints with fine threads or threads on thin shell walls. If thread slip occurs, the threads can either dilate or

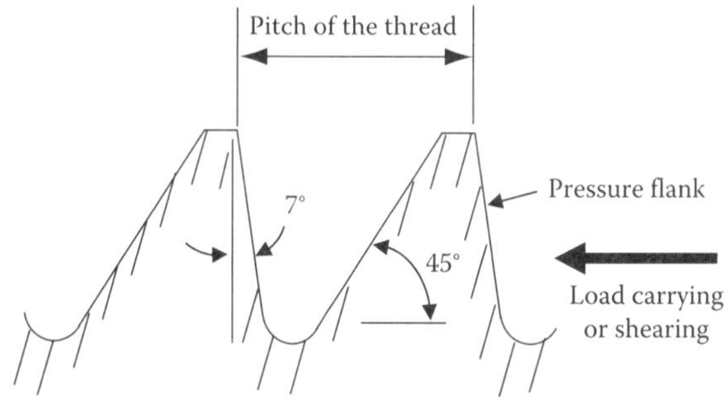

FIGURE 4.18
Depiction of a standard buttress thread.

contract elastically and the joint can pop apart with little or no apparent damage to the threads.

When we design for strength, we typically calculate the strength based on the shear area at the pitch diameter in the weaker material. This, of course, translates to half the length of engagement of the threads. This is acceptable because we usually use conservative properties and add a safety factor to account for material variations and tolerances. We must always base our calculations on the weaker material if the design is to be robust. When designing to actually fail the threads, however, we need to be more exact in our analysis and take everything such as actual material property variation and tolerancing into account or our answers will be wrong.

We will proceed in this analysis in meticulous detail, initially, as a cantilevered beam subjected to compressive and tensile stresses caused by contact forces and bending moments. This technique was first developed during the US Army's sense and destroy armor program by Dan Pangburn of Aerojet Corporation [5] and has been used by the US Army.

We consider the thread form as a short, tapered, cantilever beam and assume that failure will occur as a result of a combination of stresses and that combined bending and compressive stress precipitate the failure. This is depicted in Figure 4.19. If we examine this figure, we see that the distributed force *F* causes our beam to bend in the classical sense with the loaded flank in tension and the unloaded flank in compression about the neutral axis. We have separated an element of material out from point A in the figure. The free-body diagram of this element shows that the bending of the beam puts it in tension, while the loading on the pressure flank puts it in compression. It is this combined load that will cause failure of the material.

If we were analyzing this in a finite element code, the bending and compression would cause combined stresses and the part would fail by one of the failure criteria that were discussed earlier. However, in this case, we will use the maximum shear criteria to check for failure at some radius in the thread and will check the load at which failure occurs with the von Mises criterion at the thread roots, d_i on the male thread and d_o on the female thread. These are the diameters of the loading (i.e., the mating thread contact areas) as depicted in Figure 4.20.

For simplicity, we shall call the male thread the "bolt" (subscript 1) and the female thread the "nut" (subscript 2). The loading is further described by Figure 4.21. In this figure, the radius *r* is the plane at which the threads will shear.

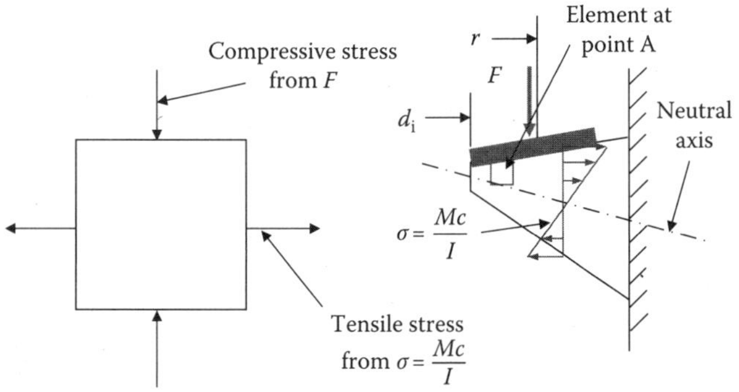

FIGURE 4.19
Depiction of a standard buttress thread.

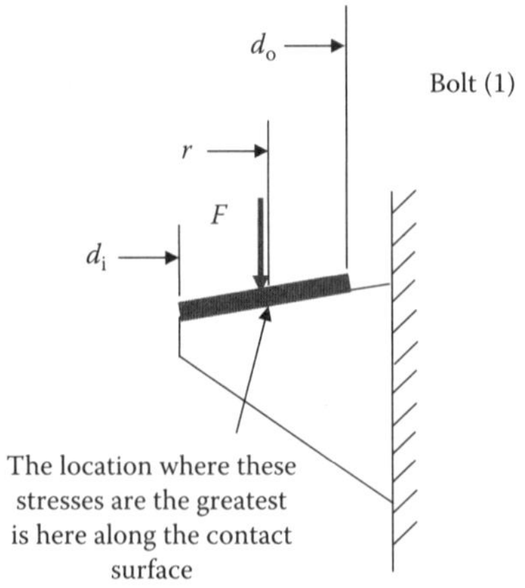

FIGURE 4.20
Definition of load radii.

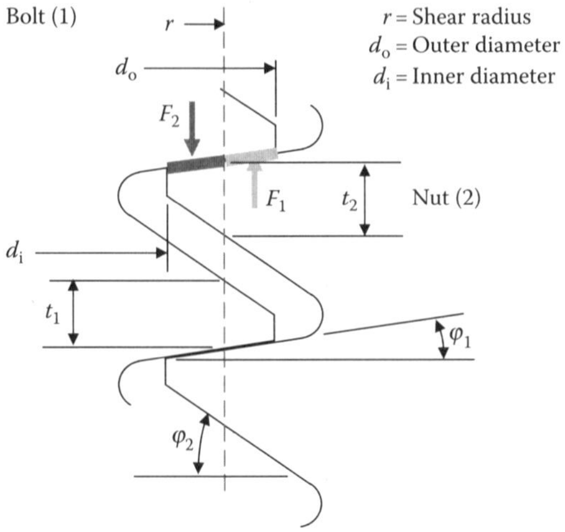

FIGURE 4.21
Loading diagram of buttress threads.

If we assume the contact is frictionless, the average normal stress is simply the total axial force F divided by the projected area A. We have assumed that the normal stress is constant over the contact area. This gives us a negative value because the stress is compressive. Figure 4.22 shows the configuration where the normal force has been termed F_4 and the thread area is A_4. Since an axial loading is what shears the threads, we need to project the components of this force along the axis of the projectile (i.e., rotate through the angle ϕ_1). This allows us to express the stress as

$$\sigma_N = \frac{-F_4}{A_4} = \frac{-\dfrac{F}{\cos\phi_1}}{\dfrac{A}{\cos\phi_1}} = -\frac{F}{A} = \sigma_v \qquad (4.112)$$

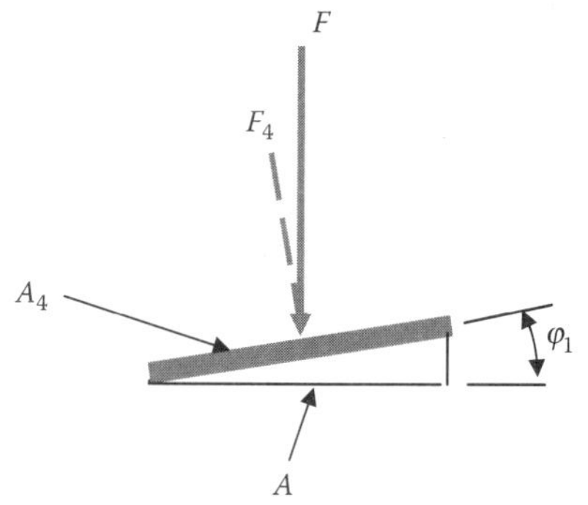

FIGURE 4.22
Loading of a thread surface.

Where σ_N and σ_v are the normal and axial stresses, respectively. By substituting the area A, we get

$$\sigma_v = -\frac{F}{\frac{\pi}{4}\left(d_o^2 - d_i^2\right)} \tag{4.113}$$

If we assume that failure takes place at a radius r, yet to be determined, the bearing force on the external thread (bolt) that produces bending in the thread is

$$F_1 = -\pi\sigma_v\left[\left(\frac{d_o}{2}\right)^2 - r^2\right] \tag{4.114}$$

Similarly, the force that produces bending in the internal thread (nut) is

$$F_2 = -\pi\sigma_v\left[r^2 - \left(\frac{d_i}{2}\right)^2\right] \tag{4.115}$$

Now the pitch diameter is defined as the location where the thickness of the thread is one-half the thread pitch. Since thread failure occurs at an assumed radius r, we need to define the thicknesses of both the male and female threads at this location.

First, recall that the thread pitch is p and then define d_{pf} as the internal (female) thread pitch diameter and d_{pm} as the external (male) thread pitch diameter. Then, t_1 and t_2 from our earlier diagram can be expressed as follows:

$$t_1 = \frac{p}{2} - \left(r - \frac{d_{pm}}{2}\right)(\tan\phi_1 + \tan\phi_2) \tag{4.116}$$

$$t_2 = \frac{p}{2} - \left(\frac{d_{pf}}{2} - f\right)(\tan\phi_1 + \tan\phi_2) \tag{4.117}$$

Then the bending stress can be calculated from simple beam theory as

$$\sigma = \frac{Mc}{I} = \frac{M\frac{t}{2}}{\frac{1}{12}(2\pi r)t^3} = \frac{3M}{\pi r t^2} \tag{4.118}$$

where c is the distance from the point of interest r to the neutral (bending) axis and I is the area moment of inertia of the cross-section. The bending stress in the external (male) thread is then

$$\sigma_1 = \frac{3F_1\left(\frac{d_o}{2} - r\right)}{2\pi r t_1^2} \tag{4.119}$$

Similarly, we can show that the bending stress in the internal (female) thread is

$$\sigma_2 = \frac{3F_2\left(r - \frac{d_i}{2}\right)}{2\pi r t_2^2} \tag{4.120}$$

In considering the failure criteria, we shall assume that the maximum shear stress in the material must not exceed 0.6 times the material strength in a tensile test. We will use the yield strength as the strength of this material because at that point in failure, the geometry of the part is changing. Experience has shown that once this begins to happen, the part is in the process of failing anyway and will not recover.

In a state of combined loading, the maximum shear stress can be found from

$$\tau_{max} = \frac{1}{2}\left|\sigma_{max} + \sigma_{min}\right| \tag{4.121}$$

This averaging can be shown to be

$$\tau_{max} = \frac{\sigma - \sigma_N}{2} = 0.6Y \tag{4.122}$$

Here we are reminded that σ_N and σ_v are therefore compressive negative numbers and Y is the yield stress in tension. The equivalent stress at failure in the male thread is then

$$Y_1 = \frac{\sigma_1 - \sigma_v}{1.2} \tag{4.123}$$

and in the female thread, it is

$$Y_2 = \frac{\sigma_2 - \sigma_v}{1.2} \tag{4.124}$$

In these equations, Y_1 and Y_2 are the yield stress in the male and female threads, respectively.

We will now combine Equations 4.123 and 4.119 as well as Equations 4.124 and 4.120 to eliminate σ_1 and σ_2, respectively. This yields

$$Y_1 = 1.25F_1\frac{\frac{1}{2}d_o - r}{\pi r t_1^2} - \frac{\sigma_v}{1.2} \tag{4.125}$$

and

$$Y_2 = 1.25F_2\frac{r - \frac{1}{2}d_i}{\pi r t_2^2} - \frac{\sigma_v}{1.2} \tag{4.126}$$

We now combine Equations 4.125 and 4.116 as well as Equations 4.126 and 4.117 to eliminate the thicknesses t_1 and t_2, respectively. This yields

$$Y_1 = 1.25F_1 \frac{\frac{1}{2}d_o - r}{\pi r\left[\frac{1}{2}p - \left(r - \frac{1}{2}d_{pm}\right)(\tan\phi_1 + \tan\phi_2)\right]^2} - \frac{\sigma_v}{1.2} \tag{4.127}$$

and

$$Y_2 = 1.25F_2 \frac{r - \frac{1}{2}d_i}{\pi r\left[\frac{1}{2}p - \left(\frac{1}{2}d_{pf} - r\right)(\tan\phi_1 + \tan\phi_2)\right]^2} - \frac{\sigma_v}{1.2} \tag{4.128}$$

We will now insert Equation 4.114 into Equation 4.127 and Equation 4.115 into Equation 4.128 to eliminate F_1 and F_2, respectively. This yields

$$Y_1 = -0.3125\sigma_v\left(d_o^2 - 4r^2\right) \frac{\frac{1}{2}d_o - r}{r\left[\frac{1}{2}p - \left(r - \frac{1}{2}d_{pm}\right)(\tan\phi_1 + \tan\phi_2)\right]^2} - \frac{\sigma_v}{1.2} \tag{4.129}$$

and

$$Y_2 = -0.3125\sigma_v\left(4r^2 - d_i^2\right) \frac{r - \frac{1}{2}d_i}{r\left[\frac{1}{2}p - \left(\frac{1}{2}d_{pf} - r\right)(\tan\phi_1 + \tan\phi_2)\right]^2} - \frac{\sigma_v}{1.2} \tag{4.130}$$

Now we must solve Equations 4.129 and 4.130 in terms of σ_v. The first of these is

$$\sigma_v = \frac{-Y_1}{G_3 + G_2 + G_1 + G_0 + \frac{1}{1.2}} \tag{4.131}$$

where

$$G_3 = \frac{0.15625d_o^3}{r\left(0.5p - r\tan\phi_1 - r\tan\phi_2 + 0.5d_{pm}\tan\phi_1 + 0.5d_{pm}\tan\phi_2\right)^2} \tag{4.132}$$

$$G_2 = \frac{-0.3125d_o^2}{\left(0.5p - r\tan\phi_1 - r\tan\phi_2 + 0.5d_{pm}\tan\phi_1 + 0.5d_{pm}\tan\phi_2\right)^2} \tag{4.133}$$

$$G_1 = \frac{-0.625rd_o}{\left(0.5p - r\tan\phi_1 - r\tan\phi_2 + 0.5d_{pm}\tan\phi_1 + 0.5d_{pm}\tan\phi_2\right)^2} \tag{4.134}$$

$$G_0 = \frac{1.25r^2}{\left(0.5p - r\tan\phi_1 - r\tan\phi_2 + 0.5d_{pm}\tan\phi_1 + 0.5d_{pm}\tan\phi_2\right)^2} \tag{4.135}$$

The second equation is

$$\sigma_v = \frac{-Y_2}{H_3 + H_2 + H_1 + H_0 + \dfrac{1}{1.2}} \tag{4.136}$$

where

$$H_3 = \frac{0.15625 d_i^3}{r\left(0.5p + r\tan\phi_1 + r\tan\phi_2 - 0.5d_{pf}\tan\phi_1 - 0.5d_{pf}\tan\phi_2\right)^2} \tag{4.137}$$

$$H_2 = \frac{-0.3125 d_i^2}{\left(0.5p + r\tan\phi_1 + r\tan\phi_2 - 0.5d_{pf}\tan\phi_1 - 0.5d_{pf}\tan\phi_2\right)^2} \tag{4.138}$$

$$H_1 = \frac{-0.625 r d_i}{\left(0.5p + r\tan\phi_1 + r\tan\phi_2 - 0.5d_{pf}\tan\phi_1 - 0.5d_{pf}\tan\phi_2\right)^2} \tag{4.139}$$

$$H_0 = \frac{1.25 r^2}{\left(0.5p + r\tan\phi_1 + r\tan\phi_2 - 0.5d_{pf}\tan\phi_1 - 0.5d_{pf}\tan\phi_2\right)^2} \tag{4.140}$$

We now solve Equation 4.113 for *F*, and we get

$$F = \frac{\pi}{4}\sigma_v\left(d_o^2 - d_i^2\right) \tag{4.141}$$

Substitution of Equation 4.131 for σv yields (for a full thread on the bolt)

$$F = \frac{\pi}{4}\left(d_o^2 - d_i^2\right)\frac{-Y_1}{G_3 + G_2 + G_1 + G_0 + \dfrac{1}{1.2}} \tag{4.142}$$

We perform a similar operation with Equation 4.136, giving us (for a full thread on the nut)

$$F = \frac{\pi}{4}\left(d_o^2 - d_i^2\right)\frac{-Y_2}{H_3 + H_2 + H_1 + H_0 + \dfrac{1}{1.2}} \tag{4.143}$$

Equations 4.142 and 4.143 now contain only two unknowns, *r* and *F*. The procedure now involves solving both Equations 4.142 and 4.143 and plotting the force *F* vs. *r*. The lowest value in either equation is then the force (and location) at which the joint will fail. It is recommended that these solutions be performed with the aid of a computerized numerical calculation program such as MathCAD.

Partial threads can have a significant effect on the failure strength of a joint. If the joint were designed to survive, it is generally best to ignore the additional strength afforded by partial threads and base the design margin on the calculation method earlier. When a joint is designed to fail, however, the lead-in and run out must be accounted for unless sufficient margin is available in the expulsion system such that two additional threads may be added to the calculation, yet still allow the joint to be overcome with ease.

4.12 Sabot Design

Sabots (French for "wooden shoe") are used in both rifled and smoothbore guns to allow a standard weapon to fire a high-density, streamlined subprojectile whose diameter is much smaller than the bore, at a velocity higher than would normally be possible if the gun were sized to the diameter of the subprojectile. Discarding sabots have been in general use since the Second World War and are still popular (in fact, an artist's rendition of a discarding sabot is illustrated on the cover of this book). They are called "discarding sabots" since they are shed from the subprojectile at the muzzle allowing it to fly unencumbered to the target.

As previously stated, velocity is proportional to the square root of the pressure achieved in the tube, the area of the bore, and the length of travel and inversely proportional to the square root of the projectile weight. In mathematical terms,

$$V \backsim \sqrt{\frac{pAL}{w_p}} \tag{4.144}$$

We can see that if the area over which the pressure is applied is much greater than the area presented at the rear of the subprojectile, a larger force would be applied to accelerate it than if it were fired at the same pressure from a bore of its own diameter. Furthermore, decreasing the launch weight of the as-fired assembly also increases the velocity. Therefore, we must design as light a sabot as feasible so that we can maintain a very dense, small diameter subprojectile (usually an armor penetrator). The combination of the full bore area; a dense, streamlined subprojectile; and a lightweight sabot has the overall effect of generating unusually high velocities, a characteristic essential for kinetic energy armor penetration.

There are many requirements for a successful sabot:

- It must seal the propellant gases behind the projectile (obturate).
- It must support the subprojectile during travel in the bore to provide stable motion (called providing a suitable wheelbase).
- It must transfer the pressure load from the propellant gases to the subprojectile.
- It must completely discard at the muzzle of the weapon without interfering with the flight of the subprojectile.
- The discarded sabot parts must also reliably fall within a danger area in front of the weapon so as not to injure troops nearby.
- It must be minimally parasitic, i.e., it must be as light as possible and remove as little energy from the subprojectile as possible.

These are formidable requirements that necessitate great ingenuity on the part of the designers.

The problem has been solved in a variety of ways. In the 1950s, designers, chiefly British, used cup- or pot-type sabots to launch APDS subprojectiles (Figure 4.23). The guns from which these munitions were fired were rifled to launch conventional full caliber, spin-stabilized rounds, and so the subprojectiles of the APDS rounds were spin-stabilized too. Such armor-defeating munitions were highly effective against the tank armor of the times, and pot-type, saboted, kinetic energy penetrators were adopted in tank cannon around the world.

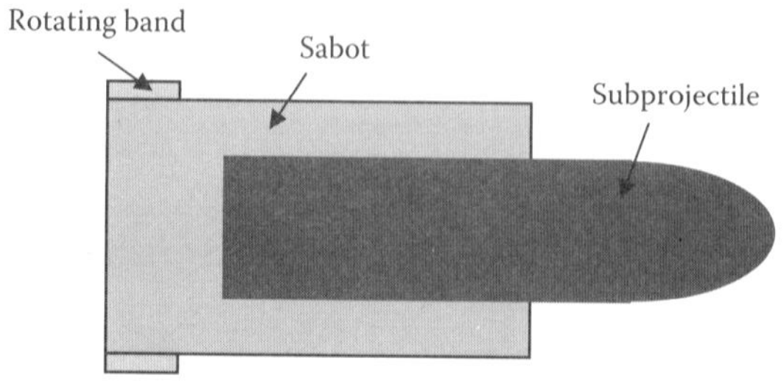

FIGURE 4.23
Simplified diagram of an APDS projectile.

Tank armor changed in the 1960s and became more difficult to penetrate with the tungsten carbide cores of the subprojectiles in use. Initially, incremental changes were made in the material of the core (sintered tungsten was used instead of sintered tungsten carbide), but it was eventually realized that longer, smaller-diameter, high-density penetrators were the answer. There are physical limits to the degree of subcalibering practical in spin-stabilized projectiles: the spin required for flight stability increases as the square of the ratio of bore to subprojectile for conventionally shaped projectiles, and it becomes nearly impossible to spin-stabilize very long projectiles. Rifling twists were increased to attempt to accommodate the APDS rounds; in one case, a 1:12 twist was tried when the normal twist would have been 1:40. In the end, APDS designs were abandoned in favor of very long, fin-stabilized penetrators (APFSDS) that used a radically different type of sabot (Figure 4.24). The guns too were changed to smoothbores; although to preserve older weapons in use, designers learned how to make fin-stabilized munitions fireable in rifled guns as well.

The basic type of sabot used with long-rod, fin-stabilized penetrators is the ring with its subvarieties: base pull, double ramp, and saddle sabots. Whereas pot sabots were essentially discarded rearward as a unit, ring sabots are segmented into three or more sections and discard radially outward at the muzzle to clear the fins that are larger in diameter than the rod. The finned subprojectile is frequently imparted with a slow spin to average out unavoidable manufacturing asymmetries during flight that could cause trajectory drift. This type of munition is now in the arsenals of all nations.

The design of the ring sabot begins with the stress analysis of the shear traction between the sabot inner diameter and the penetrator outer diameter. This analysis is crucial for determining the mass of the ring and thus the parasitic weight of the sabot. We will follow the work of Drysdale [6] throughout this development. The essential parameters of the computation are shown in Figure 4.25.

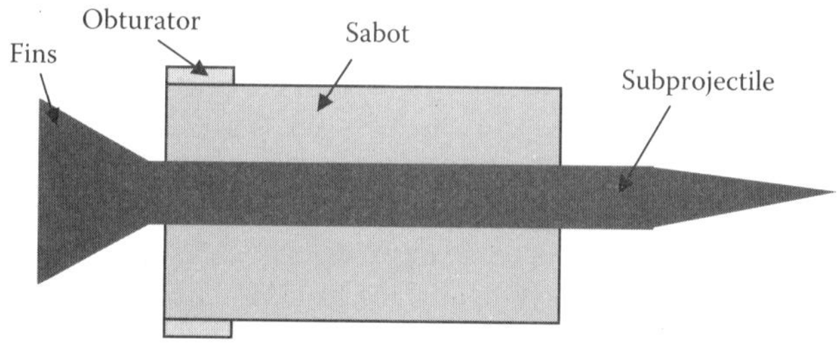

FIGURE 4.24
Simplified diagram of an armor-piercing, fin-stabilized, discarding sabot (APFSDS) projectile.

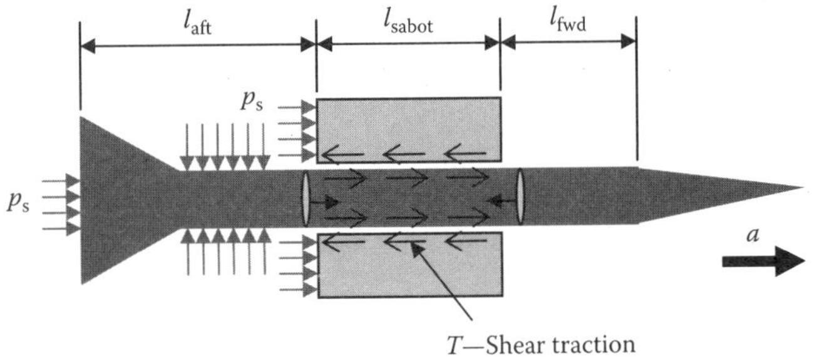

FIGURE 4.25
Free-body diagram for rings and rods.

From this free-body diagram, we can infer that

$$T = p_s\left(A - A_p\right) - m_{sabot}a \tag{4.145}$$

A reasonable estimate for the masses where the symbols are as follows:

$$m_{sabot} = \frac{1}{2}m_{subprojectile} \tag{4.146}$$

where T is the total shear traction force; A is the bore area; A_p is the area of the penetrator cross-section; m_{sabot} is the mass of the sabot; $m_{subprojectile}$ is the mass of the subprojectile; a is the projectile acceleration; p_s is the pressure on the base of the shot (note that the net pressure on the fins is zero); and σ_1 is the axial stress on the penetrator.

Because the sabot needs to be as light as possible, the material is usually much weaker than the penetrator; thus, the sabot length depends mostly on the sabot material. If the penetrator were weaker for some reason, the sabot length would depend upon that material. Thus, we can write for the surface traction

$$T_{allow} = \frac{\pi}{2}d_p l_{sabot}\tau_{allow} \tag{4.147}$$

where d_p is the diameter of the penetrator or subprojectile; T_{allow} is the allowable traction force; and τ_{allow} is the maximum shear stress allowed in the weaker material.

The shear traction is usually transmitted through matching grooves or threads. Analysis of these surfaces can be rather complicated but is similar to standard or buttress thread design practice. Given no actual data on the allowable shear stress in the material, we can use the following formulas based on the Tresca or the von Mises yield criterion:

$$\tau_{allow} = \frac{\sigma_e}{2} \tag{4.148}$$

by the Tresca criterion or

$$\tau_{allow} = \frac{1.155\sigma_e}{2} = 0.577\sigma_e \tag{4.149}$$

by the von Mises criterion. In both of these expressions, σ_e is the equivalent stress as discussed in Section 4.2. Thus, the allowable surface traction can be stated as

$$T_{allow} = K\pi d_p l_{sabot}\sigma_e, \tag{4.150}$$

where K is either 0.25 or 0.2885 dependent upon the failure criteria.

If we substitute Equation 4.150 into Equation 4.145, we can solve for the proper sabot length

$$K\pi d_p l_{sabot}\sigma_e = p_s\left(A - A_p\right) - m_{sabot}a \tag{4.151}$$

and

$$l_{sabot} = \frac{p_s}{K\pi d_p\sigma_e}\left(\frac{A}{A_p} - 1\right)A_p - \frac{m_{sabot}a}{K\pi d_p\sigma_e} \tag{4.152}$$

But

$$A_p = \frac{\pi}{4}d_p^2 \tag{4.153}$$

Then,

$$l_{sabot} = \frac{p_s}{K\pi d_p\sigma_e}\left(\frac{A}{A_p} - 1\right)A_p - \frac{m_{sabot}a}{K\pi d_p\sigma_e} \tag{4.154}$$

Now, by our earlier assumption (Equation 4.146),

$$p_s A = ma = \left(m_{sabot} + m_{subprojectile}\right)a = 3m_{sabot}a \tag{4.155}$$

Then,

$$l_{sabot} = \frac{p_s d_p}{4K\sigma_e}\left(\frac{A}{A_p} - 1\right) - \frac{m_{sabot}a}{K\pi d_p\sigma_e} \tag{4.156}$$

By multiplying and dividing the second RHS term by A_p and simplifying, we get

$$l_{sabot} = \frac{p_s d_p}{4K\sigma_e}\left(\frac{A}{A_p} - 1\right) - \frac{p_s a}{3K\pi d_p\sigma_e} \tag{4.157}$$

More generally, if the mass of the sabot is not half of the subprojectile mass, then we must use Equation 4.154 to determine the proper length.

The shape of ring sabots evolved over time from quite heavy designs to highly efficient ones. Early sabots were saddle shaped (Figure 4.26). These had points of high shear stress concentrations near the ends.

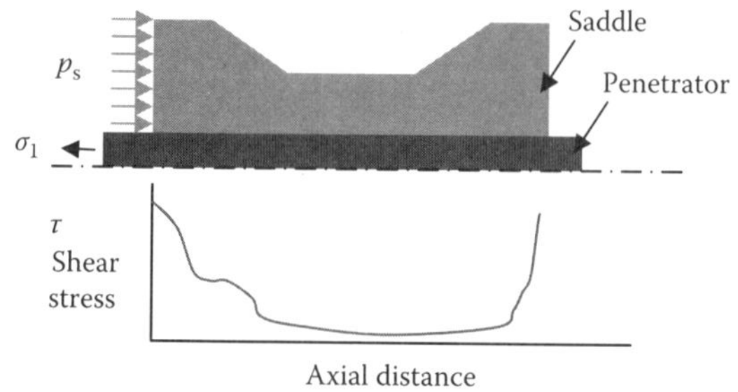

FIGURE 4.26
Shear stress variation in a saddle-type sabot.

These sabots had an excellent wheelbase (the distance between the forward and aft bourrelets), which prevented balloting in the tube and provided good accuracy. The parasitic weight, however, was high, and sufficiently high muzzle velocities were not attained.

Single- and double-ramp sabots have come into use because of the favorable weight reduction that can be obtained with this design. They utilize gun pressure to help clamp the sabot to the penetrator and have the added advantage of maintaining an almost constant shear stress between the sabot and the penetrator. The double-ramp sabot is shown in Figure 4.27.

Detailed studies have shown that a higher-order (nonlinear) curved ramp yields a constant shear stress under load. The method of solution for finding the best shape of the sabot taper depends on a free-body analysis of the sabot and the penetrator. Figures 4.28 and 4.29 show differential elements of the subprojectile and the sabot, respectively.

If we examine Figure 4.28, we see that the axial forces consist of the net internal stress $(d\sigma_{zp}/dz)\Delta z$, the inertial resistance to acceleration $\rho_p V_p a$, and the shear stress imparted by the sabot τ. Similarly, on the sabot, we have the net internal stress $(d\sigma_{zs}/dz)\Delta z$, the inertial resistance to acceleration $\rho_s V_s a$, the shear stress imparted by the subprojectile τ, and the component of pressure in the axial (z) direction. We proceed by initially finding an expression for the volume of the sabot free body. Details of this derivation are found in the study by Drysdale [6]. The incremental volume of the sabot can be shown as follows:

$$V_s = \pi\left[R_s^2(z) - R_p^2\right]\Delta z \tag{4.158}$$

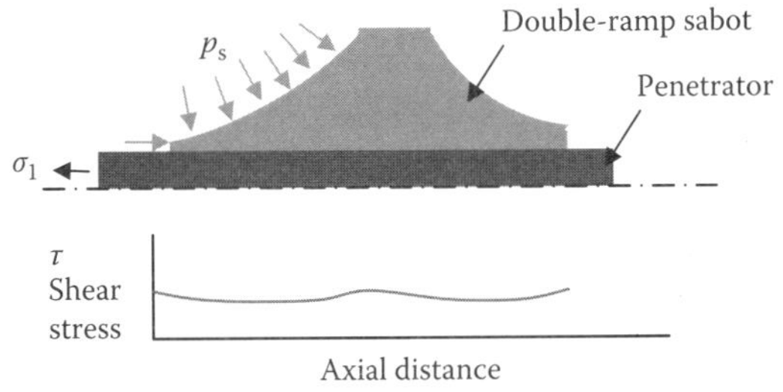

FIGURE 4.27
Shear stress variation in a double-ramp-type sabot.

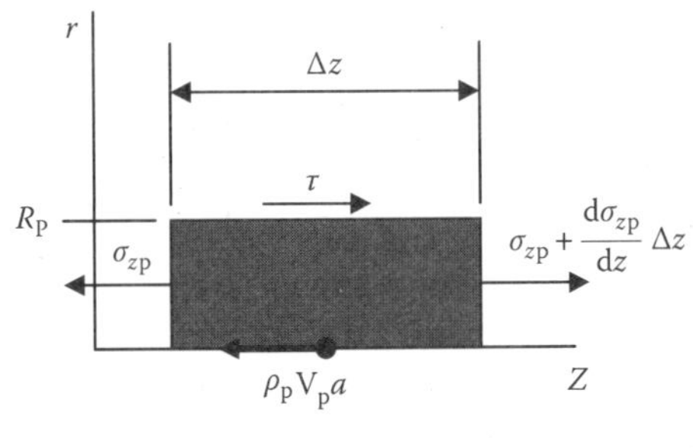

FIGURE 4.28
Differential element in a subprojectile showing forces acting.

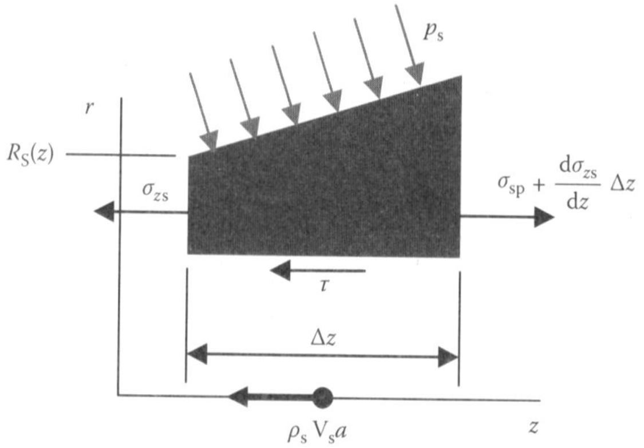

FIGURE 4.29
Differential element in a sabot showing forces acting.

We then sum the forces on the sabot in the axial direction

$$p_s\pi\left[R_s^2(z+\Delta z)-R_p^2\right]\Delta z-\sigma_{zs}\pi\left[R_s^2(z)-R_p^2\right] + \left(\sigma_{zs}+\frac{d\sigma_{zs}}{dz}\Delta z\right)\pi\left[R_s^2(z+\Delta z)R_p^2\right]-\rho_s V_s a-2\pi R_p\tau\Delta z=0 \tag{4.159}$$

After the collection of terms and simplification, we get

$$(p_s+\sigma_{zs})\frac{dR_s^2}{dz}+\left(\frac{d\sigma_{zs}}{dz}-\rho_s a\right)\left[R_s^2(z)-R_p^2\right]-2R_p\tau=0 \tag{4.160}$$

Note here that R_s and σ_{zs} are functions of z.

Next, we find σ_{zp}, assuming it is linear in z through the expression

$$\sigma_{zp}=\frac{F}{A}=\frac{1}{\pi R_p^2}\left(\rho_p V_p a-2\pi R_p\tau\Delta z\right)+\sigma_1=\frac{1}{\pi R_p^2}\left(\rho_p\pi R_p^2 a-2\pi R_p\tau\right)\Delta z+\sigma_1 \tag{4.161}$$

or

$$\sigma_{zp}=\left(\rho_p a-\frac{2\tau}{R_p}\right)\Delta z+\sigma_1 \tag{4.162}$$

where σ_1 is the axial stress in the penetrator as depicted earlier. Now we need to relate σ_{zp} to σ_{zs} by applying the assumption of strain compatibility; i.e., the strain in the sabot equals the strain in the penetrator.

We then use the appropriate elastic moduli and Poisson's ratio in Hooke's law to relate the penetrator stresses to those in the sabot

$$\varepsilon_{zs}=\frac{1}{E_s}\left[\sigma_{zs}-\nu_s(\sigma_{rs}+\sigma_{\theta s})\right]=\varepsilon_{zp}=\frac{1}{E_p}\left[\sigma_{zp}-\nu_p\left(\sigma_{rp}+\sigma_{\theta p}\right)\right] \tag{4.163}$$

Thus,

$$\sigma_{zs}=\frac{E_s}{E_p}\left[\sigma_{zp}-\nu_p\left(\sigma_{rp}+\sigma_{\theta p}\right)\right]+\nu_s(\sigma_{rs}+\sigma_{\theta s}) \tag{4.164}$$

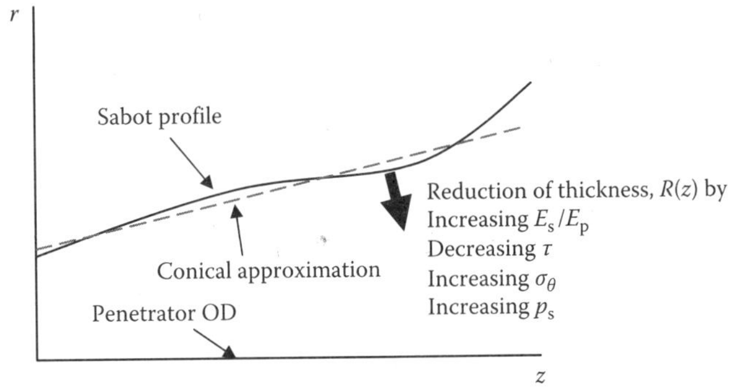

FIGURE 4.30
Sabot radial profile. (Based on analysis from Drysdale, W. H., Design of kinetic energy projectiles for structural integrity, Technical Report ARBRL-TR-02365, US Army Ballistic Research Laboratory, Aberdeen, MD, 1981.)

If we ignore the bimetallic nature of the components and assume that

$$\sigma_{rp} + \sigma_{\theta p} = \sigma_{rs} + \sigma_{\theta s} = -2p_s \tag{4.165}$$

then Equation 4.164 becomes

$$\sigma_{zs} = \frac{E_s}{E_p}\left(\sigma_{zp} + 2\nu_p p_s\right) - 2\nu_s p_s \tag{4.166}$$

These assumptions allow the integration of the differential equation for $R(z)$ producing the profile in Figure 4.30 (solid curve).

Two of the basic types of sabots are shown in Figures 4.26 and 4.27. The double ramp also incorporates a front air scoop to facilitate discard in the air stream as well as providing additional bourrelets riding surface in the tube.

A great deal of work on the effect of sabot design parameters has been accomplished at the US Army Research Laboratory (formerly the Ballistics Research Laboratory) and Picatinny Arsenal. A treatment of the effect of sabot stiffness on how clean a projectile launch is can be found in the study by Plostins et al. [7].

Problem 11

You are to design a rifled 20 mm gun system for an antivehicle application. The gun is to throw a 0.25 lbm, 0.5 in. diameter, cylindrical subprojectile at 4500 ft/s. A sabot will be placed on the outside of the subprojectile. The sabot can be any material you like and can find properties for. The sabot shall be made of three pieces, which are at least 3 in. in length. Your design must use a brass cartridge case to assist with obturation of the breech. Other assumptions and information are

1. You do not need to design the breech—assume it will hold the cartridge case in properly (in reality we can always add more threads to the design).
2. Even though there will be a slight taper on the chamber (which must be larger than the bore diameter for seating purposes), assume, for calculation purposes, that the chamber is cylindrical at its maximum diameter.
3. The tube is to be steel and assume that the yield strength is 60,000 psi (this accounts for the effect of cyclic loading). The modulus of elasticity is 29×10^6 psi. Poisson's ratio is 0.29.

4. Assume the propellant is either cylindrical or single perforated (and state your assumption).
5. Choose from the following propellants:

Propellant	Linearized Burn Rate β (in./s/psi)	Solid Density δ (lbm/in.^{3})	Adiabatic Flame Temperature T_0 (°R)	Propellant Force λ (ft-lbf/lbm)	Specific Heat Ratio γ
IMR	0.000132	0.0602	5103	327,000	1.2413
M12	0.000137	0.0600	5393	362,000	1.2326
Bullseye	0.000316	0.0590	6804	425,000	1.2523
Red Dot	0.000153	0.0593	5774	375,000	1.2400
Navy Pyro	0.000135	0.0566	4477	321,000	1.2454

6. Assume that the cartridge case is brass and use a bilinear kinematic hardening model where the brass has a modulus of elasticity of 15×10^6 psi, a local tangent modulus of 13×10^6 psi, and a yield stress of 16,000 psi (yield occurs in this material at $\varepsilon = 0.002$).
7. The weapon shall be as light as possible. The design is to proceed as follows (not necessarily in the order given):

   a. Interior ballistics design

      i. Size the chamber length and diameter.
      ii. Determine the amount of propellant needed based on your choice of the aforementioned propellants and propellant geometry (make sure it fits in the chamber).
      iii. Determine a web thickness for the propellant.
      iv. Determine the length of the gun.
      v. Determine V, p_B, and x for the projectile at peak pressure.
      vi. Determine V_c, p_{Bc}, and x_c for the projectile at charge burnout.
      vii. Determine the muzzle velocity of the projectile.

   b. Gun tube design

      i. Based on the calculations of part a, develop a pressure–distance curve to use as criteria for your gun design.
      ii. Determine the OD of the gun tube. To keep the design light as possible, use the design rules provided in this chapter and taper the tube toward the muzzle. If needed, over the chamber, you may shrink fit cylinders to build up a composite tube.
      iii. Determine the weight of your gun and comment on if it is reasonable.

   c. Cartridge case design

      i. Determine a thickness and tolerance for your cartridge case.
      ii. Determine the OD and tolerance for the cartridge case.
      iii. Decide on a tolerance for your chamber ID.
      iv. Note that for these calculations you must show that the case may be easily extracted at the limits of the tolerance.

d. Sabot design

   i. It is important that you write down all your assumptions. It is also highly likely that as you proceed further along with your design, you may come upon a situation that requires you to revisit an assumption you made earlier—this is to be expected, and it is part of the design process.

References

1. Budynas, R. G., *Advanced Strength and Applied Stress Analysis*, 2nd ed., McGraw-Hill, New York, 1999.
2. Boresi, A. P., Schmidt, R. J., and Sidebottom, O. M., *Advanced Mechanics of Materials*, 5th ed., John Wiley & Sons, New York, 1993.
3. Beer, F. P., Johnston, E. R., and DeWolf, J. T., *Mechanics of Materials*, 4th ed., McGraw-Hill, New York, 2006.
4. Montgomery, R. S., *Interaction of Copper Containing Rotating Band Metal with Gun Bores at the Environment Present in a Gun Tube*, Report AD-780-759, Watervliet Arsenal, New York, June 1974.
5. Pangburn, D., Personal communications with author December 1995 to March 2004.
6. Drysdale, W. H., *Design of Kinetic Energy Projectiles for Structural Integrity*, Technical Report ARBRL-TR-02365, US Army Ballistic Research Laboratory, Aberdeen, MD, September 1981.
7. Plostins, P., Clemins, I., Bornstein, J., and Diebler, J. E., *The Effect of Sabot Front Borerider Stiffness on the Launch Dynamics of Fin-Stabilized Kinetic Energy Ammunition*, BRL-TR-3047, US Army Ballistic Research Laboratory, Aberdeen, MD, October 1989.

Further Reading

Barber, J. R., *Intermediate Mechanics of Materials*, McGraw-Hill, New York, 2001.

Ugural, A. C., and Fenster, S. K., *Advanced Strength and Applied Elasticity*, 3rd ed., Prentice Hall, Upper Saddle River, NJ, 1995.

5

Weapon Design Practice

5.1 Fatigue and Endurance

Many parts in civil and military service are subject to fatigue. *Fatigue* is the term used for a mechanical part that undergoes cyclic loading and fails suddenly. Unlike a component that is simply overstressed and fails because the yield or ultimate strength is exceeded, a part that is subject to fatigue failure has been subjected to many small loads that stress the component below the yield strength. Damage begins to accumulate through various mechanisms such as microcrack growth or slipping along macroscopic boundaries. A simple example of fatigue is one where you take a metal paper clip and bend it 90°. After this first bend, the paper clip is still in one piece so the ultimate strength of the material was not exceeded (although it certainly has yielded). If one repeats this multiple times with the same paper clip, it will eventually break.* This failure can occur even without yielding the material.

A projectile usually undergoes one cycle of loading, so fatigue is normally a QWE issue. Gun tubes, however, undergo thousands of cycles, and fatigue is a major consideration in their design. The US design practice is to assure that a weapon shoots out before it fatigues out. What this means is that the weapon will become inaccurate because of wearing away of the rifling or the bore itself well before it fails in a sudden manner because of fatigue. This is determined by every maintenance crew by periodically checking the internal condition of the bore of the weapon. If the bore has sufficiently worn away, the tube is condemned. This condemnation is known to statistically occur after a certain number of rounds have been fired. The limit to the number of firings is compared to the design fatigue life of the weapon, and if the design was done correctly, there is sufficient margin remaining before a fatigue failure will develop.

The endurance of a material is the ability of the material to survive multiple cycles of loading. This ability of a material is graphically depicted in Figure 5.1. This figure is called an *S–N* diagram. An *S–N* diagram plots the allowable stress in the material against the number of cycles required by the designer. For example, if the designer required 10,000 cycles for a particular design using steel, it would be necessary to keep the stress below approximately 31,000 psi.

Some materials have an endurance limit. An endurance limit is the stress below which the material can withstand an infinite number of load cycles. Figure 5.1 shows that for this particular steel, the endurance limit is around 24,000 psi. Aluminums are notorious for not having an endurance limit. This means that aluminum components always have a finite fatigue life expectancy.

* This example was chosen by the author because it has been so frequently used by Dr. Jennifer Cordes of Picatinny Arsenal when she explains the nature of fatigue to new engineers or visitors.

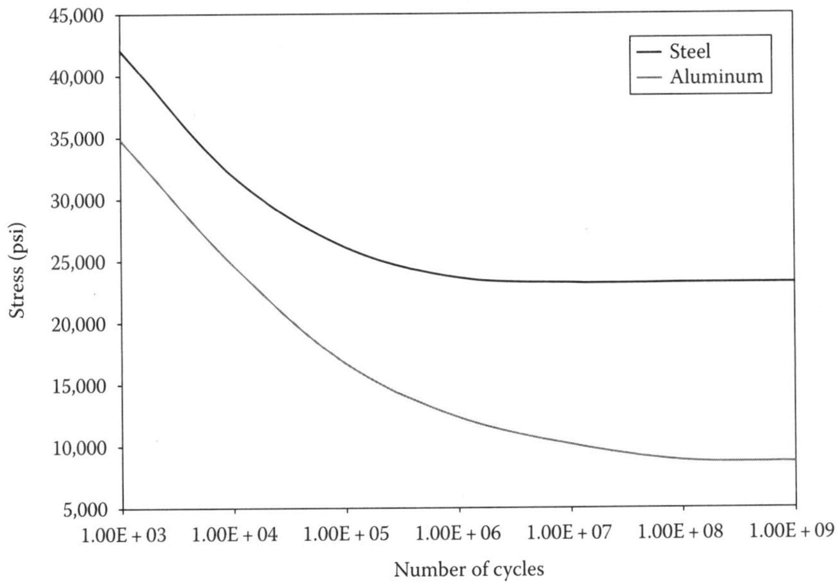

FIGURE 5.1
S–N diagram for steel and aluminum.

There are many contributing factors to the endurance of a component. Three of these factors which we have already touched upon are the material of the part, the number of loading cycles, and the stress level of each load cycle. Others are the rate of loading, rate of load reversal, the surface finish of the component, and even confidence in the endurance data used to generate the *S–N* diagram. Every reference that deals with this subject has a different twist (no pun intended) to the governing equation. The books by Deutschman et al. [1] and Norton [2] are excellent treatments of this behavior. A particularly simple approach is to define the fatigue strength of a material (i.e., the load that cannot be exceeded by any one cycle) as

$$S_n = S'_n C_R C_G C_S \tag{5.1}$$

Here S'_n is the stress in pounds per square inch read from an *S–N* diagram for the desired number of cycles; C_R is a factor that is chosen by the designer based on the reliability required in the design; C_G is a factor that is based on the rapidity of load reversal and steepness of stress gradients in the component; and C_S is a factor that accounts for the surface finish. These factors effectively reduce the allowable stress in the part (they all should be ≤ 1). Unfortunately, they are all subject to interpretation and vary with each material and even from reference to reference. Some references use additional factors as well. The best advice in the case of fatigue is for you to find a reference that has calculated fatigue in a component similar to the one you are designing and base your design on that data.

Problem 1

It is desired to construct a 75 mm gun for a pressure of 43,000 psi. The chamber diameter has been chosen to be 3.1 in. If we use AISI 4340 steel with a yield strength (S_Y) of 100,000 psi, determine the outer diameter (OD) of the weapon over the chamber. Assume that the tube is not autofrettaged and the endurance limit (S'_n) for 4340 is $0.875S_Y$ for the amount of cycles desired. Assume the following factors from our cyclic loading discussion: $C_R = 0.93$, $C_G = 0.95$, and $C_S = 0.99$. Assume the chamber is open ended as a conservative measure.

Answer: 20 in. OD will just work.

Problem 2

A shotgun is to be modified so that it can be rigidly mounted to a vehicle. The recoil force is estimated to be 800 lbf. There are two failure points: a weld on the barrel and two 10–32 screws connecting the receiver to the barrel. If we assume that each point of failure (the two screws act together) must individually take the full load, determine how many firings can be achieved using the curve for steel provided in this chapter (Figure 5.1) and the following data:

- Both materials: $C_R = 0.8$; $C_G = 0.85$
- Screws: $C_S = 0.78$; shear area = 0.019 in.^{2} each
- Welds (1/8 in. fillet): $C_S = 0.5$; shear area = 0.247 in.^{2}

Answer: The screws will survive approximately 2000 cycles; the welds will last an infinite number of cycles.

5.2 Tube Design

In the discussion of the design of conventional projectile bodies that we completed earlier, many of the concepts we introduced are now applicable, with particular modifications, to the design of gun tubes. For example, the idea of safety margins has counterparts in the design of a gun tube, but where a projectile has to withstand a single cycle of applied stress, the gun tube must remain serviceable for many cycles at stress levels very much comparable to the fired projectile.

The gun tube designer is interested in determining the structure which has the minimum weight, which usually translates to a minimum radial dimension, consistent with safely firing a projectile. The projectile designer is usually interested in determining the projectile structure of minimum weight sufficient to meet safety, reliability, and, especially, effectiveness requirements. The projectile designer needs to know the maximum pressure on the base of the moving projectile during its time in the tube, known as the single base maximum pressure. Once this single pressure-induced stress is accommodated, the designer can move on to other considerations. The tube designer, on the other hand, must know the maximum pressure exerted on the tube at every axial location in the bore as the projectile transits the tube. These are known as the station maximum pressures in tube design. We use the projectile and charge combination which applies the most stress to the weapon (usually this is the heaviest projectile and the biggest charge). These pressures are applied over and over again as the tube is cycled with each shot fired, leading to the necessity to account for and predict the fatigue failure of the design.

Finite element analysis (FEA) methods are used less frequently in gun design than in projectile design because FEA is a much more difficult method when used to predict fatigue failures. The reasons for this are that the gun launch phenomenon is highly transient, the erosion of the weapon is impossible to predict at the present time, the boundary conditions of a firing position change the dynamic response of the weapon, and, in overall gun design, there are many different parts to consider. The tried-and-true hand calculation processes developed at the Watervliet, Frankford, and Picatinny Arsenals still yield excellent, reliable weapons. But FEA will become more important as the codes develop and weight of the weapon becomes more of an issue.

Another major consideration in tube design is the degradation of material strength with temperature. The repetitive firing of a weapon with propellants burning in the chamber and in the bore generates a large amount of heat. In tube artillery or tank cannons, the temperatures developed can become high enough to begin to affect the material properties in an adverse way. In rapid-fire weapons particularly, it is absolutely critical that the degraded material strength properties be accounted for in tube and in chamber stress calculations.

There are several types of tube designs that may be encountered in service weapons: The monobloc tube is made from one piece of metal, which is not the most efficient way to construct a tube. The jacketed tube consists of separate layers or jackets built up as a composite structure; this type is mostly obsolete now and is being replaced by a process called autofrettaging or self-jacketing. The quasi-two-piece tube is formed by inserting a liner into an otherwise monobloc, pressure-containing tube; this allows for a more resilient material for the projectile to ride against and helps with the wear of the tube. British warships used a now obsolete wire-wrapped tube construction that was cheap to make, but quite inaccurate in use.

When we begin a design of a new tube, the interior ballistician computes the space-mean pressure–travel and pressure–time curves for the most stressful projectile expected to be fired at a temperature of 70°F. The maximum pressure of this curve gives the computed maximum pressure (CMP), which is the nominal pressure for the gun. However, because of the stochastic nature of a gun launch, the designer will add 2400 psi to the CMP. This is the rated maximum pressure (RMP) for the weapon. This pressure is one which cannot be exceeded by the average of the maximum pressures of a group of projectiles fired at 70°F:

$$\mathrm{RMP}\left[\frac{\mathrm{lbf}}{\mathrm{in.}^2}\right] = \mathrm{CMP}\left[\frac{\mathrm{lbf}}{\mathrm{in.}^2}\right] + 2400\left[\frac{\mathrm{lbf}}{\mathrm{in.}^2}\right] \tag{5.2}$$

After a statistically significant number of projectiles are fired out of the weapon, data are taken to validate the CMP. This experimentally determined number is the normal operating pressure for the weapon and should replace the CMP as soon as it is available and accepted.

Under service conditions, many rounds will be fired at many different operating temperatures. We define the permissible individual maximum pressure (PIMP) as the pressure which cannot be exceeded by any individual round under any service condition.

In design terms, it is calculated as 15% over the RMP:

$$\mathrm{PIMP}\left[\frac{\mathrm{lbf}}{\mathrm{in.}^2}\right] = (1.15)\mathrm{RMP}\left[\frac{\mathrm{lbf}}{\mathrm{in.}^2}\right] \tag{5.3}$$

The permissible mean maximum pressure is the pressure that cannot be exceeded by the average of all rounds fired under any service condition.

From an analysis standpoint, we need to define a pressure at which enough stress is developed (assuming tube material at 70°F) at some point in the tube so that yielding occurs; i.e., the elastic limit of the material is reached. This is the elastic strength pressure (ESP) for the tube. At higher temperatures, we must also define an ESP_{hot} to account for material strength loss at temperature. A good example of how these concepts are applied can be found in the report by Smith and Coppola [3].

When we examine the travel of the most stressful projectile down the tube, a point is reached x_{max}, where the breech pressure is at maximum $p_{B\ max}$. At this same instant, the pressure on the base of the projectile is also at a maximum (but, as we saw in the section on the Lagrange gradient, lower than the breech pressure) and will never increase beyond this value ($p_{s\,max} < p_{B\,max}$). There is a pressure gradient at every point x, between the breech of the weapon and the base of the projectile. With this in mind, it is worthy to note that the pressure at any location forward of x_{max} will never "see" a pressure higher than that acting on the base of the projectile at this point. Nevertheless, as a measure of the inbred conservatism of gun designers, we design the tube to the pressure experienced at the breech while the projectile traverses the gun. These various pressures and the gradients are shown in Figure 5.2.

Designing a gun tube requires knowledge of the stress state of the tube and a judgment of what constitutes a failure when it is under stress. For this, we turn to the von Mises criterion for failure under stress:

$$2\sigma_Y^2 = (\sigma_1 - \sigma_2)^2 + (\sigma_2 - \sigma_3)^2 + (\sigma_3 - \sigma_1)^2 \tag{5.4}$$

where σ_1 is the axial stress; σ_2 is the tangential stress; σ_3 is the radial stress; and σ_Y is the equivalent stress.

For an open-ended tube, $\sigma_1 = 0$, and Equation 5.4 becomes

$$\sigma_Y^2 = \sigma_2^2 - \sigma_2\sigma_3 + \sigma_3^2 \tag{5.5}$$

Recall the Lamé formulas for stress in a thick-walled tube:

$$\sigma_{\theta\theta} = \sigma_2 = p_i \frac{r_i^2}{r^2}\left[\frac{(r^2 + r_o^2)}{(r_o^2 - r_i^2)}\right], \quad \text{maximum at } r = r_i \tag{5.6}$$

$$\sigma_{rr} = \sigma_3 = p_i \frac{r_i^2}{r^2}\left[\frac{(r^2 - r_o^2)}{(r_o^2 - r_i^2)}\right], \quad \text{maximum at } r = r_i \tag{5.7}$$

Let us put the Lamé formulas into a more useful form by letting

$$\zeta = \frac{r_o}{r_i} > 1 \quad \text{and} \quad \zeta_x = \frac{r}{r_i} > 1 \tag{5.8}$$

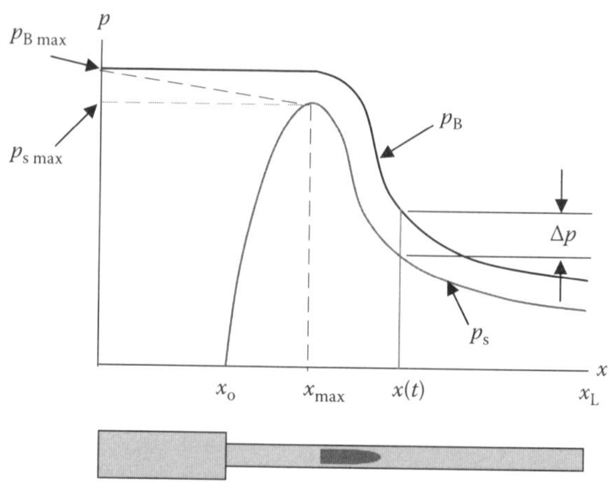

FIGURE 5.2
Pressure–distance curve for a gun tube.

then,

$$\sigma_{\theta\theta} = \sigma_2 = p_i \frac{1}{\zeta^2 - 1} \left[\frac{(\zeta_x^2 + \zeta^2)}{\zeta_x^2} \right] \tag{5.9}$$

$$\sigma_{rr} = \sigma_3 = p_i \frac{1}{\zeta^2 - 1} \left[\frac{(\zeta_x^2 - \zeta^2)}{\zeta_x^2} \right] \tag{5.10}$$

and

$$\sigma_{\theta\theta\ \max} = \sigma_{2\ \max} = p_i \frac{\zeta^2 + 1}{\zeta^2 - 1} \quad \text{at } r = r_i \tag{5.11}$$

$$\sigma_{rr\ \max} = \sigma_{3\max} = -p_i \quad \text{at } r = r_i \tag{5.12}$$

Failure is considered to have occurred when the equivalent stress σ_Y is greater than the yield strength Y of the material. If we substitute Equations 5.11 and 5.12 into Equation 5.5 and substitute Y for σ_Y, we get a solution for the ratio of internal pressure to yield strength:

$$Y^2 = \left[p_i \frac{\zeta^2 + 1}{\zeta^2 - 1} \right]^2 + \left[p_i^2 \frac{\zeta^2 + 1}{\zeta^2 - 1} \right] + p_i^2 \tag{5.13}$$

which by manipulation and expansion yields

$$\frac{Y^2}{p_i^2} = \frac{3\zeta^4 + 1}{(\zeta^2 - 1)^2} \quad \text{or} \tag{5.14}$$

$$\frac{p_i}{Y} = \frac{\zeta^2 - 1}{\sqrt{3\zeta^4 + 1}} \tag{5.15}$$

If the relationship in Equation 5.15 is plotted on a semilog plot, we see that for a monobloc tube (one which is made out of one piece) of yield strength Y and an internal pressure of $1/2$ Y, we obtain a wall thickness ratio $\zeta = 2.75$. This ratio rapidly becomes infinite as $p_i/Y \to 0.58$. This is depicted in Figure 5.3. Thus, pressure levels are restricted below this value. If we consider that a good gun steel of 180,000 psi yield strength is used, the allowable internal pressure should be kept lower than 100,000 psi. For modern, high-velocity cannons, this restriction had to be overcome and the autofrettaging process described in the following has been used with marked success.

Jackets improve the efficiency of the gun tube by utilizing more of the materials load-carrying capacity. The design concept began around 1870 and has been in use since then, but is now considered obsolete. The idea is that one can shrink fit one or more cylinders over the inner cylinder or liner so that a compressive stress is induced in the inner layers. When an internal pressure is applied, the stresses on the inner cylinders are relieved by the pressure and then put into tension as the pressure is increased. Autofrettaging (self-jacketing) rather than shrink fitting is now the process in use.

Autofrettage is a method of prestressing a tube to improve its load-carrying capability as well as its fatigue life. The procedure consists of plastically deforming the interior of the gun tube toward the outside diameter. Regions of the interior wall will now exceed the

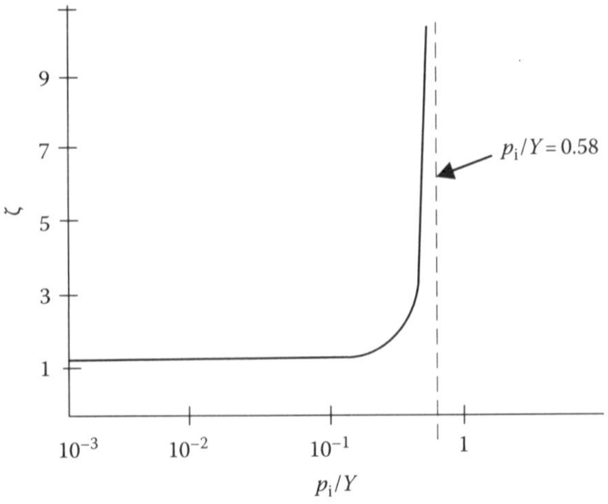

FIGURE 5.3
Wall-thickness ratio as a function of internal pressure to yield stress ratio.

yield point, but the exterior will not have yielded. When the load is removed, the outer layers of the material attempt to return to their unstressed state but cannot because of the plastically deformed portion of the wall. Thus, an equilibrium condition is attained where the outer wall regions remain in tension and the inner wall regions are in compression. The process is physically accomplished by either pressurizing the interior of the tube with water above its elastic limit or by pulling an oversized mandrel through the tube to force the yielding.

The pressure induced to autofrettage is on the order of

$$p_f = Y \ln \zeta \tag{5.16}$$

This is the pressure required to barely stress the outer wall during the process. The current practice is to keep this value below the ESP by at least 8% in a finished tube. To further ensure that the OD never goes plastic, tubes are sometimes autofrettaged in containers that act as an outer jacket during manufacture. Figures 5.3 through 5.7 illustrate the process.

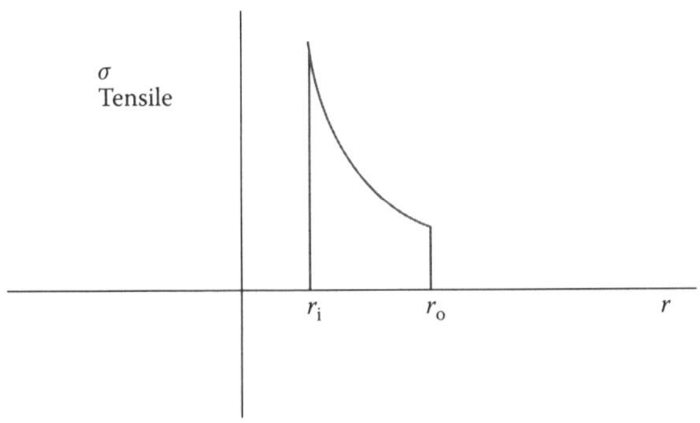

FIGURE 5.4
Stress profiles in a monolithic tube.

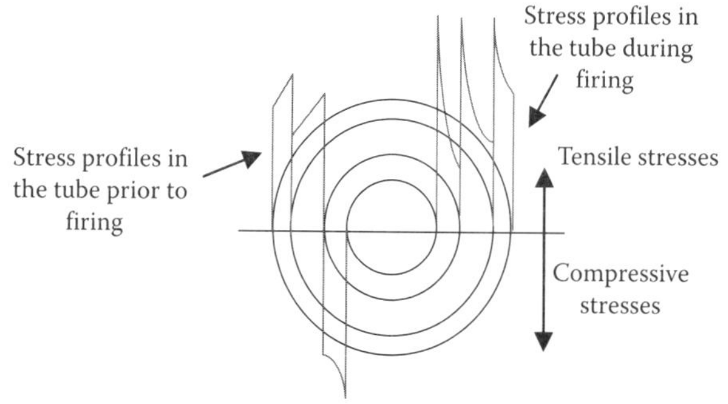

FIGURE 5.5
Stress profiles in an autofrettaged tube.

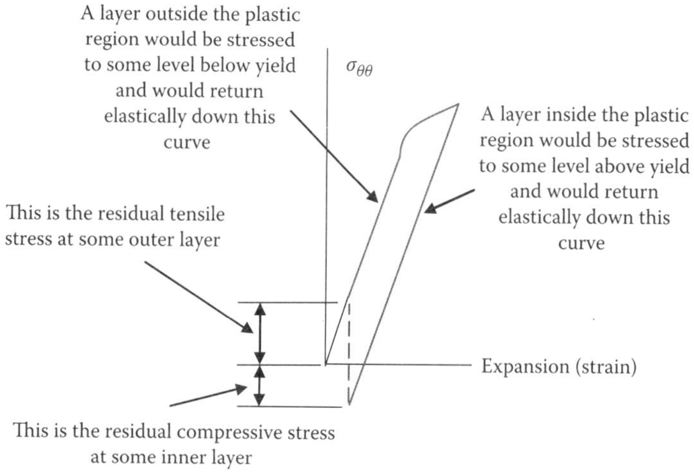

FIGURE 5.6
Hoop stress vs. strain in an autofrettaged tube.

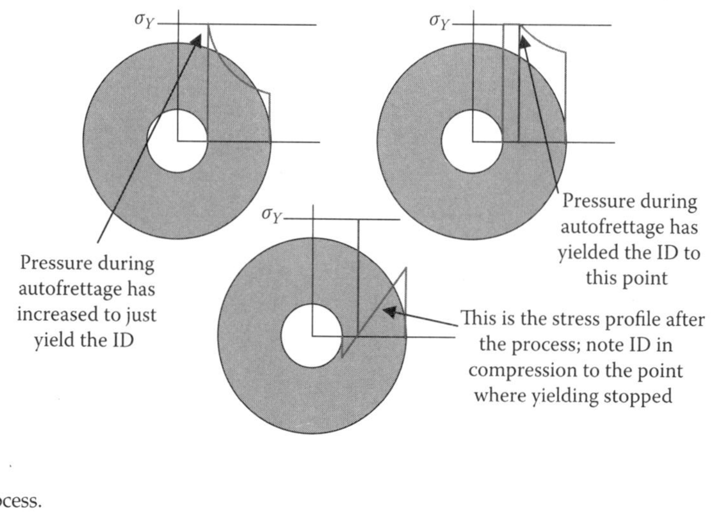

FIGURE 5.7
Autofrettage process.

Problem 3

The gun in Problem 1 is sized to a 20 in. OD. The manufacturer decides to autofrettage the weapon with 75,000 psi of hydraulic fluid. Assuming that the material behaves elastic-perfectly-plastic, approximately to what radial distance does the compressive layer extend into the tube wall?

Answer: Approximately 0.57 [in.]

5.3 Gun Dynamics

In this section, we intend to discuss how a gun behaves as a dynamic entity, during a projectile firing and immediately after the projectile exits the muzzle. We will discuss the recoil response in terms of the forces and motions and the response known as gun jump. We will not attempt to discuss recoil abatement or mounting techniques.

Recoil is generated on the gun by the reaction of its moveable parts to the impulse of the gas pressure both while the projectile is in the tube and while the propelling gases are being exhausted after the projectile exits. After projectile exits, we may simplify the actual process to one that assumes that the breech pressure decays linearly with time. This period is called the gas exhaust aftereffect and is shown in Figure 5.8.

We show the forces on the gun during the time the projectile is moving through the tube (including the forces attributable to the rifling) in Figure 5.9. Note that the projectile friction force F_{Pr} acts to tug the barrel to the right while in reaction, the acceleration of the projectile is retarded by the same magnitude of force applied to the left. Friction is normally rather small through launch with one important exception. Prior to shot start, it prevents motion until sufficient pressure builds. For the remainder of this chapter, the relations assume that

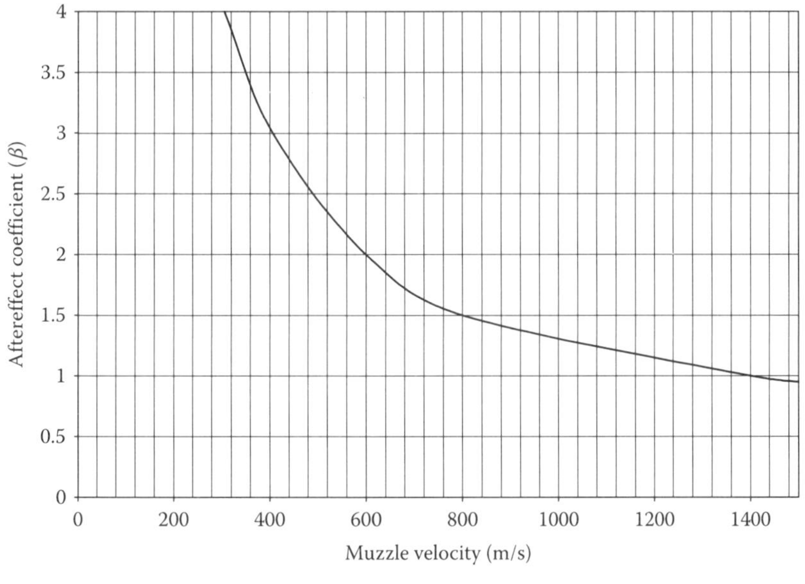

FIGURE 5.8
Aftereffect coefficient.

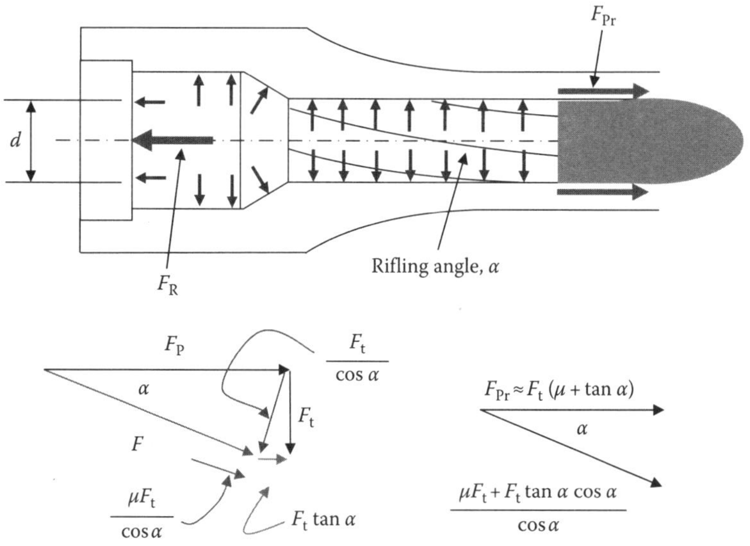

FIGURE 5.9
Forces acting on a gun tube and projectile reactions.

the shot start has occurred and the projectile is either accelerating to higher velocity or has exited the gun.

During firing, the major forces on the gun include the breech pressure p_B applied over the generally larger diameter breech area, radially directed pressures along the length of propellant gas column that are contained by the strength of the walls and vectorally cancel, and a modest forward force at the shoulder of the chamber where the local pressure vectors include a forward component as shown in Figure 5.9. It is common to approximate the local pressure at the shoulder to be equal to those at the breech. With this simplification, the magnitude of the forces acting to accelerate the cannon to the left in recoil is

$$F_R = Ap_B - F_{Pr} \tag{5.17}$$

The remaining force acting on the cannon is the reaction to rifling torque if rifling is present. This does not alter the recoil of the gun. Often a key fixed to the gun mount projects into an axial keyway groove cut along the recoil length of the OD of the cannon. During recoil, the groove slides past the stationary key acting as a cam to prevent the barrel from rotating.

The major force on the projectile remains the lower projectile base pressure p_S projected over the bore area A as developed in Equation 3.48. The net force is reduced from the pressure forces propelling the projectile to the right by the resistance force:

$$F_P = Ap_S - F_{Pr} \tag{5.18}$$

The resistance pressure is estimated as follows:

$$\begin{aligned} &\text{For smooth bores: } F_{Pr} \approx 0.01Ap_S \\ &\text{For rifled bores: } F_{Pr} \approx (\mu + \tan\alpha)F_T \end{aligned} \tag{5.19}$$

The rifling force is

$$F_T = \left(\frac{k}{d/2}\right)^2 F_P \tan\alpha \tag{5.20}$$

Here k is the radius of gyration of the projectile ($I_{zz} = wk^2$ in terms of axial moment of inertia and mass), μ is the coefficient of friction, and α is the rifling angle. For rifled bores, the interdependence on the net projectile force between Equations 5.18, 5.19, and 5.20 may be iteratively solved, or friction may neglected as small in Equation 5.20 after shot start. Our object is to find F_P.

From the Lagrange approximation for the pressure gradient, we know that

$$p_B = p_S + \left(\frac{c}{2w}\right)\left(p_S - \frac{F_{Pr}}{A}\right) \tag{5.21}$$

Students and even seasoned ballisticians can become confused by how the resistance force is introduced. The acceleration of the propellant column is the source of the pressure elevation at the breech. We know from the Lagrange assumption that the center of the propellant mass accelerates at half that of the projectile which is retarded by the friction. We must therefore decrement the projectile base pressure by the equivalent resistance pressure. The need for this is clearly seen if we consider when pressure and resistance forces are equal before shot start. In this case, the breech pressure should not yet be elevated above the projectile base pressure.

We have elected to neglect the pressure developed as the ambient gases ahead of the projectile are accelerated out of the gun. This is valid when the mass of the ambient gas to be ejected ahead of the projectile is negligibly small relative to that of the bullet. The volume of the air would be modestly less than the displacement volume of the cannon, and the density of air at standard conditions is 0.001225 kg/m^3. The displacement volume is the product of the bore area and projectile travel relative to the tube S_0 as shown in Figure 5.10:

$$\rho_{air} A S_0 \ll w \tag{5.22}$$

Now let us examine the motion of the gun during recoil. This essentially depends on the balance of momentum between the projectile and its propelling gases and the mass of the gun. Let us first write a momentum balance in the direction of fire:

$$M_{recoil} = M_{proj} + M_{prop.gas} \tag{5.23}$$

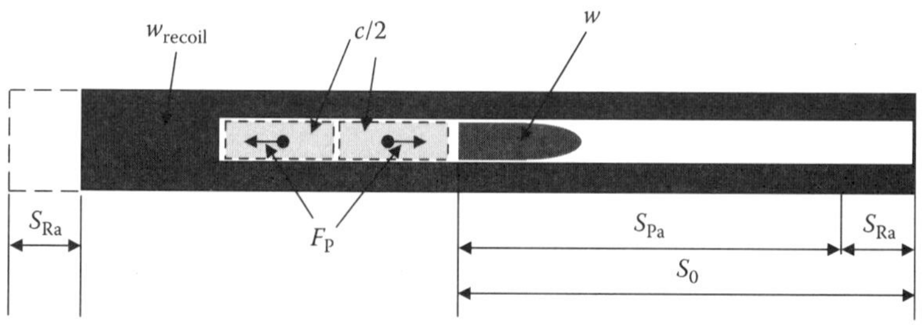

FIGURE 5.10
Diagram of gun displacements.

We recall from the Lagrange approximation for the projectile and its propelling gas that the momentum at shot exit is

$$\text{Momentum at shot exit} = \left(w + \frac{c}{2} \right) V_{\text{muzzle}} \tag{5.24}$$

Thus, at projectile exit,

$$V_{\text{Re}} = \left(\frac{w + c/2}{w_{\text{recoil}}} \right) V_{\text{muzzle}} \tag{5.25}$$

In this expression, w_{recoil} is the recoil mass of the weapon. This quantity includes all mass attached to the tube that must be moved rearward when the weapon fires such as breech closing mechanisms and sighting devices.

After the projectile has left the muzzle, the propellant gases exit at a velocity whose mass averaged value $\bar{V}$ can be approximated by

$$\bar{V} = \sqrt{V_{\text{muzzle}}^2 + \bar{c}^2} \tag{5.26}$$

where $\bar{c}$ is the average speed of sound in the propellant gases (≈ 1000 m/s or 3200 ft/s) [5]. As an example, consider using Equation 5.26 with a muzzle velocity of 3442 ft/s:

$$\bar{V} = \sqrt{(3442)^2 + (3200)^2} = 4700 \text{ ft/s} \tag{5.27}$$

This shows that old-fashioned guidance to assume $\bar{V}$ = 4700 ft/s would have been conservative as intended when muzzle velocities were generally lower than 3442 ft/s. We can define an aftereffect coefficient by

$$\bar{V} = \beta V_{\text{muzzle}} \tag{5.28}$$

Here we can estimate β from Equation 5.26.

Using β, we can write a new equation for the momentum balance:

$$w_{\text{recoil}} V_{\text{final}} = (w + \beta c) V_{\text{muzzle}} \tag{5.29}$$

This allows us to solve for the final velocity of the recoiling parts:

$$V_{\text{final}} = \frac{(w + \beta c)}{w_{\text{recoil}}} V_{\text{muzzle}} \tag{5.30}$$

During free recoil, the total distance traveled S_{Re} is the sum of the distance traveled while the projectile is in the gun S_{Ra} and the distance traveled during the gas ejection phase S_{Rn}:

$$S_{\text{Re}} = S_{\text{Ra}} + S_{\text{Rn}} \tag{5.31}$$

With no external forces acting, the common center of mass stays at rest, with half of the charge mass lumped with the gun and half with the projectile. Once again, we write the momentum balance:

$$\left(w_{\text{recoil}} + \frac{c}{2}\right)V_{\text{Re}} = \left(w + \frac{c}{2}\right)V_{\text{muzzle}} \tag{5.32}$$

By considering that the velocities, on average, are distance divided by the time, we can rewrite Equation 5.32 as a distance equation relative to an inertial reference frame before shot start:

$$\left(w_{\text{recoil}} + \frac{c}{2}\right)S_{\text{Ra}} = \left(w + \frac{c}{2}\right)S_{\text{Pa}} \tag{5.33}$$

We can see from Figure 5.10 that the recoil of the tube reduces the actual travel of the projectile because the muzzle recoils rearward prior to shot exit. The relation between the travel relative to the tube S_0, the recoil of the cannon at shot exit S_{Ra}, and the effective projectile travel relative to the inertial reference frame of the cannon at shot start S_{Pa} may be found from Equation 5.33 as

$$S_0 = S_{\text{Ra}} + S_{\text{Pa}} = S_{\text{Ra}}\left(1 + \frac{S_{\text{Pa}}}{S_{\text{Ra}}}\right) \tag{5.34}$$

We can then see that the free recoil motion of the gun while the projectile is in the tube S_{Ra} may be found from Equation 5.33 as

$$S_{\text{Ra}} = S_{\text{Pa}}\left(\frac{w + c/2}{w_{\text{recoil}} + c/2}\right) \tag{5.35}$$

And from Equation 5.34 we can show that

$$S_{\text{Ra}} = S_0\left(\frac{w + c/2}{w_{\text{recoil}} + w + c}\right) \tag{5.36}$$

We shall define the effective breech force at shot exit as

$$F_{\text{a}} = p_{\text{a}}A \tag{5.37}$$

As mentioned in Equation 5.31, further motion of the gun in free or unconstrained recoil occurs after the projectile has left the tube. It is caused by the momentum exchange of the mass of gas still exhausting from the tube after the projectile is long gone. We look for an estimate of the length of this motion S_{Rn} by examining the impulse of the gas. The duration of the tube-emptying phase t_{n} can be computed from the aftereffect impulse I_{n} by assuming that the breech pressure force linearly decreases with time (see Figure 5.11):

$$I_{\text{n}} = \frac{1}{2}F_{\text{a}}t_{\text{n}} \tag{5.38}$$

But impulse may also be defined as the change of momentum over time:

$$I_{\text{n}} = w_{\text{recoil}}(V_{\text{final}} - V_{\text{Re}}) \tag{5.39}$$

By application of Equations 5.28 and 5.30, we can show that

$$I_{\text{n}} = \left(\beta - \frac{1}{2}\right)cV_{\text{muzzle}} \tag{5.40}$$

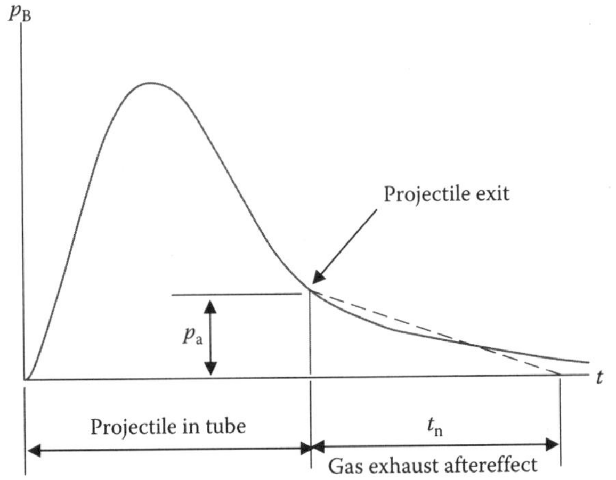

FIGURE 5.11
Breech pressure–time curve for a typical gun firing.

If we solve for t_n by inserting Equation 5.40 into Equation 5.38, we get

$$t_n = \frac{\left(\beta - \frac{1}{2}\right) c V_{muzzle}}{F_a/2} \tag{5.41}$$

Since we have assumed a linear velocity change between V_{Re} and V_{final}, we may use the trapezoidal rule of integration to compute an approximation for the remaining travel S_{Rn}:

$$S_{Rn} = t_n \frac{(V_{final} + V_{Re})}{2} \tag{5.42}$$

In this analysis, we have assumed that the weapon was in free recoil. In real weapons, this never occurs. We normally have recoil mechanisms that rely on pneumatic or hydraulic systems to slow down and finally stop the recoil within a relatively short distance. These forces need to be added to the preceding analysis to make it more accurate. The effect of a muzzle brake should be added as well. We shall discuss more about this in Chapter 6.

Let us now consider the phenomenon known as "gun jump." The axis of the gun bore, which is where the gas forces are applied, is usually not collinear with the mass center of the recoiling parts. This creates a moment couple often referred to as the "powder couple," which acts upon firing (Figure 5.12). This couple causes a rotation of the gun that usually

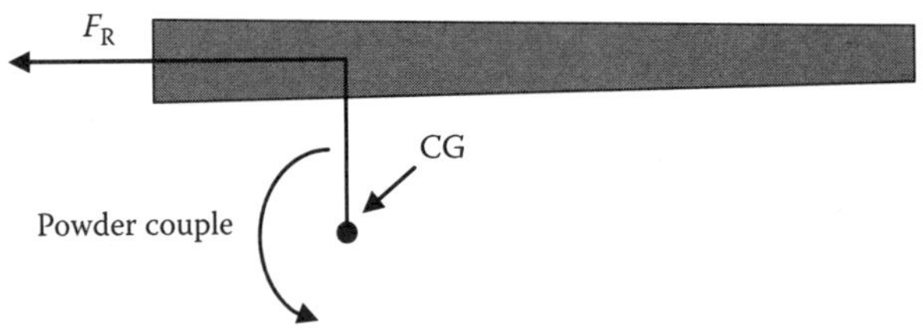

FIGURE 5.12
Powder couple illustrated.

results in muzzle rise. This contributes to projectile jump but is by no means the sole cause of it.

There are other dynamic reactions of the gun during firing. The gun is an elastic body, so that when the propelling charge is ignited, many complicated structural reactions take place. Stress and pressure waves are set up in the chamber and in the unpressurized portion of the bore, loading the tube in a highly transient fashion. Swelling and elongation occur due to pressure, the rotating band is engraved by the rifling (if present) causing local stressing of the tube, and a thermal gradient is set up. These phenomena are highly complicated, and we will not discuss them further here.

5.4 Muzzle Devices and Associated Phenomena

We use muzzle devices for three main reasons: reduce recoil, suppress flash, and decrease report. Sometimes increased accuracy results from shot to shot because of reduced weapon movement. Muzzle devices have also been devised to limit muzzle climb.

Muzzle brakes consist of surfaces placed perpendicular to the bore axis such that impinging gases exert a net forward thrust on the weapon. This thrust is accomplished through the conservation of momentum principles. Best design practice is to divert gases to the sides of the weapon because rearward diversion could affect an exposed gun crew. Downward diversion could kick up excessive debris and, without a balancing upward diversion, would strain operating gun mechanisms.

There are generally two types of muzzle brakes: closed and open. Closed brakes channel the exiting gases through fixed openings and usually have multiple baffles or ports. Open muzzle brakes generally have only one baffle and direct the gas flow to a lesser extent than closed brakes. The chief purpose of these brakes is to mitigate the recoil.

A blast deflector is similar in concept to a muzzle brake, although not designed to assist recoil as much. The purposes of the blast deflector are to direct blast away from the gun crew, minimize obscuration of the battlefield by limiting the amount of dust kicked up during the discharge of the weapon, and, in the case of small arms, limit muzzle climb. A detriment of a blast deflector is that to reduce dust, one usually needs to vent the gases upward, which tends to load the tube and to support structure of the weapon. If the weapon is already horizontal and the venting thrust has a large vertical component, this can be a substantial loading.

There are four basic types of muzzle gas deflectors (Figure 5.13). The baffle type is identical to a baffled muzzle brake with the gases vented to the sides of the piece. The perforated type, sometimes called a "pepper-pot" brake, has multiple side ports in a tubular section (none of the ports venting straight down). The T-type is the same as a single-baffle muzzle brake and, lastly, the ducted type. This latter device has a complicated array of ducting to divert the flow back and upward near the mounting trunnions. It diverts the blast load so that it is carried by the trunnions. Unfortunately, at high quadrant elevations, it ducts the blast toward the crew, which is not good.

Muzzle flash was noticed as a problem during the First World War when significant night actions were commonplace and suppression of muzzle flash became highly desirable. The study of flash has used high-speed photography and other recording devices to distinguish five types of flashes (Figure 5.14):

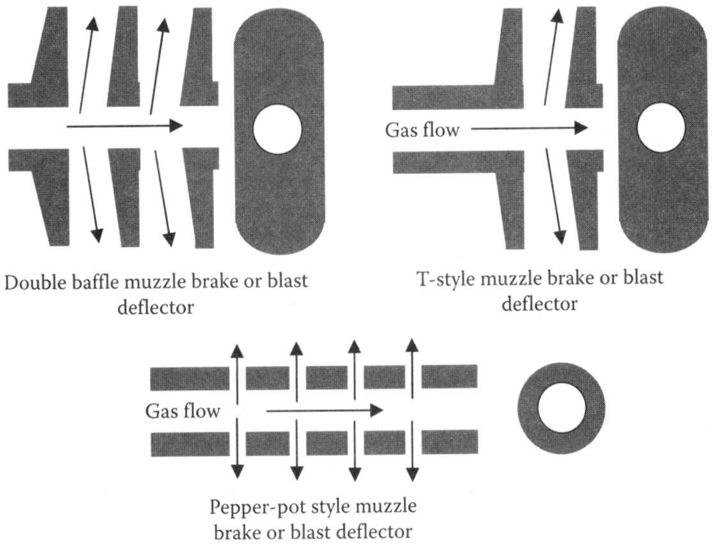

FIGURE 5.13
Typical muzzle brake or blast deflector geometry.

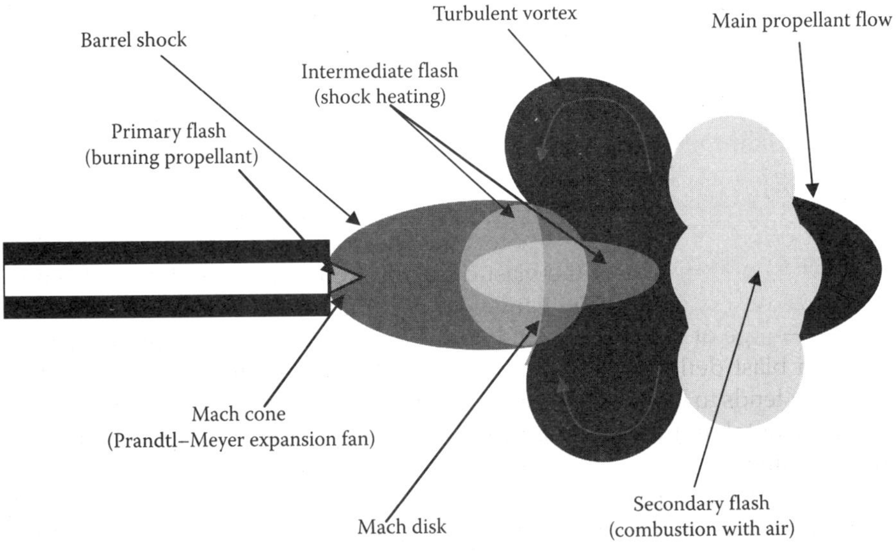

FIGURE 5.14
(See color insert.) Muzzle blast structure.

1. Preflash—This is flash caused by blowby, a condition where propellant gas leaks around the rotating band of the projectile or obturator and exits before the projectile.
2. Primary flash—This is the flash caused by any propellant solids or gases that are still burning upon muzzle exit of the projectile.
3. Muzzle glow—This is the illumination caused by gas inside the shock bottle (defined later).

4. Intermediate flash—This is the illumination caused by gas that managed to get ahead of the normal shock of muzzle gas ejection and is caused by the increased pressure and temperature of the gas as it passes through the shock front.
5. Secondary flash—This is the flash caused by the reaction of the combustion products when they enter the air (really, another, secondary oxidation reaction transpires).

Propellant additives are often used, but do not sufficiently suppress flash and, besides, add smoke. It was observed early on that muzzle brakes and blast deflectors actually suppressed flash somewhat. This has led to the development of mechanical flash suppressors.

However, the only types of muzzle flashes that can be controlled by the attachment of a mechanical flash suppressor are muzzle glow, intermediate flash, and secondary flash. These all are affected by the presence of the expanding gas shock wave. Secondary flash is the least controllable from a mechanical standpoint because the temperature of the propellant gas mixture may be so high that the shock is only an amplifying factor.

Various designs of suppressors have been developed, and they fall into two basic types (Figure 5.15):

1. Conical flash suppressors appear similar to the bell end of a trumpet and are sometimes called flash hiders.
2. Bar-type flash suppressors resemble a cage around the muzzle of the weapon. They are difficult to clean, and if they are of an open-end design, then they can get caught on objects such as clothing and vegetation during combat.

Smoke on a battlefield is disadvantageous if not generated where and when it is desired as an obscurant. In the days of black powder, it was a real problem as the battlefield became obscured for friend and foe alike. When nitrocellulose propellants were introduced, they were called "smokeless powders" because they generated much less smoke than black powder. Even with smokeless powders, large volumes of fire still produced significant quantities of smoke. An alternative would be to add chemicals to the propellant to reduce smoke, but this usually increases flash and devices that suppress flash usually increase smoke.

Smoke generated from a weapon is usually made up of a solid–liquid–gas mixture and is composed of metal or metal oxide particles from the cartridge case and its components, the

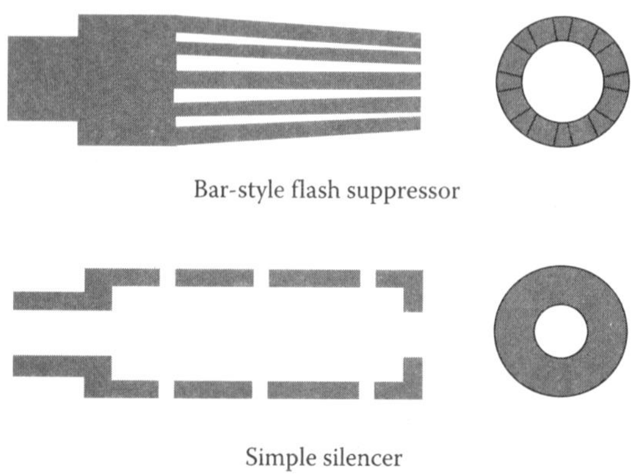

FIGURE 5.15
Typical muzzle devices.

projectile and the tube. Also present are water vapor or condensate liquid and chemical elements such as carbon, copper, lead, zinc, antimony, iron, titanium, aluminum, potassium, chlorine, sodium, and sulfur and other particulates. These components in themselves obscure vision, but they may also combine with the atmosphere to allow water vapor there to condense on the particles. Air temperature and relative humidity affect the density and longevity of the smoke as well.

Smoke suppressors are really filters that capture the solid particulates, yet allow the gaseous composition to pass through them. They are either electrostatic in nature or mechanical filters. Electrostatic types are primarily used in a laboratory environment. Mechanical types work by robbing momentum from the particles. The pores of these suppressors must be quite large so that the gas flows through them without difficulty and that only the particulates are removed.

These suppressors work by forcing the propellant gas to pass through nonstraight channels similar to pores. The impingement of the particles robs them of momentum. When the gas pressure in the suppressor (which is super caliber) becomes higher than the muzzle pressure, the gas evacuates in the opposite direction to entry, leaving the solids and liquids behind. The downside to this is that the suppressor adds weight to the tube at the muzzle end, adds cost, and requires frequent maintenance.

There are three basic types of mechanical smoke suppressors:

1. Uniform perforation spacing where no attempt to control the flow is made.
2. Increasing perforation density toward the muzzle, which allows particles that would normally build up closer to the muzzle to be spread more evenly in the device because pressure drops in the axial direction.
3. A tapered bore type which is similar to the preceding type but includes a taper that becomes smaller as one approaches the exit with a larger inner diameter near the end at the rifling. This allows some axial impingement and helps spread out the heavier particles.

Noise on the battlefield is also the subject of mitigation. Devices used to reduce noise, which are sometimes referred to as silencers, attempt to reduce the report of the weapon. Alternatively, some cannons use conical blast attenuation devices to project the noise away from the weapon and those firing it. These are geometrically the same as conical flash suppressors.

The removal of noise is important on the battlefield for several reasons. Noise affects communications, is harmful to a soldier's hearing, can reveal position, and makes covert operations difficult. Noise is related to flash and blast, and reducing one of these usually reduces noise as well. The filters used in smoke suppression generally also work well to reduce noise.

In a closed-land vehicle, ship, or aircraft, there is frequently a differential in air pressure between the interior and the exterior environment. When, after a round is fired, the breech of the weapon is opened, there is a tendency for residual propellant gas in the bore to enter the closed fighting compartment. This impairs sight and breathing, and burning particles introduced into the compartment could ignite ready ammunition. Flashback could occur when unreacted propellant gas combines with the air in the compartment similar to the events at the muzzle. The removal of residual propellant gases is a major consideration in the design of the crew compartments of fighting vehicles. In a ship mounting, ventilators are usually installed which mechanically push the muzzle gases out after shot exit. This equipment is rather large and is not practical in a land vehicle or aircraft. We design bore evacuators or bore scavengers to deal with this problem in land vehicles and aircraft.

This method is simply to mount a chamber on the outside of the tube with ports that connect directly into the tube bore. These ports are designed so that they discharge in the direction of the muzzle. When the projectile passes the open ports, gas pressure builds up in the evacuation chamber. Once shot exit occurs, the pressure in the tube eventually drops below the evacuator chamber pressure. When this occurs, the gas trapped in the evacuator rushes out of the muzzle, dragging with it the majority of the residual gas in the tube. This generates a partial vacuum so that when the breech is opened fresh air is pulled in from the compartment. If the breech is not opened for a while after firing, the vacuum dissipates, but by then, the propellant gases should have been removed. These actions are shown in Figures 5.16 through 5.19.

The phenomena of muzzle flows for which the variety of devices we have described are meant to mask or mitigate are complex and are still under active study. We will examine these flows in some detail at this point. We shall step through the muzzle exit process in the order in which the events occur.

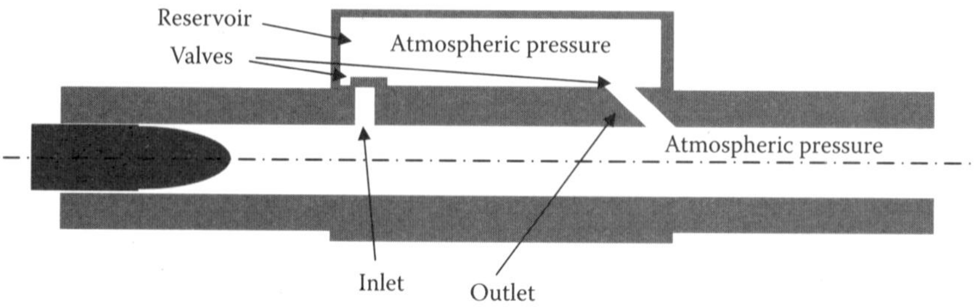

FIGURE 5.16
Projectile approaching bore evacuator.

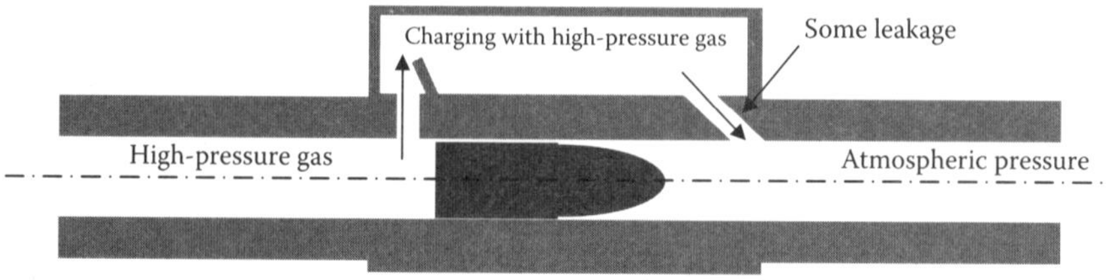

FIGURE 5.17
Bore evacuator charges with gas.

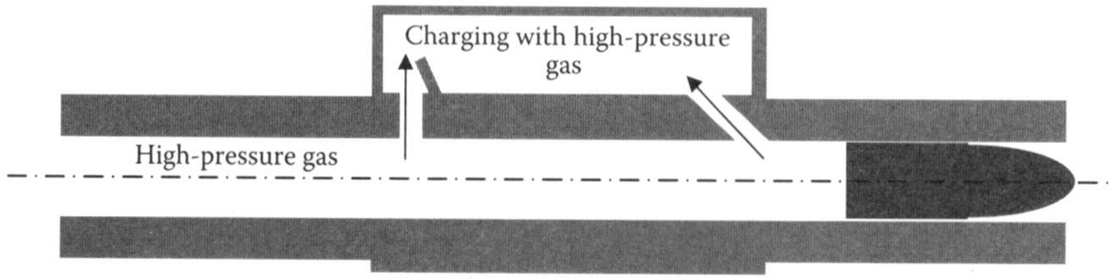

FIGURE 5.18
Bore evacuator still charges with gas.

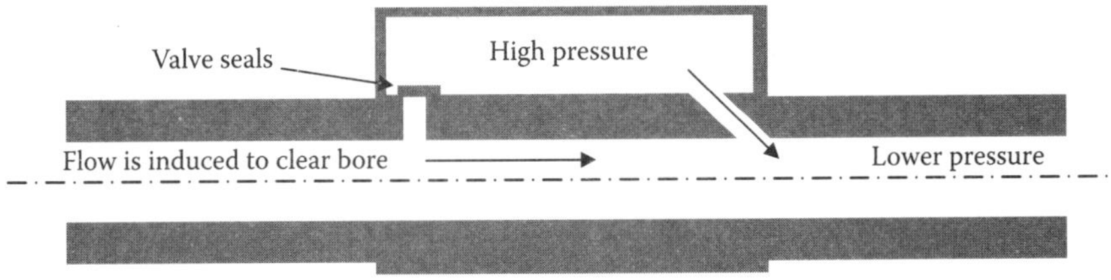

FIGURE 5.19
Projectile has exited; bore evacuator discharges inducing outflow.

As a projectile begins to move down the gun tube, it compresses the air ahead of it. The gun tube acts like a shock tube in which a near-planar shock forms. When this shock exits the muzzle, it forms a spherical shock wave as seen in Figure 5.20.

As the projectile moves faster and faster in the tube, if the velocity is low enough (that is correct, "low enough"), a second precursor will form. This precursor moves faster than the first one because it is moving into the higher-density fluid bounded by the first precursor as seen in Figure 5.21.

No projectile ever perfectly obturates because gun wear occurs; rotating bands and obturators erode; and in high-firing-rate weapons, barrel heating occurs, swelling the bore. As we know, propellants are underoxidized, and because of this, any propellant gas blowby will combine with the oxygen in the precursor flow fields and, when the temperature is high enough, react. Because this occurs before projectile exit, it is known as preflash. It can occur regardless of the presence of the precursors.

Several microseconds after the precursor shock appears, but before the projectile emerges, the so-called barrel shock and Mach cone form. This bottle-shaped structure is referred to as the shock bottle. The barrel shock is created as the higher-pressure gases being compressed by the onrushing projectile attempt to push their way into the high-pressure precursor flow

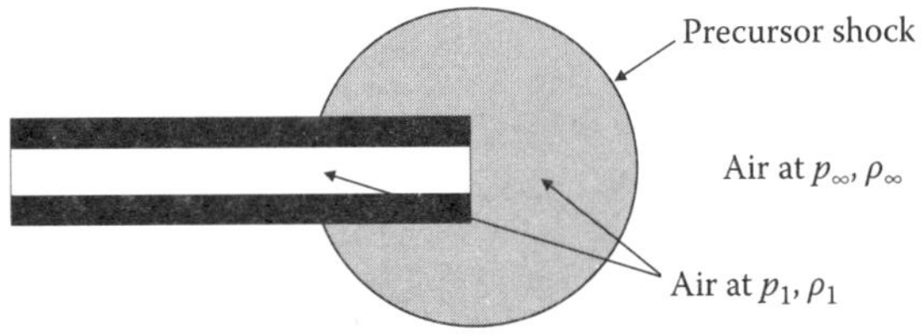

FIGURE 5.20
Precursor shock geometry.

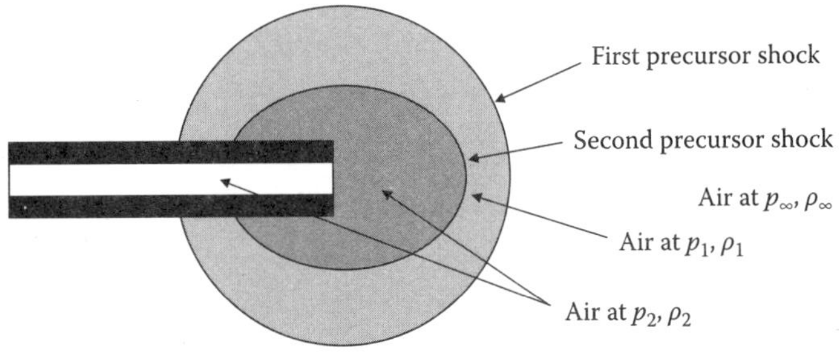

FIGURE 5.21
Second precursor shock formation.

field. One important concept to keep in mind is that pressure acts in all directions—it is a point function. Think of the precursor flow field as "pushing in" on anything that is trying to come out of the muzzle. Thus, the precursor flow field actually constrains the flow exiting the muzzle. The Mach cone is generated by the fact that the fluid jet of the gas ahead of the projectile suddenly sees that there is no more wall constraining it, and it tries to turn the 90° corner but cannot, so an expansion fan forms. This is shown in Figure 5.22.

After the formation of the shock bottle but still before shot ejection, gases are still jetting out of the muzzle. An annular vortex is formed as the gas at the center of the jet continues to rush out while gas near the outer boundary is being robbed of momentum forming a vortex. This is depicted in Figure 5.23. This vortex progresses downrange and will eventually approach the precursor shock.

When the projectile obturator uncorks from the muzzle, there is more room for high-pressure gases to escape. These gases may still be reacting and expand at a rate which

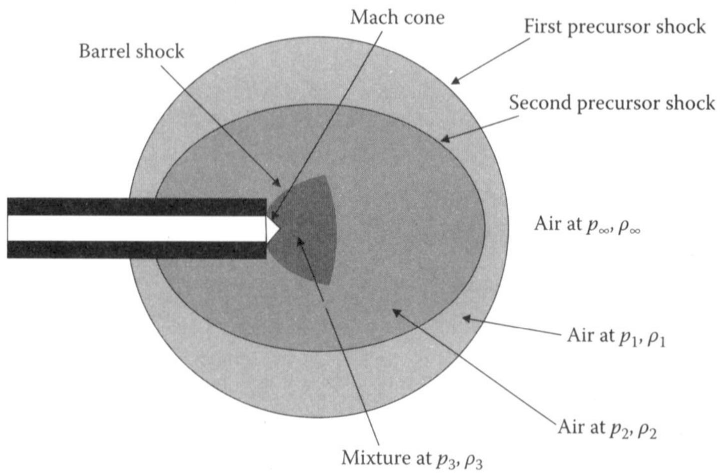

FIGURE 5.22
(See color insert.) Generation of the Mach disk.

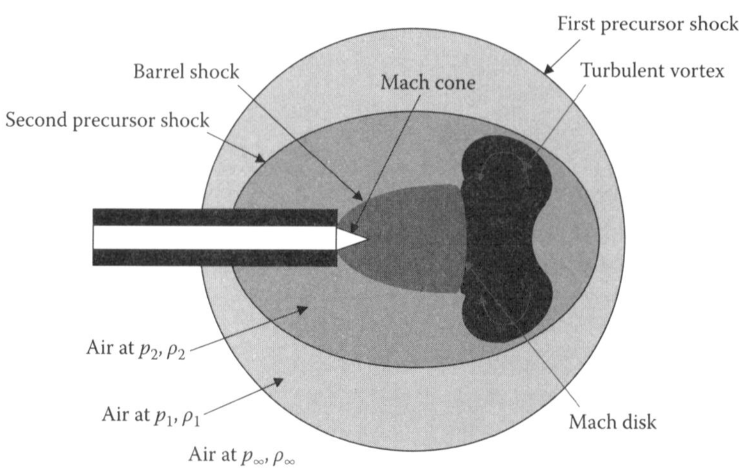

FIGURE 5.23
(See color insert.) Formation of the turbulent vortex.

results in them moving faster than the projectile. In many instances, they are supersonic with respect to the projectile and a base shock forms. The projectile may be flying backward in this flow. This propellant plume is constrained by the precursor flow field and rapidly overtakes it, since it is at a higher temperature and pressure. The result is a bulge of the propellant gases through the precursor shock both preceding and following the projectile. This is depicted in Figure 5.24.

For some time after shot exit, the flow field remains as depicted in Figure 5.24. The turbulent vortex and length of the main propellant flow increase, but the shock bottle remains fairly constant. After this phase, the propellant flow leaving the muzzle diminishes. The Mach disk retreats toward the muzzle and the shock bottle recedes. Upon completion of this process, the situation is reminiscent of effluents from a smokestack as illustrated in Figure 5.25.

Gay [4] describes the influence of the muzzle exit event on the accuracy and general motion of the projectile. This motion can be critical in direct-fire applications.

We have examined the phenomena of muzzle exit flows and the types of muzzle devices commonly used on weapons. The purpose of these devices is to affect the muzzle flow so that certain physical phenomena are altered. Research in this field is still in its infancy, and the literature abounds with theories and simulations.

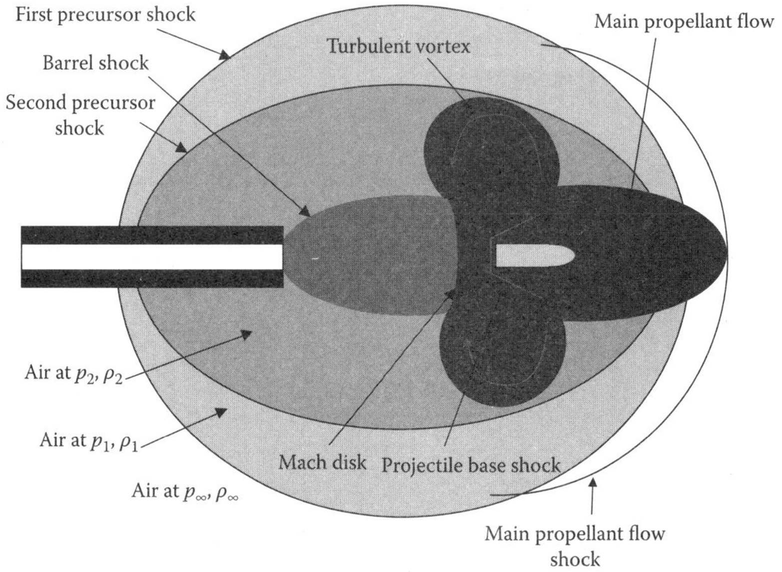

FIGURE 5.24
(See color insert.) Shock structure at shot exit.

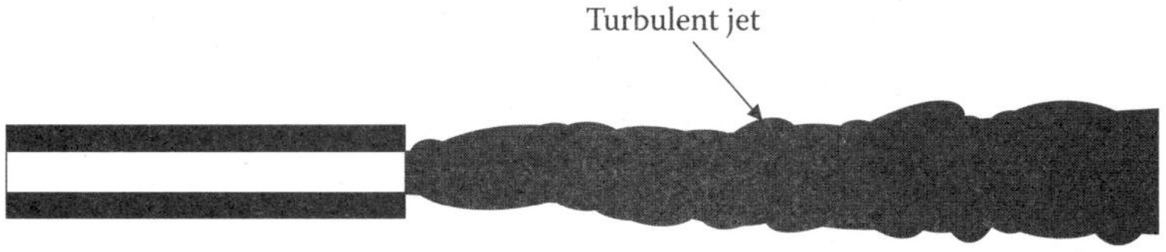

FIGURE 5.25
Turbulent jet formation.

5.5 Gas-Operated Guns

This section on gas-operated guns is located here in order to maintain continuity of the treatment. If the reader is unfamiliar with shock physics, it is recommended that they skip this section for now and proceed to read the other sections of interest, returning once the section on shock physics has been read.

Gas-operated weapons have become increasingly important in the designs of semiautomatic and fully automatic small caliber weapons. Such designs usually incorporate a so-called operating group (OG) that uses the gases generated to propel the projectile, ported through orifices in the bore, to activate the OG components that feed the next round into the chamber of the weapon. Some designs use a "direct impingement" methodology whereby the actual bolt and attachments act as the "OG." From an analysis perspective, they can be viewed using similar equations.

The governing equations, their assumptions and limitations, and the use of the Lagrange approximation to develop the equations of motions of both the projectile, whose propelling gases are being tapped to move the OG, and the OG itself, are the objects of this section. The representative symbols used in this section are described in Figure 5.26.

We will carefully describe the sequence of events during the firing of a gas-operated weapon from the instant of trigger pull of a loaded gun:

1. Propellant ignition and initial burning are constant-volume pressurization processes that occur behind the projectile until enough force is exerted to overcome the initial resistance to motion, called "shot start."
2. The projectile moves down the bore, accelerating as the propellant burns until (in theory) all of it is converted to gas, called "burnout or all-burnt." The equations describing 1 and 2 are well known from our previous work as is the projectile motion to the muzzle after burnout.
3. As the base of the projectile passes the port opening of the OG in the bore wall, gas flows into the chamber behind the piston. The burnout of the charge may not yet

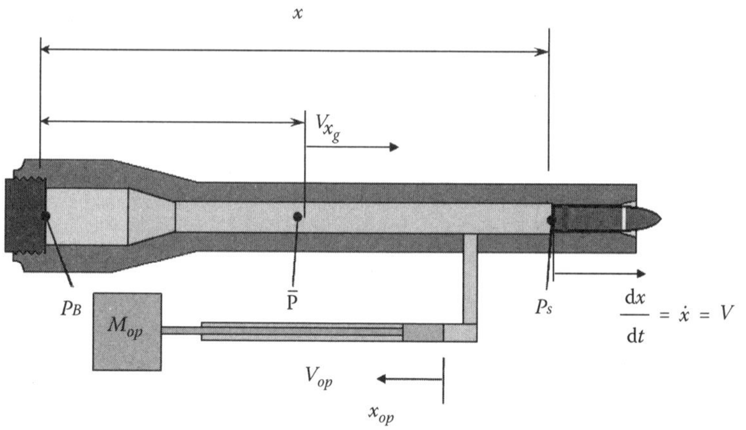

FIGURE 5.26
Depiction of a gas operating system.

have occurred, but the initial flow into the chamber will be modeled as a constant-volume event prior to the OG movement.

4. Gas removal from behind projectile fills the volume of the gas cylinder while the OG movement occurs. This movement is caused by the gas pressure, which is greater than that of a spring holding the piston and operating rod in place.
5. At some point in time, the flow into the port will "choke"; that is, only a limited mass of gas will get through the port opening even though the OG is moving and continually compressing a spring.
6. Now returning our attention to the projectile, it exits the muzzle, and a rarefaction wave passes back up the bore. When it reaches the location of the gas port, the rarefaction wave passes into the gas cylinder of the OG as well as toward the gun breech and eventually reaches the operating rod piston.
7. Once the rarefaction wave reaches the operating rod piston, the pressure drops, and after any inertial effects of the motion cease, the operating rod is accelerated back toward its original location by the compressed spring.
8. When the spring has pushed the piston all the way back to its initial location, the cycle is complete.

Before proceeding with the analysis, it would be profitable at this point to discuss some of the assumptions we are making about the equations of motion of the OG system.

The instant when the projectile passes the port is what we call t_p and the position of the projectile is x_p. At this point, we assume that the density of the gas behind the projectile is uniform or

$$\rho = \rho\left(x_g, t\right) \tag{5.43}$$

This is a bit of a courageous assumption since the gas will have to pass through the orifice, which would act as a nozzle, and the compressible effects will have to be addressed:

$$\rho(t)|_c = \frac{c}{V(t)} \tag{5.44}$$

where the subscript c refers to conditions after the charge has "burned out"; i.e., all the solid has evolved into gas. This volume V is now somewhat more complicated than before because gas will be present both in the bore proper and in the gas cylinder. Physically, the gas cylinder cannot instantaneously fill because of compressibility effects.

Regardless of the geometry of the orifice, it will behave according to the principles of compressible fluid dynamics. This is an important consideration. Once a pressure differential is established across an opening, the flow rate will continue to increase until the centerline velocity of the flow reaches the speed of sound in the gas. At this point, the flow chokes, and the mass flow cannot be exceeded irrespective of how much larger the pressure differential is. When the flow is choked, the Mach number reaches unity and can no longer increase. The good news is that propellant gases are normally at high temperatures, and this increases the speed of sound as we have seen in our earlier work. A high speed of sound implies that the velocity will be high when the Mach number is unity.

If we have a gas port area of A_{port}, then assuming ideal gas behavior (close enough with all our other assumptions), we can write

$$\dot{m}_{\text{choked}} = \left(A_{\text{port}} p_0 \sqrt{RT_0}\right)\gamma^{1/2}\left(\frac{2}{\gamma+1}\right)^{\frac{1(\gamma+1)}{2(\gamma-1)}} \tag{5.45}$$

and

$$\dot{m}_{\text{choked}} = \left(A_{\text{port}} \frac{p_0}{\sqrt{RT_0}}\right)\gamma^{1/2}\left(\frac{2}{\gamma+1}\right)^{\frac{1(\gamma+1)}{2(\gamma-1)}} \tag{5.46}$$

The pressure across the orifice that will cause choking is given by

$$p^* = p_0\left(\frac{2}{\gamma+1}\right)^{\frac{\gamma}{(\gamma-1)}} \tag{5.47}$$

The geometry of the orifice will always act to decrease the port area to some degree due to viscous effects. We handle this through the use of an estimated "discharge" coefficient C_{dscg}, giving us A_{eff} to replace A_{port}, an "effective" port area to account for the reduction in flow. This can be written as

$$A_{\text{eff}} = C_{\text{dscg}} A_{\text{port}} \tag{5.48}$$

This approach is common in fluid mechanics. The discharge coefficient varies a great deal with geometry and is determined either empirically or through computational fluid dynamics. If you have no preconceived value, experience has shown that it is acceptable to start computations with a value of C_{dscg} between 0.60 and 0.62.

To begin a calculation, the pressure distribution is needed along the gun tube from the time the projectile passes the gas port up until the projectile base leaves the muzzle. During this travel, the gas cylinder is being filled. Since the pressure varies over this period, we shall use the average base pressure on the projectile between the instant it passes the gas port and the instant it exits the muzzle. Such an average pressure is

$$\bar{p}_{\text{port}} = \frac{1}{2}\left(p_{\text{port}_{\text{ME}}} + p_{S_{\text{xp}}}\right) \tag{5.49}$$

where $\bar{p}_{\text{port}}$ is the pressure value needed, $p_{\text{port}_{\text{ME}}}$ is the pressure at the port at the moment of muzzle exit, and p_{sxp} is the pressure at the base of the projectile as it passes the port.

Using the Lagrange gradient, we can write this as

$$\bar{p}_{\text{port}} = \frac{1}{2}\left(p_{B_{\text{ME}}}\left(1-\frac{P}{L}\right)\left(1+\frac{L}{\left(1+\frac{c}{2w}\right)}\right) + p_{S_{\text{xp}}}\right) \tag{5.50}$$

As depicted in Figure 5.27, P is the distance from the bolt face to the gas port and L is the total length of travel from bolt face to the muzzle.

With an estimate of the gas pressure filling the gas cylinder, we now need to know how long this pressure is applied to the piston face. We shall assume this time to be the time it takes a rarefaction wave to move from the muzzle to the gas port against the outward gas flow. The reason this time was chosen is because the gas filling the gas port does not

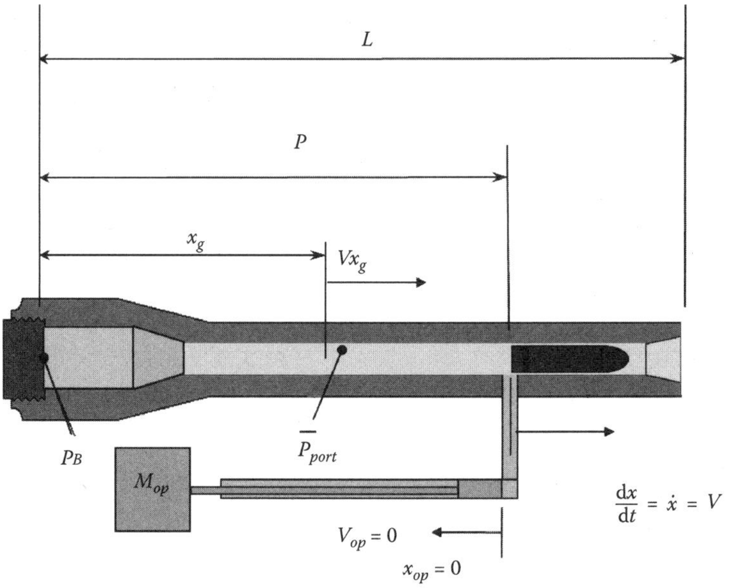

FIGURE 5.27
Depiction of a gas operating system at the instant the projectile passes the gas port.

"know" the pressure should drop until this information is communicated from the projectile exiting the muzzle. Nature handles this through a rarefaction wave.

We have previously determined in our earlier ballistics work that the gas velocity linearly decreases from the projectile velocity at the base of the projectile to zero at the breech or bolt face. So that at an orifice (say, midway down the bore), the gas velocity can be given by

$$\bar{V}_{\mathrm{g}} = \frac{1}{2} V_{\mathrm{muzzle}} \tag{5.51}$$

Usually gas ports are closer to the muzzle, and if higher accuracy is desired, a better estimate can be made based on the Lagrange gradient, but for our purposes, this will suffice.

The rarefaction wave from the muzzle propagates toward the bolt face at the speed of sound in the gas that is flowing toward the muzzle. If we assume the gun gas is an ideal gas, we can write the speed of sound as

$$a = \sqrt{\gamma R T} \tag{5.52}$$

Now this gas temperature has been reduced since the gas was generated at the adiabatic flame temperature. But we can determine an average temperature to use for the speed of sound through Equation 2.41. This temperature will be a little conservative since the actual gas temperature at charge burnout is not the adiabatic flame temperature but something lower. If we tailor Equation 2.41 to this particular situation, we obtain

$$T_{\mathrm{avg}} = T_0 \left(\frac{\bar{p}_{\mathrm{ME}}}{\bar{p}_{\mathrm{c}}} \right)^{\frac{\gamma - 1}{\gamma}} \tag{5.53}$$

When this value of the average temperature is inserted into the equation for the speed of sound (Equation 5.52), we obtain

$$a = \sqrt{\gamma R T_{avg}} \tag{5.54}$$

So the average speed of the rarefaction wave, which is moving against the gas flow, is then

$$V_{wave} = a - \frac{\bar{V}_g}{2} \tag{5.55}$$

At this speed, the rarefaction wave has to cover a distance of $L - P$ to reach the gas port from the muzzle, and the time this takes is

$$t_{gas} = \frac{L - P}{V_{wave}} \tag{5.56}$$

Now let us look at the gas cylinder. The pressure in it will depend on the rate at which gas is entering from the bore. The operating piston, having begun to move, causes the volume in the cylinder to increase. The pressure in the cylinder depends upon the rate at which the volume is increasing and the temperature of the gas in the cylinder. If we momentarily ignore the temperature change, we can write an ideal gas equation of state for the gas pressure in the cylinder as a function of time as

$$p(t) = \frac{m(t) R T_{avg}}{V_{gc}(t)} \tag{5.57}$$

The right-hand side of this equation is for the constant, choked, mass flow rate $m(t)$, so we can write

$$m(t) = m_i + \dot{m}t \tag{5.58}$$

We can tailor Equation 5.46 in terms of the updated variables at the port of pressure $\bar{p}_{port}$, the effective area of the port A_{eff}, and an average temperature T_{avg} to give us

$$\dot{m} = \left(A_{eff} \frac{\bar{p}_{port}}{\sqrt{R T_{avg}}} \right) \gamma^{1/2} \left(\frac{2}{\gamma + 1} \right)^{\frac{1(\gamma+1)}{2(\gamma-1)}} \tag{5.59}$$

If we ignore the air that was originally present in the gas cylinder or $m(0) = 0$, we can write from Equation 5.58 the mass flow $m(t)$ at any time t as

$$m(t) = \left(A_{eff} \frac{\bar{p}_{port} t}{\sqrt{R T_{avg}}} \right) \gamma^{1/2} \left(\frac{2}{\gamma + 1} \right)^{\frac{1(\gamma+1)}{2(\gamma-1)}} \tag{5.60}$$

The cylinder volume at any time t is the initial empty volume plus any additional volume created by the motion of the piston as given by

$$V_{gc}(t) = U_{gc} + A_{gc} x_{op}(t), \tag{5.61}$$

where A_{gc}, U_{gc}, $V_{gc}(t)$, and x_{op} are respectively the cylinder effective cross-sectional area, the initial empty cylinder volume, and the volume of the cylinder increased by the piston motion as it moved along the cylinder over the distance x_{op}. These are depicted at the instant the projectile passes the gas port in Figure 5.28.

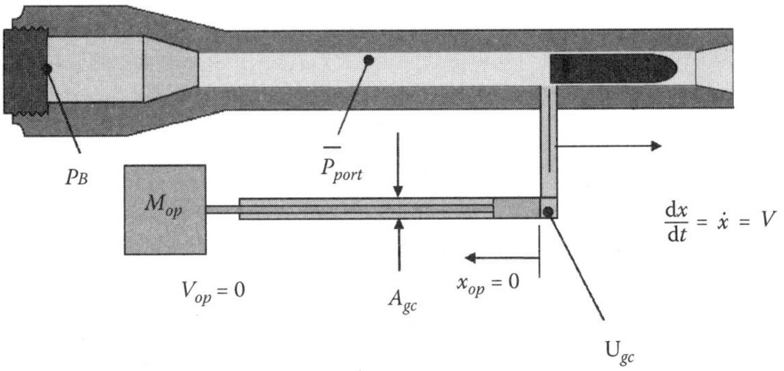

FIGURE 5.28
Cylinder geometry.

By inserting Equations 5.60 and 5.61 into Equation 5.57, we obtain

$$p(t) = \frac{\left(A_{\text{eff}}\overline{p}_{\text{port}}t\sqrt{RT_{\text{avg}}}\right)\gamma^{1/2}\left(\frac{2}{\gamma+1}\right)^{\frac{1(\gamma+1)}{2(\gamma-1)}}}{\text{U}_{\text{gc}} + A_{\text{gc}}x_{\text{op}}(t)} \tag{5.62}$$

which describes the pressure in the cylinder as a function of time. We could now get a more accurate representation of the pressure in the cylinder by letting the temperature vary by assuming adiabatic behavior through

$$T(t) = T_i\left(\frac{\text{U}_{\text{gc}}}{\text{V}_{\text{gc}}(t)}\right)^{\gamma-1} \tag{5.63}$$

Then, the pressure in the cylinder at any time t is given by

$$p(t) = \frac{\left(A_{\text{eff}}\overline{p}_{\text{port}}t\sqrt{RT_{\text{avg}}}\right)\text{U}_{\text{gc}}^{\gamma-1}\gamma^{1/2}\left(\frac{2}{\gamma+1}\right)^{\frac{1(\gamma+1)}{2(\gamma-1)}}}{\left(\text{U}_{\text{gc}} + A_{\text{gc}}x_{\text{op}}(t)\right)^{\gamma}} \tag{5.64}$$

Using these results, let us look at the motion of the OG in the gas cylinder. We will assume that there is some friction between the OG and the gas cylinder (ignoring friction is irrelevant since the equations must be numerically solved in any case, and friction would not overly complicate the work).

We begin by applying Newton's second law:

$$\left(\frac{m_{\text{op}}}{A_{\text{gc}}}\right)\left(\frac{\text{d}^2x_{\text{op}}}{\text{d}t^2}\right) - F_{\text{frict}} = \frac{\left(A_{\text{eff}}\overline{p}_{\text{port}}t\sqrt{RT_{\text{avg}}}\right)\text{U}_{\text{gc}}^{\gamma-1}\gamma^{1/2}\left(\frac{2}{\gamma+1}\right)^{\frac{1(\gamma+1)}{2(\gamma-1)}}}{\left(\text{U}_{\text{gc}} + A_{\text{gc}}x_{\text{op}}(t)\right)^{\gamma}} \tag{5.65}$$

Solving for the acceleration of the OG with the insertion of Equation 5.64 into Equation 5.65 allows us to take advantage of the pressure–time history of the OG;

$$\frac{d^2 x_{op}}{dt^2} = \frac{A_{gc}\left(A_{eff}\bar{p}_{port}t\sqrt{RT_{avg}}\right)U_{gc}^{\gamma-1}\gamma^{1/2}\left(\frac{2}{\gamma+1}\right)^{\frac{1(\gamma+1)}{2(\gamma-1)}}}{m_{op}\left(U_{gc}+A_{gc}x_{op}(t)\right)^{\gamma}} - \frac{F_{frict}}{m_{op}} \tag{5.66}$$

Equation 5.66 is a second-order, nonlinear, ordinary differential equation that must be numerically solved. The friction term can be either a constant, a function of distance, a function of velocity, or a function of both distance and velocity. This term is perhaps the most elusive of all the terms and, depending on the system design, can be very important in accurately predicting the motion.

The energy of the system, which consists of the gas cylinder and the OG (piston, springs, rods, etc.) must be conserved, and we will write a rate equation following Equation 2.48:

$$\dot{Q}_{CV} + \dot{W}_{CV} = \frac{dE_{CV}}{dt} - \dot{m}\left(h_{in} + \frac{1}{2}a^2\right) \tag{5.67}$$

If we assume the only work done is pushing the piston, accelerating the gas, or overcoming friction, we can write

$$\begin{aligned}&\dot{Q}_{CV} + m_{op}V_{op}\frac{dV_{op}}{dt} + m\left(\frac{V_{op}+a}{2}\right)\frac{1}{2}\frac{dV_{op}}{dt} - F_{frict}V_{op}\\&= \frac{d}{dt}\left(mu + \frac{1}{2}m\left(\frac{V_{op}+a}{2}\right)^2\right) - \dot{m}\left(h_1 + \frac{1}{2}a^2\right)\end{aligned} \tag{5.68}$$

If, along with these assumptions, we carry over the ideal gas behavior and use Equation 2.50, we can modify Equation 5.68 to

$$\begin{aligned}&\dot{Q}_{CV} + m_{op}V_{op}\frac{dV_{op}}{dt} + m\left(\frac{V_{op}+a}{4}\right)\frac{dV_{op}}{dt} - F_{frict}V_{op}\\&= \frac{d}{dt}\left(mu + \frac{1}{2}m\left(\frac{V_{op}+a}{2}\right)^2\right) - \dot{m}\left(C_pT_{avg} + \frac{1}{2}a^2\right)\end{aligned} \tag{5.69}$$

Performing the differentiation yields

$$\begin{aligned}&\dot{Q}_{CV} + m_{op}V_{op}\frac{dV_{op}}{dt} + m\left(\frac{V_{op}+a}{4}\right)\frac{dV_{op}}{dt} - F_{frict}V_{op}\\&= \dot{m}u(t) + m\frac{du}{dt} + \frac{1}{8}\dot{m}\left(V_{op}+a\right)^2 + \frac{1}{4}m\left(V_{op}+a\right)\frac{dV_{op}}{dt} - \dot{m}\left(C_pT_{avg} + \frac{1}{2}a^2\right)\end{aligned} \tag{5.70}$$

Finally, assuming calorically perfect behavior, by canceling terms, we can write Equation 5.70 as

$$\dot{Q}_{CV} + m_{op}V_{op}\frac{dV_{op}}{dt} - F_{frict}V_{op} = \dot{m}C_vT + mC_v\frac{dT}{dt} + \frac{1}{8}\dot{m}\left(V_{op}+a\right)^2 - \dot{m}\left(C_pT_{avg} + \frac{1}{2}a^2\right) \tag{5.71}$$

Using Equations 2.55 and 2.56, this can be written as

$$\dot{Q}_{CV} + m_{op}V_{op}\frac{dV_{op}}{dt} - F_{frict}V_{op} = \frac{\dot{m}RT}{\gamma-1} + \frac{mR}{\gamma-1}\frac{dT}{dt} + \frac{1}{8}\dot{m}\left(V_{op}+a\right)^2 - \dot{m}\left(\frac{\gamma RT_{avg}}{\gamma-1} + \frac{1}{2}a^2\right) \tag{5.72}$$

And by using Equations 5.61 and 5.63, we obtain

$$\dot{Q}_{CV} + m_{op}V_{op}\frac{dV_{op}}{dt} - F_{frict}V_{op} = \frac{\dot{m}RT_{avg}U_{gc}^{\gamma-1}}{(\gamma-1)\left(U_{gc}+A_{gc}x_{op}(t)\right)^{\gamma-1}} + \frac{mRT_{avg}U_{gc}^{\gamma-1}(1-\gamma)A_{gc}}{(\gamma-1)\left(U_{gc}+A_{gc}x_{op}(t)\right)^{\gamma}}\frac{dx_{op}}{dt} + \frac{1}{8}\dot{m}\left(V_{op}+a\right)^2 - \dot{m}\left(\frac{\gamma RT_{avg}}{\gamma-1} + \frac{1}{2}a^2\right) \tag{5.73}$$

If we write Equation 5.73 in terms of the displacement of the piston, x_{op}, and by inserting Equation 5.58, we get

$$\dot{Q}_{CV} + m_{op}\frac{dx}{dt}\frac{d^2x}{dt^2} - F_{frict}\frac{dx}{dt} = \frac{\dot{m}RT_{avg}U_{gc}^{\gamma-1}}{(\gamma-1)\left(U_{gc}+A_{gc}x\right)^{\gamma-1}} - \frac{(m_i+\dot{m}t)RT_{avg}U_{gc}^{\gamma-1}A_{gc}}{\left(U_{gc}+A_{gc}x\right)^{\gamma}}\frac{dx}{dt} + \frac{1}{8}\dot{m}\left(\frac{dx}{dt}+a\right)^2 - \dot{m}\left(\frac{\gamma RT_{avg}}{\gamma-1} + \frac{1}{2}a^2\right) \tag{5.74}$$

This we rearrange to facilitate the numerical solution:

$$m_{op}\frac{dx}{dt}\frac{d^2x}{dt^2} = -\dot{Q}_{CV} + F_{frict}\frac{dx}{dt} + \frac{\dot{m}RT_{avg}U_{gc}^{\gamma-1}}{(\gamma-1)\left(U_{gc}+A_{gc}x\right)^{\gamma-1}} - \frac{(m_i+\dot{m}t)RT_{avg}U_{gc}^{\gamma-1}A_{gc}}{\left(U_{gc}+A_{gc}x\right)^{\gamma}}\frac{dx}{dt} + \frac{1}{8}\dot{m}\left(\frac{dx}{dt}+a\right)^2 - \dot{m}\left(\frac{\gamma RT_{avg}}{\gamma-1} + \frac{1}{2}a^2\right) \tag{5.75}$$

and which we regroup as Equation 5.76, noting that the work done by the system is negative in this context:

$$\frac{d^2x}{dt^2} = \left(\frac{1}{m_{op}\dfrac{dx}{dt}}\right)\left\{-\dot{Q}_{CV} + F_{frict}\frac{dx}{dt} + \frac{\dot{m}RT_{avg}U_{gc}^{\gamma-1}}{(\gamma-1)\left(U_{gc}+A_{gc}x\right)^{\gamma-1}}\right.$$
$$\left. - \frac{(m_i+\dot{m}t)RT_{avg}U_{gc}^{\gamma-1}A_{gc}}{\left(U_{gc}+A_{gc}x\right)^{\gamma}}\frac{dx}{dt} + \frac{1}{8}\dot{m}\left(\frac{dx}{dt}+a\right)^2 - \dot{m}\left(\frac{\gamma RT_{avg}}{\gamma-1}+\frac{1}{2}a^2\right)\right\} \quad (5.76)$$

This equation is a nonlinear, second-order differential equation that we must numerically solve. When numerical solutions are compared between Equations 5.66 and 5.76, we get different answers, which is because the energy going into heating the gas actually increases the pressure. We get greater accelerations using Equation 5.76, which is not an issue because the solutions converge for the small times of the action. But without a doubt, Equation 5.76 is more accurate.

We can add a spring into Equation 5.66, where K is the spring constant and l_{und} is the undeformed spring length:

$$\frac{d^2x_{op}}{dt^2} = \frac{A_{gc}\left(A_{eff}\overline{p}_{port}t\sqrt{RT_{avg}}\right)U_{gc}^{\gamma-1}\gamma^{1/2}\left(\dfrac{2}{\gamma+1}\right)^{\frac{1(\gamma+1)}{2(\gamma-1)}}}{\left(m_{op}+\dfrac{m_{spring}}{3}\right)\left(U_{gc}+A_{gc}x_{op}(t)\right)^{\gamma}} - \frac{F_{frict}}{\left(m_{op}+\dfrac{m_{spring}}{3}\right)}$$
$$-\frac{K\left\{x_{op}(t)+l_{und}\right\}}{\left(m_{op}+\dfrac{m_{spring}}{3}\right)} \quad (5.77)$$

And similarly, Equation 5.76 can be modified to incorporate a spring:

$$\left(m_{op}+\frac{m_{spring}}{4}\right)\frac{dx}{dt}\frac{d^2x}{dt^2} = -\dot{Q}_{CV} + F_{frict}\frac{dx}{dt} + \frac{\dot{m}RT_{avg}U_{gc}^{\gamma-1}}{(\gamma-1)\left(U_{gc}+A_{gc}x\right)^{\gamma}} - K\left(x+l_{und}\right)\frac{dx}{dt}$$
$$-\frac{(m_i+\dot{m}t)RT_{avg}U_{gc}^{\gamma-1}A_{gc}}{\left(U_{gc}+A_{gc}x\right)^{\gamma}}\frac{dx}{dt} + \frac{1}{8}\dot{m}\left(\frac{dx}{dt}+a\right)^2 - \dot{m}\left(\frac{\gamma RT_{avg}}{\gamma-1}+\frac{1}{2}a^2\right) \quad (5.78)$$

The time the pressure acts on the piston is of great interest to the gun designer since the rate of fire depends on this time. While we have the interior ballistic tools to exactly calculate how long it takes the projectile to leave the muzzle from when it just passes the orifice, we can estimate this time by

$$t_{me} = \frac{L-P}{V_{muzzle}} \quad (5.79)$$

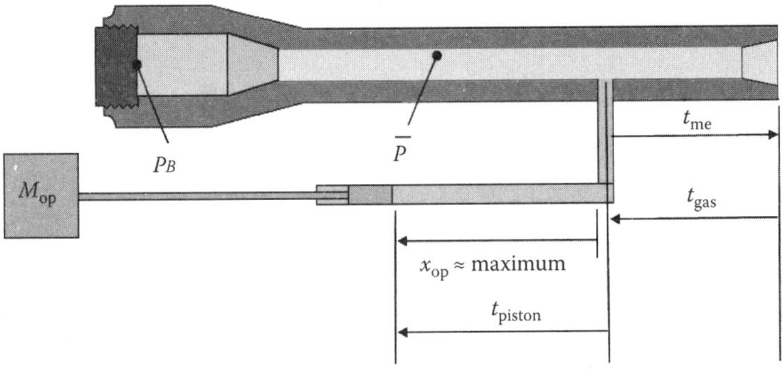

FIGURE 5.29
Timeline superimposed on drawing of the system at the instant the pressure begins to decay.

where t_{me} is the time to muzzle exit. Then, the total time the gas is filling the gas cylinder is given by

$$t_{tot} = t_{me} + t_{gas} \tag{5.80}$$

where t_{tot} is the sum of the time the projectile takes to reach the muzzle from the gas port and the time it takes for the rarefaction wave to cut off flow at the cylinder orifice. If we now use this time to determine the position of the piston and assume the speed of sound in the gas cylinder is far greater than that of the operating rod, we can estimate how long it takes the rarefaction wave to reach the piston. This can estimated through

$$t_{piston} = \frac{x(t_{tot})}{\sqrt{\gamma R T_{avg}}} \tag{5.81}$$

Based on this value, an estimate of when the pressure is no longer acting on the piston can be obtained from

$$t_{end} = t_{tot} + t_{piston} = t_{me} + t_{gas} + t_{piston} \tag{5.82}$$

Although this equation is actually the time the head of the rarefaction wave reaches the piston (which means the pressure is just starting to drop), it is a good approximation for the point at which the spring begins to push the operating rod forward. Knowing that time, the position of the OG can be estimated. This timeline is superimposed on a drawing of the operating rod at the instant the pressure has begun to decay in Figure 5.29.

While some of the equations appear complicated, solutions can be reached through numerical methods which yield details of value to the weapon designer. Details of the design of the individual weapon can greatly affect the solutions. Exact models are possible but involve coupled fluid–solid codes that support kinematics as well.

Gun Dynamics Nomenclature

c	Charge mass
d	Bore diameter
F	Force
I	Impulse
I_{zz}	Polar moment of inertia of projectile
k	Radius of gyration of projectile
M	Momentum
p_B	Breech pressure
p_S	Base pressure on projectile
S	Distance
V	Velocity
w	Projectile mass
w_{recoil}	Mass of recoiling parts
α	Rifling angle
β	Aftereffect coefficient
μ	Coefficient of friction

Subscripts on *F*, *S*, *V*, *M*, and *I*

a	Time when projectile exits the muzzle
n	Aftereffect phase
o	Total distance in tube
P	Projectile base
Pa	Reaction at the base when the projectile exits the muzzle
Pr	Reaction on gun caused by projectile rotating band
Ps	Reaction at the projectile base (essentially the same as P)
R	Reaction at breech
Ra	Reaction at breech until projectile exits muzzle
Re	Reaction at end of recoil
Rn	Reaction at breech from when projectile exits to the end of the aftereffect
t	Tangential direction

References

1. Deutschman, A. D., Michels, W. J., and Wilson, C. E., *Machine Design, Theory, and Practice*, Macmillan., New York, 1975.
2. Norton, R. L., *Machine Design, an Integrated Approach*, 3rd ed., Pearson–Prentice Hall, Upper Saddle River, NJ, 2006.
3. Smith, D. C., and Coppola, E. E., *Safe Maximum Pressure Determination for the M829E3/M256 Cannon Qualification Program, US Army ARDEC* Technical Report ARCCB-TR-02013, Benet Laboratories, Watervliet, New York, September 2002.
4. Gay, H. P., *On the Motion of a Projectile as It Leaves the Muzzle*, Technical Note No. 1425 AD-801–974, USA BRL, Aberdeen Proving Ground, MD, August 1961.
5. Germershausen, R. et al., *Handbook on Weaponry*, 2nd English ed., Rheinmetall, Düsseldorf, 1982.

Further Reading

Klingenberg, G., and Heimerl, J. M., Gun muzzle blast and flash, *Progress in Astronautics and Aeronautics*, Vol. 139, American Institute of Astronautics and Aeronautics, Washington, DC, 1992.

US Army Materiel Command Headquarters, *Gun Tubes, US Army Engineering Design Handbook AMCP-706–252*, US Army Research Office, Washington, DC, February 1964.

US Army Materiel Command Headquarters, *Muzzle Devices, US Army Engineering Design Handbook AMCP-706–251*, US Army Research Office, Washington, DC, May 1968.

6

Recoil Arresting and Recoilless Guns

The recoil momentum endowed during gun launch can be problematic to manage. In this chapter, we consider how recoil is arrested by most mounted cannon and introduce two guns that manage recoil differently.

The baseplate-mounted trench mortar is a very light cannon, limited to high-trajectory fire.

The recoilless gun is the focus of this chapter as it provides an enduring means to enable large caliber guns to be fired from lightweight platforms.

6.1 Recoil Arresting

Most guns employ a recoil arresting system within a gun mount that provides space for the recoil motion of the cannon in the opposite direction to that of the projectile. We may consider a simplified recoil brake beneath the cannon as grounded to an inertial reference frame as shown in Figure 6.1. Like Figure 5.12, this shifts the center of gravity below the centerline of the barrel. As the cannon recoils to the left, the piston of the recoil rod travels through the fixed recoil brake housing. Real gun mounts also include springs to later return the cannon to its firing position that are not shown.

By combining Equations 5.24 and 5.30, we have an expression for the recoil momentum imparted by firing a gun:

$$M_{\text{recoil}} = (w + \beta c)V_{\text{muzzle}} \tag{6.1}$$

This momentum endows the cannon with recoil velocity as we developed in Equation 5.31. This allows us to solve for the kinetic energy of free recoil as

$$\text{KE}_{\text{freerecoil}} = \frac{M^2_{\text{recoil}}}{2w_{\text{recoil}}} \tag{6.2}$$

We can see from this that the mass of the recoiling parts plays a critical role in attenuating the recoil energy. It is useful to consider an idealized recoil arresting force F_{R} as a constant force applied until the recoiling parts are brought to rest after traversing some distance behind the initial firing position. This ideal is approached in practice through the use of variable orifice hydraulic brakes. Examination of Figure 6.1 reveals that the orifice constricts as the cannon recoils further to the left. The increasing constriction tends to increase the force while the slowing of the cannon tends to reduce it.

The total distance is the final recoil stroke S_{Rf}. The purpose of the recoil system is to apply much smaller forces to the firing platform than the peak recoil force delivered by the propellant gases at the breech as quantified in Equation 5.17. As the designed recoil arresting forces are made smaller, the required final recoil stroke becomes larger. As the design recoil strokes become large relative to the displacements at shot exit and after tube

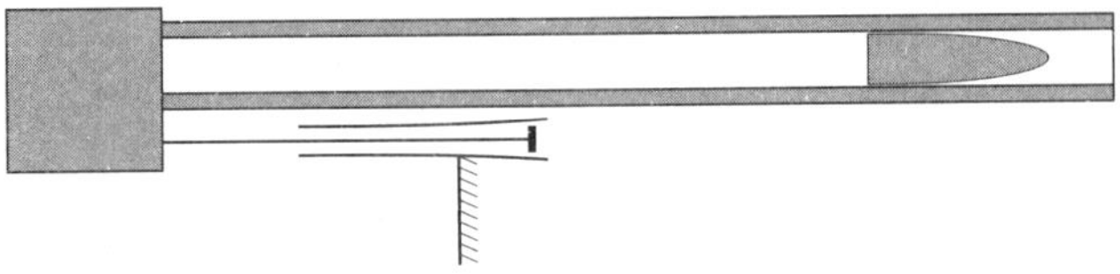

FIGURE 6.1
Notional recoiling cannon.

emptying (Equations 5.37 and 5.42), we may continue with the free recoil approximation. Here, we make the approximation that the entire firing momentum, including the after effect, is instantaneously imparted. We may then analyze relationship between the idealized force and stroke using the free recoil approximation.

$$F_R S_{Rf} = \frac{M_{recoil}^2}{2w_{recoil}} \tag{6.3}$$

We may consider space-limited applications, such as a gun turret, with S_{Rf} fixed. In this case, the required recoil arresting force is inversely proportional to the mass of the recoiling parts. Alternatively, we might consider the case of a carriage that can tolerate only a limited maximum recoil force. Here, the required recoil stroke is inversely proportional to the mass of the recoiling parts. These two scenarios and Equation 6.3 provide a potent disincentive to reducing the mass of gun tubes and breeches.

In real weapons, recoil arresting forces will commence from the outset of the firing, and they are not held constant. This topic of retarded recoil and sizing a variable orifice recoil brake is beyond the scope of the text, but the reader is referred to the text by Hayes [1]. Hayes includes a description of the French 75 mm introduced in 1898, which is widely considered the first modern recoiling artillery cannon and remained in use through World War II.

6.2 Muzzle Brakes

Muzzle brakes are shown in Figure 5.13. These devices divert the flow of propellant from the muzzle to reduce the gas momentum contribution to recoil. A figure of merit is the gas performance index of the muzzle brake Γ. The merit of a muzzle brake must consider the reduction of the total momentum without the muzzle brake M_{recoil} as computed in Equation 6.1 to the level attained with the muzzle brake M_w. Here we assume that the muzzle velocity is left unaffected by the muzzle brake. By Equation 5.24, this means that the change in recoil achieved by the muzzle brake may be solely attributed to changes in the propellant momentum. The performance index is defined as the ratio of the magnitude of the recoil reduction divided by the original propellant momentum without a muzzle brake [2]:

$$\Gamma \equiv \frac{M_{recoil} - M_w}{M_{recoil} - M_{proj}} = \frac{M_{recoil} - M_w}{\beta c V_{muzzle}} \tag{6.4}$$

We may now rearrange and use Equation 6.1 to solve for the total momentum with a muzzle brake as

$$M_w = M_{recoil} - \Gamma \beta c V_{muzzle} = (w + (1 - \Gamma)\beta c) V_{muzzle} \tag{6.5}$$

The case of $\Gamma = 1$ implies that the muzzle brake has eliminated all the gas momentum. A performance index of $\Gamma = 1/2$ would reduce half of the gas momentum. Disadvantages of muzzle brakes with high indexes include elevated blast near the gun crew, the sudden application of large forces to the end of the cannon, and the weight often associated with the structure required to endure these forces. We see from Equation 6.2 that the kinetic energy of recoil increases with the square of total momentum. This amplifies the utility of momentum reduction to ease the burden of recoil arresting.

We may now introduce the two kinds of guns that get around the recoil mass challenge.

6.3 Trench Mortar

The first approach to getting around the recoil mass is commonly employed by trench mortars. These are described in the introduction as man-portable cannon than cannot fire a flat trajectory at all. The reason is simple. Most mortars are fired off baseplates that couple the recoil forces directly to the soil beneath them. A ball socket interface between the breech of the cannon and plate allows traverse and elevation of the cannon. They are limited to fire at elevations no less than 45°. The base plates are typically circular and employ vertical projections or grousers that bite into the soil to better hold the baseplate in place. If the firing elevation were made a little bit too shallow, the baseplate would shift during firing and introduce accuracy errors. If fired far too low, rapid disemplacement of the cannon and baseplate would ensue presenting a hazard.

A notional baseplate-mounted mortar is shown in Figure 6.2. These are generally muzzle loaded weapons and rest on a lightweight bipod support that is not shown.

Mortars are typically low-pressure cannon. The reason can be readily understood by reflecting on the required wall ratio of Equation 5.15. As the operating pressure becomes

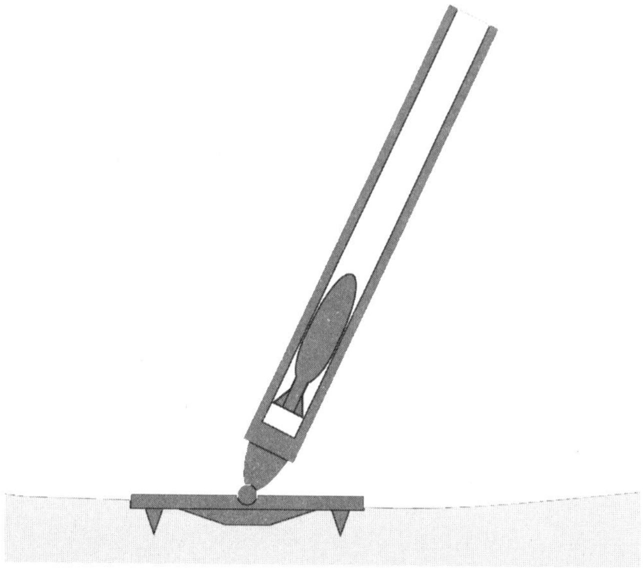

FIGURE 6.2
Baseplate-mounted trench mortar cannon.

small relative to the yield strength of the cannon, the required outer radius becomes close to the inner radius of the tube. As a result, the load sharing becomes more nearly uniform through the thin wall. This nearly uniform stress is more structurally efficient and allows for a lighter cannon. This is an important attribute for a man-portable cannon.

The shells fired by trench mortars are also typically subsonic and intended for relatively short ranges.

6.4 Recoilless Guns

A second way around the recoil mass challenge is to eliminate the recoil. For a brief period following World War I, this was done by ganging two cannons together that shared a common chamber. One cannon would fire the projectile at the intended target. The other cannon sharing a common centerline would fire a countermass projectile in the opposite direction. This is known as a Davis gun in tribute to its inventor, Commander Cleland Davis [3]. The guns were mounted to slow-moving flying boat aircraft and were manned by external gunners in the free airstream at the nose of the aircraft. They could manually point the cannon and fire over the bow of the flying boat at submarines below. Although it is successful at eliminating recoil, the Davis gun is cumbersome to reload, the two cannons are rather heavy, and the firing of two projectiles leads to heavy ammunition.

The Soviets were the first to implement a new kind of recoilless gun. By 1937, they had fully developed no fewer than three systems including 37, 76, and 305 mm guns [4]. The 305 mm gun was impressive in that it delivered a 250 kg shell to a range of 16 km. The first significant use of this technology was by the Germans in their airborne invasion of Crete in 1941. These new guns did not use a countermass projectile to eliminate recoil. Instead, they allowed propellant to escape from the gun chamber through a rearward-directed nozzle. This produces thrust in much the same manner as rocket motor. The configuration is shown in Figure 6.3.

For a given nozzle discharge area, the thrust generated may be balanced to that required to eliminate gun recoil by imposing a flow constriction between the chamber and nozzle. The minimum area of the constriction is known as the throat of the nozzle. A larger passage will allow more propellant to pass through the nozzle increasing thrust. In contrast to the Davis gun, the nozzle provides for very high discharge velocities. This reduces the quantity of countermass that must be exhausted rearward reducing the weight of the ammunition relative to the Davis gun. The nozzles may be achieved with short expansion nozzle cones in place of the heavy second cannon of the Davis gun. This renders the weapon to be compact and lightweight. For the same reasons as the mortar, relatively low-operating pressure is often preferred to allow the mass efficiency of a thin-walled cannon.

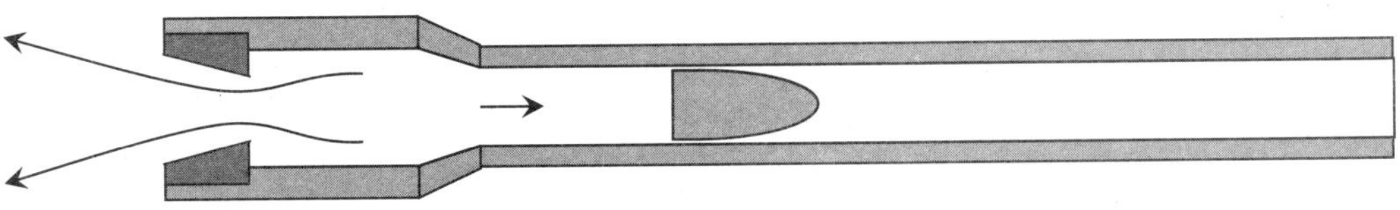

FIGURE 6.3
Recoilless gun with central nozzle through the breech.

The nozzle is a form of a de Laval nozzle, with the convergent portion upstream of the throat and the divergent portion downstream. This convergence and expansion is reflected in the particle path trajectories shown in Figure 6.3. The early Russian and German guns used a cartridge case with a plastic disk that would be unsupported by the closed breechblock with a central hole down the middle. The disk would fulfill its role as a cartridge case describe in Section 4.8 as a container to protect the propellant charge and to prove structurally robust during handing, shipping, stowage, and loading. It would serve as a case seal during the early ignition of the propellant and then blow out of the gun leaving in its absence the throat of the nozzle. This is known as a central nozzle design. Clearly, unburnt solid propellant would tend to leak unimpeded through the nozzle, reducing the efficiency of the system.

An American innovation during World War II was the use of an annular nozzle. Rather than rupture the rear of the case, it was fully supported like a closed breech gun. Much of the span of the camber was larger in diameter than the cartridge case. And the case wall was perforated with a large number of small holes. A thin plastic liner inside the case served to protect the propellant and would tear open early during the ignition of the propellant. The propellant was then free to radially flow through the case wall into the larger chamber. To exit the gun, the propellant flowed through four cut out regions forming the throats of four small nozzles between the threads of the breechblock and cannon. This is shown in Figure 6.4. The net area of all the case wall perforations was kept larger than that of the four throats to prevent large pressure differences from developing that could bulge the cartridge case and hinder extraction.

An advantage of this approach is that propellant gas passes through the perforations more readily than unburned solid propellant. This reduces, but does not eliminate, leakage of solid propellant through the nozzle. A second advantage is that the four narrower nozzle throats allow a shorter expansion section to serve as the nozzle making for a more compact breech interface. A third advantage is that during operation, the chamber volume is significantly larger than that within the cartridge case alone. This allows the loading of a larger propelling charge in the volume of the cartridge without resulting in higher than desired operating pressure. As a result, the ammunition is less bulky for the perforated case design.

Relative to closed breech guns, recoilless guns use much more propellant and generate a new danger area behind the gun. The larger guns have widely been replaced by guided missiles that have more than tripled the accuracy range. The depth of the danger area behind the developmental 120 mm recoilless gun that lost the heavy antiarmor mission to the tube-launched, optically tracked, wire-guided TOW missile reached 80 ft [3]. Firing the launch rocket motor of the TOW missile more than doubled the depth of the danger area beyond 160 ft. We will learn in the next section how to assess the increased propellant consumption. Back blast assessment is beyond the scope of this text. An extensive discussion of the back blast hazard is available in the handbook by the US Army Materiel Command Headquarters [3].

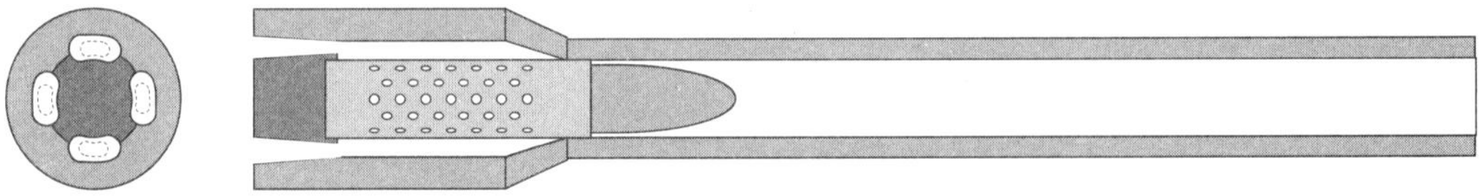

FIGURE 6.4
Recoilless gun with an annular nozzle between the threads of a breechblock.

6.5 Interior Ballistics of Recoilless Guns

The interior ballistic process at work in a recoilless gun may be considered a hybrid between that of a solid propellant rocket motor and a closed breech gun. Zero dimensional models have been developed for recoilless guns [4], although they are subject to greater loss of fidelity than for closed breech guns. We will discuss this fidelity issue later.

Let us simplify the recoilless gun into two separate devices: a closed breech gun firing to the right and a solid propellant rocket directed to the left. If we impose a common area and pressure between the breech of the gun and the chamber of the rocket, we now have a coupled pair of zero-dimensional models of the system. Since the pressure and area are common, the need for a breechblock and end closure to the combustion chamber vanishes, and we may return to the common chamber of the recoilless gun of Figure 6.3. For instructional purposes, it is useful to first consider them as separate devices.

Recoilless operation occurs when the total momentum developed by both devices cancels. By combining Equations 5.24 and 5.30, we already have an expression for the recoil momentum for the gun side model:

$$M_{\text{recoil}} = (w + \beta c) V_{\text{muzzle}} \tag{6.6}$$

We may use any desired gun model to design the propelling charge such as the one introduced in Section 3.2 and develop the required pressure distance curve for the gun, as shown in Figure 6.5.

We may then consider a portion of the same propelling charge granulation to burn within the rocket chamber. The rocket nozzle may now be defined by its throat area and expansion ratio. Typical recoilless gun contractions achieve a throat area of about 30% less than the bore area. The expansion ratio is equal to the ratio of the nozzle exit area to that of its throat. Many recoilless guns have expansion ratios of nearly 2.

For the zero-dimensional rocket model, we could use the adiabatic flame temperature of solid propellant combustion and combine this with the common pressure imposed by the breech pressure of a gun model to define the gas generation rate and reservoir state of the nozzle gases. Then we could apply the isentropic relations of Equation 2.61 and the Lagrangian frame of the first law of thermodynamics of Equation 2.42 to compute gas velocity developed as pressure is lost through the nozzle. We may anticipate a problem with this model. The breech pressure is modestly higher than the average gun pressure as developed in Equation 3.59. As a consequence, the rocket chamber will burnout before the gun, and it will prove impossible to maintain pressure balance after this occurs. We also need an approximation to accommodate the leakage of unburnt propellant.

A less complicated approach for the rocket model is to borrow the concept of effective nozzle exhaust velocity V_{nozzle} from rocket design practice [5]. If we further simplify and consider an average effective exhaust velocity for the duration of the rocket burn, we may treat the developed thrust in exact analogy to how the muzzle gas momentum was treated using a mean exit velocity in Equation 5.27. Using this simplifying approach, the unburned

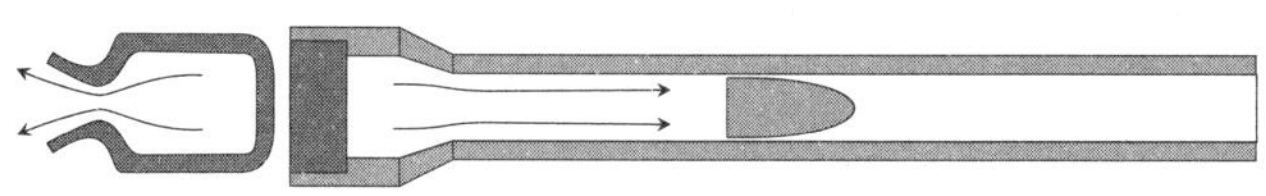

FIGURE 6.5
Zero-dimensional models of a closed breech gun balanced by a rocket motor.

propellant is accounted for. All else the same, the effective exhaust velocity will decline as more propellant is discharged unburnt. The total momentum developed by the propelling charge mass dedicated to the nozzle c_{nozzle}, is

$$M_{nozzle} = c_{nozzle} V_{nozzle} \tag{6.7}$$

Typical solid propellant rockets may have effective exhaust velocities near 8000 ft/s. Close combat missiles alternatives more comparable to recoilless guns tend to be under 6000 ft/s. The rocket community often prefers to divide the effective nozzle velocity by the gravitational acceleration of the earth to define the weight specific impulse:

$$I_{sp} = \frac{V_{nozzle}}{g} \tag{6.8}$$

This results in nearly 250 s for an exhaust velocity of 8000 ft/s.

Three factors reduce nozzle performance for the recoilless gun. First, rockets need to endure only one firing, so they often use high-adiabatic-flame-temperature propellants that are too erosive for guns. This relegates guns to lower-energy and, therefore, lower-effective-exhaust-velocity propellants. Second, recoilless guns leak in the range of one-fifth to one-third of the propellant unburned through the nozzle. Third, the focus on compact designs prefers recoilless expansion nozzles to be short with relatively low expansion. This leaves 5000 ft/s and 155 s to be more representative for the effective nozzle exhaust of recoilless guns. The total recoilless charge may now be estimated to cancel the recoil between the two zero-dimensional models:

$$c_{recoilless} = \left(\frac{V_{muzzle}}{V_{nozzle}}\right) w + c + \left(\frac{V_{muzzle}}{V_{nozzle}}\right) \beta c \tag{6.9}$$

The first term on the right-hand side of Equation 6.9 represents the propellant expelled through the nozzle to counter the recoil of the projectile alone. It is the largest term for low-velocity recoilless guns. The second term is the charge required to drive the projectile down the bore of the gun as well as propelling those gases that follow the projectile down the gun. The third term is the smallest for low-velocity guns. It is the additional propellant propelled through the nozzle to counter the forward momentum of those gases following the projectile down the bore.

In the 1970s, Aviars Celmiņš [6], an interior ballistician in the Ballistic Research Laboratory, developed an early one-dimensional ballistics model for the launch bore of recoilless guns. He relegated zero-order models to be coarse approximations that will be in error on the order of 20% with respect to pressure. This leaves models like the one presented here useful only for quick estimates reinforced by experimental checks. We already identified one problem for such models arising after the burnout of the propelling charge. Celmiņš further showed that because of the proximity of the nozzle to the chamber, the chamber pressure falls more swiftly after burnout than would occur in a closed breech gun. As a consequence, the chamber pressure and density fall resulting in some of the bore gases reversing their forward speed to empty into the chamber. As a result, recoilless guns do not maintain recoil balance throughout the entire launch. During burning earlier in the cycle, they may err to introduce small levels of forward momentum to accommodate the loss of thrust later in the cycle after burnout. Concern over this nonideal behavior should be tempered by the reality that recoilless guns work well in practice over a wide variety of calibers and velocities.

For our purposes, we now know that the aftereffect coefficient of Equation 3.26 is overstated. The recoilless gun problem later will show that this uncertainty leads to small changes in the ballistics. Using $\beta = 1/2$ is likely a reasonable approximation [7].

One means to reduce the imbalanced thrust effect of burnout on recoilless ballistics would be to delay it, perhaps until shot exit. Upon reflection, we may consider that the leakage of unburnt propellant can occur only prior to burnout. Delaying burnout may be anticipated to exacerbate leakage of unburned propellant. Therefore, this effect is likely best left tolerated.

Problem 1

You are asked to analyze the mass of the charge require to fire a 100 mm recoilless system. A 15 lb projectile is to be fired at a muzzle velocity of 1500 ft/s. The propellant properties are listed in the following. Using Equation 3.150, you may assume a typical ballistic efficiency of 33% to estimate *c*, the amount of charge allocated to the gun side of the recoilless gun. You may consider two cases. First, the aftereffect is computed using Equation 5.28. For the second case, use $\beta = 1/2$.

- Specific heat ratio $\gamma = 1.25$
- Propellant force $\lambda = 375{,}000$ ft-lbf/lbm
- Effective nozzle exhaust velocity $V_{\text{nozzle}} = 5000$ ft/s

Answer: $c = 1.06$ lb

$c_{\text{recoilless}} = 5.95$ or 5.72 lb, assuming the full aftereffect increased the charge by only 4%

Problem 2

We want to examine the ballistic efficiency of recoilless guns as a function of velocity. We know that the ballistic efficiency of Equation 3.150 should decline at higher velocities because more kinetic energy must be invested in the propellant gas column to keep up with the shot as quantified in Equation 3.20. Let us suppose we know that the efficiency for the close breech approximation to the gun and propellant of Problem 1 is 33% at 500 m/s and declines to 25% at 1500 m/s. A linear fit between these two points is

$$\varepsilon_{\text{cb}} = 0.37 - 80V/(1{,}000{,}000 \text{ m/s}) \tag{6.10}$$

Using the approximation of Equation 6.9, please plot the ballistic efficiency using Equation 3.150 for the total propelling charge of a recoilless gun between 500 and 1500 m/s. You may use $\beta = 1/2$. What is the highest ballistic efficiency and at what muzzle velocity does it occur?

Answer: The highest efficiency is nearly 10.2% near 4655 ft/s.

References

1. 1. Hayes, T. J., *Elements of Ordnance*, Wiley, Hoboken, NJ, 1938.
2. 2. Germershausen, R. et al., *Handbook on Weaponry*, 2nd English ed., Rheinmetall GmbH, Düsseldorf, 1982.

3. 3. US Army Materiel Command Headquarters, Recoilless rifle weapon systems, *US Army Engineering Design Handbook AMCP-706-238,* US Army Research Office, Durham, NC, January 1976.
4. 4. Tsygankov. I. S., and Sosulin, Y. A., *Orudiye, Minomet, Boyevaya Mashina,* (Gun, Mortar, Combat Equipment), Moscow, 1980.
5. 5. Sutton, G., and Biblarz, O., *Rocket Propulsion Elements,* 8th ed., Wiley, Hoboken, NJ, 2010.
6. 6. Celmiņš, A., Mathematical modeling of recoilless rifle interior ballistics, *Interior Ballistics of Guns,* Edited by Herman Krier and Martin Summerfield, American Institute of Aeronautics and Astronautics, Washington, DC, 1979.
7. 7. Hirschfelder, J. O., Kershner, R. B., Curtiss, C. F., and Johnson, R. E., *Interior Ballistics of Recoilless Guns,* National Defense Research Committee Armor and Ordnance Report A-215, September 1943.

Section II

Exterior Ballistics

7

Introductory Concepts

7.1 Definitions

We will begin with the simplest case, the consideration of the projectile as a point mass flying in a vacuum with only the force due to gravity acting on it. Then, we will proceed to introduce the force due to the pressure of the air, but still considering the projectile as a mass concentrated at a point. Finally, we will consider the projectile as a three-dimensional body acted upon by the air, its spin, and gravity. In the final sections of this part of the chapter, we shall examine the complex motions arising from the coupling of projectile dynamics and aeromechanical forces. Our object will be to examine the conditions necessary for a precise, predictable, satisfactory trajectory enabling the projectile to fulfill its terminal ballistic utility.

Since this text is intended to have a broad scope, some of the material is not derived in detail. The reader is encouraged to seek the more detailed treatments in the references noted throughout each section.

Many of the principles and terms concerned with fluid mechanics required for the understanding of interior ballistics were introduced in Section 2.7. These principles will be extended in this section with a view toward an exterior ballistician—commonly called an aeroballistician.

We shall first examine the elements of a trajectory as depicted in Figure 7.1. These terms are commonly used in the military by gunners and researchers alike. Although most of the symbols and terms in this figure are self-explanatory, some require comment. First is the so-called map range. This is the range to the target that the gunner would see if he or she were to plan firing using a map. The base of the trajectory is quite important and is defined as being level in a plane with the firing point. Gunners of large-caliber weapons and mortars take great care in assuring that the sights on the weapon are leveled in the direction depicted as well as the plane out of the paper.

Since larger ordnance fires over extensive ranges, it is common to assume that the origin of the trajectory is coincident with the ground beneath the artillery piece. The line of site and angle of site (yes, they are spelled that way in much of the literature) are what the gunner uses to aim at the target. As you can see, they only assist in the determination of the pointing of the weapon and the relative height of the target.

An important feature of this diagram is the line of departure. You have probably noticed that it is not collinear with the elevation of the weapon (i.e., where the bore is pointed).

The reality is that a projectile almost never leaves the bore of a gun aligned with the bore—we shall discuss this in detail later. For now, we will simply state that this is due to the dynamics of the projectile and gun as well as aerodynamic effects. It should be noted that Figure 7.1 is drawn as two dimensional. The out-of-plane angular position of the

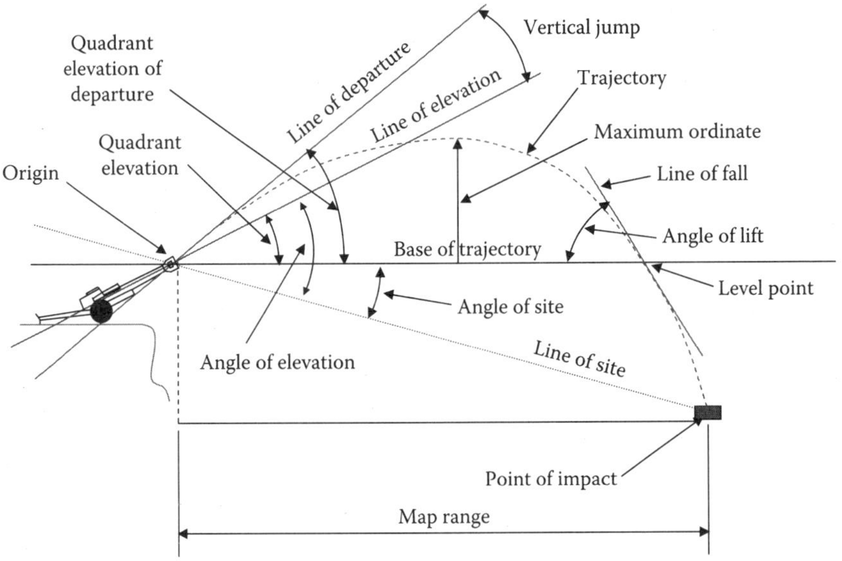

FIGURE 7.1
Elements of a trajectory.

projectile at muzzle exit is known as lateral or azimuthal jump. This will vectorially combine with the vertical jump that is depicted to give a resultant jump vector.

The angle of lift and line of fall are defined for the level point; however, it is common to see these used at the target even though, officially, these quantities at the target are called angle of impact and line of impact (sometimes shot line).

The aerodynamics and ballistics literature are quite diverse, and the terminology is far from consistent. This has particular significance in the coordinate systems used to define the equations of motion. In this text, we shall use the coordinate system by McCoy [1] as depicted in Figure 7.2. The primary difference between this scheme and those of, say, Murphy [2], McShane et al. [3], Nicolaides [4], and Nielsen [5] is that the y-axis is deemed to be positive pointing up, with the z-axis as positive to the right as opposed to the z-axis down and y-axis to the right. This makes sense to the authors with up being a more intuitive positive direction. The only issues (and some people consider them significant) with this scheme are that, first, the nice right-handed naming convention of the aerodynamic

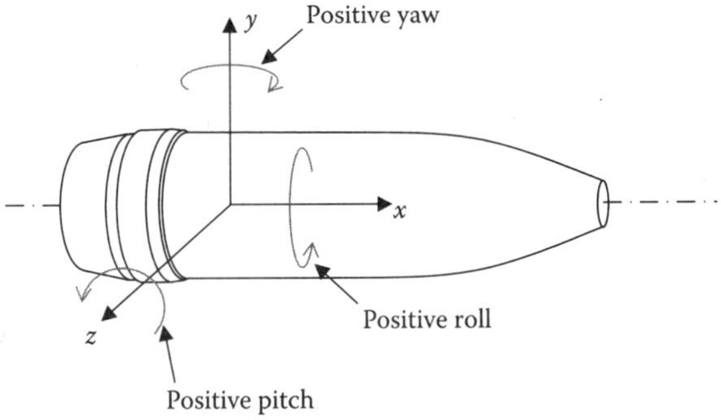

FIGURE 7.2
Definition of projectile coordinates.

coefficients is disturbed (as we shall see later x–y–z corresponds to l–n–m not l–m–n as one would normally like) and, second, what would normally be a positive rotation in the y-direction (i.e., nose left) is defined as negative—we shall handle this when we define the associated equations.

We shall now define some terminology and, more importantly, the forces, moments, and associated coefficients that are used throughout this part of the chapter. It is important for the reader to recognize that these force, moment, and coefficient definitions are by no means an all-inclusive collection. Occurrences of additional forces or moments at times require additional definitions—e.g., control deflections. We shall adhere to the broad scope of this chapter by including only what is necessary for a basic understanding of ballistics.

We mentioned the yaw and pitch of the projectile earlier in this section. The projectile geometry in an arbitrary state of yaw is depicted in Figure 7.3. This illustration shows the projectile yawed and pitched to some angle α_t relative to the velocity vector. The illustration also shows the trajectory which is defined as the curve traced out by the velocity vector. Thus, the velocity vector is everywhere tangent to the trajectory curve. The inset shows the decomposition of the angle between the projectile axis of symmetry $\mathbf{x}(\overline{OB})$ and the velocity vector $\mathbf{V}(\overline{OA})$. We first measure the sideslip angle $\beta(\angle AOC)$ and then measure the yaw angle $\alpha(\angle COB)$ from the axis of symmetry $\mathbf{x}$ to side $\overline{OC} = V\cos\beta$. The side $\overline{BC}$ of right triangle OBC then has a value of $V\cos\beta\sin\alpha$. The resulting angle $\angle AOB$ is defined as the total yaw angle α_t, and, in triangle, AOB, where side $\overline{AB} = V\sin\alpha_t$. It should be noted that triangle ABC with sides $V\sin\alpha_t$, $V\cos\beta\sin\alpha$, and $V\sin\beta$ is also a right triangle.

Most projectiles have at least a trigonal symmetry. This is symmetry about three planes through the projectile long axis, 120° apart. Because of symmetry, it is common to vectorially combine the yaw and pitch of the projectile into one term which we simply call total yaw α_t. All our coefficients will be based on this total yaw. Later, when we discuss advanced topics, it will be necessary to once again separate them.

An examination of Figure 7.3 shows that we can relate the total yaw to α and β through

$$\sin\alpha_t = \sqrt{\sin^2\beta + \cos^2\beta\sin^2\alpha} \tag{7.1}$$

The drag on a projectile is the force exerted on it by the medium through which it is moving, usually air. Since the drag is generated by the motion of the projectile through the air, it is naturally directed opposite to the velocity vector as illustrated in Figure 7.4.

There are, in general, two types of drag: pressure drag and skin friction drag. This is because nature can act on the surface area of the projectile in only two ways: normal to the

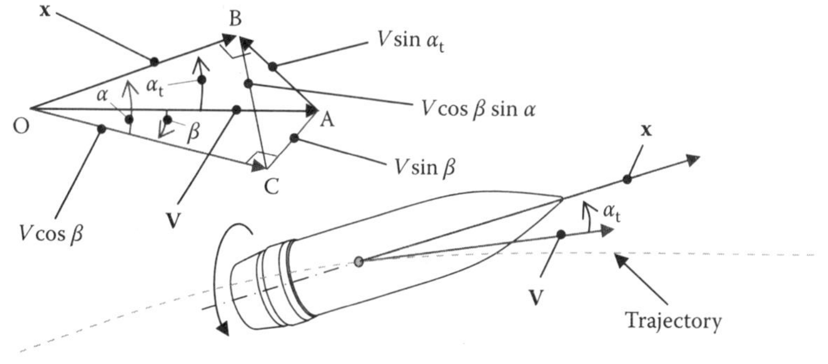

FIGURE 7.3
Generalized yaw of a projectile.

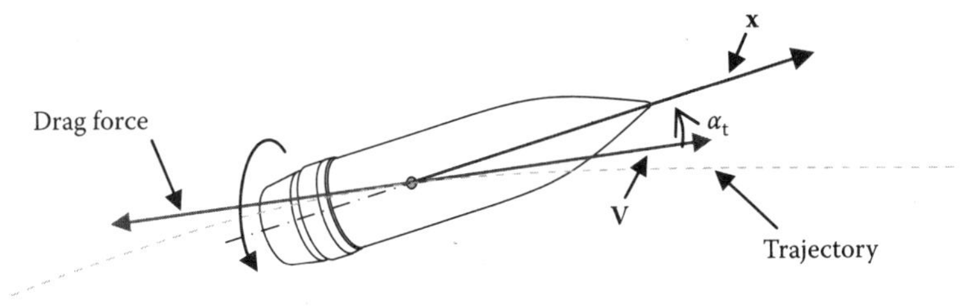

FIGURE 7.4
Drag of a projectile.

surface and along it. A third type of drag called wave drag is a form of pressure drag generated by a shock wave formed when the local velocity along the surface of the projectile reaches Mach 1. We will discuss drag in further detail later, but in all cases, it is convenient to lump the effect of the drag into one coefficient called the drag coefficient. The drag force is defined in terms of this drag coefficient as

$$\text{Drag force} = F_D = \frac{1}{2}\rho S C_D \mathbf{V} V = \frac{1}{2}\rho V^2 S C_D \tag{7.2}$$

Equation 7.2 shows two forms of the defining expression for drag force, vector and scalar. We shall define all our forces and moments in this way because; although we will initially examine the scalar forms, it will be necessary later to use the vector forms. For now, knowing that the drag force is opposite to the velocity vector and that it is a scalar quantity, as depicted on the far right-hand side (RHS) of Equation 7.2, is sufficient.

Like many of the coefficients we shall discuss, the drag coefficient can be a complicated function of the yaw angle. In a more general form, we can write the drag coefficient as the sum of a linear part and a yaw-dependent term:

$$C_D = C_{D_0} + C_{D_{\delta^2}} \delta^2 \tag{7.3}$$

Here δ is the total yaw defined as

$$\delta = \sin \alpha_t \tag{7.4}$$

The first term on the RHS is the linear part of the drag coefficient, known as the zero-yaw drag coefficient, while the second term is known as the yaw drag coefficient. The reason that there is no intermediate term is that for a symmetric body, the drag has to be the same whether the body is angled at, say +5° or −5°. This is discussed more elegantly by Murphy [2].

We shall see later that the drag coefficient varies with Mach number in a complex manner.

Dynamic pressure is a quantity defined as $1/2\rho V^2$, where ρ is the density of a fluid that an object is immersed in and V is the velocity of the fluid relative to the object. It is simply the physical reaction of the fluid when trying to force an object through it and occurs so often that it has been given its own name. This dynamic pressure is multiplied by a reference area S. It is always important to know what reference area is used in the definition of the coefficients. In every case we shall examine, this reference area is based on the projectile circular cross section. Also, as we shall soon see, moments require a length scale as well. In all these instances, we shall use the projectile diameter as the reference length.

When a projectile spins in a medium, the viscous interaction of the medium and the projectile surface is such that the projectile will spin down throughout the flight. This phenomenon is accounted for by a moment applied to the projectile called the spin-damping moment. It is defined as

$$\text{Spin-damping moment} = M_{l_p} = \frac{1}{2}\rho V^2 Sd\left(\frac{pd}{V}\right)C_{l_p} \quad (7.5)$$

This moment is directed opposite to the spin vector **p** of the projectile as depicted in Figure 7.5, and the tendency is for the projectile to spin down; thus, there is no negative sign in Equation 7.5, because the vector handles the decay. One needs to note that the figure is drawn for a right-hand twist. If a left-hand twist were involved, the spin vector **p** and the spin-damping moment vector would be reversed.

Some projectiles have fins or jets which impart a roll torque to the projectile such that the spin rate increases. This rolling moment is depicted in Figure 7.6 and defined through

$$\text{Rolling moment} = M_{l_\delta} = \frac{1}{2}\rho V^2 Sd\delta_F C_{l_\delta} \quad (7.6)$$

In this expression, δ_F is a cant angle provided to the fins to generate the lift required to sustain rotation.

Lift is defined as the aerodynamic force which acts orthogonal to the velocity vector. This is depicted in Figure 7.7. The lift force can be defined in both scalar and vector notations as

$$\text{Lift force} = F_L = \frac{1}{2}\rho SC_{L_\alpha}[\mathbf{V} \times (\mathbf{x} \times \mathbf{V})] = \frac{1}{2}\rho V^2 SC_{L_\alpha}\delta \quad (7.7)$$

The lift force coefficient can be written in its nonlinear form as

$$C_{L_\alpha} = C_{L_{\alpha_0}} + C_{L_{\alpha_2}}\delta^2 \quad (7.8)$$

With a symmetric projectile, we must note that if there is no angle of attack (i.e., $\delta = 0$), then there is no lift. This is obvious even for the linear case since δ appears in Equation 7.7. Some authors prefer to work in coordinates other than those we are utilizing here. In those cases, expressions such as C_x and C_N are used for drag and lift, respectively. In these cases,

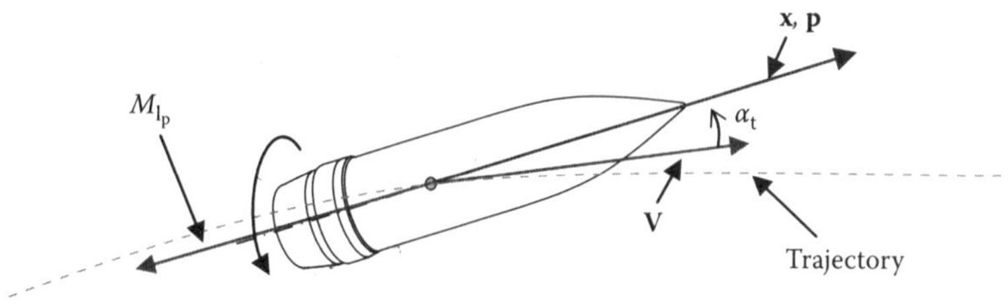

FIGURE 7.5
Spin damping of a projectile.

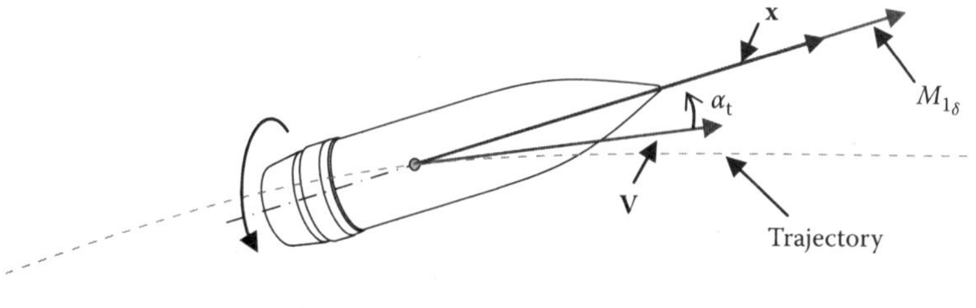

FIGURE 7.6
Roll moment of a projectile.

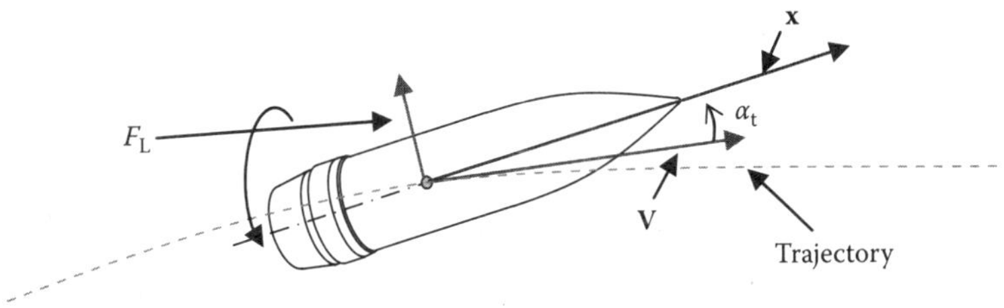

FIGURE 7.7
Lift vector of a projectile.

it is important that proper transformations are used to change the coefficients. An example of this is provided by McCoy [1].

At this point, we must discuss two quantities known as center of pressure (CP) and center of gravity (CG) (sometimes called center of mass). The CG is the location on the projectile where all the mass can be concentrated, so that for an analysis, the gravitational vector will operate at this point. The CP is the point through which a vector can be drawn, i.e., the resultant of all infinitesimal pressure forces acting on the projectile. For most projectiles that are spin-stabilized, the CP is ahead of the CG, and the reverse is true with fin- or drag-stabilized projectiles. Figure 7.8 is an illustration of this.

The separation of the CP and CG gives rise to an overturning moment in all projectiles (Figure 7.9). As we shall see later, this moment is destabilizing for spin-stabilized projectiles (which is why they must be spun) and stabilizing for fin-stabilized projectiles. The overturning moment (sometimes called the pitching moment) is defined as

$$\text{Overturning moment} = M_\alpha = \frac{1}{2}\rho S d V C_{M_\alpha}(\mathbf{V} \times \mathbf{x}) = \frac{1}{2}\rho V^2 S d C_{M_\alpha}\delta \tag{7.9}$$

We can see from Equation 7.9 that this moment is a function of the angle of attack and because of the cross product, a positive overturning moment (nose up) is oriented along the positive z-axis.

The overturning moment coefficient can be written in a nonlinear form similar to the lift and drag forces as

$$C_{M_\alpha} = C_{M_{\alpha_0}} + C_{M_{\alpha_2}}\delta^2 \tag{7.10}$$

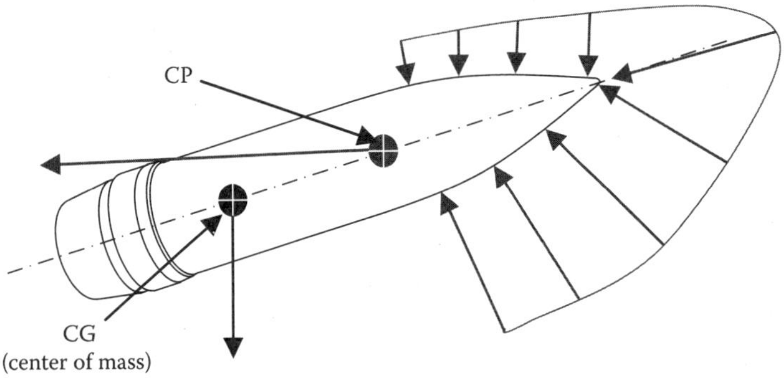

FIGURE 7.8
CG and CP illustrated.

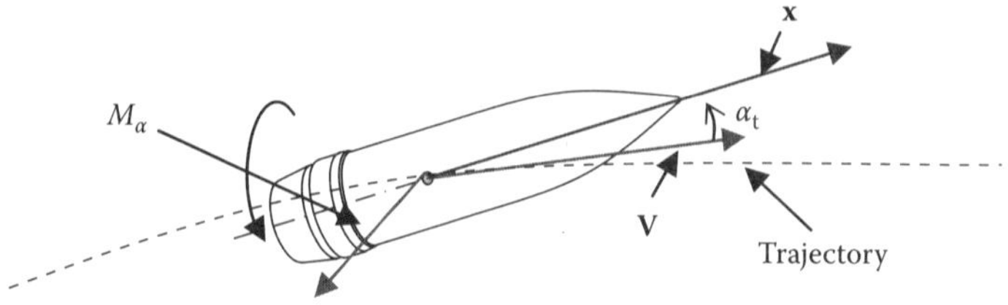

FIGURE 7.9
Overturning moment vector of a projectile.

When a body of circular cross section is immersed in a flow-field perpendicular to its axis and is spun about its axis, a force known as the Magnus force is developed [6]. This force comes about because on one side of the body, the free stream velocity of the flow is added to the velocity of the surface, while on the other side, the free stream velocity is reduced by the surface velocity. On the basis of Bernoulli's equation (Equation 7.11), we see that along the body surface streamline, the pressure must be higher on the side with the lower velocity [7]:

$$\frac{p}{\rho} + \frac{1}{2}V^2 + z = \text{constant} \tag{7.11}$$

This results in a side force on the body as illustrated in Figure 7.10.

This might not seem like a big deal because a projectile almost never flies sideways, but if we consider a projectile in a crosswind or, more importantly, one that is yawed, we see that this side component can somewhat contribute to the aerodynamic loading. For all practical purposes, however, if a projectile is not yawed in flight, then there is no Magnus force. The Magnus force is defined for our purposes as

$$\text{Magnus force} = F_{N_{P\alpha}} = \frac{1}{2}\rho SV\left(\frac{\omega d}{V}\right)C_{N_{P\alpha}}(\mathbf{V}\times\mathbf{x}) = \frac{1}{2}\rho V^2 S\left(\frac{\omega d}{V}\right)C_{N_{P\alpha}}\delta \tag{7.12}$$

The Magnus force coefficient can be written in a nonlinear form in the same manner as Equation 7.10, which we will not repeat (Figure 7.11).

In many cases, the Magnus force is small and is usually neglected with respect to the other forces acting on the projectile. In contrast, the moment developed because of this force is considerable. We define the Magnus moment as

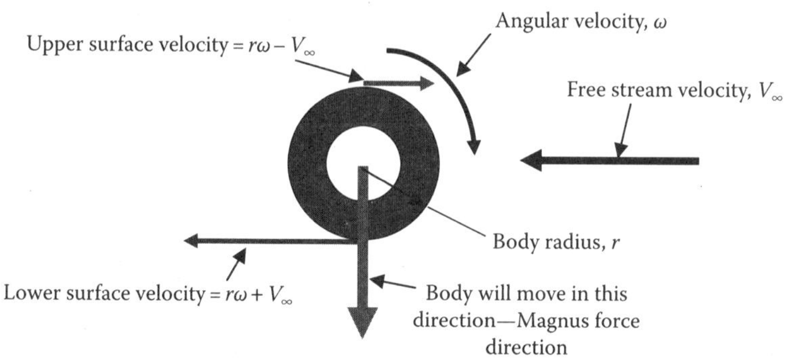

FIGURE 7.10
Magnus effect on a projectile.

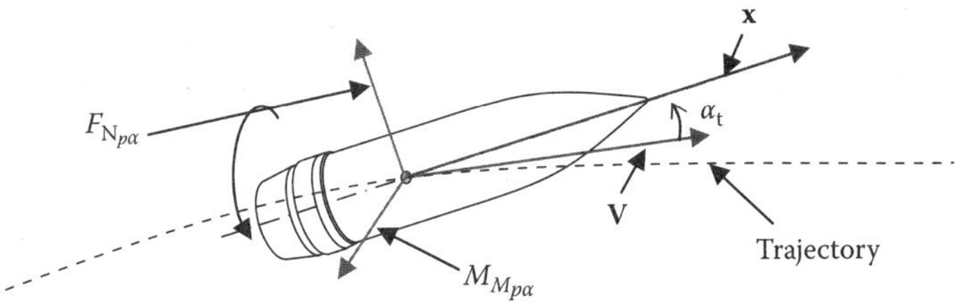

FIGURE 7.11
Magnus force and moment on a projectile.

$$\text{Magnus moment} = M_{M_{P\alpha}} = \frac{1}{2}\rho VSd\left(\frac{\omega d}{V}\right)C_{M_{P\alpha}}\left[\mathbf{x}\times(\mathbf{V}\times\mathbf{x})\right] = \frac{1}{2}\rho V^2Sd\left(\frac{\omega d}{V}\right)C_{M_{P\alpha}}\delta \quad (7.13)$$

The Magnus moment significantly contributes to the stability of the projectile and will be discussed in detail later. The Magnus moment coefficient can be written as a nonlinear term in the same way as all our other coefficients.

The CP for the lift force and the CP for the Magnus force are usually not the same; thus, the moments will act through differing moment arms. The reason for this is the different physics that give rise to the different phenomena. These change during flight as well since the yaw of a projectile changes as it moves downrange.

Pitch damping is the tendency of a projectile to cease its pitching motion due to air resistance. It is usually more difficult to visualize for someone new to the field. It is relatively simple to think about a right circular cylinder mounted in a fixture with its spin axis held by a frictionless bearing on each end. If we spin the projectile, it will slow down because of the sticking of the fluid to the surface and the resultant viscous action (remember the bearings are magically frictionless). If we mount the projectile such that the bearing is transverse to the long axis and spin it, we will still have the viscous action slowing the projectile down; however, this will be overwhelmed by the pressure forces that retard the motion, and the projectile will spin down much faster. This combination of forces is called pitch damping. For projectiles, we can define the pitch-damping force as

$$\text{Pitch-damping forces} = \mathbf{F}_{N_{q+\dot{\alpha}}} = \frac{1}{2}\rho VSd\left(\frac{d\mathbf{x}}{dt}\right)C_{N_q} + \frac{1}{2}\rho VSdC_{N_{\dot{\alpha}}}\left(\frac{d\mathbf{x}}{dt} - \frac{d\mathbf{l}}{dt}\right) \quad (7.14)$$

Or, in scalar terms,

$$\text{Pitch-damping forces} = F_{N_{q+\dot{\alpha}}} = \frac{1}{2}\rho V^2S\left[\left(\frac{q_t d}{V}\right)C_{N_q} + \left(\frac{\dot{\alpha}_t d}{V}\right)C_{N_{\dot{\alpha}}}\right] \quad (7.15)$$

In Equation 7.15, we have defined the total pitching motion q_t and the total rate of change of angle of attack α_t as

$$q_t = \sqrt{q^2 + r^2} \quad \text{and} \quad \dot{\alpha}_t = \frac{d\alpha_t}{dt} \quad (7.16)$$

We note here that this pitch damping comes about through two motions. The first motion is brought about through the pitching rate q, while the second is developed because of the resistance to the changing angle of attack. This is described in eloquent detail by Nielsen [5]. The simplest way of depicting this is to assume a sinusoidal motion of a projectile along its flight path. With this assumption, Figure 7.12 shows what motions would result if only q was present and contrasts this with motion if only $\dot{\alpha}$ were present.

It is generally difficult to separate q and $\dot{\alpha}$ in experimental flight data. For this reason, the two coefficients are almost always written as a sum and recorded in the literature as such. With assumptions on the yawing motion of the projectile and the practice of combining coefficients, as previously described in Equations 7.14 and 7.15, they can be combined as detailed by McCoy [1] into

$$\begin{aligned}\text{Pitch-damping force } &= F_{N_{q+\dot{\alpha}}} = \frac{1}{2}\rho V S d\left(C_{N_q} + C_{N_{\dot{\alpha}}}\right)\frac{d\mathbf{x}}{dt} \\ &= \frac{1}{2}\rho V^2 S d\left(\frac{q_t d}{V}\right)\left(C_{N_q} + C_{N_{\dot{\alpha}}}\right)\end{aligned} \tag{7.17}$$

The pitch-damping force is, like the Magnus force, generally neglected because it is small with respect to the other forces such as lift and drag. The moment caused by this pitch damping is frequently significant (Figure 7.13). It can be mathematically described as follows:

$$\text{Pitch-damping moment} = \mathbf{M}_{M_{q+\dot{\alpha}}} = \frac{1}{2}\rho V S d^2\left(\mathbf{x} \times \frac{d\mathbf{x}}{dt}\right)C_{M_q} + \frac{1}{2}\rho V S d^2 C_{M_{\dot{\alpha}}}\left[\left(\mathbf{x} \times \frac{d\mathbf{x}}{dt}\right) - \left(\mathbf{x} \times \frac{d\mathbf{l}}{dt}\right)\right] \tag{7.18}$$

In scalar form, we can write

$$\text{Pitch-damping moment} = M_{M_{q+\dot{\alpha}}} = \frac{1}{2}\rho V^2 S d\left[\left(\frac{q_t d}{V}\right)C_{M_q} + \left(\frac{\dot{\alpha}_t d}{V}\right)C_{M_{\dot{\alpha}}}\right] \tag{7.19}$$

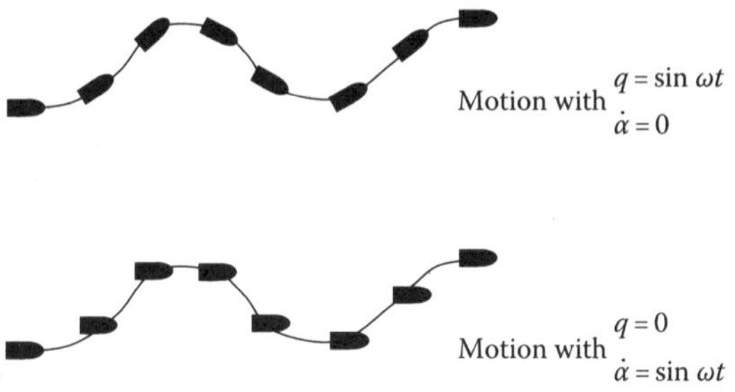

FIGURE 7.12
Pictorial description of q and $\dot{\alpha}$.

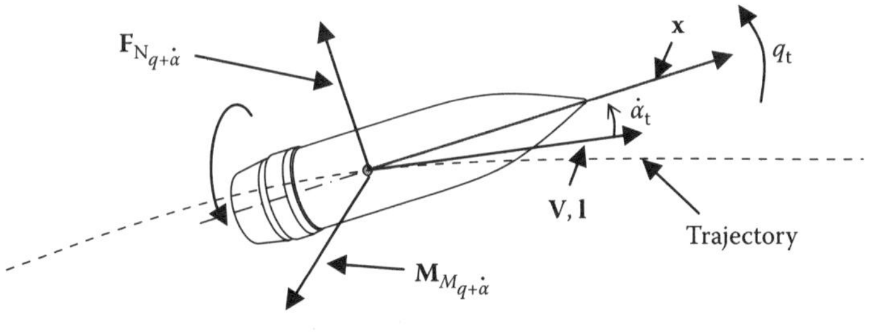

FIGURE 7.13
Pitch-damping force and moment on a projectile.

These can be simplified as per McCoy [1] into

$$\text{Pitch-damping moment} = \mathrm{M}_{M_{q+\dot{\alpha}}} = \frac{1}{2}\rho V S d^2 \left(C_{M_q} + C_{M_{\dot{\alpha}}}\right)\left(\mathbf{x} \times \frac{d\mathbf{x}}{dt}\right) \tag{7.20}$$

$$\text{Pitch-damping moment} = M_{M_{q+\dot{\alpha}}} = \frac{1}{2}\rho V^2 S d\left(\frac{q_t d}{V}\right)\left[C_{M_q} + C_{M_{\dot{\alpha}}}\right] \tag{7.21}$$

At certain times and in some special cases, there are other combinations of forces and moments, and therefore, additional coefficients require attention. We will not go any further here as this chapter is meant to be most general.

We now have the basic terms defined that we shall use in our study of exterior ballistics.

7.2 Development of Total Yaw Angle from Orthogonal Measurements

In this section, we shall discuss how to obtain total yaw angle from two orthogonal measurements obtained in a range firing. We shall assume that the cameras are perfectly orthogonal. Any errors in this assumption have to be accounted for in the data reduction. This procedure will not be covered because it will differ based on test configurations.

Assume that, through whatever data collection, we have been able to determine the two vectors **V** and **x** as depicted in Figure 7.3. We shall assume a test range inertial coordinate system similar to that used in Figure 7.2, where **X** points downrange, **Y** is up, and **Z** is to the right as viewed from the rear of the weapon. This is depicted in Figure 7.14.

At this point, if we have the vectors determined in some manner, we can simply determine the total yaw angle directly from

$$\alpha_t = \cos^{-1}\left(\frac{\mathbf{x} \cdot \mathbf{V}}{|\mathbf{x}||\mathbf{V}|}\right) \tag{7.22}$$

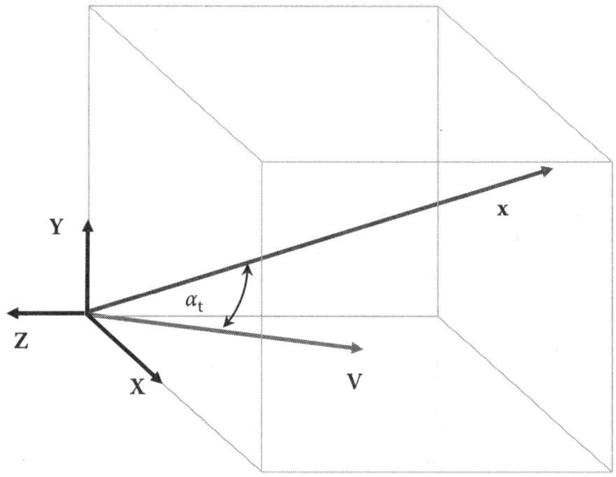

FIGURE 7.14
Depiction of initial data from test range.

We can take this further, determining the pitch and yaw angles, α and β, respectively as depicted in Figure 7.3, through the following procedure. First, we need to rotate both the velocity vector and the position vector through an angle such that the velocity vector is aligned with the **X**-axis. To begin, let us write the velocity and position vectors in terms of their components in the **X**, **Y**, **Z** system:

$$\mathbf{V} = V_X\mathbf{i} + V_Y\mathbf{j} + V_Z\mathbf{k} \tag{7.23}$$

$$\mathbf{x} = x_X\mathbf{i} + x_Y\mathbf{j} + x_Z\mathbf{k} \tag{7.24}$$

We shall now define the projection of the vectors (which are still vectors) by using subscripts for the plane in which they lie. For instance, $\mathbf{V}_{XY}$ is the projection of the velocity vector **V** into the **X–Y** plane. This is illustrated in Figure 7.15. We shall call this angle γ_{YX}, and it is defined as

$$\gamma_{YX} = \sin^{-1}\left(\frac{V_Z}{|\mathbf{V}|}\right) = \sin^{-1}\left(\frac{V_Z}{\sqrt{V_X^2 + V_Y^2 + V_Z^2}}\right) \tag{7.25}$$

Similarly, we can define γ_{ZX} as follows

$$\gamma_{ZX} = \cos^{-1}\left(\frac{V_Y}{|\mathbf{V}|}\right) = \cos^{-1}\left(\frac{V_Y}{\sqrt{V_X^2 + V_Y^2 + V_Z^2}}\right) \tag{7.26}$$

This is illustrated in Figure 7.16.

It is important to perform the rotation about the **Y**-axis first as it preserves the angle between **V** and the **X–Z** plane. The proper rotation, however, is determined by the angle we shall call δ_{ZX} as illustrated in Figure 7.17.

We can write this angle as

$$\delta_{ZX} = \tan^{-1}\left(\frac{V_Z}{V_X}\right) \tag{7.27}$$

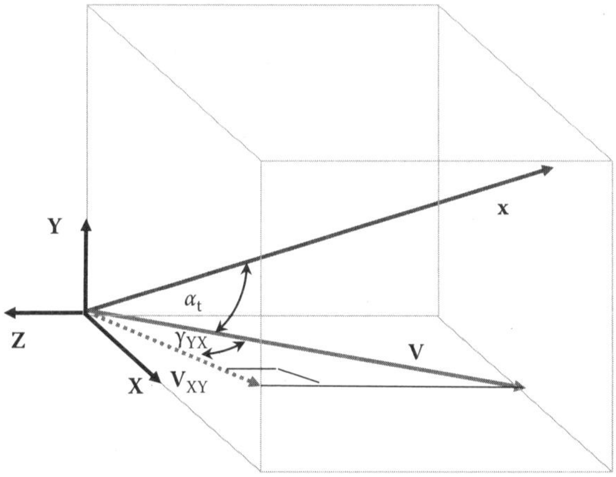

FIGURE 7.15
Example of a projection of **V** onto the **X–Y** plane.

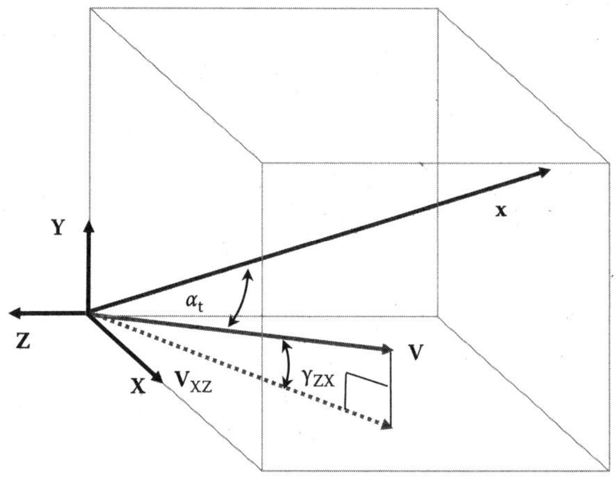

FIGURE 7.16
Example of a projection of **V** onto the **X–Z** plane.

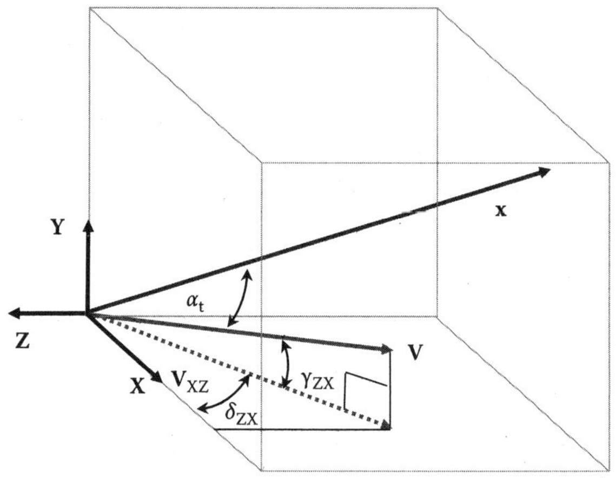

FIGURE 7.17
Rotation of **V** about the **Y**-axis.

Rotating **V** through this angle can be written as

$$\mathbf{V}' = \begin{bmatrix} \cos\delta_{ZX} & 0 & \sin\delta_{ZX} \\ 0 & 1 & 0 \\ -\sin\delta_{ZX} & 0 & \cos\delta_{ZX} \end{bmatrix} \begin{bmatrix} V_X \\ V_Y \\ V_Z \end{bmatrix} \tag{7.28}$$

We now follow the same procedure to rotate about the **Z**-axis but this time using **V**′ as the vector:

$$\mathbf{V}'' = \begin{bmatrix} \cos\delta_{ZX} & \sin\delta_{ZX} & 0 \\ -\sin\delta_{ZX} & \cos\delta_{ZX} & 0 \\ 0 & 0 & 1 \end{bmatrix} \begin{bmatrix} V'_X \\ V'_Y \\ V'_Z \end{bmatrix} \tag{7.29}$$

Now our velocity vector will be aligned with the **X**-axis, and if we perform the exact same rotation on the projectile axis vector, we will have preserved the total yaw angle, and the situation will be as illustrated in Figure 7.18. The equation is

$$\mathbf{x}'' = \begin{bmatrix} \cos\delta_{ZX} & \sin\delta_{ZX} & 0 \\ -\sin\delta_{ZX} & \cos\delta_{ZX} & 0 \\ 0 & 0 & 1 \end{bmatrix} \begin{bmatrix} \cos\delta_{ZX} & 0 & \sin\delta_{ZX} \\ 0 & 1 & 0 \\ -\sin\delta_{ZX} & 0 & \cos\delta_{ZX} \end{bmatrix} \begin{bmatrix} x_X \\ x_Y \\ x_Z \end{bmatrix} \tag{7.30}$$

So we can now extract α and β from these equations:

$$\alpha = \sin^{-1}\left(\frac{x''_Y}{|\mathbf{x}''|}\right) = \sin^{-1}\left(\frac{x''_Y}{\sqrt{x''^2_X + x''^2_Y + x''^2_Z}}\right) \tag{7.31}$$

$$\beta = \tan^{-1}\left(\frac{x''_Z}{x''_X}\right) \tag{7.32}$$

These equations will be consistent with Equation 7.1.

Problem 1

In a test range, a 0.50-caliber M33 ball projectile has a velocity vector which is at an angle of 10° to the horizontal (assume zero azimuth) with a velocity of 3013 ft/s. The initial pitch and yaw angles are 1.030° and 1.263°, respectively. The initial rotational rate of change of the axial unit vector ($d\mathbf{x}/dt$) is provided in the following. If the projectile has the following coefficients at this particular instant, draw the situation and determine the following:

1. Velocity vector (ft/s)
2. Projectile axial unit vector (**x**)
3. Drag force vector (lbf)
4. Spin-damping moment vector (lbf-in)
5. Overturning moment vector (lbf-in)
6. Magnus moment vector (lbf-in)

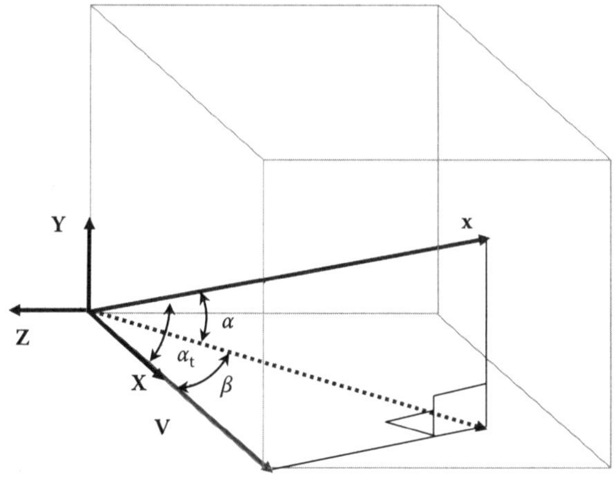

FIGURE 7.18
Completed rotation of both **V** and **x** about the **Y**- and **Z**-axes.

Assume the weapon has a right-hand twist.
Projectile information:

$$
\begin{array}{llll}
C_D = 0.2938 & \left(C_{M_q} + C_{M_{\dot{\alpha}}}\right) = -5.5 & I_P = 7.85\left[g-cm^2\right] \\
C_{M_\alpha} = 2.88 & \left(C_{N_q} + C_{N_{\dot{\alpha}}}\right) = 0.004 & I_T = 74.5\left[g-cm^2\right] \\
C_{L_\alpha} = 2.69 & C_{M_{p\alpha}} = 0.05 & m = 42.02\left[g\right] \\
C_{N_{p\alpha}} = -0.01 & & \omega = 15,404\left[\dfrac{rad}{s}\right] \\
C_{l_p} = 0.003 & \rho = 0.0751\left[\dfrac{lbm}{ft^3}\right] &
\end{array}
$$

$$
\frac{d\mathbf{x}}{dt} = \{0.405\mathbf{e}_1 - 1.963\mathbf{e}_2 - 0.981\mathbf{e}_3\}\left[\frac{rad}{s}\right]
$$

Please supply all answers in an inertial coordinate system labeled 1, 2, and 3 with 1 being along the downrange direction and 3 being to the right side. Treat all missing coefficients as equal to zero. It is very important that you *draw* the situation. This will have a great deal of influence in obtaining the correct answer.

References

1. McCoy, R. L., *Modern Exterior Ballistics*, Schiffer Military History, Atglen, PA, 1999.
2. Murphy, C. H., *Free Flight Motion of Symmetric Missiles*, Ballistic Research Laboratory Report No. 1216, Ballistic Research Laboratory, Aberdeen Proving Ground, MD, 1963.
3. McShane, E. J., Kelley, J. L., and Reno, F. V., *Exterior Ballistics*, University of Denver Press, Denver, CO, 1953.
4. Nicolaides, J. D., *On the Free Flight Motion of Missiles Having Slight Configurational Asymmetries*, Ballistic Research Laboratory Report No. 858, Ballistic Research Laboratory, Aberdeen Proving Ground, MD, 1953.
5. Nielsen, J. N., *Missile Aerodynamics*, AIAA reprint, American Institute of Aeronautics and Astronautics, Reston, VA, 1988.
6. White, F. M., *Fluid Mechanics*, 5th ed., McGraw-Hill, New York, 2003.
7. Fox, R. W. and McDonald, A. T., *Introduction to Fluid Mechanics*, 4th ed., John Wiley & Sons, New York, 1992.

Further Reading

Bull, G. V., and Murphy, C. H., *Paris Kanonen—The Paris Guns (Wilhelmgeschutze) and Project HARP*, Verlag E.S. Mittler & Sohn, Herford und Bonn, 1988.

8

Dynamics Review

Throughout the study of exterior ballistics, dynamics play a great role in the flight of the projectile. The Coriolis effect in long-range trajectories or drag changes due to the precessional and nutational motion of the projectile are just two examples of the effect of projectile body dynamics on flight. We will find that at least a cursory review of dynamics is essential to the understanding of projectile motion. Analyzing dynamics of projectile flight is best approached through the use of vectors, and we will begin our review with their study.

A vector is defined as a quantity with a magnitude and a direction. Two vectors are considered equal if both their magnitude and direction are identical. However, this does not mean that they have to originate at the same point; i.e., a translation has no effect on whether vectors are equal. A scalar is simply a numerical quantity (a magnitude). When a scalar and a vector are multiplied (in any order), they form a vector. Thus, we can define any vector as a scalar magnitude multiplied by a vector of unit length (a unit vector) in the proper direction (Figure 8.1):

$$\mathbf{A} = A\mathbf{e}_A \tag{8.1}$$

A vector can be written as the sum of its scalar magnitude in each individual coordinate direction times a unit vector in that particular direction:

$$\mathbf{A} = A_x\mathbf{i} + A_y\mathbf{j} + A_z\mathbf{k} \tag{8.2}$$

The magnitude of the vector is defined as

$$A = |\mathbf{A}| = \sqrt{A_x^2 + A_y^2 + A_z^2} \tag{8.3}$$

Vectors may be added together in any order by summing the individual components in each direction. This is the commutative property:

$$\mathbf{A} + \mathbf{B} = \mathbf{B} + \mathbf{A} = (A_x + B_x)\mathbf{i} + \left(A_y + B_y\right)\mathbf{j} + (A_z + B_z)\mathbf{k} \tag{8.4}$$

The following is also true when adding more than one vector together:

$$(\mathbf{A} + \mathbf{B}) + \mathbf{C} = \mathbf{A} + (\mathbf{B} + \mathbf{C}) \tag{8.5}$$

Equation 8.5 represents the associative property of vectors. In all of the preceding expressions, note that **i**, **j**, and **k** are the unit vectors in the x-, y-, and z-coordinate directions, respectively.

Multiplication of vectors can occur in two different ways—each applicable to particular situations. Consider the two vectors **A** and **B** shown in Figure 8.2; we define the scalar product or dot product as

$$\mathbf{A} \cdot \mathbf{B} = A \cdot B \cdot \cos\theta \tag{8.6}$$

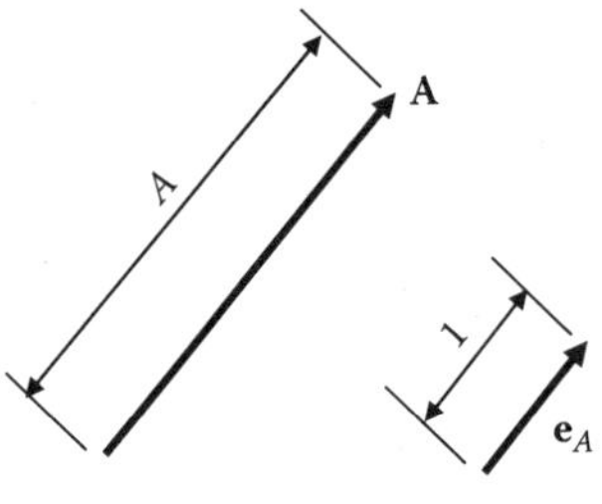

FIGURE 8.1
Vector and associated unit vector.

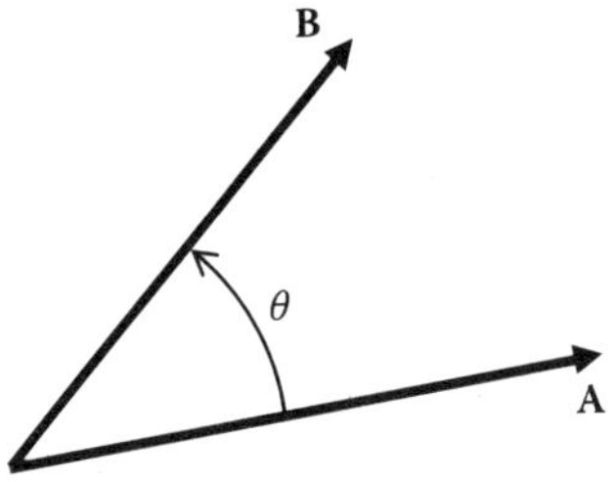

FIGURE 8.2
Vector pair illustrated.

Both the commutative and associative laws of multiplication apply to the dot product:

$$\mathbf{A} \cdot \mathbf{B} = \mathbf{B} \cdot \mathbf{A} \tag{8.7}$$

$$(\mathbf{A} + \mathbf{B}) \cdot \mathbf{C} = \mathbf{A} \cdot \mathbf{C} + \mathbf{B} \cdot \mathbf{C} \tag{8.8}$$

The dot product of two vectors is given by

$$\mathbf{A} \cdot \mathbf{B} = \left(A_x\mathbf{i} + A_y\mathbf{j} + A_z\mathbf{k}\right) \cdot \left(B_x\mathbf{i} + B_y\mathbf{j} + B_z\mathbf{k}\right) \tag{8.9}$$

This equation when expanded is

$$\mathbf{A} \cdot \mathbf{B} = A_xB_x\mathbf{i} \cdot \mathbf{i} + A_xB_y\mathbf{i} \cdot \mathbf{j} + A_xB_z\mathbf{i} \cdot \mathbf{k} + A_yB_x\mathbf{j} \cdot \mathbf{i} + A_yB_y\mathbf{j} \cdot \mathbf{j} + A_yB_z\mathbf{j} \cdot \mathbf{k} + A_zB_x\mathbf{k} \cdot \mathbf{i} + A_zB_y\mathbf{k} \cdot \mathbf{j} + A_zB_z\mathbf{k} \cdot \mathbf{k} \tag{8.10}$$

However, since the unit vectors are orthogonal and the dot product of two orthogonal vectors is identically zero, while the dot product of parallel vectors is unity as follows from

$$\mathbf{i} \cdot \mathbf{i} = \mathbf{j} \cdot \mathbf{j} = \mathbf{k} \cdot \mathbf{k} = 1 \cdot 1 \cdot \cos(0^\circ) = 1$$

$$\mathbf{i} \cdot \mathbf{j} = \mathbf{j} \cdot \mathbf{i} = \mathbf{j} \cdot \mathbf{k} = \mathbf{k} \cdot \mathbf{j} = \mathbf{i} \cdot \mathbf{k} = \mathbf{k} \cdot \mathbf{i} = 1 \cdot 1 \cdot \cos(90^\circ) = 0$$

Therefore, we can write

$$\mathbf{A} \cdot \mathbf{B} = A_xB_x + A_yB_y + A_zB_z \tag{8.11}$$

The second type of vector multiplication is the vector or cross product, which is defined as

$$\mathbf{A} \times \mathbf{B} = A \cdot B \cdot \sin\theta\mathbf{e}_n \tag{8.12}$$

Here $\mathbf{e}_n$ is a unit vector normal to the plane made by vectors **A** and **B**. This is depicted in Figure 8.3. The cross product does not obey the commutative property because

$$\mathbf{A} \times \mathbf{B} = -\mathbf{B} \times \mathbf{A} \tag{8.13}$$

The distributive property, however, does apply to the cross product. Thus,

$$(\mathbf{A} + \mathbf{B}) \times \mathbf{C} = \mathbf{A} \times \mathbf{C} + \mathbf{B} \times \mathbf{C} \tag{8.14}$$

The cross product of two vectors is given by

$$\mathbf{A} \times \mathbf{B} = \left(A_x\mathbf{i} + A_y\mathbf{j} + A_z\mathbf{k}\right) \times \left(B_x\mathbf{i} + B_y\mathbf{j} + B_z\mathbf{k}\right) \tag{8.15}$$

When expanded, Equation 8.15 can be written as

$$\begin{aligned}\mathbf{A} \times \mathbf{B} = {} & A_xB_x\mathbf{i} \times \mathbf{i} + A_xB_y\mathbf{i} \times \mathbf{j} + A_xB_z\mathbf{i} \times \mathbf{k} + A_yB_x\mathbf{j} \times \mathbf{i} + A_yB_y\mathbf{j} \times \mathbf{j} + A_yB_z\mathbf{j} \times \mathbf{k} \\ & + A_zB_x\mathbf{k} \times \mathbf{i} + A_zB_y\mathbf{k} \times \mathbf{j} + A_zB_z\mathbf{k} \times \mathbf{k}\end{aligned} \tag{8.16}$$

Since the unit vectors are orthogonal,

$$\mathbf{i} \times \mathbf{i} = \mathbf{j} \times \mathbf{j} = \mathbf{k} \times \mathbf{k} = 1 \cdot 1 \cdot \sin(0^\circ)\mathbf{e}_n = 0 \quad \text{and} \quad \mathbf{i} \times \mathbf{j} = 1 \cdot 1 \cdot \sin(90^\circ) = \mathbf{e}_n$$

But, since we have a right-handed coordinate system, by the right-hand rule, the normal to **i** and **j** is the unit vector **k**; thus, $\mathbf{i} \times \mathbf{j} = \mathbf{k}$. We can also invoke Equation 8.13 to get $\mathbf{j} \times \mathbf{i} = -\mathbf{i} \times \mathbf{j} = -\mathbf{k}$. We can carry this logic further to show that $\mathbf{j} \times \mathbf{k} = \mathbf{i}$ or $\mathbf{k} \times \mathbf{j} = -\mathbf{i}$ and $\mathbf{i} \times \mathbf{k} = -\mathbf{j}$ or $\mathbf{k} \times \mathbf{i} = \mathbf{j}$. Thus, we can rewrite Equation 8.16 as

$$\mathbf{A} \times \mathbf{B} = \left(A_yB_z - A_zB_y\right)\mathbf{i} + (A_zB_x - A_xB_z)\mathbf{j} + \left(A_xB_y - A_yB_x\right)\mathbf{k} \tag{8.17}$$

Equation 8.17 is the following determinant expanded by its minors:

$$\mathbf{A} \times \mathbf{B} = \begin{vmatrix} \mathbf{i} & \mathbf{j} & \mathbf{k} \\ A_x & A_y & A_z \\ B_x & B_y & B_z \end{vmatrix} \tag{8.18}$$

We will proceed next to the calculus of vectors. Let us consider a vector **A** dependent upon a scalar variable u, as shown in Figure 8.4. Then, $\mathbf{A} + \Delta\mathbf{A}$ corresponds to $u + \Delta u$ and we can write for its derivative

$$\frac{d\mathbf{A}}{du} = \lim_{\Delta u \to 0} \frac{\Delta\mathbf{A}}{\Delta u} \tag{8.19}$$

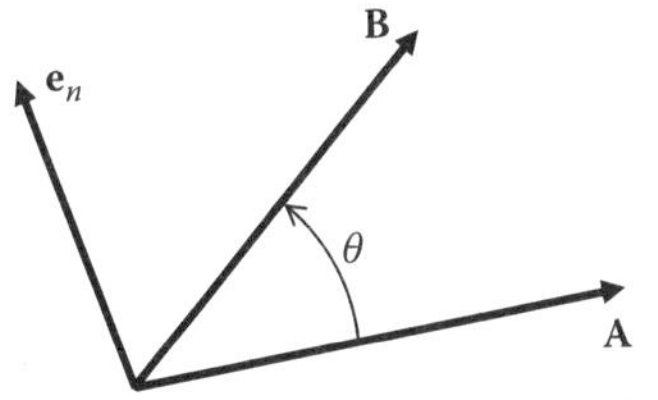

FIGURE 8.3
Vector cross product normal unit vector.

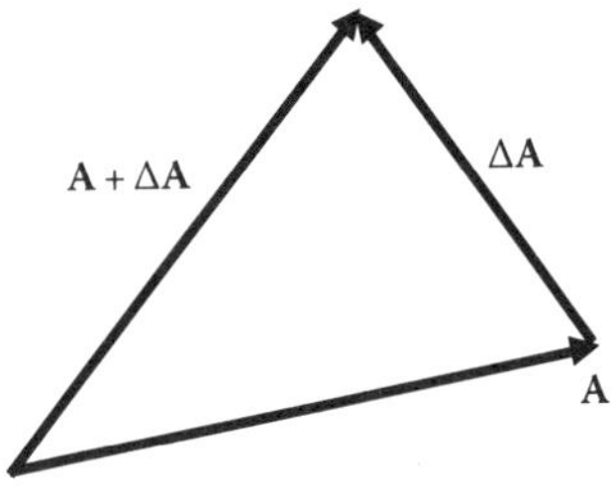

FIGURE 8.4
Vector sum illustrated.

Differentiation is distributive so that

$$\frac{\mathrm{d}(\mathbf{A}+\mathbf{B})}{\mathrm{d}u}=\frac{\mathrm{d}\mathbf{A}}{\mathrm{d}u}+\frac{\mathrm{d}\mathbf{B}}{\mathrm{d}u} \tag{8.20}$$

The chain rule also applies for scalar and vector products so that

$$\frac{\mathrm{d}}{\mathrm{d}u}(g\mathbf{A})=\frac{\mathrm{d}g}{\mathrm{d}u}\mathbf{A}+g\frac{\mathrm{d}\mathbf{A}}{\mathrm{d}u} \tag{8.21}$$

$$\frac{\mathrm{d}}{\mathrm{d}u}(\mathbf{A}\cdot\mathbf{B})=\frac{\mathrm{d}\mathbf{A}}{\mathrm{d}u}\cdot\mathbf{B}+\mathbf{A}\cdot\frac{\mathrm{d}\mathbf{B}}{\mathrm{d}u} \tag{8.22}$$

$$\frac{\mathrm{d}}{\mathrm{d}u}(\mathbf{A}\times\mathbf{B})=\frac{\mathrm{d}\mathbf{A}}{\mathrm{d}u}\times\mathbf{B}+\mathbf{A}\times\frac{\mathrm{d}\mathbf{B}}{\mathrm{d}u} \tag{8.23}$$

Consider a vector **A** dependent upon time t. If we take its derivative with respect to time, we get

$$\frac{\mathrm{d}\mathbf{A}}{\mathrm{d}t}=\frac{\mathrm{d}A_x}{\mathrm{d}t}\mathbf{i}+\frac{\mathrm{d}A_y}{\mathrm{d}t}\mathbf{j}+\frac{\mathrm{d}A_z}{\mathrm{d}t}\mathbf{k}+A_x\frac{\mathrm{d}\mathbf{i}}{\mathrm{d}t}+A_y\frac{\mathrm{d}\mathbf{j}}{\mathrm{d}t}+A_z\frac{\mathrm{d}\mathbf{k}}{\mathrm{d}t} \tag{8.24}$$

If the coordinate system is inertial (i.e., it does not move), we can write

$$\frac{\mathrm{d}\mathbf{A}}{\mathrm{d}t}=\frac{\mathrm{d}A_x}{\mathrm{d}t}\mathbf{i}+\frac{\mathrm{d}A_y}{\mathrm{d}t}\mathbf{j}+\frac{\mathrm{d}A_z}{\mathrm{d}t}\mathbf{k} \tag{8.25}$$

If the coordinate system is moving (like on a rotating earth), the rate of change terms for the unit vectors cannot be neglected. This gives rise to what we call "Coriolis terms," as we shall discuss later.

We will now examine the kinematics of a particle. Kinematics is the study of the motion of particles and rigid bodies without regard to the forces which generate the motion. Particle kinematics assumes that a point can represent the body. The rotations of the particle itself are neglected making this a three-degree-of-freedom model. If we have the inertial reference frames x, y, and z, the position of a particle P is defined by a position vector **r** drawn from the origin to the particle as shown in Figure 8.5.

If the particle P moves along a trajectory T, its instantaneous velocity is always in a direction tangent to the trajectory and its magnitude is the speed at which it moves along the curve. Thus, the tip of this vector **r** traces out the trajectory (Figure 8.6) and the velocity **v** is defined as the time rate of change of **r**, written as

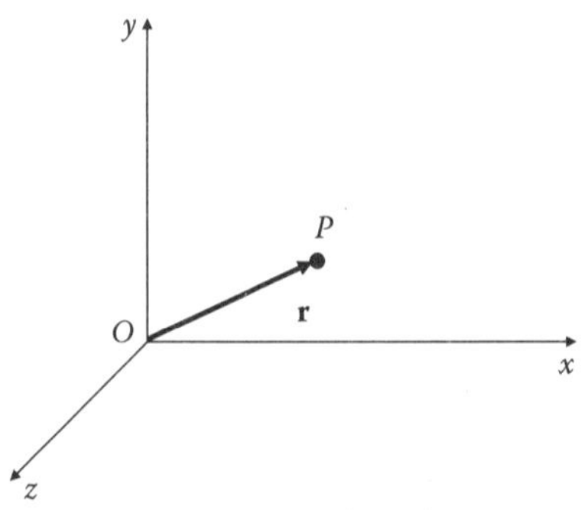

FIGURE 8.5
Position vector.

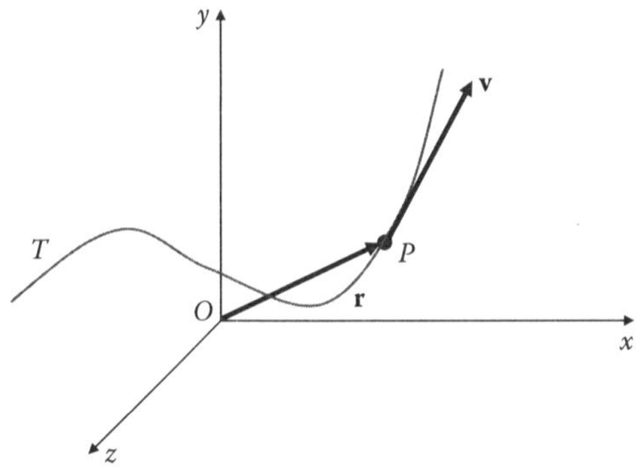

FIGURE 8.6
Trajectory curve.

$$\mathbf{v} = \frac{\mathrm{d}\mathbf{r}}{\mathrm{d}t} \tag{8.26}$$

If we were to take the velocity vector at every instant of time and fix its tail to the origin of an inertial coordinate system, then the curve traced out by its tip would be called a hodograph (Figure 8.7) and the velocity of the tip would be the time rate of change of velocity or the acceleration **a**. Thus, we can write

$$\mathbf{a} = \frac{\mathrm{d}\mathbf{v}}{\mathrm{d}t} = \frac{\mathrm{d}^2\mathbf{r}}{\mathrm{d}t^2} \tag{8.27}$$

Now, if we examine the particle as moving in two dimensions only, we can break its motion up into two components, one parallel to and one perpendicular to the position vector **r** (Figure 8.8). The position vector written in this coordinate system is given by

$$\mathbf{r} = r\mathbf{e}_r \tag{8.28}$$

So from our definition for the velocity in Equation 8.26, we get

$$\mathbf{v} = \frac{\mathrm{d}\mathbf{r}}{\mathrm{d}t} = \frac{\mathrm{d}r}{\mathrm{d}t}\mathbf{e}_r + r\frac{\mathrm{d}\mathbf{e}_r}{\mathrm{d}t} \tag{8.29}$$

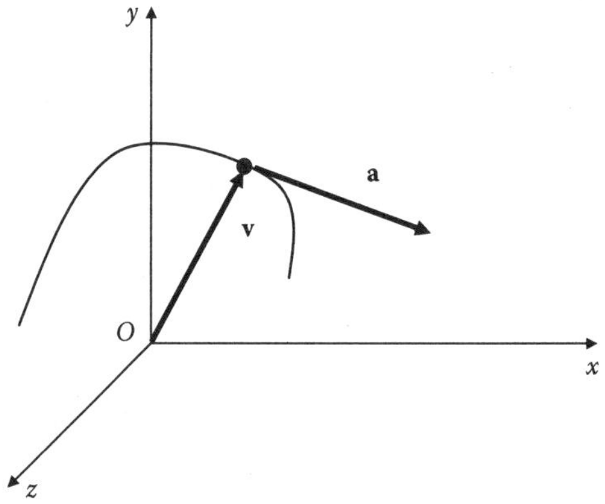

FIGURE 8.7
Hodograph.

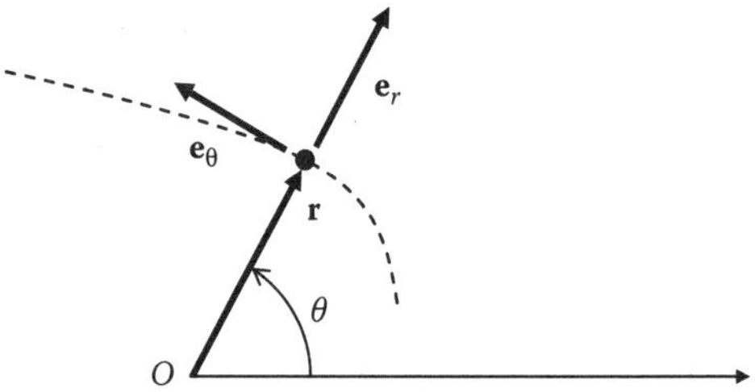

FIGURE 8.8
Differentiation of a vector through use of tangential and radial unit vectors.

Since $\mathbf{e}_r$ is a unit vector (its magnitude is a constant = 1), the only thing that changes with time is its direction.

This introduces the concept of curvilinear motion with radial coordinates (r, θ). The direction is defined by the angle θ. For a small change in the angle θ, we can write

$$\frac{d\mathbf{e}_r}{dt} = \lim_{\Delta t \to 0} \frac{\Delta \mathbf{e}_r}{\Delta t} \tag{8.30}$$

However, we can observe that for small angles

$$\Delta \mathbf{e}_r = |\mathbf{e}_r| \sin(\Delta\theta) = (1)\sin(\Delta\theta) \approx \Delta\theta \tag{8.31}$$

We also see from Figure 8.9 that $\Delta \mathbf{e}_r$ acts in the $\mathbf{e}_\theta$ direction; thus,

$$\Delta \mathbf{e}_r \approx \Delta\theta \mathbf{e}_\theta \tag{8.32}$$

Then, returning to Equation 8.30 we can write

$$\frac{d\mathbf{e}_r}{dt} = \mathbf{e}_\theta \lim_{\Delta t \to 0} \frac{\Delta\theta}{\Delta t} = \frac{d\theta}{dt}\mathbf{e}_\theta \tag{8.33}$$

Now, we can insert Equation 8.33 into Equation 8.29 to get the desired relation for the velocity:

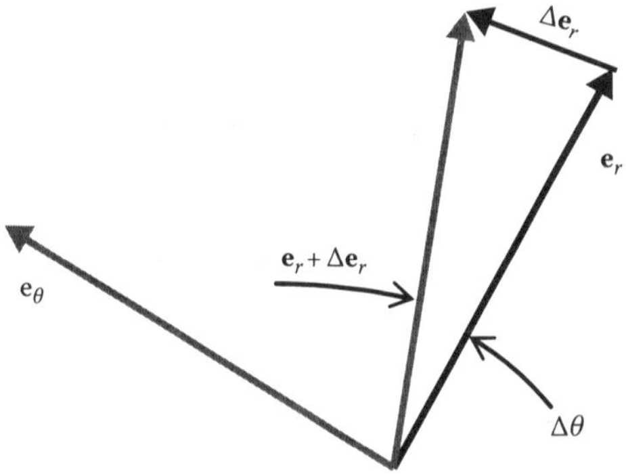

FIGURE 8.9
Rotation of the radial unit vector.

$$\mathbf{v} = \frac{\mathrm{d}\mathbf{r}}{\mathrm{d}t} = \frac{\mathrm{d}r}{\mathrm{d}t}\mathbf{e}_r + r\frac{\mathrm{d}\theta}{\mathrm{d}t}\mathbf{e}_\theta \tag{8.34}$$

The first term on the right-hand side of Equation 8.34 is the radial velocity, and the second term is the tangential velocity; these are sometimes denoted as v_r and v_θ, respectively. The magnitude of the velocity is given by

$$v = |\mathbf{v}| = \sqrt{v_r^2 + v_\theta^2} = \sqrt{\left(\frac{\mathrm{d}r}{\mathrm{d}t}\right)^2 + \left(r\frac{\mathrm{d}\theta}{\mathrm{d}t}\right)^2} \tag{8.35}$$

To obtain the acceleration in curvilinear coordinates, we need to take the time derivative of Equation 8.34 as follows:

$$\mathbf{a} = \frac{\mathrm{d}\mathbf{v}}{\mathrm{d}t} = \frac{\mathrm{d}}{\mathrm{d}t}\left(\frac{\mathrm{d}r}{\mathrm{d}t}\mathbf{e}_r + r\frac{\mathrm{d}\theta}{\mathrm{d}t}\mathbf{e}_\theta\right) = \frac{\mathrm{d}^2r}{\mathrm{d}t^2}\mathbf{e}_r + \frac{\mathrm{d}r}{\mathrm{d}t}\frac{\mathrm{d}\mathbf{e}_r}{\mathrm{d}t} + \frac{\mathrm{d}r}{\mathrm{d}t}\frac{\mathrm{d}\theta}{\mathrm{d}t}\mathbf{e}_\theta + r\frac{\mathrm{d}^2\theta}{\mathrm{d}^2t}\mathbf{e}_\theta + r\frac{\mathrm{d}\theta}{\mathrm{d}t}\frac{\mathrm{d}\mathbf{e}_\theta}{\mathrm{d}t} \tag{8.36}$$

We have already solved for the derivative of $\mathbf{e}_r$ with respect to time; now, in a similar manner, we will find the derivative of the tangential component. Again, since $\mathbf{e}_\theta$ is a unit vector, the only thing that changes with time is its direction. This direction is again defined by the angle θ, so for a small change in the angle θ, we can write

$$\frac{\mathrm{d}\mathbf{e}_\theta}{\mathrm{d}t} = \lim_{\Delta t \to 0}\frac{\Delta\mathbf{e}_\theta}{\Delta t} \tag{8.37}$$

But we see again that for small angles

$$\Delta\mathbf{e}_\theta = |\mathbf{e}_\theta|\sin(\Delta\theta) = (1)\sin(\Delta\theta) \approx \Delta\theta \tag{8.38}$$

which acts in the negative $\mathbf{e}_r$ direction as depicted in Figure 8.10. If

$$\Delta\mathbf{e}_\theta \approx -\Delta\theta\mathbf{e}_r \tag{8.39}$$

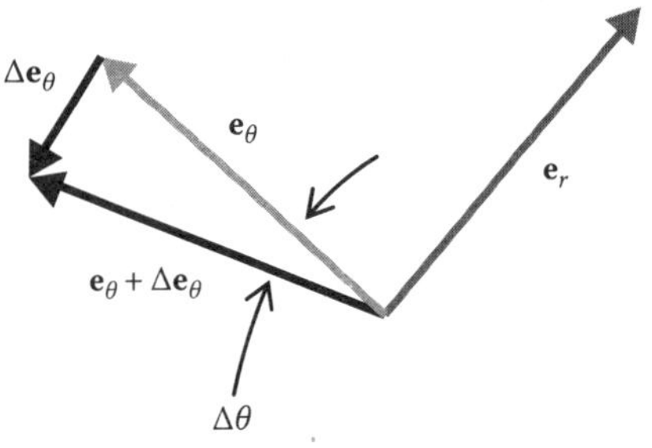

FIGURE 8.10
Rotation of the tangential unit vector.

then we can write

$$\frac{d\mathbf{e}_\theta}{dt} = -\mathbf{e}_r \lim_{\Delta t \to 0} \frac{\Delta\theta}{\Delta t} = -\frac{d\theta}{dt} e_r \tag{8.40}$$

Insertion of Equations 8.33 and 8.40 into Equation 8.36 yields

$$\mathbf{a} = \frac{d^2 r}{dt^2}\mathbf{e}_r + \frac{dr}{dt}\frac{d\theta}{dt}\mathbf{e}_\theta + \frac{dr}{dt}\frac{d\theta}{dt}\mathbf{e}_\theta + r\frac{d^2\theta}{d^2 t}\mathbf{e}_\theta - r\left(\frac{d\theta}{dt}\right)^2 \mathbf{e}_r \tag{8.41}$$

Rearranging and combining like terms gives us

$$\mathbf{a} = \left[\frac{d^2 r}{dt^2} - r\left(\frac{d\theta}{dt}\right)^2\right]\mathbf{e}_r + \left(r\frac{d^2\theta}{d^2 t} + 2\frac{dr}{dt}\frac{d\theta}{dt}\right)\mathbf{e}_\theta \tag{8.42}$$

Each of these terms has a specific name and meaning in the dynamics of a body:

$$\frac{d^2 r}{dt^2} = \text{radial acceleration}$$

$$r\left(\frac{d\theta}{dt}\right)^2 = \text{centripetal acceleration}$$

$$\frac{d^2\theta}{dt^2} = \text{angular acceleration}$$

$$2\frac{dr}{dt}\frac{d\theta}{dt} = \text{Coriolis acceleration}$$

To move on in our study, we need to examine the planar kinematics of a rigid body. First, we will examine a pure translation where we have a body-fixed coordinate system moving relative to our inertial coordinate system. Here we note from Figure 8.11 that by vector addition, we obtain

$$\mathbf{r}_B = \mathbf{r}_A + \mathbf{r}_{B/A} \tag{8.43}$$

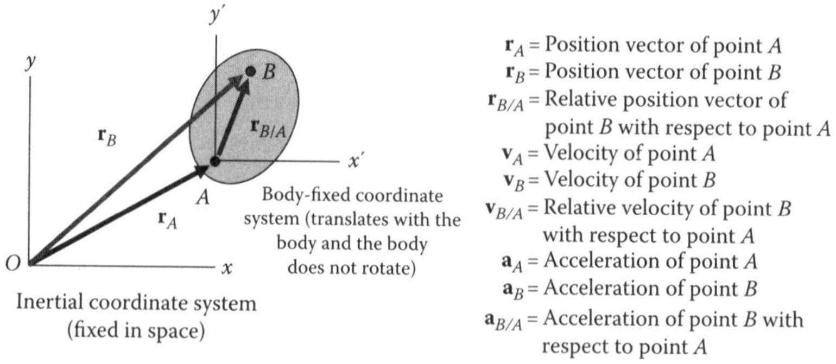

FIGURE 8.11
Definition of vectors associated with rigid body translational motion.

To determine the velocity of point B, which is under a pure translation, we have to differentiate Equation 8.43 to get

$$\mathbf{v}_B = \frac{d\mathbf{r}_B}{dt} = \frac{d\mathbf{r}_A}{dt} + \frac{d\mathbf{r}_{B/A}}{dt} \tag{8.44}$$

We know, however, that since this is a pure translation (no rotation), $d\mathbf{r}_{B/A}/dt = 0$ and $d\mathbf{r}_A/dt = \mathbf{v}_A$. Thus, for a pure translation,

$$\mathbf{v}_B = \mathbf{v}_A \tag{8.45}$$

If we differentiate Equation 8.45, we get the acceleration of a point during a pure translation as

$$\mathbf{a}_B = \frac{d\mathbf{v}_B}{dt} = \frac{d\mathbf{v}_A}{dt} = \mathbf{a}_A \tag{8.46}$$

We will now examine the rotation of a body fixed in space. Let us define the angular velocity ω as the time rate of change of angular position θ; thus,

$$\omega = \frac{d\theta}{dt} \tag{8.47}$$

Let us further define the angular acceleration α as the time rate of change of angular velocity, or

$$\alpha = \frac{d\omega}{dt} = \frac{d^2\theta}{dt^2} \tag{8.48}$$

The angular velocity ω and the angular acceleration α are depicted in Figure 8.12. We will now look at the rotation in terms of the vector kinematic equations for point P. We first examine the velocity whose direction we specify by the right-hand rule. Then, let us define the position of point P by the position vector $\mathbf{r}$ as shown in Figure 8.13. Now we can write the velocity $\mathbf{v}$ in terms of radial and circumferential components as we discussed earlier.

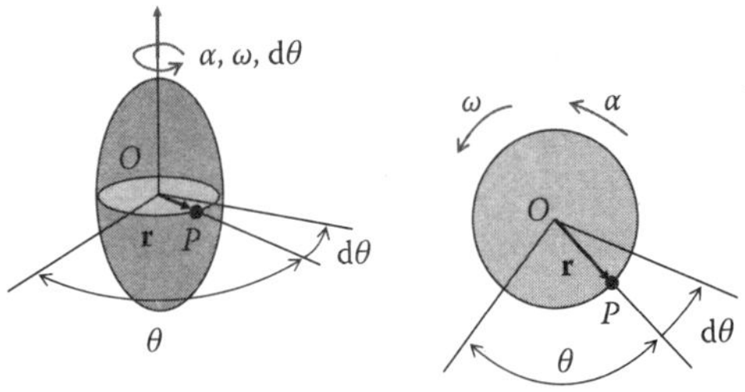

FIGURE 8.12
Example of rigid body rotation. On the left is the body rotating in space. On the right is a view of this same body looking down the axis toward O.

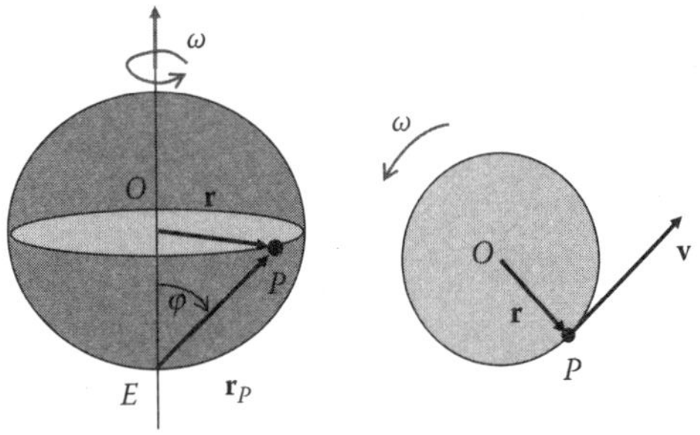

FIGURE 8.13
Rigid body rotation. On the left is the body rotating at angular velocity ω. On the right is a view of this same body looking down the axis toward O.

Thus, from Equation 8.34 we have

$$\mathbf{v} = \frac{d\mathbf{r}}{dt} = \frac{dr}{dt}\mathbf{e}_r + r\frac{d\theta}{dt}\mathbf{e}_\theta \tag{8.49}$$

But, since this is a rigid body, $dr/dt = 0$, so we get

$$\mathbf{v} = r\frac{d\theta}{dt}\mathbf{e}_\theta = r\omega\mathbf{e}_\theta \tag{8.50}$$

If we were instead to draw the position vector from a more general location such as point E in Figure 8.13, we see that

$$r = r_P \sin\phi \tag{8.51}$$

If we substitute Equation 8.51 into Equation 8.50, we see a form we have derived earlier:

$$v = r_P\omega \sin\phi\mathbf{e}_\theta \tag{8.52}$$

This can be written in vector form if we invoke Equation 8.12. Thus, we have

$$\mathbf{v} = \boldsymbol{\omega} \times \mathbf{r}_P \tag{8.53}$$

We now will examine the acceleration whose direction is once more specified by the right-hand rule. We shall define the position of point P by the position vector $\mathbf{r}$ as shown in

Figure 8.14. We can then write the acceleration **a** in terms of radial and circumferential components as discussed earlier.

We need to recall Equation 8.42 and note that dr/dt = 0. This leaves us with

$$\mathbf{a} = \left[-r\left(\frac{d\theta}{dt}\right)^2\right]\mathbf{e}_r + \left(r\frac{d^2\theta}{dt^2}\right)\mathbf{e}_\theta \tag{8.54}$$

Here we need to note that the first term is negative because it acts in the negative radial direction. Now, from our previous definitions, we can rewrite Equation 8.54 as

$$\mathbf{a} = \left(-r\omega^2\right)\mathbf{e}_r + (r\alpha)\mathbf{e}_\theta \tag{8.55}$$

We can define the normal and tangential components of acceleration as

$$\mathbf{a}_n = -r\boldsymbol{\omega}^2\mathbf{e}_r \quad \text{and} \quad \mathbf{a}_t = r\alpha\mathbf{e}_\theta \tag{8.56}$$

where $\mathbf{a}_n$ is the normal (centrifugal) acceleration and $\mathbf{a}_t$ is the tangential acceleration.

We can differentiate Equation 8.53 to obtain the more general result

$$\mathbf{a} = \frac{d\mathbf{v}}{dt} = \frac{d\boldsymbol{\omega}}{dt} \times \mathbf{r}_P + \boldsymbol{\omega}\frac{d\mathbf{r}_P}{dt} \tag{8.57}$$

If we insert Equations 8.48 and 8.53 into Equation 8.57, we get the general vector form for the acceleration of a rigid body rotating in an inertial coordinate system:

$$\mathbf{a} = \boldsymbol{\alpha} \times \mathbf{r}_P + \boldsymbol{\omega} \times (\boldsymbol{\omega} \times \mathbf{r}_P) \tag{8.58}$$

To move closer toward a more general treatment, we shall now derive the kinematic equations for the plane motion of a rigid body by using a translating coordinate system (the body is free to rotate). We can break down any planar motion of the rigid body into a translation and a rotation about some point. Let us choose point A in Figure 8.15 to be a location about which the body rotates. Equation 8.43 is still valid, but d**r**/dt no longer equals zero.

Thus, from Equation 8.44, we have

$$\mathbf{v}_B = \frac{d\mathbf{r}_B}{dt} = \frac{d\mathbf{r}_A}{dt} + \frac{d\mathbf{r}_{B/A}}{dt} \tag{8.59}$$

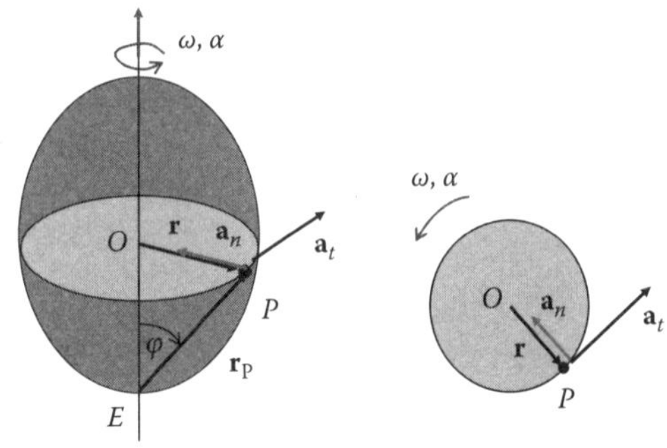

FIGURE 8.14
Rigid body rotation with acceleration. On the left is the body rotating at angular velocity ω and accelerating with angular acceleration α. On the right is a view of this same body looking down the axis toward O.

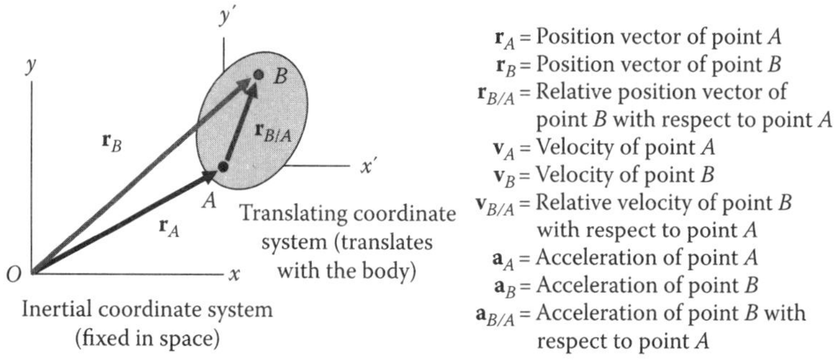

FIGURE 8.15
Definition of vectors associated with rigid body translational and rotational motions.

Not only is $\mathrm{d}\mathbf{r}_{B/A}/\mathrm{d}t \neq 0$ but also, since we chose point A as one about which rotation will take place, $\mathrm{d}\mathbf{r}_A/\mathrm{d}t$ is a pure translation and $\mathrm{d}\mathbf{r}_{B/A}/\mathrm{d}t$ is a pure rotation about point A. Thus, we can write Equation 8.59 in terms of the velocities as

$$\mathbf{v}_B = \mathbf{v}_A + \mathbf{v}_{B/A} \tag{8.60}$$

We saw earlier that for a pure rotation, we can write the velocity in the form of Equation 8.53. Thus, we have

$$\mathbf{v}_{B/A} = \boldsymbol{\omega} \times \mathbf{r}_{B/A} \tag{8.61}$$

If we substitute Equation 8.61 into Equation 8.60, we obtain the vector equation for the planar motion of a rigid body in which our coordinate system translates with a point in the body but does not rotate with the body:

$$\mathbf{v}_B = \mathbf{v}_A + \boldsymbol{\omega} \times \mathbf{r}_{B/A} \tag{8.62}$$

To obtain the acceleration of point B, we need to differentiate Equation 8.60; thus,

$$\mathbf{a}_B = \frac{\mathrm{d}\mathbf{v}_B}{\mathrm{d}t} = \frac{\mathrm{d}\mathbf{v}_A}{\mathrm{d}t} + \frac{\mathrm{d}\mathbf{v}_{B/A}}{\mathrm{d}t} \tag{8.63}$$

Let us examine this equation term by term. The first term is straightforward, showing that the acceleration of the translation is simply the linear acceleration of our chosen reference point:

$$\frac{\mathrm{d}\mathbf{v}_A}{\mathrm{d}t} = \mathbf{a}_A \tag{8.64}$$

The second term is differentiated as follows:

$$\frac{\mathrm{d}\mathbf{v}_{B/A}}{\mathrm{d}t} = \frac{\mathrm{d}}{\mathrm{d}t}\left(\boldsymbol{\omega} \times \mathbf{r}_{B/A}\right) = \frac{\mathrm{d}\boldsymbol{\omega}}{\mathrm{d}t} \times \mathbf{r}_{B/A} + \boldsymbol{\omega} \times \frac{\mathrm{d}\mathbf{r}_{B/A}}{\mathrm{d}t} \tag{8.65}$$

We again need to call upon Equations 8.48 and 8.53 to get Equation 8.65 into a more general form:

$$\frac{\mathrm{d}\mathbf{v}_{B/A}}{\mathrm{d}t} = \boldsymbol{\alpha} \times \mathbf{r}_{B/A} + \boldsymbol{\omega} \times \left(\boldsymbol{\omega} \times \mathbf{r}_{B/A}\right) \tag{8.66}$$

Inserting Equations 8.64 and 8.66 into Equation 8.63 yields the kinematic equation for the acceleration of a rigid body in a coordinate system that translates with the body. The coordinate system in this case moves with the body but does not rotate, allowing the body to rotate relative to the moving coordinate system:

$$\mathbf{a}_B = \mathbf{a}_A + \boldsymbol{\alpha} \times \mathbf{r}_{B/A} + \boldsymbol{\omega} \times \left(\boldsymbol{\omega} \times \mathbf{r}_{B/A}\right) \tag{8.67}$$

We shall now derive the kinematic equations for plane motion of a rigid body by using a translating and rotating coordinate system (the body is capable of movement in both coordinate systems). We choose point A in Figure 8.16 to be a location from which we want to measure the motion of the body. At the instant considered, point A has a position $\mathbf{r}_A$, a velocity $\mathbf{v}_A$, and an acceleration $\mathbf{a}_A$, while the x–y-axes (and the body) are rotating with angular velocity $\boldsymbol{\omega}$ and accelerating with angular acceleration $\boldsymbol{\alpha}$.

Equation 8.43 is still valid for determination of the position vectors. For the velocity of point B, we shall use the form

$$\mathbf{v}_B = \mathbf{v}_A + \frac{d\mathbf{r}_{B/A}}{dt} \tag{8.68}$$

If we examine our earlier derivations under the curvilinear motion and replace $\mathbf{e}_r$ with $\mathbf{i}$ and $\mathbf{e}_\theta$ with $\mathbf{j}$ ($\mathbf{i}$ and $\mathbf{j}$ represent the unit vectors in our x–y coordinate system), we obtain the following relations:

$$\frac{d\mathbf{i}}{dt} = \frac{d\theta}{dt}\mathbf{j} = \omega\mathbf{j} \tag{8.69}$$

$$\frac{d\mathbf{j}}{dt} = \frac{d\theta}{dt}(-\mathbf{i}) = -\omega\mathbf{i} \tag{8.70}$$

Using the definition of the cross product and noting they are orthogonal, Equations 8.69 and 8.70 can be rewritten as

$$\frac{d\mathbf{i}}{dt} = \boldsymbol{\omega} \times \mathbf{i} \tag{8.71}$$

$$\frac{d\mathbf{j}}{dt} = \boldsymbol{\omega} \times \mathbf{j} \tag{8.72}$$

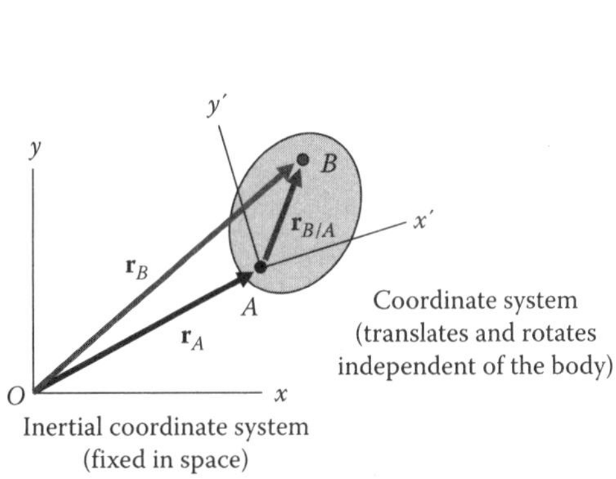

FIGURE 8.16
Definition of vectors associated with rigid body translational and rotational motions including a rotating local coordinate system.

If we look at the body and our moving coordinate system as shown in Figure 8.17, we see that if the body translates and rotates relative to our x–y-axes, we can write, in light of Equation 8.62,

$$\frac{d\mathbf{r}_{B/A}}{dt} = \left(\mathbf{v}_{B/A}\right)_{xyz} + \boldsymbol{\omega} \times \mathbf{r}_{B/A} \tag{8.73}$$

Substituting this into Equation 8.68 yields the general relation for the velocity of a point in a rigid body as seen from an arbitrary coordinate system:

$$\mathbf{v}_B = \mathbf{v}_A + \left(\mathbf{v}_{B/A}\right)_{xyz} + \boldsymbol{\omega} \times \mathbf{r}_{B/A} \tag{8.74}$$

To obtain the acceleration of our point B, we need to differentiate Equation 8.74 with respect to time; thus,

$$\mathbf{a}_B = \frac{d\mathbf{v}_A}{dt} + \frac{d(\mathbf{v}_{B/A})_{xyz}}{dt} + \frac{d\boldsymbol{\omega}}{dt} \times \mathbf{r}_{B/A} + \boldsymbol{\omega} \times \frac{d\mathbf{r}_{B/A}}{dt} \tag{8.75}$$

Let us again move term by term through Equation 8.75. The first term is

$$\frac{d\mathbf{v}_A}{dt} = \mathbf{a}_A \tag{8.76}$$

Since our point A does not rotate, the acceleration of its translation is simply this linear acceleration.

The second term is differentiated by first breaking up $\mathbf{v}_{B/A}$ into its components along the x and y-axes of our moving frame. Since we are looking only at motion on the x–y plane, the z component is nonexistent:

$$\left(\mathbf{v}_{B/A}\right)_{xyz} = \left(v_{B/A}\right)_x \mathbf{i} + \left(v_{B/A}\right)_y \mathbf{j} \tag{8.77}$$

This allows us to write the second term as

$$\frac{d\left(\mathbf{v}_{B/A}\right)_{xyz}}{dt} = \frac{d\left(\mathbf{v}_{B/A}\right)_x}{dt}\mathbf{i} + \frac{d\left(\mathbf{v}_{B/A}\right)_y}{dt}\mathbf{j} + \left(v_{B/A}\right)_x \frac{d\mathbf{i}}{dt} + \left(v_{B/A}\right)_y \frac{d\mathbf{j}}{dt} \tag{8.78}$$

The first pair of terms in Equation 8.78 is made up of the acceleration components of point B relative to point A as seen by an observer moving with the coordinate system at point A.

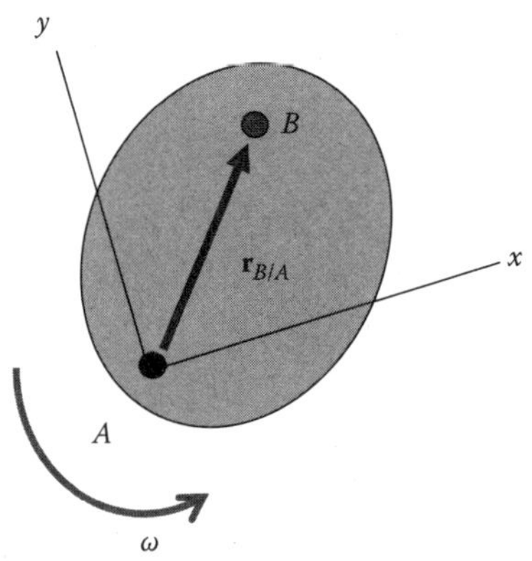

FIGURE 8.17
Body rotating in a moving coordinate system.

TABLE 8.1

Vector Terms Used in Equations 8.74 and 8.81

Variable	Definition
$\mathbf{r}_{B/A}$	Relative position vector of point B with respect to point A
$\mathbf{v}_A$	Velocity of point A in the inertial coordinate system
$\mathbf{v}_B$	Velocity of point B in the inertial coordinate system
$(\mathbf{v}_{B/A})_{xyz}$	Relative velocity of point B with respect to point A in the x–y–z coordinate system
$\mathbf{a}_A$	Acceleration of point A in the inertial coordinate system
$\mathbf{a}_B$	Acceleration of point B in the inertial coordinate system
$(\mathbf{a}_{B/A})_{xyz}$	Acceleration of point B with respect to point A in the x–y–z coordinate system
$\boldsymbol{\omega}$	Angular velocity of the x–y–z coordinate system measured from the inertial coordinate system
$\boldsymbol{\alpha}$	Angular acceleration of the x–y–z coordinate system measured from the inertial coordinate system

The second pair of terms of Equation 8.78 can be rewritten as the cross product of the angular velocity of the x–y coordinate system and the velocity vector of point B relative to point A. Thus, we can write

$$\frac{\mathrm{d}\left(\mathbf{v}_{B/A}\right)_{xyz}}{\mathrm{d}t} = \left(\mathbf{a}_{B/A}\right)_{xyz} + \boldsymbol{\omega} \times \left(\mathbf{v}_{B/A}\right)_{xyz} \tag{8.79}$$

Returning now to Equation 8.75, in the third term, we simply rewrite the term $\mathrm{d}\boldsymbol{\omega}/\mathrm{d}t$ as $\boldsymbol{\alpha}$. Finally, for the last term of Equation 8.75, we use Equation 8.73 to obtain

$$\mathbf{a}_B = \mathbf{a}_A + \left(\mathbf{a}_{B/A}\right)_{xyz} + \boldsymbol{\omega} \times \left(\mathbf{v}_{B/A}\right)_{xyz} + \boldsymbol{\alpha} \times \mathbf{r}_{B/A} + \boldsymbol{\omega} \times \left(\mathbf{v}_{B/A}\right)_{xyz} + \boldsymbol{\omega} \times \left(\boldsymbol{\omega} \times \mathbf{r}_{B/A}\right) \tag{8.80}$$

After a slight rearrangement and combination of like terms, we have the general kinematic equation for acceleration of point B:

$$\mathbf{a}_B = \mathbf{a}_A + \boldsymbol{\alpha} \times \mathbf{r}_{B/A} + \boldsymbol{\omega} \times \left(\boldsymbol{\omega} \times \mathbf{r}_{B/A}\right) + 2\boldsymbol{\omega} \times \left(\mathbf{v}_{B/A}\right)_{xyz} + \left(\mathbf{a}_{B/A}\right)_{xyz} \tag{8.81}$$

It is important to review each of the terms in Equations 8.74 and 8.81. First, let us review the generalized velocity and acceleration equations we derived:

$$\mathbf{v}_B = \mathbf{v}_A + \left(\mathbf{v}_{B/A}\right)_{xyz} + \boldsymbol{\omega} \times \mathbf{r}_{B/A} \tag{8.74}$$

$$\mathbf{a}_B = \mathbf{a}_A + \boldsymbol{\alpha} \times \mathbf{r}_{B/A} + \boldsymbol{\omega} \times \left(\omega \times \mathbf{r}_{B/A}\right) + 2\boldsymbol{\omega} \times \left(\mathbf{v}_{B/A}\right)_{xyz} + \left(\mathbf{a}_{B/A}\right)_{xyz} \tag{8.81}$$

The terms in these equations have meanings as listed in Table 8.1.

As one can imagine the addition of the third dimension in these equations adds significant complexity to the expressions although the basic principles remain the same. The interested reader is referred to the book by Greenwood [1] or any similar text on dynamics to familiarize themselves with the three-dimensional uses of these equations.

Problem 1

A 155 mm projectile is in flight at its maximum ordinate. At this instant in time, the nose of the projectile is pointing along (and spinning about) the unit vector:

$$\mathbf{x} = \left(0.998\mathbf{e}_1 + 0.030\mathbf{e}_2 + 0.056\mathbf{e}_3\right)$$

The projectile velocity vector is

$$\mathbf{V} = (1199\mathbf{e}_1 + 0\mathbf{e}_2 + 49\mathbf{e}_3)\left[\frac{\text{ft}}{\text{s}}\right]$$

In both of these cases, $\mathbf{e}_1$, $\mathbf{e}_2$, and $\mathbf{e}_3$ are unit vectors on the x, y, and z planes, respectively. Also, at this location, the air density, spin rate, and projectile mass are as follows:

$$\rho = 0.052\left[\frac{\text{lbm}}{\text{ft}^3}\right], \quad p = \omega = 150\left[\frac{\text{rev}}{\text{s}}\right], \quad \text{and} \quad m = 100\,[\text{lbm}]$$

The projectile characteristics are assumed to be as follows:

$$C_D = 0.29$$

$$C_{M_\alpha} = 3.0 \quad \left(C_{M_q} + C_{M_{\dot\alpha}}\right) = -10.2$$

$$C_{L_\alpha} = 2.12 \quad \left(C_{N_q} + C_{N_{\dot\alpha}}\right) = 0.002$$

$$C_{N_{p\alpha}} = -0.010 = -10.2 \quad C_{M_{p\alpha}} = 0.51$$

$$C_{l_p} = -0.015$$

Please answer the following questions:

1. Draw the situation.
2. Determine the drag force vector.
   Answer: $F_D = (-68.39\mathbf{e}_1 - 2.80\mathbf{e}_3)$ [lbf]
3. Determine the lift force vector.
   Answer: $F_L = (-0.310\mathbf{e}_1 + 14.980\mathbf{e}_2 + 7.600\mathbf{e}_3)$ [lbf]
4. Determine the overturning moment vector.
   Answer: $M_M = (-0.441\mathbf{e}_1 - 5.468\mathbf{e}_2 + 10.783\mathbf{e}_3)$ [ft-lbf]
5. Determine the magnus moment vector.
   Answer: $M_{M_{p\alpha}} = (0.043\mathbf{e}_1 - 0.732\mathbf{e}_2 - 0.369\mathbf{e}_3)$ [ft-lbf]

Reference

1. Greenwood, D. T., *Principles of Dynamics*, 2nd ed., Prentice Hall, Englewood Cliffs, NJ, 1988.

Further Reading

Beer, F. P., and Johnston, E. R., *Vector Mechanics for Engineers—Statics and Dynamics*, 7th ed., McGraw-Hill, New York, 2004.

Colley, S. J., *Vector Calculus*, Prentice Hall, Upper Saddle River, NJ, 1998.

Hibbeler, R. C., *Engineering Mechanics—Statics and Dynamics*, 7th ed., Prentice Hall, Englewood Cliffs, NJ, 1995.

O'Neil, P. V., *Advanced Engineering Mathematics*, 2nd ed., Wadsworth, Belmont, CA, 1986.

9

Trajectories

Now that the basics of the terminology and the dynamic equations have been presented, we shall begin to look at their uses in the form of prediction of trajectories. The aeroballistician is usually faced with one of two problems: "If I want to hit a target at position x, to what elevation (and perhaps with how much propelling charge) do I have to elevate the weapon?" or "My weapon is elevated to elevation x and I expect muzzle velocity y—where is the projectile going to end up?"

To approach this in a logical and easily understandable fashion, we shall begin with a great many simplifying assumptions, relieving these as we progress. Each section builds upon the previous one so that we recommend even seasoned veterans progress in numerical order.

Initially, we will look only at the effect that gravity imposes on the projectile, a vacuum trajectory, so that even the air is removed from our area of concern, thus neutralizing the fluid mechanics for a while. As we progress, we shall add in the atmosphere but neglect dynamics, atmospheric perturbations, and earth rotation. One by one, we shall continually step up the complexity until, finally, we shall introduce the full six degree-of-freedom (6 DOF) equations.

One might initially think that these simplified models have no practical use, but are merely educational stepping-stones. Nothing could be further from the truth. In many instances, some of the complications only slightly affect the solution, and a ballistician is well placed to assume them away. Some of these common situations will be pointed out as they arise.

9.1 Vacuum Trajectory

In this section, we will make two broad assumptions: first, that the projectile mass is concentrated at a point (which allows us to neglect body dynamics affected by mass distribution) and, second, that the only force acting on the projectile is that due to the acceleration of gravity (this allows us to neglect the rather complicated fluid dynamic effects when a solid body moves through a fluid). With these assumptions, the two governing differential equations of motion are

$$m\ddot{x} = 0 \tag{9.1}$$

$$\ddot{y} = -g \tag{9.2}$$

The solutions to these equations, found by integrating with respect to time, are

$$x = V_0 t \cos\phi_0 \tag{9.3}$$

$$y = V_0 t \sin\phi_0 - \frac{1}{2} g t^2 \tag{9.4}$$

A sketch of this simplified trajectory is seen in Figure 9.1. Notice that unlike the generalized trajectory shown previously in Figure 7.1, the terminal point is on the line $y = 0$, and the entire trajectory is on the x–y plane.

But from Equations 9.3 and 9.4, we can write

$$\frac{y}{x} = \tan\phi_0 - \frac{gt}{2V_0 \cos\phi_0} \tag{9.5}$$

By solving Equation 9.3 for time t, we get

$$t = \frac{x}{V_0 \cos\phi_0} \tag{9.6}$$

We can then write Equation 9.5 putting y in terms of x only. Therefore,

$$y = x\tan\phi_0 - \frac{gx^2}{2V_0^2 \cos^2\phi_0} \tag{9.7}$$

This equation is in the form of a parabola in x- and y-coordinates; the path the projectile will follow in a vacuum. Solving for the range x when $y = 0$ gives

$$x\left(2V_0^2 \cos\phi_0 \sin\phi_0 - gx\right) = 0 \tag{9.8}$$

This says that either $x = 0$ (the trivial solution) or

$$x = \frac{2V_0^2}{g} \cos\phi_0 \sin\phi_0 = \frac{V_0^2}{g} \sin 2\phi_0 \tag{9.9}$$

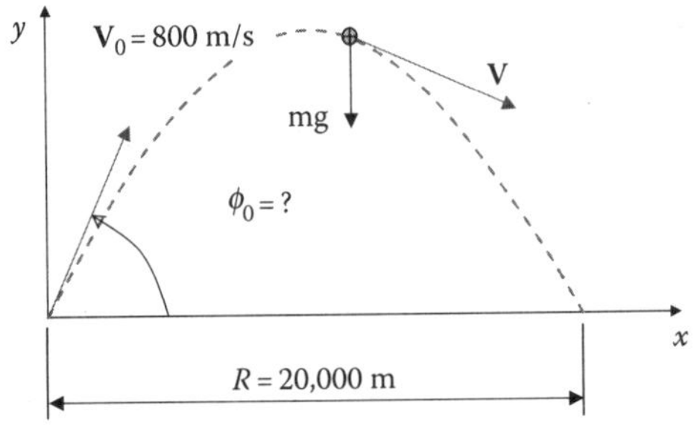

FIGURE 9.1
Vacuum trajectory.

Because the trajectory is a parabola, the maximum range is attained at

$$\phi_0 = \frac{\pi}{4} \quad (9.10)$$

since

$$\sin\frac{\pi}{2} = 1 \quad (9.11)$$

When we substitute this into Equation 9.9, the maximum range can be found to be

$$x_{max} = \frac{V_0^2}{g} \quad (9.12)$$

The maximum ordinate of the trajectory is at half the maximum range and is

$$y_{max} = \frac{V_0^2}{4g} \quad (9.13)$$

If we differentiate Equation 9.9 with respect to ϕ_0 and set this equal to 0, we can prove Equation 9.10 as shown next:

$$\frac{dx_{max}}{d\phi_0} = \frac{2V_0^2}{g}\cos 2\phi_0 = 0 \quad (9.14)$$

This gives the launch angle for maximum range in a vacuum as $\pi/4$.

Except for the maximum range, there are two quadrant elevations (QEs) that will allow a projectile to impact at a given distance. We will designate the second QE with a caret "^." Its existence is due to the identity

$$\sin\phi = \sin(180° - \phi) \quad (9.15)$$

Then,

$$\sin 2\phi_0 = \sin 2(90° - \phi_0) \quad (9.16)$$

Thus,

$$\hat{\phi}_0 = 90° - \phi_0 = 90° - \frac{1}{2}\sin^{-1}\left(\frac{gR}{V_0^2}\right) \quad (9.17)$$

where x has been replaced by the range R. The maximum ordinate is achieved when the y-component of the velocity is 0. By differentiating Equation 9.4 with respect to time and setting the result equal to 0, we get

$$V_0\sin\phi_0 - gt_s = 0 \quad (9.18)$$

or

$$t_s = \frac{V_0\sin\phi}{g} \quad (9.19)$$

Substituting this into Equation 9.4 gives

$$y_s = V_0 \sin\phi_0\left(\frac{V_0 \sin\phi_0}{g}\right) - \frac{1}{2}g\left(\frac{V_0 \sin\phi_0}{g}\right)^2 = \frac{1}{2}\frac{V_0^2 \sin^2\phi_0}{g} \quad (9.20)$$

If we note that at impact, the y-coordinate is zero, we can find the time of flight (TOF) to impact with Equation 9.4:

$$0 = V_0 t_I \sin\phi_0 - \frac{1}{2}g t_I^2 \quad (9.21)$$

or

$$t_I = \frac{2V_0 \sin\phi_0}{g} \quad (9.22)$$

This is double the time to the maximum ordinate and the trajectory in a vacuum is symmetrical about this ordinate. Further evidence of the symmetry may be seen by examining the angle of fall. If we differentiate Equation 9.7 with respect to x and substitute the value of x we found at impact in Equation 9.9 in the differentiated result, we see that

$$\left.\frac{dy}{dx}\right|_I = \tan\phi_0 - \frac{\sin 2\phi_0}{\cos^2\phi_0} \quad (9.23)$$

But

$$\sin 2\phi_0 = 2\sin\phi_0 \cos\phi_0 \quad (9.24)$$

Therefore,

$$\tan\phi_I = \tan\phi_0 - 2\frac{\sin\phi_0}{\cos\phi_0} = -\tan\phi_0 \quad (9.25)$$

Thus, in a vacuum trajectory, the projectile impacts at the mirror image of the angle it had when it was launched.

For any given launch velocity V_0, the maximum range in a vacuum is achieved with an initial launch angle of 45°. To reach any range shorter than the maximum, there are two launch elevations, one greater than 45°, and the other less, a high angle and a low angle of fire. The trajectory envelope is a curve that bounds all possible trajectories that attempt to reach all ranges from zero to the maximum range possible for the given launch velocity [1]. We shall now mathematically describe this curve.

We know from Equation 9.7 that

$$y = x\tan\phi_0 - \frac{gx^2}{2V_0^2\cos^2\phi_0} \quad (9.7)$$

This can also be written as

$$y = x\tan\phi_0 - \frac{gx^2}{2V_0^2}\sec^2\phi_0 \quad (9.26)$$

If we make use of the trigonometric identity $\sec^2\phi = 1 + \tan^2\phi$, we can, with substitution and manipulation, write

$$\tan^2\phi_0 - \frac{2V_0^2}{gx}\tan\phi_0 + \frac{2V_0^2}{gx^2}y + 1 = 0 \tag{9.27}$$

Equation 9.27 is quadratic in ϕ_0 and, as such, when solved, yields two roots that correspond to the two elevations that achieve the same range as discussed earlier.

The exceptions to this are when the range is zero or the range is maximum. These conditions yield a repeated root. The other instances a repeated root occurs are whenever the trajectory touches the trajectory envelope. This occurs only once at any given elevation. If the roots of this equation are complex conjugates, the range in question cannot be achieved with the given muzzle velocity. We can solve for all the double roots to obtain the equation of the trajectory envelope.

We proceed by first completing the square in Equation 9.27 noting that

$$\tan^2\phi_0 - \frac{2V_0^2}{gx}\tan\phi_0 + \left(\frac{V_0^2}{gx}\right)^2 = \left(\tan\phi_0 - \frac{V_0^2}{gx}\right)^2 \tag{9.28}$$

By adding and subtracting a term, $(V_0^2/gx)^2$, to Equation 9.27, we complete the square of a part of the equation and can operate on the remainder of it:

$$\tan^2\phi_0 - \frac{2V_0^2}{gx}\tan\phi_0 + \left(\frac{V_0^2}{gx}\right)^2 - \left(\frac{V_0^2}{gx}\right)^2 + \frac{2V_0^2}{gx^2}y + 1 = 0 \tag{9.29}$$

Breaking apart Equation 9.29 into two terms and setting each equal to zero gives us from Equation 9.28

$$\tan^2\phi_0 - \frac{2V_0^2}{gx}\tan\phi_0 + \left(\frac{V_0^2}{gx}\right)^2 = \left(\tan\phi_0 - \frac{V_0^2}{gx}\right)^2 = 0 \tag{9.30}$$

and

$$\frac{2V_0^2}{gx^2}y - \left(\frac{V_0^2}{gx}\right)^2 + 1 = 0 \tag{9.31}$$

The double root in Equation 9.30 occurs when

$$\tan\phi_0 = \frac{V_0^2}{gx_e} \tag{9.32}$$

Where $x_e = x$ on the envelope curve. We can also pursue the equation for the envelope curve more directly:

$$\frac{2V_0^2}{gx_e^2}y_e - \left(\frac{V_0^2}{gx_e}\right)^2 + 1 = 0 \tag{9.33}$$

Where y_e is the y-coordinate on the envelope curve. Equation 9.33 can be further rearranged to yield the final equation of the trajectory envelope:

$$y_e = \frac{1}{2}\frac{V_0^2}{g} - \frac{gx_e^2}{2V_0^2} \tag{9.34}$$

A typical trajectory envelope is illustrated in Figure 9.2.

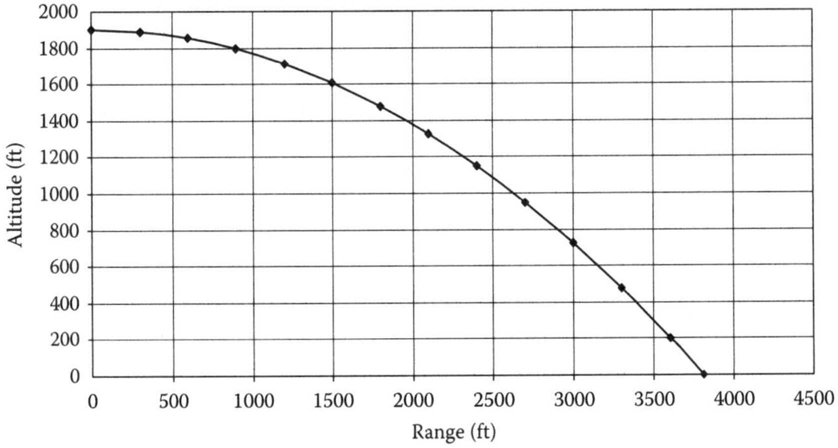

FIGURE 9.2
Trajectory envelope.

To move to a different subject in the study of the vacuum trajectory, when the trajectory of the projectile is relatively flat, certain simplifying assumptions may be made, which allow the equations of motion to be solved with greater ease. In particular, if we rewrite Equation 9.7 as

$$y = x \tan \phi_0 - \frac{gx^2}{2V_0^2} \sec^2 \phi_0 \tag{9.35}$$

We now take its derivative with respect to ϕ_0, and we get

$$\frac{dy}{d\phi_0} = x \sec^2 \phi_0 - \frac{gx^2}{V_0^2} \tan \phi_0 \sec^2 \phi_0 = x\left(1 - \frac{gx}{V_0^2} \tan \phi_0\right) \sec^2 \phi_0 \tag{9.36}$$

Now because

$$\sec^2 \phi_0 = 1 + \tan^2 \phi_0$$

and if $\tan^2 \phi_0 \ll 1$, then $\sec^2 \phi_0 \approx 1$. This occurs when $\phi_0 < 5°$. This is the requirement for what is commonly called the flat fire approximation to be valid. We can then translate Equations 9.35 and 9.36 into

$$y \approx x \tan \phi_0 - \frac{gx^2}{2V_0^2} \tag{9.37}$$

and

$$\frac{dy}{d\phi_0} \approx x\left(1 - \frac{gx}{V_0^2} \tan \phi_0\right) \tag{9.38}$$

Equation 9.38 can be even further simplified for short ranges if $-\frac{gx}{V_0^2}\tan\phi_0 \ll 1$, then

$$\frac{dy}{d\phi_0} \approx x \tag{9.39}$$

This is sometimes known as the rigid trajectory because the trajectory appears to rotate rigidly with the elevation angle. The vertical error that arises from the use of the flat fire approximation in a vacuum trajectory is

$$\varepsilon_y = \frac{gx^2}{2V_0^2}\tan^2\phi_0 \tag{9.40}$$

Which, as is readily seen, states that as the range or launch angle increases, the error increases. Flat fire is characteristic of the engagements experienced with high-powered, high-velocity tank cannons where initial launch angles for direct-fire ranges of several kilometers are less than 5°. Elevation changes to correct fire are measured in fractions of a degree (known as mils) as well. One mil is equal to 1/6400 of a circle.

Problem 1

A target is located at 20 km. The projectile muzzle velocity is 800 m/s. Assuming a vacuum trajectory, at what QE should one set the weapon to hit the target?
Answer: $\phi_0 = 158.7$ [mil]

Problem 2

The enemy in the aforementioned problem is very smart and has located his unit on the reverse slope of a hill that is 3,000 m in height with its peak located 18,000 m from your firing position. Assuming that the target is at the same level as you (just behind the hill), determine a firing solution (QE, if there is one) to hit him assuming a vacuum trajectory.
Answer: It can be hit. You find the initial QE.

Problem 3

The US pattern 1917 (M1917) Enfield rifle was the most numerous rifle used by our troops in the First World War. It was an easier rifle to manufacture than the M1903 Springfield (even though the Springfield was officially the US Army's service rifle), and the troops liked its accuracy better. In fact, the famous Sergeant Alvin York was actually armed (there is still argument over this today) with an Enfield, not a Springfield as is commonly believed, when he single-handedly captured over 100 German soldiers in the Argonne Forest in 1918. The pattern 1917 used the standard M1 30-'06 cartridge in US service. The bullet had a mass of 174 grains (a grain is a common unit of measure in small arms ammunition and is defined as 1/7000 lbm) and a diameter of 0.308 in. This cartridge–rifle combination has a muzzle velocity of 2800 ft/s. Assuming a vacuum trajectory,

1. Determine the angle in degrees to set the sights on the rifle (i.e., the QE) if the target is level with the firer and at a range of 200 yards.
   Answer: $\phi_0 = 0.0705°$

2. If the target is at the same horizontal range but 20 yards higher, and the firer does not adjust the sights, how much higher or lower will the bullet strike?

   Answer: $y_{miss} = -0.0125$ [in.]

Problem 4

You are asked to create a rough safety fan for a maximum range test at Yuma Proving Ground. The test consists of a US M198 155 mm howitzer firing an M549 projectile at maximum charge with rocket off. The projectile weighs 96 lbm. The muzzle velocity is 880 m/s.

1. Using a vacuum trajectory, calculate and plot the trajectory envelope for the test.

   Answer: $R = 78940$ [m]

2. Determine the longest TOF of the projectile.

   Answer: $t_I = 179.4$ [s]

Problem 5

It is desired to use a 105 mm battery to set up a line of illumination candles over an enemy defensive position. The enemy is positioned in depth of about 2 km. Assuming that four weapons are available and the candle array is to be placed such that the closest candle is 9 km away from the gun position (assume all guns occupy the same point in space) and there is one candle every 500 m (i.e., four candles total at 9.0, 9.5, 10.0, and 10.5 km), determine the QE for each weapon, the time set on each fuze, and the time and order of fire of each gun such that all the expulsion events simultaneously occur at an altitude of 750 m over the heads of the enemy. Assume a muzzle velocity of 600 m/s and a projectile mass of 30 lbm. Use a vacuum trajectory for the purposes of this problem. (Note that this would *never* be acceptable in actual practice.)

Could the same result be achieved with less than four guns? Show why or why not with a calculation.

Problem 6

The fire control problem: During the age of battleships, it was essential that a fire control computer be installed on the ships. From 1915 onwards, these were mechanical devices by which one could put in the course and speed of their own ship as well as the estimated course and speed of the target ship. These data and the expected muzzle velocity of the projectile (as well as TOF) allowed the guns to be aimed where the enemy ship would be when the shells landed. To get a feel for the magnitude of the problem, we are going to examine it in a greatly simplified form assuming a vacuum trajectory. Given the data that follow, and assuming a vacuum trajectory, provide a firing angle off the bow, elevation, and timing to fire each of four guns so that a pattern is created to hit if the target veers 10° to the port or starboard of its present course. A hit can be assumed to occur if the shell lands on the point where the enemy ship will actually be or if it lands in its "danger space." Because of the trajectory of the shells, a hit will occur if the trajectory passes over the target and lands within 100 yards behind it—the ship creates a "shadow" or danger space. An example pattern might look like the one drawn in the figure, but feel free to create your own if it meets the aforementioned criteria. The shells should all be fired at the same time. Assume that your speed estimate of the target is exact. Also assume the four guns are mounted in two pairs so that you have only two azimuths to work, with but the elevations can be independently varied. The target is about 400 ft long, so some error in azimuth is acceptable—but the error should be quantified. Plot the impact points and target position at the time of impact. Remember that your ship is moving!

The weapons are British 12 in. mark IX naval guns with a muzzle velocity of 2800 ft/s and a maximum elevation of 20°.

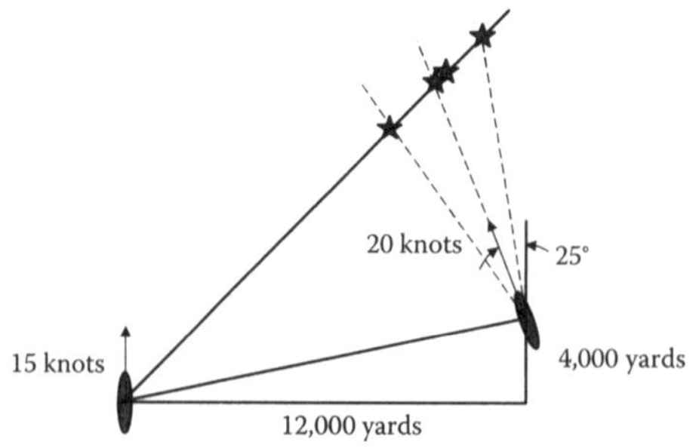

9.2 Simple Air Trajectory (Flat Fire)

As we progress in our study of exterior ballistics, we now introduce the concept of drag by substituting air for the medium through which our point mass projectile flies. We do this so that projectile dynamics do not enter yet into the equations of motion. We are essentially still dealing with a spherical, nonrotating cannonball. Furthermore, to simplify the mathematics, we will insist on a flat fire trajectory, with launch angles below about 6°. A flat fire trajectory is depicted in Figure 9.3. The methods and equations we will develop were used in the 1950s for direct-fire calculations over relatively short ranges [1].

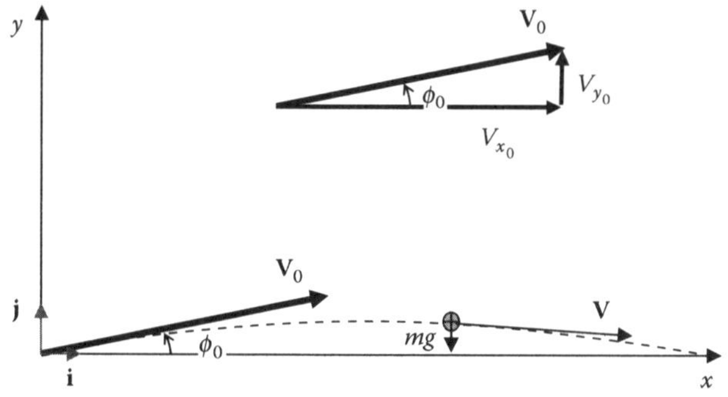

FIGURE 9.3
Flat fire trajectory.

We begin with Newton's second law in an inertial reference frame and use vector representations (boldface) where appropriate:

$$\mathbf{F} = m\mathbf{a} \tag{9.41}$$

$$m\frac{d\mathbf{V}}{dt} = \Sigma\mathbf{F} + m\mathbf{g} \tag{9.42}$$

Where m is the projectile mass; $\mathbf{V}$ is the projectile velocity vector; and t is the time; $\mathbf{a} = \frac{d\mathbf{v}}{dt}$ = vector acceleration; $\Sigma\mathbf{F}$ is the vector sum of all aerodynamic forces; and $\mathbf{g}$ is the vector acceleration due to gravity.

The inertial reference frame allows us to neglect the Coriolis acceleration, which is the result of the earth's rotation. Since there is no angle of yaw, the lift and drag forces due to yaw and the Magnus force due to spin are also negligibly small. These will be discussed in detail later. Thus, only the projectile drag forces (base, wave, and skin friction) are working to slow the projectile down, and gravity is pulling it toward the earth. The aerodynamic drag force acting on the projectile is then given by

$$F_D = -\frac{1}{2}\rho S C_D \mathbf{V} V = \frac{1}{2}\rho V^2 S C_D \tag{9.43}$$

Where C_D is the drag coefficient, introduced earlier. Another coefficient in common use in ballistics is the ballistic coefficient C, which is defined as

$$C = \frac{m}{d^2} \tag{9.44}$$

Where m and d are the mass and diameter of the projectile, respectively. As a matter of convenience, we also define

$$\hat{C}_D^* = \frac{\rho S C_D}{2m} = \frac{\rho\pi}{8}\frac{C_D}{C} \tag{9.45}$$

This allows us to save a little energy in typing since this combination of parameters appears so often. It is known as a starred coefficient [2]. Equation 9.45 stems from the fact that S, the frontal area of the projectile, is

$$S = \frac{\pi d^2}{4} \tag{9.46}$$

We can combine Equations 9.42 and 9.45 and divide by the mass to get an expression for the time rate of change of velocity (acceleration)

$$\frac{d\mathbf{V}}{dt} = -\frac{1}{2m}\rho S C_D \mathbf{V} V + \mathbf{g} \tag{9.47}$$

The negative sign was placed in front of the force (Equation 9.43) because the drag always opposes the velocity vector (otherwise, it is called thrust). We can separate the velocity, acceleration, and gravitational vectors into components along the coordinate axes so that they will be convenient to work with:

$$\frac{d\mathbf{V}}{dt} = \dot{V}_x\mathbf{i} + \dot{V}_y\mathbf{j} + \dot{V}_z\mathbf{k} + \left(V_x\dot{\mathbf{i}} + V_y\dot{\mathbf{j}} + V_z\dot{\mathbf{k}}\right) \tag{9.48}$$

and

$$\mathbf{g} = -g\mathbf{j} \tag{9.49}$$

But because we are in an inertial frame, $\dot{\mathbf{i}} = \dot{\mathbf{j}} = \dot{\mathbf{k}} = 0$, and therefore,

$$\frac{d\mathbf{V}}{dt} = \dot{V}_x\mathbf{i} + \dot{V}_y\mathbf{j} + \dot{V}_z\mathbf{k} \tag{9.50}$$

If we break Equation 9.47 into its components, we get three coupled, ordinary, nonlinear differential equations

$$\dot{V}_x = -\hat{C}_D^* V V_x \tag{9.51}$$

$$\dot{V}_y = -\hat{C}_D^* V V_y - g \tag{9.52}$$

$$\dot{V}_z = -\hat{C}_D^* V V_z \tag{9.53}$$

The equation that couples Equations 9.51 through 9.53 is

$$V = \sqrt{V_x^2 + V_y^2 + V_z^2} \tag{9.54}$$

We can linearize these equations by making a few assumptions. First, let us assume that there is no crosswind, so $V_z = 0$, and if we further constrain the ratio of the vertical velocity to the horizontal velocity to $|V_y/V_x| = \tan\phi < 0.1$, then V and V_x are within 0.5% of each other, and we have constrained the launch and fall angles to be less than 5.7°, the angles introduced in the preceding section for the flat fire approximation.

So with the assumptions that $V = V_x$ and $V_z = 0$, we can develop Equations 9.51 through 9.53 into

$$\dot{V}_x = -\hat{C}_D^* V_x^2 \tag{9.55}$$

$$\dot{V}_y = -\hat{C}_D^* V_x V_y - g \tag{9.56}$$

$$\dot{V}_z = 0 \tag{9.57}$$

These differential equations use time as the independent variable. It is often convenient to use distance as the independent variable. By making a common transformation of variables to allow distance along the trajectory to be the independent variable instead of time, we can improve our ability to work with these expressions. Performing the transformation results in equations of the form

$$V_x V_x' = -\hat{C}_D^* V_x^2 \tag{9.58}$$

$$V_x V_y' = -\hat{C}_D^* V_x V_y - g \tag{9.59}$$

where the prime denotes differentiation with respect to distance.

By dividing both equations by V_x, we obtain Equations 9.60 and 9.61 that use the downrange distance x as the independent variable. For these equations, an analytic solution does exist:

$$V_x' = -\hat{C}_D^* V_x \tag{9.60}$$

$$V_y' = -\hat{C}_D^* V_y - \frac{g}{V_x} \tag{9.61}$$

Equation 9.60 can be integrated by separation of variables as

$$V_x = V_{x_0} \exp\left(-\hat{C}_D^* \int_0^x dx_1 \right) \tag{9.62}$$

In this equation and future equations, we use a variable x_i or t_i as a dummy variable of integration. Equation 9.61 can also be solved by quadrature methods since it is of the form

$$\frac{dV_y}{dx} + \hat{C}_D^* V_y = -\frac{g}{V_x} \tag{9.63}$$

Equation 9.61 can be solved for V_y for initial conditions at $x = 0$, $t = 0$, and $V_y = V_{y_0}$ as

$$V_y = \left\{ \exp\left(-\int_0^x \hat{C}_D^* \, dx_1 \right) \right\} \left[V_{y_0} - \int_0^x \left(\frac{g}{V_x} \right) \left\{ \exp\left(\int_0^{x_2} \hat{C}_D^* \, dx_1 \right) \right\} dx_2 \right] \tag{9.64}$$

If we take the ratio of the x- and y-velocity components, we can obtain the relation for the angle the velocity vector makes with the horizontal. This can be shown to be

$$\tan\phi = \left[\tan\phi_0 - \frac{1}{V_x} \int_0^x \left(\frac{g}{V_x} \right) \left\{ \exp\left(\int_0^{x_2} \hat{C}_D^* \, dx_1 \right) \right\} dx_2 \right] \tag{9.65}$$

Where ϕ_0 is the initial launch angle.

To complete our study of the flat fire trajectory, we need to find the elements of it, i.e., the x and y values along it, from launch to termination. To do this, we must integrate over time the velocities we have found in Equations 9.62 and 9.64, which we had earlier transformed into distance variables. We know by definition

$$y = \int_0^t V_y \, dt \quad \text{and} \quad x = \int_0^t V_x \, dt \tag{9.66}$$

By substituting Equations 9.62 and 9.64 into each of the equations in Equation 9.66 in turn and performing the integrations, we can show that with the initial conditions of $x = 0$, $t = 0$, and $y = y_0$,

$$y = t \exp\left(\int_0^x \hat{C}_D^* dx_1 \right) \left[V_{y_0} - \int_0^x \left(\frac{g}{V_x} \right) \left\{ \exp\left(\int_0^{x_2} \hat{C}_D^* \, dx_1 \right) \right\} dx_2 \right] + y_0 \tag{9.67}$$

Now we can find t from $V_x = dx/dt$. By separating the variables and substituting Equation 9.62 for V_x, we get

$$t = \frac{1}{V_{x_0}} \int_0^x \exp\left(\int_0^{x_2} \hat{C}_D^* \, dx_1 \right) dx_2 \tag{9.68}$$

Through a somewhat tedious set of algebraic substitutions and manipulations that are contained in the book by McCoy [1], we can arrive at our desired equation in x and y; the launch angle ϕ_0; the dummy range variables x_1, x_2, and x_3; the initial launch velocity V_{x_0}; the initial ordinate y_0; and the drag coefficient $\hat{C}_D^*$:

$$y = y_0 + x\tan\phi_0 - \frac{gx^2}{2V_{x_0}^2}\left[\frac{2}{x^2}\int_0^x\int_0^{x_3}\left\{\exp\left(2\int_0^{x_2}\hat{C}_D^*\,dx_1\right)\right\}dx_2dx_3\right] \tag{9.69}$$

The disadvantage of these equations is that the variation of drag coefficient has to be simple to evaluate the integrals. Since the drag coefficient does not vary in a simple manner with Mach number, this makes the analytic solutions inaccurate and difficult to accomplish. Figure 9.4 depicts a typical drag curve that varies with Mach number. One can see from this figure that there is no simple analytic solution to this variation. With computer power nowadays, we usually solve or approximate the exact solutions numerically, doing the quadratures by breaking the area under the curve into quadrilaterals and summing the areas.

To integrate these equations analytically, we will examine three forms of the drag coefficient:

1. Constant C_D that is useful for the subsonic flight regime, $M < 1$
2. C_D inversely proportional to the Mach number that is characteristic of the high-supersonic flight regime, $M \gg 1$
3. C_D inversely proportional to the square root of the Mach number that is useful in the low-supersonic flight regime, $M \geq 1$

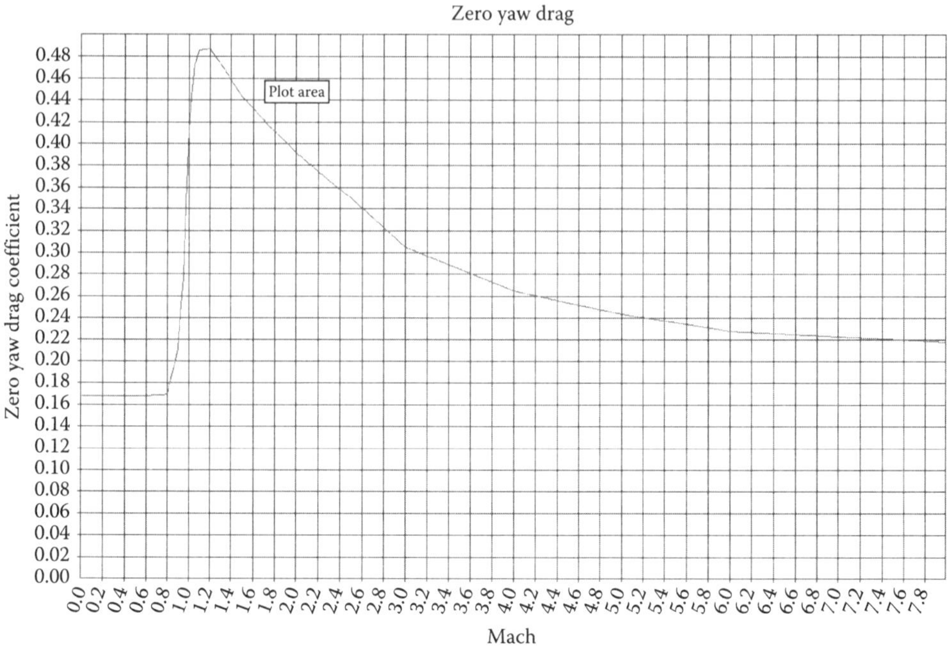

FIGURE 9.4
Drag coefficient vs. Mach number for a typical projectile.

First, we will examine the case of a constant drag coefficient. If we examine Figure 9.4, we can see that this would be a useful approximation for our projectile behavior if the launch velocity was, say, between Mach 0.8 and 0. We assume that the drag force varies with the square of the velocity (the drag coefficient was the drag force divided by the dynamic pressure $\frac{1}{2}\rho V^2$), and we set the drag coefficient equal to a constant K_1. We shall use terminology consistent with that of McCoy [1] so that

$$\hat{C}_D^* = \frac{\rho S}{2m} C_D = \frac{\rho S}{2m} K_1 = k_1 \tag{9.70}$$

Which we then substitute in Equation 9.62:

$$V_x = V_{x_0} \exp\left(-k_1 \int_0^x dx_1\right) = V_{x_0} \exp(-k_1 x) \tag{9.71}$$

We can find t by substituting Equation 9.70 into Equation 9.68 to give

$$t = \frac{1}{V_{x_0}} \int_0^x \exp\left(\int_0^{x_2} k_1 dx_1\right) dx_2 = \frac{1}{V_{x_0}} \int_0^x \exp(k_1 x_2) dx_2 \tag{9.72}$$

or

$$t = \frac{1}{V_{x_0} k_1} (\exp[k_1 x] - \exp[0]) = \frac{1}{V_{x_0} k_1} (\exp[k_1 x] - 1) \tag{9.73}$$

Noting that from Equation 9.62,

$$\exp\left(-\int_0^x \hat{C}_D^* dx_1\right) = \frac{V_x}{V_{x_0}} \quad \text{and} \quad \exp\left(\int_0^x \hat{C}_D^* dx_1\right) = \frac{V_{x_0}}{V_x} \tag{9.74}$$

and noting that $(V_{y_0}/V_{x_0}) = \tan\phi_0$, we can show through manipulation [1] that

$$V_y = V_x \left[\tan\phi_0 - \frac{gt}{V_{x_0}}\left(1 + \frac{V_{x_0} k_1 t}{2}\right)\right] \tag{9.75}$$

To find the angle of fall ϕ as a function of range x and the instantaneous velocity at x, V_x, we solve Equation 9.71 for k_1:

$$k_1 = \frac{1}{x} \ln\left(\frac{V_{x_0}}{V_x}\right) \tag{9.76}$$

We now substitute Equation 9.76 into Equation 9.73 for t. Taking the result and recalling that $\tan\phi = V_y/V_x$ for any x, we use this new equation for t and transform Equation 9.65 into

$$\tan\phi = \tan\phi_0 - \frac{gt}{V_{x_0}}\left[1 + \frac{V_{x_0} t}{2} \frac{1}{x} \ln\left(\frac{V_{x_0}}{V_x}\right)\right] \tag{9.77}$$

Finally, to find the altitude y at any point along the trajectory as a function of the range and the velocity at that range, we transform Equation 9.65 with the constant drag coefficient k_1, use the new equation for t that we derived earlier, and after manipulation arrive at

$$y = y_0 + x\tan\phi_0 - \frac{g}{2}\left[\frac{x}{V_{x_0}}\frac{1}{\ln\left(\frac{V_{x_0}}{V_x}\right)}\right]^2\left[\frac{1}{2}\left(\frac{V_{x_0}}{V_x}-1\right)^2+\left(\frac{V_{x_0}}{V_x}-1\right)-\ln\left(\frac{V_{x_0}}{V_x}\right)\right] \quad (9.78)$$

A constant drag coefficient is useful when analyzing low-subsonic projectiles since most of them have nearly constant drag coefficients. Also, projectiles at hypersonic speeds (usually described as a Mach number greater than five) can be analyzed with this assumption (look again at Figure 9.4). Essentially, we are linearizing the problem when we do this.

Our next effort will be to examine a nonconstant drag coefficient, one varying as the inverse of the Mach number. In this case, we assume that the drag force varies linearly with the velocity (because the drag coefficient is the drag force divided by the dynamic pressure, $\frac{1}{2}\rho V^2$, and when we divide by the Mach number, we essentially divide by the velocity times a constant). Now we set the drag coefficient equal to K_2/M, then

$$C_D = \frac{K_2}{M} \quad (9.79)$$

and

$$\hat{C}_D^* = \frac{\rho S}{2m}C_D = \frac{\rho S}{2m}\frac{K_2}{M} \quad (9.80)$$

Recall that the Mach number is V/a, where a is the speed of sound in air. Then, for our flat fire approximation, we can define a constant k_2 such that

$$k_2 = \frac{\rho S}{2m}K_2 a \quad (9.81)$$

Then,

$$\hat{C}_D^* = \frac{\rho S}{2m}\frac{K_2 a}{V_x} = \frac{k_2}{V_x} \quad (9.82)$$

From Equations 9.55 and 9.56, we see that

$$\dot{V}_x = -k_2 V_x \quad (9.83)$$

and

$$\dot{V}_y = -k_2 V_y - g \quad (9.84)$$

Also from Equation 9.60, we see that

$$V'_x = -\hat{C}_D^* V_x \quad (9.85)$$

Using these three equations and proceeding in the same fashion as we did with the constant C_D, we can derive equations for the x- and y-velocities, the TOF to any range x, the angle of fall ϕ; and the trajectory ordinate at any range. These equations are (details in the book by McCoy [1])

$$V_x = V_{x_0}\exp(-k_2 t) \quad \text{(in terms of } t\text{)} \quad (9.86)$$

$$V_y = \left(V_{y_0} + \frac{g}{k_2}\right)\exp(-k_2 t) - \frac{g}{k_2} \quad \text{(in terms of } t\text{)} \quad (9.87)$$

$$t = \frac{\dfrac{x}{V_{x_0}} \ln\left(\dfrac{V_{x_0}}{V_x}\right)}{\left(1 - \dfrac{V_x}{V_{x_0}}\right)} \tag{9.88}$$

$$\tan\phi = \tan\phi_0 + \frac{gx}{V_{x_0}^2}\left(\frac{1 - \dfrac{V_{x_0}}{V_x}}{1 - \dfrac{V_x}{V_{x_0}}}\right) \tag{9.89}$$

$$y = y_0 + x\tan\phi_0 - \left(\frac{gt^2}{\ln\dfrac{V_{x_0}}{V_x}}\right)\left(1 - \frac{\left(1 - \dfrac{V_x}{V_{x_0}}\right)}{\ln\left(\dfrac{V_{x_0}}{V_x}\right)}\right) \tag{9.90}$$

These relations, for C_D proportional to $1/M$, are useful in the analysis of high-supersonic projectiles such as kinetic energy armor penetrators where $2.5 < M < \sim 5$.

For the case where the drag coefficient varies as the $\sqrt{M}$, we assume that the drag varies with velocity to the 3/2 power and we set the drag coefficient equal to $K_3/\sqrt{M}$; then,

$$\hat{C}_D^* = \frac{\rho S}{2m}\frac{K_3}{\sqrt{M}} \tag{9.91}$$

Since

$$\sqrt{M} = \sqrt{\frac{V}{a}} \tag{9.92}$$

We can define a new constant as

$$k_3 = \frac{\rho S}{2m} K_3\sqrt{a} \tag{9.93}$$

We can then write (assuming $V = V_x$)

$$\hat{C}_D^* = \frac{\rho S}{2m} K_3 \sqrt{\frac{a}{V_x}} = \frac{k_3}{\sqrt{V_x}} \tag{9.94}$$

Proceeding as we did in the earlier two cases, we can derive V_x, V_y, t, ϕ, and the ordinate y. The details of the derivations are again available in the book by McCoy [1]:

$$V_x = \frac{4V_{x_0}}{\left(k_3\sqrt{V_{x_0}}t + 2\right)} \quad \text{(in terms of } t\text{)} \tag{9.95}$$

$$V_y = -\frac{g}{\left(k_3\sqrt{V_{x_0}}t+2\right)}\left(\frac{k_3^2 V_{x_0} t^3}{3} + 2k_3\sqrt{V_{x_0}}t^2 + 4t\right) + \frac{4V_{y_0}}{\left(k_3\sqrt{V_{x_0}}t+2\right)^2} \quad \text{(in terms of } t\text{)} \tag{9.96}$$

$$t = \frac{x}{V_{x_0}}\sqrt{\frac{V_{x_0}}{V_x}} \tag{9.97}$$

$$\tan\phi = \tan\phi_0 - \frac{gt}{V_{x_0}}\left[\frac{1}{3}\left(\frac{V_{x_0}}{V_x} + \sqrt{\frac{V_{x_0}}{V_x}} + 1\right)\right] \tag{9.98}$$

$$y = y_0 + x\tan\phi_0 - \frac{1}{2}gt^2\left[\frac{1}{3}\left(1 + 2\sqrt{\frac{V_x}{V_{x_0}}}\right)\right] \tag{9.99}$$

These last equations (Equations 9.95 through 9.99) are useful for flight in the low- to moderate-supersonic regime, $1 < M < 2.5$.

In summary, we have derived the equations of motion assuming a flat fire trajectory. We use them when the angle of departure and angle of fall are both below 5.7°. We have solved them with three drag assumptions:

1. A constant drag coefficient that is useful in the subsonic and hypersonic regimes and can be used over short distances in all Mach regimes
2. A drag coefficient inversely proportional to the Mach number that is useful in the high-supersonic regime
3. A drag coefficient inversely proportional to the square root of the Mach number that is useful in the low-supersonic flight regime

Problem 7

The French infantry rifle model 1886 called the Lebel was their standard weapon from 1886 into the First World War and even saw limited use in the Second World War. You can see this 51 in. long monster in any movie involving the French Foreign Legion. It used an 8 mm cartridge called the balle D with a bullet mass of 198 grains and a diameter of 0.319 in. This cartridge–rifle combination has a muzzle velocity of 2296 ft/s. Assuming flat fire with $K_3 = 0.5$ and using standard sea level met data ($\rho = 0.0751$ lbm/ft^3; $a = 1120$ ft/s)

1. Create a table containing range (yards), impact velocity (feet per second), TOF (seconds), initial QE angle (minutes), and angle at impact (minutes) in 200-yard increments out to 1000 yards.
2. If an infantryman is looking at a target at 2000 yards, what angle will the sight have relative to the tube assuming they used standard met in the design?

   Answer: About 10.3°

Comment on the validity of this method with respect to question 2.

Problem 8

British 0.303 in. ball ammunition is to be fired in an Mk.1 Maxim machine gun. The mass of the bullets is 175 grains. When used in this weapon, it has a muzzle velocity of 1820 ft/s. Assuming flat fire with $K_3 = 0.5$ and using standard sea level met data ($\rho = 0.0751$ lbm/ft^3; $a = 1120$ ft/s)

1. Create a table containing range (yards), impact velocity (feet per second), TOF (seconds), initial QE angle (minutes), and angle at impact (minutes) in 200-yard increments out to 1000 yards.
2. The weapon was used by British units assigned to bolster the Italians in the Alps during the First World War (Italy came in on the Allied side because they wanted the Tyrol region from Austria more than they wanted the Nice region from France). At an altitude of 3000 ft, how much higher or lower will a bullet fired from this weapon impact a level target if the sights are set using the sea level conditions given earlier and the target is at 600 yards? At this altitude, assume the density and temperature of the atmosphere are $\rho = 0.0551$ lbm/ft^3 and $T = 20°$F.

   Answer: $y = 3.078$ [ft] (too high)

Problem 9

The main armament in the Italian M13-40 during the Second World War was a 47 mm/32 caliber weapon designed and built by the Ansaldo Arms company. The most effective antitank projectile it carried was an armor-piercing, ballistic capped round, which had a muzzle velocity of 2060 ft/s. With this particular projectile–weapon combination, the assumption of constant drag coefficient seems to yield reasonable results. The k_1 value for this case is 0.00025 [1/m]. Using the flat fire, point mass trajectory create a table of range (yards), velocity (feet per second), initial QE (minutes), and impact QE (minutes) out to 1000 yards in 200-yard increments.

Problem 10

A US 37 mm projectile is fired with a muzzle velocity of 2600 [ft/s]. The projectile weighs 1.61 lbm. Assuming $K_2 = 0.841$ [unitless] and using standard sea level met data ($\rho = 0.0751$ lbm/ft^3; $a = 1120$ ft/s; $R = 1716$ [ft lbf/slug R])

1. Determine the drag coefficient C_D and drag force on the projectile if the projectile is fired in still air.

   Answer: $F_D = 33.04$ [lbf]

2. Create a table containing range (yards), impact velocity (feet per second), TOF (seconds), initial QE angle (minutes), and angle at impact (minutes) in 100-yard increments out to 800 yards.
3. If this weapon is used at an increased altitude and assuming that the density and temperature of the atmosphere are $\rho = 0.060$ lbm/ft^3 and $T = 30°$F, respectively, how much higher or lower will the weapon have to be aimed to hit a target at 800 yards?

   Answer: The weapon must be aimed 0.28 mil or 0.98 min lower.

Problem 11

You are asked to create a rough safety fan for at maximum range test at Yuma Proving Ground. The test consists of an experimental 155 mm howitzer projectile at a severe overpressure charge. The projectile weighs 106 lbm. The muzzle velocity is 980 m/s.

1. Using a vacuum trajectory calculate and plot the trajectory envelope for the test.
2. Determine the longest TOF of the projectile.

If we assume an average constant drag coefficient (on the way up) of $C_D = 0.5/M$ and the projectile were fired vertically, what would the maximum ordinate be? Assume standard sea level met data ($\rho = 0.0751$ lbm/ft^3; $a = 1120$ ft/s; $R = 1716$ [ft·lbf/slug·R]).
Hint: Look at the flat fire assumptions and rederive the equation of motion assuming there never is an x-velocity.

Problem 12

The drag of a sphere below Mach number of 0.5 is well approximated by $C_D = AM^2 + BM + C$, where $A = 0.0262$, $B = 0.0456$, $C = 0.4666$, and M is the Mach number. If we would like to analyze the motion of a cannonball fired from a demiculverin circa 1646, which fired a spherical shot weighing 9 lbm and of 4.5 in. diameter, please develop the equation of motion for V_x (only) as a function of downrange distance based on the flat fire assumptions. Then, using standard sea level met data ($\rho = 0.0751$ lbm/ft^3; $a = 1120$ ft/s; R=1716 [ft·lbf/slug·R]), determine V_x at a range of 1800 yards (the reported maximum range of the weapon). Assume a muzzle velocity of 550 ft/s.

Problem 13

A French 240 mm projectile is fired from a model 1873 cannon with a muzzle velocity of 440 [m/s]. The projectile mass is 144 kg. Assuming $K_1 = 0.55$ [unitless] and using standard sea level met data ($\rho = 0.0751$ lbm/ft^3; $a = 1120$ ft/s; $R = 1716$ [ft·lbf/slug·R])

1. Create a table containing range (meters), impact velocity (meters per second), TOF (seconds), initial quadrant elevation angle (degrees), and angle at impact (degrees) in 600 m increments out to 7800 m.
2. Experiments were conducted in France in 1884 to determine the effect of rotating and locations on the range of this projectile. These experiments were reported on by Breger in *Notes on the Construction of Ordnance No. 27*, in June 10, 1884. The farther rearward the rotating band was placed, the greater the projectile yaw at muzzle. The same QE that provided a range of 7800 m actually only made 7688 m with a (assume constant) yaw angle of 4°. Assume that the 7800 m range was achieved with zero initial yaw. With this information, calculate the yaw drag coefficient $C_{D_{\delta^2}}$.

Problem 14

It is desired to develop a close protection system using a 0.50 caliber machine gun. The muzzle velocity of the weapon is 2950 ft/s. The projectile is an M8 bullet with a diameter of 0.50 in. and a mass of 0.09257 lbm. Since the projectile interception mission is to occur relatively close to the firing platform, we can assume the projectile behaves according to the flat fire assumption with a constant drag coefficient of $K_1 = 0.45$ [unitless]. Assuming the flat fire assumption is valid for the trajectory and the intercept point is to be 20 ft above the ground, develop a firing table for the intercept out to 300 yards in 50-yard increments. Assume standard sea level met data (ρ = 0.0751 lbm/ft^3; a = 1,120 ft/s; R = 1716 [ft·lbf/slug·R]).

Create a table containing range (yards), impact velocity (feet per second), TOF (seconds), and initial quadrant elevation angle (degrees) in 50-yard increments out to 300 yards. Comment on the accuracy of your answer.

Problem 15

If we were to assume that the flat fire conditions were to hold in a coordinate system that were elevated to an angle θ, rederive the differential equations in time and space coordinates—*do not* solve them, just put them in the form of Equations 9.60 and 9.61. Also, write the transformation equations between the normal and slant velocity (V_n and V_s) and V_x and V_y.

Problem 16

One of the problems with hit-to-kill close protection systems is the ability to accurately point the weapon and fire in a timely fashion at a very small target. Assume that we must impact a sphere 4 in. in diameter at the ranges developed in Problem 14. For each range, assuming perfect timing as well as perfect projectile tracking, determine the allowable tolerance in QE to impact the target.

Problem 17

Let us now assume that the pointing against the incoming round in Problem 14 is absolutely perfect. Determine the tolerance in lock time (officially the time from pulling the trigger to weapon firing—but we will assume it is to muzzle exit) to hit the target at the conditions of Problem 14. Assume that the incoming projectile is moving at 300 ft/s. How does this change if the velocity estimate is ±20 ft/s?

Problem 18

The main armament of the last prewar US Heavy Cruisers (known as the tin-clads) was an 8 in./55 caliber weapon. The effective range of this weapon was 30,000 yards at an elevation of 40°43′. During the Second World War, there were many night actions in the Pacific where these weapons were used at an extremely short range (less than 10,000 yards). You are asked to create a firing table for this weapon at the short ranges. The muzzle velocity of the weapon/projectile/propellant combination is 2500 ft/s. The projectile is an armor-piercing, capped round with a diameter of 8 in. and a mass of 335 lbm. Since the range is short, we can assume the projectile behaves according to the flat fire assumption with a drag coefficient inversely proportional to the Mach number of K_2 = 0.62 [unitless] (note that this is not really a great fit for this projectile). Assuming the flat fire assumption is valid for the trajectory, develop a firing table for the system to 10,000 yards in 1,000-yard increments. Assume standard sea level met data (ρ = 0.0751 lbm/ft^3; a = 1,120 ft/s; R = 1716 [ft·lbf/slug · R]).

Create a table containing range (yards), impact velocity (feet per second), TOF (seconds), and initial quadrant elevation angle (degrees) in 1,000-yard increments out to 10,000 yards.

Problem 19

The analog fire control computers installed on board ships during the Second World War were amazing devices. The inputs required were course and speed of the firing ship, estimated range to the target and course, and speed of the target. Inaccuracies in the target course and speed estimates were compensated for by generating a "ladder"; this was a shell pattern that was a linear array using as many guns as were available in one "salvo." The ideal result was that the target would end up directly in the middle of the salvo and be "straddled." Because of the relatively flat projection of the shells, being straddled usually guaranteed that the target was hit by at least one projectile. The shorter the range to a target, the larger the "danger space" offered and the better the chance of a hit. In this problem, you assume the role of the fire control computer. The weapons available are nine 8 in. /55 caliber guns with the ballistic performance from Problem 18. Your ship is moving due north at 30 knots (one knot is one nautical mile per hour or 2000 yards per 3600 s). At the instant of fire, the enemy ship is dead ahead of your ship traveling at 35 knots on course 090 (see the following figure) at a range of 8000 yards. Ignore the effect of the launch platform motion on the drag (only) of the projectile.

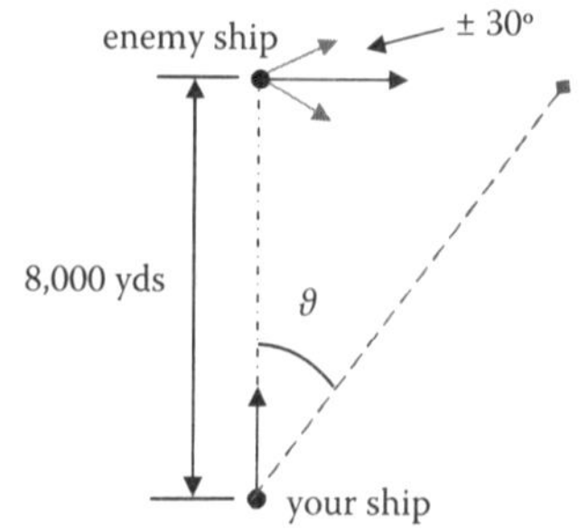

1. Determine the firing solution assuming both ships continue straight ahead (QE and relative angle to the bow of your ship) for one shell to impact the enemy. (Because of the flat trajectory, it is good to aim so the shell falls a little behind the enemy ship.)

   Hint: Remember the projectiles are leaving from a ship that is moving!

2. Perform the same calculation if the enemy turns 30° toward the firing ship and 30° away from the firing ship.
3. Using a method of your choosing, examine the sensitivity of the fire control problem to (a) an incorrect firing ship speed, (b) an incorrect target ship course, and (c) an incorrect target ship speed. Quantify their relative importance.

Problem 20

The US 7.62 mm Ball M80 (projectile diameter = 0.308 in.; mass m = 147 grains) is fired in a test range. Based on data given next, estimate the coefficient C_D. Assume the projectile is fired with a muzzle velocity of 2810 ft/s, under standard sea level met conditions (ρ = 0.0751 lbm/ft^3, a = 1120 ft/s). Justify your answer by explaining why you chose the appropriate drag model. Validate your answer with an appropriate calculation.

Range (yards)	V_0 (ft/s)	V_x (ft/s)
400	2810	1960
500	2810	1765
600	2810	1580

Problem 21

Many times, all the data we need for a projectile are not provided to us and we have to extract the information from different sources. You are given the following information about a British 2 pounder projectile [3]:

- Projectile diameter: 40 mm
- Projectile weight: 2.375 lbm
- Muzzle velocity: 2600 ft/s
- Armor penetration as a function of distance

Distance (yards)	Thickness Perforated (mm)
100	55
500	47
1000	37
1500	27

In terminal ballistics, we will find that, based on some work by Zener and Holloman in 1942, the penetration of this type of projectile is proportional to the velocity as follows:

$$V \sim t\sqrt{\frac{d}{m}},$$

where t is the target thickness; d is the projectile diameter; and m is the projectile mass. With only this information at your disposal,

1. Determine the best drag model for this projectile.
2. Generate the proper coefficient from the data.
3. Create a table of range (yards), velocity (feet per second), TOF (seconds), launch angle (minutes), and impact angle (minutes) if the projectile is fired with no wind at each position.

Please justify your answer.

Problem 22

A Hornady 0.308 in. diameter 208 grain Amax bullet is to be analyzed. Based on data collected by Litz [4], it has the following drag characteristics:

V (ft/s)	C_D
1500	0.354
2000	0.306
2500	0.274
3000	0.250

If it is fired with a muzzle velocity of 2780 ft/s, assuming standard sea level met data (ρ = 0.0751 lbm/ft^3; a = 1120 ft/s),

1. Determine a proper drag model to use over this range. Justify your answer by comparing numbers from all three drag models that were introduced.

2. Assuming $K_3 = 0.409$, create a table of range (yards), velocity (feet per second), TOF (seconds), launch angle (minutes), impact angle (minutes), and drift (yards) if the projectile is fired with no wind every 100 yards out to 600 yards.
3. Determine the deflection in inches assuming the projectile experiences a headwind of 20 ft/s for the entire flight. Tabulate every 100 yards out to 600 yards.
4. Determine the deflection in inches assuming the projectile experiences a crosswind (left or right—your choice) of 20 ft/s for the entire flight. Tabulate every 100 yards out to 600 yards.

Problem 23

Many precision shooters develop specific propellant loads for competition shooting for a given propellant type, projectile, barrel, cartridge, and primer combination. The strategy for optimizing a propellant loading is to incrementally increase the propelling charge with the weapon aimed at the same location and determine the charge at which the projectiles "group" the most. This method, called a Creighton Audette ladder test, finds the so-called "sweet spot" of the gun. The projectiles group due to a combination of barrel motion owing to recoil and vibration. To illustrate this point, let us assume that we have a perfectly rigid (i.e., no vibration) 22 in. barrel that fires the projectile of Problem 22. Assume a range of 300 yards. Assume that the projectile velocity during barrel transit can be approximated as ½ of the muzzle velocity. Also, assume that the recoil is fairly constant and causes a constant upward rotation of the muzzle of 3.3 rad/s. For the following 20 velocities, determine the muzzle velocity (which would correspond to a propellant load) that is in the sweet spot. Please plot impact height vs. muzzle velocity.

Hint: The muzzle rise will affect the initial launch angle, but the flatter trajectory will compensate to some degree. Although not done in practice, for this calculation, start at the high muzzle velocity and track the bullet impact heights as you lower the velocity.

V_0 (ft/s)
2780
2775
2770
2765
2760
2755
2750
2745
2740
2735
2730
2725
2720
2715
2710
2705
2700
2695
2690
2685

Problem 24

Normally on a fin-stabilized projectile, the spin damping due to the body is much smaller than that due to the fins. A 155 mm projectile weighs 101 lbm and is designed so that it leaves the muzzle of the weapon at 600 m/s and spinning at 220 Hz. After 0.5 s, fins are deployed. At this instant in time, the spin rate is 197 Hz and the velocity is 500 m/s. After another 0.5 s, the projectile has achieved its steady state spin rate of 12 Hz and is at a velocity of 400 m/s. The changes in polar moment associated with the fin deployment are provided next (be sure to think about which one is before and which is after). Determine the following:

1. The drag coefficient of the both flight configurations
2. The spin-damping coefficient for the body (only) and the fins (only)
3. The roll coefficient (C_{1_δ}) of the fins assuming a 1° cant

Assume

$$\rho = 0.076\left[\frac{\text{lbm}}{\text{ft}^3}\right] \quad a \approx 330\left[\frac{\text{m}}{\text{s}}\right] \quad I_{P_A} = 10.1\left[\text{lbm}\cdot\text{ft}^2\right] \quad I_{P_B} = 10.8\left[\text{lbm}\cdot\text{ft}^2\right]$$

It is important in this problem, since it is pretty open ended, that you list all your assumptions and present your answer in a manner appropriate to those assumptions.

Problem 25

In a test range, a 0.50 caliber M33 ball projectile is fired at an elevation of 10° with a muzzle velocity of 3013 ft/s. The initial pitch and yaw angles are 1.030° and 1.263°, respectively. The initial pitch and yaw rates are 2 rad/s nose down and 1 rad/s nose left, respectively. If the projectile has the following coefficients at this particular instant, write the acceleration vector and the angular momentum vector.

Please ignore the Coriolis acceleration and assume the weapon has a right-hand twist.

- Projectile information:

$$C_D = 0.2938 \quad \left(C_{M_q} + C_{M_{\dot\alpha}}\right) = -5.5 \quad I_P = 7.85\left[\text{g-cm}^2\right]$$

$$C_{M_\alpha} = 2.88 \quad \left(C_{N_q} + C_{N_{\dot\alpha}}\right) = 0.004 \quad I_T = 74.5\left[\text{g-cm}^2\right]$$

$$C_{L_\alpha} = 2.69 \quad C_{M_{p\alpha}} = 0.05 \quad m = 42.02\left[\text{g}\right]$$

$$C_{N_{p\alpha}} = -0.01 \quad \rho = 0.0751\left[\frac{\text{lbm}}{\text{ft}^3}\right] \quad p = 15{,}404\left[\frac{\text{rad}}{\text{s}}\right]$$

Please supply all answers in an inertial coordinate system labeled 1, 2, and 3, with 1 being along the downrange direction and 3 being to the right side. Treat all missing coefficients as equal to zero. It is very important that you *draw* the situation.

Problem 26

The projectile given in Problem 25 is fired off of a fast attack boat chasing down some pirates. All the conditions in Problem 25 are identical except for the mounting of the weapon and its movement. The weapon is mounted on the starboard (right) side of the boat. At the instant of firing, the boat is moving at 40 knots in a straight line. The boat is rolling at 6 Hz. in a counterclockwise direction as viewed from behind. The boat is pitching into a swell so that the bow is dropping at 3 Hz. The gunner is slewing the weapon toward the bow at 1.5 rad/s while simultaneously lowering the muzzle at 2 rad/s. Assuming, at the instant of muzzle exit, the weapon is pointed directly to starboard and has the same elevation as in the test firing (i.e., 10°), find the values for the acceleration and angular momentum as was done in Problem 25. Comment on the results. For a proper comparison, use the same coordinate system as in Problem 25 but now with the 1 direction pointing to starboard and the 3 direction pointing to the stern of the boat. Once again the drawings are important.

Problem 27

In a test range, a modified 105 mm M1 projectile is fired at an elevation of 7° with a muzzle velocity of 1022 ft/s. The initial pitch and yaw angles are 1.0° and 1.5°, respectively. The initial pitch and yaw rates are 3 rad/s nose down and 2 rad/s nose left, respectively. If the projectile has the following coefficients at this particular instant, write the acceleration vector and the angular momentum vector.
Please ignore the Coriolis acceleration and assume the weapon has a right-hand twist.

- Projectile information:

$$C_D = 0.131 \qquad \left(C_{M_q} + C_{M\dot{\alpha}}\right) = -8.7 \qquad I_P = 0.547\ [\text{lbm-ft}^2]$$

$$C_{D_{\delta^2}} = 4.20 \qquad \left(C_{N_q} + C_{N_{\dot{\alpha}}}\right) \approx 0 \qquad I_T = 5.377\ [\text{lbm-ft}^2]$$

$$C_{M_\alpha} = 4.30 \qquad C_{M_{p\alpha}} = -0.892 \qquad m = 32.1\ [\text{lbm}]$$

$$C_{L_\alpha} = 1.65 \qquad C_{l_p} = -0.028 \qquad p = 932 \left[\frac{\text{rad}}{\text{s}}\right]$$

$$C_{N_{p\alpha}} = -0.55 \qquad \rho = 0.0751 \left[\frac{\text{lbm}}{\text{ft}^3}\right]$$

Please supply all answers in an inertial coordinate system labeled 1, 2, 3 with 1 being along the downrange direction and 3 being to the right side. Treat all missing coefficients as equal to zero. It is very important that you *draw* the situation. This will have a great deal of influence in obtaining the correct answer

Problem 28

A modified 105 mm M1 projectile is fired downward at an angle of −45° from the horizontal from an aircraft moving horizontally at 200 knots with a muzzle velocity of 1022 ft/s. The initial pitch and yaw angles are 1.0° and 1.5°, respectively. The initial pitch and yaw rates are 3.5 rad/s nose down and 2.5 rad/s nose left, respectively.

1. If the projectile is fired off the right side of the aircraft and has the following coefficients at this particular instant, write the acceleration vector and the angular momentum vector.
2. Write the acceleration vector and the angular momentum vector assuming everything is the same except now the projectile is fired off the left side of the aircraft.
3. Comment on the differences between parts 1 and 2.

Please ignore the Coriolis acceleration and assume the weapon has a right-hand twist.

- Projectile information:

$$C_D = 0.131 \qquad \left(C_{M_q} + C_{M\dot{\alpha}}\right) = -8.7 \qquad I_P = 0.547\ [\text{lbm-ft}^2]$$

$$C_{D_{\delta^2}} = 4.20 \qquad \left(C_{N_q} + C_{N_\alpha}\right) \approx 0 \qquad I_T = 5.377\ [\text{lbm-ft}^2]$$

$$C_{M_\alpha} = 4.30 \qquad C_{M_{p\alpha}} = -0.892 \qquad m = 32.1\ [\text{lbm}]$$

$$C_{L_\alpha} = 1.65 \qquad C_{l_p} = -0.028 \qquad p = 932 \left[\frac{\text{rad}}{\text{s}}\right]$$

$$C_{N_{p\alpha}} = -0.55 \qquad \rho = 0.0751 \left[\frac{\text{lbm}}{\text{ft}^3}\right]$$

Please supply all answers in an inertial coordinate system labeled 1, 2, and 3, with 1 being along the downrange direction and 3 being to the right side of the gun looking from the breech (be careful as this will change between parts 1 and 2). Treat all missing coefficients as equal to zero.

9.3 Wind Effects on a Simple Air Trajectory

We continue with our study of a point mass projectile model by adding a further complication to its flat fire trajectory—a crosswind or a range wind, as dynamic atmospheric phenomena. In the basic equations, we have neglected any change in air density with a change in altitude since the effect is small. We also have assumed the equations could be solved in closed form. We want to be able to solve them with winds that are both constant and variable along the flight path.

We begin with a modified version of Equation 9.47 by using vector notation

$$\frac{d\mathbf{V}}{dt} = -\frac{\rho S C_D}{2m}\tilde{V}(\mathbf{V}-\mathbf{W}) + \mathbf{g} \tag{9.100}$$

or

$$\frac{d\mathbf{V}}{dt} = -\hat{C}_D^*\tilde{V}(\mathbf{V}-\mathbf{W}) + \mathbf{g} \tag{9.101}$$

Where m is the projectile mass; $\mathbf{W}$ is the wind velocity vector; $\mathbf{V}$ is the projectile velocity vector; $\mathbf{g}$ is the vector acceleration due to gravity; t is the time; ρ is the air density; $\mathbf{a} = \frac{d\mathbf{V}}{dt}$ is the vector acceleration; S is the projectile reference area; $\hat{C}_D^* = \frac{\rho S C_D}{2m}$; and C_D is the dimensionless drag coefficient.

In the aforementioned equations, we have replaced the velocity vector $\mathbf{V}$ by the vector $(\mathbf{V} - \mathbf{W})$ because drag measurements are made relative to the air stream not relative to the ground. We have also replaced the scalar velocity (the speed) with

$$\tilde{V} = |\mathbf{V} - \mathbf{W}|$$

This is the scalar difference of the projectile and wind velocities. A diagram of the problem is shown in Figure 9.5.

We can resolve **V**, **W**, and **g** into components along the coordinate axes as follows:

$$\mathbf{V} = V_x\mathbf{i} + V_y\mathbf{j} + V_z\mathbf{k} \tag{9.102}$$

$$\mathbf{W} = W_x\mathbf{i} + W_y\mathbf{j} + W_z\mathbf{k} \tag{9.103}$$

$$\mathbf{g} = -g\,\mathbf{j} \tag{9.104}$$

Note that

$$\tilde{V}^2 = |\mathbf{V} - \mathbf{W}| \cdot |\mathbf{V} - \mathbf{W}|$$

and

$$|\mathbf{V}| = \sqrt{V_x^2 + V_y^2 + V_z^2}$$

This leads us to

$$|\mathbf{V} - \mathbf{W}| \cdot |\mathbf{V} - \mathbf{W}| = (V_x - W_x)^2 + (V_y - W_y)^2 + (V_z - W_z)^2$$

Then,

$$\tilde{V} = \sqrt{(V_x - W_x)^2 + (V_y - W_y)^2 + (V_z - W_z)^2} \tag{9.105}$$

If we insert Equations 9.102 through 9.104 into Equation 9.101, we get

$$\frac{d\mathbf{V}}{dt} = \left[-\hat{C}_D^*\tilde{V}(V_x - W_x)\right]\mathbf{i} + \left[-\hat{C}_D^*\tilde{V}\,(V_y - W_y) - g\right]\mathbf{j} + \left[-\hat{C}_D^*\tilde{V}(V_z - W_z)\right]\mathbf{k}$$

We can separate this vector equation into its three scalar components:

$$\dot{V}_x = \frac{dV_x}{dt} = -\hat{C}_D^*\tilde{V}(V_x - W_x) \tag{9.106}$$

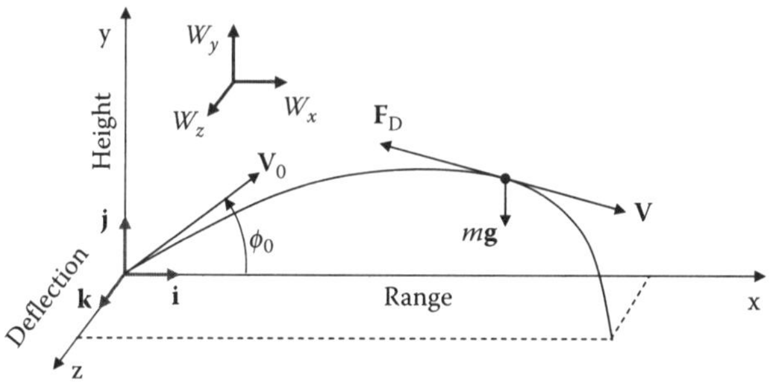

FIGURE 9.5
Coordinate system for projectile launch including wind effects.

$$\dot{V}_y = \frac{dV_y}{dt} = -\hat{C}_D^* \tilde{V}(V_y - W_y) - g \tag{9.107}$$

$$\dot{V}_z = \frac{dV_z}{dt} = -\hat{C}_D^* \tilde{V}(V_z - W_z) \tag{9.108}$$

Equations 9.106 through 9.108 are the exact equations for a point mass trajectory of a projectile acted upon by gravity, wind, and aerodynamic drag. They are first-order, nonlinear, coupled, ordinary differential equations that are coupled through Equation 9.105. The nonlinearity, as previously discussed, creates difficulties when we attempt to analytically solve these expressions. We can solve the exact equations only by using numerical methods. This will necessitate making the simplifying assumption of flat fire, which will allow us to solve them in closed form. We can alter Equation 9.105 by multiplying by the fraction $(V_x - W_x)/(V_x - W_x) = 1$ and then simplifying to get

$$\tilde{V} = (V_x - W_x)\sqrt{1 + \varepsilon_y^2 + \varepsilon_z^2} \tag{9.109}$$

Where

$$\varepsilon_y = \frac{(V_y - W_y)}{(V_x - W_x)} \tag{9.110}$$

and

$$\varepsilon_z = \frac{(V_z - W_z)}{(V_x - W_x)} \tag{9.111}$$

Using the binomial expansion of the form

$$\sqrt{1 + \zeta} = \left[1 + \frac{1}{2}\zeta - \frac{1}{8}\zeta^2 + \cdots\right]$$

We can operate on the radical of Equation 9.109, arriving at

$$\tilde{V} = (V_x - W_x)\left[1 + \frac{1}{2}\left(\varepsilon_y^2 + \varepsilon_z^2\right) - \frac{1}{8}\left(\varepsilon_y^2 + \varepsilon_z^2\right)^2 + \cdots\right] \tag{9.112}$$

Because the velocity of a projectile is usually much greater than winds of even hurricane force, we can assume that

$$|W_x|,\ |W_y|,\ \text{ and }\ |W_z| \ll V_x \tag{9.113}$$

and

$$\varepsilon_y^2\ \text{ and }\ \varepsilon_z^2 \ll 1 \tag{9.114}$$

If we look at the first inequality of Equation 9.114 and consider the assumptions of Equation 9.113, we find that

$$\varepsilon_y = \frac{V_y}{V_x} - \frac{W_y}{W_x} \approx \frac{V_y}{V_x} = \tan\phi \ll 1$$

This was the approximation developed around a similar binomial expansion in Section 9.2. We recall that this relation restricted us to $V_y/V_x < 0.1$, which, by squaring, results in the requirement that winds be at least two orders of magnitude smaller than the velocity V_x and this is easily the case. The second inequality of Equation 9.114 is also satisfied if W_z and V_z are comparable in size from

$$\varepsilon_z = \frac{(V_z - W_z)}{(V_x - W_x)} = \frac{(V_z - W_z)}{(V_x)} \ll 1$$

All this results in $\tilde{V}$ and $(V_x - W_x)$ being within about 1% of each other. So if $\tilde{V} \approx (V_x - W_x)$, we can rewrite Equations 9.106 through 9.108 as

$$\dot{V}_x = \frac{dV_x}{dt} = -\hat{C}_D^*(V_x - W_x)^2 \tag{9.115}$$

$$\dot{V}_y = \frac{dV_y}{dt} = -\hat{C}_D^*(V_x - W_x)(V_y - W_y) - g \tag{9.116}$$

$$\dot{V}_z = \frac{dV_z}{dt} = -\hat{C}_D^*(V_x - W_x)(V_z - W_z) \tag{9.117}$$

Updrafts and downdrafts are usually so small (and usually have the same effect as a crosswind for reasons we shall later describe) that we neglect them completely. Thus, we shall set W_y equal to zero from now on. We will now look first at the effect where only a crosswind is present (i.e., where $W_x = W_y = 0$) and then examine the effect of a headwind or tailwind. If we make this substitution into Equations 9.115 through 9.117, we obtain

$$\dot{V}_x = \frac{dV_x}{dt} = -\hat{C}_D^* V_x^2 \tag{9.118}$$

$$\dot{V}_y = \frac{dV_y}{dt} = -\hat{C}_D^* V_x V_y - g \tag{9.119}$$

$$\dot{V}_z = \frac{dV_z}{dt} = -\hat{C}_D^* V_x (V_z - W_z) \tag{9.120}$$

Equations 9.118 and 9.119 are identical to Equations 9.55 and 9.56 from our earlier work in the zero wind case. If we now change from time to space variables, as we did in the zero wind case, and recall that $dt/dx = 1/(dx/dt)$, then we arrive at the equations as follows:

$$V_x' = -\hat{C}_D^* V_x \tag{9.121}$$

$$V_y' = -\hat{C}_D^* V_y - \frac{g}{V_x} \tag{9.122}$$

$$V_z' = -\hat{C}_D^*(V_z - W_z) \tag{9.123}$$

Once again, the prime symbol represents differentiation with respect to x, and Equations 9.121 and 9.122 are identical to those developed for the zero wind case. Now we have already solved differential Equations 9.121 and 9.122 under their previous guise with the result of

$$V_x = V_{x_0} \exp\left(-\int_0^x \hat{C}_D^* \, dx_1\right) \tag{9.124}$$

$$V_y = \left\{\exp\left(-\int_0^x \hat{C}_D^* \, dx_1\right)\right\}\left[V_{y_0} - \int_0^x \left(\frac{g}{V_x}\right)\left\{\exp\left(\int_0^{x_2} \hat{C}_D^* \, dx_1\right)\right\} dx_2\right] \tag{9.125}$$

Equation 9.123 is somewhat more difficult to solve. It is a first-order, linear differential equation of the form $y' + P(x)y = Q$, whose solution, after the necessary integrations and substitution of initial conditions that at $x = 0$ and $V_z = 0$, is

$$V_z = \left\{\exp\left(-\int_0^x \hat{C}_D^* \, dx_1\right)\right\}\int_0^x \hat{C}_D^* W_z \left\{\exp\left(\int_0^x \hat{C}_D^* \, dx_1\right)\right\} dx_2 \tag{9.126}$$

From Equation 9.124, we see that the exponential is V_x/V_{x_0}, and this can be directly inserted into Equation 9.126. Also, if we assume that W_z is a constant, it can be removed from the integral to give

$$V_z = \frac{V_x}{V_{x_0}} W_z \int_0^x \hat{C}_D^* \exp\left(\int_0^x \hat{C}_D^* dx_1\right) dx_2 \tag{9.127}$$

We can integrate $\int_0^x \hat{C}_D^* \exp\left(\int_0^x \hat{C}_D^* \, dx_1\right) dx_2$ by parts in Equation 9.127 to yield

$$\int_0^x \hat{C}_D^* \exp\left(\int_0^x \hat{C}_D^* dx_1\right) dx_2 = \exp\left(\int_0^x \hat{C}_D^* dx_1\right)\int_0^x \hat{C}_D^* dx_2 - \int_0^x \left(\int_0^x \hat{C}_D^* dx_2\right) \hat{C}_D^* \exp\left(\int_0^x \hat{C}_D^* dx_1\right) dx_1 \tag{9.128}$$

The integral of the last term can be solved through a series of substitutions and evaluations at the limits to yield

$$\int_0^x \hat{C}_D^* \exp\left(\int_0^x \hat{C}_D^* dx_1\right) = \exp\left(\int_0^x \hat{C}_D^* dx_1\right) - 1 \tag{9.129}$$

If this is inserted into Equation 9.127, the result is

$$V_z = \frac{V_x}{V_{x_0}} W_z \left[\exp\left(\int_0^x \hat{C}_D^* \, dx_1\right) - 1\right] \tag{9.130}$$

We can further manipulate Equation 9.130 by inserting the value of the exponential from Equation 9.124. In doing so, we get

$$V_z = \frac{V_x}{V_{x_0}} W_z \left(\frac{V_{x_0}}{V_x} - 1 \right) = W_z \left(1 - \frac{V_x}{V_{x_0}} \right) \tag{9.131}$$

Since $0 < V_x < V_{x_0}, V_z$ always has to be less than the wind speed W_z. Thus, W_z is an upper bound on V_z.

If we examine the deflection due to a constant crosswind, we can write

$$\begin{aligned} z &= \int_0^t V_z \mathrm{d}t = W_z \int_0^t \left(1 - \frac{V_x}{V_{x_0}} \right) \mathrm{d}t = W_z \left(t|_0^t - \frac{1}{V_{x_0}} \int_0^t V_x \mathrm{d}t \right) \\ z &= W_z \left(t - \frac{x}{V_{x_0}} \right) \end{aligned} \tag{9.132}$$

Equation 9.132 is known as the lag rule for predicting crosswind effects. It is an exact solution for a constant crosswind. The quantity in the brackets is known as the lag time because a projectile in a real atmosphere would take longer to reach the same range than one fired in a vacuum.

Another interesting point is seen from examination of Equation 9.131. If V_x is always equal to the initial x-velocity, no matter how hard the wind blows, the projectile will not be affected. Thus, a rocket motor that maintains the initial x-velocity could make the projectile insensitive to wind, a concept called automet. Note also that if the thrust is greater than the initial velocity, the projectile will actually move into the wind.

We consider next the effect of a variable crosswind. A simple way to model this effect on a projectile is to superimpose solutions for constant crosswinds over incremental distances and piece the resultant trajectory together. This technique of superposition works only with linear phenomena. However, since Equation 9.132 is linear in x and t, we can apply this method. An alternative approach would be to apply Equation 9.132 in a piecewise fashion using the information from the previous calculation in the subsequent one. To do this, we shall rewrite Equation 9.132 as a difference equation

$$\Delta z_i = W_{z_i} \left[\Delta t_i - \frac{\Delta x_i}{V_{x_{i0}}} \right] \tag{9.133}$$

Where Δz_i is the distance traveled in the z-direction from time $i - 1$ to the time i; Δx_i is the distance traveled in the x-direction from time $i - 1$ to the time i; Δt_i is the time between time $i - 1$ to the time i; W_{z_i} is the constant crosswind acting on the projectile between time $i - 1$ and time i; and $V_{x_{i0}}$ is the x-velocity at time $i - 1$.

We can rewrite Equation 9.133 as

$$z_i - z_{i-1} = W_{z_i} \left[(t_i - t_{i-1}) - \frac{(x_i - x_{i-1})}{V_{x_{i0}}} \right] \tag{9.134}$$

To use this method, one must first tabulate t, V_x, and x as described earlier and then perform the calculation for z at each interval. With some modifications, a forward difference technique can also be used. These tedious calculations are best done with a computer program for small intervals of time.

We will now examine the effects of a constant range wind, both head on and a tailwind. We do this by comparing the effects to a flat fire, no-wind flight and will determine the effects on TOF, impact, and velocity at impact.

We make the initial assumption that there is no crosswind, i.e., $W_y = W_z = 0$, and insert this into Equations 9.115 through 9.117, the component differential equations for a point-mass, flat fire trajectory:

$$\dot{V}_x = \frac{\mathrm{d}V_x}{\mathrm{d}t} = -\hat{C}_\mathrm{D}^*(V_x - W_x)^2 \tag{9.135}$$

$$\dot{V}_y = \frac{\mathrm{d}V_y}{\mathrm{d}t} = -\hat{C}_\mathrm{D}^*(V_x - W_x)V_y - g \tag{9.136}$$

$$\dot{V}_z = \frac{\mathrm{d}V_z}{\mathrm{d}t} = -\hat{C}_\mathrm{D}^*(V_x - W_x)V_z \tag{9.137}$$

Because there is no crosswind, Equation 9.137 reduces to

$$V_z = 0$$

By a change of time to space variables and various algebraic manipulations, we can change Equation 9.135 to

$$V_x' + \hat{C}_\mathrm{D}^* V_x = \hat{C}_\mathrm{D}^* W_x\left(2 - \frac{W_x}{V_x}\right) \tag{9.138}$$

Similarly, we do the same to Equation 9.136 and arrive at a distance equation in a y-variable only:

$$V_y' + \hat{C}_\mathrm{D}^*\left(1 - \frac{W_x}{V_x}\right)V_y = \frac{-g}{V_x} \tag{9.139}$$

Recall from our earlier discussion that the wind speed is about two orders of magnitude smaller than the projectile velocity, so mathematically, we can express this condition as

$$\frac{W_x}{V_x} \le 0.01$$

We can then rewrite Equations 9.138 and 9.139, allowing them to be equalities as follows:

$$V_x' + \hat{C}_\mathrm{D}^* V_x = 2\hat{C}_\mathrm{D}^* W_x \tag{9.140}$$

and

$$V_y' + \hat{C}_\mathrm{D}^* V_y = -\frac{g}{V_x} \tag{9.141}$$

We can then solve these equations for V_x and V_y.

As we saw in the earlier solutions for the constant crosswind, with appropriate integrations, algebraic manipulation, and the insertion of the initial condition that at $x = 0$ and $V_x = V_{x_0}$, we see that

$$V_x = \left\{\exp\left(-\int_0^x \hat{C}_\mathrm{D}^*\,\mathrm{d}x_1\right)\right\}\int_0^x 2\hat{C}_\mathrm{D}^* W_x\left\{\exp\left(\int_0^x \hat{C}_\mathrm{D}^*\,\mathrm{d}x_1\right)\right\}\mathrm{d}x_2 + V_{x_0}\left\{\exp\left(-\int_0^x \hat{C}_\mathrm{D}^*\,\mathrm{d}x_1\right)\right\} \tag{9.142}$$

From Equation 9.129, recall that

$$\int_0^x \hat{C}_D^* \exp\left(\int_0^x \hat{C}_D^* \, dx_1\right) = \exp\left(\int_0^x \hat{C}_D^* \, dx_1\right) - 1$$

Using this fact and by substituting it into Equation 9.142, factoring the result, and considering that W_x is constant, we arrive at

$$V_x = V_{x_0}\left\{\exp\left(-\int_0^x \hat{C}_D^* \, dx_1\right)\right\} + 2W_x\left[1 - \left\{\exp\left(-\int_0^x \hat{C}_D^* \, dx_1\right)\right\}\right] \tag{9.143}$$

The first term on the right-hand side (RHS) of Equation 9.143 is simply the velocity decay caused by drag of the projectile. The second term is the effect of the range wind on it. If we examine Equation 9.124, which was an analysis for a firing in the absence of range wind, the first term of Equation 9.143 represents V_x, the x-velocity with no wind. The second term, when we substitute for the exponential, then represents the effect of the range wind on the flight. Thus, we can see the range wind effects shown as the variable of interest with a tilde (~) in the following:

$$\tilde{V}_x = V_x + 2W_x\left[1 - \frac{V_x}{V_{x_0}}\right] \tag{9.144}$$

This equation shows that at any time t, a tailwind (i.e., one blowing in the positive x-direction) has the effect of increasing the velocity (relative to the ground), while the opposite is true of a headwind. This is important because if we had a table of velocities vs. range for the no-wind case, we could then tabulate the effect of range wind.

If we now look at the y-velocity, we can operate on Equation 9.141 with the initial condition that at $x = 0$, $V_y = V_{y_0}$. This provides us with the solution of the space variable equation

$$V_y = -\exp\left(-\int_0^x \hat{C}_D^* \, dx_1\right)\int_0^x \frac{g}{V_x}\exp\left(\int_0^x \hat{C}_D^* \, dx_1\right)dx_2 + V_{y_0}\exp\left(-\int_0^x \hat{C}_D^* \, dx_1\right) \tag{9.145}$$

At this point, we can introduce the range wind by inserting Equation 9.143 for V_x, arriving at

$$V_y = -g\exp\left(\int_0^x \hat{C}_D^* dx_1\right)\int_0^x \frac{\exp\left(\int_0^x \hat{C}_D^* dx_1\right)}{V_{x_0}\exp\left(\int_0^x \hat{C}_D^* dx_1\right) + 2W_x\left[1 - \exp\left(-\int_0^x \hat{C}_D^* dx_1\right)\right]}dx_2 + V_{y_0}\exp\left(-\int_0^x \hat{C}_D^* dx_1\right) \tag{9.146}$$

This rather complicated integral can be simplified somewhat; however, another approach [1] to the problem that makes use of the no-wind method previously used in Equation 9.144 simplifies things even further. This is seen as

$$\tilde{V}_y = -\exp\left(-\int_0^x \hat{C}_D^* \, dx_1\right)\left[\int_0^x \frac{g}{\tilde{V}_x} \exp\left(\int_0^x \hat{C}_D^* \, dx_1\right) dx_2 + \frac{V_{y_0}}{V_{x_0}} V_{x_0} \exp\left(-\int_0^x \hat{C}_D^* \, dx_1\right)\right] \tag{9.147}$$

$$\text{or} \quad \text{since } \frac{V_{y_0}}{V_{x_0}} = \tan\phi_0,$$

$$\tilde{V}_y = -g\exp\left(-\int_0^x \hat{C}_D^* \, dx_1\right)\int_0^x \frac{1}{\tilde{V}_x} \exp\left(\int_0^x \hat{C}_D^* \, dx_1\right) dx_2 + V_x \tan\phi_0 \tag{9.148}$$

Further use of $\tilde{V}_x$ and some algebraic manipulation gives

$$\tilde{V}_y = V_x \tan\phi_0 - gV_x \int_0^x \frac{1}{V_x \tilde{V}_x} dx_2 \tag{9.149}$$

Recalling Equation 9.144, we can rewrite the denominator of the integral as

$$V_x \tilde{V}_x = V_x^2 + 2W_x\left(V_x - \frac{V_x^2}{V_{x_0}}\right) = V_x^2\left(1 + \frac{2W_x}{V_x} - \frac{2W_x}{V_{x_0}}\right) \tag{9.150}$$

If we again use the fact that the wind velocity is at least two orders of magnitude smaller than the projectile velocity, the last two terms in the product on the RHS vanish, leaving

$$V_x \tilde{V}_x \approx V_x^2 \tag{9.151}$$

Then, we can rewrite Equation 9.149 as

$$\tilde{V}_y = V_x \tan\phi_0 - gV_x \int_0^x \frac{1}{V_x^2} dx_2 \tag{9.152}$$

This equation has exactly the same form as the flat fire equation for V_y. Hence, we can say that for a flat fire trajectory, with a small range wind compared to the projectile velocity, the vertical component of velocity is not appreciably affected.

We can now turn our attention to the TOF of a projectile with a constant range wind by first defining an average downrange velocity following the procedure by McCoy [1] as

$$\tilde{V}_{x_{avg}} = \frac{1}{x}\int_0^x \tilde{V}_x \, dx_1 \tag{9.153}$$

For $\tilde{V}_x$, we substitute Equation 9.144 giving

$$\tilde{V}_{x_{avg}} = \frac{1}{x}\int_0^x \left(V_x + 2W_x - 2W_x \frac{V_x}{V_{x0}}\right) dx_1$$

By performing the integration on the second term of the integrand and rewriting, we get

$$\tilde{V}_{x_{avg}} = \frac{1}{x}\left(2W_x x + \int_0^x V_x dx_1 - 2\frac{W_x}{V_{x_0}}\int_0^x V_x dx_1\right) \tag{9.154}$$

Rearranging Equation 9.154 and knowing that the velocity averaging also applies to the no-wind case, i.e.,

$$V_{x_{avg}} = \frac{1}{x}\int_0^x V_x \, dx_1 \tag{9.155}$$

We get

$$\tilde{V}_{x_{avg}} = V_{x_{avg}} + 2W_x\left(1 - \frac{V_{x_{avg}}}{V_{x_0}}\right) \tag{9.156}$$

The TOF can be expressed as the range divided by the average velocity for the case of either no range wind or with range wind included. Thus, we can write

$$t = \frac{R}{V_{x_{avg}}} \tag{9.157}$$

or

$$\tilde{t} = \frac{R}{\tilde{V}_{x_{avg}}} \tag{9.158}$$

By taking the reciprocal of Equation 9.158, performing judicious substitutions, gathering terms, and finally taking the reciprocal of the result, we can write

$$\tilde{t} = \left[\frac{t}{1 + 2W_x\left(\frac{t}{R} - \frac{1}{V_{x_0}}\right)}\right] \tag{9.159}$$

This shows, as we would expect, that a tailwind (W_x positive) reduces the TOF, while a headwind (W_x negative) increases it.

Let us summarize what we have done for crosswinds and range winds. We modified the flat fire equations to account for crosswind and range wind. We use them when the angle of departure and angle of fall are both below 5.7°. We solved the crosswind equations assuming constant and variable crosswinds and introduced the classic lag rule. With variable crosswinds, we saw that it is fairly accurate to piece the trajectory together using locally constant values for the crosswind. We have solved the range wind equations assuming only constant range wind. We could treat variable range wind in a manner similar to variable crosswinds, but the difference in results is usually not worth the added effort. For range wind, we used a solution technique that compared the velocities, positions, and time to the no-wind case.

Problem 29

A US 37-mm AP projectile is fired with a muzzle velocity of 2600 [ft/s]. The projectile weighs 1.61 lbm. Assuming flat fire with $K_2 = 0.841$ [unitless] and using standard sea level met data ($\rho = 0.0751$ lbm/ft^3; $a = 1120$ ft/s)

1. Create a table containing range (yards), impact velocity (feet per second), TOF (seconds), initial QE angle (minutes), and angle at impact (minutes) in 200-yard increments out to 1000 yards assuming no-wind effects.
   Answer: At 1000 yards, $V = 1837$ [ft/s].

2. Determine the deflection of the projectile with a 20 mi/h crosswind blowing from left to right as viewed from behind the weapon.
   Answer: At 1000 yards, $z = 6.217$ [ft].

3. Determine the impact velocity, change in TOF and how high the projectile will hit if fired at the same QEs with a 20 mi/h tailwind and no crosswind.
   Answer: The projectile will hit 1.402 in. higher than expected.

Problem 30

A British 12 in. projectile has a K_3 of 0.8 and a weight of 850 lbm. If it is fired at an initial QE of 130 mil with a muzzle velocity of 2800 ft/s,

1. Create a table of range (yards), altitude (yards), velocity (feet per second), TOF (seconds), inclination angle (degrees), and drift (yards) if the projectile is fired with no wind.
2. Repeat part 1 if the projectile is fired with a headwind of 25 ft/s for the first 3000 yards of flight and a crosswind (left or right—your choice) of 35 ft/s for the remainder of the flight. Tabulate every 1000 yards with the impact location as the last entry in the table.

Problem 31

The US 0.30 caliber Ball M2 (projectile diameter = 0.308 in.) was the standard infantry rifle cartridge in WWII. Based on data collected by McCoy [1], it is a 150 grain, flat-based Spitzer shape with a K_3 of 0.491 from Mach 1.2 to 3. If it is fired with a muzzle velocity of 2780 ft/s, assuming standard sea level met data ($\rho = 0.0751$ lbm/ft^3; $a = 1120$ ft/s):

1. Create a table of range (yards), velocity (feet per second), TOF (seconds), launch angle (minutes), impact angle (minutes), and drift (yards) if the projectile is fired with no wind every 100 yards out to 600 yards.
2. Determine the deflection in inches assuming the projectile experiences a headwind of 20 ft/s for the entire flight. Tabulate every 100 yards out to 600 yards.
3. Determine the deflection in inches assuming the projectile experiences a crosswind (left or right—your choice) of 20 ft/s for the entire flight. Tabulate every 100 yards out to 600 yards.

Problem 32

Precision shooters are always in search of "tight groupings," that is, the grouping of the impact points of the projectiles at a given range. We shall examine the rifle and projectile combination given in Problem 31 and determine the effect each of several parameters has on the precision. Vary each of the following parameters individually by 1% (up and down) from Problem 31 and determine the miss distance in inches for a 200-yard range. Please carry answers to five significant figures as a baseline of comparison:

1. Muzzle velocity
2. Projectile mass
3. Drag coefficient (C_D)
4. Air density
5. Air temperature (in Rankine)
6. Launch angle of departure (we will assume this is due to weapon or shooter motion)
7. Headwind (0 ± 0.2 ft/s)
8. Crosswind (0 ± 0.2 ft/s)
9. Choose any two of these and vary them together—what is the result? Is the answer simply a linear superposition of the two individual errors? Why or Why not? Is this true for all of the parameters? Can you draw any conclusions from looking at all of the results?

9.4 Generalized Point Mass Trajectory

In keeping with our practice of introducing ever-increasing complexity into our theory, we will now remove most of the restrictions of the earlier work. We will examine the effects of an unrestricted launch angle and make the high-angle fire of mortars and howitzers amenable to trajectory analysis. We still reserve for later study the effects on flight of a three-dimensional body whose shape, physical properties, and motions add a significant level of complexity to trajectory analysis.

The aerodynamic behavior of a projectile can be examined from three separate viewpoints: motion affected only by the acceleration of gravity and the initial velocity (vacuum trajectory); motion affected by gravity, initial velocity, and aerodynamic drag (point mass trajectory); motion affected by the shape, physical properties, and dynamics (which actually manifests itself as changing drag) of the projectile, as well as gravity and launch conditions. We will concentrate on the second viewpoint in this section.

In the equations that follow, we are assuming that the projectile is still a cannonball with all of its mass concentrated at one point. This allows us to continue to neglect the rigid body kinematics that would be present in a distributed mass. However, we shall include wind effects, earth rotational effects, and, therefore three-dimensional motion. As stated, flat fire restrictions are removed so that the analysis is applicable to all launch angles.

We begin with the same set of equations of motion, except for the addition of a term for the Coriolis force $m\mathbf{\Lambda}$:

$$\mathbf{F} = m\mathbf{a}, \tag{9.160}$$

$$m\frac{\mathrm{d}\mathbf{V}}{\mathrm{d}t} = \Sigma\mathbf{F} + m\mathbf{g} + m\Lambda \tag{9.161}$$

Where m is the projectile mass; $\mathbf{V}$ is the velocity vector; t is the time; $\mathbf{a} = \frac{\mathrm{d}\mathbf{V}}{\mathrm{d}t}$ is the vector acceleration; $\Sigma\mathbf{F}$ is the vector sum of all aerodynamic forces; $\mathbf{g}$ is the vector acceleration due to gravity; and $\mathbf{\Lambda}$ is the vector Coriolis acceleration due to rotation of the earth.

We also recall from our earlier work with wind effects

$$\frac{\mathrm{d}\mathbf{V}}{\mathrm{d}t} = -\hat{C}_{\mathrm{D}}^{*}\tilde{V}(\mathbf{V} - \mathbf{W}) + \mathbf{g} \tag{9.162}$$

Where $\mathbf{W}$ is the wind velocity vector and $\hat{C}_{\mathrm{D}}^{*} = \rho S C_{\mathrm{D}}/2m$. In the aforementioned equations, we have replaced the velocity vector $\mathbf{V}$ by the vector $(\mathbf{V} - \mathbf{W})$ because drag measurements are made relative to the air stream, not relative to the ground. We have also again replaced the scalar velocity (the speed) with $\tilde{V} = |\mathbf{V} - \mathbf{W}|$, which is the scalar difference of the projectile and wind velocities. A diagram of this is shown in Figure 9.6.

Without repeating the entire procedure, it can be shown that we may separate Equation 9.162 into individual components to obtain the differential equations for a point mass:

$$\dot{V}_x = \frac{\mathrm{d}V_x}{\mathrm{d}t} = -\hat{C}_{\mathrm{D}}^{*}\tilde{V}(V_x - W_x) \tag{9.163}$$

$$\dot{V}_y = \frac{\mathrm{d}V_y}{\mathrm{d}t} = -\hat{C}_{\mathrm{D}}^{*}\tilde{V}(V_y - W_y) - g \tag{9.164}$$

$$\dot{V}_z = \frac{\mathrm{d}V_z}{\mathrm{d}t} = -\hat{C}_{\mathrm{D}}^{*}\tilde{V}(V_z - W_z) \tag{9.165}$$

The scalar velocity $\tilde{V}$ is again

$$\tilde{V} = \sqrt{(V_x - W_x)^2 + (V_y - W_y)^2 + (V_z - W_z)^2} \tag{9.166}$$

In all the aforementioned equations, the wind velocity is variable and is considered positive when it blows in the positive direction of one of the coordinate axes. Equations

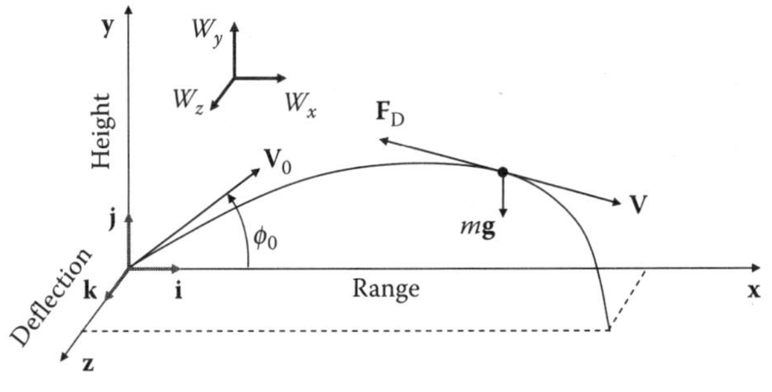

FIGURE 9.6
Generalized point mass trajectory.

9.163 through 9.165 are nonlinear, coupled differential equations which are the exact solution to Newton's laws governing the motion of a projectile affected by wind, gravity, and aerodynamic drag. These equations are coupled through Equation 9.166. Now, as we did in our discussion of flat fire, we would like to evaluate Equations 9.163 through 9.165 by using the downrange distance x as the independent variable. To do this, we simply note that for each of the time derivatives, we can write

$$\dot{V}_x = \frac{dV_x}{dt} = \frac{dV_x}{dt}\frac{dt}{dx}\frac{dx}{dt} = V_x\frac{dV_x}{dx} = V_xV_x' \tag{9.167}$$

And similarly

$$\dot{V}_y = V_xV_y' \tag{9.168}$$

$$\dot{V}_z = V_xV_z' \tag{9.169}$$

We can now write the three equations of motion with x as the independent variable as follows:

$$V_x' = \frac{1}{V_x}\dot{V}_x = -\hat{C}_D^*\left(\frac{\tilde{V}}{V_x}\right)(V_x - W_x) \tag{9.170}$$

$$V_y' = \frac{1}{V_x}\dot{V}_y = -\hat{C}_D^*\left(\frac{\tilde{V}}{V_x}\right)(V_y - W_y) - \left(\frac{g}{V_x}\right) \tag{9.171}$$

$$V_z' = \frac{1}{V_x}\dot{V}_z = -\hat{C}_D^*\left(\frac{\tilde{V}}{V_x}\right)(V_z - W_z) \tag{9.172}$$

As we noted earlier, the vertical component of the wind W_y is usually extremely small and will be neglected in further treatment. Further, as we mentioned, these equations are impossible to solve in closed form, and we must resort to numerical methods for their solution.

Without the restrictions of flat fire, projectiles fired at high angles of departure may traverse the atmosphere to great altitudes. In their flight, they encounter air temperatures and pressures that constantly change. These changes must be accounted for in the numerical computations to adequately solve the trajectory. Hence, knowledge of the standard atmosphere must serve as input to the calculations. There are two standards in common use: Army Standard Metrology and the International Civil Aviation Organization (ICAO) atmosphere. ICAO atmosphere is the most used of the two. Temperature and pressure vs. altitude are shown for the ICAO model in Figure 9.7. These atmospheric models are usually incorporated into ballistics codes.

Now, to become familiar with the physics of the Coriolis acceleration which was brought to its final form by Gaspard de Coriolis in 1835, we will study the effects of the rotation of the earth on a flat fire, vacuum trajectory example. We do this because we will be able to see these effects without resorting to a computer for calculation. The effect is really due to the fact that the firing point and target are located on the rotating earth; thus when the projectile lands, the earth has rotated through an angle and has thus moved the target. Figure 9.8 shows the geometry of the earth, the latitude of the firing site, and the orientation of the axes.

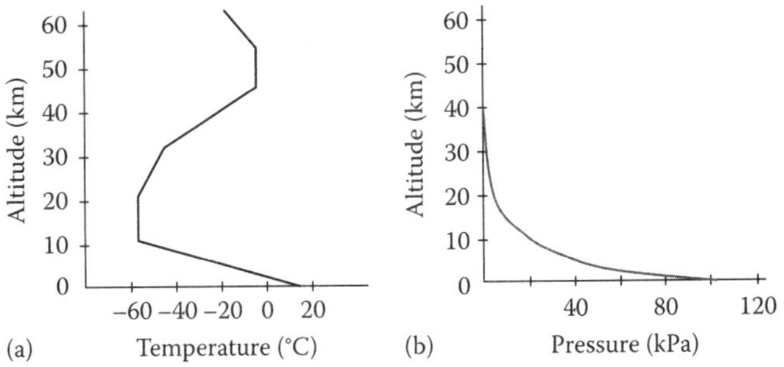

FIGURE 9.7
ICAO models for (a) atmospheric temperature and (b) pressure.

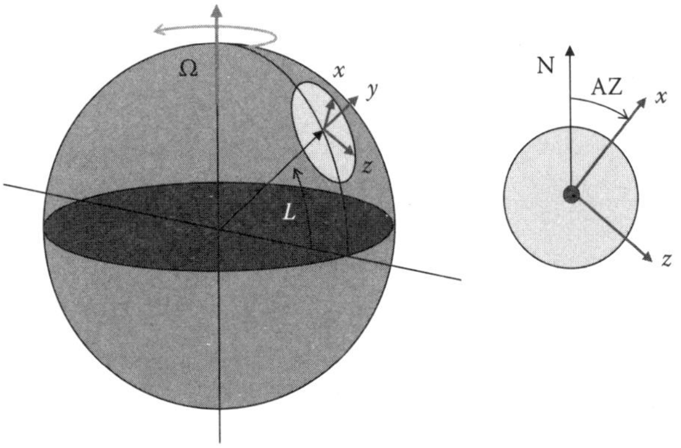

FIGURE 9.8
Angles used for coriolis acceleration calculations. Picture on the right represents a map of the corresponding area on the globe.

The Coriolis acceleration is defined as

$$2\boldsymbol{\omega} \times \left(\mathbf{v}_{\mathrm{B/A}}\right)_{xyz} = 2\Omega \times (\mathbf{v})_{xyz} \tag{9.173}$$

We have written Equation 9.173 in this way because the angular velocity we are considering is that of the earth and our projectile velocity is relative to our firing position (and therefore the earth), which moves with the x–y–z coordinate system. For this equation to be useful to us, we have to write the angular velocity Ω of the earth in terms of our x–y–z coordinate system. We will see that this acceleration is independent of the projectile weight but dependent upon its velocity. From Figure 9.8, we see that we can readily define Ω in terms of our moving coordinate system as

$$\Omega = \Omega\cos L\cos \mathrm{AZ}\mathbf{i} + \Omega\sin L\mathbf{j} - \Omega\cos L\sin \mathrm{AZ}\mathbf{k} \tag{9.174}$$

If we also note that $\mathbf{v}$ is defined as

$$(\mathbf{v})_{xyz} = V_x\mathbf{i} + V_y\mathbf{j} + V_z\mathbf{k} \tag{9.175}$$

Then inserting Equations 9.174 and 9.175 into Equation 9.173 gives us

$$2\Omega \times (\mathbf{v})_{xyz} = 2\Omega \begin{bmatrix} (V_z \sin L + V_y \cos L \sin \mathrm{AZ})\mathbf{i} \\ (-V_z \cos L \cos \mathrm{AZ} - V_x \cos L \sin \mathrm{AZ})\mathbf{j} \\ (V_y \cos L \cos \mathrm{AZ} - V_x \sin L)\mathbf{k} \end{bmatrix} \tag{9.176}$$

We will write the Coriolis acceleration in terms of a d'Alembert force (i.e., the negative of what we have in Equation 9.176), so we shall define the Coriolis term in our equation of motion (Equation 9.161) as

$$\mathbf{\Lambda} = -2\Omega \times (\mathbf{v})_{xyz} = 2\Omega \begin{bmatrix} (-V_y \cos L \sin \mathrm{AZ} - V_z \sin L)\mathbf{i} \\ (V_x \cos L \cos \mathrm{AZ} + V_z \cos L \cos \mathrm{AZ})\mathbf{j} \\ (V_x \sin L - V_y \cos L \cos \mathrm{AZ})\mathbf{k} \end{bmatrix} \tag{9.177}$$

Here we need to define the following variables:

- $\mathbf{\Lambda}$ is the vector Coriolis acceleration.
- Ω is the angular velocity of the earth about its polar axis = 0.00007292 (rad/s).
- L is the latitude of the firing site, positive in the northern hemisphere, negative in the southern.
- AZ is the azimuth angle of fire, measured clockwise from north.
- V_x, V_y, and V_z are the velocities in the x, y, z directions, respectively, positive along the positive coordinate axes.

Now that we have defined some terminology, we shall examine the effect that the Coriolis acceleration has on a vacuum trajectory. While this is stretching the vacuum trajectory much beyond its usefulness in ballistics, we remind the reader that the purpose is to demonstrate the physics that result from Coriolis effects. We begin by recalling Equation 9.161:

$$m\frac{d\mathbf{V}}{dt} = \Sigma\mathbf{F} + m\mathbf{g} + m\mathbf{\Lambda} \tag{9.161}$$

Now, since this is a vacuum trajectory, the force term on the RHS is zero, and we can divide by the mass m to obtain the vector equation for a vacuum trajectory

$$\frac{d\mathbf{V}}{dt} = \mathbf{g} + \Lambda \tag{9.178}$$

Rewriting Equation 9.178 in terms of its vector components gives (note that the g term appears only in Equation 9.180)

$$\frac{dV_x}{dt} = 2\Omega\left(-V_y \cos L \sin \mathrm{AZ} - V_z \sin L\right) \tag{9.179}$$

$$\frac{dV_y}{dt} = 2\Omega(-V_x \cos L \sin \mathrm{AZ} + V_z \cos L \cos \mathrm{AZ}) - g \tag{9.180}$$

$$\frac{dV_z}{dt} = 2\Omega\left(V_x \sin L - V_y \cos L \cos \mathrm{AZ}\right) \tag{9.181}$$

We shall now provide examples of the effect. These examples are based on the work of McCoy and can be found in his book [1]. Let us first consider a purely vertical firing (i.e., $V_x = V_z = 0$). One may initially consider this a trivial example, but for test purposes, we occasionally do fire vertically. And, by the way, as we will see, what goes up does not come straight down. Let us also choose due east as positive x, so AZ = 90°. With these assumptions, Equations 9.179 through 9.181 become

$$\frac{dV_x}{dt} = -2\Omega V_y \cos L \tag{9.182}$$

$$\frac{dV_y}{dt} = -g \tag{9.183}$$

$$\frac{dV_z}{dt} = 0 \tag{9.184}$$

These equations are well behaved and no longer coupled, so we can solve them independently. We shall integrate Equation 9.182 by first rewriting it, then integrating it:

$$\frac{dV_x}{dt} = -2\Omega \frac{dy}{dt} \cos L \tag{9.185}$$

$$V_x = -2\Omega y \cos L + C \tag{9.186}$$

To determine C, we know that at $y = y_0$, $V_x = 0$ so we can write

$$V_x = -2\Omega y \cos L + 2\Omega y_0 \cos L = -2\Omega \cos L(y - y_0) \tag{9.187}$$

If you recall our coordinate system, this means a projectile fired straight up will drift to the west and one fired (or dropped) straight down will drift to the east. Now we will integrate Equation 9.183 to get

$$V_y = -gt + C \tag{9.188}$$

Again, solving for the constant by inserting the initial conditions that at $t = 0$, $V_y = V_{y_0}$, we get

$$V_y = V_{y_0} - gt \tag{9.189}$$

Now we shall rewrite and integrate Equation 9.189 a second time making use of the fact that at $t = 0$, $y = y_0$, to obtain

$$y = V_{y_0} t - \frac{1}{2} g t^2 + y_0 \tag{9.190}$$

We can now insert Equation 9.190 into Equation 9.187 and rewrite it as

$$\frac{dx}{dt} = -2\Omega \cos L\left(y_0 + V_{y_0} t - \frac{1}{2} g t^2 - y_0 \right) = -2\Omega \cos L\left(V_{y_0} t - \frac{1}{2} g t^2 \right) \tag{9.191}$$

This can be integrated using the initial conditions that at $t = 0$, $x = 0$ to give

$$x = -\Omega \cos L\left(V_{y_0} t^2 - \frac{1}{3} g t^3 \right) \tag{9.192}$$

Let us now look at the special case of a bomb dropped from a given height with $V_{y_0} = 0$, and let $y = 0$. If we know the altitude from which we are dropping the bomb, we can determine its TOF from Equation 9.190; thus,

$$y_0 = \frac{1}{2}gt^2 \tag{9.193}$$

or

$$t = \sqrt{\frac{2y_0}{g}} \tag{9.194}$$

If we insert this into Equation 9.192, we get

$$x = \frac{1}{3}g\Omega\cos L\left(\frac{2y_0}{g}\right)^{3/2} \tag{9.195}$$

This says that since we are on the positive x-axis, the bomb will drift to the east. This drift would be greatest at the equator and zero at the poles.

Another example that uses the vacuum trajectory analysis is a projectile that is fired vertically upward with velocity V_{y_0}. We can find the time to maximum ordinate from Equation 9.189 knowing that at maximum ordinate $V_y = 0$:

$$t = \frac{V_{y_0}}{g} \tag{9.196}$$

If we insert this value of t into Equation 9.192, we get

$$x = -\frac{2}{3}\Omega g t^3 \cos L \tag{9.197}$$

The time to maximum ordinate can be put in terms of the height at maximum ordinate y_s through Equation 9.190:

$$y_s = \frac{1}{2}gt^2 \tag{9.198}$$

Therefore, the time to maximum ordinate is

$$t_s = \sqrt{\frac{2y_s}{g}} \tag{9.199}$$

And therefore, the Coriolis-caused displacement at maximum ordinate along the x-axis is found by inserting Equation 9.199 into Equation 9.197, giving

$$x_s = -\frac{4}{3}\Omega\sqrt{\frac{2y_s^3}{g}}\cos L \tag{9.200}$$

Lastly, we can apply the Coriolis analysis to the vacuum trajectory, flat fire situation and determine a correction for the acceleration in that case. We begin by making the usual assumptions for the flat fire trajectory of

$$V_y \ll V_x \quad \text{and} \quad V_z \ll V_x \tag{9.201}$$

We substitute these into Equations 9.179 through 9.181, yielding

$$\frac{dV_x}{dt} \approx 0 \tag{9.202}$$

$$\frac{dV_y}{dt} \approx 2\Omega V_x \cos L \sin AZ - g \tag{9.203}$$

$$\frac{dV_z}{dt} \approx 2\Omega V_x \sin L \tag{9.204}$$

The solution of Equation 9.202 with the initial conditions of $V_x = V_{x_0}$ at $t = 0$ yields

$$V_x \approx V_{x_0} \tag{9.205}$$

The solution of Equation 9.203, after insertion of Equation 9.205 with the initial conditions of $V_y = V_{y_0}$ at $t = 0$ and integrating, yields

$$V_y \approx V_{y_0} - gt\left[1 - \left(\frac{2\Omega V_{x_0}}{g}\right)\cos L \sin \mathrm{AZ}\right] \tag{9.206}$$

The solution of Equation 9.204 with the initial conditions of $V_z = 0$ at $t = 0$ yields the following after insertion of Equation 9.205 and integration:

$$V_z \approx 2\Omega V_{x_0} t \sin L \tag{9.207}$$

If we now integrate Equations 9.205 through 9.207 subject to $x = 0$, $y = y_0$, and $z = 0$ at $t = 0$ to get the displacements in the x, y, and z directions, we get

$$x \approx V_{x_0} t \tag{9.208}$$

$$y \approx y_0 + V_{y_0} t - \frac{gt^2}{2}\left[1 - \left(\frac{2\Omega V_{x_0}}{g}\right)\cos L \sin \mathrm{AZ}\right] \tag{9.209}$$

$$z = \Omega V_{x_0} t^2 \sin L \tag{9.210}$$

If we want to parameterize Equations 9.209 and 9.210 in terms of the downrange distance x, we can rewrite Equation 9.208 as

$$t \approx \frac{x}{V_{x_0}} \tag{9.211}$$

We can then insert this value of time into Equations 9.209 and 9.210 to obtain

$$y \approx y_0 + \frac{V_{y_0}}{V_{x_0}} x - \frac{gx^2}{2V_{x_0}^2}\left[1 - \left(\frac{2\Omega V_{x_0}}{g}\right)\cos L \sin \mathrm{AZ}\right] \tag{9.212}$$

and

$$z \approx \frac{\Omega x^2}{V_{x_0}} \sin L \tag{9.213}$$

Equation 9.212 was rearranged in the following form (including the substitution of tan ϕ_0) for comparison with Equation 9.37 and also modified to include a y_0:

$$y \approx y_0 + x \tan \phi_0 - \frac{gx^2}{2V_0^2} \tag{9.37}$$

$$y \approx y_0 + x \tan \phi_0 - \frac{gx^2}{2V_{x_0}^2}\left[1 - \left(\frac{2\Omega V_{x_0}}{g}\right)\cos L \sin \mathrm{AZ}\right] \tag{9.214}$$

From this comparison, we see that the incorporation of the Coriolis acceleration in the flat fire vacuum trajectory manifests itself in a modification to the gravitational term. Thus, as defined by McCoy [1], we can define a Coriolis factor f_C as

$$f_C = \left[1 - \left(\frac{2\Omega V_{x_0}}{g}\right)\cos L \sin \mathrm{AZ}\right] \tag{9.215}$$

and we could rewrite Equation 9.214 as

$$y \approx y_0 + x \tan \phi_0 - f_C \frac{gx^2}{2V_{x_0}^2} \tag{9.216}$$

If we closely look at Equation 9.214, we note several things: the value of cos L is anywhere between 0 and 1 for all possible latitudes; thus, if we were firing due north or due south, there would be no effect on the vertical component of impact; if we fired due east (AZ = 90°), the Coriolis effect essentially weakens the gravity term and the bullet would hit high; a due west firing would strike low; and the maximum effect on gravity is to alter it by 1.8%. Since sin L fluctuates between +1 and −1, the drift, the z-component, will vary right or left depending on the hemisphere where the firing occurs.

Now that the physics of the Coriolis effect is understood, the only difference when applied to the nonvacuum point mass trajectory is the fact that the velocity is changing with time due to drag. This is best handled numerically and will not be covered here.

In summary, for the generalized point mass trajectory, we included drag but ignored the dynamic effects of the projectile on drag. We described the origins of the Coriolis acceleration acting on a projectile. The physics was demonstrated through the vacuum trajectory and further examined with the flat fire assumptions. The incorporation of the Coriolis

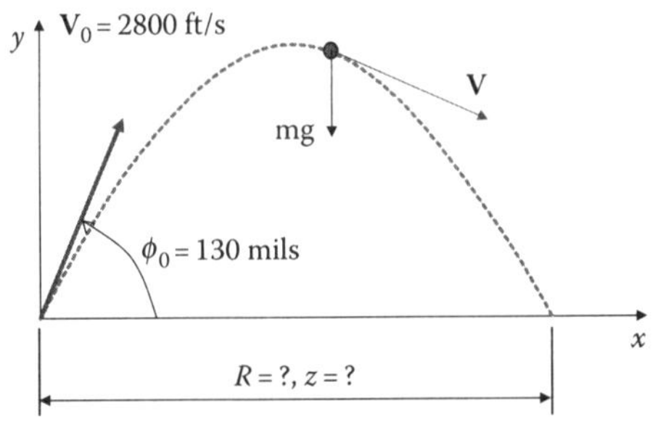

FIGURE 9.9
Graphical representation of long-range fire for Problem 33.

acceleration into the generalized point mass assumption is only affected by the variation of velocity over the trajectory and best handled numerically.

Problem 33

A projectile fired from a British 12 in. Mark IX naval gun had a muzzle velocity of 2800 ft/s and was fired at a QE of 130 mil (Figure 9.9). Assuming a vacuum trajectory, at what deflection would the shot hit the ground?

Assume the firing is taking place at 50° south latitude and the round is being fired due north.

Answer: $z = -75.8$ [ft]

9.5 Six Degree-of-Freedom (6 DOF) Trajectory

In keeping with our plan of increasing the complexity of our analyses to more closely approach the physical realities of projectile flight, we will now consider the projectile as a distributed mass. Since projectiles are relatively stiff structures, a 6 DOF model can adequately represent its position and attitude at any time. Each degree of freedom is tied to a coordinate necessary to completely describe the position of a body.

While this model is necessarily more complex than anything we have studied so far, the underlying physical principles remain the same. In the following work, we will use vectors (boldface, nonitalicized letters) in many of the derivations. We continue to do this because of the brevity and elegance of the notation.

In the equations that follow, we assume that the projectile is a rigid body of finite length with its mass distributed based on its geometry. This allows us to account for the effect of projectile attitude on drag and allows the full dynamics to come into play. We shall use direction cosines with respect to the projectile axis of symmetry (and thus a coordinate system with unit vectors **i**, **j**, and **k** that translates with the center of gravity (CG) but does not rotate and remains aligned with the projectile axis) as opposed to Eulerian angles (angles that are measured relative to the inertial coordinate system). This is illustrated in Figure 9.10.

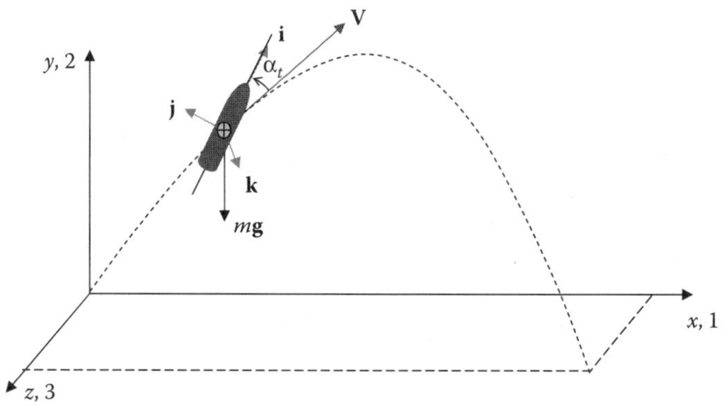

FIGURE 9.10
Coordinate system for 6 DOF model.

Once again, we restate the equations of motion, which, for generality, includes a term for rocket propulsion of the projectile. However, because this force is usually assumed to be aligned with the longitudinal axis of the projectile, its effect on the motions we will study is uncoupled from the other motions and may be added in afterward. Consequently, we will ignore it in our further work:

$$\mathbf{F} = m\mathbf{a} \tag{9.217}$$

$$m\frac{d\mathbf{V}}{dt} = \Sigma\mathbf{F} + m\mathbf{g} + m\mathbf{\Lambda} + \Sigma\mathbf{R}_\mathrm{T} \tag{9.218}$$

Where m is the projectile mass; $\mathbf{V}$ is the projectile velocity vector; t is the time; $\mathbf{a} = \frac{d\mathbf{V}}{dt}$ is the vector acceleration; $\Sigma\mathbf{F}$ is the vector sum of all aerodynamic forces; $\mathbf{g}$ is the vector acceleration due to gravity; $\mathbf{\Lambda}$ is the vector Coriolis acceleration due to rotation of the earth; and $\Sigma\mathbf{R}_\mathrm{T}$ is the vector sum of all rocket thrust forces (to be ignored).

We can also write the equation for the conservation of angular momentum as

$$\frac{d\mathbf{H}}{dt} = \Sigma\mathbf{M} + \Sigma\mathbf{R}_\mathrm{M} \tag{9.219}$$

Where $\mathbf{H}$ is the vector angular momentum of the projectile; $\Sigma\mathbf{M}$ is the vector sum of all aerodynamic moments; and $\Sigma\mathbf{R}_\mathrm{M}$ is the vector sum of all rocket thrust moments (to be ignored).

Because the projectile is assumed to be symmetric, every axis transverse to the longitudinal axis through the CG is a principal axis of inertia. The longitudinal axis itself is also, of course, a principal axis of inertia. The definition of the inertia tensor, which we will use, is

$$\mathbf{I} = \begin{bmatrix} I_{xx} & -I_{xy} & -I_{xz} \\ -I_{yx} & I_{yy} & -I_{yz} \\ -I_{zx} & -I_{zy} & I_{zz} \end{bmatrix} \tag{9.220}$$

Here the diagonal terms are called the moments of inertia and the off-diagonal terms are called the products of inertia. We know that there is a rotation that can be applied to this tensor such that the off-diagonal elements go to zero. In this orientation, the axes are said to be principal axes of inertia, and the tensor is written as

$$\mathbf{I} = \begin{bmatrix} I_x & 0 & 0 \\ 0 & I_y & 0 \\ 0 & 0 & I_z \end{bmatrix} \tag{9.221}$$

In our coordinate system, we shall define the unit vectors, **i**, **j**, and **k** so that they all lie along the principal axes of the projectile. Because of this unique situation, the total angular momentum of the projectile can be expressed as the sum of two vectors: the angular momentum about **i** and the angular momentum about any axis perpendicular to **i** through the CG. Since the **i**-axis is what we usually call the polar axis, we will denote the polar moment of inertia as I_P. With the symmetry of the projectile, the other moments of inertia about axes perpendicular to **i** are known as the transverse moments of inertia $I_y = I_z = I_T$.

We can then rewrite the inertia tensor as

$$\mathbf{I} = \begin{bmatrix} I_P & 0 & 0 \\ 0 & I_T & 0 \\ 0 & 0 & I_T \end{bmatrix} \tag{9.222}$$

If a projectile is spinning at spin rate p, the angular momentum about the polar axis is defined as

$$\mathbf{H}_P = I_P p \mathbf{i} \tag{9.223}$$

The angular momentum about any transverse axis is defined as

$$\mathbf{H}_T = I_T \left(\mathbf{i} \times \frac{d\mathbf{i}}{dt} \right) \tag{9.224}$$

With this, we can write the total momentum vector as

$$\mathbf{H} = I_P p \mathbf{i} + I_T \left(\mathbf{i} \times \frac{d\mathbf{i}}{dt} \right) \tag{9.225}$$

By defining a specific angular momentum $\mathbf{h} = \mathbf{H}/I_T$, we can write

$$\mathbf{h} = \frac{I_P p}{I_T} \mathbf{i} + \left(\mathbf{i} \times \frac{d\mathbf{i}}{dt} \right) \tag{9.226}$$

If we take the derivative of Equation 9.226 with respect to time, we get

$$\frac{d\mathbf{h}}{dt} = \frac{I_P}{I_T} \dot{p} \mathbf{i} + \frac{I_P p}{I_T} \frac{d\mathbf{i}}{dt} + \left(\frac{d\mathbf{i}}{dt} \times \frac{d\mathbf{i}}{dt} \right) + \left(\mathbf{i} \times \frac{d^2\mathbf{i}}{dt^2} \right) \tag{9.227}$$

Since the cross product of a vector with itself is zero, we get

$$\frac{d\mathbf{h}}{dt} = \frac{I_P}{I_T} \dot{p} \mathbf{i} + \frac{I_P p}{I_T} \frac{d\mathbf{i}}{dt} + \left(\mathbf{i} \times \frac{d^2\mathbf{i}}{dt^2} \right) \tag{9.228}$$

In anticipation of a later need, we shall take the dot product and cross product of vector **h** with unit vector **i** to get

$$\mathbf{h} \cdot \mathbf{i} = \left[\frac{I_P p}{I_T} \mathbf{i} + \left(\mathbf{i} \times \frac{d\mathbf{i}}{dt} \right) \right] \cdot \mathbf{i} = \frac{I_P p}{I_T} \tag{9.229}$$

$$\mathbf{h} \times \mathbf{i} = \left[\frac{I_P p}{I_T} \mathbf{i} + \left(\mathbf{i} \times \frac{d\mathbf{i}}{dt} \right) \right] \times \mathbf{i} = \frac{d\mathbf{i}}{dt} \tag{9.230}$$

In Equations 9.229 and 9.230, we have used the orthogonality properties of vectors as follows:

$$\mathbf{i} \cdot \mathbf{i} = 1$$

$$\mathbf{i} \cdot \mathbf{j} = \mathbf{i} \cdot \mathbf{k} = 0$$

$$\mathbf{i} \times \mathbf{i} = 0$$

$$\mathbf{i} \times \mathbf{j} = \mathbf{k} \rightarrow (\mathbf{i} \times \mathbf{j}) \times \mathbf{i} = (\mathbf{k}) \times \mathbf{i} = \mathbf{j} \therefore (\mathbf{i} \times \mathbf{j}) \times \mathbf{i} = \mathbf{j}$$

We will now examine all the forces and then the moments acting on the projectile and combine them into Equations 9.218 and 9.219. We have discussed all these items in Chapter 7, so we shall simply refresh their meanings briefly and move on. The first force acting on the projectile is the drag force, which acts opposite to the velocity vector, so we have

$$\text{Drag force} = \mathbf{F}_{\mathrm{D}} = -\frac{1}{2}\rho S C_{\mathrm{D}} \mathbf{V} V \tag{9.231}$$

The second force is the lift force, which we modified for our coordinate system as follows:

$$\text{Lift force} = \mathbf{F}_{\mathrm{L}} = -\frac{1}{2}\rho S C_{\mathrm{L}_\alpha}[\mathbf{V} \times (\mathbf{i} \times \mathbf{V})] \tag{9.232}$$

This equation contains a vector triple product in it that we replace with the relationship from vector algebra

$$\mathbf{A} \times (\mathbf{B} \times \mathbf{C}) = (\mathbf{A} \cdot \mathbf{C})\mathbf{B} - (\mathbf{A} \cdot \mathbf{B})\mathbf{C} \tag{9.233}$$

Which, for the product in Equation 9.232, can be written as

$$\mathbf{V} \times (\mathbf{i} \times \mathbf{V}) = V^2\mathbf{i} - (\mathbf{V} \cdot \mathbf{i})\mathbf{V}$$

When inserted into Equation 9.232, we have

$$\text{Lift force} = \mathbf{F}_{\mathrm{L}} = \frac{1}{2}\rho S C_{\mathrm{L}_\alpha}\left[V^2\mathbf{i} - (\mathbf{V} \cdot \mathbf{i})\mathbf{V}\right] \tag{9.234}$$

The next force is the Magnus force, brought on by the spin or roll of the projectile and taken from Equation 7.12:

$$\text{Magnus force} = \mathbf{F}_{\mathrm{M}} = \frac{1}{2}\rho S V\left(\frac{pd}{V}\right) C_{\mathrm{N}_{p\alpha}}(\mathbf{V} \times \mathbf{i}) \tag{9.235}$$

However, from Equation 9.229, we know that $p = (I_{\mathrm{T}}/I_{\mathrm{P}})(\mathbf{h} \cdot \mathbf{i})$ and, from vector algebra, $\mathbf{V} \times \mathbf{i} = -\mathbf{i} \times \mathbf{V}$. Then, we can manipulate Equation 9.235 to the form

$$\text{Magnus force} = \mathbf{F}_{\mathrm{M}} = -\frac{1}{2}\rho S d C_{\mathrm{N}_{p\alpha}}\left(\frac{I_{\mathrm{T}}}{I_{\mathrm{P}}}\right)(\mathbf{h} \cdot \mathbf{i})(\mathbf{i} \times \mathbf{V}) \tag{9.236}$$

Next, we need to include the pitch-damping force from Equation 7.14, where we will write $\mathbf{v}'$ as the unit vector along the velocity vector:

$$\text{Pitch-damping force} = \frac{1}{2}\rho V S d\left(\frac{\mathrm{d}\mathbf{i}}{\mathrm{d}t}\right) C_{\mathrm{N}_q} + \frac{1}{2}\rho V S d C_{\mathrm{N}_{\dot{\alpha}}}\left(\frac{\mathrm{d}\mathbf{i}}{\mathrm{d}t} - \frac{\mathrm{d}\mathbf{v}'}{\mathrm{d}t}\right) \tag{9.237}$$

If we assume $d\mathbf{v}'/dt = d\mathbf{i}/dt$ (this means that the rate at which the velocity vector is rotating to follow the curve of the trajectory is much smaller than the rate at which the axis of the projectile is moving) and we include Equation 9.230, we get a relation similar to Equation 7.17:

$$\text{Pitch-damping force} = \frac{1}{2}\rho VSdC_{N_q}\left(\frac{d\mathbf{i}}{dt}\right) + \frac{1}{2}\rho VSdC_{N_{\dot{\alpha}}}\left(\frac{d\mathbf{i}}{dt}\right) \tag{9.238}$$

or by Equation 9.230,

$$\text{Pitch-damping force} = \frac{1}{2}\rho VSd\left(C_{N_q} + C_{N_{\dot{\alpha}}}\right)(\mathbf{h}\times\mathbf{i}) \tag{9.239}$$

With all our forces now expressed in terms of our defined coefficients, we can divide Equation 9.218, omitting the rocket motor, by the projectile mass and inserting the coefficients to give

$$\begin{aligned}\frac{d\mathbf{V}}{dt} = &-\frac{\rho VSC_D}{2m}\mathbf{V} + \frac{\rho SC_{L_\alpha}}{2m}[V^2\mathbf{i} - (\mathbf{V}\cdot\mathbf{i})\mathbf{V}] - \frac{\rho SdC_{N_{p\alpha}}}{2m}\left(\frac{I_T}{I_P}\right)(\mathbf{h}\cdot\mathbf{i})(\mathbf{i}\times\mathbf{V}) \\ &+ \frac{\rho VSd\left(C_{N_q} + C_{N_{\dot{\alpha}}}\right)}{2m}(\mathbf{h}\times\mathbf{i}) + \mathbf{g} + \Lambda\end{aligned} \tag{9.240}$$

We will now examine the moments involved in Equation 9.219, the first of which is the spin-damping moment written as

$$\text{Spin-damping moment} = \mathbf{M}_S = \frac{1}{2}\rho V^2Sd\left(\frac{pd}{V}\right)C_{l_p}\mathbf{i} \tag{9.241}$$

However, if we again insert Equation 9.229 into the aforementioned equation, we get

$$\text{Spin-damping moment} = \mathbf{M}_S = \frac{1}{2}\rho VSd^2C_{l_p}\frac{I_T}{I_P}(\mathbf{h}\cdot\mathbf{i})\mathbf{i} \tag{9.242}$$

The rolling moment comes from Equation 7.6 and is

$$\text{Rolling moment} = \mathbf{M}_R = \frac{1}{2}\rho V^2Sd\delta_F C_{l_\delta}\mathbf{i} \tag{9.243}$$

The overturning moment can be written from Equation 7.9 as

$$\text{Overturning moment} = \mathbf{M}_\alpha = \frac{1}{2}\rho SdVC_{M_\alpha}(\mathbf{V}\times\mathbf{i}) \tag{9.244}$$

The Magnus moment can be written from Equation 7.13, and, by using the relations of Equations 9.233 and 9.236, we get

$$\text{Magnus moment} = \mathbf{M}_{p\alpha} = \frac{1}{2}\rho Sd^2C_{M_{p\alpha}}\frac{I_T}{I_P}(\mathbf{h}\cdot\mathbf{i})[\mathbf{V} - (\mathbf{V}\cdot\mathbf{i})\mathbf{i}] \tag{9.245}$$

We can obtain the pitch-damping moment by rewriting Equation 6.19 as well as using the relation of Equation 9.229 to get

$$\text{Pitch-damping moment} = \mathbf{M}_q = \frac{1}{2}\rho VSd^2\left(C_{M_q} + C_{M_{\dot{\alpha}}}\right)[\mathbf{h} - (\mathbf{h}\cdot\mathbf{i})\mathbf{i}] \tag{9.246}$$

We can now place all these relations into Equation 9.219, again omitting the rocket term, to yield

$$\frac{d\mathbf{H}}{dt} = \mathbf{M}_S + \mathbf{M}_R + \mathbf{M}_\alpha + \mathbf{M}_{p\alpha} + \mathbf{M}_q \tag{9.247}$$

Equation 9.247 can be changed to a more desirable form by dividing by I_T, which yields

$$\frac{d\mathbf{h}}{dt} = \frac{\mathbf{M}_S}{I_T} + \frac{\mathbf{M}_R}{I_T} + \frac{\mathbf{M}_\alpha}{I_T} + \frac{\mathbf{M}_{p\alpha}}{I_T} + \frac{\mathbf{M}_q}{I_T} \tag{9.248}$$

This, in turn, may be rewritten by inserting the various moment equations derived earlier as

$$\begin{aligned}\frac{d\mathbf{h}}{dt} &= \frac{\rho V S d^2 C_{l_P}}{2I_P}(\mathbf{h}\cdot\mathbf{i})\mathbf{i} + \frac{\rho V^2 S d \delta_F C_{l_\delta}}{2I_T}\mathbf{i} + \frac{\rho V S d C_{M_\alpha}}{2I_T}(\mathbf{V}\times\mathbf{i}) \\ &+ \frac{\rho S d^2 C_{M_{p\alpha}}}{2I_P}(\mathbf{h}\cdot\mathbf{i})[\mathbf{V}-(\mathbf{V}\cdot\mathbf{i})\mathbf{i}] + \frac{\rho V S d^2\left(C_{M_q}+C_{M_{\dot\alpha}}\right)}{2I_T}[\mathbf{h}-(\mathbf{h}\cdot\mathbf{i})\mathbf{i}]\end{aligned} \tag{9.249}$$

Note that the equations of motion are highly coupled to one another, and the reason we call the model a 6 DOF is readily apparent. When we break the equations up into their individual components, we have six equations and six unknowns (x, y, z, p, α, and β). Let us recall that the x-, y-, *and* z-axes are axes fixed to the earth, independent of the projectile, while **i** is the unit vector along the axis of symmetry of the projectile and has components along the x, y, z earth axes, and p, α, and β are the spin rate, pitch, and yaw angles, respectively. For convenience and clarity, and to facilitate analysis, we will relabel the x, y, z unit vectors (normally **i, j, and k**) as $\mathbf{e}_1$, $\mathbf{e}_2$, and $\mathbf{e}_3$, respectively, letting the subscripts denote the x-, y-, z-axes in that order (see Figure 9.10). Then, in terms of components in the earth-fixed system,

$$\mathbf{h} = h_1\mathbf{e}_1 + h_2\mathbf{e}_2 + h_3\mathbf{e}_3 \tag{9.250}$$

$$\mathbf{i} = i_1\mathbf{e}_1 + i_2\mathbf{e}_2 + i_3\mathbf{e}_3 \tag{9.251}$$

$$\mathbf{V} = V_1\mathbf{e}_1 + V_2\mathbf{e}_2 + V_3\mathbf{e}_3 \tag{9.252}$$

$$\mathbf{W} = W_1\mathbf{e}_1 + W_2\mathbf{e}_2 + W_3\mathbf{e}_3 \tag{9.253}$$

We shall also define

$$\mathbf{v} = \mathbf{V} - \mathbf{W} = (V_1 - W_1)\mathbf{e}_1 + (V_2 - W_2)\mathbf{e}_2 + (V_3 - W_3)\mathbf{e}_3 \tag{9.254}$$

and further defining

$$v_1 = (V_1 - W_1), \quad v_2 = (V_2 - W_2), \quad v_3 = (V_3 - W_3) \tag{9.255}$$

and

$$v = \sqrt{v_1^2 + v_2^2 + v_3^2} \tag{9.256}$$

We can insert the definition for **v** in place of **V** in Equations 9.245 and 9.249 to yield

$$\frac{d\mathbf{V}}{dt} = -\frac{\rho v S C_D}{2m}\mathbf{v} + \frac{\rho S C_{L_\alpha}}{2m}[v^2\mathbf{i} - (\mathbf{v}\cdot\mathbf{i})\mathbf{v}] - \frac{\rho S d C_{N_{p\alpha}}}{2m}\left(\frac{I_T}{I_P}\right)(\mathbf{h}\cdot\mathbf{i})(\mathbf{i}\times\mathbf{v}) + \frac{\rho v S d(C_{N_q} + C_{N_{\dot\alpha}})}{2m}(\mathbf{h}\times\mathbf{i}) + \mathbf{g} + \boldsymbol{\Lambda} \tag{9.257}$$

and

$$\frac{d\mathbf{h}}{dt} = \frac{\rho v S d^2 C_{lp}}{2I_P}(\mathbf{h}\cdot\mathbf{i})\mathbf{i} + \frac{\rho v^2 S d \delta_F C_{l\delta}}{2I_T}\mathbf{i} + \frac{\rho v S d C_{M_\alpha}}{2I_T}(\mathbf{v}\times\mathbf{i}) + \frac{\rho S d^2 C_{M_{p\alpha}}}{2I_P}(\mathbf{h}\cdot\mathbf{i})[\mathbf{v} - (\mathbf{v}\cdot\mathbf{i})\mathbf{i}] + \frac{\rho v S d^2\left(C_{M_q} + C_{M_{\dot\alpha}}\right)}{2I_T}[\mathbf{h} - (\mathbf{h}\cdot\mathbf{i})\mathbf{i}] \tag{9.258}$$

Our goal is to examine Equations 9.257 and 9.258 and break each into three equations, one for each coordinate direction (actually for the acceleration in each coordinate direction). But before we attempt to break Equations 9.257 and 9.258 into their components, it will be best to solve for some of the vector quantities that occur in them. Beginning with the second term of Equation 9.257, we can, with appropriate vector multiplication, obtain

$$(\mathbf{v}\cdot\mathbf{i})\mathbf{v} = \left(v_1^2 i_1 + v_1 v_2 i_2 + v_1 v_3 i_3\right)\mathbf{e}_1 + \left(v_1 v_2 i_1 + v_2^2 i_2 + v_2 v_3 i_3\right)\mathbf{e}_2 + \left(v_1 v_3 i_1 + v_2 v_3 i_2 + v_3^2 i_3\right)\mathbf{e}_3 \tag{9.259}$$

Another useful relation is

$$(\mathbf{v}\cdot\mathbf{i}) = (v_1\mathbf{e}_1 + v_2\mathbf{e}_2 + v_3\mathbf{e}_3)\cdot(i_1\mathbf{e}_1 + i_2\mathbf{e}_2 + i_3\mathbf{e}_3) \tag{9.260}$$

or

$$(\mathbf{v}\cdot\mathbf{i}) = v_1 i_1 + v_2 i_2 + v_3 i_3 \tag{9.261}$$

However, we can show that

$$\cos\alpha_t = \frac{(\mathbf{v}\cdot\mathbf{i})}{v} = \frac{v_1 i_1 + v_2 i_2 + v_3 i_3}{v} \tag{9.262}$$

The next relations in Equation 9.257 are

$$(\mathbf{h}\cdot\mathbf{i}) = h_1 i_1 + h_2 i_2 + h_3 i_3 \tag{9.263}$$

and

$$(\mathbf{i}\times\mathbf{v}) = \begin{vmatrix} \mathbf{e}_1 & \mathbf{e}_2 & \mathbf{e}_3 \\ i_1 & i_2 & i_3 \\ v_1 & v_2 & v_3 \end{vmatrix} = (i_2 v_3 - i_3 v_2)\mathbf{e}_1 + (i_3 v_1 - i_1 v_3)\mathbf{e}_2 + (i_1 v_2 - i_2 v_1)\mathbf{e}_3 \tag{9.264}$$

Now, also in Equation 9.257 is the term

$$(\mathbf{h}\cdot\mathbf{i})(\mathbf{i}\times\mathbf{v}) = (h_1 i_1 + h_2 i_2 + h_3 i_3)[(i_2 v_3 - i_3 v_2)\mathbf{e}_1 + (i_3 v_1 - i_1 v_3)\mathbf{e}_2 + (i_1 v_2 - i_2 v_1)\mathbf{e}_3] \tag{9.265}$$

But with the fact that $(\mathbf{h} \cdot \mathbf{i}) = I_P p / I_T$ as shown earlier, then we can write

$$(\mathbf{h} \cdot \mathbf{i})(\mathbf{i} \times \mathbf{v}) = \frac{I_P p}{I_T}(i_2 v_3 - i_3 v_2)\mathbf{e}_1 + \frac{I_P p}{I_T}(i_3 v_1 - i_1 v_3)\mathbf{e}_2 + \frac{I_P p}{I_T}(i_1 v_2 - i_2 v_1)\mathbf{e}_3 \tag{9.266}$$

Another relation we have to deal with is

$$(\mathbf{h} \times \mathbf{i}) = \begin{vmatrix} \mathbf{e}_1 & \mathbf{e}_2 & \mathbf{e}_3 \\ h_1 & h_2 & h_3 \\ i_1 & i_2 & i_3 \end{vmatrix} = (h_2 i_3 - h_3 i_2)\mathbf{e}_1 + (h_3 i_1 - h_1 i_3)\mathbf{e}_2 + (h_1 i_2 - h_2 i_1)\mathbf{e}_3 \tag{9.267}$$

The next relation we generate with the help of Equation 9.229 is

$$(\mathbf{h} \cdot \mathbf{i})\mathbf{i} = \frac{I_P p}{I_T} i_1 \mathbf{e}_1 + \frac{I_P p}{I_T} i_2 \mathbf{e}_2 + \frac{I_P p}{I_T} i_3 \mathbf{e}_3 \tag{9.268}$$

In Equation 9.258, we need

$$(\mathbf{v} \times \mathbf{i}) = \begin{vmatrix} \mathbf{e}_1 & \mathbf{e}_2 & \mathbf{e}_3 \\ v_1 & v_2 & v_3 \\ i_1 & i_2 & i_3 \end{vmatrix} = (i_3 v_2 - i_2 v_3)\mathbf{e}_1 + (i_1 v_3 - i_3 v_1)\mathbf{e}_2 + (i_2 v_1 - i_1 v_2)\mathbf{e}_3 \tag{9.269}$$

And finally in Equation 9.258,

$$(\mathbf{v} \cdot \mathbf{i})\mathbf{i} = (v_1 i_1 + v_2 i_2 + v_3 i_3)(i_1 \mathbf{e}_1 + i_2 \mathbf{e}_2 + i_3 \mathbf{e}_3) \tag{9.270}$$

or

$$\begin{aligned}(\mathbf{v} \cdot \mathbf{i})\mathbf{i} = {}& \left(v_1 i_1^2 + v_2 i_1 i_2 + v_3 i_1 i_3\right)\mathbf{e}_1 + \left(v_1 i_1 i_2 + v_2 i_2^2 + v_3 i_2 i_3\right)\mathbf{e}_2 \\ & + \left(v_1 i_1 i_3 + v_2 i_2 i_3 + v_3 i_3^2\right)\mathbf{e}_3\end{aligned} \tag{9.271}$$

Let us now look at Equation 9.257 with all the vector quantities broken into their components

$$\begin{aligned}\frac{dV_1}{dt}\mathbf{e}_1 + \frac{dV_2}{dt}\mathbf{e}_2 + \frac{dV_3}{dt}\mathbf{e}_3 = {}& -\frac{\rho v S C_D}{2m}(v_1\mathbf{e}_1 + v_2\mathbf{e}_2 + v_3\mathbf{e}_3) \\ & + \frac{\rho S C_{L_\alpha}}{2m}[v^2 i_1 \mathbf{e}_1 + v^2 i_2 \mathbf{e}_2 + v^2 i_3 \mathbf{e}_3 - v\cos\alpha_t(v_1\mathbf{e}_1 + v_2\mathbf{e}_2 + v_3\mathbf{e}_3)] \\ & - \frac{\rho S d C_{N_{p\alpha}}}{2m}\left(\frac{I_T}{I_P}\right)\left(\frac{I_P p}{I_T}\right)[(v_3 i_2 - v_2 i_3)\mathbf{e}_1 + (v_1 i_3 - v_3 i_1)\mathbf{e}_2 + (v_2 i_1 - v_1 i_2)\mathbf{e}_3] \\ & + \frac{\rho v S d\left(C_{N_q} + C_{N_{\dot\alpha}}\right)}{2m}[(h_2 i_3 - h_3 i_2)\mathbf{e}_1 + (h_3 i_1 - h_1 i_3)\mathbf{e}_2 + (h_1 i_2 - h_2 i_1)\mathbf{e}_3] \\ & + g_1\mathbf{e}_1 + g_2\mathbf{e}_2 + g_3\mathbf{e}_3 + \Lambda_1\mathbf{e}_1 + \Lambda_2\mathbf{e}_2 + \Lambda_3\mathbf{e}_3\end{aligned} \tag{9.272}$$

Similarly, let us perform the same operation on Equation 9.258

$$\frac{dh_1}{dt}\mathbf{e}_1+\frac{dh_2}{dt}\mathbf{e}_2+\frac{dh_3}{dt}\mathbf{e}_3=\frac{\rho v S d^2 C_{l_p}}{2I_P}\left(\frac{I_P p}{I_T}\right)(i_1\mathbf{e}_1+i_2\mathbf{e}_2+i_3\mathbf{e}_3)$$
$$+\frac{\rho v^2 S d\delta_F C_{l_\delta}}{2I_T}(i_1\mathbf{e}_1+i_2\mathbf{e}_2+i_3\mathbf{e}_3)+\frac{\rho v S d C_{M_\alpha}}{2I_T}\left[(v_2 i_3-v_3 i_2)\mathbf{e}_1+(v_3 i_1-v_1 i_3)\mathbf{e}_2+(v_1 i_2-v_2 i_1)\mathbf{e}_3\right]$$
$$+\frac{\rho S d^2 C_{M_{p\alpha}}}{2I_P}\left(\frac{I_P p}{I_T}\right)\left[(v_1\mathbf{e}_1+v_2\mathbf{e}_2+v_3\mathbf{e}_3)-v\cos\alpha_t(i_1\mathbf{e}_1+i_2\mathbf{e}_2+i_3\mathbf{e}_3)\right]$$
$$+\frac{\rho v S d^2\left(C_{M_q}+C_{M_{\dot\alpha}}\right)}{2I_T}\left[(h_1\mathbf{e}_1+h_2\mathbf{e}_2+h_3\mathbf{e}_3)-\left(\frac{I_P p}{I_T}\right)(i_1\mathbf{e}_1+i_2\mathbf{e}_2+i_3\mathbf{e}_3)\right] \tag{9.273}$$

We will first operate on Equation 9.272 by collecting all the terms with the unit vectors $\mathbf{e}_1$, then $\mathbf{e}_2$ and $\mathbf{e}_3$ and by putting them into the equations for linear and angular momentum:

$$\frac{dV_1}{dt}=-\frac{\rho v S C_D}{2m}v_1+\frac{p S C_{L_\alpha}}{2m}\left[v^2 i_1-v v_1\cos\alpha_t\right]-\frac{\rho S d C_{N_{p\alpha}} p}{2m}(v_3 i_2-v_2 i_3)$$
$$+\frac{\rho v S d\left(C_{N_q}+C_{N_{\dot\alpha}}\right)}{2m}(h_2 i_3-h_3 i_2)+g_1+\Lambda_1 \tag{9.274}$$

$$\frac{dV_2}{dt}=-\frac{\rho v S C_D}{2m}v_2+\frac{p S C_{L_\alpha}}{2m}\left[v^2 i_2-v v_2\cos\alpha_t\right]-\frac{\rho S d C_{N_{p\alpha}} p}{2m}(v_1 i_3-v_3 i_1)$$
$$+\frac{\rho v S d\left(C_{N_q}+C_{N_{\dot\alpha}}\right)}{2m}(h_3 i_1-h_1 i_3)+g_2+\Lambda_2 \tag{9.275}$$

$$\frac{dV_3}{dt}=-\frac{\rho v S C_D}{2m}v_3+\frac{p S C_{L_\alpha}}{2m}\left[v^2 i_3-v v_3\cos\alpha_t\right]-\frac{\rho S d C_{N_{p\alpha}} p}{2m}(v_2 i_1-v_1 i_2)$$
$$+\frac{\rho v S d\left(C_{N_q}+C_{N_{\dot\alpha}}\right)}{2m}(h_1 i_2-h_2 i_1)+g_3+\Lambda_3 \tag{9.276}$$

Next is Equation 9.273, where the same procedure will be followed:

$$\frac{dh_1}{dt}=\frac{\rho v S d^2 C_{l_P} p}{2I_T}i_1+\frac{\rho v^2 S d\delta_F C_{l_\delta}}{2I_T}i_1+\frac{\rho v S d C_{M_\alpha}}{2I_T}(v_2 i_3-v_3 i_2)$$
$$+\frac{\rho S d^2 C_{M_{p\alpha}} p}{2I_T}[v_1-v i_1\cos\alpha_t]+\frac{\rho v S d^2\left(C_{M_q}+C_{M_{\dot\alpha}}\right)}{2I_T}\left[h_1-\left(\frac{I_P p}{I_T}\right)i_1\right] \tag{9.277}$$

$$\frac{dh_2}{dt} = \frac{\rho v S d^2 C_{l_P} p}{2I_T} i_2 + \frac{\rho v^2 S d \delta_F C_{l_\delta}}{2I_T} i_2 + \frac{\rho v S d C_{M_\alpha}}{2I_T} (v_3 i_1 - v_1 i_3)$$
$$+ \frac{\rho S d^2 C_{M_{p\alpha}} p}{2I_T} [v_2 - v i_2 \cos\alpha_t] + \frac{\rho v S d^2 \left(C_{M_q} + C_{M_{\dot\alpha}}\right)}{2I_T} \left[h_2 - \left(\frac{I_P p}{I_T}\right) i_2\right] \quad (9.278)$$

$$\frac{dh_3}{dt} = \frac{\rho v S d^2 C_{l_P} p}{2I_T} i_3 + \frac{\rho v^2 S d \delta_F C_{l_\delta}}{2I_T} i_3 + \frac{\rho v S d C_{M_\alpha}}{2I_T} (v_1 i_2 - v_2 i_1)$$
$$+ \frac{\rho S d^2 C_{M_{p\alpha}} p}{2I_T} [v_3 - v i_3 \cos\alpha_t] + \frac{\rho v S d^2 \left(C_{M_q} + C_{M_{\dot\alpha}}\right)}{2I_T} \left[h_3 - \left(\frac{I_P p}{I_T}\right) i_3\right] \quad (9.279)$$

We can simplify Equations 9.274 through 9.279 considerably by defining the following coefficients:

$$\tilde{C}_D = \frac{\rho v S C_D}{2m} \qquad \tilde{C}_{l_p} = \frac{\rho v S d^2 C_{l_p} p}{2I_T}$$

$$\tilde{C}_{L_\alpha} = \frac{\rho S C_{L_\alpha}}{2m} \qquad \tilde{C}_{l_\delta} = \frac{\rho v^2 S d \delta_F C_{l_\delta}}{2I_T}$$

$$\tilde{C}_{N_{p\alpha}} = \frac{\rho S d C_{N_{p\alpha}} p}{2m} \qquad \tilde{C}_{M_\alpha} = \frac{\rho v S d C_{M_\alpha}}{2I_T}$$

$$\tilde{C}_{N_q} = \frac{\rho v S d \left(C_{N_q} + C_{N_{\dot\alpha}}\right)}{2m} \qquad \tilde{C}_{M_{p\alpha}} = \frac{\rho S d^2 C_{M_{p\alpha}} p}{2I_T}$$

$$\tilde{C}_{M_q} = \frac{\rho v S d^2 \left(C_{M_q} + C_{M_\alpha}\right)}{2I_T}$$

With these coefficients, we can write Equations 9.274 through 9.276 in a more compact form:

$$\frac{dV_1}{dt} = -\tilde{C}_D v_1 + \tilde{C}_{L_\alpha}\left(v^2 i_1 - v v_1 \cos\alpha_t\right) - \tilde{C}_{N_{p\alpha}}(v_3 i_2 - v_2 i_3) + \tilde{C}_{N_q}(h_2 i_3 - h_3 i_2) + g_1 + \Lambda_1 \quad (9.280)$$

$$\frac{dV_2}{dt} = -\tilde{C}_D v_2 + \tilde{C}_{L_\alpha}\left(v^2 i_2 - v v_2 \cos\alpha_t\right) - \tilde{C}_{N_{p\alpha}}(v_1 i_3 - v_3 i_1) + \tilde{C}_{N_q}(h_3 i_1 - h_1 i_3) + g_2 + \Lambda_2 \quad (9.281)$$

$$\frac{dV_3}{dt} = -\tilde{C}_D v_3 + \tilde{C}_{L_\alpha}\left(v^2 i_3 - v v_3 \cos\alpha_t\right) - \tilde{C}_{N_{p\alpha}}(v_2 i_1 - v_1 i_2) + \tilde{C}_{N_q}(h_1 i_2 - h_2 i_1) + g_3 + \Lambda_3 \quad (9.282)$$

We can do the same with Equations 9.277 through 9.279:

$$\frac{dh_1}{dt} = \left(\tilde{C}_{l_P} + \tilde{C}_{l_\delta}\right) i_1 + \tilde{C}_{M_\alpha}(v_2 i_3 - v_3 i_2) + \tilde{C}_{M_{p\alpha}}(v_1 - v i_1 \cos\alpha_t) + \tilde{C}_{M_q}\left[h_1 - \left(\frac{I_P p}{I_T}\right) i_1\right] \quad (9.283)$$

$$\frac{\mathrm{d}h_2}{\mathrm{d}t}=\left(\tilde{C}_{l_P}+\tilde{C}_{l_\delta}\right)i_2+\tilde{C}_{M_\alpha}(v_3i_1-v_1i_3)+\tilde{C}_{M_{p\alpha}}(v_2-vi_2\cos\alpha_t)+\tilde{C}_{M_q}\left[h_2-\left(\frac{I_Pp}{I_T}\right)i_2\right] \tag{9.284}$$

$$\frac{\mathrm{d}h_3}{\mathrm{d}t}=\left(\tilde{C}_{l_P}+\tilde{C}_{l_\delta}\right)i_3+\tilde{C}_{M_\alpha}(v_1i_2-v_2i_1)+\tilde{C}_{M_{p\alpha}}(v_3-vi_3\cos\alpha_t)+\tilde{C}_{M_q}\left[h_3-\left(\frac{I_Pp}{I_T}\right)i_3\right] \tag{9.285}$$

Now that we have the six coupled equations for our six accelerations, we would like to determine the position of the projectile in space and time. We do this by creating a vector **X** to the center of mass of the projectile. If we note that $\mathbf{X} = [x\mathbf{e}_1 + y\mathbf{e}_2 + z\mathbf{e}_3]$ in the earth-fixed coordinate system, then we can break the individual components into

$$x = x_0 + \int_0^t V_1\,\mathrm{d}t \tag{9.286}$$

$$y = y_0 + \int_0^t V_2\,\mathrm{d}t \tag{9.287}$$

$$z = z_0 + \int_0^t V_3\,\mathrm{d}t \tag{9.288}$$

Recognize that when firing a long-range weapon, we usually do so with grid coordinates on a map of the earth. A map is, in theory, created by peeling the geometry off a globe. Thus, the coordinates and distances are correct in the downrange and cross-range directions (x and z). However, the altitude y has to be corrected for the curvature of the earth. This is depicted with the applicable equations in Figure 9.11. A similar rotation occurs with the gravity vector as depicted in Figure 9.12.

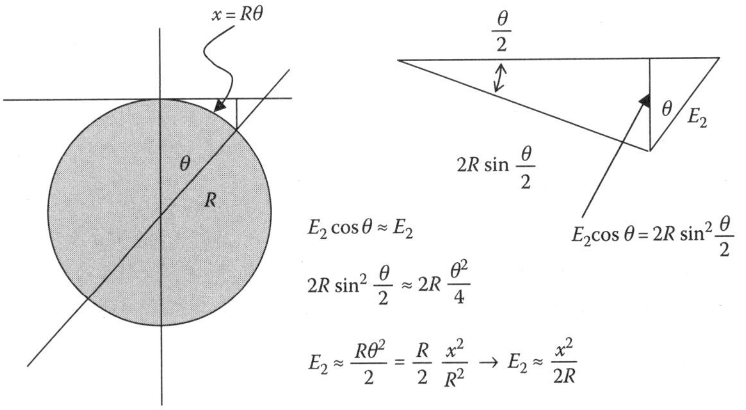

FIGURE 9.11
Altitude error over long trajectories.

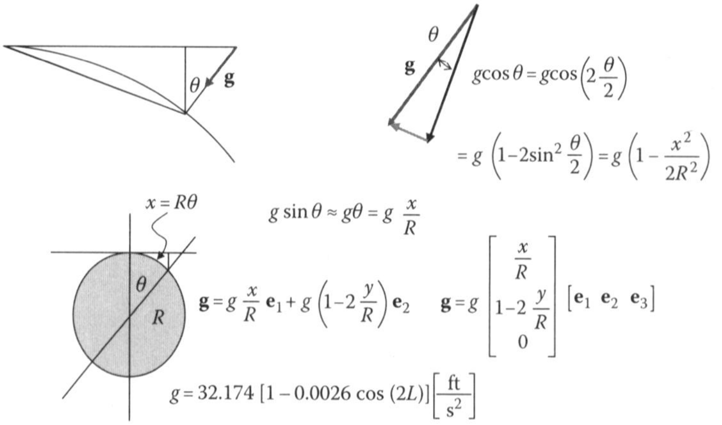

FIGURE 9.12
Rotation of the gravity vector due to earth curvature and associated equations.

With this relationship, we can write the projectile position vector in earth coordinates as

$$\mathbf{E} \approx [E_1\mathbf{e}_1 + E_2\mathbf{e}_2 + E_3\mathbf{e}_3] = \left[x\mathbf{e}_1 + \left(y + \frac{x^2}{2R} \right)\mathbf{e}_2 + z\mathbf{e}_3 \right] = \begin{bmatrix} x \\ y + \dfrac{x^2}{2R} \\ z \end{bmatrix} [\mathbf{e}_1 \quad \mathbf{e}_3 \quad \mathbf{e}_3] \tag{9.289}$$

Where R is the average radius of the earth, taken to be 6,951,844 yards or 6,356,766 m. The use of earth coordinates is recommended at ranges beyond about 2000 yards (at 2000 yards, there is a 10.36 in. difference in height [1]). Furthermore, the acceleration of gravity varies with altitude (and, in fact, latitude and longitude as well), and we need to consider this.

To complete the equations of motion, we must consider the form of the Coriolis acceleration vector. We have discussed this extensively previously so we shall simply write the components of this vector as

$$\Lambda_1 = 2\Omega(-V_2 \cos L \sin \mathrm{AZ} - V_3 \sin L) \tag{9.290}$$

$$\Lambda_2 = 2\Omega(V_1 \cos L \sin \mathrm{AZ} + V_3 \cos L \cos \mathrm{AZ}) \tag{9.291}$$

$$\Lambda_3 = 2\Omega(V_1 \sin L - V_2 \cos L \cos \mathrm{AZ}) \tag{9.292}$$

or as a vector

$$\boldsymbol{\Lambda} = \begin{bmatrix} \Lambda_1 \\ \Lambda_2 \\ \Lambda_3 \end{bmatrix} [\mathbf{e}_1 \ \mathbf{e}_2 \ \mathbf{e}_3] = 2\Omega \begin{bmatrix} -V_2 \cos L \sin \mathrm{AZ} - V_3 \sin L \\ V_1 \cos L \sin \mathrm{AZ} + V_3 \cos L \cos \mathrm{AZ} \\ V_1 \sin L - V_2 \cos L \cos \mathrm{AZ} \end{bmatrix} [\mathbf{e}_1 \ \mathbf{e}_2 \ \mathbf{e}_3] \tag{9.293}$$

We now have the differential equations of motion but need initial conditions to solve them. Let us examine the projectile at the instant of muzzle exit without worrying about how it attained its state of motion there (this is the job of the interior ballistician). We shall define the initial tube angle in azimuth and elevation as θ_0 and ϕ_0, respectively. Then, our initial velocity vector can be defined as

$$\mathbf{V}_0 = \begin{bmatrix} \Lambda_1 \\ \Lambda_2 \\ \Lambda_3 \end{bmatrix} [\mathbf{e}_1\ \mathbf{e}_2\ \mathbf{e}_3] = V_0 \begin{bmatrix} \cos\phi_0 \cos\theta_0 \\ \sin\phi_0 \cos\theta_0 \\ \sin\theta_0 \end{bmatrix} [\mathbf{e}_1\ \mathbf{e}_2\ \mathbf{e}_3] \tag{9.294}$$

And, if we also take the wind into account, we have

$$\mathbf{v}_0 = \mathbf{V}_0 - \mathbf{W}_0 = \begin{bmatrix} v_{1_0} \\ v_{2_0} \\ v_{3_0} \end{bmatrix} [\mathbf{e}_1\ \mathbf{e}_2\ \mathbf{e}_3] = \begin{bmatrix} V_{1_0} - W_{1_0} \\ V_{2_0} - W_{2_0} \\ V_{3_0} - W_{3_0} \end{bmatrix} [\mathbf{e}_1\ \mathbf{e}_2\ \mathbf{e}_3] \tag{9.295}$$

Here the usual relationships for these vectors apply. These are

$$V_0 = \sqrt{V_{1_0}^2 + V_{2_0}^2 + V_{3_0}^2} \tag{9.296}$$

$$v_0 = \sqrt{v_{1_0}^2 + v_{2_0}^2 + v_{3_0}^2} \tag{9.297}$$

The initial orientations of the body-fixed unit vectors in the earth-fixed system are

$$\mathbf{i}_0 = i_{1_0}\mathbf{e}_1 + i_{2_0}\mathbf{e}_2 + i_{3_0}\mathbf{e}_3 = \begin{bmatrix} i_{1_0} \\ i_{2_0} \\ i_{3_0} \end{bmatrix} [\mathbf{e}_1\ \mathbf{e}_2\ \mathbf{e}_3] = \begin{bmatrix} \cos(\phi_0 + \alpha_0)\cos(\theta_0 + \beta_0) \\ \sin(\phi_0 + \alpha_0)\cos(\theta_0 + \beta_0) \\ \sin(\theta_0 + \beta_0) \end{bmatrix} [\mathbf{e}_1\ \mathbf{e}_2\ \mathbf{e}_3] \tag{9.298}$$

$$\mathbf{j}_0 = j_{1_0}\mathbf{e}_1 + j_{2_0}\mathbf{e}_2 + j_{3_0}\mathbf{e}_3 = \begin{bmatrix} j_{1_0} \\ j_{2_0} \\ j_{3_0} \end{bmatrix} [\mathbf{e}_1\ \mathbf{e}_2\ \mathbf{e}_3]$$

$$= \frac{1}{\sqrt{Q}} \begin{bmatrix} -\cos^2(\theta_0 + \beta_0)\sin(\phi_0 + \alpha_0)\cos(\phi_0 + \alpha_0) \\ \cos^2(\theta_0 + \beta_0)\cos^2(\phi_0 + \alpha_0) + \sin^2(\theta_0 + \beta_0) \\ -\sin(\theta_0 + \beta_0)\cos(\theta_0 + \beta_0)\sin(\phi_0 + \alpha_0) \end{bmatrix} [\mathbf{e}_1\ \mathbf{e}_2\ \mathbf{e}_3] \tag{9.299}$$

$$\mathbf{k}_0 = k_{1_0}\mathbf{e}_1 + k_{2_0}\mathbf{e}_2 + k_{3_0}\mathbf{e}_3 = \begin{bmatrix} k_{1_0} \\ k_{2_0} \\ k_{3_0} \end{bmatrix} [\mathbf{e}_1\ \mathbf{e}_2\ \mathbf{e}_3] = \frac{1}{\sqrt{Q}} \begin{bmatrix} -\sin(\theta_0 + \beta_0) \\ 0 \\ \cos(\theta_0 + \beta_0)\cos(\phi_0 + \alpha_0) \end{bmatrix} [\mathbf{e}_1\ \mathbf{e}_2\ \mathbf{e}_3] \tag{9.300}$$

In the aforementioned equations, α_0 and β_0 are the initial pitch and yaw angles, respectively, of the projectile. Thus, they directly add to the weapon azimuth and elevation angles. We define quantity Q following McCoy [1] as

$$Q = \sin^2(\theta_0 + \beta_0) + \cos^2(\theta_0 + \beta_0)\cos^2(\phi_0 + \alpha_0) \tag{9.301}$$

If we now consider the rotation $(\boldsymbol{\omega})_{ijk}$ of the projectile about its axis of symmetry (thus relative to the **i–j–k** triad) and we define an arbitrary initial projectile rotation as

$$(\boldsymbol{\omega}_0)_{ijk} = \omega_{i_0}\mathbf{i}_0 + \omega_{j_0}\mathbf{j}_0 + \omega_{k_0}\mathbf{k}_0 \tag{9.302}$$

Here this initial angular velocity is dependent upon the initial orientation of the unit vector i_0. Then, the initial velocity of the unit vector can be written as follows:

$$\frac{d\mathbf{i}_0}{dt} = (\boldsymbol{\omega}_0)_{ijk} \times \mathbf{i}_0 = \begin{vmatrix} \mathbf{i}_0 & \mathbf{j}_0 & \mathbf{k}_0 \\ \omega_{i_0} & \omega_{j_0} & \omega_{k_0} \\ 1 & 0 & 0 \end{vmatrix} = \left[\omega_{k_0}\mathbf{j}_0 - \omega_{j_0}\mathbf{k}_0\right] = \begin{bmatrix} \dot{i}_{1_0} \\ \dot{i}_{2_0} \\ \dot{i}_{3_0} \end{bmatrix} [\,\mathbf{e}_1 \ \mathbf{e}_2 \ \mathbf{e}_3\,]$$

$$= \begin{bmatrix} \omega_{k_0} j_{1_0} - \omega_{j_0} k_{1_0} \\ \omega_{k_0} j_{2_0} - \omega_{j_0} k_{2_0} \\ \omega_{k_0} j_{3_0} - \omega_{j_0} k_{3_0} \end{bmatrix} [\,\mathbf{e}_1 \ \mathbf{e}_2 \ \mathbf{e}_3\,] \tag{9.303}$$

Note that Equation 9.303 is a tensor equation. Tensors are higher-order vectors but can be treated the same. If we insert the results of Equations 9.299 and 9.300 into the aforementioned equation, we get

$$\dot{i}_{1_0} = \frac{1}{\sqrt{Q}} = \left[\,\omega_{j_0} \sin(\theta_0 + \beta_0) - \omega_{k_0} \cos^2(\theta_0 + \beta_0) \sin(\phi_0 + \alpha_0) \cos(\phi_0 + \alpha_0)\,\right] \tag{9.304}$$

$$\dot{i}_{2_0} = \frac{1}{\sqrt{Q}} = \left[\omega_{k_0} \cos^2(\theta_0 + \beta_0) \cos^2(\phi_0 + \alpha_0) + \omega_{k_0} \sin^2(\theta_0 + \beta_0)\right] \tag{9.305}$$

$$\dot{i}_{3_0} = \frac{1}{\sqrt{Q}} \left[-\omega_{j_0} \cos(\theta_0 + \beta_0) \cos(\phi_0 + \alpha_0) - \omega_{k_0} \sin(\theta_0 + \beta_0) \cos(\theta_0 + \beta_0) \sin(\phi_0 + \alpha_0)\right] \tag{9.306}$$

Continuing with our statement of the initial conditions, a positive pitch rotates the nose of the projectile upward and a positive yaw rotates the nose to the left as viewed from the rear. The initial value of the modified angular momentum vector is given by

$$\mathbf{h}_0 = \frac{I_P p_0}{I_T} \mathbf{i}_0 + \left(\mathbf{i}_0 \times \frac{d\mathbf{i}_0}{dt} \right) \tag{9.307}$$

We can rewrite $d\mathbf{i}_0/dt$ as

$$\frac{d\mathbf{i}_0}{dt} = \dot{i}_{1_0}\mathbf{e}_1 + \dot{i}_{2_0}\mathbf{e}_2 + \dot{i}_{3_0}\mathbf{e}_3 \tag{9.308}$$

Which then allows us to write

$$\mathbf{i}_0 \times \frac{d\mathbf{i}_0}{dt} = \begin{vmatrix} \mathbf{e}_1 & \mathbf{e}_2 & \mathbf{e}_3 \\ i_{1_0} & i_{2_0} & i_{3_0} \\ \dot{i}_{1_0} & \dot{i}_{2_0} & \dot{i}_{3_0} \end{vmatrix} = \left(i_{2_0}\dot{i}_{3_0} - i_{3_0}\dot{i}_{2_0}\right)\mathbf{e}_1 + \left(i_{3_0}\dot{i}_{1_0} - i_{1_0}\dot{i}_{3_0}\right)\mathbf{e}_2 + \left(i_{1_0}\dot{i}_{2_0} - i_{2_0}\dot{i}_{1_0}\right)\mathbf{e}_3 \tag{9.309}$$

We can then incorporate Equation 9.309 into Equation 9.307 to yield

$$\mathbf{h}_0 = \begin{bmatrix} h_{1_0} \\ h_{2_0} \\ h_{3_0} \end{bmatrix} [\,\mathbf{e}_1 \ \mathbf{e}_2 \ \mathbf{e}_3\,] = \begin{bmatrix} \dfrac{I_P p_0}{I_T} i_{1_0} + i_{2_0}\dot{i}_{3_0} - i_{3_0}\dot{i}_{2_0} \\ \dfrac{I_P p_0}{I_T} i_{2_0} + i_{3_0}\dot{i}_{1_0} - i_{1_0}\dot{i}_{3_0} \\ \dfrac{I_P p_0}{I_T} i_{3_0} + i_{1_0}\dot{i}_{2_0} - i_{2_0}\dot{i}_{1_0} \end{bmatrix} [\,\mathbf{e}_1 \ \mathbf{e}_2 \ \mathbf{e}_3\,] \tag{9.310}$$

Here the initial value of the spin rate p_0 is determined by the axial velocity and the twist rate n (in calibers per revolution) of the weapon through $p_0 = 2\pi V_0/nd$. We have thus completed all the initial conditions necessary to perform the calculation.

As we will discuss in a later section, the motion of the motion can be characterized as epicyclic. The tip of a vector drawn from the CG of the projectile to the nose will trace out a curve that contains two cyclic modes: a fast mode, known as nutation, and a slow mode, known as precession. If the round is stable, these modes will eventually damp down to near zero, leaving only some movement because of nonlinear forces and moments. We shall explore this more later.

Some other terms come up in the succeeding sections that require definitions. Since they are essential to the understanding of trajectories, we will define them now.

The yaw of a projectile of repose is the yaw created by the action of gravity on the projectile as it attempts to follow its trajectory curve. As stated earlier, the nose of the projectile is usually above the trajectory. There is then a net aerodynamic force through the CP which wants to rotate the nose up. With a right-hand spinning projectile, this results in a yaw of the nose to the right. This is called the yaw of repose.

Failure to trail is a situation that arises when the base of the projectile does not follow the nose (it flies base first after maximum ordinate). This is depicted in Figure 9.13.

The trail angle is the quadrant elevation angle (particular to a gun, projectile, and charge combination) above which the projectile will not turn over and will fail to trail.

We can summarize this section by saying that for a rigid projectile, the 6 DOF model is as accurate as one can get to the trajectories. If the model yields an inaccurate answer, the problem is usually a wrong assumption in the metrology, initial conditions, or projectile mass properties. Lastly, the only practical method of solving these equations is by numerical methods, and, with the speed of computers today, the codes run very efficiently and quickly. This last statement makes it difficult to generate meaningful problems for the interested reader. We have endeavored to create useful exercises by stipulating a large number of conditions and requiring the reader to examine the accelerations of the projectile at a point in space.

Problem 34

A British bomber is flying at a speed of 200 mph in still air. If the 0.303 in. machine guns are fired sideways, calculate the axial acceleration vector and the angular acceleration vector acting on the projectile through use of the 6 DOF equations if the projectile is

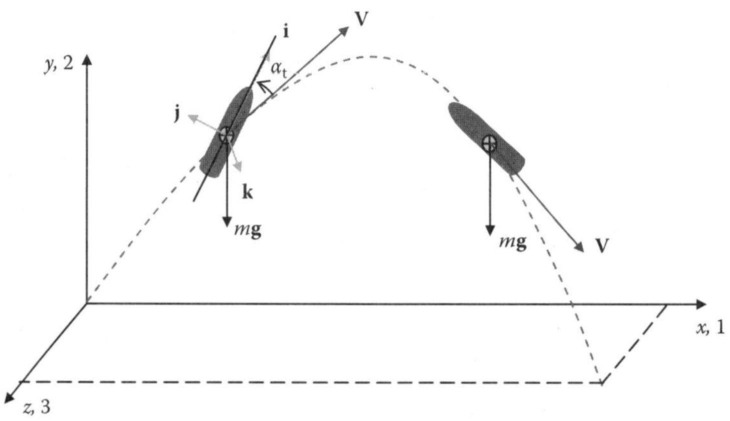

FIGURE 9.13
Projectile that has failed to trail.

1. Fired to the right
   Answer:

$$\mathbf{a} = [-1376\mathbf{e}_1 + 5.89\mathbf{e}_2 - 1184\mathbf{e}_3]\left[\frac{\text{ft}}{\text{s}}\right]$$

$$\frac{d\mathbf{h}}{dt} = [39.85\mathbf{e}_1 - 35,922\mathbf{e}_2 - 2.45\mathbf{e}_3]\left[\frac{\text{rad}}{\text{s}^2}\right]$$

2. Fired to the left
   Answer:

$$\mathbf{a} = [-1376\mathbf{e}_1 + 70.29\mathbf{e}_2 - 1184\mathbf{e}_3]\left[\frac{\text{ft}}{\text{s}^2}\right]$$

$$\frac{d\mathbf{h}}{dt} = [39.85\mathbf{e}_1 + 35,922\mathbf{e}_2 + 2.45\mathbf{e}_3]\left[\frac{\text{rad}}{\text{s}^2}\right]$$

3. Discuss the effect of the angular momentum on the projectile nose (which way does it tip?)

Please ignore the Coriolis acceleration; assume there is no yaw at muzzle exit and assume a muzzle velocity of 2440 ft/s; the weapon has a right-hand twist.

- Projectile information:

$$C_{D_0} = 0.35 \qquad \left(C_{M_q} + C_{M_{\dot{\alpha}}}\right) = -16.2 \qquad I_P = 0.00026\,[\text{lbm-in.}^2]$$

$$C_{D_{\delta^2}} = 3.46 \qquad \left(C_{N_q} + C_{N_{\dot{\alpha}}}\right) = 0.003 \qquad I_T = 0.00258\,[\text{lbm-in.}^2]$$

$$C_{M_\alpha} = 2.36 \qquad C_{M_{p\alpha}} = 0.02 \qquad m = 0.025\,[\text{lbm}]$$

$$C_{L_\alpha} = 2.81$$

$$C_{N_{p\alpha}} = -0.67 \qquad \rho = 0.060\left[\frac{\text{lbm}}{\text{ft}^3}\right] \qquad p = 2033\left[\frac{\text{rev}}{\text{s}}\right].$$

Please supply all answers in an inertial coordinate system labeled 1, 2, and 3, with 1 being along the aircraft flight path and 3 being off the right side of the plane. Treat all missing coefficients as equal to zero.

Problem 35

One of the interesting aspects of the forces acting on a projectile occurs as the projectile leaves an aircraft sideways. This problem is encountered all the time in the AC-130 gunship. Let us examine a 105 mm HE projectile being fired into a city from both the top of a building and from the AC-130 in flight. The velocity of the projectile is 1510 ft/s. With the information provided,

1. Calculate the total acceleration vector for both cases.
2. Comment on the differences.

Positional information:north.

- 33.5° north latitude.
- Azimuth of velocity vector: 80° true.
- Angle of velocity vector to horizontal: −10°.
- Wind is calm.
- $\alpha = +2°$ (nose up); $\beta = -1.5°$ (nose to the left looking downrange).
- The projectile nose is rotating to the right of the velocity vector at 0.5 rad/s.
- The aircraft is flying at 300 mph to the north.

Projectile information:

$$C_{D_0} = 0.39 \quad \left(C_{M_q} + C_{M_{\dot{\alpha}}}\right) = -6.5 \quad I_P = 0.547\,[\text{lbm-ft}^2]$$

$$C_{D_{\delta^2}} = 8.0 \quad \left(C_{N_q} + C_{N_{\dot{\alpha}}}\right) = 0.005 \quad I_T = 5.377\,[\text{lbm-ft}^2]$$

$$C_{M_\alpha} = 3.80 \quad C_{M_{p\alpha}} = 0.05 \quad m = 32.1\,[\text{lbm}]$$

$$C_{L_\alpha} = 1.9$$

$$C_{N_{p\alpha}} = -0.01 \quad \rho = 0.060\left[\frac{\text{lbm}}{\text{ft}^3}\right] \quad p = 200\left[\frac{\text{rev}}{\text{s}}\right].$$

Please supply all answers in an inertial coordinate system labeled 1, 2, and 3, with 1 being due north and 3 being due east.

Problem 36

The Paris gun was built by Germany in the First World War to shell Paris from 75 mi away. The weapon was a 210 mm diameter bore with the shells preengraved to compensate for wear of the tube. During firing of this weapon, all things such as wind effects and Coriolis had to be accounted for (they really could have used a good 6 DOF model and a computer). Write the acceleration vector for this projectile at an instant in its trajectory when the velocity (relative to the ground) is 2500 ft/s and the following conditions apply (please note that there is no rocket motor):

- Positional information:

  48.75° north latitude.
  Azimuth of velocity vector: 300° true.
  Angle of velocity vector to horizontal: +10°.
  Wind is blowing at 20 mph due south and horizontal.
  $\alpha = 1°$; $\beta = 1.5°$.
  The projectile nose is rotating up at 2 rad/s.

- Projectile information:

$$C_D = 0.28 \quad \left(C_{M_q} + C_{M_{\dot{\alpha}}}\right) = -16.5 \quad I_P = 19.13\,[\text{lbm-ft}^2]$$

$$C_{M_\alpha} = 3.50 \quad \left(C_{N_q} + C_{N_{\dot{\alpha}}}\right) = 0.005 \quad I_T = 66.40\,[\text{lbm-ft}^2]$$

$$C_{L_\alpha} = 2.50 \quad C_{M_{p\alpha}} = 0.55 \quad m = 220\,[\text{lbm}]$$

$$C_{N_{p\alpha}} = -0.02 \quad \rho = 0.060\left[\frac{\text{lbm}}{\text{ft}^3}\right] \quad p = 150\left[\frac{\text{rev}}{\text{s}}\right].$$

Note that the aforementioned numbers are guesses at the characteristics of the projectile; they do not represent the performance of the real projectile as no data are available from any source researched.

Please supply all answers in an inertial coordinate system labeled 1, 2, and 3, with 1 being due west and 3 being due north.

Answer:

$$\text{The linear acceleration vector is } \mathbf{a} = \{-85\mathbf{e}_1 - 33\mathbf{e}_2 - 35\mathbf{e}_3\}\left[\frac{\text{ft}}{\text{s}^2}\right]$$

$$\text{The angular acceleration vector is } \frac{d\mathbf{h}}{dt} = \{-14\mathbf{e}_1 - 42\mathbf{e}_2 + 38\mathbf{e}_3\}\left[\frac{\text{rad}}{\text{s}^2}\right]$$

Problem 37

An AC-130 is flying at a speed of 200 mph in still air. If the 105 mm weapon is fired sideways, calculate the axial acceleration vector and the angular acceleration vector acting on the projectile through use of the 6 DOF equations if the projectile is fired to the left. Discuss the effect of the angular momentum on the projectile nose (which way does it want to tip?)

Please ignore the Coriolis acceleration; assume there is no yaw at muzzle exit, and assume a muzzle velocity of 1500 ft/s; the weapon has a right-hand twist.

- Projectile information:

$$C_{D_0} = 0.39 \qquad \left(C_{M_q} + C_{M_{\dot{\alpha}}}\right) = -6.5 \qquad I_P = 0.547\ [\text{lbm-ft}^2]$$

$$C_{D_{\delta^2}} = 8.0 \qquad \left(C_{N_q} + C_{N_{\dot{\alpha}}}\right) = 0.005 \qquad I_T = 5.377\ [\text{lbm-ft}^2]$$

$$C_{M_\alpha} = 3.80 \qquad C_{M_{p\alpha}} = 0.05 \qquad m = 32.1\ [\text{lbm}]$$

$$C_{L_\alpha} = 1.9$$

$$C_{N_{p\alpha}} = -0.01 \qquad \rho = 0.060\left[\frac{\text{lbm}}{\text{ft}^3}\right] \qquad p = 220\left[\frac{\text{rev}}{\text{s}}\right].$$

Please supply all answers in an inertial coordinate system labeled 1, 2, 3, with 1 being along the aircraft flight path and 3 being off the right side of the plane. Treat all missing coefficients as equal to zero.

Problem 38

An F-86 Saber jet flying with a speed of 600 mph. The aircraft is pulling up out of a dive and the pilot is experiencing 2.5 g's. The pilot fires his 0.50 caliber machine guns at a ground target during the pull out. At the instant the plane makes an angle of 30° to the ground, a projectile leaves the muzzle of the weapon with a muzzle velocity of 1800 ft/s, 3° to the right of the bore axis and is rotating to the left at 2 rad/s in the plane of the gun, relative to the gun. Also the projectile is pitched up 1.5° and yawed left 1° to the line of fire. Ignoring Coriolis effects, write the vectors that completely define the initial conditions of the projectile so that you could use them in a 6 DOF model. *Do not* solve the equations of motion—just write the initial conditions. Assume coordinate directions as follows:

1. Horizontal direction in which the aircraft is flying
2. Straight up
3. To the right as viewed from the rear of the aircraft

Problem 39

In the study of ballistics, we normally neglect Magnus forces as small, and only the moments generated by these forces are significant. In low-velocity projectiles, however, Magnus forces may be considerable. Consider a baseball thrown with spin toward a batter. The ball weighs 5 oz. and is 2.9 in. in diameter. The distance from the pitcher's mound to the home plate is 60 ft 6 inches. Assume the ball is thrown at 70 mph and that there is no wind. The release point is 4 ft above the ground with an upward angle of 4°. We need to determine how much spin is required to move the ball 2 ft to the left as viewed from the pitcher. Please perform the following calculations:

1. Assuming constant drag and Magnus force coefficients, develop the equations of motion for the ball assuming flat fire is valid and the spin axis remains vertical for the entire flight.
2. Determine the spin rate and direction to cause the desired motion.
3. Determine the final velocity, TOF, and height of the ball.
4. Determine if the assumption of flat fire really was valid.

Please list all your assumptions. The following coefficients may be assumed:

$$C_D = 0.52; \quad \rho = 0.076 \left[\frac{\text{lbm}}{\text{ft}^3}\right]; \quad C_{N_{p\alpha}} = 0.09$$

9.6 Modified Point Mass Trajectory

The equations of the 6 DOF model, when fully developed, described an epicyclical motion with fast (nutational) and slow (precessional) modes that (hopefully) would damp out early on, allowing the projectile to assume a yaw of repose for the remainder of the flight. This yaw of repose, which remains nearly constant, we assume will account for most of the drag induced by the yawing of the projectile. If we can simplify the 6 DOF model, which is computationally expensive to run, by accounting only for the yaw of repose, we could get a model that will allow the projectile to drift the proper amount and still be quite accurate.

Once again, following McCoy [1], we will make the point mass assumption in the equations that follow. Recall that because of this assumption, the projectile is essentially represented as a cannonball with all its mass concentrated at one point. We shall then add some details, which will account for the yawing of the projectile, by assuming the projectile yaw is relatively constant or varies little with time compared to the steady state yaw angle. This assumption is usually valid except in high-angle fire situations.

We begin with the usual equations of motion and Newton's second law

$$\mathbf{F} = m\mathbf{a} \tag{9.311}$$

Particularized as

$$m\frac{d\mathbf{V}}{dt} = \Sigma\mathbf{F} + m\mathbf{g} + m\mathbf{\Lambda} \tag{9.312}$$

Here the variables are the same as we described in the 6 DOF section. In the aforementioned equations, we replace the velocity vector **V** by the vector (**V** – **W**) because drag measurements are made relative to the air stream n relative to the ground. We will also again replace the scalar velocity (the speed) with the difference between the projectile and wind velocities:

$$\mathbf{v} = \mathbf{V} - \mathbf{W} \rightarrow \tilde{V} = |\mathbf{V} - \mathbf{W}| \tag{9.313}$$

The diagram of the problem is shown in Figure 9.14.

From our work on 6 DOF model, recall Equations 9.257 and 9.258, which we rewrite here neglecting the pitch-damping and rocket forces:

$$\frac{d\mathbf{V}}{dt} = -\frac{\rho v S C_D}{2m}\mathbf{v} + \frac{\rho S C_{L\alpha}}{2m}[v^2\mathbf{i} - (\mathbf{v}\cdot\mathbf{i})\mathbf{v}] - \frac{\rho S d C_{N_{p\alpha}}}{2m}\left(\frac{I_T}{I_P}\right)(\mathbf{h}\cdot\mathbf{i})(\mathbf{i}\times\mathbf{v}) + \mathbf{g} + \mathbf{\Lambda} \tag{9.314}$$

$$\begin{aligned}\frac{d\mathbf{h}}{dt} &= \frac{\rho v S d^2 C_{l_p}}{2I_P}(\mathbf{h}\cdot\mathbf{i})\mathbf{i} + \frac{\rho v^2 S d \delta_F C_{l_\delta}}{2I_T}\mathbf{i} + \frac{\rho v S d C_{M_\alpha}}{2I_T}(\mathbf{v}\times\mathbf{i}) \\ &+ \frac{\rho S d^2 C_{M_{p\alpha}}}{2I_P}(\mathbf{h}\cdot\mathbf{i})[\mathbf{v} - (\mathbf{v}\cdot\mathbf{i})\mathbf{i}] + \frac{\rho v S d^2\left(C_{M_q} + C_{M_{\dot\alpha}}\right)}{2I_P}[\mathbf{h} - (\mathbf{h}\cdot\mathbf{i})\mathbf{i}]\end{aligned} \tag{9.315}$$

Recall also from 6 DOF model (Equation 9.228) that introduces the polar and transverse moments of inertia that we can write

$$\frac{d\mathbf{h}}{dt} = \frac{I_P}{I_T}\dot{p}\mathbf{i} + \frac{I_P p}{I_T}\frac{d\mathbf{i}}{dt} + \left(\mathbf{i}\times\frac{d^2\mathbf{i}}{dt^2}\right) \tag{9.316}$$

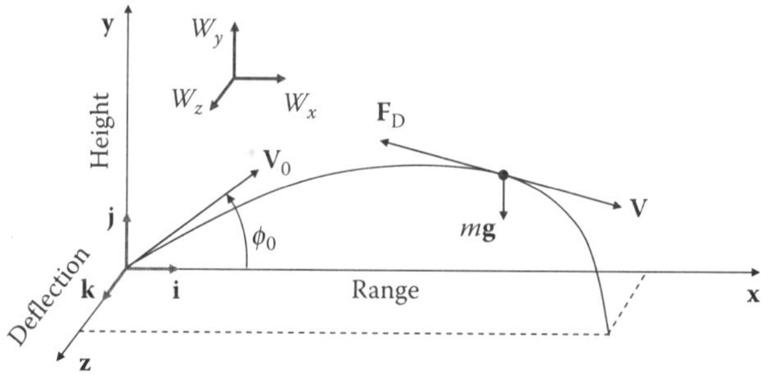

FIGURE 9.14
Modified point mass trajectory.

Equations 9.314 and 9.315 may be simplified by introducing the tilde (~) coefficients:

$$\tilde{C}_D = \frac{\rho v S C_D}{2m} \qquad \tilde{C}_{l_p} = \frac{\rho v S d^2 C_{l_p} p}{2I_T}$$

$$\tilde{C}_{L_\alpha} = \frac{\rho S C_{L_\alpha}}{2m} \qquad \tilde{C}_{l_\delta} = \frac{\rho v^2 S d \delta_F C_{l_\delta}}{2I_T}$$

$$\tilde{C}_{N_{p\alpha}} = \frac{\rho S d C_{N_{p\alpha}} p}{2m} \qquad \tilde{C}_{M_\alpha} = \frac{\rho v S d C_{M_\alpha}}{2I_T}$$

$$\tilde{C}_{N_q} = \frac{\rho v S d \left(C_{N_q} + C_{N_{\dot{\alpha}}}\right)}{2m} \qquad \tilde{C}_{M_{p\alpha}} = \frac{\rho S d^2 C_{M_{p\alpha}} p}{2I_T}$$

$$\tilde{C}_{M_q} = \frac{\rho v S d^2 \left(C_{M_q} + C_{M_{\dot{\alpha}}}\right)}{2I_T}$$

Using this notation, the modified equations are written as

$$\frac{d\mathbf{V}}{dt} = -\tilde{C}_D \mathbf{v} + \tilde{C}_{L_\alpha}\left[v^2\mathbf{i} - (\mathbf{v}\cdot\mathbf{i})\mathbf{v}\right] - \tilde{C}_{N_{p\alpha}}\left(\frac{I_T}{I_P}\right)(\mathbf{h}\cdot\mathbf{i})(\mathbf{i}\times\mathbf{v}) + \mathbf{g} + \mathbf{\Lambda}, \tag{9.317}$$

$$\frac{d\mathbf{h}}{dt} = \left(\tilde{C}_{l_p} + \tilde{C}_{l_\delta}\right)\mathbf{i} + \tilde{C}_{M_\alpha}(\mathbf{v}\times\mathbf{i}) + \tilde{C}_{M_{p\alpha}}(\mathbf{h}\cdot\mathbf{i})[\mathbf{v} - (\mathbf{v}\cdot\mathbf{i})\mathbf{i}] + \tilde{C}_{M_q}[\mathbf{h} - (\mathbf{h}\cdot\mathbf{i})\mathbf{i}] \tag{9.318}$$

or alternatively,

$$\frac{d\mathbf{V}}{dt} = -\tilde{C}_D \mathbf{v} + \tilde{C}_{L_\alpha}[\mathbf{v}\times(\mathbf{i}\times\mathbf{v})] - \tilde{C}_{N_{p\alpha}}(\mathbf{v}\times\mathbf{i}) + \mathbf{g} + \mathbf{\Lambda}, \tag{9.319}$$

$$\frac{d\mathbf{h}}{dt} = (\tilde{C}_{l_p} + \tilde{C}_{l_\delta})\mathbf{i} + \tilde{C}_{M_\alpha}(\mathbf{v}\times\mathbf{i}) + \tilde{C}_{M_{p\alpha}}(\mathbf{h}\cdot\mathbf{i})[\mathbf{v} - (\mathbf{v}\cdot\mathbf{i})\mathbf{i}] + \tilde{C}_{M_q}[\mathbf{h} - (\mathbf{h}\cdot\mathbf{i})\mathbf{i}] \tag{9.320}$$

We have shown earlier that since the unit vector **i** is always perpendicular to its derivative d**i**/dt, the dot product of **i** and d**i**/dt is identically zero. We shall combine Equations 9.316 and 9.320 to yield

$$\frac{I_P}{I_T}\dot{p}\mathbf{i} + \frac{I_P}{I_T}p\frac{d\mathbf{i}}{dt} + \left(\mathbf{i}\times\frac{d^2\mathbf{i}}{dt^2}\right) = \left(\tilde{C}_{l_p} + \tilde{C}_{l_\delta}\right)\mathbf{i} + \tilde{C}_{M_\alpha}(\mathbf{v}\times\mathbf{i}) + \tilde{C}_{M_{p\alpha}}[\mathbf{i}\times(\mathbf{v}\times\mathbf{i})] + \tilde{C}_{M_q}\left[\mathbf{i}\times\frac{d\mathbf{i}}{dt}\right] \tag{9.321}$$

We will now take the dot product of **i** with Equation 9.321 to yield

$$\frac{I_P}{I_T}\dot{p} = \left(\tilde{C}_{l_p} + \tilde{C}_{l_\delta}\right) \rightarrow \frac{dp}{dt} = \frac{I_T}{I_P}\left(\tilde{C}_{l_p} + \tilde{C}_{l_\delta}\right) \tag{9.322}$$

Here use has been made of the facts that a cross product results in a vector that is orthogonal to both of the original vectors and that the dot product of orthogonal vectors is identically zero. These relationships are written in mathematical terms here:

$$\mathbf{i}\cdot\frac{d\mathbf{i}}{dt} = \mathbf{i}\cdot(\omega\times\mathbf{i}) = 0 \quad \mathbf{i}\cdot\left(\mathbf{i}\times\frac{d\mathbf{i}}{dt}\right) = 0$$

$$\mathbf{i}\cdot\left[\mathbf{i}\times(\mathbf{v}\times\mathbf{i})\right] = 0 \quad \mathbf{i}\cdot(\mathbf{i}\times\mathbf{v}) = \mathbf{i}\cdot(\mathbf{v}\times\mathbf{i}) = 0.$$

Equation 9.322 has an important consequence; for a rotationally symmetric projectile, the spin is decoupled from the yawing motion. Now, if we substitute Equation 9.322 into Equation 9.321, we get

$$\begin{aligned}&\left(\tilde{C}_{l_p}+\tilde{C}_{l_\delta}\right)\mathbf{i}+\frac{I_P}{I_T}p\frac{d\mathbf{i}}{dt}+\left(\mathbf{i}\times\frac{d^2\mathbf{i}}{dt^2}\right)\\&\quad=\left(\tilde{C}_{l_p}+\tilde{C}_{l_\delta}\right)\mathbf{i}+\tilde{C}_{M_\alpha}(\mathbf{v}\times\mathbf{i})+\tilde{C}_{M_{p\alpha}}[\mathbf{i}\times(\mathbf{v}\times\mathbf{i})]+\tilde{C}_{M_q}\left[\mathbf{i}\times\frac{d\mathbf{i}}{dt}\right]\end{aligned}\tag{9.323}$$

or

$$\frac{I_P}{I_T}p\frac{d\mathbf{i}}{dt}+\left(\mathbf{i}\times\frac{d^2\mathbf{i}}{dt^2}\right)=\tilde{C}_{M_\alpha}(\mathbf{v}\times\mathbf{i})+\tilde{C}_{M_{p\alpha}}[\mathbf{i}\times(\mathbf{v}\times\mathbf{i})]+\tilde{C}_{M_q}\left[\mathbf{i}\times\frac{d\mathbf{i}}{dt}\right]\tag{9.324}$$

With Equations 9.319, 9.322, and 9.324, we have merely restated our 6 DOF model. Murphy [2] formulated the differential equation of motion as a second-order equation in terms of complex variables and solved it. The particular solution was the (relatively) constant yaw of repose, and the complementary solution was the transient epicyclic motion. In the modified point mass approach, we extract the particular solution and ignore the transient motion, instead concentrating on the yaw of repose, the drift, and the effect of the yaw drag. We assume that the epicyclic pitching and yawing motion are negligible everywhere along the trajectory, i.e., in many instances, reasonable, since it should damp early in the trajectory and thus contributes little to the drift. We proceed by defining another unit vector triad in the same sense as our **i–j–k** triad. Instead of it being aligned with the geometric axis of the projectile, we align it with the velocity vector and utilize **l–n–m** as the principal directions. We can then define **l** (Figure 9.15) as

$$\mathbf{l}=\frac{\mathbf{v}}{|\mathbf{v}|}\tag{9.325}$$

and formally define our vector yaw of repose as

$$\boldsymbol{\alpha}_\mathbf{R}=\mathbf{l}\times(\mathbf{i}\times\mathbf{l})\tag{9.326}$$

But we know that, where α_t is the total angle of attack,

$$\mathbf{l}\times(\mathbf{i}\times\mathbf{l})=(1)^2\mathbf{i}-(\mathbf{l}\cdot\mathbf{i})\mathbf{l}=\mathbf{i}-(1)(1)\cos\alpha_t\mathbf{l}\tag{9.327}$$

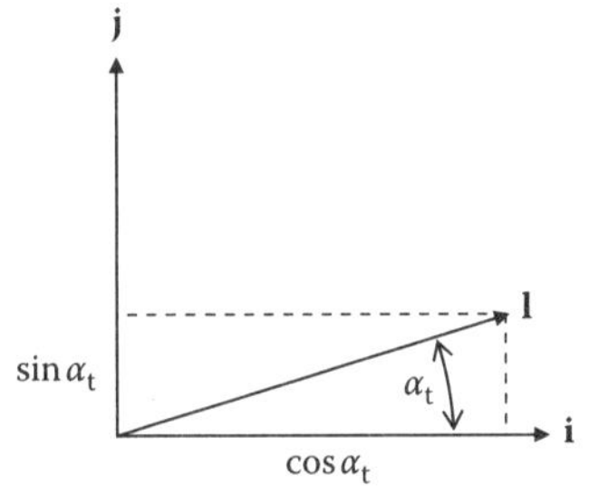

FIGURE 9.15
Projectile axial unit vector **l** illustrated.

so that in terms of α_t

$$\boldsymbol{\alpha}_R = \mathbf{l} \times (\mathbf{i} \times \mathbf{l}) = \mathbf{i} - (\cos\,\alpha_t)\mathbf{l} \tag{9.328}$$

For simplicity [1,2], if we choose the plane that **l** lies in to be the plane that **j** lies in as well, we can write Equation 9.328 as

$$\boldsymbol{\alpha}_R = \mathbf{i} - \cos\,\alpha_t(\cos\alpha_t\mathbf{i} + \sin\alpha_t\mathbf{j}) = \left(1 - \cos^2\alpha_t\right)\mathbf{i} + \sin\alpha_t\cos\,\alpha_t\mathbf{j}$$

then

$$\alpha_R = \sqrt{(1 - \cos^2\alpha_t)^2 + \cos^2\alpha_t\,\sin^2\alpha_t} = \sqrt{\sin^4\alpha_t + \cos^2\alpha_t\,\sin^2\alpha_t}$$

and

$$\alpha_R = \sin\alpha_t\sqrt{\sin^2\alpha_t + \cos^2\alpha_t} = \sin\alpha_t$$

We shall now differentiate Equation 9.328 with respect to time:

$$\frac{d\boldsymbol{\alpha}_R}{dt} = \frac{d\mathbf{i}}{dt} - (\cos\,\alpha_t)\frac{d\mathbf{l}}{dt} + \sin\alpha_t\mathbf{l} \tag{9.329}$$

We have made the assumption early in this analysis that the yaw of repose is relatively constant; thus, $(d\boldsymbol{\alpha}_R/dt) \approx 0$. We also note that for a small yaw angle, $\sin\,\alpha_R \approx 0 \ll \cos\,\alpha_R$. If we incorporate these approximations into Equation 9.329, we get

$$\frac{d\mathbf{i}}{dt} = (\cos\alpha_t)\frac{d\mathbf{l}}{dt} \tag{9.330}$$

Taking the time derivative of Equation 9.330 yields

$$\frac{d^2\mathbf{i}}{dt^2} = (\cos\,\alpha_t)\frac{d^2\mathbf{l}}{dt^2} - (\sin\,\alpha_t)\frac{d\mathbf{l}}{dt} \tag{9.331}$$

But the small angle approximation still applies so that

$$\frac{d^2\mathbf{i}}{dt^2} = (\cos\alpha_t)\frac{d^2\mathbf{l}}{dt^2} \tag{9.332}$$

We can also solve Equation 9.328 for **i** to get

$$\mathbf{i} = \boldsymbol{\alpha}_R + (\cos\alpha_t)\mathbf{l} \tag{9.333}$$

Since **v** and **l** are parallel and the cross products of parallel vectors are zero, we can write

$$\mathbf{i} \times \mathbf{v} = [\boldsymbol{\alpha}_R + (\cos\,\alpha_t)\mathbf{l}] \times \mathbf{v} = \boldsymbol{\alpha}_R \times \mathbf{v} + (\cos\,\boldsymbol{\alpha}_t)\mathbf{l} \times \mathbf{v} = \boldsymbol{\alpha}_R \times \mathbf{v} \tag{9.334}$$

and

$$\mathbf{v} \times (\mathbf{i} \times \mathbf{v}) = \mathbf{v} \times (\boldsymbol{\alpha}_R \times \mathbf{v}) = v^2\boldsymbol{\alpha}_R - (\mathbf{v} \cdot \boldsymbol{\alpha}_R)\mathbf{v} = v^2\boldsymbol{\alpha}_R \tag{9.335}$$

Also, in a similar fashion operating on Equation 9.334, we can write

$$\mathbf{v} \times \mathbf{i} = \mathbf{v} \times [\boldsymbol{\alpha}_R + (\cos\,\boldsymbol{\alpha}_t)\mathbf{l}] = \mathbf{v} \times \boldsymbol{\alpha}_R + \mathbf{v} \times (\cos\,\alpha_t)\mathbf{l} = \mathbf{v} \times \boldsymbol{\alpha}_R \tag{9.336}$$

We can also show that

$$\mathbf{i} \times (\mathbf{v} \times \mathbf{i}) = v \cos \alpha_t \boldsymbol{\alpha}_R + v(\sin^2 \alpha_t)\mathbf{l} \tag{9.337}$$

We now have relations in Equations 9.330, 9.332, and 9.333 for $\mathbf{i}$, $\mathrm{d}\mathbf{i}/\mathrm{d}t$, and $\mathrm{d}^2\mathbf{i}/\mathrm{d}t^2$, respectively, and can substitute them into Equations 9.319 and 9.324 to eliminate $\mathbf{i}$. We shall start with Equation 9.319 and noting that

$$\mathbf{v} \times \boldsymbol{\alpha}_R = (v\mathbf{l}) \times \boldsymbol{\alpha}_R = v(\mathbf{l} \times \boldsymbol{\alpha}_R) \tag{9.338}$$

We get

$$\frac{\mathrm{d}\mathbf{V}}{\mathrm{d}t} = -\tilde{C}_D \mathbf{v} + \tilde{C}_{L_\alpha} v^2 \boldsymbol{\alpha}_R + \tilde{C}_{N_{p\alpha}} v(\mathbf{l} \times \boldsymbol{\alpha}_R) + \mathbf{g} + \boldsymbol{\Lambda} \tag{9.339}$$

It is also worth noting that

$$\frac{\mathrm{d}\mathbf{V}}{\mathrm{d}t} = \frac{\mathrm{d}}{\mathrm{d}t} V\mathbf{l} = \frac{\mathrm{d}V}{\mathrm{d}t}\mathbf{l} + V\frac{\mathrm{d}\mathbf{l}}{\mathrm{d}t} = \dot{V}\mathbf{l} + V\frac{\mathrm{d}\mathbf{l}}{\mathrm{d}t} \tag{9.340}$$

We will now attack each term of Equation 9.324, but first we define

$$\gamma = (\mathbf{l} \cdot \mathbf{i}) = \cos \alpha_t \tag{9.341}$$

Then, this and the succeeding relations follow as

$$\frac{I_P}{I_T} p \frac{\mathrm{d}\mathbf{i}}{\mathrm{d}t} \approx \frac{I_P}{I_T} p \cos \alpha_t \frac{\mathrm{d}\mathbf{l}}{\mathrm{d}t} = \gamma \frac{I_P}{I_T} p \frac{\mathrm{d}\mathbf{l}}{\mathrm{d}t} \tag{9.342}$$

$$\left(\mathbf{i} \times \frac{\mathrm{d}^2\mathbf{i}}{\mathrm{d}t^2}\right) = \gamma\left(\boldsymbol{\alpha}_R \times \frac{\mathrm{d}^2\mathbf{l}}{\mathrm{d}t^2}\right) + \gamma^2\left(\mathbf{l} \times \frac{\mathrm{d}^2\mathbf{l}}{\mathrm{d}t^2}\right) \tag{9.343}$$

$$\left[\mathbf{i} \times \frac{\mathrm{d}\mathbf{i}}{\mathrm{d}t}\right] = \gamma\left(\boldsymbol{\alpha}_R \times \frac{\mathrm{d}\mathbf{l}}{\mathrm{d}t}\right) + \gamma^2\left(\mathbf{l} \times \frac{\mathrm{d}\mathbf{l}}{\mathrm{d}t}\right) \tag{9.344}$$

Combining the terms of Equation 9.324, we get

$$\begin{aligned}\gamma \frac{I_P}{I_T} p \frac{\mathrm{d}\mathbf{l}}{\mathrm{d}t} + \gamma\left(\boldsymbol{\alpha}_R \times \frac{\mathrm{d}^2\mathbf{l}}{\mathrm{d}t^2}\right) + \gamma^2\left(\mathbf{l} \times \frac{\mathrm{d}^2\mathbf{l}}{\mathrm{d}t^2}\right) &= \tilde{C}_{M_\alpha} v(\mathbf{l} \times \boldsymbol{\alpha}_R) + \tilde{C}_{M_{p\alpha}}[v\gamma\boldsymbol{\alpha}_R + v\sin^2\alpha_t \mathbf{l}] \\ &\quad + \tilde{C}_{M_q}\left[\gamma\left(\boldsymbol{\alpha}_R \times \frac{\mathrm{d}\mathbf{l}}{\mathrm{d}t}\right) + \gamma^2\left(\mathbf{l} \times \frac{\mathrm{d}\mathbf{l}}{\mathrm{d}t}\right)\right]\end{aligned} \tag{9.345}$$

At this point, we continue with our simplifying assumptions and neglect the Coriolis term in comparison with the gravitational acceleration term and neglect the $\sin^2\alpha_t$ in comparison to γ. Thus, we can rewrite Equations 9.339 (including the relation of Equation 9.340) and 9.345 as

$$\dot{V}\mathbf{l} + V\frac{\mathrm{d}\mathbf{l}}{\mathrm{d}t} = -\tilde{C}_D \mathbf{v} + \tilde{C}_{L_\alpha} v^2 \boldsymbol{\alpha}_R + \tilde{C}_{N_{p\alpha}} v(\mathbf{l} \times \boldsymbol{\alpha}_R) + \mathbf{g} \tag{9.346}$$

and

$$\gamma\frac{I_{\mathrm{P}}}{I_{\mathrm{T}}}p\frac{\mathrm{d}\mathbf{l}}{\mathrm{d}t}+\gamma\left(\boldsymbol{\alpha}_{\mathrm{R}}\times\frac{\mathrm{d}^2\mathbf{l}}{\mathrm{d}t^2}\right)+\gamma^2\left(\mathbf{l}\times\frac{\mathrm{d}^2\mathbf{l}}{\mathrm{d}t^2}\right)=\tilde{C}_{\mathrm{M}_\alpha}v(\mathbf{l}\times\boldsymbol{\alpha}_{\mathrm{R}})+\tilde{C}_{\mathrm{M}_{p\alpha}}v\gamma\boldsymbol{\alpha}_{\mathrm{R}}+\tilde{C}_{\mathrm{M}_q}\left[\gamma\left(\boldsymbol{\alpha}_{\mathrm{R}}\times\frac{\mathrm{d}\mathbf{l}}{\mathrm{d}t}\right)+\gamma^2\left(\mathbf{l}\times\frac{\mathrm{d}\mathbf{l}}{\mathrm{d}t}\right)\right] \tag{9.347}$$

We shall now take the vector cross product of **l** with Equations 9.346 and 9.347 and observe how each term behaves: first, the left-hand side (LHS),

$$\mathbf{l}\times\left(\dot{V}\mathbf{l}+V\frac{\mathrm{d}\mathbf{l}}{\mathrm{d}t}\right)=\mathbf{l}\times\dot{V}\mathbf{l}+\mathbf{l}\times V\frac{\mathrm{d}\mathbf{l}}{\mathrm{d}t}=0+V\left(\mathbf{l}\times\frac{\mathrm{d}\mathbf{l}}{\mathrm{d}t}\right) \tag{9.348}$$

Then each term on the RHS of Equation 9.346:

$$\mathbf{l}\times\left(-\tilde{C}_{\mathrm{D}}\mathbf{v}\right)=\mathbf{l}\times\left(-\tilde{C}_{\mathrm{D}}v\mathbf{l}\right)=-\tilde{C}_{\mathrm{D}}v(\mathbf{l}\times\mathbf{l})=0 \tag{9.349}$$

$$\mathbf{l}\times\tilde{C}_{\mathrm{L}_\alpha}v^2\boldsymbol{\alpha}_{\mathrm{R}}=\tilde{C}_{\mathrm{L}_\alpha}v^2(\mathbf{l}\times\boldsymbol{\alpha}_{\mathrm{R}}) \tag{9.350}$$

$$\begin{aligned}\mathbf{l}\times\left(-\tilde{C}_{\mathrm{N}_{p\alpha}}v\right)(\mathbf{l}\times\boldsymbol{\alpha}_{\mathrm{R}})&=-\tilde{C}_{\mathrm{N}_{p\alpha}}v[\mathbf{l}\times(\mathbf{l}\times\boldsymbol{\alpha}_{\mathrm{R}})]=\tilde{C}_{\mathrm{N}_{p\alpha}}v[\mathbf{l}\times(\boldsymbol{\alpha}_{\mathrm{R}}\times\mathbf{l})]\\&=\tilde{C}_{\mathrm{N}_{p\alpha}}v[\alpha_{\mathrm{R}}-(\mathbf{l}\cdot\boldsymbol{\alpha}_{\mathrm{R}})\mathbf{l}]\end{aligned} \tag{9.351}$$

and

$$\mathbf{l}\times\mathbf{g}=\mathbf{l}\times\mathbf{g} \tag{9.352}$$

Continuing on Equation 9.347, first the LHS,

$$\mathbf{l}\times\gamma\frac{I_{\mathrm{P}}}{I_{\mathrm{T}}}p\frac{\mathrm{d}\mathbf{l}}{\mathrm{d}t}=\gamma\frac{I_{\mathrm{P}}}{I_{\mathrm{T}}}p\left(\mathbf{l}\times\frac{\mathrm{d}\mathbf{l}}{\mathrm{d}t}\right) \tag{9.353}$$

$$\mathbf{l}\times\gamma\left(\boldsymbol{\alpha}_{\mathrm{R}}\times\frac{\mathrm{d}^2\mathbf{l}}{\mathrm{d}t^2}\right)=\gamma\,\mathbf{l}\times\left(\boldsymbol{\alpha}_{\mathrm{R}}\times\frac{\mathrm{d}^2\mathbf{l}}{\mathrm{d}t^2}\right)=\gamma\left[\left(\mathbf{l}\cdot\frac{\mathrm{d}^2\mathbf{l}}{\mathrm{d}t^2}\right)\boldsymbol{\alpha}_{\mathrm{R}}-(\mathbf{l}\cdot\boldsymbol{\alpha}_{\mathrm{R}})\frac{\mathrm{d}^2\mathbf{l}}{\mathrm{d}t^2}\right] \tag{9.354}$$

$$\mathbf{l}\times\gamma^2\left(\mathbf{l}\times\frac{\mathrm{d}^2\mathbf{l}}{\mathrm{d}t^2}\right)=\gamma^2\left[\mathbf{l}\times\left(\mathbf{l}\times\frac{\mathrm{d}^2\mathbf{l}}{\mathrm{d}t^2}\right)\right]=\gamma^2\left[\left(\mathbf{l}\cdot\frac{\mathrm{d}^2\mathbf{l}}{\mathrm{d}t^2}\right)\mathbf{l}-\frac{\mathrm{d}^2\mathbf{l}}{\mathrm{d}t^2}\right] \tag{9.355}$$

Then each term on the RHS of Equation 9.347:

$$\mathbf{l}\times\tilde{C}_{\mathrm{M}_\alpha}v(\mathbf{l}\times\boldsymbol{\alpha}_{\mathrm{R}})=\tilde{C}_{\mathrm{M}_\alpha}v[\mathbf{l}\times(\mathbf{l}\times\boldsymbol{\alpha}_{\mathrm{R}})]=-\tilde{C}_{\mathrm{M}_\alpha}v[\boldsymbol{\alpha}_{\mathrm{R}}-(\mathbf{l}\cdot\boldsymbol{\alpha}_{\mathrm{R}})\mathbf{l}] \tag{9.356}$$

$$\mathbf{l}\times\tilde{C}_{\mathrm{M}_{p\alpha}}v\gamma\boldsymbol{\alpha}_{\mathrm{R}}=\tilde{C}_{\mathrm{M}_{p\alpha}}v\gamma(\mathbf{l}\times\boldsymbol{\alpha}_{\mathrm{R}}) \tag{9.357}$$

continuing

$$\mathbf{l}\times\tilde{C}_{\mathrm{M_q}}\gamma\left(\boldsymbol{\alpha}_{\mathrm{R}}\times\frac{\mathrm{d}\mathbf{l}}{\mathrm{d}t}\right)=\tilde{C}_{\mathrm{M_q}}\gamma\left[\mathbf{l}\times\left(\boldsymbol{\alpha}_{\mathrm{R}}\times\frac{\mathrm{d}\mathbf{l}}{\mathrm{d}t}\right)\right] \tag{9.358}$$

$$\mathbf{l}\times\tilde{C}_{M_q}\gamma\left(\boldsymbol{\alpha}_R\times\frac{d\mathbf{l}}{dt}\right)=\tilde{C}_{M_q}\gamma\left[\left(\mathbf{l}\cdot\frac{d\mathbf{l}}{dt}\right)\boldsymbol{\alpha}_R-(\mathbf{l}\cdot\boldsymbol{\alpha}_R)\frac{d\mathbf{l}}{dt}\right]=\tilde{C}_{M_q}\gamma(\mathbf{l}\cdot\boldsymbol{\alpha}_R)\frac{d\mathbf{l}}{dt} \tag{9.359}$$

and finally

$$\mathbf{l}\times\tilde{C}_{M_q}\gamma^2\left(\mathbf{l}\times\frac{d\mathbf{l}}{dt}\right)=\tilde{C}_{M_q}\gamma^2\left[\mathbf{l}\times\left(\mathbf{l}\times\frac{d\mathbf{l}}{dt}\right)\right]=\tilde{C}_{M_q}\gamma^2\left[0-(\mathbf{l}\cdot\mathbf{l})\frac{d\mathbf{l}}{dt}\right]=-\tilde{C}_{M_q}\gamma^2\frac{d\mathbf{l}}{dt} \tag{9.360}$$

We will now insert Equations 9.348 through 9.352 into Equation 9.346 to get

$$V\left(\mathbf{l}\times\frac{d\mathbf{l}}{dt}\right)=\tilde{C}_{L_\alpha}v^2(\mathbf{l}\times\boldsymbol{\alpha}_R)-\tilde{C}_{N_{p\alpha}}v[\boldsymbol{\alpha}_R-(\mathbf{l}\cdot\boldsymbol{\alpha}_R)\mathbf{l}]+(\mathbf{l}\times\mathbf{g}) \tag{9.361}$$

We then do the same with Equations 9.353 through 9.360, inserting them into Equation 9.347, yielding

$$\begin{aligned}&\gamma\frac{I_P}{I_T}p\left(\mathbf{l}\times\frac{d\mathbf{l}}{dt}\right)+\gamma\left[\left(\mathbf{l}\cdot\frac{d^2\mathbf{l}}{dt^2}\right)\boldsymbol{\alpha}_R-(\mathbf{l}\cdot\boldsymbol{\alpha}_R)\frac{d^2\mathbf{l}}{dt^2}\right]+\gamma^2\left[\left(\mathbf{l}\cdot\frac{d^2\mathbf{l}}{dt^2}\right)\mathbf{l}-\frac{d^2\mathbf{l}}{dt^2}\right]\\&=-\tilde{C}_{M_\alpha}v[\boldsymbol{\alpha}_R-(\mathbf{l}\cdot\boldsymbol{\alpha}_R)\mathbf{l}]+\tilde{C}_{M_{p\alpha}}v\gamma(\mathbf{l}\times\boldsymbol{\alpha}_R)-\tilde{C}_{M_q}\left[\gamma(\mathbf{l}\cdot\boldsymbol{\alpha}_R)\frac{d\mathbf{l}}{dt}+\gamma^2\frac{d\mathbf{l}}{dt}\right]\end{aligned} \tag{9.362}$$

We now have a pair of equations that essentially comprise two vector variables: the yaw of repose $\boldsymbol{\alpha}_R$ and the vector $\mathbf{l}\times\boldsymbol{\alpha}_R$. Cumbersome as it may seem, this is a linear system that can be readily solved through the use of matrices and their determinants to yield $\boldsymbol{\alpha}_R$ in terms of $\mathbf{l}$ and the other scalars and coefficients. The solution for $\boldsymbol{\alpha}_R$ after much manipulation is

$$\boldsymbol{\alpha}_R=\frac{\begin{array}{c}-\tilde{C}_{M_{p\alpha}}v\gamma V\left[\left(\mathbf{l}\times\frac{d\mathbf{l}}{dt}\right)-(\mathbf{l}\times\mathbf{g})\right]\\+\tilde{C}_{L_\alpha}v^2\gamma^2\left[\left(\mathbf{l}\cdot\frac{d^2\mathbf{l}}{dt^2}\right)\mathbf{l}-\frac{d^2\mathbf{l}}{dt^2}\right]+\tilde{C}_{L_\alpha}v^2\gamma\frac{I_P}{I_T}p\left(\mathbf{l}\times\frac{d\mathbf{l}}{dt}\right)+\tilde{C}_{L_\alpha}\tilde{C}_{M_q}v^2\gamma^2\frac{d\mathbf{l}}{dt}\end{array}}{\tilde{C}_{N_{p\alpha}}\tilde{C}_{M_{p\alpha}}v^2\gamma-\tilde{C}_{L_\alpha}v^2\gamma\left(\mathbf{l}\cdot\frac{d^2\mathbf{l}}{dt^2}\right)+\tilde{C}_{L_\alpha}\tilde{C}_{M_\alpha}v^3} \tag{9.363}$$

We can further simplify this expression by returning to Equation 9.346 and taking the dot product of it with $\mathbf{l}$. Through vector algebra and the use of the fact that $(\mathbf{l}\cdot\boldsymbol{\alpha}_R)=0$, we see that

$$\dot{V}=-\tilde{C}_Dv+(\mathbf{l}\cdot\mathbf{g})=-\frac{\rho SC_Dv^2}{2m}+(\mathbf{l}\cdot\mathbf{g}) \tag{9.364}$$

When this is substituted back into Equation 9.346, we get

$$-\tilde{C}_D\mathbf{v}+(\mathbf{l}\cdot\mathbf{g})\mathbf{l}+V\frac{d\mathbf{l}}{dt}=-\tilde{C}_D\mathbf{v}+\tilde{C}_{L_\alpha}v^2\boldsymbol{\alpha}_R+\tilde{C}_{N_{p\alpha}}v(\mathbf{l}\times\boldsymbol{\alpha}_R)+\mathbf{g} \tag{9.365}$$

Which can be rewritten as

$$V\frac{d\mathbf{l}}{dt} = \tilde{C}_{L_\alpha} v^2 \boldsymbol{\alpha}_R + \tilde{C}_{N_{p\alpha}} v(\mathbf{l} \times \boldsymbol{\alpha}_R) + \mathbf{g} - (\mathbf{l} \cdot \mathbf{g})\mathbf{l} \tag{9.366}$$

and by neglecting the Magnus force term as it is small and noting that $\mathbf{l} \times (\mathbf{g} \times \mathbf{l}) = \mathbf{g} - (\mathbf{l} \cdot \mathbf{g})\mathbf{l}$, we get

$$V\frac{d\mathbf{l}}{dt} = \tilde{C}_{L_\alpha} v^2 \boldsymbol{\alpha}_R + [\mathbf{l} \times (\mathbf{g} \times \mathbf{l})] \tag{9.367}$$

Remember that we wish to find a useful form with which we can calculate the quasi-steady state yaw of repose as shown in Equation 9.363. That equation encompasses different vector functions and time derivatives of **l**, but our ultimate goal is to find expressions that only involve the measurable quantities of the aeroballistic coefficients, spin, gravity, and velocity and not the unit vector, **l**. To do this, although it may seem a devious process, we begin by taking the time derivative of Equation 9.367, getting

$$\dot{V}\frac{d\mathbf{l}}{dt} + V\frac{d^2\mathbf{l}}{dt^2} = 0 + \frac{d}{dt}[\mathbf{l} \times (\mathbf{g} \times \mathbf{l})] \tag{9.368}$$

We use Equation 9.364 and substitute it in Equation 9.368, arriving at

$$-\tilde{C}_D v\frac{d\mathbf{l}}{dt} + (\mathbf{l} \cdot \mathbf{g})\frac{d\mathbf{l}}{dt} + V\frac{d^2\mathbf{l}}{dt^2} = \frac{d\mathbf{g}}{dt} - (\mathbf{l} \cdot \mathbf{g})\frac{d\mathbf{l}}{dt} - \left(\frac{d\mathbf{l}}{dt} \cdot \mathbf{g}\right)\mathbf{l} - \left(\mathbf{l} \cdot \frac{d\mathbf{g}}{dt}\right)\mathbf{l} \tag{9.369}$$

Noticing that $d\mathbf{g}/dt = 0$ (if not, we really will have problems), we can rewrite this as

$$V\frac{d^2\mathbf{l}}{dt^2} = -2(\mathbf{l} \cdot \mathbf{g})\frac{d\mathbf{l}}{dt} - \left(\frac{d\mathbf{l}}{dt} \cdot \mathbf{g}\right)\mathbf{l} + \tilde{C}_D v\frac{d\mathbf{l}}{dt} \tag{9.370}$$

If we examine the RHS of Equation 9.370 term by term and realize from Equation 9.367 that $d\mathbf{l}/dt$ can be solved for, then

$$-2(\mathbf{l} \cdot \mathbf{g})\frac{d\mathbf{l}}{dt} = -\frac{2}{V}[(\mathbf{l} \cdot \mathbf{g})\tilde{C}_{L_\alpha} v^2 \boldsymbol{\alpha}_R + (\mathbf{l} \cdot \mathbf{g})\mathbf{g} - (\mathbf{l} \cdot \mathbf{g})^2\mathbf{l}] \tag{9.371}$$

also,

$$\left(\frac{d\mathbf{l}}{dt} \cdot \mathbf{g}\right)\mathbf{l} = \frac{1}{V}\left[\tilde{C}_{L_\alpha} v^2(\boldsymbol{\alpha}_R \cdot \mathbf{g}) + g^2 - (\mathbf{l} \cdot \mathbf{g})^2\right]\mathbf{l} \tag{9.372}$$

and

$$\tilde{C}_D v\frac{d\mathbf{l}}{dt} = \frac{\tilde{C}_D v}{V}[\tilde{C}_{L_\alpha} v^2 \boldsymbol{\alpha}_R + \mathbf{g} - (\mathbf{l} \cdot \mathbf{g})\mathbf{l}] \tag{9.373}$$

Putting these expressions back into Equation 9.370 complicates things considerably, viz.,

$$V\frac{d^2\mathbf{l}}{dt^2} = -\frac{2}{V}[\tilde{C}_{L_\alpha} v^2(\mathbf{l} \cdot \mathbf{g})\boldsymbol{\alpha}_R + (\mathbf{l} \cdot \mathbf{g})\mathbf{g} - (\mathbf{l} \cdot \mathbf{g})^2\mathbf{l}] - \frac{1}{V}[\tilde{C}_{L_\alpha} v^2(\boldsymbol{\alpha}_R \cdot \mathbf{g})\mathbf{l} + g^2\mathbf{l} - (\mathbf{l} \cdot \mathbf{g})^2\mathbf{l}] + \tilde{C}_D v\frac{d\mathbf{l}}{dt} \tag{9.374}$$

If we assume that the terms containing our modified lift $\tilde{C}_{L_\alpha}$ and drag $\tilde{C}_D$ coefficients are either small with respect to the other terms or cancel one another, the third term in Equation 9.374 disappears, and the others simplify to eventually give

$$V^2 \frac{d^2\mathbf{l}}{dt^2} = \left[3(\mathbf{l}\cdot\mathbf{g})^2 - g^2\right]\mathbf{l} - 2(\mathbf{l}\cdot\mathbf{g})\mathbf{g} \tag{9.375}$$

Recalling that the triad (**l, n, m**) are unit vectors, with **l** in the direction of the velocity vector, **v**, we can write $\mathbf{l} = \mathbf{v}/v'$ and can transform Equation 9.367 into

$$\frac{d\mathbf{l}}{dt} = \frac{1}{V}\left\{\tilde{C}_{L_\alpha} v^2 \boldsymbol{\alpha}_R + \frac{1}{v^2}\left[\mathbf{v}\times(\mathbf{g}\times\mathbf{v})\right]\right\} \tag{9.376}$$

Likewise, our Equation 9.374, by dividing both sides by V, can be transformed to

$$\begin{aligned}\frac{d^2\mathbf{l}}{dt^2} = &-\frac{2}{V^2}\left[\tilde{C}_{L_\alpha} v(\mathbf{v}\cdot\mathbf{g})\boldsymbol{\alpha}_R + \frac{1}{v}(\mathbf{v}\cdot\mathbf{g})\,\mathbf{g} - \frac{1}{v^3}(\mathbf{v}\cdot\mathbf{g})^2\mathbf{v}\right] \\ &-\frac{1}{V^2}\left[\tilde{C}_{L_\alpha} v(\boldsymbol{\alpha}_R\cdot\mathbf{g})\mathbf{v} + \frac{g^2}{v}\mathbf{v} - \frac{1}{v^3}(\mathbf{v}\cdot\mathbf{g})^2\mathbf{v}\right] + \frac{\tilde{C}_D v}{V}\frac{d\mathbf{l}}{dt}\end{aligned} \tag{9.377}$$

Now, if we examine each term of Equation 9.363, we will have to deal with such terms as $\mathbf{l}\times\frac{d\mathbf{l}}{dt}$, $\mathbf{l}\cdot\frac{d^2\mathbf{l}}{dt^2}$, and $\left(\mathbf{l}\cdot\frac{d^2\mathbf{l}}{dt^2}\right)\mathbf{l} - \frac{d^2\mathbf{l}}{dt^2}$. We can perform all the substitutions of these terms with what we have shown in Equations 9.376 and 9.377, and then examine the resulting Equation 9.363 for terms that can be neglected for comparative magnitudes. When all the algebra is completed, the resulting equation, applicable to spinning projectiles [1], is

$$\boldsymbol{\alpha}_R = \frac{-2I_P p\left(\mathbf{v}\times\frac{d\mathbf{V}}{dt}\right)}{\rho S d v^4 C_{M_\alpha}} \tag{9.378}$$

For nonspinning projectiles, we can simplify Equation 9.363 by removing the spin terms and neglecting terms of small magnitude. The resulting expression is

$$\boldsymbol{\alpha}_R = \frac{\tilde{C}_{M_q}\gamma^2\left[\mathbf{v}\times(\mathbf{v}\times\mathbf{g})\right]}{\tilde{C}_{M_\alpha} V v^3} \tag{9.379}$$

The vector mechanics work out so that when there is a positive overturning moment (statically unstable projectile), the yaw of repose vector points to the right for a right-hand spin. The yaw of repose for a statically stable nonspinning projectile is such that the nose points slightly above the trajectory. Either Equation 9.378 or 9.379 can be inserted into Equation 9.361 and numerically integrated simultaneously with Equation 9.362 to yield the velocity and position at any time. This forms the basis of the modified point mass method.

Problem 40

The Paris gun was built by Germany in the First World War to shell Paris from 75 miles away. The weapon was a 210 mm diameter bore with the shells preengraved to account for wear of the tube. During firing of this weapon, all things such as wind effects and Coriolis had to be accounted for. When the United States entered the war, the doughboys (the nickname for American troops) were to take the St. Mihiel salient where the gun was located. We shall assume that the Germans have turned the gun to fire on the Americans. The projectile is at some point in space defined later. To demonstrate your knowledge of the modified point mass equations,

1. Draw the situation.
2. Calculate the vector yaw of repose for this projectile using Equation 9.326.
   Answer: $\boldsymbol{\alpha}_R = [0.008\mathbf{e}_2 + 0.002\mathbf{e}_3]$ [rad].
3. Write the acceleration vector for this projectile using Equation 9.339 at the instant in its trajectory when the velocity (relative to the ground) is 2100 ft/s and the conditions given next apply.

Note: You do not need all of the information given next. It is provided to you so you can compare the differences in formulations with the 6 DOF model.

Answer:

$$\frac{d\mathbf{V}}{dt} = [-61\mathbf{e}_1 - 29\mathbf{e}_2 - 10\mathbf{e}_3]\left[\frac{\text{ft}}{\text{s}^2}\right]$$

4. Why do we not need to obtain the angular acceleration vector dh/d*t*?

Positional information:

- 48° north latitude.
- Azimuth of velocity vector: 190° true.
- Angle of velocity vector to horizontal: +1°.
- Wind is blowing at 15 mph due south and horizontal.
- $\alpha = 0.5°$; $\beta = 0.25°$.
- The projectile nose is rotating down at 1 rad/s.

Projectile information:

$C_D = 0.28$ $\quad \left(C_{M_q} + C_{M_{\dot\alpha}}\right) = -16.5$ $\quad I_P = 19.13\ [\text{lbm-ft}^2]$

$C_{M_\alpha} = 3.50$ $\quad \left(C_{N_q} + C_{N_{\dot\alpha}}\right) = 0.005$ $\quad I_T = 66.40\ [\text{lbm-ft}^2]$

$C_{L_\alpha} = 2.50$ $\quad C_{M_{p\alpha}} = 0.55$ $\quad m = 220\ [\text{lbm}]$

$C_{N_{p\alpha}} = -0.02$

$C_{1_P} = -0.01$ $\quad \rho = 0.060\left[\frac{\text{lbm}}{\text{ft}^3}\right]$ $\quad p = 130\left[\frac{\text{rev}}{\text{s}}\right].$

Supply all answers in an inertial coordinate system labeled 1, 2, and 3, with 1 being due south and 3 being due west.

Problem 41
If we were to use a modified point mass assumption for both of the cases cited in Problem 35,

1. Calculate the vector yaw of repose for both cases.

   Answer: $\boldsymbol{\alpha}_R = [0.0003\mathbf{e}_1 + 0.0060\mathbf{e}_2 - 0.0009\mathbf{e}_3]$ for the building and $\boldsymbol{\alpha}_R = [-0.0009\mathbf{e}_1 - 0.0007\mathbf{e}_2 - 0.0000\mathbf{e}_3]$ for the gunship.
2. Draw and explain what this vector represents.
3. Comment on whether this model is applicable for each case and why.

9.7 Probability of First Round Hit

The theories describing the characteristics of exterior ballistic trajectories that we have just completed, while increasingly complex, do not deal with the problems of real-life gunnery. What an artillery commander or a tank commander wants to know is what the probability will be that the first round fired will land where desired. This is less of a problem for the indirect fire mode, where a spotter can call in corrective advice to the gunner, than it is in the direct fire mode, where the urgency of a first round hit may be a life-and-death matter. When you fire your weapon at a visible enemy, you may be subject to immediate return fire. So a first round hit with high probability of target destruction is of primary interest to gunner and commander.

Predicted fire is what we are talking about, particularly tank gun vs. tank target, and the variables of the combat situation are what govern these probabilities. The advent of smart, even autonomous munitions has, of course, reduced the problems of predicted fire, but at great monetary expense and the subjection of such munitions to countermeasures that negate their usefulness. The study of predicted fire is still a worthwhile exercise.

Firing real projectiles from real weapon systems involves interactions among many variables. For example, the location (terrain) of the launching platform; the condition of the gun, its age, wear history, and mounting method; aiming procedures including fire control; condition of the crew including fatigue, experience, fear of attack; weather conditions, including temperature, humidity, air density, and pressure; target location relative to the launching platform and target size; and not least, the characteristics of the projectile itself.

Leaving these generalities, let us examine two of the areas where hit probabilities are generated: the proving ground and troop training locations. In the proving ground, great care can be taken to reduce, and even eliminate, some of the variables. For the ammunition designer, this is the opportunity to assess how precisely the design functions round to round: the gunner can take all the time he/she needs to aim the weapon—no one is shooting back at him. The weapon can be new or in the early part of its wear life; the propellant temperature can be closely controlled, reducing temperature effects on muzzle velocity variations. In some cases, special mounts can be built so that even large caliber cannon can be fired from what amounts to a bench rest similar to ones used with small arms. The object of such care is to find out the round-to-round precision (sometimes called dispersion) of the ammunition. This is not its accuracy, but is an important component of what is known as "first round hit probability."

The group (usually five or more projectiles made as identical as possible) fired under such controlled conditions clusters about an average point called the "center of impact" or CI. It is the predictability and reproducibility of the CI that assures the gunner and tank commander that when the first round of an engagement is fired, it will go where it is intended. Rarely are second and third rounds fired at the same target in combat.

The normal course of a development program sees multiple cycles of design, manufacture of test samples, and test and evaluation until a stable design is achieved. These cycles are usually followed by a series of confirmatory tests that evaluate the performance of the stable design under varying conditions of temperature, gun wear, range performance, and a host of other tests. By normalizing the gun superelevations of each group after firing, it is possible then to compare the CIs of the groups. At the proving ground, the azimuth or left-right setting of the gun would be closely maintained as a constant. This comparison speaks to the accuracy of the design, that is, the reproducibility of the CIs under varying conditions of proving ground firings. The precision (dispersion) of the design would also be found under these variations.

How would this scatter of the CIs relate to accuracy under real-world combat situations? What could we then say about the probability of first round hit (PFRH) under combat conditions? To answer these questions, serious studies have been conducted in probabilistic terms about the conditions to be found around the world in combat in what are termed *quasi-combat* conditions [5]. Extensive computer programs have been devised, as well, more recently, to compute PFRHs for designs of direct fire munitions under quasi-combat conditions [6]. Much of these theories and computations are dependent on the fire control system of the vehicle and data collections of conditions of terrain and atmospherics around the world. The original work in this area was done by H. Brodkin of the Pittman–Dunn Laboratory at the now-defunct Frankford Arsenal in 1958 and remains classified.

As shown by Christman [7], the PFRH is dependent on certain fixed biases, variable biases, and random errors. Hit probability computed this way is heavily dependent, as it should be, on both the type of fire control system employed and the flight properties of the projectiles being judged. For the predicted fire hit probabilities Christman computed in 1959, laser range finding with its very small ranging errors was not in use, neither were the short times of flight of the fin-stabilized, small-diameter but highly dense, armor-defeating projectiles (armor-piercing fin-stabilized discarding sabot). They were not yet invented.

The superelevation of the gun, that is, the amount of elevation above the line of sight to the target that the gunner must use to lay the weapon, is greatly reduced when both the flight velocity is increased and the drag of the projectile is reduced with its consequently shortened TOF. These factors have greatly aided the improved PRFH with the latest tank cannon and fire control systems now in use. It is nevertheless instructive to show the error sources considered in a stationary-to stationary engagement of a tank to a vertical 2.3 × 2.3 m target. Such computations are now carried out on computer programs similar to those shown in the report by Bunn [5].

Some error sources are tabulated in Tables 9.1 and 9.2 [7].

As can be seen from Tables 9.1 and 9.2, there are many error sources that affect PFRH in a quasi-combat situation. Some of these require explanation. For example,

- *Zeroing*: The weapon is zeroed by firing a small group usually at 1200 m range, in a noncombat situation, recording the CI of the group and adjusting the optical fire control system so that it has a basis from which range-affected variables can be set. Once zeroed, the other fixed and variable biases and random errors remain to determine the PFRH at the time.
- *Cant*: This error, caused by the unevenness of terrain, causes a tilt of the trunnion of the weapon in both the roll and pitch attitudes and, thus, a consequent error in superelevation (gun elevation above boresight) dictated by the fire control.
- *Drift*: This error is usually the result of the Coriolis effect, that is, the effect of the rotation of the earth under the trajectory causing a deviation in the CI. This is

TABLE 9.1

Horizontal Errors on First Rounds—Quasi-Combat Conditions

Fixed biases
Drift
Fire control
Jump, mean
Parallax
Variable biases
Cant
Fire control
Jump variation
Wind, cross
Zeroing
Cant
Fire control
Group center of impact
Jump variation
Observation of CI
Wind, cross
Random errors
Round-to-round dispersion
Laying error

dependent on the latitude of the firing and target sites as well as the hemisphere (northern vs. southern hemispheres reverse the direction of the deviation).

- *Fire control and parallax*: These errors are the result of the slop in the mechanical linkages of the controls and the small offset in the alignment of the optical sight with the axis of the gun tube.
- *Cross wind*: This is a variable bias usually affecting the firing of different groups on different occasions
- *Laying error*: This is the result of random performance of gunners sighting in on targets and laying the gun from different directions occasion to occasion.
- *Range estimation*: This is a bias which is fixed for laser ranging but range dependent for optical ranging.
- *Round-to-round dispersion*: This is a random error dependent on the type of ammunition being fired. It can also vary from lot to lot in the same ammunition type due to muzzle velocity variations shot to shot in the lot.
- *Jump*: This is what is left over in the error budget when all other errors are accounted for.

The test procedure by the US Army Armor and Engineer Board [6] meticulously lays out the testing procedures used in qualifying ammunition for service use. Appendix A provides a nomogram for computing PFRH. It also provides techniques for measuring some of the error sources mentioned.

TABLE 9.2

Vertical Errors on First Rounds—Quasi-Battle Conditions

Fixed biases
Fire control
Jump
Parallax (at 0 range)
Range estimation
Variable biases
Jump variation
Cant
Fire control
Muzzle velocity (lot to lot)
Range estimation
Zeroing
Fire control
Cant
Group CI
Jump
Muzzle velocity (lot to lot)
Observation of CI
Range estimation
Random errors
Round-to-round dispersion
Laying error

We must emphasize that this discussion only points out the complexity of the problem of computing PFRH. If the model by Bunn [5] is used to run such computations, it requires considerable detailed data input, which for modern tank, fire control, and ammunition designs are not available to general audiences. Furthermore, recall that the discussion was only for stationary-to-stationary situations where the opposing tanks are firing head on. Targets moving must provide lead data. Firing from moving vehicles, with stabilized turrets as in the US M1 Abrams tank, adds further error sources inherent in the stabilization system but whose advanced computers and fire control systems may eliminate some of the other errors.

References

1. McCoy, R. L., *Modern Exterior Ballistics*, Schiffer Military History, Atglen, PA, 1999.
2. Murphy, C. H., *Free Flight Motion of Symmetric Missiles*, Ballistic Research Laboratory Report No. 1216, Ballistic Research Laboratory, Aberdeen Proving Ground, MD, 1963.
3. Jentz, T. L., *Tank Combat in North Africa, the Opening Rounds*, Schiffer Military History, Atglen, PA, 1998.
4. Litz, B., *Applied Ballistics for Long Range Shooting*, 2nd ed., Applied Ballistics, Cedar Springs, MI, 2011.

5. Bunn, F. L., *The Tank Accuracy Model*, US Army Research Laboratory Report No. ARL-MR-A8, US Army Research Laboratory, Aberdeen Proving Ground, MD, 1993 (DTIC AD 262812).
6. US Army Armor and Engineer Board, *Common Service Test Procedure, Round-to-Round Dispersion*, US Army Armor and Engineer Board , Fort Knox, KY, 1970 (DTIC AD 872085).
7. Christman, E. C., *The Effect of System Design Characteristics on First Round Hit Probability of Tank Fired Projectiles*, Ballistic Research Laboratory Report No. 1192, Ballistic Research Laboratory, Aberdeen Proving Ground, MD, 1959 (DTIC AD 316221).

Further Reading

Capecelatro, A., and Salzarulo, L., *Quantitative Physics for Scientists and Engineers—Mechanics*, Auric Associates, Newark, NJ, 1980.

Sears, F. W., Zemanski, M. W., and Young, H. D., *University Physics*, 6th ed., Addison-Wesley, Reading, MA, 1982.

10

Linearized Aeroballistics

The aeroballistics topics discussed so far have built up to where the reader has an appreciation for the techniques required to analyze projectile motion to a great degree of accuracy. The culmination of this study was the development of the equations for a six degree-of-freedom (6-DOF) model, which accurately describes the motion of a rigid body through air. With a 6-DOF model in hand, the aeroballistician can examine the effects of a *given* configuration. The word *given* was italicized for emphasis because the aeroballistician must know the configuration properties before he or she analyzes the projectile. The implications of this are that without other tools to determine what needs to be changed in a design to alter the projectile behavior, one must simply guess at a new configuration, determine the aerodynamic coefficients, and reanalyze. This process can be very inefficient. The solution to this problem is to develop a theory that can be used to quickly determine what must be changed in a projectile to alter its flight behavior, make the changes, and reassess. This will be the topic for the remainder of this section.

Linearized theory was (at least in the opinion of the authors) refined to an exceptional degree by Murphy [1] in 1963. Other authors before and since [2–4] have developed similar theories, and a good description of these can be found in the book by McCoy [5]. Linearized theory is so named because the aerodynamic coefficients are assumed to be linear functions of the angle of attack. In other words,

$$F_{\mathrm{j}} \propto C_{\mathrm{j}0}(\sin \alpha_t) \quad \text{or} \quad M_{\mathrm{j}} \propto C_{\mathrm{j}0}d(\sin \alpha_t) \tag{10.1}$$

where the subscript j indicates any parameter of interest as introduced earlier. There are good points and bad points (as always) with this technique. The good news is that the mathematics become simple enough to determine quantities of interest extremely quickly and find means of changing the flight characteristics of a projectile quickly. The bad news is that the use of linear coefficients prevents us from duplicating some motions that occur frequently enough in projectile flight to warrant the inclusion of their nonlinear brethren—and the math becomes complicated to boot.

We will continue the practice of using the definitions of the appropriate vectors and scalars based on the book by McCoy [5]. The choice is somewhat arbitrary, but for several years now, the authors have used this lucid work as a supplementary textbook, and it is a matter of convenience. Our coordinate system is defined as in Figure 10.1.

The aerodynamic coefficients introduced in the beginning of this chapter were written for both forces and moments as

$$F_{\mathrm{j}} = \frac{1}{2}\rho V^2 S C_{\mathrm{j}} \tag{10.2}$$

$$M_{\mathrm{j}} = \frac{1}{2}\rho V^2 S d C_{\mathrm{j}} \tag{10.3}$$

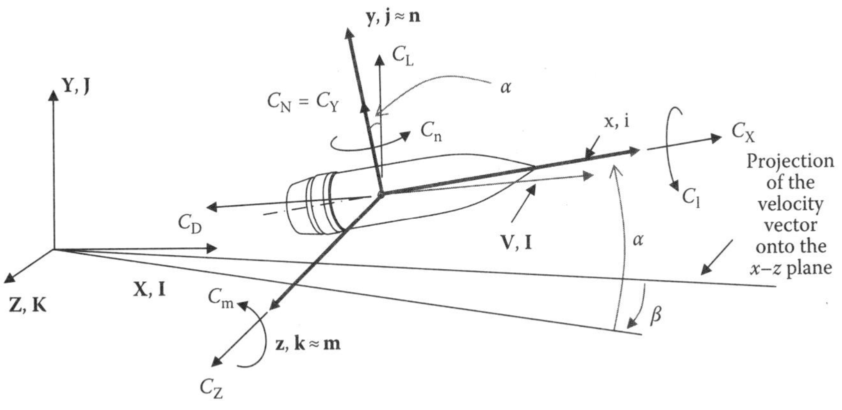

FIGURE 10.1
Coordinate system for projectile aerodynamic coefficients.

We have also defined the angular rates of the projectile as

$$p = \text{Roll (spin) rate} \tag{10.4}$$

$$q = \text{Pitch rate} \tag{10.5}$$

$$r = \text{Yaw rate} \tag{10.6}$$

The projectile angular position with respect to the velocity vector was given by

$$\alpha = \text{Angle of attack} \tag{10.7}$$

$$\beta = \text{Angle of sideslip} \tag{10.8}$$

The aerodynamic coefficients are functions of the rates expressed in Equations 10.4 through 10.6, as well as angular positions expressed in Equations 10.7 and 10.8. Additionally, these coefficients are also functions of the time rate of change of α and β that do not normally coincide with q and r. Thus, we can write

$$C_j = C_j(\alpha, \beta, \dot{\alpha}, \dot{\beta}, p, q, r) \tag{10.9}$$

With this nomenclature, any coefficient can be normally expressed as a series expansion in the seven variables

$$C_j = C_{j0} + C_{j_\alpha}\alpha + C_{j_\beta}\beta + C_{j_{\dot{\alpha}}}\left(\frac{\dot{\alpha}d}{V}\right) + C_{j_{\dot{\beta}}}\left(\frac{\dot{\beta}d}{V}\right) + C_{j_p}\left(\frac{pd}{V}\right) + C_{j_q}\left(\frac{qd}{V}\right) + C_{j_r}\left(\frac{rd}{V}\right) + \cdots \tag{10.10}$$

In Equation 10.10, we have included the terms in parentheses to maintain the nondimensional characteristics of the coefficient. We can see that this expansion results in a large number of terms that must be carried. Seldom in aeroballistics do we require terms in this expression beyond the second order, but they can be included if data are available. When we discuss linear aeroballistics, we are limiting ourselves to the eight terms displayed in Equation 10.10.

The linearization implies that

$$C_{j_k} = \left.\frac{\partial C_j}{\partial k}\right|_{k=0} \tag{10.11}$$

Further simplifications will be made as we progress which will assist us in tackling the mathematics. We shall make use in this section of starred coefficients. These coefficients are defined in terms of their unstarred counterparts as

$$C^*_{j_k} = \frac{\rho S d}{2m} C_{j_k} \tag{10.12}$$

10.1 Linearized Pitching and Yawing Motions

In the beginning of this chapter, we discussed terminology that allowed us to describe the pitching and yawing motion of a projectile. Because of the symmetry of typical projectiles, we combined pitch and yaw into a total yaw without any explanation. In this section, we will discuss the two motions separately and then formally make the assumptions that allowed us to combine them. This approach was formulated by Murphy [1], and what follows is basically that development with the coordinate system altered to fit our needs.

If we have a projectile as depicted in Figure 10.1 and allow it only to move in a truly pitching motion, we can look down the z-axis and we would see what is depicted in Figure 10.2. Some interesting observations can be made from this figure. First, we see that the velocity vector **V** and the associated unit vector **I** are pitched up at angle ϕ to the earth-fixed coordinate system. The projectile is actually pointed above this angle by the pitch angle α. The vector along which the projectile is pointed is the geometric axis unit vector **x** and the spin (principal) axis unit vector **i**. In an axially symmetric projectile, these are identical. We can see that through a rigid body rotation, this forces the unit vectors of the transverse geometric axis **y** and transverse principal axis **j** to be rotated from the earth-fixed **Y**-axis through an angle of $\phi + \alpha$. If we assume that the projectile is constrained to pitch only, then the time rate of change of this total angle is q and the yaw angle and yaw rate are equal to zero as depicted in the figure.

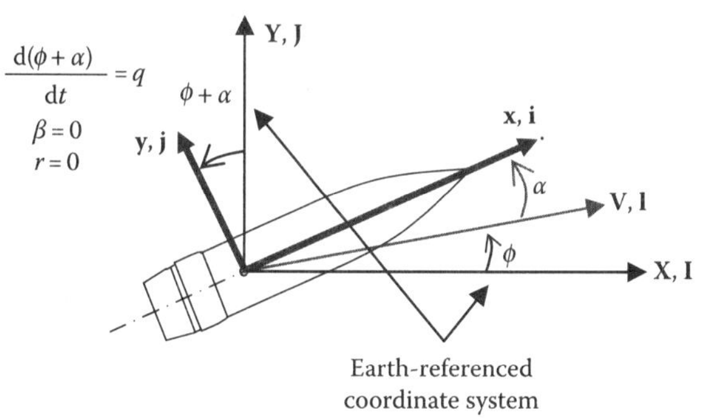

FIGURE 10.2
Projectile in a pure pitching motion.

In a similar manner, we can constrain our projectile to motion in the yaw plane only, which is depicted in Figure 10.3. In this case, the velocity and its associated unit vector are yawed with respect to the earth-fixed coordinate system by angle θ. The projectile geometric axis as well as the principal axis is yawed at angle β with respect to the velocity vector. This results in a rotation of the transverse principle axis from the earth-fixed coordinate system of $\theta + \beta$ as depicted in the illustration. The rate of change of this total angle is the yaw rate r, and because of our constraints, there is no pitching motion as identified in the figure.

We shall now develop the equations of motion for each of these two specialized cases with the purpose of combining them in the end. For the purpose of this development, we shall define the force in the **Y**- and **Z**-directions using force coefficients C_Y and C_Z, respectively.

If we examine our projectile constrained to a pitching motion only, we can define the force coefficient as

$$C_Y = C_{Y_0} + C_{Y_\alpha}\alpha + C_{Y_{\dot{\alpha}}}\left(\frac{\dot{\alpha}d}{V}\right) + C_{Y_q}\left(\frac{qd}{V}\right) \tag{10.13}$$

where we have restricted ourselves to the linear coefficients. We can see that this pitching motion causes a force in the **Y**-direction that is affected by angle of attack, rate of change of angle of attack, and pitching rate. An item worthy of note is that for a perfectly symmetrical projectile, C_{Y_0} would be zero. It is included here for completeness and can be present if an asymmetry exists.

The corresponding moment for pitching motion only is given by

$$C_m = C_{m_0} + C_{m_\alpha}\alpha + C_{m_{\dot{\alpha}}}\left(\frac{\dot{\alpha}d}{V}\right) + C_{m_q}\left(\frac{qd}{V}\right) \tag{10.14}$$

where the same comments about the nondimensionalization and C_{m_0} apply as well.

Now we will examine the equations of motion. The force and moment equations are given by

$$\mathbf{F} = m\mathbf{a} \tag{10.15}$$

$$\mathbf{M} = I\dot{\boldsymbol{\alpha}} \tag{10.16}$$

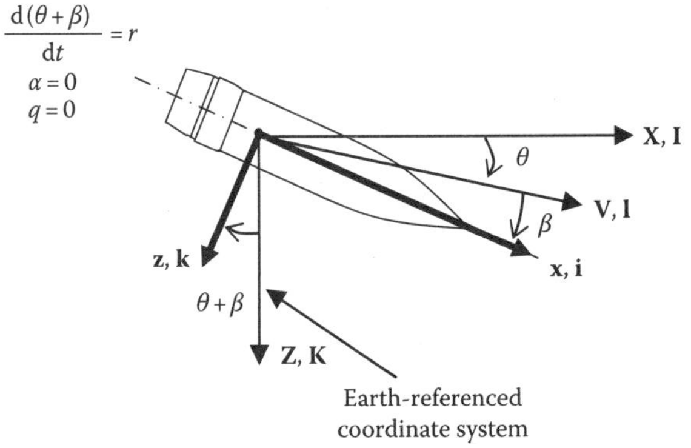

FIGURE 10.3
Projectile in a pure yawing motion.

If we define angles $\hat{\phi}$ and $\hat{\theta}$ as

$$\hat{\phi} = \phi + \alpha \tag{10.17}$$

$$\hat{\theta} = \theta + \beta \tag{10.18}$$

then our scalar equations of a projectile in flight exhibiting pure pitching motion are

$$m\frac{dV_x}{dt} \approx m\frac{dV}{dt} = F_x \tag{10.19}$$

$$m\frac{d^2Y}{dt^2} = F_Y \cos\hat{\phi} + F_x \sin\hat{\phi} - mg \tag{10.20}$$

$$\frac{d^2\hat{\phi}}{dt^2} = \frac{M_Z}{I_Z} \tag{10.21}$$

If we examine a small time of the projectile flight, we can assume constant velocity. If we further limit the pitching motion to small angles, we can assume

$$F_x = -F_D \approx 0 \tag{10.22}$$

$$\cos\hat{\phi} \approx 1 \tag{10.23}$$

$$\sin\hat{\phi} \approx \hat{\phi} \tag{10.24}$$

These assumptions can be used in Equations 10.19 and 10.20 to yield

$$m\frac{dV}{dt} = 0 \tag{10.25}$$

$$m\frac{d^2Y}{dt^2} = F_Y - mg \tag{10.26}$$

We know that

$$F_Y = \frac{1}{2}\rho V^2 S C_Y \tag{10.27}$$

If we then substitute Equations 10.13 and 10.27 into Equation 10.26, we obtain

$$m\frac{d^2Y}{dt^2} = \frac{1}{2}\rho V^2 S\left[C_{Y_0} + C_{Y_\alpha}\alpha + C_{Y_{\dot{\alpha}}}\left(\frac{\dot{\alpha}d}{V}\right) + C_{Y_q}\left(\frac{qd}{V}\right)\right] - mg \tag{10.28}$$

or, using our definition of starred coefficients, we have

$$\frac{d^2Y}{dt^2} = \frac{V^2}{d}\left[C^*_{Y_0} + C^*_{Y_\alpha}\alpha + C^*_{Y_\alpha}\left(\frac{\dot{\alpha}d}{V}\right) + C^*_{Y_q}\left(\frac{qd}{V}\right)\right] - g \tag{10.29}$$

Equation 10.29 can be combined with Equation 10.21 to develop a single equation for projectile motion. With this, the dynamic equation for the pure pitching motion of a projectile can then be described as

$$\ddot{\alpha} + \hat{H}_{1d}\dot{\alpha} - \hat{M}_1\alpha = \hat{A}_1 + \hat{G}_d \tag{10.30}$$

This linear, second-order differential equation with constant coefficients was established by Murphy [1] and modified here (the terms with the d subscript) to account for the assumption of zero drag. In this expression, we identify the coefficients as follows:

$$\hat{H}_{1d} = -\left[C^*_{Y_\alpha} + \frac{1}{k_Z^2}\left(C^*_{m_q} + C^*_{m_{\dot{\alpha}}}\right)\right]\left(\frac{V}{d}\right) \tag{10.31}$$

$$\hat{M}_1 = \left(\frac{1}{k_Z^2} C^*_{m_\alpha}\right)\left(\frac{V}{d}\right)^2 \tag{10.32}$$

$$\hat{A}_1 = \left(\frac{1}{k_Z^2} C^*_{m_0}\right)\left(\frac{V}{d}\right)^2 \tag{10.33}$$

$$\hat{G}_d = -\left(\frac{1}{k_Z^2} C^*_{m_q}\right)\left(\frac{g}{d}\right) \tag{10.34}$$

$$k_Z^2 = \frac{I_Z}{md^2} \tag{10.35}$$

If we include drag (and, thus, ignore Equation 10.22) yet leave all the other assumptions in place, we obtain a result identical to that of Murphy [1]. This results in Equations 10.30, 10.31, and 10.34 being modified to

$$\ddot{\alpha} + \hat{H}_1\dot{\alpha} - \hat{M}_1\alpha = \hat{A}_1 + \hat{G} \tag{10.36}$$

$$\hat{H}_1 = -\left[C^*_{Y_\alpha} + C^*_D + \frac{1}{k_Z^2}\left(C^*_{m_q} + C^*_{m_{\dot{\alpha}}}\right)\right]\left(\frac{V}{d}\right) \tag{10.37}$$

$$\hat{G} = -\left(\frac{1}{k_Z^2} C^*_{m_q} - C^*_D\right)\left(\frac{g}{d}\right) \tag{10.38}$$

It is more convenient to examine the differential Equations 10.30 and 10.36 with dimensionless distance (defined as *s/d*) instead of time as the independent variable. The time derivatives of dimensionless distance can then be written as

$$\frac{\mathrm{d}s}{\mathrm{d}t} = \left(\frac{V}{d}\right) \tag{10.39}$$

and

$$\frac{\mathrm{d}^2 s}{\mathrm{d}t^2} = \left(\frac{\dot{V}}{d}\right) \tag{10.40}$$

With this, we can use the relations

$$\frac{\mathrm{d}}{\mathrm{d}t}() = \frac{\mathrm{d}}{\mathrm{d}s}()\frac{\mathrm{d}s}{\mathrm{d}t} = \left(\frac{V}{d}\right)\frac{\mathrm{d}}{\mathrm{d}s}() \tag{10.41}$$

$$\frac{\mathrm{d}^2}{\mathrm{d}t^2}() = \left(\frac{V}{d}\right)^2 \frac{\mathrm{d}^2}{\mathrm{d}s^2}() + \left(\frac{\dot{V}}{d}\right)\frac{\mathrm{d}}{\mathrm{d}s}() \tag{10.42}$$

to rewrite Equations 10.30 and 10.36, respectively, as

$$\alpha'' + H_{1d}\alpha' - M_1\alpha = A_1 + G_d \tag{10.43}$$

$$\alpha'' + H_1\alpha' - M_1\alpha = A_1 + G \tag{10.44}$$

The coefficients in these equations are given by

$$H_{1d} = -\left[C^*_{Y_\alpha} + \frac{1}{k_Z^2}\left(C^*_{m_q} + C^*_{m_{\dot\alpha}}\right)\right] \tag{10.45}$$

$$H_1 = -\left[C^*_{Y_\alpha} + 2C^*_D + \frac{1}{k_Z^2}\left(C^*_{m_q} + C^*_{m_{\dot\alpha}}\right)\right] \tag{10.46}$$

$$M_1 = \frac{1}{k_Z^2}C^*_{m_\alpha} \tag{10.47}$$

$$A_1 = \frac{1}{k_Z^2}C^*_{m_0} \tag{10.48}$$

$$G_d = -\left(\frac{1}{k_Z^2}C^*_{m_q}\right)\left(\frac{gd}{V_0^2}\right) \tag{10.49}$$

$$G = -\left(\frac{1}{k_Z^2}C^*_{m_q} - C^*_D\right)\left(\frac{gd}{V_0^2}\right) \tag{10.50}$$

where V_0 is the muzzle (or a reference) velocity of the projectile.

A similar procedure can be followed to define motion constrained to the yaw plane only. This gives the result (details covered in the report by Murphy [1]) of

$$\beta'' + H_{2d}\beta' - M_2\beta = A_2 \tag{10.51}$$

$$\beta'' + H_2\beta' - M_2\beta = A_2 \tag{10.52}$$

The coefficients in these equations are given by

$$H_{2d} = -\left[C^*_{Z_\beta} + \frac{1}{k_Y^2}\left(C^*_{n_r} + C^*_{n_{\dot\beta}}\right)\right] \tag{10.53}$$

$$H_2 = -\left[C^*_{Z_\beta} + 2C^*_D\frac{1}{k_Y^2}\left(C^*_{n_r} + C^*_{n_{\dot\beta}}\right)\right] \tag{10.54}$$

$$M_2 = \frac{1}{k_Y^2}C^*_{m_\beta} \tag{10.55}$$

$$A_2 = \frac{1}{k_Y^2}C^*_{n_0} \tag{10.56}$$

$$k_Y^2 = \frac{I_Y}{md^2} \tag{10.57}$$

Assuming a projectile is axially symmetric implies that any plane orthogonal to the polar axis is a principal axis. This forces the two transverse moments of inertia to be equal and, with an assumption of small yaw, allows us to write

$$I_T = I_y = I_z \approx I_Y = I_Z \tag{10.58}$$

This symmetry also allows us to equate the pitch and yaw coefficients. Thus, we define

$$C_{N_\alpha} \equiv C_{Y_\alpha} = C_{Z_\beta} \tag{10.59}$$

$$C_{M_q} \equiv C_{m_q} = C_{n_r} \tag{10.60}$$

$$C_{M_\alpha} \equiv C_{m_\alpha} = C_{n_\beta} \tag{10.61}$$

$$C_{M_{\dot\alpha}} \equiv C_{m_{\dot\alpha}} = C_{n_{\dot\beta}} \tag{10.62}$$

Complex numbers are commonly used to define pitch and yaw angles. This is extremely convenient because it allows us to collapse two differential equations into one. We shall define the complex yaw angle as ξ, which shall be thus defined

$$\xi \equiv \alpha + i\beta \tag{10.63}$$

This definition allows one to look downrange as a projectile flies along a trajectory and visualize the imaginary part of the equation affecting the yaw of the projectile and the real part of the equation as affecting pitch. This is illustrated in Figure 10.4. In this figure, the origin is the trajectory of the projectile looking downrange.

The two differential equations of motion Equations 10.44 and 10.52 can then be combined by first multiplying Equation 10.52 by the imaginary number i and adding them together. This results in

$$\xi'' + H\xi' - M\xi = A + G\left(\frac{V_0}{V}\right)^2 \tag{10.64}$$

The coefficients in this equation are given by

$$H = -\left[C^*_{N_\alpha} + 2C^*_D + \frac{1}{k_T^2}\left(C^*_{M_q} + C^*_{M_{\dot\alpha}}\right)\right] \tag{10.65}$$

$$M = \frac{1}{k_T^2} C^*_{M_\alpha} \tag{10.66}$$

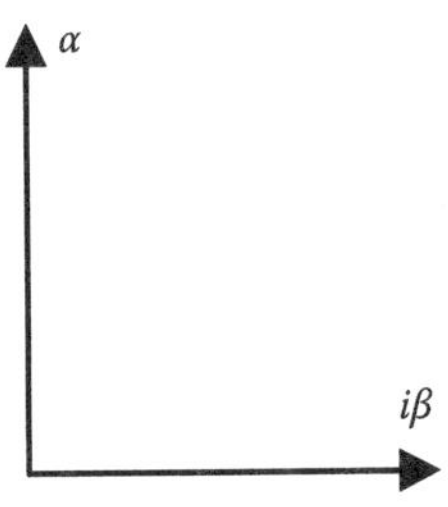

FIGURE 10.4
Complex yaw plane.

$$A = \frac{1}{k_{\mathrm{T}}^2}\left(C_{\mathrm{m}_0}^* + iC_{\mathrm{n}_0}^*\right) \tag{10.67}$$

$$G = -\left(\frac{1}{k_{\mathrm{T}}^2} C_{\mathrm{M}_q}^* - C_{\mathrm{D}}^*\right)\left(\frac{gd}{V_0^2}\right) \tag{10.68}$$

The solution to Equation 10.64 can be found for a nonspinning projectile to be [1]

$$\xi = K_1 \exp(i\psi_1) + K_2 \exp(i\psi_2) + K_3 \exp(i\psi_{30}) + \xi_{\mathrm{g}} \tag{10.69}$$

In this equation, each term K_{j} is known as an arm to be subsequently described. Mathematically, we can express these terms as

$$K_{\mathrm{j}} = K_{\mathrm{j}_0} \exp\left(\lambda_{\mathrm{j}} s\right) \tag{10.70}$$

Here we see that each arm is a function of its initial value (that occurring at the muzzle of the gun) and an exponential damping term. The exponential damping term decides whether the amplitude of the motion will decay, grow, or remain constant. The damping terms are given by [1,5]

$$\lambda_1 = \lambda_2 = -\frac{1}{2} H \tag{10.71}$$

The exponential terms in Equation 10.69 contain phase angles ψ_{j}. These phase angles represent the instantaneous angle that each arm makes with the imaginary axis. These can be written in terms of their initial value and a turning frequency as

$$\psi_{\mathrm{j}} = \psi_{\mathrm{j}_0} + \psi_{\mathrm{j}}' s \tag{10.72}$$

The turning frequencies are given for a nonspinning projectile by [1,5]

$$\psi_1' = -\psi_2' = \sqrt{-M} \tag{10.73}$$

The third term on the right-hand side (RHS) of Equation 10.69 is the so-called trim arm. This is a measure of the amount that a fin-stabilized projectile will trim (i.e., fly with constant pitch or yaw) during flight. It is given by

$$K_3 \exp(i\psi_{30}) = -\frac{i\left(C_{\mathrm{m}_0} + iC_{\mathrm{n}_0}\right)}{C_{\mathrm{M}_\alpha}} \tag{10.74}$$

The fourth term on the RHS of Equation 10.69 is the yaw caused by interaction of the projectile with the gravity vector, sometimes called the yaw of repose. It is defined as

$$\xi_{\mathrm{g}} = \frac{i\left(C_{\mathrm{M}_q} - k_{\mathrm{T}}^2 C_{\mathrm{D}}\right)\left(\frac{gd}{V^2}\right)}{C_{\mathrm{M}_\alpha}} \tag{10.75}$$

To visualize the physical meaning of Equation 10.69, we shall imagine we have a projectile and we are looking downrange along the trajectory such that the complex plane lies perpendicular to the trajectory curve. Our projectile will be at some arbitrary yaw angle. This is depicted in Figure 10.5. We need to note that the arms usually do not point to the nose of the projectile; they point to the symmetry axis; however, it is easiest to visualize the situation by scaling them to point to the nose. Imagine that we follow the projectile depicted in Figure 10.5 as it traverses the trajectory. We would see the nose motion swirling around. Throughout this time, we would also see the length of each of the arms changing (growing,

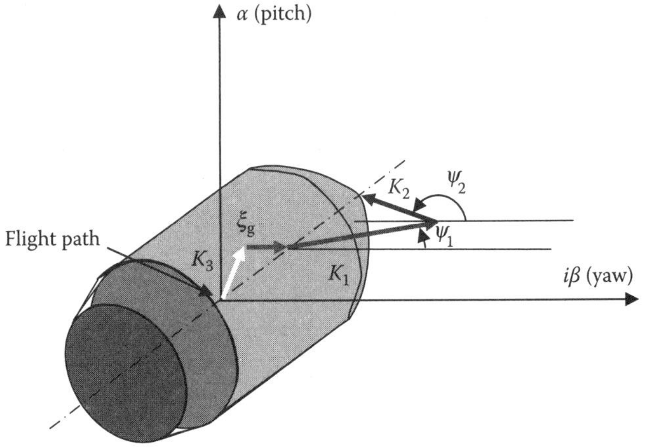

FIGURE 10.5
Example of tricyclic arms.

decaying, or remaining the same) as dictated by Equation 10.69. Additionally, we would see the arms rotating around their respective origins at rates described by Equation 10.72. All through this time, our viewpoint would be changing because we have our gaze fixed on the complex plane, and it is rotating into the paper because of the curvature of the trajectory.

In the development of Equation 10.64 and its solution Equation 10.69, the spin of the projectile was neglected. Because of this, these equations are specific to fin- or drag-stabilized projectiles that have relatively small spin rates. Murphy [1], Vaughn [2], and McCoy [5] developed the equation of motion for spinning projectiles in exactly the same manner. The results essentially incorporate the third angular component known as the roll or spin. The differential equation for a spinning projectile is given by

$$\xi'' + (H - iP)\xi' - (M + iPT)\xi = -iPG \tag{10.76}$$

In this formulation, we can utilize axial symmetry and, thus, define our coefficients as follows:

$$H = C_{L_\alpha}^* - C_D^* - \frac{1}{k_T^2}\left(C_{M_q}^* + C_{M_{\dot\alpha}}^*\right) \tag{10.77}$$

$$M = \frac{1}{k_T^2} C_{M_\alpha}^* \tag{10.78}$$

$$T = C_{L_\alpha}^* + \frac{1}{k_P^2} C_{M_{p\alpha}}^* \tag{10.79}$$

$$G = \frac{gd}{V_0^2} \tag{10.80}$$

$$P = \left(\frac{I_P}{I_T}\right)\left(\frac{pd}{V}\right) \tag{10.81}$$

The solution to Equation 10.76 is

$$\xi = K_{10} \exp[\lambda_1 s] \exp\left[i\left(\psi_{10} + \psi_1' s\right)\right] + K_{20} \exp[\lambda_2 s] \exp\left[i\left(\psi_{20} + \psi_2' s\right)\right] + \xi_g \tag{10.82}$$

This equation is essentially the same form as Equation 10.69 except for the deletion of the trim arm. It is also noteworthy that we have expanded the slow and fast arm terms and exponents to display their exponential behavior. The expression is also commonly written as

$$\xi = K_1 \exp(i\psi_1) + K_2 \exp(i\psi_2) + \xi_g \tag{10.83}$$

where the definitions of Equations 10.70 and 10.71 apply. The λ_j terms are known as the exponential damping coefficients, and the ψ_j terms are the precessional and nutation frequencies of the projectile. These are commonly defined as a complex pair where

$$\lambda_{1,2} + i\psi_{1,2} = \frac{1}{2}\left[-H + iP \pm \sqrt{4M + H^2 - P^2 + 2iP(2T - H)}\right] \tag{10.84}$$

As a parting note, we need to discuss the behavior of the fast and slow arms and the associated motion that they undergo. For a nonspinning projectile, we shall examine Equation 10.73. In this expression, the sign of M is important. For a nonspinning projectile, M is negative. That tells us that the arms turn in opposite directions with K_1 being positive (clockwise) and K_2 negative (counterclockwise). This is depicted in Figure 10.6.

Likewise for a spinning projectile, we need to examine the derivative with respect to s of Equation 10.84. In this case, we would find that

$$\psi_{1,2}' = \frac{1}{2}\left(P \pm \sqrt{P^2 - 4M}\right) \tag{10.85}$$

Here we shall see in the following section that for stability, this must result in a solution that has no imaginary part. So both values of the root will have the same sign; thus, the two arms turn in the same direction as shown in Figure 10.7.

Initial conditions that are present when the projectile leaves the muzzle of the weapon are important as our starting point for the values of the fast and slow arms. These can even cause drastically different flight behavior when nonlinear coefficients are introduced later. The initial sizes of the fast and slow arms can be expressed as functions of the precession and nutation rates, the damping exponents, and the initial yaw and yaw rates [5] as

$$K_{10} \exp(i\psi_{10}) = \frac{\xi_0' - (\lambda_2 + i\psi_2')\xi_0}{\lambda_1 - \lambda_2 + i\left(\psi_1' - \psi_2'\right)} \tag{10.86}$$

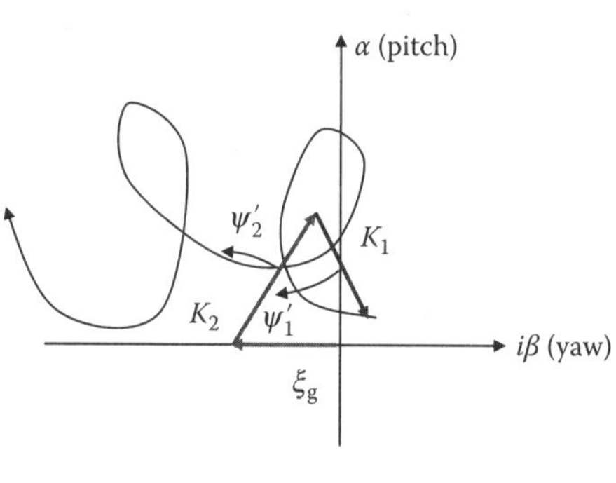

FIGURE 10.6
Example of tricyclic arms for fin-stabilized projectile.

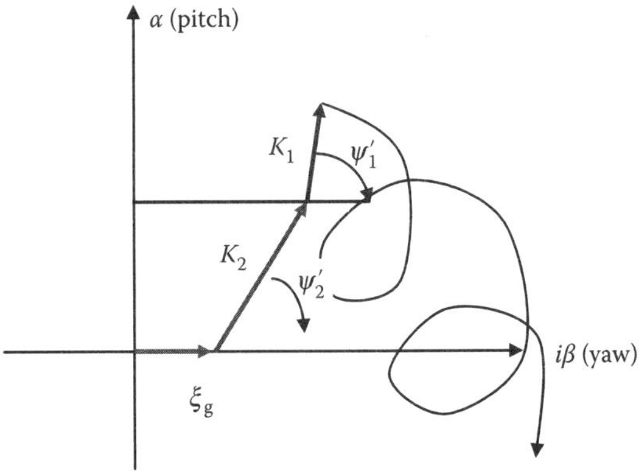

FIGURE 10.7
Example of tricyclic arms for spin-stabilized projectiles.

$$K_{20}\exp(i\psi_{20}) = \frac{\xi_0' - (\lambda_1 + i\psi_1')\xi_0}{\lambda_2 - \lambda_1 + i(\psi_2' - \psi_1')} \tag{10.87}$$

Because the damping exponents are usually an order of magnitude or more smaller than the precession and nutation rates, these equations can be simplified to

$$K_{10}\exp(i\psi_{10}) = \frac{i\xi_0' + \psi_2'\xi_0}{\psi_2' - \psi_1'} \tag{10.88}$$

$$K_{20}\exp(i\psi_{20}) = \frac{i\xi_0' + \psi_1'\xi_0}{\psi_1' - \psi_2'} \tag{10.89}$$

These equations are important because they allow one to determine the initial amplitudes of the arms given an assumed or measured initial yaw, yaw rate, and muzzle exit conditions for a known projectile geometry.

The expressions introduced in this section are the basis for stability criterion to be established next. In the next section, we shall discuss the behavior of these equations and use them to define stability criteria for a projectile.

Problem 1
A 155 mm M549A1 projectile has the following properties and initial conditions:

$$C_D = 0.3$$

$$C_{L_\alpha} = 0.13 \qquad \rho = 0.0751\left[\frac{\text{lbm}}{\text{ft}^3}\right] \qquad I_P = 505.5\,[\text{lbm-in.}^2]$$

$$C_{M_\alpha} = 4.28$$

$$C_{l_p} = -0.024 \qquad d = 155\,[\text{mm}] \qquad I_T = 6610\,[\text{lbm-in.}^2]$$

$$C_{M_q} + C_{M_{\dot{\alpha}}} = -26 \qquad V_{\text{muzzle}} = 3000\left[\frac{\text{ft}}{\text{s}}\right] \qquad m = 96\,[\text{lbm}]$$

$$C_{M_{p\alpha}} = 0.876$$

At an instant in time after launch when

$$p = 220\,[\text{Hz}]$$

$$\phi = \delta = 4^\circ$$

$$V = 1764\left[\frac{\text{ft}}{\text{s}}\right]$$

determine

1. The yaw of repose
   Answer:

$$\beta_R = 0.00172\ [\text{rad}]$$

2. The precessional frequency in hertz
   Answer:

$$\frac{\mathrm{d}\psi_2}{\mathrm{d}t} = 1.9\,[\text{Hz}]$$

3. The nutational frequency in hertz
   Answer:

$$\frac{\mathrm{d}\psi_1}{\mathrm{d}t} = 14.9\,[\text{Hz}]$$

Problem 2
For the projectile given in Chapter 9, Problem 37, determine the precessional and nutational frequencies in hertz.

10.2 Gyroscopic and Dynamic Stabilities

In the previous section, we developed a pair of equations and their solutions using linear aeroballistic coefficients that allow us to examine the motion of a projectile in pitch, yaw, and roll. These equations will now be examined in detail so that we can establish criteria for a stable projectile. In so doing, we will examine some interesting characteristics of motion, which will be displayed as curves in the complex plane.

We shall repeat the equations and their solutions here for ease of reference but leave the coefficient definitions in Section 10.1 to preserve space. The governing equations are as follows:

- For a nonspinning or slowly spinning projectile,

$$\xi'' + H\xi' - M\xi = A + G\left(\frac{V_0}{V}\right)^2 \tag{10.90}$$

with the solution

$$\xi = K_{10}\exp[\lambda_1 s]\exp\left[i\left(\psi_{10} + \psi_1' s\right)\right] + K_{20}\exp[\lambda_2 s]\exp\left[i\left(\psi_{20} + \psi_2' s\right)\right]$$
$$+ K_3\exp[i\psi_{30}] + \xi_g \tag{10.91}$$

- For a spinning projectile,

$$\xi'' + (H - iP)\xi' - (M + iPT)\xi = -iPG \tag{10.92}$$

with the solution

$$\xi = K_{10}\exp[\lambda_1 s]\exp\left[i\left(\psi_{10} + \psi_1' s\right)\right] + K_{20}\exp[\lambda_2 s]\exp\left[i\left(\psi_{20} + \psi_2' s\right)\right] + \xi_g \tag{10.93}$$

For our general development of stability, we shall focus on Equation 10.92 and its solution (Equation 10.93), since the trim term in Equation 10.91 can be easily dealt with separately.

If we examine Equation 10.93, we can readily see that nasty things can happen to us mathematically because of the exponential terms. Since K_{10} and K_{20} are constants (they are the initial magnitudes of the fast and slow arms, respectively), we can focus on the exponential terms that they are multiplied by as a means of determining whether they will grow, shrink, or remain the same.

We shall consider the exponential functions of ψ and ψ' first using the fast arm terms as examples. The term ψ_{10} is a constant and will be ignored. This leaves the term $\psi_1 s$, which is multiplied by i in the exponent. If ψ_1' is purely real, then, when multiplied by i, it becomes purely imaginary in the exponent (because s must be real), the solution is oscillatory, and this will cause the fast arm to increase and decrease in amplitude (i.e., oscillate), neither increasing nor decreasing beyond the established limits of oscillation. This would be a gyroscopically stable projectile. If it has an imaginary component, then, when multiplied by i in the exponent, the solution has a real part. This real part will be multiplied by s and continue to grow throughout the flight as s continually increases. This would result in a gyroscopically unstable projectile.

The question to answer at this point is, "What governs whether the exponents have real or imaginary parts?" This can be answered by the examination of a version of Equation 10.84, whereby all aerodynamic forces and moments are ignored except for the largest (pitching) moment. This has been shown [1,5] to result in a governing equation of

$$\xi'' - iP\xi' - M\xi = -iPG \tag{10.94}$$

with the solution

$$\xi = K_1\exp\left[i\left(\psi_{10} + \psi_1' s\right)\right] + K_2\exp\left[i\left(\psi_{20} + \psi_2' s\right)\right] + \xi_g \tag{10.95}$$

resulting in

$$\psi_{1,2}' = \frac{1}{2}\left(P \pm \sqrt{P^2 - 4M}\right) \tag{10.96}$$

where the subscripts 1 and 2 represent the fast and slow arms, respectively.

Using Equation 10.96, we recall that for a gyroscopically stable projectile, ψ' must be real; therefore, for gyroscopic stability, we require that

$$\left(P^2 - 4M\right) > 0 \tag{10.97}$$

This expression has some interesting implications. If we look back at the definition of our parameter M in Equation 10.66, we see that it is dependent upon the pitching moment coefficient. This happens to always be negative for a fin-stabilized projectile since the fins impart a restoring moment. Unless there is some unique drag device, this moment is positive in a non-fin-stabilized projectile. Because of this, a fin-stabilized projectile is always gyroscopically stable because P^2 must be positive. However, a non-fin-stabilized projectile must have a spin sufficient to make $P^2 > 4M$. We therefore define a statically stable projectile as one in which $M < 0$. With this definition, a statically stable projectile is always gyroscopically stable.

Gyroscopic stability is a necessary but not sufficient condition for a stable projectile. The second condition required is that of dynamic stability. Let us once again examine Equation 10.93, but this time, we shall assume that we have a gyroscopically stable projectile. This means that the exponential terms containing ψ' decay or remain constant, leaving the terms containing λ as potentially destabilizing. We can readily see that since these are multiplied by the downrange distance s, they must be negative to assure that the fast and slow arms decay in magnitude. With this, we shall define a dynamically stable projectile as one in which both λs are negative throughout the flight. Recall that we calculate λ as the real part of Equation 10.84. For convenience, we shall express them directly as

$$\lambda_{1,2} = -\frac{1}{2}\left[H \mp \frac{P(2T - H)}{\sqrt{P^2 - 4M}}\right] \tag{10.98}$$

It should be noted here that, as is common in ballistics, there are always exceptions to any rule. Some successful projectiles have been fielded where instability occurs for a very short time in a flight or in a range where a certain projectile will never be fired. Of course, it is always best to avoid these situations, but sometimes the lack of design space makes it unavoidable. In these instances, the rational examination of the instability is necessary and should be well documented.

We have mathematically shown how we define stability and the parameters that affect stability. Sometimes, it is desirable to quantify how stable a projectile is. We do this through use of a gyroscopic and dynamic stability factors. We define the gyroscopic stability factor as

$$S_g = \frac{P^2}{4M} \tag{10.99}$$

Here, with our earlier discussion, $S_g > 1$ to assure gyroscopic stability. In a similar fashion, we can define a dynamic stability factor as

$$S_d = \frac{2T}{H} \tag{10.100}$$

where for a symmetric projectile to be deemed stable, whether spinning or nonspinning, we require

$$\frac{1}{S_g} < S_d(2 - S_d) \tag{10.101}$$

For a statically stable projectile, we require that $0 < S_d < 2$ for dynamic stability. This leads to an interesting condition where one can spin a statically stable projectile too fast, resulting in instability. This condition translated to dimensionless spin rate is given by

$$P < \sqrt{\frac{4M}{S_d(2 - S_d)}} \tag{10.102}$$

for a statically stable projectile.

It is interesting to combine Equations 10.96 and 10.98 in various ways writing them in terms of the dimensionless parameters *P*, *M*, *H*, and *T*. The details of this can be found in the books by Murphy [1] and McCoy [5] with the following results:

$$P = \psi_1' + \psi_2' \tag{10.103}$$

$$M = \psi_1'\psi_2' - \lambda_1\lambda_2 \tag{10.104}$$

$$H = -(\lambda_1 + \lambda_2) \tag{10.105}$$

$$PT = -\left(\psi_1'\lambda_1 + \psi_2'\lambda_2\right) \tag{10.106}$$

If we again examine Equation 10.91 or 10.93, we see that the magnitude of the precessional and nutational arms is highly dependent upon initial conditions. Without going into details (which are described quite well in the book by McCoy), we can express these initial conditions in terms of the complex angle of attack and damping parameters as

$$K_{10}\exp[i\psi_{10}] = \frac{\xi_0' - (\lambda_2 + i\psi_2')\xi_0}{\lambda_1 - \lambda_2 + i\left(\psi_1' - \psi_2'\right)} \tag{10.107}$$

$$K_{20}\exp[i\psi_{20}] = \frac{\xi_0' - (\lambda_1 + i\psi_1')\xi_0}{\lambda_2 - \lambda_1 + i\left(\psi_2' - \psi_1'\right)} \tag{10.108}$$

In these equations, ξ_0 and ξ_0' are the initial complex yaw and yaw rates, respectively. These parameters are determined by measurements as the projectile leaves the gun tube or are assumed values.

We now have solid criteria by which we can determine whether a projectile will be stable or not. These developments have been made assuming that the projectile aerodynamic coefficients behave in a linear fashion. As such, a projectile is either stable or it is not. This stability, even with our linear model, will change during the flight based on Mach number and angle of attack. We will discuss in a later section how a nonlinearity can help or hurt matters. The true power of these equations is that they can tell us which coefficients need to be altered to affect stability. This can be used in instances where we want to change a physical configuration to make a projectile "drop out of the sky" or design a round such that it damps more quickly and can thus fly with lower drag. Other uses for these equations allow for tweaking the flight characteristics for better flight behavior in general.

Problem 3

Up until the late 1960s, many US and foreign ships carried the Bofors 40 mm gun as a general light support weapon. Originally designed as an antiaircraft weapon, this gun served as an antitank weapon if the situation required it, and its high rate of fire made it quite successful as an antipersonnel weapon. Assuming the properties of the system given next,

1. Calculate the gyroscopic stability factor at the beginning and at the end of the flight assuming a terminal velocity of 2450 ft/s and the spin rate is 10% lower than the initial value.
   Answer: At the beginning of flight, $S_g = 5.814$.
2. Is the projectile stable throughout the flight?
   Answer: Yes.

3. Assuming that this is the longest time of flight for the projectile, at what spin rate will the projectile become unstable?
   Answer:

$$p_{\text{unstable}} < 1887 \left[\frac{\text{rad}}{s}\right]$$

4. Where will the instability occur?
   Answer: At the muzzle of the weapon.

Projectile and weapon information:

$$C_{M\alpha} = 3.10 \qquad \rho = 0.067 \left[\frac{\text{lbm}}{\text{ft}^3}\right] \qquad I_P = 1.231 \left[\text{lbm-in.}^2\right]$$

$$C_{l_p} = -0.011 \qquad d = 40\,[\text{mm}] \qquad I_T = 6.263 \left[\text{lbm-in.}^2\right]$$

$$V_{\text{muzzle}} = 2850 \left[\frac{\text{ft}}{\text{s}}\right] \qquad m = 1.985\,[\text{lbm}] \qquad n = \frac{1}{30}\left[\frac{\text{rev}}{\text{cal}}\right]$$

Please note that this weapon actually has a progressive twist, but when faced with this situation, you only need the muzzle velocity and the twist at the muzzle to calculate initial spin.

Problem 4

For the projectile described in Problem 35 of Chapter 9,

1. Determine the precessional damping exponent.
   Answer: $\lambda_2 = -0.0000608$
2. Determine the nutational damping exponent.
   Answer: $\lambda_1 = -0.001135$
3. With 1 and 2, which of these modes will damp out?
   Answer: Both
4. Determine the dynamic stability factor S_d.
   Answer: $S_d = 0.74$
5. Consider a cargo projectile with identical properties to our projectile in Chapter 9, Problem 35. The designer did not secure the cargo well enough so that the cargo fails to spin completely up during gun launch in a worn tube. When this happens, immediately after muzzle exit, the round spins down (and the cargo spins up a little more) so that the projectile finally reaches a spin rate of 100 Hz. The velocity is unaffected.

   a. Determine the gyroscopic stability factor for each of the two situations.
      Answer: $S_g = 3.192$ and $S_g = 0.659$.
   b. Will both projectiles fly properly? Why or why not?
      Answer: No, the second projectile will tumble.

Problem 5

A 155 mm HE projectile is fired from a cannon. The muzzle velocity of the projectile is 800 m/s and the twist of the rifling is 1:20. The projectile and filler properties are given next. Assuming the aerodynamic forces and moments are negligible and that the projectile is dynamically stable,

1. Determine the initial spin rate of the complete projectile.
   Answer:

$$p_{\text{muzzle}} = 1621.5 \left[\frac{\text{rad}}{s}\right]$$

2. Determine the spin rate of the projectile in flight assuming the fill does not spin up in the bore and both shell and fill come into dynamic equilibrium.
   Answer:

$$p_{\text{total}} = 1259.2 \left[\frac{\text{rad}}{s}\right]$$

3. Determine the gyroscopic stability factors for 1 and 2.
   Answer:

$$S_g = 16.54 \quad \text{and} \quad S_g = 9.97$$

4. Is the projectile stable in 1 and 2?
   Answer: Yes.

Projectile and weapon information:

$$C_{M\alpha} = 1.07 \qquad \rho = 0.067 \left[\frac{\text{lbm}}{\text{ft}^3}\right] \qquad I_{P_{total}} = 555 \left[\text{lbm-in.}^2\right]$$

$$C_{l_p} = -0.012 \qquad I_{T_{\text{total}}} = 3335 \left[\text{lbm-in.}^2\right]$$

$$I_{P_{\text{shell}}} = 431 \left[\text{lbm-in.}^2\right] \qquad d = 155\,[\text{mm}] \qquad m = 106\,[\text{lbm}]$$

$$I_{P_{\text{fill}}} = 124 \left[\text{lbm-in.}^2\right] \qquad V_{\text{muzzle}} = 800 \left[\frac{\text{m}}{s}\right] \qquad n = \frac{1}{20} \left[\frac{\text{rev}}{\text{cal}}\right]$$

Problem 6
For the projectile given in Problem 5, determine the nutation and precession frequencies in hertz.
Answer:

$$\frac{d\psi_1}{dt} = 42.05\,[\text{Hz}] \quad \text{and} \quad \frac{d\psi_2}{dt} = 0.67\,[\text{Hz}]$$

Problem 7
Assume the projectile in Problem 1 has slipped its rotating band, and the spin at the same instant in time is 130 Hz. Is the projectile stable?
Answer: No.

Problem 8
What is the minimum spin (Hz) required to stabilize the projectile in Problem 1?
Answer: $p_{\text{min}} = 140$ [Hz].

Problem 9
For the projectile given in Chapter 9, Problem 37, determine the minimum spin rate for stability. (*Hint*: Remember when a projectile is least stable.)

Problem 10

A right circular cylinder is to be horizontally fired for an impact test. If the cylinder is made of steel ($\rho = 0.283$ lbm/in.^{3}) and it is 0.5 in. in diameter and 0.75 in. long,

1. Determine the spin rate required to stabilize the projectile (if it can be stabilized).
2. Comment on the preceding answer—what is dominant in the problem?
3. Determine the precessional and nutational frequencies of the projectile at this spin rate in hertz.

The projectile properties are provided next:

$$C_D = 0.4$$
$$C_{L_\alpha} = 0.18 \qquad \rho_{air} = 0.0751 \left[\frac{\text{lbm}}{\text{ft}^3}\right]$$
$$C_{M_\alpha} = 8.0 \qquad d = 0.5\,[\text{in}]$$
$$C_{l_p} = -0.02 \qquad l = 0.75\,[\text{in}]$$
$$C_{M_q} + C_{M_{\dot{\alpha}}} = -29 \qquad V_{muzzle} = 6000 \left[\frac{\text{ft}}{\text{s}}\right]$$
$$C_{M_{p\alpha}} = 0.92$$

Problem 11

Modifications are made to a 155 mm M483A1 projectile so that it has the following properties and initial conditions:

$$C_D = 0.2$$
$$C_{L_\alpha} = 1.975 \qquad \rho = 0.0751 \left[\frac{\text{lbm}}{\text{ft}^3}\right] \qquad I_P = 537.1\,[\text{lbm-in.}^2]$$
$$C_{M_\alpha} = 4.573 \qquad d = 155\,[\text{mm}] \qquad I_T = 5753\,[\text{lbm-in.}^2]$$
$$C_{l_p} = -0.0285$$
$$C_{M_q} + C_{M_{\dot{\alpha}}} = -15.2 \qquad V_{muzzle} = 2900 \left[\frac{\text{ft}}{\text{s}}\right] \qquad m = 103\,[\text{lbm}]$$
$$C_{M_{p\alpha}} = 1.20$$

At an instant in time after launch when

$$p = 100\,[\text{Hz}]$$
$$\phi = \delta = 2°$$
$$V = 1000 \left[\frac{\text{ft}}{\text{s}}\right]$$

determine

1. If the projectile is stable
2. The precessional frequency in hertz
3. The nutational frequency in hertz

Problem 12

What is the minimum spin (Hz) required to stabilize the projectile in Problem 11?

Problem 13

For the projectile given in Chapter 9, Problem 25, determine the precessional and nutational frequencies in hertz. Determine the minimum spin rate for the projectile to be stable.

Problem 14

For the projectile given in Chapter 9, Problem 27, determine the precessional and nutational frequencies in hertz. Determine the minimum spin rate for the projectile to be stable.

Problem 15

For the projectile given in Chapter 9, Problem 28, determine the precessional and nutational frequencies in hertz. Determine the minimum spin rate for the projectile to be stable. Does it differ depending on which side of the aircraft it is fired from? Calculate it for both cases.

10.3 Yaw of Repose

In Section 10.1, we introduced the yaw of repose for a projectile and defined it in Equation 10.75 using the symbol ξ_g. The subscript g was used to denote that this quantity comes about through the action of gravity on the projectile. In terms of our dimensionless parameters, we can rewrite Equation 10.75 as

$$\xi_g = \frac{PG}{M + iPT} \tag{10.109}$$

A qualitative look at this expression leads to some extremely interesting results. First and foremost is that the spin rate directly affects the yaw. The greater the spin (and therefore, the larger the value of P), the greater the yaw of repose is.

The second useful item to note is that the more abrupt the trajectory curve is, the greater the yaw of repose is. In fact, if we look at the term G; it is linear in the cosine of angle of attack ϕ. For details of this form that G takes, the reader is referred to the book by McCoy [5]. Thus, when the projectile approaches maximum ordinate, the yaw of repose should be a maximum given that all the other parameters remain constant. Because of the decay of the other terms, the result is that the yaw of repose is usually a maximum shortly before or after reaching maximum ordinate.

The sign of the yaw of repose is important. In our convention, the term P is positive for a right-hand twist. Thus, a positive value of ξ_g causes the projectile to nose over to the right. Note that there can also be a significant pitch component to this quantity; this is easily seen as the real part of Equation 10.109.

If we examine a plot of pitch (α) vs. yaw (β) for a British 14 in. projectile in Figure 10.8, we can imagine the yaw of repose as the vector pointing to the right (viewed from the rear) to the center of the precessional path similar to Figure 10.7. We can see that the magnitude as well as the direction of this vector change as the projectile moves downrange. In Figure 10.8, the projectile was analyzed using the PRODAS software and was fired with a muzzle velocity of 2483 ft/s, spin rate of 71 Hz corresponding to a 1:30 twist with an initial pitch angle of 0.1°. There was no initial yaw or pitch/yaw rate. This projectile has progressed through only one and one-half yaw cycles (about 1.7 s) when the analysis was stopped to yield a nice clear illustration.

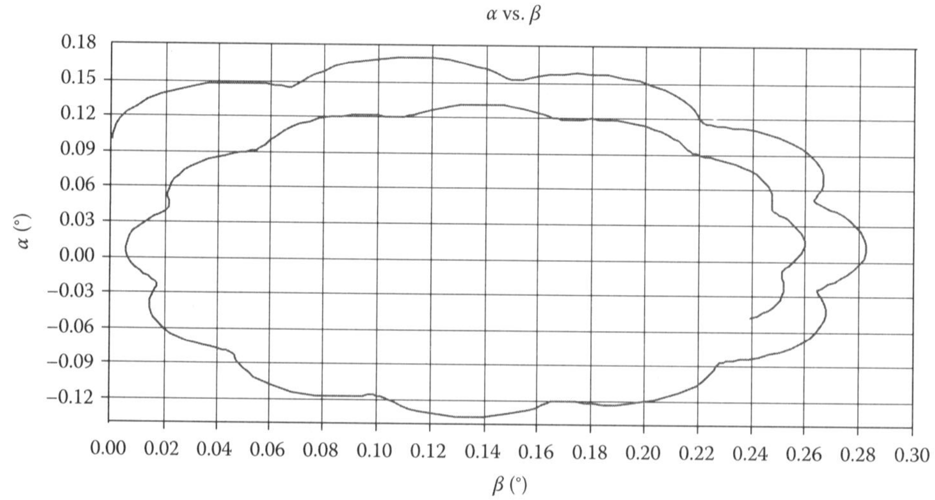

FIGURE 10.8
Pitching and yawing motion for a British 14 in. Mk I projectile fired at 2483 ft/s with a 0.1° initial pitch angle.

10.4 Roll Resonance

Until this point, we have assumed that the projectiles under study have been axially symmetric. This rarely happens in practice because of manufacturing tolerances in a given projectile design. In Chapter 11, we shall discuss the means of handling a slight mass asymmetry. In this section, we shall discuss the implications of a geometric (including slight mass) asymmetry as applied to a fin-stabilized projectile.

Fin asymmetries commonly occur when a finned projectile is manufactured or can be the result of damage owing to rough handling. In the field of explosively formed penetrators, which are normally drag- or fin-stabilized, inconsistencies can (and usually do) arise due to the explosive formation process. In either case, this effect may be coupled with some mass asymmetry as well.

In Equation 10.74, the trim arm was introduced, which would force a statically stable projectile to fly with an angle of attack. It is for this reason that all fin- and drag-stabilized projectiles are designed to slightly roll to increase accuracy. One can see from the way that this equation was written that there is no change in the orientation of K_{30}. It was fixed, oriented at the initial angle ψ_{30}.

To begin our assessment of this specific type of asymmetry, we shall start with the governing equation for a spin-stabilized projectile (Equation 10.76) because the roll is going to play a part. We shall alter the RHS to incorporate a forcing term representing the lifting force and moment that is caused by the asymmetry (say, e.g., a bent fin). We shall write this in such a way that the direction of the applied force and moment rotates with the projectile:

$$\xi'' + (H - iP)\xi' - (M + iPT)\xi = -iA_3 \exp(i\psi) \quad (10.110)$$

where

$$A_3 = \left(\frac{\rho S d}{2m}\right)\left(\frac{1}{k_T^2}\right)\left(C_{m_0} + iC_{n_0}\right) + (\psi' - P)\left(C_{Z_0} + iC_{Y_0}\right) \quad (10.111)$$

$$\psi' = \frac{pd}{V}, \quad \text{dimensionless turning rate} \tag{10.112}$$

$$\psi = \int_0^s \psi' ds, \quad \text{dimensionless distance} \tag{10.113}$$

This development was put forth in the studies by Murphy [1], Nicolaides [2], and McCoy [5]. If we look closely at these equations, we see that the forcing function A_3 rotates with the projectile.

If we solve Equation 10.110, assuming a solution for the particular part of

$$\xi_p = K_3 \exp[i(\psi + \psi_0)] \tag{10.114}$$

where ψ_0 is some arbitrary angle that contains the plane of the asymmetry, we obtain a general solution for a constant roll rate of

$$\xi = K_1 \exp[i\psi_1] + K_2 \exp[i\psi_2] + K_3 \exp[i(\psi + \psi_0)] \tag{10.115}$$

and, after inserting the initial conditions, say, of $\psi_0 = 0$, we obtain

$$K_3 = \frac{-iA_3}{\psi'^2 - P\psi' + M - i(\psi' H - PT)} \tag{10.116}$$

This is the expression for the yaw component caused by a lift force and corresponding moment constrained to rotate at the projectile spin rate. If the spin rate is zero, the orientation of this lift force will be fixed and the projectile will drift more and more in that direction. This is not desirable from an accuracy standpoint so we must have some spin.

The denominator in Equation 10.116 is normally dominated by its real part because H and the product PT are small by comparison. However, much like a resonance in a spring mass system, if the roll frequency ever approaches either one of the precession or nutation frequencies (and remains there for some time), the denominator in Equation 10.114 approaches zero and the yaw becomes very large [1,5]. This usually occurs when the nutational frequency is approached and is called roll resonance or spin pitch resonance [1]. Since projectiles are usually changing spin rate throughout their flight, this is only a problem if there is a slow change of spin rate when the frequencies are close.

Another way of looking at this is to imagine a projectile where this asymmetry is present. Since the asymmetry is at the same frequency as the nutation rate, every time the projectile is at the outer limit of its motion, it gets kicked a little further, similar to pushing a child on a swing. This disturbance grows as long as the two motions stay coupled (i.e., at the same frequency); however, if they became out of phase, the problem would correct itself.

An example of roll resonance is depicted in Figure 10.9. In this case, an explosively formed penetrator (EFP) was the device under test. Keep in mind that only the total angle of attack is measured here so the yawing motion is not constrained to a single plane. We see that as the EFP approached a spin rate of ~300 rad/s, it locked in and flew very far off of the target.

Problem 16

The roll resonance of a projectile occurs when the spin rate approaches a "forcing" frequency of the projectile. This occurs more frequently in fin-stabilized projectiles than in spin-stabilized projectiles, because in the latter, the spin rate is usually quite high in order to maintain stability, the overturning moment is positive, and these forcing functions unless they are intentional—like thrusters are usually due to asymmetries (like bent fins)—are

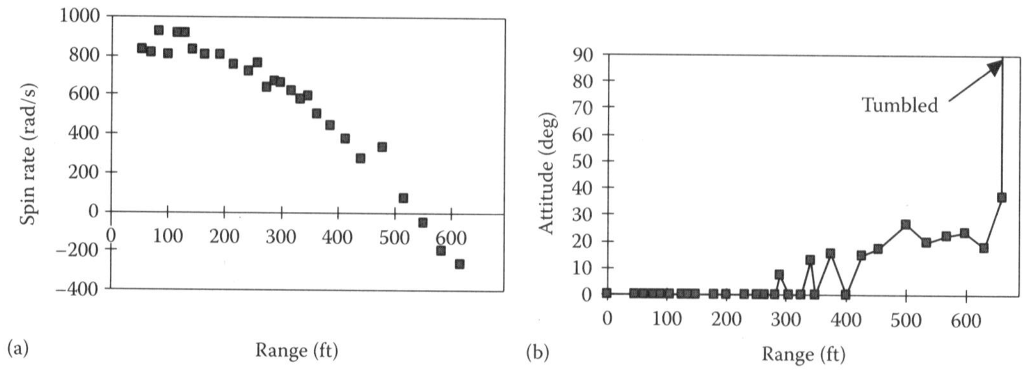

FIGURE 10.9
EFP experiencing roll resonance. (Courtesy of Eric Volkmann, Alliant Techsystems, Hopkins, MN.)

usually small. If we examine the projectile of Chapter 9, Problem 37, instead as a fin-stabilized projectile, we can write the equation for the pointing direction as

$$\xi = K_1 \exp[i\psi_1] + K_2 \exp[i\psi_2] + K_3 \exp[i(\psi + \psi_0)] \tag{10.117}$$

where the subscripts 1 and 2 represent the fast and slow modes, respectively. The third term is the forcing function where we can define

$$\psi' = \frac{pd}{V} \tag{10.118}$$

and, after inserting the initial conditions, say, of $\psi_0 = 0$, we obtain

$$K_3 = \frac{-iA_3}{\psi'^2 - P\psi' + M - i(\psi' H - PT)} \tag{10.119}$$

Based on what you know about the behavior of imaginary numbers and Equations 10.118 and 10.119, determine the spin rate at which catastrophic yaw will occur. Use the aerodynamic properties of the projectile from Chapter 9, Problem 37, but assume that the overturning moment is the negative of what was provided (we are essentially faking a fin-stabilized version). You may assume that the velocity stays constant at 750 ft/s.

References

1. Murphy, C. H., *Free Flight Motion of Symmetric Missiles*, Ballistic Research Laboratory Report No. 1216, Ballistic Research Laboratory, Aberdeen Proving Ground, MD, 1963.
2. Vaughn, H., *A Detailed Development of the Tricyclic Theory*, Sandia National Laboratories Report No. SC-M-67–2933, Sandia National Laboratories, Albuquerque, NM, February 1968.
3. McShane, E. J., Kelley, J. L., and Reno, F. V., *Exterior Ballistics*, University of Denver Press, Denver, CO, 1953.
4. Nicolaides, J. D., *On the Free Flight Motion of Missiles Having Slight Configurational Asymmetries*, Ballistic Research Laboratory Report No. 858, Ballistic Research Laboratory, Aberdeen Proving Ground, MD, 1953.
5. McCoy, R. L., *Modern Exterior Ballistics*, Schiffer Military History, Atglen, PA, 1999.

11

Mass Asymmetries

Until this point we have assumed that the projectile has been an axially symmetric body. This allowed us to simplify the equations of motion considerably. Projectiles are rarely axially symmetric. The asymmetry usually comes about through manufacturing tolerances, damage due to rough handling, and cargo slippage, or more recently, they are simply designed that way. The purpose of this section is simply to introduce the geometry of mass asymmetries, which will be introduced into the equations of motion for the projectile in later sections.

Mass asymmetries come in two categories: static imbalance and dynamic imbalance. In a static imbalance, the center of gravity (CG) of the projectile is not located on the geometric axis of symmetry. The geometric axis of symmetry can be defined by imagining a projectile with the same exterior dimensions as the unbalanced projectile but of uniform density. The symmetry axis would then be centrally located in the body of revolution (i.e., a perfectly axially symmetric body). In a statically imbalanced projectile, this axis would be shifted to pass through the CG but remain parallel to the geometric axis. This is illustrated in Figure 11.1.

A dynamically imbalanced projectile also has a CG that is offset from the geometric axis of symmetry. In this case, however, the mass distribution is such that the principal axis of inertia resides as some angle to the geometric axis as well. This is illustrated in Figure 11.2.

Whether a projectile is statically or dynamically imbalanced, we shall define the plane in which the CG offset is located relative to some reference plane (we shall arbitrarily use the x–y plane as the reference, which we have defined in earlier sections) using the symbol Φ. This is illustrated in Figure 11.3 as viewed from the rear of the projectile.

The effect of these mass asymmetries on projectile flight can dramatically affect accuracy, especially in direct fire systems. Consider a projectile with an imbalance in the gun tube. While in the tube, the projectile is constrained to rotate about the tube geometric axis. If we idealize this situation to say that the tube is perfectly straight, inflexible, and fits the projectile snugly, we can further state that the projectile is constrained to rotate about its own geometric axis. Note that there is a wealth of literature dedicated to the real situation (e.g., [1–10]).

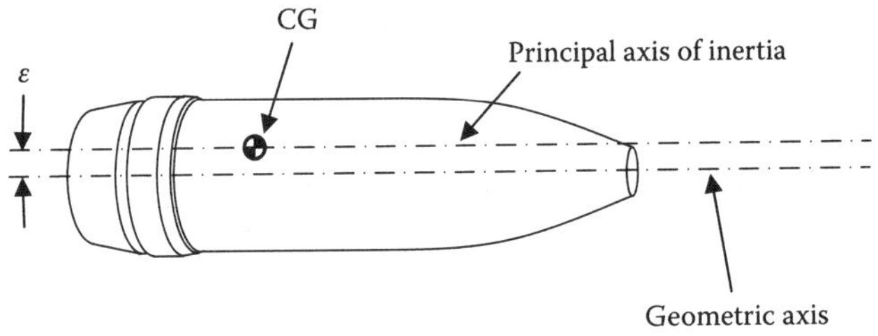

FIGURE 11.1
Statically imbalanced projectile.

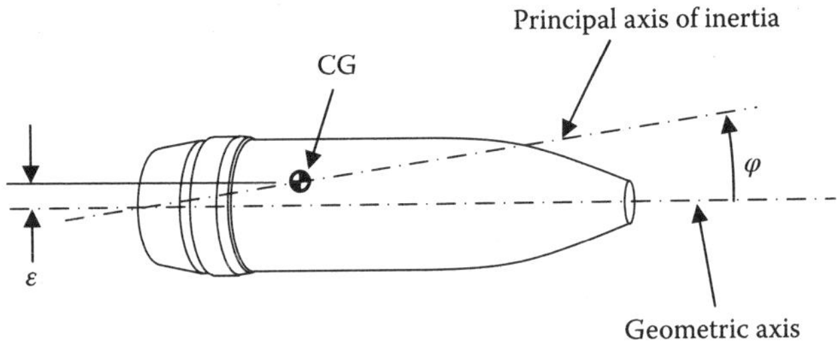

FIGURE 11.2
Dynamically imbalanced projectile.

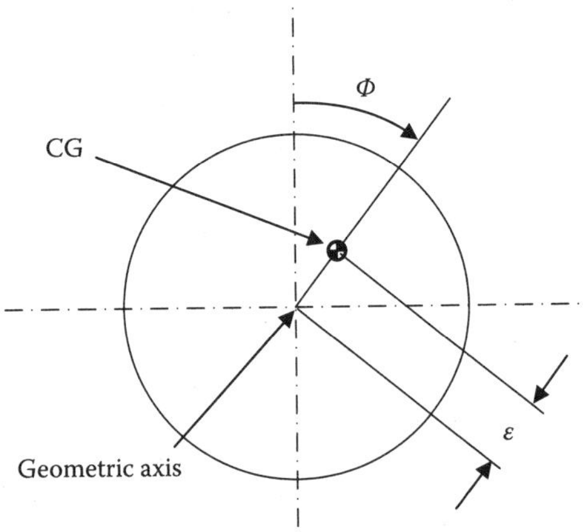

FIGURE 11.3
CG offset viewed from rear of projectile.

References

1. Berger, M. P., Position and form of bands for projectiles, *Memoires Militaires et Scientifiques*, Publies par le Department de la Marine, Translated by LT C.C. Morrison, Notes on the Construction of Ordnance, Washington, DC, June 10, 1884.
2. Kent, R. H., and Hitchcock, H. P., *Comparison of Predicted and Observed Yaw in Front of the Muzzle of a 12' Gun*, Report No. 990 AD-116-140, US Ballistic Research Laboratory, Aberdeen Proving Ground, MD, July 1956.
3. Kent, R. H., and McShane, E. J., *An Elementary Treatment of the Motion of a Spinning Projectile about Its Center of Gravity*, Report No. 459 AD-491-943, US Ballistic Research Laboratory, Aberdeen Proving Ground, MD, April 1944.
4. Heppner, L. D., *Setback and Spin for Artillery, Mortar, Recoilless Rifle and Tank Ammunition*, Report No. DPS-2611, US Ballistic Research Laboratory, Aberdeen Proving Ground, MD, January 1968.
5. Gay, H. P., and Elder, A. S., *The Lateral Motion of a Tank Gun and Its Effect on the Accuracy of Fire*, Report No. 1070 AD-217-657, US Ballistic Research Laboratory, Aberdeen Proving Ground, MD, March 1959.
6. Kirkendall, R. D., *The Yawing Motions of Projectiles in the Bore*, Technical Note No. 1739 AD-878-327-L, US Ballistic Research Laboratory, Aberdeen Proving Ground, MD, September 1970.
7. Zaroodny, S. J., *On Jump due to Muzzle Disturbances*, Report No. 703 AD-805-876, US Ballistic Research Laboratory, Aberdeen Proving Ground, MD, June 1949.
8. Gay, H. P., *On the Motion of a Projectile as It Leaves the Muzzle*, Technical Note No. 1425 AD-801-974, US Ballistic Research Laboratory, Aberdeen Proving Ground, MD, August 1961.
9. Sterne, T. E., *On Jump due to Bore Clearance*, Report No. 491 AD-491-938, US Ballistic Research Laboratory, Aberdeen Proving Ground, MD, September 1944.
10. Line, L. E., *The Erosion of Guns at the Muzzle*, NRDC Report No. A-357, OSRD Report No. 6322, National Defense Research Committee, Office of Scientific Research and Development, Washington, DC, November 1945.

12

Lateral Throwoff

Earlier in the book, we stated that projectiles rarely leave the tube with their velocity vectors aligned with the geometric axis of the gun tube. This chapter and Chapter 13 describe this behavior. The result of this behavior is weapon inaccuracy, and it must be well understood by the practicing ballistician because, although it is not practical to completely eliminate the behavior, we would like to reduce it to acceptable levels. The first component of this behavior is known as lateral throwoff. It is a dynamic response of the projectile to either a static or a dynamic imbalance and will now be described in detail.

If we imagine a projectile with a mass asymmetry as depicted in Figure 11.3, we can imagine the spinning motion as viewed from the rear. If we ignore the axial velocity by simply spinning the projectile at a high rate, say, between two flexible supports on a test stand, we would see a wobble develop as a result of the centrifugal action on the center of mass. All the time the projectile is being spun up in the gun, the tube walls and stiffness of the supporting members prevent this wobble (to the extent the clearances allow) from developing. At the instant the projectile is free from the constraints of the tube we expect it to become affected by the centrifugal loading due to spin. This is known as lateral throwoff because the effect is to fling the projectile in a direction off the tube centerline.

We can use the analogy of a vacuum trajectory to examine the lateral throwoff effect generated by either a static or a dynamic imbalance. Consider the projectile asymmetry from Figure 11.3. If we examine the projectile over a short period of flight, ignoring gravity as well as assuming no drag because of the vacuum assumption, we would see the dynamic forces acting on the projectile as depicted in Figure 12.1. In this figure, the only force acting is the centrifugal force due to spin. This dynamic action will result in the force vector changing direction, although since there is no angular acceleration or deceleration, it maintains a constant magnitude. It is worth noting that we have resorted to our complex plane in this example, as it is convenient to use in our development. At the instant, in time depicted here, we can break the force into a component in the y-direction and one in the iz-direction.

We are not necessarily concerned with the force acting on the center of gravity (CG) per se. We want to see where the projectile moves because of this force. To accomplish this, we need to use Newton's second law. We know that

$$F_r = ma_r \tag{12.1}$$

This is the centripetal force. The centrifugal force would be equal but opposite in sign. From dynamics [1], we recall that

$$a_r = -rp^2 \tag{12.2}$$

In the case we are considering here, we see that

$$r = \varepsilon \quad \text{and} \quad \Phi = pt \tag{12.3}$$

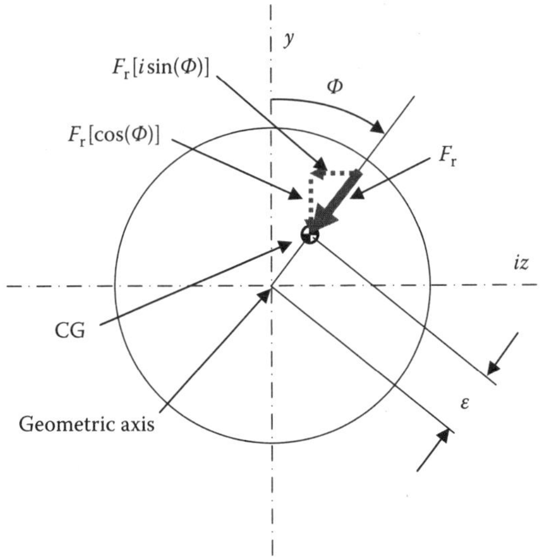

FIGURE 12.1
Dynamic force acting on a statically or dynamically imbalanced projectile.

With this, we can write the magnitude of the force as

$$F_r = m\varepsilon p^2 \tag{12.4}$$

and the centripetal acceleration on the complex plane as

$$a = -\frac{F_r}{m}[\cos(pt) + i\sin(pt)] = -\varepsilon p^2[\cos(pt) + i\sin(pt)] \tag{12.5}$$

The complex velocity can therefore be expressed as

$$V = -\varepsilon p^2 \int_0^t [\cos(pt) + i\sin(pt)]\mathrm{d}t \tag{12.6}$$

Evaluating the integral and assuming that as the projectile leaves the muzzle, we have an initial orientation of the mass asymmetry of $\Phi = \Phi_0$ yields

$$V = -\varepsilon p[\sin(pt + \Phi_0) - i\cos(pt + \Phi_0)] = \varepsilon p[-\sin(pt + \Phi_0) + i\cos(pt + \Phi_0)] \tag{12.7}$$

To see how much lateral movement has developed, we can again integrate

$$r = \varepsilon p \int_0^t [-\sin(pt + \Phi_0) + i\cos(pt + \Phi_0)]\mathrm{d}t \tag{12.8}$$

The evaluation of which yields

$$r = \varepsilon[\cos(pt + \Phi_0) + i\sin(pt + \Phi_0)] \tag{12.9}$$

As an example, if we were only concerned with motion in the cross-range direction, we could state

$$z = \mathrm{Im}\{\varepsilon[\cos(pt + \Phi_0) + i\sin(pt + \Phi_0)]\} = \varepsilon\sin(pt + \Phi_0) \tag{12.10}$$

To apply numbers to this example, let us consider a projectile that weighs 100 lbm and is spinning at a rate of 270 Hz. We shall assume that the projectile has a CG offset of 0.25 in. If this were the case, the velocity in the *z*-direction as well as the motion for the first 4 s of flight can be seen in Figures 12.2 and 12.3. Here we have assumed that the CG offset has emerged from the weapon at the 12 o'clock position.

The most interesting observation between the figures is that for this arbitrary emergence of the CG offset, we see that the projectile would like to move laterally to the right for a right-hand spin. This is commonly known as drift. Just to put things into perspective, the muzzle velocity consistent with the 270 Hz spin rate is about 2750 ft/s, so the projectile would only have gone about 0.4 ft to the right after it traversed 11,000 ft downrange.

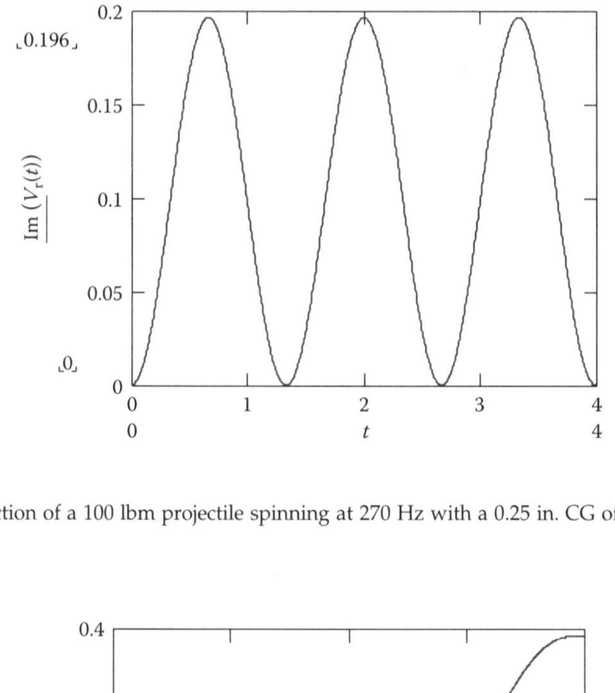

FIGURE 12.2
Velocity in the *z*-direction of a 100 lbm projectile spinning at 270 Hz with a 0.25 in. CG offset.

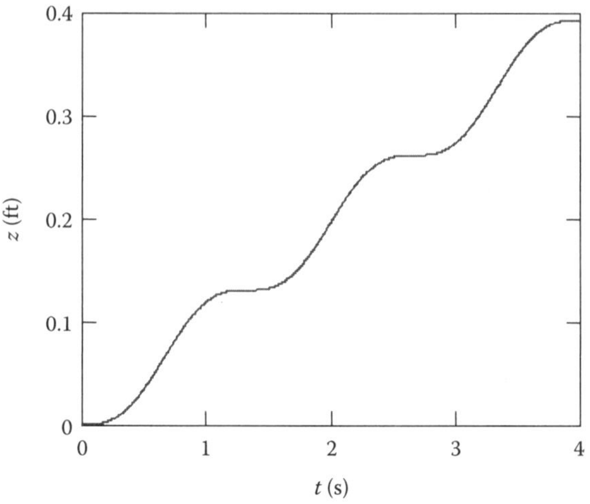

FIGURE 12.3
Displacement in the *z*-direction of a 100 lbm projectile spinning at 270 Hz with a 0.25 in. CG offset.

We must always bear in mind that this example was an idealized situation. In the case of a real projectile, there are other forces acting which complicate the motion; however, it is instructive to look at simplifications such as this to see the phenomenon at work. We will now move on to examine the dynamic behavior in terms of the equations of motion of a projectile from statically imbalanced and dynamically imbalanced projectiles. We shall see how this affects lateral throwoff.

12.1 Static Imbalance

In Figure 11.1, we saw the effect on the principal axis of a static imbalance. Although this rarely happens in production (imbalances are usually of the dynamic type), it can happen and presents an interesting case. We shall follow the analysis procedure documented by McCoy [2] in the development, correcting terms to fit our coordinate system.

If we examine the velocity of the center of mass of the projectile as it leaves the gun tube, we see a scene as depicted in Figure 12.4. If the projectile is constrained as it continues down the gun tube, the motion of the CG would resemble a spiral or helix similar to a thread on a bolt, except that the pitch of the helix would continue to increase as the axial velocity increases. We could express this mathematically using a cylindrical set of coordinates with x indicating the axial distance, r indicating the radius of the CG from the centerline, and Φ indicating the angular position from the vertical plane. If we assume that the tube is straight, then the axial component of the velocity vector will be constrained along the tube and our unit vector $\mathbf{l}$ will adequately describe the direction. We shall use the unit vectors $\mathbf{e}_r$ and $\mathbf{e}_\Phi$ to represent the radial and angular positions, respectively. If we use our instantaneous spin rate p as defined in Equation 12.3, we can write the tangential component of velocity as

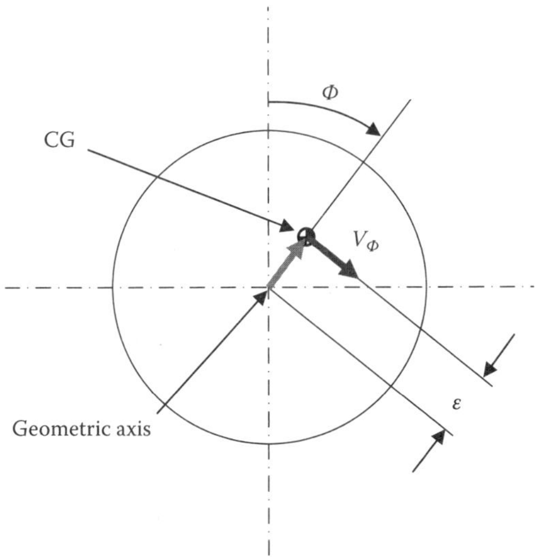

FIGURE 12.4
Velocity of the CG of a statically imbalanced projectile.

$$V_\Phi = rp\mathbf{e}_\Phi = \varepsilon p\mathbf{e}_\Phi \tag{12.11}$$

The axial velocity is simply

$$V_x = V\mathbf{l} \tag{12.12}$$

Then, the velocity vector could be written in cylindrical coordinates as

$$V = V\mathbf{l} + \varepsilon p\mathbf{e}_\Phi \tag{12.13}$$

Or, if we like to remain in Cartesian coordinates, we can combine Equation 12.13 with Equation 12.7 to yield

$$\mathbf{V} = V\mathbf{l} + \varepsilon p(-\sin(pt + \Phi_0)\mathbf{n} + i\cos(pt + \Phi_0)\mathbf{m}) \tag{12.14}$$

These Cartesian coordinates are useful when we want to write the velocity vector at the muzzle of the weapon. The lateral throwoff caused by a static imbalance can be described as the tangent of the angle of the projectile CG as it exits. For small angles (usually the case), this is approximately the angle itself in radians. With this, we can define the lateral throwoff at the muzzle owing to a static imbalance as

$$T_L = \frac{\varepsilon p_0}{V_0}[-\sin(\Phi_0) + i\cos(\Phi_0)] \tag{12.15}$$

where we have used $t = 0$ at the muzzle and specified the spin rate and muzzle velocity. We must keep in mind that this is an angular measure for small angles or, more precisely, a tangent of an angle. We can use the relationship

$$i\exp(i\theta) = -\sin\theta + i\cos\theta \tag{12.16}$$

to write

$$T_L = i\frac{\varepsilon p_0}{V_0}\exp(i\Phi_0) \tag{12.17}$$

If the projectile has a rotating band that forces it to spin based on the rifling twist, this expression can be written in terms of the projectile diameter and twist rate as well. This is extremely straightforward and left as an exercise for the reader.

12.2 Dynamic Imbalance

The diagram of Figure 11.2 represents the most common case of a projectile asymmetry, a dynamic imbalance. In this case, a lateral throwoff effect as described in Section 12.1 will result as the projectile leaves the muzzle of the gun, and there will be significant flight dynamic effects as the projectile moves downrange. Usually, this mass asymmetry is small and can be treated as a small amount of mass removed from or added to a projectile at a point defined by a radial set of coordinates from the CG. We shall use the former approach following the development of McCoy [2]. This is depicted in Figure 12.5. Figure 12.6 depicts how this removed mass is oriented relative to the CG offset in the radial direction.

The development put forward by McCoy [2] assumes that the removed mass is much smaller than the overall mass of the projectile. As established earlier, we will use two orthogonal coordinate systems. The first is our **i–j–k** triad, which is oriented along the

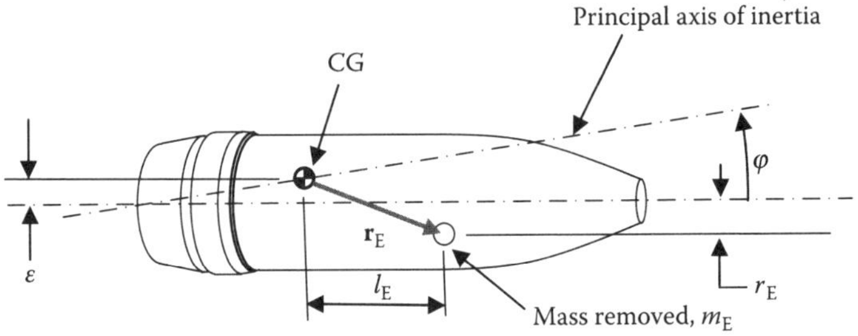

FIGURE 12.5
Dynamically imbalanced projectile with mass removed.

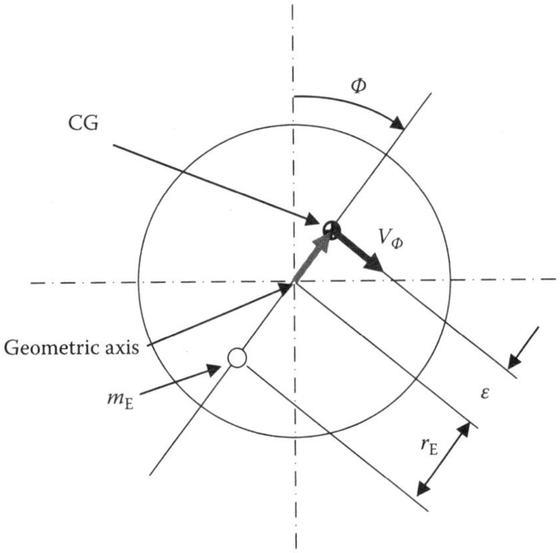

FIGURE 12.6
Velocity of the CG of a dynamically imbalanced projectile.

projectile axis as depicted in Figure 10.1. This coordinate system does not roll with the projectile. We shall also make use of a second nonrolling coordinate system using the *l–n–m* system depicted in the same figure. In this case, the coordinate system is oriented along the velocity vector. The coordinate systems are related to one another, assuming small yaw angles, through the relationships

$$\mathbf{i} = \gamma\mathbf{l} + \alpha\mathbf{n} + \beta\mathbf{m} \tag{12.18}$$

$$\mathbf{j} = -\alpha\mathbf{l} + \mathbf{n} \tag{12.19}$$

$$\mathbf{k} = -\beta\mathbf{l} + \mathbf{m} \tag{12.20}$$

where α is the pitch angle, β is the yaw angle, and γ is defined as

$$\gamma = \cos\alpha\cos\beta \approx 1 \tag{12.21}$$

The angular momentum of the projectile is the vector sum of all the angular momenta and is closely approximated by

$$\mathbf{H} = I_P p\mathbf{i} + I_T\left(\mathbf{i}\times\frac{\mathrm{d}\mathbf{i}}{\mathrm{d}t}\right) - m_E\left(\mathbf{r}_E \times \mathbf{v}_E\right) \tag{12.22}$$

Here $\mathbf{H}$ is the total angular momentum and $\mathbf{v}_E$ is the velocity of the removed mass.

This velocity can be broken into two components, one owing to the rotation about the spin axis and the other owing to the yawing motion of the projectile as follows:

$$\mathbf{v}_E = p(\mathbf{i}\times\mathbf{r}_E) + \left[\left(\mathbf{i}\times\frac{\mathrm{d}\mathbf{i}}{\mathrm{d}t}\right)\times\mathbf{r}_E\right] \tag{12.23}$$

Then, inserting this relationship into Equation 12.22 and combining terms gives us, after the utilization of the vector triple product,

$$\begin{aligned}\mathbf{H} &= \left(I_P - m_E r_E^2\right)p\mathbf{i} + \mathbf{I}_T\left(\mathbf{i}\times\frac{\mathrm{d}\mathbf{i}}{\mathrm{d}t}\right) - m_E \\ &\times\left[-p(\mathbf{r}_E\cdot\mathbf{i})\mathbf{r}_E - \left(\mathbf{r}_E\times\frac{\mathrm{d}\mathbf{i}}{\mathrm{d}t}\right)(\mathbf{r}_E\times\mathbf{i}) + (\mathbf{r}_E\cdot\mathbf{i})\left(\mathbf{r}_E\times\frac{\mathrm{d}\mathbf{i}}{\mathrm{d}t}\right)\right]\end{aligned} \tag{12.24}$$

If we examine Figure 12.5, we see that

$$(\mathbf{r}_E\cdot\mathbf{i}) = l_E \tag{12.25}$$

And we note that for a spin-stabilized projectile, the yaw rate $\mathrm{d}\mathbf{i}/\mathrm{d}t$ is much smaller than the spin rate p, so we can eliminate terms in Equation 12.24 to yield

$$\mathbf{H} \approx I_P p\mathbf{i} + I_T\left(\mathbf{i}\times\frac{\mathrm{d}\mathbf{i}}{\mathrm{d}t}\right) + m_E p l_E \mathbf{r}_E \tag{12.26}$$

We can express the mass asymmetry vector $\mathbf{r}_E$ in terms of the projectile geometric axes as

$$\mathbf{r}_E = l_E\mathbf{i} + \mathbf{r}_E\cos\Phi\mathbf{j} + r_E\sin\Phi\mathbf{k} \tag{12.27}$$

This can be expressed in our coordinate system attached to the velocity vector through the relationships in Equations 12.18 through 12.20 as

$$\mathbf{r}_E = (l_E\gamma - r_E\alpha\cos\Phi - r_E\beta\sin\Phi)\mathbf{l} + (l_E\alpha + r_E\cos\Phi)\mathbf{n} + (l_E\beta + r_E\sin\Phi)\mathbf{m} \tag{12.28}$$

We can simplify this expression somewhat if we use the fact that both α and β are much smaller than γ. In this case, the expression would simplify to

$$\mathbf{r}_E = (l_E\gamma)\mathbf{l} + (l_E\alpha + r_E\cos\Phi)\mathbf{n} + (l_E\beta + r_E\sin\Phi)\mathbf{m} \tag{12.29}$$

We can take the derivative of Equation 12.29 by using the fact that the coordinate system is effectively not rotating to write

$$\frac{\mathrm{d}\mathbf{r}_E}{\mathrm{d}t} = (l_E\dot{\gamma})\mathbf{l} + (l_E\dot{\alpha} - r_E p\sin\Phi)\mathbf{n} + \left(l_E\dot{\beta} + r_E p\cos\Phi\right)\mathbf{m} \tag{12.30}$$

Here we have used the fact that

$$\frac{\mathrm{d}\Phi}{\mathrm{d}t} = p \tag{12.31}$$

As in our previous analyses, we shall consider a short period of flight. By doing this, we can neglect all forces and moments except the pitching (overturning) moment. This allows us to equate the rate of change of angular momentum to the applied pitching moment:

$$\frac{\mathrm{d}\mathbf{H}}{\mathrm{d}t} \approx I_{\mathrm{P}} p \frac{\mathrm{d}\mathbf{i}}{\mathrm{d}t} + I_{\mathrm{T}}\left(i \times \frac{\mathrm{d}^2\mathbf{i}}{\mathrm{d}t^2}\right) + m_{\mathrm{E}} p l_{\mathrm{E}} \frac{\mathrm{d}\mathbf{r}_{\mathrm{E}}}{\mathrm{d}t} = mC^*_{\mathrm{M}_\alpha} V^2 (\mathbf{1} \times \mathbf{i}) \tag{12.32}$$

This expression can be written as a set of three equations in terms of each component as follows:

$$I_P p\dot{\gamma} + I_T\left(\alpha \frac{\mathrm{d}^2\beta}{\mathrm{d}t^2} - \beta \frac{\mathrm{d}^2\alpha}{\mathrm{d}t^2}\right) + m_E p l_E^2 \dot{\gamma} = 0 \tag{12.33}$$

$$I_{\mathrm{P}} p\dot{\alpha} + I_{\mathrm{T}}\left(\beta \frac{\mathrm{d}^2\gamma}{\mathrm{d}t^2} - \gamma \frac{\mathrm{d}^2\beta}{\mathrm{d}t^2}\right) + m_{\mathrm{E}} p l_{\mathrm{E}}\left(l_{\mathrm{E}}\dot{\alpha} - r_{\mathrm{E}} p \sin\Phi\right) = -mC^*_{\mathrm{M}_\alpha} V^2 \beta \tag{12.34}$$

$$I_{\mathrm{P}} p\dot{\beta} + I_{\mathrm{T}}\left(\gamma \frac{\mathrm{d}^2\alpha}{\mathrm{d}t^2} - \alpha \frac{\mathrm{d}^2\gamma}{\mathrm{d}t^2}\right) + m_{\mathrm{E}} p l_{\mathrm{E}}\left(l_{\mathrm{E}}\dot{\beta} + r_{\mathrm{E}} p \cos\Phi\right) = mC^*_{\mathrm{M}_\alpha} V^2 \alpha \tag{12.35}$$

The details of this are provided by McCoy [2]. If we change the temporal derivatives into spatial derivatives along a dimensionless downrange distance s and define the following:

$$P = \left(\frac{I_{\mathrm{P}}}{I_{\mathrm{T}}}\right)\left(\frac{pd}{V}\right) \tag{12.36}$$

$$M = \frac{md^2}{I_{\mathrm{T}}} C^*_{\mathrm{M}_\alpha} \tag{12.37}$$

$$I_{\mathrm{E}} = m_{\mathrm{E}} r_{\mathrm{E}} l_{\mathrm{E}} \tag{12.38}$$

we can rewrite Equations 12.33 through 12.35 as

$$P\left(1 + \frac{m_{\mathrm{E}} l_{\mathrm{E}}^2}{I_{\mathrm{T}}}\right)\gamma' + \alpha\beta'' - \beta\alpha'' = 0 \tag{12.39}$$

$$P\left(1 + \frac{m_{\mathrm{E}} l_{\mathrm{E}}^2}{I_{\mathrm{T}}}\right)\alpha' + \beta\gamma'' - \gamma\beta'' + M\beta - \left(\frac{I_{\mathrm{E}} I_{\mathrm{T}}}{I_{\mathrm{P}}^2}\right)P^2 \sin\Phi = 0 \tag{12.40}$$

$$P\left(1 + \frac{m_{\mathrm{E}} l_{\mathrm{E}}^2}{I_{\mathrm{T}}}\right)\beta' + \gamma\alpha'' - \alpha\gamma'' - M\alpha + \left(\frac{I_{\mathrm{E}} I_{\mathrm{T}}}{I_{\mathrm{P}}^2}\right)P^2 \cos\Phi = 0 \tag{12.41}$$

Here the primed quantities are differentiated with respect to s. With small yaw as well as classical size assumptions (see the book by McCoy [2] and the report by Murphy [3]), we can neglect several of these terms because they are either products of small numbers or summed with a much larger number. This results in Equation 12.39 vanishing altogether and the other two transforming into

$$P\alpha' - \beta'' + M\beta = \left(\frac{I_E I_T}{I_P^2}\right) P^2 \sin\Phi \tag{12.42}$$

$$P\beta' + \alpha'' - M\alpha = -\left(\frac{I_E I_T}{I_P^2}\right) P^2 \cos\Phi \tag{12.43}$$

If we now multiply Equation 12.42 by $-i$ and add it to Equation 12.43, we obtain

$$(\alpha'' + i\beta'') + P(\beta' - i\alpha') - M(\alpha + i\beta) = -\left(\frac{I_E I_T}{I_P^2}\right) P^2 (\cos\Phi + i\sin\Phi) \tag{12.44}$$

If we invoke our definition of complex yaw angle, we can write this as

$$\xi'' - iP\xi - M\xi = \left(\frac{I_E I_T}{I_P^2}\right) P^2 \exp(i\Phi) \tag{12.45}$$

The solution to this differential equation was discussed in Section 10.1. The difference here is that the forcing term on the right-hand side is somewhat different. If we use a solution written as

$$\xi = K_1 \exp(i\psi_1) + K_2 \exp(i\psi_2) + K_4 \exp(i\Phi) \tag{12.46}$$

where K_1 and K_2 are the solutions to the homogeneous part of the equation and K_4 is our new term which depends on the spin rate and the mass asymmetry we can solve for the magnitude of the trim arm caused by the asymmetry. If we solve Equation 12.46 for the particular solution, we find that this new trim arm caused by the mass asymmetry is given by

$$K_4 = \frac{I_E}{I_T - I_P + \left(\dfrac{I_P^2 M}{I_T P}\right)} \tag{12.47}$$

The third term in the denominator is usually very small so this term has been approximated (see the book by McCoy [2] and the report by Murphy [4]).

$$K_4 \approx \frac{I_E}{I_T - I_P} \tag{12.48}$$

This trim arm due to a mass asymmetry is usually small.

Throughout this development, I_P and I_T have been used as the moments of inertia even though, in the purest sense, the mass asymmetry removes the axially symmetric properties of the projectile. For most cases, it is sufficient to use these quantities based on an axially symmetric projectile.

References

1. Greenwood, D. T., *Principles of Dynamics*, 2nd ed., Prentice Hall, Englewood Cliffs, NJ, 1988.
2. McCoy, R. L., *Modern Exterior Ballistics*, Schiffer Military History, Atglen, PA, 1999.
3. Murphy, C. H., *Free Flight Motion of Symmetric Missiles*, Ballistic Research Laboratory Report No. 1216, Ballistic Research Laboratory, Aberdeen Proving Ground, MD, 1963.
4. Murphy, C. H., *Yaw Induction by Means of Asymmetric Mass Distributions*, Ballistic Research Laboratory Memorandum Report No. 2669, Ballistic Research Laboratory, Aberdeen Proving Ground, MD, 1976.

13

Swerve Motion

Following our procedure of slowly introducing complexity into the description of projectile behavior, we shall now develop equations to characterize the remainder of what is known in general as swerve motion. We saw in Chapter 12 that a mass asymmetry can cause projectile motion transverse to the original line of fire even in a vacuum. We stated in that section that a dynamic projectile imbalance was more common than a static imbalance but either can actually occur.

Chapter 6 explains many aspects of projectile behavior that arise due to the presence of the air stream. All the coefficients were functions of the angle of the attack observed by the projectile relative to that air stream. If we examine how a statically or dynamically imbalanced projectile would behave as viewed from above the trajectory curve based on its spin, we would see motion as depicted in Figures 13.1 and 13.2. We must keep in mind that the motion in these figures is greatly exaggerated for ease of viewing.

We can imagine, by looking at these figures, that the aerodynamic forces would be considerable because even in the case of the statically imbalanced projectile, motion laterally across the trajectory will manifest itself in an angle of attack and therefore affect the flight characteristics.

In this chapter, we shall describe and evaluate the aerodynamic forces that arise from this behavior and include them in our equations of motion for projectile flight. We shall also include the effect of configurational asymmetries such as bent fins or damaged form, because these will result in similar behavior even without the mass asymmetry present. In fact, to a varying degree, every projectile has a combination of both form and mass asymmetries present.

13.1 Aerodynamic Jump

McCoy [1] showed that the equation of motion for the point mass solution plus swerving motion is given by

$$\frac{d^2y}{ds^2} + i\frac{d^2z}{ds^2} = C_{L_\alpha}^{*}\xi - \frac{gd}{V_0^2}\exp(2C_D^{*}s) \tag{13.1}$$

with a solution of

$$y + iz = (y_0 + iz_0) + \left(\left.\frac{dy}{ds}\right|_0 + i\left.\frac{dz}{ds}\right|_0\right)s + C_{L_\alpha}^{*} I_L - \frac{s^2 gd}{2V_0^2}\left[\frac{\exp(2C_D^{*}s) - 2C_D^{*}s - 1}{2(C_D^{*}s)^2}\right] \tag{13.2}$$

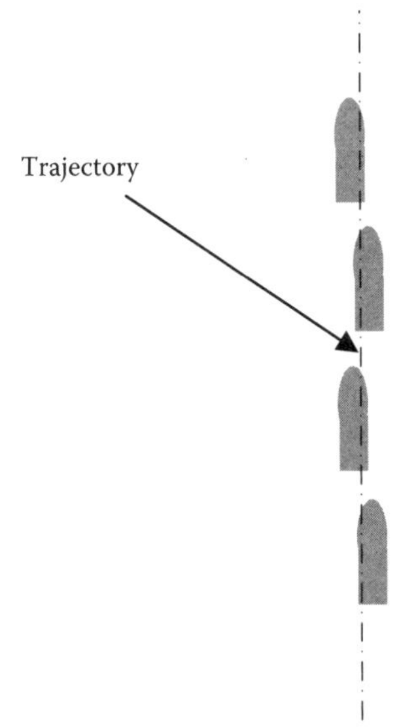

FIGURE 13.1
Motion of a statically imbalanced projectile.

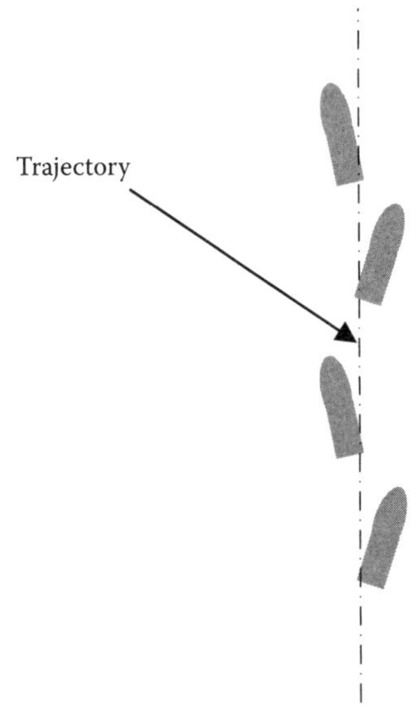

FIGURE 13.2
Motion of a dynamically imbalanced projectile.

where

$$I_L = \int_0^s \int_0^{s_1} \xi \, ds_1 \, ds_2 \tag{13.3}$$

Here we have used s_1 and s_2 as dummy variables representing integrations with respect to s. Equation 13.2 describes the position of the projectile in a direction perpendicular to the trajectory based on flat fire point mass assumptions.

Equation 10.72 was developed as a solution for ξ. McCoy [1] showed that the solution to the double integral of Equation 13.3 can be obtained by substitution of Equation 10.72 into Equation 13.3, resulting in

$$\begin{aligned} I_L = &-\left[\left(\frac{\lambda_1 - i\psi_1'}{\lambda_1^2 + \psi_1'^2}\right) K_{10} \exp(i\psi_{10}) + \left(\frac{\lambda_2 - i\psi_1'}{\lambda_2^2 + \psi_2'^2}\right) K_{20} \exp(i\psi_{20})\right] s \\ &+ (R_{11} - iR_{12}) K_{10} \exp(i\psi_{10}) \left\{\exp\left[(\lambda_1 - i\psi_1')s\right] - 1\right\} \\ &+ (R_{21} - iR_{22}) K_{20} \exp(i\psi_{20}) \left\{\exp\left[(\lambda_2 - i\psi_2')s\right] - 1\right\} + i\frac{PG_0}{M} s^2 \left[\frac{\exp(2C_{D}^{*}s) - 2C_{D}^{*}s - 1}{(2C_{D}^{*}s)^2}\right] \end{aligned} \tag{13.4}$$

Here we have used

$$G_0 = \frac{gd}{V_0^2} \tag{13.5}$$

$$R_{11} = \frac{\lambda_1^2 - \psi_1'^2}{(\lambda_1^2 + \psi_1'^2)^2} \tag{13.6}$$

$$R_{12} = \frac{2\lambda_1 \psi_1'}{(\lambda_1^2 + \psi_1'^2)^2} \tag{13.7}$$

$$R_{21} = \frac{\lambda_2^2 - \psi_2'^2}{(\lambda_2^2 + \psi_2'^2)^2} \tag{13.8}$$

$$R_{22} = \frac{2\lambda_2 \psi_2'}{(\lambda_2^2 + \psi_2'^2)^2} \tag{13.9}$$

If we make the assumption that $\lambda_{1,2}^2 \ll \psi_{1,2}'^2$, the aforementioned parameters become

$$R_{11} \approx -\frac{1}{\psi_1'^2}, \quad R_{12} \approx 0, \quad R_{21} \approx -\frac{1}{\psi_2'^2}, \quad \text{and} \quad R_{22} \approx 0 \tag{13.10}$$

Inserting these assumptions into Equation 13.4 yields the following result:

$$I_L = i\left(\frac{1}{\psi_1'}K_{10}\exp(i\psi_{10}) + \frac{1}{\psi_2'}K_{20}\exp(i\psi_{20})\right)s - \frac{1}{\psi_1'^2}K_{10}\exp(i\psi_{10})\left\{\exp\left[(\lambda_1 + i\psi_1')s\right] - 1\right\}$$
$$- \frac{1}{\psi_2'^2}K_{20}\exp(i\psi_{20})\left\{\exp\left[(\lambda_2 + i\psi_2')s\right] - 1\right\} + i\frac{PG_0}{M}s^2\left[\frac{\exp(2C_D^* s) - 2C_D^* s - 1}{(2C_D^* s)^2}\right] \tag{13.11}$$

This result is important because it depicts the three components of swerve motion. The first term on the right-hand side is called the aerodynamic jump J_A, and it is what we will examine for the remainder of this section. The second two terms are the epicyclic swerve S_E and will be discussed in Section 13.2. The third term is called drift D_R and will be discussed in Section 13.3. To keep things simple, we will restate the aerodynamic jump as

$$J_A = iC_{L_\alpha}^*\left(\frac{1}{\psi_1'}K_{10}\exp(i\psi_{10}) + \frac{1}{\psi_2'}K_{20}\exp(i\psi_{20})\right) \tag{13.12}$$

We should note a few things about Equation 13.12. First, we must keep in mind that in Equation 13.11, this aerodynamic jump term is multiplied by a downrange distance s, implying that it is actually an angular measure (for small angles). A second observation is that the aerodynamic jump is completely dependent upon the initial conditions of the projectile and how these couple in with the fast and slow arm turning rates.

If we insert our approximated initial fast and slow arm amplitudes from Equations 10.78 and 10.79 into Equation 13.12, we obtain

$$J_A = iC_{L_\alpha}^*\left[\frac{-i\xi_0' - \psi_2'\xi_0'}{\psi_1'(\psi_1' - \psi_2')} + \frac{i\xi_0' + \psi_1'\xi_0'}{\psi_2'(\psi_1' - \psi_2')}\right] \tag{13.13}$$

This can be rewritten [1] as

$$J_A = k_T^2\left(\frac{C_{L_\alpha}}{C_{M_\alpha}}\right)(iP\xi_0 - \xi_0') \tag{13.14}$$

This result shows that by knowing the projectile mass properties and launch conditions, we can determine to what angle a projectile will "jump." We can envision this jump effect as shown in Figure 13.3.

While we have said a great deal mathematically about aerodynamic jump, we have not really described the physics behind it. Because of the presence of aerodynamic lift on the projectile, there is a strong influence of angle of attack on the resultant motion. We saw earlier that a projectile, through purely dynamic means, can yaw because of either spin or some geometric asymmetry. When this happens, the aerodynamic forces change, either improving or worsening the situation. This interaction of the aerodynamic forces with the projectile manifests itself in the jump angle as depicted in Figure 13.3.

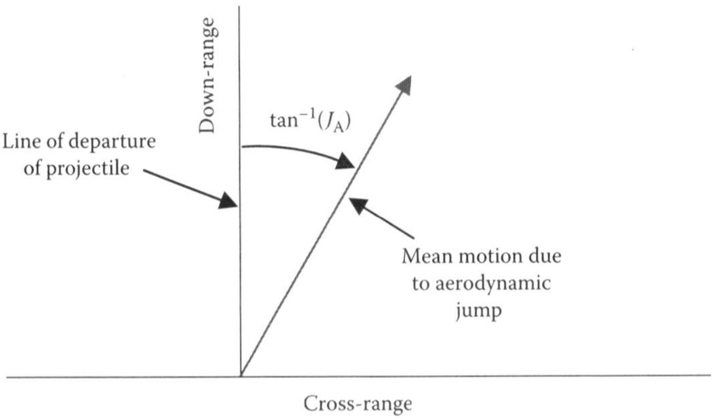

FIGURE 13.3
Graphical representation of aerodynamic jump.

13.2 Epicyclic Swerve

The second two terms in Equation 13.11 describe the epicyclic swerve of a projectile. We can define this parameter specifically [1] as

$$S_E = -C^*_{L_\alpha}\left(\frac{1}{\psi_1'^2}K_{20}\exp(i\psi_{20})\left\{\exp\left[\left(\lambda_1 + i\psi_1'\right)s\right]-1\right\}\right.$$
$$\left.+\frac{1}{\psi_2'^2}K_{20}\exp(i\psi_{20})\left\{\exp\left[\left(\lambda_2 + i\psi_2'\right)s\right]-1\right\}\right) \tag{13.15}$$

McCoy [1] showed that this equation can be put into a more useful form through the use of the relation

$$C^*_{L_\alpha} = k_T^2\left(\frac{C_{L_\alpha}}{C_{M_\alpha}}\right)\psi_1'\psi_2' \tag{13.16}$$

If we insert Equation 13.16 into Equation 13.15, we obtain

$$S_E = -k_T^2\left(\frac{C_{L_\alpha}}{C_{M_\alpha}}\right)\left(\frac{\psi_2'}{\psi_1'}K_{10}\exp(i\psi_{10})\left\{\exp\left[\left(\lambda_1 + i\psi_1'\right)s\right]-1\right\}\right.$$
$$\left.+\frac{\psi_1'}{\psi_2'}K_{20}\exp(i\psi_{20})\left\{\exp\left[\left(\lambda_2 + i\psi_2'\right)s\right]-1\right\}\right) \tag{13.17}$$

Following McCoy, we shall examine two special cases of this equation. The first is where we have a projectile that is nonspinning (statically stable) and the second is a spin-stabilized projectile with a good gyroscopic stability (measured at muzzle exit) of at least 1.5.

For the nonspinning projectile, the following conditions apply:

$$M < 0, \quad P = 0, \quad \text{and} \quad \psi_2' = -\psi_1' \tag{13.18}$$

If we insert these conditions into Equation 13.17, we get

$$S_{E_{\text{non-spin}}} = k_T^2\left(\frac{C_{L_\alpha}}{C_{M_\alpha}}\right)\Big(K_{10}\exp(i\psi_{10})\left\{\exp\left[(\lambda_1 + i\psi_1')s\right]-1\right\} + K_{20}\exp(i\psi_{20})\left\{\exp\left[(\lambda_2 + i\psi_2')s\right]-1\right\}\Big) \tag{13.19}$$

Now we can invoke the fact that the spin is equal to zero and insert Equation 10.59, in which we shall neglect the trim and yaw of repose, into Equation 13.19 to yield

$$S_{E_{\text{non-spin}}} = k_T^2\left(\frac{C_{L_\alpha}}{C_{M_\alpha}}\right)(\xi - \xi_0) \tag{13.20}$$

This relationship will produce a motion in exactly the same manner as the aerodynamic jump developed in Section 13.1. It essentially couples the yawing motion of the projectile to the swerving motion. Both will thus damp together, and the more yaw, the greater the epicyclic swerve.

If we examine the spin-stabilized projectile, we can write

$$M > 0 \quad \text{and} \quad \psi_1'^2 \gg \psi_2'^2 \tag{13.21}$$

With the aforementioned mathematical statements, McCoy [1] stated that an excellent approximation of Equation 13.15 for a spinning projectile is

$$S_{E_{\text{spin}}} = -k_T^2\left(\frac{C_{L_\alpha}}{C_{M_\alpha}}\right)\left(\frac{\psi_1'}{\psi_2'}\right)\Big(K_{20}\exp(i\psi_{20})\left\{\exp\left[(\lambda_2 + i\psi_2')s\right]-1\right\}\Big) \tag{13.22}$$

An interesting comparison may be drawn between the epicyclic swerving behavior of a spinning projectile and a nonspinning projectile. If we compare Equations 13.22 and 13.20, we see that in the latter, the yawing motion and swerve are locked together and operate on the same fixed plane. This is because the lift generated by the motion never rotates. In a spin-stabilized projectile, the lift vector is always rotating; thus, the center of mass of the projectile will move in a helical manner around the flight path. Furthermore, the motion will be locked to the rate of turning of the slow arm and will damp or increase as the slow arm does.

13.3 Drift

The last term in Equation 13.11 describes the drift of a projectile. We can define this parameter specifically [1] as

$$D_R = i\frac{PG_0}{M}s^2\left[\frac{\exp(2C_D^*s) - 2C_D^*s - 1}{(2C_D^*s)^2}\right] \tag{13.23}$$

If we expand the term in brackets in a power series, we can rewrite this equation as

$$D_R = i\left(\frac{PG_0}{M}\right)s^2\left[1+\frac{2}{3}(C_D^*s)+\frac{1}{3}(C_D^*s)^2+\cdots\right] \tag{13.24}$$

The examination of the drift equation in this form has some advantages. First, we can see that if a projectile has no spin, $P = 0$ and there is no drift. If we look at a fin- or drag-stabilized projectile where $M < 0$, we see that the projectile will drift in the direction opposite to the spin. That is, a left-hand spin will produce a right-hand drift and vice versa. In a statically unstable (spin-stabilized) projectile where $M > 0$, we see that the projectile will drift in the same direction as the spin. It must be noted that this drift is very small compared to the other swerve components as well as Coriolis drift. In fact, to even measure it, some researchers [1] have fired two projectiles simultaneously out of side-by-side gun barrels with both left- and right-hand twists to remove Coriolis and wind drift components, which would equally affect each.

The interested reader should consult the book by McCoy [1] for further information on this topic.

Reference

1. McCoy, R. L., *Modern Exterior Ballistics*, Schiffer Military History, Atglen, PA, 1999.

14

Nonlinear Aeroballistics

Until this point, we have concerned ourselves with linear behavior of the aerodynamic coefficients only. This is very convenient for direct fire projectiles and projectiles that fly with very little yaw. It had the benefit of allowing us to make a black or white decision with regard to projectile stability as well—the projectile was either stable or not. In real systems, several of the coefficients are only linear over a small range of angles of attack. This can be either helpful or hurtful to a particular design.

Limit-cycle motion is motion that develops over time in a projectile, whereby the angle of attack of projectiles grows until a certain (sometimes rather large) angle is achieved. As the angle of attack increases (or some other parameter such as the air density changes), the coefficients change so that the projectile will actually become stable at some large angle of attack. At first, this may seem like it is a desirable quality in a projectile; however, range is sacrificed due to the larger drag generally associated with this large yaw. Some systems have been unwittingly fielded in this condition, and it was only after a large number of firings in the field that this was determined to be an issue.

This nonlinear behavior arises out of the interaction between the air and the surfaces of the projectile. It is a rather complicated mechanism that can arise (many times in a discontinuous manner) from boundary layer separation, fin masking, vortex shedding, etc. All of which are fluid dynamic phenomenon. This is and continues to be a challenging area of aeroballistic research, where experimental, theoretical, and computational techniques are pushed to the limit of their usefulness.

The next two sections will look at this behavior to some degree of detail; however, because of space constraints, the reader is encouraged to consult the literature for more detailed mathematical and theoretical treatment.

14.1 Nonlinear Forces and Moments

In general, we can divide nonlinear forces and moments into two categories: geometric and aerodynamic nonlinearities. The geometric nonlinearities arise from the cosine terms in the equations of motion that were eliminated when we assumed a small yaw angle. This small angle assumption is generally valid for most projectiles in flight. If a projectile is flying with a large yaw, the cosine terms must be retained and the resulting equations are more difficult to solve. Since this behavior is usually designed out of projectiles, we shall focus on the second type of nonlinearity, the aerodynamic nonlinearity.

The aerodynamic nonlinearity can exist even at angles of attack that are consistent with the small yaw assumption. They arise due to the fluid–mechanic interaction of the air with the solid projectile body. This interaction can consist of phenomena such as vortex shedding, separation, and shock interactions.

The most dominant force acting on the body is the drag force. In all our previous discussions, we have stated that the forces that arise due to other sources are small, and that is still true for the case of nonlinearities; however, the moments caused by these other forces cannot be neglected. We can define a nonlinear drag coefficient as

$$C_D = C_{D_0} + C_{D_{\delta^2}} \delta^2 + \cdots \tag{14.1}$$

In this equation, the first term on the right-hand side is the zero-yaw–drag coefficient and the second term is the yaw–drag coefficient. More coefficients can be added, but typically the expression is truncated at the yaw–drag term.

There are essentially two common ways of determining the yaw–drag coefficient: experimentally or computationally. Experimental evaluation is more common, although recent advances in computational fluid dynamics (CFD) [1] have shown that it is possible to extract coefficients directly from analyses. In either case, the overall drag coefficient at multiple angles of attack is determined from either a direct force measurement (in the case of a wind tunnel or CFD model) or the velocity decay (in a free flight test), and the results are plotted as C_D vs. angle of attack. The slope of the resulting line (hopefully, it is a line) is then the yaw–drag coefficient, and the y-intercept is the zero-yaw–drag coefficient. In the case of a free flight firing where the projectile is dragging down continuously, Murphy [2] and McCoy [3] suggested an averaging scheme that has been successfully demonstrated based on a great deal of experience.

The aforementioned technique is known as a quasi-linear approach because it defines a linear function that is a solution to a nonlinear equation. The same approach is used to determine the nonlinear moments, which are generally assumed to have the same form as Equation 14.1.

In general, both the zero-yaw–drag coefficient and the yaw–drag coefficient are positive values. In the case of the pitching or overturning moment of a spin-stabilized projectile, the zero-yaw overturning moment coefficient is positive while the cubic overturning moment coefficient is negative [3]. This condition can have some interesting effects on stability as summarized by McCoy [3].

The overall equation of motion that includes all the nonlinear terms that is equivalent to our linear equation (Equation 10.76) with the gravitational term neglected is

$$\xi'' + \left(H_0 + H_2\delta^2 - iP\right)\xi' - \left[M_0 + M_2\delta^2 + iP\left(T_0 + T_2\delta^2\right)\right]\xi = 0 \tag{14.2}$$

We can define our coefficients as follows:

$$H_0 = \frac{\rho S d}{2m}\left[C_{L_{\alpha_0}} - C_{D_0} - \frac{1}{k_T^2}(C_{M_q} + C_{M_{\dot{\alpha}}})_0\right] \tag{14.3}$$

$$H_2 = \frac{\rho S d}{2m}\left[C_{L_{\alpha_2}} - C_{D_2} - \frac{1}{k_T^2}(C_{M_q} + C_{M_{\dot{\alpha}}})_2\right] \tag{14.4}$$

$$M_0 = \frac{\rho S d}{2m}\frac{1}{k_T^2} C_{M_{\alpha_0}} \tag{14.5}$$

$$M_2 = \frac{\rho S d}{2m}\frac{1}{k_T^2} C_{M_{\alpha_2}} \tag{14.6}$$

$$T_0 = \frac{\rho S d}{2m}\left(C_{L_{\alpha_0}} + \frac{1}{k_P^2} C_{M_{P\alpha_0}}\right) \tag{14.7}$$

$$T_2 = \frac{\rho S d}{2m}\left(C_{L_{\alpha_2}} + \frac{1}{k_P^2} C_{M_{P\alpha_2}}\right) \tag{14.8}$$

$$P = \left(\frac{I_P}{I_T}\right)\left(\frac{pd}{V}\right) \tag{14.9}$$

The solution to Equation 14.2 is

$$\xi = K_{10}\exp[\lambda_1 s]\exp\left[i\left(\psi_{10} + \psi_1' s\right)\right] + K_{20}\exp[\lambda_2 s]\exp\left[i\left(\psi_{20} + \psi_2' s\right)\right] \tag{14.10}$$

where again, we are reminded that the gravitational term has been neglected. McCoy [3] has written expressions for the coefficients in terms of the damping exponents and turning rates as follows:

$$\delta_{e1}^2 = K_1^2 + 2K_2^2 \tag{14.11}$$

$$\delta_{e2}^2 = K_2^2 + 2K_1^2 \tag{14.12}$$

$$\psi_1' + \psi_2' = P + M_2\left(\frac{K_1^2 + K_2^2}{\psi_1' - \psi_2'}\right) \approx P \tag{14.13}$$

$$\delta_e^2 = \frac{\psi_1'\delta_{e2}^2 - \psi_2'\delta_{e1}^2}{\psi_1' - \psi_2'} \tag{14.14}$$

$$\psi_1'\psi_2' = M_0 + M_2\delta_e^2 \tag{14.15}$$

$$\lambda_1 = \frac{-H_0\psi_1' + P\left(T_0 - T_2\delta_{e1}^2\right) - H_2\left[\psi_1'\left(K_1^2 + K_2^2\right) + \psi_1' K_2^2\right]}{\psi_1' - \psi_2'} \tag{14.16}$$

$$\lambda_2 = \frac{-H_0\psi_2' + P\left(T_0 - T_2\delta_{e2}^2\right) - H_2\left[\psi_2'\left(K_1^2 + K_2^2\right) + \psi_1' K_1^2\right]}{\psi_1' - \psi_2'} \tag{14.17}$$

In terms of some of these parameters, McCoy [3] derived a form for the nonlinear lift coefficient as

$$C_{L_\alpha} = C_{L_{\alpha_0}} + C_{L_{\alpha_2}}\left(\frac{\psi_2'^2\delta_{e1}^2 K_1 e^{i\psi_1} + \psi_1'^2\delta_{e2}^2 K_2 e^{i\psi_2}}{\psi_2'^2 K_1 e^{i\psi_1} + \psi_1'^2 K_2 e^{i\psi_2}}\right) \tag{14.18}$$

As is readily apparent, these expressions are significantly more complex than their linear cousins. Because of this, they are generally solved using numerical schemes. The interested reader is referred to the studies by Murphy [4–7], Platus [8], Tobak et al. [9], Seginer and Rosenwasser [10], Platou [11], and Cohen et al. [12] for a more detailed treatment as well as examples of this behavior.

Problem 1

If the projectile in Problem 1 of Chapter 10 happens to be flying at a limit-cycle yaw of 4° with a spin rate of 130 Hz and a velocity of 1764 ft/s, what would the nonlinear pitching moment have to be for the projectile to be marginally stable?

Hints:

1. Assume all the other coefficients are linear.
2. Recall the definition of the nonlinear pitching moment (you have the linear part in Problem 1 of Chapter 10).
   Answer: $C_{M_{\alpha 2}} = -119.079$

14.2 Bilinear and Trilinear Moments

We have discussed nonlinear forces and moments and their implications in the previous section. At this point, we shall turn our attention to nonlinear moments in which the cubic behavior itself can be described by a bilinear or trilinear curve. This is evident when the cubic coefficient is plotted vs. yaw angle. A bilinear coefficient would have two different linear slopes, while a trilinear moment would have three. This is quite useful since many experimental data can be fitted using these curves. In particular, we shall examine the Magnus moment and its implications because this is the dominant moment in spin-stabilized projectile flight behavior [3].

If we are examining projectile flight data, it is often tempting to fit a higher-order polynomial curve to deal with the nonlinearity. This is usually not advisable since the abrupt changes in behavior at certain angles of attack are caused by fluid–solid interactions such as boundary layer separation and vortex shedding.

To describe the behavior of projectiles with nonlinear Magnus moment coefficients, we shall use two examples: one with a linear cubic Magnus moment and one with a bilinear Magnus moment. We are interested in two things: first, the effect of initial conditions on projectile stability and, second, limit-cycle motion.

In the excellent treatment by McCoy [3], for illustrative purposes, the author suggested assuming a linear pitch-damping moment with a cubic Magnus moment coefficient. This will force H_2 to be zero and allow Equations 14.16 and 14.17 to be written as

$$\lambda_1 = \frac{-H_0\psi_1' + P\left(T_0 + T_2\delta_{e1}^2\right)}{\psi_1' - \psi_2'} \tag{14.19}$$

$$\lambda_2 = \frac{-H_0\psi_2' + P\left(T_0 + T_2\delta_{e2}^2\right)}{\psi_1' - \psi_2'} \tag{14.20}$$

We can put these equations into the form

$$\lambda_1 = \lambda_{1_0} + \lambda_{1_2}\delta_{e1}^2 \tag{14.21}$$

$$\lambda_2 = \lambda_{2_0} + \lambda_{2_2}\delta_{e2}^2 \tag{14.22}$$

where we can define

$$\lambda_{1_0} = \frac{-H_0\psi_1' + PT_0}{\psi_1' - \psi_2'} \tag{14.23}$$

$$\lambda_{2_0} = \frac{-H_0\psi_2' + PT_0}{\psi_1' - \psi_2'} \tag{14.24}$$

$$\lambda_{1_2} = -\lambda_{2_2} = \frac{PT_2}{\psi_1' - \psi_2'} \tag{14.25}$$

With these expressions, we can draw plots of damping coefficients vs. yaw angle in a manner similar to the coefficients.

At this juncture, we need to recall that these damping exponents will decrease the yaw of their particular mode if they are negative and increase the yaw if they are positive. Thus, negative values are stabilizing, and positive values are destabilizing. As a simple example, let us look at a projectile that has a linear cubic Magnus moment. In analyzing this projectile, we create two plots of damping coefficient vs. yaw. These are depicted as in Figures 14.1 and 14.2.

In Figure 14.1, we can see that the fast mode damping coefficient is negative for all yaw angles of interest (if the projectile is flying at an angle above 11°, we probably have a problem). Thus, the fast mode will always damp for this projectile. Examination of Figure 14.1 reveals that as long as the projectiles yaw angle is below 5.73°, the slow arm will damp to zero (recall that the yaw angle is equal to sin(δ)); above this angle, it will grow without bound. Although this angle is fairly large for a projectile, there have been instances documented where a slowly launched missile was stable when fired from one side of a fast

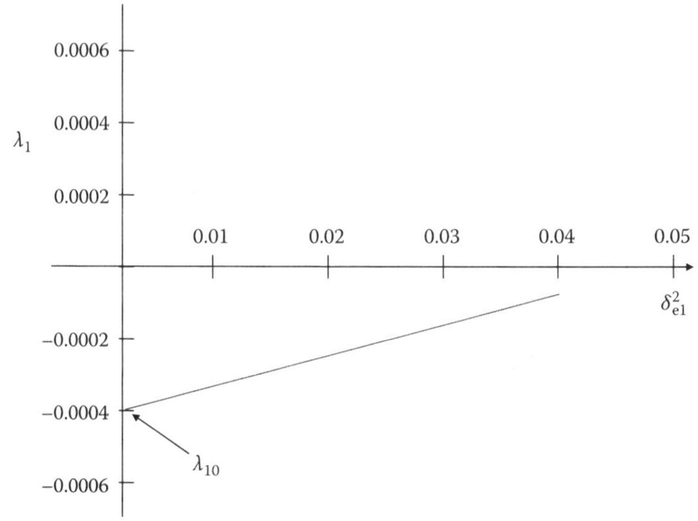

FIGURE 14.1
Plot of fast mode damping coefficient vs. yaw for linear cubic fast mode.

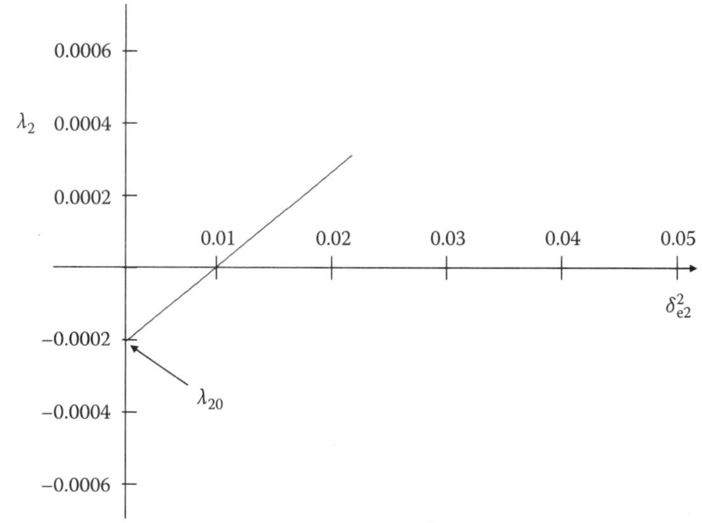

FIGURE 14.2
Plot of slow mode damping coefficient vs. yaw for linear cubic slow mode.

warship, but unstable when launched from the other [2,8]. The instability was caused by the vector addition of the ships own speed with the launch velocity.

Figures 14.3 and 14.4 show the fast and slow damping exponents for a projectile with bilinear cubic Magnus moment behavior. This is an interesting example because it illustrates how a projectile can enter into limit-cycle motion. Limit-cycle motion is motion in which the projectile cones in a predictable manner about the velocity vector.

If we examine Figure 14.3, we see that, similar to our earlier case, the fast arm damping coefficient is everywhere negative. Because of this, the fast mode will always damp to zero. The interesting part of the story is shown in Figure 14.4. Here, we see that for small angles,

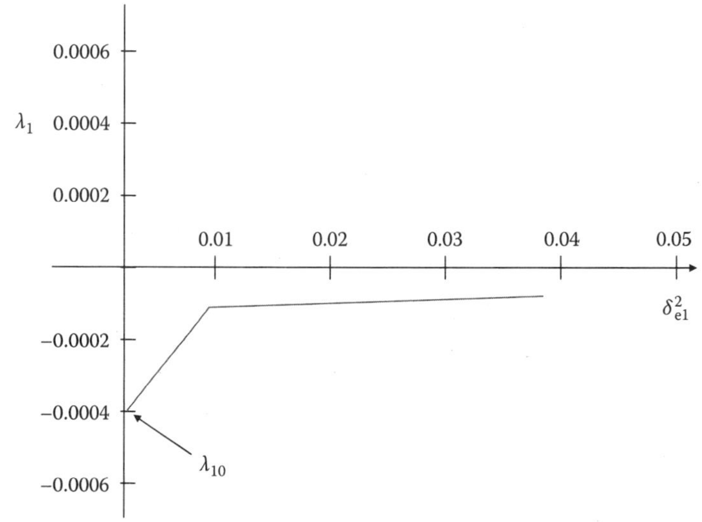

FIGURE 14.3
Plot of fast mode damping coefficient vs. yaw for bilinear cubic fast mode.

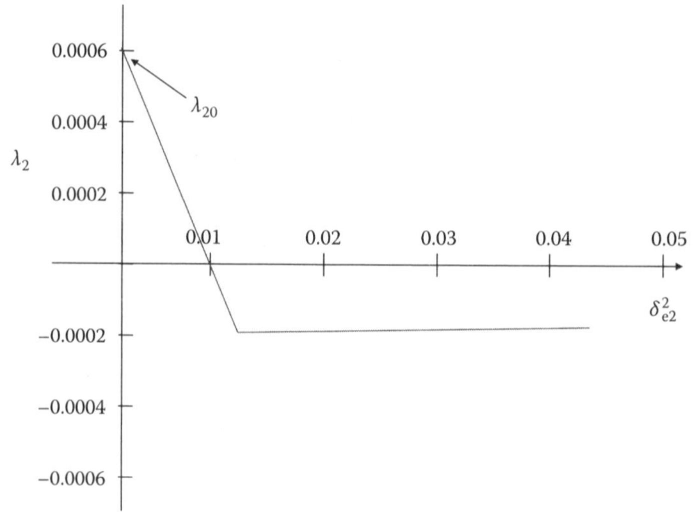

FIGURE 14.4
Plot of slow mode damping coefficient vs. yaw for bilinear cubic slow mode.

the projectiles slow arm will continue to grow because the damping exponent is positive. Once the amplitude of the motion grows beyond 5.74°, the sign of the coefficient changes driving the motion back to zero. However, the motion cannot be driven all the way back to zero because as soon as the angle decreases below 5.74°, the now positive damping coefficient will again cause it to increase. The result will be a projectile that cones about the velocity vector at a 5.74° angle.

These examples assumed that the velocity of the projectile has had no effect on the exponents. We must always keep in mind that there are many interrelated phenomena that affect these coefficients—the real world is a complicated place. This discussion should provide you with a feel for the physics of the projectile behavior.

References

1. DeSpirito, J., and Heavey, K. R., *CFD Computation of Magnus Moment and Roll Damping Moment of a Spinning Projectile*, AIAA Paper No. 2004-4713, American Institute of Aeronautics and Astronautics, New York, August 2004.
2. Murphy, C. H., *Free Flight Motion of Symmetric Missiles*, Ballistic Research Laboratory Report No. 1216, Ballistic Research Laboratory, Aberdeen Proving Ground, MD, 1963.
3. McCoy, R. L., *Modern Exterior Ballistics*, Schiffer Military History, Atglen, PA, 1999.
4. Murphy, C. H., *Data Reduction for the Free Flight Spark Ranges*, Report No. 900, US Ballistic Research Laboratory, Aberdeen Proving Ground, MD, February 1954.
5. Murphy, C. H., *Limit Cycles for Non-spinning Statically Stable Symmetric Missiles*, Report No. 1071, US Ballistic Research Laboratory, Aberdeen Proving Ground, MD, March 1959.
6. Murphy, C. H., *The Measurement of Non-linear Forces and Moments by Means of Free Flight Tests*, Report No. 974, US Ballistic Research Laboratory, Aberdeen Proving Ground, MD, February 1956.
7. Murphy, C. H., *Advances in the Dynamic Analysis of Range Data*, Memorandum Report No. 1270, US Ballistic Research Laboratory, Aberdeen Proving Ground, MD, May 1960.

8. Platus, D. H., *Dynamic Instability of Finned Missiles Caused by Unequal Effectiveness of Windward and Leeward Fins*, AIAA Paper No. 70–206, American Institute of Aeronautics and Astronautics, New York, January 1970.
9. Tobak, M., Schiff, L. B., and Peterson, V. L., Aerodynamics of bodies of revolution in coning motion, *AIAA Journal*, 7(1), January 1969, 95–99.
10. Seginer, A., and Rosenwasser, I., *Magnus Effect on Spinning Transonic Finned Missiles*, AIAA Paper No. 83–2146, American Institute of Aeronautics and Astronautics, New York, August 1983.
11. Platou, A. S., Magnus characteristics of finned and nonfinned projectiles, *AIAA Journal*, 3(1), January 1965, 83–90.
12. Cohen, C. J., Clare, T. A., and Stevens, F. L., *Analysis of the Non-linear Rolling Motion of Finned Missiles*, AIAA Paper No. 72-980, American Institute of Aeronautics and Astronautics, New York, September 1972.

Section III

Terminal Ballistics

15

Introductory Concepts

Terminal ballistics is the regime that the projectile enters at the conclusion of its flight. It has been delivered into its flight by the interior ballistician, pursued and guided through its flight by the exterior ballistician, and now, at its target, becomes the responsibility of the terminal ballistician. The basic objective of firing the projectile is to defeat some type of target, and we will study the widely varying phenomena of terminal effects that are the tools of the terminal ballistician. These end effects are dependent on the design and mission of the projectile. The most common of the missions are as follows: fragmentation of the projectile body by its cargo of high explosives; penetration or perforation of the target by the application of kinetic or chemical energy; blast at the target area delivered by the chemical energy of the explosive cargo; and the dispersal of the cargo for lethal or other missions, e.g., smoke, illumination, and propaganda dispersal.

Since most terminal ballistic phenomena involve the generation and effects of stress waves in solids, we will spend some time examining the details of this field. We must gain some knowledge of terminal ballistic terminology to be able to study the theories of kinetic energy penetration of solid targets; detonation, deflagration, and burning of energetic materials; the fundamentals of shaped charges; fragmentation theories; blast effects; and lethality with the study of wound ballistics.

We shall begin by introducing some concepts that we shall use throughout our study of this field.

In examination of penetration theories, we need to consider the following items: What constitutes defeat of the target? What is the source of the data for which we have to create a theory? Does the theory track with respect to momentum balance or energy balance? How many empirically derived constants are there in the model (this tells us how universal the theory will be)? What simplifications and assumptions were made?

Penetration is defined as an event during which a projectile creates a discontinuity in the original surface of the target. Perforation requires that after projectile or its remnants are removed, light may be seen through the target. Since penetration is a somewhat stochastic event, we need to define some statistical parameters. V_{10} is the velocity at which a given projectile will defeat a given target 10% of the time. V_{50} is the velocity at which a given projectile will defeat a given target 50% of the time, and V_{90} is the velocity at which a given projectile will defeat a given target 90% of the time. These quantities are depicted in Figure 15.1.

The 50% penetration velocity is commonly used as both an experimental measurement and a production check. The following procedure illustrates its usage in an experiment. The reader should refer to Figure 15.2 to illustrate the meaning. First, we should estimate V_{50} through a calculation. Once this is accomplished, we fire a projectile with a V_s as close to V_{50} as we can achieve. Let us say that the velocity of this experimental firing is a bit over our estimate (at 1 in Figure 15.2). Assuming shot 1 only partially penetrated, we considerably increase the velocity, and let us say that we achieve complete penetration at 2 in the figure. We now assume that V_{50} is midway between 1 and 2. We now would attempt to fire at the velocity halfway between 1 and 2 (at 3), and, say, we get complete penetration. We would

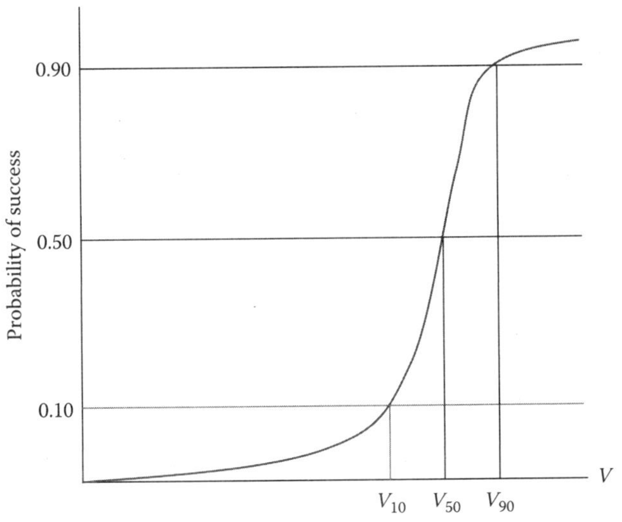

FIGURE 15.1
Statistical velocities defined.

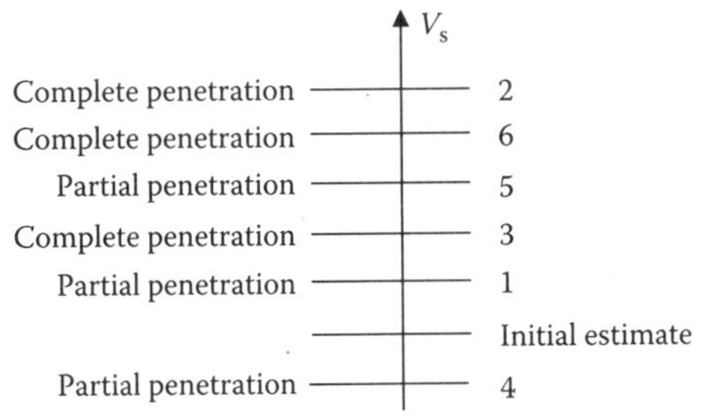

FIGURE 15.2
Illustration of the V_{50} experimental procedure.

next lower the velocity to get a partial penetration, say, at 4, and then we would increase it to get a complete penetration (but let us say that we get only a partial penetration at 5). We would then have to increase the next shot velocity to 6.

We would continue the aforementioned procedure, commonly known as an up and down test, until we obtained three complete penetrations and three partial penetrations with the difference between the highest and the lowest velocities in the set less than 200 ft/s. At that point, we would calculate the experimental V_{50} from

$$V_{50} = \frac{\sum_{i=1}^{6} V_i}{6} \tag{15.1}$$

The limit velocity V_l (sometimes called the ballistic limit when referring to the armor) is the velocity below which a given projectile will not defeat a given target. The technique for determining it was invented by the US Army Ballistic Research Laboratory, Aberdeen, Maryland. The object is to fire a few projectiles that achieve complete penetration, measuring the residual velocity through the use of flash X-rays, and then generate a curve as shown in Figure 15.3. We then plot the residual velocity after penetration vs. the striking

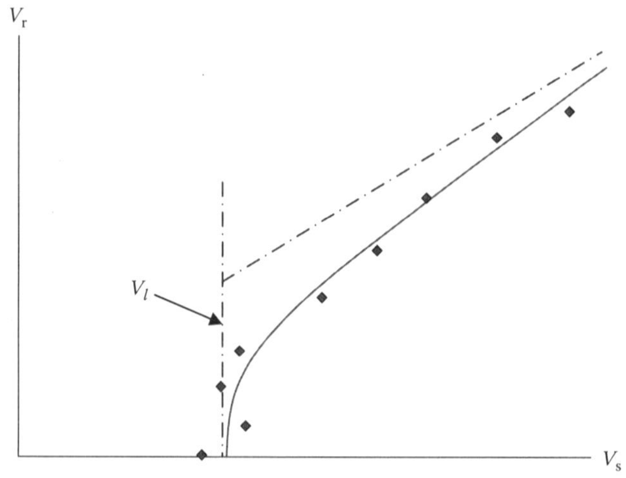

FIGURE 15.3
Limit velocity illustrated.

velocity. Usually, there will be a lower limit that develops below which the armor is not penetrated or the projectile gets stuck in the armor.

From experimental evidence, we know that the following factors affect the limit velocity: material hardness, yaw at impact, projectile density, projectile nose shape, and length-to-diameter ratio of the projectile. For the material hardness, in general, the harder the target, the higher V_{50} becomes, while the harder the penetrator, the lower V_{50} becomes, and there is more residual penetrator. With respect to yaw at impact, the more yaw, the greater chance for breakup or ricochet and the higher V_{50} becomes. With projectile density, we find that the denser the projectile is, the lower V_{50} becomes. A blunter nose translates, in general, to a higher V_{50}. If the target is significantly overmatched, however, the nose shape has negligible effect. The length-to-diameter ratio can go either way, and a great deal depends on the obliquity of impact.

We will now introduce some concepts, which we shall use in our examination of penetration events.

16

Penetration Theories

16.1 Penetration and Perforation of Metals

Metals are by and large the most common target of medium- to large-caliber projectiles. Although small-caliber ammunition is generally used against soft targets, there are times when even they are called upon to penetrate metal objects. This section will discuss several models of penetration into two of the most common metals: steel and aluminum. While these formulas are not exactly perfect for other metals, usually a metal will behave like one or the other.

Projectiles may impact metallic targets under a wide range of velocities. The nature of the target material is such that different velocities must be handled using somewhat different techniques. At very low velocities (<250 m/s), the penetration is usually coupled to the overall structural dynamics of the target. Responses are on the order of 1 ms. As the impact velocity increases (500–2000 m/s), the local behavior of the target (and sometimes penetrator) material dominates the problem. This local zone is approximately two to three projectile diameters from the center of impact. With further increases in velocity (2000–3000 m/s), the high pressures involved allow the materials to be modeled as fluids in the early stages of impact. At impact speeds greater than 12,000 m/s, energy exchange occurs at such a high rate that some of the colliding material will vaporize. This energy exchange must be accounted for. We will not treat this last case as it is beyond the normal scope of military applications.

A typical sequence of events that occur during a projectile impact is developed here [1]. Given that a projectile strikes a target, compressive waves propagate into both the projectile and the target. Relief waves propagate inward from the lateral free surfaces of the penetrator, cross at the centerline, and generate a high tensile stress. If the impact were normal, we would have a two-dimensional stress state. If the impact were oblique, bending stresses will be generated in the penetrator. When the compressive wave reached the free surface of the target, it would rebound as a tensile wave. The target may fracture at this point as will be seen in Section 18.3. The projectile may change direction if it perforates (usually toward the normal of the target surface).

Because of the differences in target behavior based on the proximity of the distal surface, we must categorize targets into four broad groups. A semi-infinite target is one where there is no influence of distal boundary on penetration. A thick target is one in which the boundary influences penetration after the projectile is some distance into the target. An intermediate-thickness target is a target where the boundaries exert influence throughout the impact. Finally, a thin target is one in which stress or deformation gradients are negligible throughout the thickness.

There are several methods by which a target will fail when subjected to an impact. The major variables are the target and penetrator material properties, the impact velocity, the projectile shape (especially the ogive), the geometry of the target supporting structure, and the dimensions of the projectile and target.

The failure modes of the target are depicted in Figure 16.1. They will now be described. Spalling is very common and is the result of wave reflection from the rear face of the plate. It is common for materials stronger in compression than in tension. Scabbing is similar to spalling, but the fracture predominantly results from large plate deformation, which begins a crack at a local inhomogeneity. These failure mechanisms will be expounded upon in Section 18.3. Brittle fracture usually occurs in weak and lower-density targets. Radial cracking is common in ceramic-type materials where the tensile strength is lower than the compressive strength, but it does occur in some steel armor. Plugging occurs in materials that are fairly ductile and usually when the projectile impact velocity is very close to the ballistic limit. Petaling occurs when the radial and circumferential stresses are high and the projectile impact velocity is close to the ballistic limit.

Because of the very high loading rates and correspondingly high temperatures, we need to describe some phenomena that occur during penetration events. Terms such as these occur throughout the literature, so it is good to understand what they mean.

The concept of adiabatic shearing is encountered in impacts where a plug has been formed. On initial impact, a local ring of intense shear is generated. Since this occurs very quickly (on the order of microseconds), the target does not have sufficient time to build up any motion. Locally intense heat is generated. Because of the time scale and a large deformation rate, the heat cannot be conducted away. Since the material properties are weaker at this high temperature, the material tends to yield readily and flow plastically. The process then feeds on itself. Finally, a plug is formed and breaks free. If the minimum

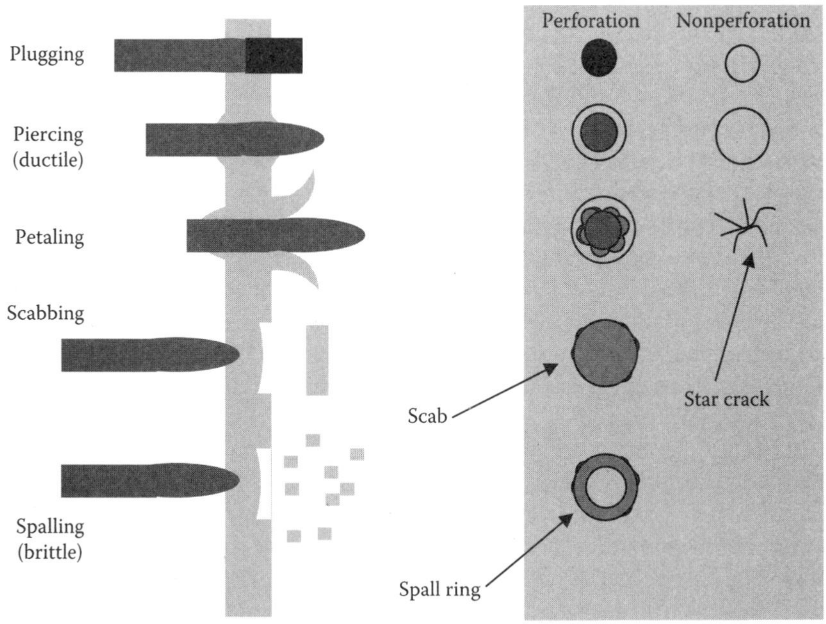

FIGURE 16.1
Target failure modes.

perforation velocity is exceeded by more than about 5–10%, the plug will usually break up. Blunt noses on projectiles tend to increase the propensity to fail a target by adiabatic shear.

Hydrodynamic erosion is an important concept in terminal ballistics. Metal cutting tools such as water jets or soft metal penetrators and shaped charge jets can defeat a target by hydrodynamic erosion. During hydrodynamic erosion, the penetrator material forces the target material aside in a manner similar to a punch being pushed into the target material except that the hole will be larger. This phenomenon usually occurs at impact velocities over 1000 m/s. The deposition of the penetrator material on the walls of the hole is an indication that this failure mechanism played a part in the penetration.

The hydrodynamic transition velocity is the velocity below which the projectile and target act as essentially elastic bodies and above which both target and projectile can be treated as fluids. This concept is illustrated by the penetration sequence of Zukas et al. [1]. For all penetration velocities, the target material is radially accelerated away from the axis of penetration. At low velocities, elastic strain keeps the target material in contact with the penetrator. At high velocities, the material is thrown away from the projectile, so that the hole becomes bigger than the projectile diameter. The radial acceleration of the material is greatest at the tip of the projectile. At the hydrodynamic transition velocity, the tip of the penetrator deforms laterally. The projectile tip becomes spherically blunted and forms a stable shape that penetrates the target for the remainder of the event. The transition velocity inversely varies with the tip radius. Hydrodynamic transition velocity is possibly related to the rate of rod erosion and plastic wave propagation.

Shear banding is a form of adiabatic shearing in which layers of material in a like state of shear tend to form. There are discontinuities in stress and strain instead of a gradual increase in shear strain near the disturbed region. Uranium and tungsten tend to display this phenomenon. Normal material models used in finite element codes do not show this effect. A model that includes thermal softening is required.

The analytical models in use today to solve these types of problems can be organized into three broad categories: empirical or quasi-analytical, approximate analytical, and numerical. In empirical or quasi-analytical models, algebraic equations are developed from large amounts of experimental data. These models are generally curve fits (results based). They usually do not incorporate physics and tend to be configuration dependent. An approximate analytical model attempts to examine the physics of a particular aspect of the penetration process or failure mechanism such as petaling and plugging. The mathematics becomes tractable because we must make simplifying assumptions. They are usually limited to particular situations. Numerical models usually attempt to solve the full equations of continuum mechanics by using finite difference or finite element techniques. This is the most general method. The problem with numerical models is that good material models are required, and this can be expensive.

Most analytical models can consider only one damage mechanism (like plugging or fracture) or conservation law before they become mathematically intractable. Some allow as many as two mechanisms. The approach is to make simplifying assumptions. Typical assumptions are to assume localized influence where the projectile is influenced only by a small region of the target; to ignore rigid body motions; to ignore thermal, friction, shock heating, and any material behavioral changes due to these mechanisms, and to assume that the target is initially stress free, etc. One important thing to recognize is that a complicated model does not necessarily yield a more accurate answer.

The perforation of finite-thickness plates in which plugging is the predominant penetration mode is divided into three stages. In the first stage, locally, the material ahead of the projectile is compressed and the mass is added to the projectile (i.e., the projectile decelerates somewhat and the added mass accelerates). In the second stage, more material is accelerated but shearing is occurring on the surface area of the plug. In the third stage, the plug has completely sheared out and both the plug and the projectile move with the same velocity. If this model is used for an oblique impact, one must use the line-of-sight thickness. At velocities from 1200 to 5000 m/s, the model used usually involves hydrodynamic erosion of the projectile tip as the first stage as well. This can be followed by both plugging and further tip erosion. In the third stage, we usually consider the projectile to be completely eroded and the plug is ejected from the armor [1].

Some models account for the flexibility of the target. This is usually required as the impact velocities approach the limit velocity. In this case, a significant amount of energy is consumed in both elastically and plastically bending the target plate.

We shall examine the underlying assumptions in a few penetration theories before moving on to detailed examination of the theories themselves. Theories that are derived from a momentum balance are typically used for thin plates. These theories can be used with minor modifications when the target petals. They usually require that the projectile remains intact.

Theories that are derived from an energy balance are typically used for thick and moderately thick plates. With moderately thick targets, plugging can occur. Thick plates are usually defeated by a piercing phenomenon that also has distinct phases. The first phase is a radial displacement of the target material. Sometimes, there is plugging at this stage. This stage is followed by plastic flow and yielding of the target. The target material may well be able to be treated like a fluid during this phase.

Many empirically based predictive relationships are based on energy approaches. A particularly popular model takes the form of

$$E = kd^m t^n \tag{16.1}$$

where we have

$$m + n \approx 3 \tag{16.2}$$

In these equations, E is the perforation energy; d is the projectile diameter; t is the plate thickness; and k is an empirically derived constant (see Figure 16.2).

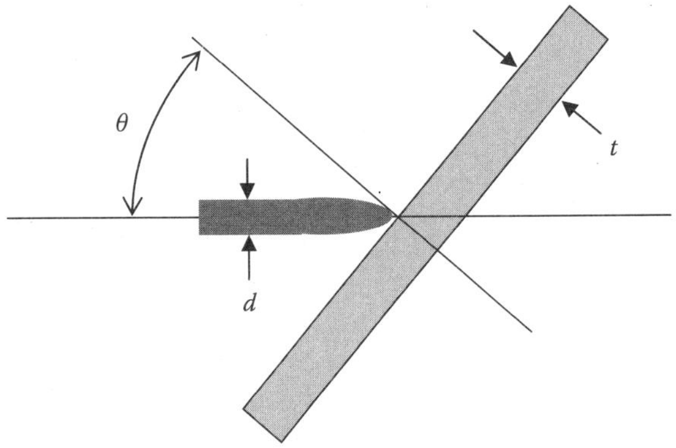

FIGURE 16.2
Projectile impact problem illustrated.

If we let $m = 1.5$ and $n = 1.4$, we get the famous DeMarre formula for normal impact. If we would like to include an angle of obliquity in the aforementioned formula, it is common practice to use

$$E = kd^m t^n \sec^p \theta \tag{16.3}$$

Here p is an experimental parameter based on the projectile–armor combination and θ is the angle of obliquity measured from the normal to the plate. Sometimes, the armor fabrication process will affect the penetration. In this case, there is a function called the figure of merit, where the perforation velocity of the armor is compared to that of mild steel:

$$\text{FOM} = \frac{V_l}{V_{l_{\text{mild steel}}}} \tag{16.4}$$

Note that in Equation 16.4, the velocity used does not necessarily have to be the limit velocity. Another useful relationship commonly employed by the projectile designer is

$$E_{\text{perf}} = \frac{1}{2} m V_{\text{perf}}^2 \tag{16.5}$$

Inserting Equation 16.3 into Equation 16.5 yields

$$V_{\text{perf}}^2 = 2k \frac{d^m t^n}{m} \sec^p \theta \tag{16.6}$$

Now taking the square root and assimilating terms, we get

$$V_{\text{perf}} = k \sqrt{\frac{d^m t^n}{m}} \sec^{j(\theta)} \theta \tag{16.7}$$

In 1886, DeMarre developed a famous formula for the penetration of a plate given a normal impact:

$$\frac{mV^2}{d^3} = \alpha \frac{t^{1.4}}{d^{1.5}} \tag{16.8}$$

where m is the penetrator mass; V is the impact velocity; d is the diameter of the projectile; and t is the plate thickness, with α being an empirically derived constant.

As a word of caution, many of these formulas are dangerous because of the units in the empirically derived constant; it is commonplace to see centimeter–gram–second (CGS) units in these formulas as well. Over time, many have modified the DeMarre formula and used it in the form

$$\frac{mV^2}{d^3} = \alpha \left(\frac{t}{d} \right)^\beta \tag{16.9}$$

Here β is an empirically derived constant as well. In the aforementioned form, the DeMarre formula is used when considering a normal impact. Some researchers have extended its use to include an oblique impact, and it would then take the following form:

$$\frac{mV^2}{d^3} = \alpha \left[\frac{tg(\theta)}{d} \right]^\beta \tag{16.10}$$

where $g(\theta)$ is a function of the angle of obliquity and is most often taken as sec θ.

We sometimes define the specific limit energy (SLE) as

$$\frac{mV_l^2}{d^3} \equiv \text{SLE} \tag{16.11}$$

Hans Bethe, a physicist at Cornell University in 1941, determined that for piercing-type problems (i.e., thick plate perforation where a hole is laterally or radially widened by the penetrator), the constant β should be equal to 1, thus yielding

$$mV_l^2 \sim td^2 \tag{16.12}$$

Around the same time (1942), Zener and Holloman from Watertown Arsenal came up with a formula for use when plugging or petaling is the predominant penetration mode. They stated that in this case, β should equal to 2, thus yielding

$$mV_l^2 \sim t^2 d \tag{16.13}$$

In 1943, Curtis and Taub attempted to modify the DeMarre formula to account for a mode change during the penetration event. In a thick plate, the mode changes at some point from a piercing to a plugging at the rear surface. This results in a decrease in energy consumed per unit path length, so the DeMarre formula had to be further modified to

$$\frac{mV_l^2}{d^3} = \alpha\left(\frac{t}{d} + \gamma\right) \tag{16.14}$$

Here α and γ are constants and $\gamma < 0$. If we define t' as depicted in Figure 16.3, then γ is a quadratic function of t'. Also $t' \sim d$ and is the distance after the mode changes.

S. Jacobson, working at the Picatinny Arsenal in New Jersey, further refined the concept that there is a different energy relationship for each of the two modes. For plugging, this is

$$E_{\text{plug}} = \text{force} \cdot \text{distance} \approx \pi d t Y_s \cdot t \tag{16.15}$$

where Y_s is the shear yield strength of the material. For the piercing mode, we have

$$E_{\text{piercing}} = Y_{\text{flow}} \cdot \text{V} \approx \frac{\pi d^2}{4} t Y_{\text{flow}} \tag{16.16}$$

where V is the volume of the plug and Y_{flow} is the flow or plastic yield stress of the target material.

We can rewrite Equations 16.15 and 16.16 as

$$E_{\text{plug}} = k_{\text{plug}} d^3 \left(\frac{t}{d}\right)^2 Y_s \tag{16.17}$$

$$E_{\text{piercing}} = k_{\text{piercing}} d^3 \left(\frac{t}{d}\right) Y_{\text{flow}} \tag{16.18}$$

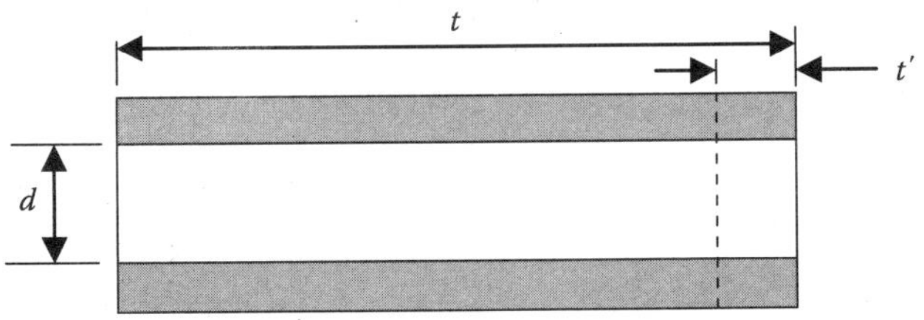

FIGURE 16.3
Section of a target plate that defines t and t'.

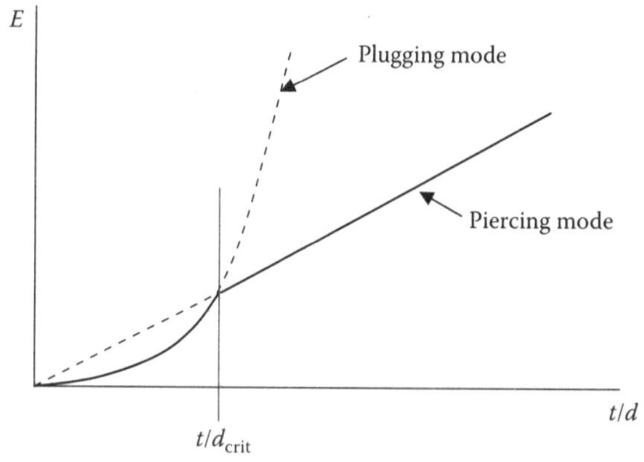

FIGURE 16.4
Energy in penetration modes based on the model of Jacobson.

If we graph both expressions, we obtain a plot as illustrated in Figure 16.4. To obtain t/d_{crit}, we solve Equations 16.17 and 16.18 where

$$E_{\text{plug}} = E_{\text{piercing}} \tag{16.19}$$

using the relations that

$$Y_s \approx 0.6Y_{\text{flow}} \tag{16.20}$$

$$k_{\text{piercing}} = \frac{\pi}{4} \quad \text{and} \quad k_{\text{plug}} = \pi \tag{16.21}$$

Then, combining Equations 16.17 and 16.18, we get

$$\frac{\pi}{4}d^3\left(\frac{t}{d}\right)Y_{\text{flow}} = \pi d^3\left(\frac{t}{d}\right)^2 0.6Y_{\text{flow}} \rightarrow \left(\frac{t}{d}\right)_{\text{crit}} = 0.42 \tag{16.22}$$

This value of t/d is the point where the mode of penetration changes from plugging to piercing. Thus, against targets whose thicknesses are such that an attack by a penetrator whose t/d ratio is greater than 0.42, we can expect that the penetration mode will be piercing; otherwise, plugging is to be expected.

Lambert and Zukas proposed a model in 1982 while working at the Ballistic Research Laboratory to cover more general cases of penetration. If we examine Equation 16.14, we can see that as the plate thickness goes to zero, the residual velocity should approach the striking velocity and the limit velocity should approach zero. Expressed mathematically, we require that

$$\lim_{t\to 0} V_l \rightarrow 0 \tag{16.23}$$

However, if we look at Equation 16.14, we note that if $V_l = 0$ and $t = 0$, it requires the product $\gamma\alpha$ to equal zero, which is not physically possible. Therefore, the Lambert model replaces γ by $[\exp(-t/d) - 1]$ as

$$\frac{mV_l^2}{d^3} = \alpha\left[\frac{t}{d} + \exp\left(-\frac{t}{d}\right) - 1\right] \tag{16.24}$$

This forces

$$V_l = 0 \quad \text{at} \quad t = 0 \tag{16.25}$$

and

$$V_l = \infty \quad \text{at} \quad t = \infty \tag{16.26}$$

Since the penetrator volume is proportional to d^2l and since there should be a dependence on this volume in the SLE, we want to keep the dimension of diameter cubed in Equation 16.24; thus, we shall write

$$d^3 \to d^{3-c}l^c = d^2 l\left(\frac{l}{d}\right)^{c-1} = d^3\left(\frac{l}{d}\right)^c \tag{16.27}$$

where c is a constant. We can then incorporate this into Equation 16.24 as

$$\frac{mV_l^2}{d^3} = \left(\frac{l}{d}\right)^c \alpha\left[\frac{t}{d} + \exp\left(-\frac{t}{d}\right) - 1\right] \tag{16.28}$$

Next, we will include obliquity effects by adding in the angle of obliquity θ through replacement of t by $t \sec^k \theta$. In this case, if $k = 1$, we have the true path length through the armor plate (line-of-sight thickness). We shall define

$$z = \frac{t}{d}\sec^k\theta \tag{16.29}$$

We can now rewrite Equation 16.29 as

$$\frac{mV_l^2}{d^3} = \alpha\left(\frac{l}{d}\right)^c\left[\frac{t}{d}\sec^k\theta + \exp\left(-\frac{t}{d}\sec^k\theta\right) - 1\right] \tag{16.30}$$

If we solve Equation 16.30 for the limit velocity, we obtain

$$V_l = \sqrt{\alpha\left(\frac{l}{d}\right)^c\left[\frac{t}{d}\sec^k\theta + \exp\left(-\frac{t}{d}\sec^k\theta\right) - 1\right]\frac{d^3}{m}} \tag{16.31}$$

The Lambert model was used to examine the firing of 200 long rods into rolled homogeneous armor (RHA). The test conditions were as follows:

$$0.5 \le m\,[\text{g}] \le 3630 \quad 0.6 \le t\,[\text{cm}] \le 15$$

$$0.2 \le d\,[\text{cm}] \le 0.5 \quad 0° \le \theta \le 60°$$

$$4 \le \frac{l}{d} \le 30 \qquad 7.8 \le \rho\left[\frac{\text{g}}{\text{cm}^3}\right] \le 19.0$$

A least-squares fit of the results yielded the following: $\alpha = (4000)^2$, $c = 0.3$, and $k = 0.75$. If we insert these into Equation 16.31, we get

$$V_l = \left(\frac{l}{d}\right)^{0.15}(4000)\sqrt{\frac{d^3}{m}\left[\frac{t}{d}\sec^{0.75}\theta + \exp\left(-\frac{t}{d}\sec^{0.75}\theta\right) - 1\right]}\left[\frac{\text{m}}{\text{s}}\right] \tag{16.32}$$

Please note the CGS units. The authors suggest that the model is applicable where $t/d > 1.5$. Also, we must note that nose geometry has a significant influence for $t/d < 1.0$. RHA or good-quality steel is the target (the specific properties are unimportant).

One measure of lethal effects once a projectile has perforated the target material is the residual velocity. V_r is the symbol for the residual velocity of the penetrator. That is the velocity that the penetrator moves with once it perforates the target. Mathematically, it is defined in the Lambert model as

$$V_r = \begin{Bmatrix} 0, & 0 \le V_s \le V \\ a\left(V_s^p - V_l^p\right)^{\frac{1}{p}}, & V_s > V_l \end{Bmatrix} \tag{16.33}$$

If we assume that V_s is large so that the absorption of momentum by the target is negligible, then the momentum balance can be written in terms of identifiable penetrator mass and velocity (m_r and V_r) and the large quantity of unidentifiable target and penetrator ejecta with each particle m_i having a particular velocity V_i. Thus, the momentum balance is

$$m_r V_r + \sum_{i=1}^{n} m_i V_i \to m_s V_s \text{ as } V_s \to \infty \tag{16.34}$$

Even though Equation 16.34 is mathematically satisfying, in practice, it is usually difficult to measure the mass and velocity of all the fragments, so most of the $m_i V_i$ will remain unknown.

We shall now consider a general case of impact as illustrated in Figure 16.5. Here we shall let m' be the mass of the ejecta. We can then write

$$m' = \rho \frac{\pi}{4} d^3 z \tag{16.35}$$

where

$$z = \frac{t}{d} \sec^{0.75}\theta \tag{16.36}$$

therefore,

$$m' = \rho \frac{\pi}{4} d^3 \left(\frac{t}{d} \sec^{0.75}\theta \right) = \rho \frac{\pi}{4} d^2 t \sec^{0.75}\theta \tag{16.37}$$

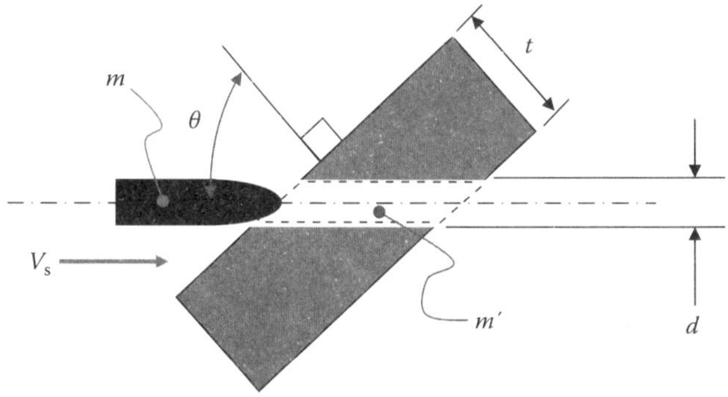

FIGURE 16.5
General case of projectile impact.

If we now assume that

$$\sum_{i=1}^{n} m_i V_i = hm' V_r \tag{16.38}$$

This is equivalent to stating that m' is the mass of material pushed ahead of the penetrator. This mass, m' is ejected with speed V_r (plugging theory), and the total momentum of the ejecta jumble is proportional to $m'V_r$. We can also write, in the limiting case, that the residual momentum approaches the initial momentum or, mathematically,

$$\frac{M_r}{M} \to 1 \tag{16.39}$$

If we substitute Equation 16.38 into Equation 16.34, we get

$$mV_r + hm'V_r \to m_s V_s \quad \text{as} \quad V_s \to \infty \tag{16.40}$$

which can be rearranged to yield

$$\frac{V_r}{V_s} \to \left(\frac{m_s}{m_r + hm'} \right) \quad \text{as} \quad V_s \to \infty \tag{16.41}$$

We know that if penetration occurred, Equation 16.33 applies, so we have

$$V_r = a\left(V_s^p - V_l^p\right)^{\frac{1}{p}} \tag{16.42}$$

We can divide Equation 16.42 by V_s to get

$$\frac{V_r}{V_s} = a\left[1 - \left(\frac{V_l}{V_s}\right)^p\right]^{\frac{1}{p}} \tag{16.43}$$

which means that as V_s approaches infinity, the second term in the parentheses approaches zero or

$$\frac{V_r}{V_s} \to a \quad \text{as} \quad V_s \to \infty \tag{16.44}$$

This is illustrated in Figure 16.6.

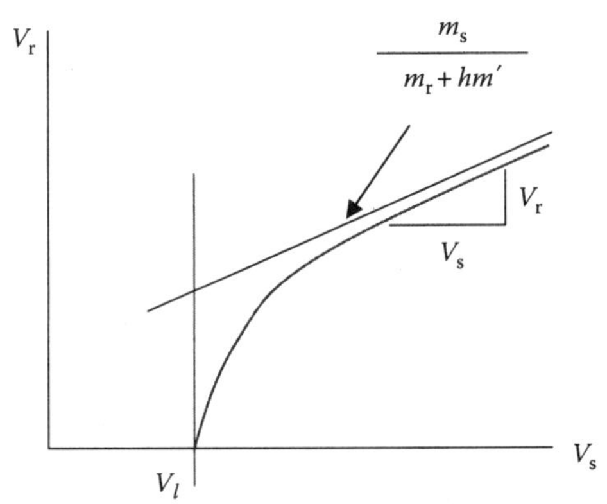

FIGURE 16.6
Asymptote on limit velocity.

If we look at Equations 16.44 and 16.41, we see that

$$a = \left(\frac{m_s}{m_r + hm'} \right) \tag{16.45}$$

Furthermore, we can assume in the plugging mode that the mass of the penetrator does not significantly change during penetration, so we get $m_s = m_r = m$. We can then write Equation 16.45 as

$$a = \left(\frac{m}{m + hm'} \right) \tag{16.46}$$

There is empirical evidence that suggests that $h \approx 1/3$, so we can write

$$a = \left(\frac{m}{m + \frac{1}{3}m'} \right) \tag{16.47}$$

If we assume that the penetrator remains intact throughout the perforation event, we can write

$$\mathrm{KE_{impact}} = \mathrm{KE_{limit}} + \mathrm{KE_{residual}} \tag{16.48}$$

This can also be expressed as

$$V_r^2 \sim V_s^2 - V_l^2 \rightarrow V_r \sim \left(V_s^2 - V_l^2\right)^{\frac{1}{2}} \tag{16.49}$$

which, if written as

$$V_r = a\left(V_s^2 - V_l^2\right)^{\frac{1}{2}} \tag{16.50}$$

would say that $p = 2$. If we looked at momentum, we would get

$$V_s \sim V_l + V_r \tag{16.51}$$

which could be written as

$$V_s = a(V_l + V_r) \tag{16.52}$$

Equation 16.52 implies that for a momentum balance, $p = 1$. Thus, it is clear that the value for p should fall between 1 and 2. Lambert accounted for this by choosing

$$p = 2 + z = 2 + \frac{t}{d}\sec^{0.75}\theta \tag{16.53}$$

where both p and z monotonically grow as

$$\frac{t}{d} \rightarrow \infty \quad \text{and/or} \quad \theta \rightarrow \frac{\pi}{2} \tag{16.54}$$

also,

$$p \rightarrow 2 \quad \text{as} \quad t \rightarrow 0 \tag{16.55}$$

Lambert also found that a better empirical fit was obtained if he let

$$p = 2 + \frac{z}{3} = 2 + \frac{t}{3d}\sec^{0.75}\theta \tag{16.56}$$

A numerical model for penetration was proposed by A. Tate to determine penetration of metals [2]. The base equation for this model is

$$\frac{1}{2}\rho_P(V_i - u)^2 + Y_P = \frac{1}{2}\rho_t u^2 + R_t \tag{16.57}$$

where ρ_p is the density of the projectile material; ρ_t is the density of the target material; V_i is the impact velocity; u is the instantaneous projectile velocity; and Y_p and R_t are the ballistic resistances of the projectile and target, respectively, defined as

$$Y_P = 1.7\sigma_P \tag{16.58}$$

$$R_t = \sigma_t\left[\frac{2}{3} + \ln\left(0.57\frac{E_t}{\sigma_t}\right)\right] \tag{16.59}$$

where σ_P is the yield strength of the projectile material; σ_t is the yield strength of the target material; and E_t is the modulus of elasticity of the target material.

The way the Tate model is used is to numerically integrate Equation 16.57 until the velocity goes to zero or perforation occurs. When the projectile stops, a second integration determined the depth of penetration. When perforation occurs, the value of u is the residual velocity. Tate states that the accuracy of this method is within 20%. One of the downsides of the model is that it does not handle oblique impacts, but it can at least be altered by the line-of-sight thickness.

If a penetrator hits a target at a great enough angle, it may ricochet. The ricochet process can be described as follows. During impact, both the projectile and the target are elastically compressed. When this energy is released, it will change the motion of the projectile. Deformations because of resisting force of the target will change the direction of the penetrator. Rotating moments are generated by internal forces acting within the projectile.

In general, thin plates do not allow ricochet except at extreme angles of attack. Tate produced a ricochet formula for the critical ricochet angle (oblique impacts at angles greater than this will ricochet):

$$\tan^3\beta > \frac{2}{3}\frac{\rho_P V^2}{Y_P}\left(\frac{L^2 + d^2}{Ld}\right)\left[1 + \left(\frac{\rho_P}{\rho_t}\right)^{\frac{1}{2}}\right] \tag{16.60}$$

Here Y_p is a characteristic strength usually taken as the Hugoniot elastic limit (described in Section 18.3), the subscripts p and t are projectile and target, respectively, and L, d, and V are the length, diameter, and velocity of the penetrator, respectively.

As vehicles become lighter, aluminum is being used more and more as armor. It is therefore necessary to determine the penetration capabilities of projectiles into aluminum.

Aluminum behaves a little differently than steel during penetration by ogival projectiles in its tendency to be pierced rather than to develop plugs. A penetrator is usually of significantly greater density than the target in most cases. One significant difference is the evidence of a layer of aluminum with an altered microstructure on the penetrated surface. This indicates a melt layer that is believed to assist in penetration.

A simple model of projectile penetration into aluminum was put forward by Forrestal et al. in 1992 [3]. A distinct advantage of this model is its simplicity. A possible disadvantage is that the empirical nature is not universal. Even though the study was specifically performed with 7075-T651 targets, it yields a fairly good representation of aluminum penetration.

The model assumes normal impact of the projectile and that the projectile is rigid. This may, at first, seem to be a restrictive assumption, but the method provides reasonable estimates for slightly yawed projectiles if the angle is below about 5° and possibly further.

We first define the caliber radius head as

$$\psi = \frac{s}{d} \tag{16.61}$$

where d is the diameter of the projectile; ψ is the caliber radius head; and s is the ogive radius.

We can also define the nose length as

$$l = \frac{d}{2}\sqrt{4\psi - 1} \tag{16.62}$$

This geometry is illustrated in Figure 16.7.

We shall say that the resistance force of the aluminum target on the penetrator in this case will have two components: one normal to the surface (normal stresses) and one tangential to the surface (shear stresses and friction). If we lump the shear stress in with the stress owing to friction and furthermore assume that the tangential stress is proportional to normal stress, we can write

$$\sigma_t = \mu\sigma_n \tag{16.63}$$

where σ_t is the tangential stress; σ_n is the normal stress; and μ is the proportionality constant (coefficient of sliding friction).

Forrestal et al. [4] developed a formula for the axial force on an ogival nose:

$$F_z = 2\pi s \int_{\theta_0}^{\pi/2} \left\{ \left[\sin\theta - \left(\frac{s - \frac{d}{2}}{s} \right) \right] (\cos\theta + \mu\sin\theta) \right\} \sigma_n(V_z, \theta)\, d\theta \tag{16.64}$$

where

$$\theta_0 = \sin^{-1}\left(\frac{s - \frac{d}{2}}{s} \right) \tag{16.65}$$

Here V_z is the instantaneous velocity during penetration. The stress function $\sigma_n(V_z, \theta)$ is assumed to be similar to that of a spherically symmetric expanding cavity (defined later). If we let V be the constant velocity at which the tip of the projectile radially expands the hole, then we can write the radial stress at the cavity surface as

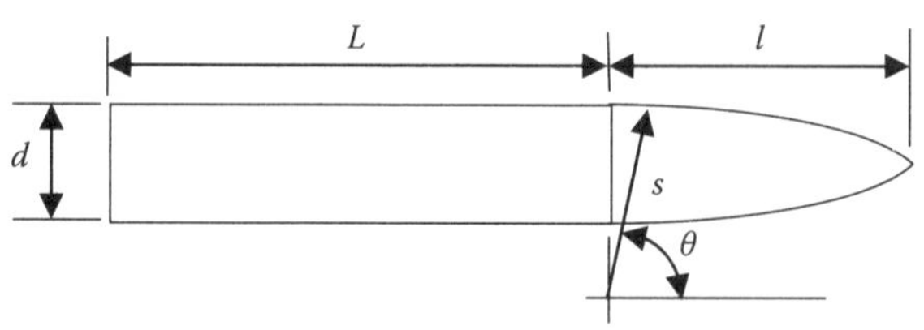

FIGURE 16.7
Ogival penetrator for the model of Forrestal et al. [3].

$$\frac{\sigma_r}{Y} = A + B\left(\sqrt{\frac{\rho_t}{Y}}V\right)^2 \tag{16.66}$$

where σ_r is the radial stress; Y is the material yield stress; ρ_t is the target density; and A and B are constants defined as

$$A = \frac{2}{3}\left[1 + \left(\frac{2E}{3Y}\right)^n I\right] \tag{16.67}$$

$$B = \frac{3}{2} \tag{16.68}$$

where

$$I = \int_0^{1-\left(\frac{3Y}{2E}\right)} \frac{(-\ln x)^n}{1-x}\,dx \tag{16.69}$$

In these expressions, E is Young's modulus and n is the strain hardening exponent (assumes power law strain hardening). For an assumed incompressible 7075-T651 aluminum, Forrestal et al. [3] provided $I = 3.896$ and $A = 4.609$.

Empirically, curve fitting the stress–strain curves (thus including compressibility) for 7075-T651 yielded slightly different results with $A = 4.418$ and $B = 1.068$.

To approximate the normal stress on the ogive, we can replace the spherically symmetric velocity V in Equation 16.66 with $V_z \cos\theta$; then, we have

$$\frac{\sigma_n(V_z,\theta)}{Y} = A + B\left(\sqrt{\frac{\rho_t}{Y}}V_z\cos\theta\right)^2 \tag{16.70}$$

If we insert Equation 16.70 into Equation 16.64, we obtain

$$F_z = 2\pi s Y \int_{\theta_0}^{\pi/2} \left\{\left[\sin\theta - \left(\frac{s-\frac{d}{2}}{s}\right)\right](\cos\theta + \mu\sin\theta)\right\}\left(A + B\frac{\rho_t}{Y}V_z^2\cos^2\theta\right)d\theta \tag{16.71}$$

Now we integrate to obtain

$$F_z = \frac{\pi d^2}{4}Y\left(\alpha + \beta\frac{\rho_t V_z^2}{Y}\right) \tag{16.72}$$

where

$$\alpha = A\left[1 + 4\mu\psi^2\left(\frac{\pi}{2} - \theta_0\right) - \mu(2\psi - 1)\sqrt{4\psi - 1}\right] \tag{16.73}$$

$$\beta = B\left[\frac{8\psi - 1}{24\psi^2} + \mu\psi^2\left(\frac{\pi}{2} - \theta_0\right) - \frac{\mu(2\psi - 1)(6\psi^2 + 4\psi - 1)\sqrt{4\psi - 1}}{24\psi^2}\right] \tag{16.74}$$

Now that we have an expression for force as a function of velocity, we need to come up with how this varies during penetration.

We can write Newton's second law as

$$-F_z = m\frac{dV_z}{dt} \tag{16.75}$$

We can convert this time integral to a distance integral and rewrite it as follows:

$$-F_z = mV_z\frac{dV_z}{dz} \tag{16.76}$$

One can write the mass of our projectile in terms of the parameters we have already described. The mass of the cylindrical section of the projectile is

$$m_{\text{cylinder}} = \rho_p\frac{\pi d^2}{4}L \tag{16.77}$$

We can write the mass of the ogive as

$$m_{\text{ogive}} = \rho_p\frac{\pi d^3}{8}k \tag{16.78}$$

where

$$k = \left(4\psi^2 - \frac{4\psi}{3} + \frac{1}{3}\right)\sqrt{4\psi - 1} - 4\psi^2(2\psi - 1)\sin^{-1}\left[\frac{\sqrt{4\psi - 1}}{2\psi}\right] \tag{16.79}$$

Now the total mass of the projectile is

$$m = m_{\text{cylinder}} + m_{\text{ogive}} = \rho_p\frac{\pi d^2}{4}\left(L + \frac{kd}{2}\right) \tag{16.80}$$

If we insert Equations 16.80 and 16.72 into Equation 16.76, we get, after some rearrangement,

$$-dz = \rho_p\left(L + \frac{kd}{2}\right)\frac{V_z}{\alpha Y + \beta\rho_t V_z^2}dV_z \tag{16.81}$$

This can be integrated as

$$-\int_0^P dz = \rho_p\left(L + \frac{kd}{2}\right)\int_{V_0}^{0}\frac{V_z}{\alpha Y + \beta\rho_t V_z^2}dV_z \tag{16.82}$$

The result of this integration is

$$P = \frac{1}{2\beta}\left(\frac{\rho_p}{\rho_t}\right)\left(L + \frac{kd}{2}\right)\ln\left[1 + \left(\frac{\beta}{\alpha}\right)\left(\frac{\rho_t V_0^2}{Y}\right)\right] \tag{16.83}$$

where P is the final penetration depth and V_0 is the impact velocity.

If the penetration depth P is greater than the target thickness, perforation will occur.

When this is the case, it is useful to be able to calculate the residual velocity of the penetrator, which we do by integrating Equation 16.82 with different limits of integration:

$$-\int_0^T dz = \rho_p\left(L + \frac{kd}{2}\right)\int_{V_0}^{V_r}\frac{V_z}{\alpha Y + \beta\rho_t V_z^2}dV_z \tag{16.84}$$

where T is the target thickness. Performing the integration yields

$$V_r = \sqrt{\left(\frac{\alpha Y}{\beta\rho_t} + V_0^2\right)\exp\left[-\frac{2\beta\rho_t T}{\rho_p\left(L + \frac{kd}{2}\right)}\right] - \frac{\alpha Y}{\beta\rho_t}} \quad (16.85)$$

This model has proven to be fairly accurate (within 15%) once the coefficients have been tuned. It is fairly sensitive to the friction coefficient μ incorporated in both α and β, which Forrestal et al. [4] suggested should be between 0 and 0.06.

Problem 1

A German 280 mm armor-piercing projectile weighs 666 lbm and is about 34 in. in length. It strikes a British warship in the 1/2 in. thick vertical side plating at an angle of 12° from the horizontal along the following path depicted. The initial impact velocity is 2000 ft/s. Determine the residual velocity of the shell after passing through each compartment and how far through the ship it will go (i.e., in which compartment will it stop).

Assume the density of the armor plate to be $\rho = 0.283$ lbm/in.^{3}

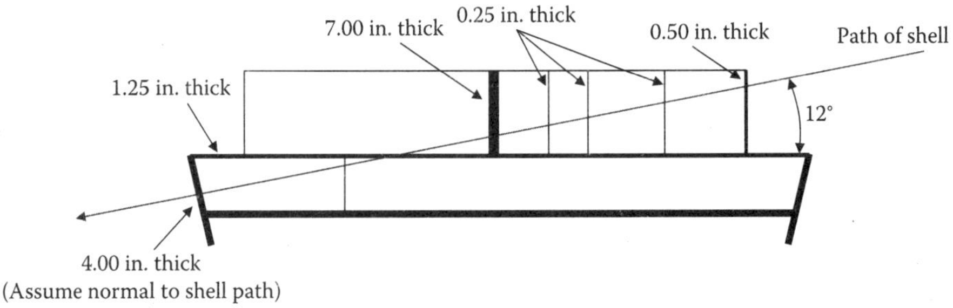

Answer: The projectile is arrested by the 1.25 in. deck.

Problem 2

An explosively formed penetrator impacts a 4 in. thick RHA plate at a velocity of 1500 m/s. The penetrator parameters are given later. Determine if the penetrator will perforate the target by using the Lambert/Zukas model given

1. A normal impact

   Answer: $V_l = 1299 \left[\frac{\text{m}}{\text{s}}\right]$; yes

2. An impact at 30° obliquity

   Answer: $V_l = 1389 \left[\frac{\text{m}}{\text{s}}\right]$; yes

Penetrator information:

$$l = 95\,[\text{mm}] \quad m = 1.25\,[\text{lbm}]$$

$$d = 22\,[\text{mm}] \quad V_s = 1500\left[\frac{\text{m}}{\text{s}}\right]$$

Problem 3

A German 7.5 cm Gr 34A1 projectile is fired at a 2 in. thick armor plate at a 30° obliquity. The impact velocity is 400 m/s. The penetrator parameters are given later.

1. Determine whether the penetration mode will be plugging or piercing through use of the Jacobson model for a normal impact.
   Answer: Piercing

2. Determine if the penetrator will perforate the armor through the use of the Lambert model.
   Answer: No

3. Comment on the validity of the model.

Penetrator information:

$$l = 39\,[\text{cm}] \quad m = 1.25\,[\text{kg}]$$

$$d = 7.5\,[\text{cm}] \quad V_s = 400\left[\frac{\text{m}}{\text{s}}\right]$$

Problem 4

A Japanese 20 mm projectile with the properties given later impacts the 1/2 in. thick aluminum armor plate on the rear gun mount of a US plane at 30° obliquity. If the projectile and the armor have the properties given later,

1. Determine how deep the projectile will penetrate into the armor (assume $\mu = 0.03$).
   Answer: $P = 53.1$ [mm] = 2.09 [in.]

2. If the projectile perforates the armor, determine its residual velocity.
   Answer: $V_r = 423\left[\frac{\text{m}}{\text{s}}\right]$

Estimated penetrator information:

$$s = 40\,[\text{mm}] \quad m = 128\,[\text{g}]$$

$$d = 20\,[\text{mm}] \quad V_s = 500\left[\frac{\text{m}}{\text{s}}\right] \quad \rho_p = 0.283\left[\frac{\text{lbm}}{\text{in.}^3}\right]$$

$$L = 60\,[\text{mm}]$$

Estimated armor information:

$$A = 4.418Y = 39{,}000\,[\text{psi}] \quad \rho_t = 0.098\left[\frac{\text{lbm}}{\text{in.}^3}\right]$$

$$B = 1.068$$

Problem 5

You are asked to design a "slug butt" for a test gun. The requirement for the device is to stop a 106 lbm steel projectile moving at 1000 m/s. You have available a pile of 2 in. thick RHA plates that you can stack together. Note that stacking plates together results in slightly worse performance than that of a solid plate—but we can account for this by adding some extra material to provide "margin." Normally slug butts are constructed at an angle to deflect ricochets downward and improve ballistic performance, but this decreases the "target" area. Given the information provided earlier and later, determine the number of plates you need to prevent a penetration and determine the angle of the slug butt—do not use an angle greater than 40° as the target area will be too small. List all assumptions and comment on your design.

Projectile information:

$$l = 40\ [\text{in.}] \qquad m = 106\ [\text{lbm}]$$

$$d = 155\ [\text{mm}]$$

$$\rho_p = 0.283\left[\frac{\text{lbm}}{\text{in.}^3}\right] \qquad V_s = 1000\left[\frac{\text{m}}{\text{s}}\right]$$

Armor information:

$$\rho_s = 0.283\left[\frac{\text{lbm}}{\text{in.}^3}\right] \quad \text{available in 2 in. thick plates}$$

Problem 6

A German 305 mm armor-piercing projectile weighs 894 lbm and is about 35.2 in. in length. It strikes a British warship in the 3 1/4 in. thick turret crown at an angle of 20° from the horizontal along the path depicted later. The initial impact velocity is 1800 ft/s. Determine the residual velocity of the shell after passing through each compartment and how far through the ship it will go (i.e., in which compartment will it stop).

Assume the density of the armor plate to be $\rho = 0.283$ lbm/in.^{3}

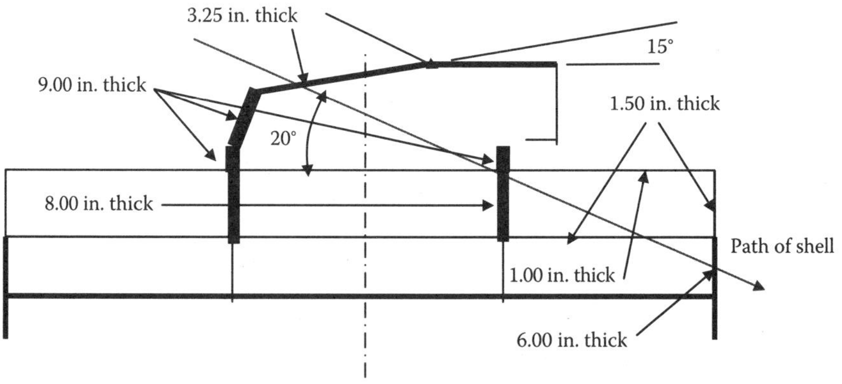

Problem 7

A German 280 mm armor-piercing projectile weighs 666 lbm and is about 34 in. in length. It strikes a British warship in the 0.43 in. thick vertical side plating at an angle of 10° from the horizontal along the path depicted later. The initial impact velocity is 1900 ft/s. Determine the residual velocity of the shell after passing through each compartment and how far through the ship it will go (i.e., in which compartment will it stop).

Assume the density of the armor plate to be $\rho = 0.283$ lbm/in.^{3}

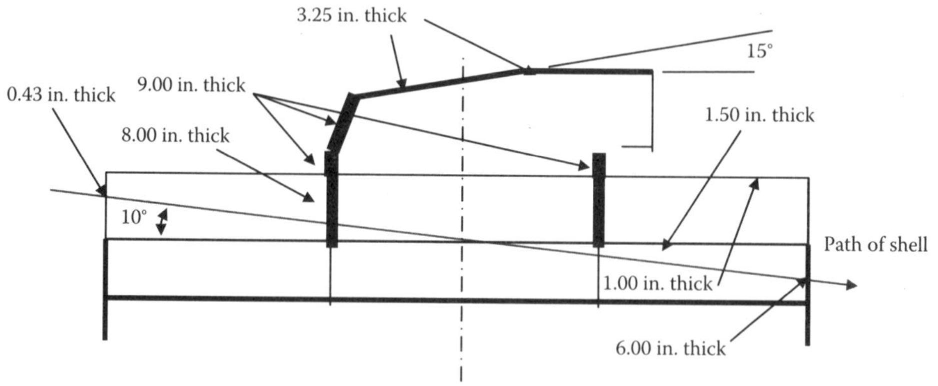

Problem 8

A British short-magazine Lee–Enfield is fired at a sniper plate across no-man's-land in World War I (WWI) (because of the static nature of the fighting, snipers in the opposing lines fired through small holes in thick metal plates to minimize exposure). The projectile has a mass of 175 grains, and the projectile is 1.1 in. long and made of lead (ignoring the copper jacket). The diameter is 0.310 in. The range to the target is 200 yards, so we can assume an impact velocity of 1512 ft/s. The angle of impact is 12.85 min from the normal. Using the Lambert–Zukas model, determine the thickness of armor plate that the projectile will penetrate (i.e., obtain V_{50}).

Assume the density of the armor plate to be $\rho_t = 0.283$ lbm/in.^{3}. Assume the density of lead to be $\rho_p = 0.407$ lbm/in.^{3}.

Problem 9

Using the same information in Problem 8, determine how deep a projectile will penetrate into a 1 in. thick steel sniper plate assuming the impact is normal this time and the impact velocity is 1800 ft/s. Use the Tate formula. The additional target and bullet properties are as follows:

$$\sigma_p = 10{,}000 \left[\frac{\text{lbf}}{\text{in.}^2}\right]$$

$$\sigma_t = 36{,}000 \left[\frac{\text{lbf}}{\text{in.}^2}\right] \quad E_t = 29{,}000{,}000 \left[\frac{\text{lbf}}{\text{in.}^2}\right]$$

The Tate formula shows the energy balance between the projectile and the target. When solved for the penetration depth, the following equation results:

$$P = \frac{\rho_p}{Y_p} \int_{v_c}^{V_i} u(v) l(v) dv$$

In this formula, $u(v)$ is the instantaneous velocity of the base of the hole, v is the instantaneous velocity of the projectile, and $l(v)$ is the instantaneous projectile length. The limits of integration are between the impact velocity V_i and a cutoff velocity v_c that depends on whether the projectile is stronger than the target or not. For this case, we can use

$$v_c = \sqrt{2\frac{R_t - Y_p}{\rho_p}} \quad \text{for } R_t > Y_p$$

The other terms are as follows:

$$u(v) = \left(\frac{1}{1-\gamma^2}\right)\left(v - \gamma\sqrt{v^2 + A}\right)\gamma = \sqrt{\frac{\rho_t}{\vec{\rho}_p}}$$

$$A = \frac{2(R_t - T_p)(1-\gamma^2)}{\rho_t}$$

$$\frac{l(v)}{L} = \left(\frac{v + \sqrt{v^2 + A}}{V_i + \sqrt{V_i + A}}\right)^{\frac{R_t - Y_p}{\gamma Y_p}} \exp\left\{\frac{\gamma\rho_p}{2(1-\gamma^2)Y_p}\left[\left(v\sqrt{v^2 + A} - \gamma v^2\right) - \left(V_i\sqrt{V_i^2 + A} - \gamma V_i^2\right)\right]\right\}$$

where L is the initial length of the projectile. Comment on the results.

Problem 10

At the end of WWI, the German navy surrendered to the British at Scapa Flow (a large anchorage in northern Scotland). When it appeared that surrender negotiations were breaking down, the German sailors opened the sea cocks (valves in the bottom of the ships) and sank most of their ships within sight of astounded British onlookers. In the 1920s, the British raised what ships they could and used them as targets to assess the penetration and bursting characteristic of their heavy shell [5]. One such test was against the side armor of SMS. *Baden*, the largest warship built by Germany during the war. We would like to examine two impacts against the armor of this vessel. In each case, the projectile data are given next:

Projectile information [6]:

$$l = 66\,[\text{in.}]$$

$$d = 15\,[\text{in.}]$$

$$s = 90\,[\text{in.}] \qquad m = 1938\,[\text{lbm}]$$

$$\rho_p = 0.283\left[\frac{\text{lbm}}{\text{in.}^3}\right] \qquad V_s = 1550\left[\frac{\text{ft}}{\text{s}}\right]$$

Armor information:

$$\rho_s = 0.283\left[\frac{\text{lbm}}{\text{in.}^3}\right]$$

For each of the following situations, determine the residual velocity of the projectile (if any):

1. Gundeck	2. B – turretroof
$\theta = 16.5\,[°]$	$\theta = 75.25\,[°]$
$t = 1.187\,[\text{in.}]$	$t = 4\,[\text{in.}]$
$\rho_t = 0.283\left[\frac{\text{lbm}}{\text{in.}^3}\right]$	$\rho_t = 0.283\left[\frac{\text{lbm}}{\text{in.}^3}\right]$

16.2 Penetration and Perforation of Concrete

Concrete-penetrating munitions have always been important in the military arsenal. Bunkers, buildings, and walls are used as cover by an enemy, and it is required to perforate the structure and deliver some type of lethal or nonlethal effect behind the obstruction.

Concrete comes in a variety of forms that have variable strengths, reinforcement geometries, and material properties owing to curing. Each of these forms behaves somewhat differently when impacted by a projectile. There is some evidence that once the impact velocity of a projectile is great enough, one can ignore reinforcement, and only the concrete strength becomes important. As a consequence of the high compressive strength of concrete relative to its tensile strength, it tends to readily spall.

A relatively simple model of projectile penetration into concrete was put forward by Forrestal et al. in 1994 [7]. This model has an advantage in its simplicity. But a slight disadvantage is that its empirical nature makes its global applicability somewhat limited. We shall use this model as a fairly good representation of concrete penetration physics. The model assumes normal impact of the projectile and that the projectile is rigid. This may seem to be restrictive assumptions; however, the method provides reasonable estimates for slightly yawed projectiles if the angle is below about 5° based on this author's own work.

The point of departure is the determination of the force on the nose of the projectile that is defined in a manner similar to a fluid mechanics analysis as

$$F = \frac{\pi d^2}{4}\left(\tau_0 A + NB\rho V^2\right) \tag{16.86}$$

with N defined as

$$N = \frac{8\,\psi - 1}{24\,\psi^2} \tag{16.87}$$

In these equations, the projectile properties are as follows (see Figure 16.7): d is the diameter of the projectile; ψ is the caliber radius head, defined in Equation 16.88; V is the projectile velocity (assuming rigid body motion); and s (used in Equation 16.88) is the ogive radius.

The caliber radius head is defined as

$$\psi = \frac{s}{d} \tag{16.88}$$

The target properties used in Equation 16.86 are as follows: ρ is the density of the target, the product $\tau_0 A$ is a shear strength parameter obtained from a triaxial strength test, and B is a compressive strength parameter. In this model, the parameters are set as

$$B = 1 \tag{16.89}$$

$$\tau_0 A = Sf_c' \tag{16.90}$$

Here S is a dimensionless empirical constant that depends upon the unconfined compressive strength f_c'. If we define the instantaneous depth of penetration as z, we find that for $z > 2d$, we can write

$$F = \frac{\pi d^2}{4}\left(Sf_c' + N\dot{\rho}V^2\right), \quad z > 2d \tag{16.91}$$

This equation is valid for deep penetration depths. For depths less than two projectile diameters, the penetration process is affected by surface cratering. Beyond two projectile diameters, the hole caused by the projectile will be approximately equal to the projectile diameter. This is known as the tunnel region. We shall define the penetration depth as P.

Below two projectile diameters, the damage to the concrete will, in general, be a conical taper called the crater. This is illustrated in Figure 16.8.

In the surface crater region, the force on the projectile nose is proportional to the penetration depth or, mathematically,

$$F = cz, \quad 0 < z < 2d \tag{16.92}$$

Here c is a constant, which we will soon define.

If we begin with Newton's second law, we see that

$$F = ma = m\frac{d^2z}{dt^2} \tag{16.93}$$

Here m is the mass of the projectile. Since we know the force acting on the projectile will tend to slow it down, we can equate Equations 16.92 and 16.93:

$$m\frac{d^2z}{dt^2} = -cz \tag{16.94}$$

We can rewrite Equation 16.94 as

$$\frac{d^2z}{dt^2} = -\omega^2 z \tag{16.95}$$

where we have defined

$$\omega^2 = \frac{c}{m} \tag{16.96}$$

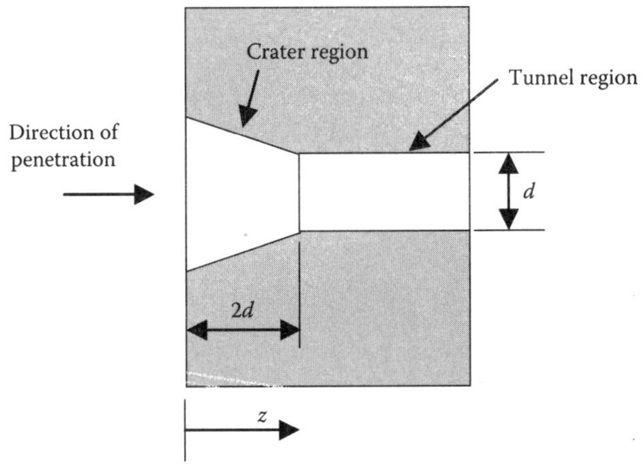

FIGURE 16.8
Illustration of a concrete penetration.

If we assume a solution of the form

$$z = A_1 \sin \omega t \tag{16.97}$$

we can write

$$\frac{\mathrm{d}z}{\mathrm{d}t} = A_1 \omega \cos \omega t \tag{16.98}$$

$$\frac{\mathrm{d}^2 z}{\mathrm{d}t^2} = -A_1 \omega^2 \sin \omega t \tag{16.99}$$

Our initial conditions are such that at $t = 0$, $\mathrm{d}z/\mathrm{d}t = V_s$, where V_s is our striking velocity, so

$$V_s = A_1 \omega \rightarrow A_1 = \frac{V_s}{\omega} \tag{16.100}$$

Then, we have for $z < 2d$

$$z = \frac{V_s}{\omega} \sin \omega t \tag{16.101}$$

$$\frac{\mathrm{d}z}{\mathrm{d}t} = V_s \cos \omega t \tag{16.102}$$

$$\frac{\mathrm{d}^2 z}{\mathrm{d}t^2} = -\omega V_s \sin \omega t \tag{16.103}$$

We now use a compatibility condition that at $z = 2d$, both Equations 16.103 and 16.91 must yield the same answer. We shall call the time it takes the projectile to reach $2d$, t_1, and the velocity at that point will be V_1; thus, at $z = 2d$, we have

$$F|_{t=t_1} = \frac{\pi d^2}{4}\left(Sf_c' + N\rho V_1^2\right), \quad z = 2d \tag{16.104}$$

$$\frac{\mathrm{d}^2 z}{\mathrm{d}t^2} = -\omega V_s \sin \omega t_1, \quad z = 2d \tag{16.105}$$

Since $F = ma$, we can combine the aforementioned equations to write

$$m\omega V_s \sin \omega t_1 = \frac{\pi d^2}{4}\left(Sf_c' + N\rho V_1^2\right), \quad z = 2d \tag{16.106}$$

Also at $t = t_1$, Equations 16.101 and 16.102 can be written as

$$2d = \frac{V_s}{\omega} \sin \omega t_1 \tag{16.107}$$

$$V_1 = V_s \cos \omega t_1 \tag{16.108}$$

We now rearrange Equation 16.108 to

$$V_s = \frac{\omega 2d}{\cos \omega t_1} \tag{16.109}$$

Now insert Equation 16.109 into Equation 16.106, giving us

$$m\omega^2 2d = \frac{\pi d^2}{4}\left(Sf_c' + N\rho V_1^2\right) \tag{16.110}$$

And if we make use of Equation 16.96, we can obtain c as

$$c = \frac{\pi d}{8}\left(Sf_c' + N\rho V_1^2\right) \tag{16.111}$$

We now need to find V_1, which we do by squaring Equations 16.107 and 16.108 and adding them, resulting in

$$V_1^2 + \frac{4cd^2}{m} = V_s^2 \sin^2 \omega t_1 + V_s^2 \cos^2 \omega t_1 \tag{16.112}$$

Making use of a trigonometric identity and rearranging brings us to

$$c = \frac{m}{4d^2}\left(V_s^2 - V_1^2\right) \tag{16.113}$$

If we now equate Equations 16.111 and 16.113, we get

$$V_1^2 = \frac{2mV_s^2 - \pi d^3 Sf_c'}{2m + \pi d^3 N\rho} \tag{16.114}$$

Once we have V_1 and c, the determination of the time t_1 is simply found through the use of Equation 16.108:

$$t_1 = \frac{1}{\omega}\cos^{-1}\left(\frac{V_1}{V_s}\right) = \sqrt{\frac{m}{c}}\cos^{-1}\left(\frac{V_1}{V_s}\right) \tag{16.115}$$

To summarize the analysis procedure for the crater region, we must first find V_1 through the use of Equation 16.114, then we find c through the use of Equation 16.113, and finally, we find t_1 through use of Equation 16.115.

If V goes to zero before time t_1 is reached, the projectile never penetrates deeper than the crater region and our analysis would be complete. The depth of penetration in this case would be found from Equation 16.102:

$$V = 0 = V_s \cos \omega t \tag{16.116}$$

This would occur when

$$\omega t = \frac{\pi}{2} \rightarrow t = \frac{\pi}{2}\sqrt{\frac{m}{c}} \rightarrow \sin \omega t = 1 \tag{16.117}$$

If we insert this result into Equation 16.101, we obtain the achieved depth of penetration P:

$$P = V_s\sqrt{\frac{m}{c}} \tag{16.118}$$

The striking velocity that would make this true would be determined from Equation 16.114 with V_1 set equal to zero. So for a projectile to stop before creating a tunnel, the velocity is given by

$$V_{s_{\text{Notunnel}}} \leq \sqrt{\frac{2\pi d^3 S f_c'}{2m}} \tag{16.119}$$

If the projectile penetrates beyond two diameters into the concrete, it will enter the so-called tunnel region. When the projectile continues into the tunnel region, there is a change in the governing equation as discussed earlier. To determine the depth of penetration, we begin by combining Equations 16.91 and 16.93 to obtain

$$m\frac{\mathrm{d}^2 z}{\mathrm{d}t^2} = \frac{\pi d^2}{4}\left(S f_c' + N\rho V^2\right), \quad 2d < z < P \tag{16.120}$$

We can transform our independent variable from time to distance, and we can write

$$mV\frac{\mathrm{d}V}{\mathrm{d}z} = \frac{\pi d^2}{4}\left(S f_c' + N\rho V^2\right), \quad 2d < z < P \tag{16.121}$$

We rewrite Equation 16.121 as follows:

$$\frac{\mathrm{d}V}{\mathrm{d}z} = \frac{\pi d^2}{4m}\left(\frac{S f_c'}{V} + N\rho V\right) \tag{16.122}$$

Now we integrate it from V_1 to zero and $2d$ to P, so we can write

$$\int_{V_1}^{0}\left(\int_{2d}^{P}\frac{\mathrm{d}V}{\mathrm{d}z}\mathrm{d}z\right)\mathrm{d}V = \int_{V_1}^{0}\left[\int_{2d}^{P}\frac{\pi d^2}{4m}\left(\frac{S f_c'}{V} + N\rho V\right)\mathrm{d}z\right]\mathrm{d}V \tag{16.123}$$

which results in

$$P = \frac{2m}{\pi d^2 N\rho}\ln\left(1 + \frac{N\rho V_1^2}{S f_c'}\right) - 2d, \quad 2d < P \tag{16.124}$$

If we have determined through the use of Equations 16.113 through 16.115 that a projectile will penetrate beyond the tunnel region, we can write a procedure to determine the depth of penetration as follows. First, calculate V_1, c, and t_1 as described earlier for the crater region. Then, calculate P from Equation 16.124. If this is greater than the concrete thickness, the projectile will perforate. If not, the projectile will penetrate to depth P. It would be good to see if a spall thickness is created (as will be described in Section 18.3) by the impact, and if this is the case, we could add the spall thickness to P, and perforation may still result (although with low residual velocity).

Since the dimensionless parameter S is obtained or experimentally verified, it would be nice to know how close we came to our estimate by direct calculation. If one had an experiment where a given projectile penetrated to depth P, we can back calculate S as follows. We start with Equation 16.124 and rearrange thusly

$$S = \frac{N\rho V_1^2}{f_c'}\frac{1}{\left\{\exp\left[(P - 2d)\dfrac{\pi d^2 N\rho}{2m}\right] - 1\right\}} \tag{16.125}$$

In an experiment, we are usually given the striking velocity, so we want to replace V_1 in this equation with V_s, so we use Equation 16.114:

$$S = \frac{N\rho(2mV_s^2 - \pi d^3 S f_c')}{f_c'(2m + \pi d^3 N\rho)} \frac{1}{\left\{\exp\left[(P-2d)\frac{\pi d^2 N\rho}{2m}\right] - 1\right\}} \tag{16.126}$$

which can be simplified to

$$S = \frac{N\rho V_s^2\left(1 - \frac{\pi d^3 S f_c'}{2mV_s^2}\right)}{f_c'\left(1 + \frac{\pi d^3 N\rho}{2m}\right)} \frac{1}{\left\{\exp\left[(P-2d)\frac{\pi d^2 N\rho}{2m}\right] - 1\right\}} \tag{16.127}$$

With this equation, one can find S if you know the striking velocity and the concrete strength. Forrestal et al. [7] calibrated this equation with several experiments. A reproduction of their chart is shown in Figure 16.9 with the addition of upper and lower bounds based on their data. The equation used to determine S given the unconfined compressive strength f_c' is

$$S = 93.48 f_c'^{-0.5603} \tag{16.128}$$

Here recall that f_c' is in megapascals and S is dimensionless. Bounding equations are shown in Figure 16.9. These equations were obtained through use of a curve fit routine.

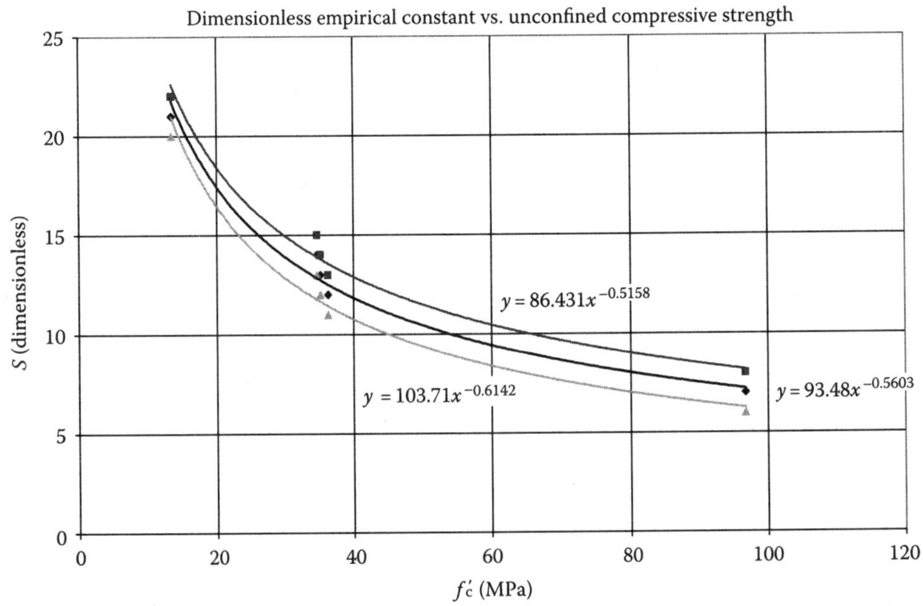

FIGURE 16.9
Determination of dimensionless parameter S for Forrestal et al. [8] concrete penetration model. (Reproduced from Forrestal, M. J. et al., *International Journal Of Impact Engineering*, 15, 395–405, 1994.)

Problem 11

A 0.50-caliber projectile is fired at an extremely thick concrete wall of 2100 psi unconfined compressive strength and density of 0.084 lbm/in.^{3}. It strikes with no obliquity and a 2000 ft/s velocity. How far does it penetrate?

Answer: $P = 20.4$ [cm]

Projectile information:

$$s = 63.50\ [\text{mm}] \quad m = 662\ [\text{grains}]$$

$$d = 12.70\ [\text{mm}] \quad V_s = 2000\left[\frac{\text{ft}}{\text{s}}\right]$$

Problem 12

Areal density is a measure of the mass of 1 in.^{2} (or sometimes a square foot or meter) of a material used as armor. Given a threat projectile with the properties provided later, find the armor solution with the lowest areal density that prevents perforation if your choices are steel with density 0.283 lbm/in.^{3}, concrete of 2500 psi unconfined compressive strength and density of 0.084 lbm/in.^{3}, and aluminum with the following properties:

$$A = 4.418$$

$$B = 1.068$$

$$Y = 39{,}000\ [\text{psi}]$$

$$\rho_t = 0.098\left[\frac{\text{lbm}}{\text{in.}^3}\right]$$

Assume the projectile strikes with 15° of obliquity.

Penetrator information:

$$s = 35\ [\text{mm}]$$

$$d = 12\ [\text{mm}]$$

$$L = 15\ [\text{mm}]$$

$$m = 0.03\ [\text{kg}]$$

$$V_s = 600\left[\frac{\text{m}}{\text{s}}\right]$$

$$\rho_p = 0.283\left[\frac{\text{lbm}}{\text{in.}^3}\right]$$

Problem 13

We would like to compare defenses against the projectile in Problem 10. Assuming this projectile impacts a steel plate at 0° obliquity, do the following:

1. Determine the thickness of the armor required to prevent penetration
2. Determine the thickness of 1500 psi unconfined compressive strength concrete required to do the same, assuming that the concrete does not spall—assume a density of 0.084 lbm/in.^{3} for the concrete—note that this particular shell has a

secant ogive so as an estimate for the purposes of this problem—divide the resultant ogive length by 2—please note that this is simply a guess and not based on physics.
3. Comment on the validity of the answer to part 2.

16.3 Penetration and Perforation of Soils

In recent times, the penetration of soils has gained importance in the terminal ballistic field. Enemy strong points have been encountered below a soil layer. Land mines need to be defeated below various types of soils as well. It is therefore necessary to determine the penetration capabilities of projectiles into soils with the intention of defeating a buried target.

As a reasonable approach to determine soil penetration, we shall use the method of Forrestal and Luk [8]. While other approaches exist, this rather simple procedure is excellent for introducing the physics of the problem.

Soils widely vary in their behavior under penetration loadings. Because the behavior is somewhat complicated, more parameters are needed to describe a soil than a material such as a metal. The first thing we have to realize is that soil can be in a state where the density is less than its locked density. The locked density is where the soil behaves like a solid or fluid in compression (i.e., its states are defined by a hydrostat). We therefore need to introduce two densities: ρ_0, its initial density, and ρ^*, its locked density. We also need to define η^*, its locked volumetric strain. Here we define η^* as

$$\eta^* = 1 - \frac{\rho_0}{\rho^*} \tag{16.129}$$

Two typical models used for soils come directly from our failure theories of structures. They are the Tresca (maximum shear stress) theory and the Mohr–Coulomb theory of failure. Both of these were introduced in Section 4.2. Here we shall use a combination of the two: a Mohr–Coulomb yield criterion with a Tresca flow rule. For the Tresca criterion, once a shear stress failure level is achieved, the material strength is not increased with increasing load. With the Mohr–Coulomb criterion, the yield stress in the material increases with compressive load. The combination of the two allows the material to resist more load as compression is applied up to a point, then further increase in the compressive loading will not affect the material strength.

Similar to the aluminum penetration model, we again define the caliber radius head as

$$\psi = \frac{s}{d} \tag{16.130}$$

where d is the diameter of the projectile; ψ is the caliber radius head; and s is the ogive radius.

We again define nose length as

$$l = \frac{d}{2}\sqrt{4\psi - 1} \tag{16.131}$$

The method considers the resistance force of the soil on the penetrator to have two components: a normal force (normal stresses) and a tangential force (shear stresses and friction). If we lump the shear stress in with the stress owing to friction and furthermore assume that the tangential stress is proportional to normal stress, we can again write

$$\sigma_t = \mu \sigma_n \tag{16.132}$$

where σ_t is the tangential stress; σ_n is the normal stress; and μ is the proportionality constant (a coefficient of sliding friction).

Forrestal et al. [4] developed a formula for the axial force on an ogival nose that we introduced in Section 15.1 and we again use here:

$$F_z = 2\pi s \int_{\theta_0}^{\pi/2} \left\{ \left[\sin\theta - \left(\frac{s - \frac{d}{2}}{s} \right) \right] (\cos\theta + \mu \sin\theta) \right\} \sigma_n (V_z, \theta) \, d\theta \tag{16.133}$$

where

$$\theta_0 = \sin^{-1}\left(\frac{s - \frac{d}{2}}{s} \right) = \sin^{-1}\left(\frac{2\psi - 1}{2\psi} \right) \tag{16.134}$$

Here V_z is the instantaneous velocity during penetration. The stress function $\sigma_n(V_z, \theta)$ is assumed to be similar to that of a spherically symmetric expanding cavity.

At this point, we are going to depart from the mathematics to look at the penetration event in a qualitative manner. Let us assume that we are at some axial location in the ogive of the projectile and we are looking in the direction of penetration at time t. What we would see is illustrated in Figure 16.10. The projectile would be opening a cavity at a rate that we shall call V_t. The plastic zone would be expanding at some rate ct. Here c is the speed of the

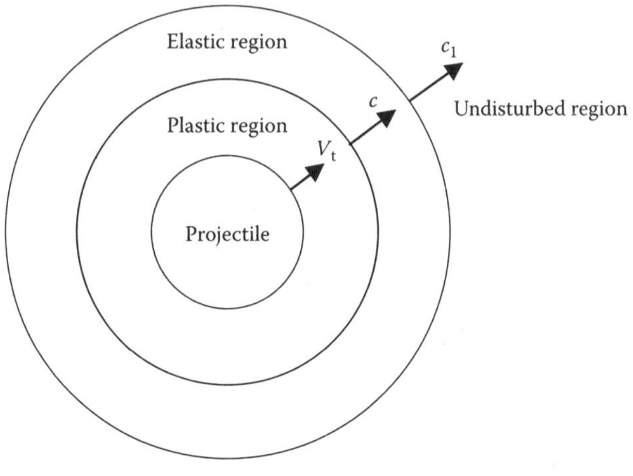

FIGURE 16.10
Elastic and plastic compression zones at a section of an ogive penetrating into soil looking in the direction of penetration.

plastic wave (dependent upon the Hugoniot jump conditions to be discussed in Section 18.1). The elastic zone would be expanding at a rate $c_1 t$. Here c_1 is the speed of the dilatational wave in the material. It can be shown that V and c are related (as per the shock theory that will be discussed in Chapter 18), and we can define a parameter γ as

$$\gamma = \frac{V}{c} = \left[\left(1+\frac{\tau_c}{2E}\right)^3 - (1-\eta^*)\right]^{\frac{1}{3}} \tag{16.135}$$

where E is Young's modulus. With the aforementioned physics, Forrestal and Luk [8] derived material response models for each of the three failure models we have discussed earlier. For the detailed derivation, the interested reader is referred to that paper. The basic idea was to have a general function for the force acting on the projectile nose that we can integrate using Newton's second law to obtain the velocity and penetration distance as a function of time.

If we insert expressions that relate the radial expansion velocity of the cavity V to the projectile penetration velocity V_z, we can put the expression for the retarding force in the form

$$F_z = \alpha_s + \beta_s V_z^2 \tag{16.136}$$

where

$$\alpha_s = \frac{\pi d^2}{4}\tau_c A\left[1+4\mu\psi^2\left(\frac{\pi}{2}-\theta_0\right) - \mu(2\psi-1)\sqrt{(4\psi-1)}\right] \tag{16.137}$$

$$\beta_s = \frac{\pi d^2}{4}\rho_0 B\left[\frac{8\psi-1}{24\psi^2} + \mu\psi^2\left(\frac{\pi}{2}-\theta_0\right) - \frac{\mu(2\psi-1)(6\psi^2+4\psi-1)\sqrt{4\psi-1}}{24\psi^2}\right] \tag{16.138}$$

Here the coefficients A and B are dependent upon the material model used for the soil.

Recall that the definition for the Tresca criterion implies that once a material reaches its state of maximum shear stress, it begins to plastically deform and cannot support any more load. For a soil that behaves in a Tresca-type manner, we have

$$A = \frac{2}{3}\left\{1 - \ln\left[\frac{\left(1+\frac{\tau_c}{2E}\right)^3 - (1-\eta^*)}{\left(1+\frac{\tau_c}{2E}\right)^3}\right]\right\} \tag{16.139}$$

$$B = \frac{3}{2(1-\eta^*)} + \frac{\frac{3\tau_c}{E} + \eta^*\left(1-\frac{3\tau_c}{2E}\right)^2}{\left[\left(1+\frac{\tau_c}{2E}\right)^3 - (1-\eta^*)\right]^{\frac{2}{3}}} - \frac{\left[\left(1+\frac{\tau_c}{2E}\right)^3 - (1-\eta^*)\right]^{\frac{1}{3}}}{2\left(1+\frac{\tau_c}{2E}\right)^4}\left[1+\frac{3\left(1+\frac{\tau_c}{2E}\right)^3}{(1-\eta^*)}\right] \tag{16.140}$$

Recall that the definition for the Mohr–Coulomb criterion implies that as the compressive forces increase, it becomes harder to have the material fail in shear. For a soil that behaves in a Mohr–Coulomb-type manner, we have

$$A = \frac{1}{\alpha}\left(\frac{1+\dfrac{\tau_c}{2E}}{\gamma}\right)^{2\alpha} - \frac{1}{\lambda} \tag{16.141}$$

$$B = \frac{3}{(1-\eta^*)(1-2\alpha)(2-\alpha)}$$

$$+\frac{1}{\gamma^2}\left(\frac{1+\dfrac{\tau_c}{2E}}{\gamma}\right)^{2\alpha}\left\{\frac{3\tau_c}{E}+\eta^*\left(1-\frac{3\tau_c}{2E}\right)^2 - \frac{\gamma^3\left[2(1-\eta^*)(2-\alpha)+3\gamma^3\right]}{(1-\eta^*)(1-2\alpha)(2-\alpha)\left(1+\dfrac{\tau_c}{2E}\right)^4}\right\} \tag{16.142}$$

Also note that the Tresca criterion behaves the same as the Mohr–Coulomb criterion with $\lambda = 0$. We define

$$\alpha = \frac{3\lambda}{3+2\lambda} \tag{16.143}$$

Because of a singularity in the governing equations, there is a special set of equations for the Mohr–Coulomb criterion when we have $\lambda = 3/4$. In this case,

$$A = 2\left(\frac{1+\dfrac{\tau_c}{2E}}{\gamma}\right) - \frac{4}{3} \tag{16.144}$$

$$B = \frac{-2\ln\gamma}{(1-\eta^*)} + \frac{\left(1+\dfrac{\tau_c}{2E}\right)\left[\dfrac{3\tau_c}{E}+\eta^*\left(1-\dfrac{3\tau_c}{2E}\right)^2\right]}{\gamma^3} - \frac{2}{3}\left[\frac{1}{\left(1+\dfrac{\tau_c}{2E}\right)^3} - \frac{3\ln\left(1+\dfrac{\tau_c}{2E}\right)}{(1-\eta^*)}\right] \tag{16.145}$$

When a material behaves according to the model that combines both Mohr–Coulomb and Tresca behaviors, things become slightly more complicated. Parameters *A* and *B* will be dependent upon the rate of loading. One must keep in mind that this failure criterion implies that up to some stress level, the material will have improved resistance to compressive loading because of the internal friction of the grains, and after a limit load is reached (τ_m), the material simply yields regardless of load. Thus, we can consider three velocity regimes: $V < V_{min}$, where the yielding is completely Mohr–Coulomb behavior; $V_{min} < V < V_{max}$, where the yielding closest to the projectile is by Tresca criterion and the yielding near the elastic–plastic interface is according to the Mohr–Coulomb criterion; and $V > V_{max}$, where the entire yield region is according to the Tresca model. We shall consider each of these cases.

If $V < V_{min}$, we stated that the yielding is completely according to the Mohr–Coulomb model. Thus, Equations 16.141 through 16.145 apply. The equation required to determine V_{min} is

$$V_{\min} = \sqrt{\frac{\tau_c}{\alpha \rho_0 B}\left[\frac{\tau_m}{\tau_c} - \left(\frac{1+\frac{\tau_c}{2E}}{\gamma}\right)^{2\alpha}\right]} \tag{16.146}$$

Recall that τ_m is the stress level at which the material behaves according to the Tresca model. We shall discuss how we determine V shortly.

If $V_{\min} < V < V_{\max}$, the zone of yielding material has two subzones: a zone next to the projectile that behaves according to the Tresca model and a zone next to the elastic region that behaves according to the Mohr–Coulomb model. We shall first write the equation for $V_{\max}$:

$$V_{\max} = \sqrt{\frac{\tau_c \gamma^2}{\rho_0\left[\frac{3\tau_c}{E} + \eta^*\left(1-\frac{3\tau_c}{2E}\right)^2\right]}\left[\frac{1}{\alpha}\left(\frac{\tau_m}{\tau_c}\right) - \frac{1}{\lambda} - \frac{2}{3}\right]} \tag{16.147}$$

If we define a coordinate ξ that varies from 0 at the projectile surface to 1 at the elastic–plastic interface, we can determine a coordinate ξ_m, where the yield behavior changes from Tresca to Mohr–Coulomb. Unfortunately, this crossover point has to be numerically solved with the equation that follows:

$$\frac{\left(1+\frac{\tau_c}{2E}\right)^{2\alpha}}{\left[(1-\eta^*)\xi_m^3+\gamma^3\right]^{\frac{2\alpha}{3}}} + \frac{\alpha\rho_0 V^2}{\tau_c \gamma^2}\left\{\frac{\gamma^3\left[2(1-\eta^*)(2-\alpha)\xi_m^3+3\gamma^3\right]}{(1-\eta^*)(1-2\alpha)(2-\alpha)\left[(1-\eta^*)\xi_m^3+\gamma^3\right]^{\frac{4}{3}}}\right\}$$
$$+\frac{\alpha\rho_0 V^2\left(1+\frac{\tau_c}{2E}\right)^{2\alpha}}{\tau_c\gamma^2\left[(1-\eta^*)\xi_m^3+\gamma^3\right]^{\frac{2\alpha}{3}}}\left\{\frac{3\tau_c}{E}+\eta^*\left(1-\frac{3\tau_c}{2E}\right)^2 - \frac{\gamma^3\left[2(1-\eta^*)(2-\alpha)+3\gamma^3\right]}{(1-\eta^*)(1-2\alpha)(2-\alpha)\left(1+\frac{\tau_c}{2E}\right)^4}\right\} - \frac{\tau_m}{\tau_c} = 0 \tag{16.148}$$

Keep in mind here that we know all the information (including V) and we are solving for ξ_m. A good math code will generally solve this equation quickly.

Once we have ξ_m, then A and B are given at the projectile surface (Tresca) by

$$A = \frac{1}{\alpha}\left(\frac{\tau_m}{\tau_c}\right) - \frac{1}{\lambda} + \frac{2}{3}\left(\frac{\tau_m}{\tau_c}\right)\ln\left[1+(1-\eta^*)\left(\frac{\xi_m}{\gamma}\right)^3\right] \tag{16.149}$$

$$B = \frac{1}{2(1-\eta^*)}\left\{3 - \frac{3+4(1-\eta^*)\left(\frac{\xi_m}{\gamma}\right)^3}{\left[1+(1-\eta^*)\left(\frac{\xi_m}{\gamma}\right)^3\right]^{\frac{4}{3}}}\right\} \tag{16.150}$$

These equations account for the fact that the yielding is Mohr–Coulomb outside of $\xi = \xi_m$.

If $V > V_{\max}$, the yielding is completely according to the Tresca model. Thus, A and B are given by

$$A = \frac{2}{3} - 2\left(\frac{\tau_m}{\tau_c}\right)\ln\left[\frac{\gamma}{\left(1+\frac{\tau_c}{2E}\right)}\right] \tag{16.151}$$

$$B = \frac{3}{2(1-\eta^*)} + \frac{\frac{3\tau_c}{E} + \eta^*\left(1-\frac{3\tau_c}{2E}\right)^2}{\gamma^2} - \frac{\gamma}{2\left(1+\frac{\tau_c}{2E}\right)^4}\left[1+\frac{3\left(1+\frac{\tau_c}{2E}\right)^3}{(1-\eta^*)}\right] \tag{16.152}$$

To approximate the normal stress on the ogive, we can replace the spherically symmetric velocity V in our previous equations with $V_z \cos\theta$. We can write an equation for the normal stress function on the ogive [6] as

$$\sigma_n(V_z, \theta) = \tau_c A + \rho_0 B[V_z \cos\theta]^2 \tag{16.153}$$

We can write Newton's second law as

$$-F_z = m\frac{dV_z}{dt} \tag{16.154}$$

We can then convert this time integral to a distance integral as before to yield

$$-F_z = mV_z\frac{dV_z}{dz} \tag{16.155}$$

If we substitute Equation 15.136 into the aforementioned equation and integrate, we get an equation for the acceleration, velocity, and depth of penetration, respectively, as a function of time:

$$a = -\frac{\frac{\alpha_s}{m}}{\cos^2\left[\tan^{-1}\left(\sqrt{\frac{\beta_s}{\alpha_s}}V_0\right) - \frac{t\sqrt{\alpha_s\beta_s}}{m}\right]} \tag{16.156}$$

$$V_z = \sqrt{\frac{\alpha_s}{\beta_s}}\tan\left[\tan^{-1}\left(\sqrt{\frac{\beta_s}{\alpha_s}}V_0\right) - \frac{t\sqrt{\alpha_s\beta_s}}{m}\right] \tag{16.157}$$

$$z = \frac{m}{\beta_s}\ln\left\{\frac{\cos\left[\tan^{-1}\left(\sqrt{\frac{\beta_s}{\alpha_s}}V_0\right) - \frac{t\sqrt{\alpha_s\beta_s}}{m}\right]}{\cos\left[\tan^{-1}\left(\sqrt{\frac{\beta_s}{\alpha_s}}V_0\right)\right]}\right\} \tag{16.158}$$

If we determine the distance where the velocity of the projectile slows to zero, we obtain the depth of penetration as

$$P = \frac{m}{2\beta_s}\ln\left(1+\frac{\beta_s V_0^2}{\alpha_s}\right) \tag{16.159}$$

where P is the final penetration depth and V_0 is the impact velocity.

So now that we have developed penetration formulas for soils, what do we do with them? The use of models such as this one, as nice as it is, usually carries with it some practical issues. A detailed model like this requires detailed material properties that, in practice, one rarely has. It will usually require a test or two to calibrate it. Forrestal and Luk [8] suggested using a value of 0.13 for η^*. The authors claim the model is relatively insensitive to it. The model was derived for normal penetration, but the authors claim good results up to impact yaw angles of 30°. In this case, they used the line-of-sight penetration depth. As one might expect, the accuracy of this particular model significantly varies with the properties of the soil. Rocks, roots, and soil layers further complicate everything. Nevertheless, the model is an excellent tool and describes the physics of soil penetrations well. This is a highly active area of current research.

Problem 14

A 0.50-caliber projectile is fired at a soil berm with properties established later. How far does it penetrate?

Answer: $P = 84.9$ [cm]

Projectile information:

$$s = 63.50\,[\text{mm}] \quad m = 662\,[\text{grains}]$$

$$d = 12.70\,[\text{mm}] \quad V_s = 2000\left[\frac{\text{ft}}{\text{s}}\right]$$

Soil information (assume Mohr–Coulomb behavior):

$$\text{Initial density } \rho_0 = 1860\left[\frac{\text{kg}}{\text{m}^3}\right] \quad \mu = 0.1$$

$$\text{Locked density } \rho* = 2125\left[\frac{\text{kg}}{\text{m}^3}\right] \quad \lambda = 0.33$$

$$\tau_c = 1500\left[\frac{\text{lbf}}{\text{in.}^2}\right] \quad E = 2\times 10^7\left[\frac{\text{lbf}}{\text{in.}^2}\right]$$

$$\tau_m = 2500\left[\frac{\text{lbf}}{\text{in.}^2}\right]$$

16.4 Penetration and Perforation of Ceramics

The desire to decrease the weight of vehicles coupled with constant improvements in manufacture has increased interest in the use of ceramics as armor. The design of an armored vehicle using ceramics requires an understanding of their behavior under impact loads. The advantages of ceramic armor are its relatively low density, high hardness, and high compressive strength. The disadvantages are that ceramics are usually brittle; they have low tensile strength that when coupled with high compressive strength can be a problem from a spallation standpoint; they allow the protection to be degraded in a multihit situation; and they are somewhat expensive. Their complex structural behavior makes them difficult to model, although this is a disadvantage only to the designers.

The response of a ceramic to penetration is unique among those of all the other materials discussed in this chapter. The material behaves differently depending on the radial confinement and whether it is backed or not. For reasons such as spallation, they are usually backed by a fiber-reinforced composite, plastic, elastomer, or metal plate. If a ceramic is not backed, it will most likely spall when subjected to a high-shock load. This spallation can be analyzed by the techniques we will discuss in Chapter 18 on shock theory.

Although not exhaustive, this is a list of common ceramics currently either in use or being studied for armor applications:

- Boron carbide (B_4C)
- Silicon carbide (SiC)
- Titanium diboride (TiB_2)
- Aluminum nitride (AlN)
- Alumina (Al_2O_3)

Historically, terra cotta (ceramic) armor has been found in Chinese tombs dating from 400 BC. Before WWI, the practice of placing coal bunkers around magazines to take advantage of comminution, a phenomenon that we shall discuss shortly, is another example of energy that is mirrored in ceramics.

If a ceramic is backed, one can take advantage of its high compressive strength to resist penetration. This will cause the tip of the penetrator to deform. Large stresses then build up in the penetrator. If the striking velocity is low enough, the penetrator will break up or ricochet. This process is called interface defeat or infinite dwell. If the penetrator survives the initial impact, the ceramic begins to fail. This process is complicated, which is why it is difficult to model, but it is key to understanding the behavior and utilizing the ceramic to the maximum extent possible.

The ceramic penetration process was documented by Cheeseman [9]. After an initial dwell and several reflections of the shocks and rarefactions, the following events occur and will either continue to perforation or stop when the penetration is arrested. Initially, tensile cracks appear near the penetrator forming circular rings. These cracks propagate along the principal stress planes that are usually 25–75° from the surface normal.

Once the cracks reach the distal boundary, they coalesce into conical form. At this point, if the ceramic was not backed, a plug would be ejected and the material would be perforated. If the plate is backed, then at the time when the conoid is formed, the stress is circumferentially redistributed and radial cracks appear. After the appearance of radial cracks, lateral cracking in the plane of the impact surface forms. This process is illustrated in Figure 16.11. With backing material present that holds the ceramic plug in place, the material has nowhere to go, so microcracking begins. This pulverizes the ceramic material. This is known as the comminuted zone. The process of comminution and the sand-like character of the comminuted material erode the penetrator at a rapid rate. The powdered material continually gets in the way of the penetrator. This material radially flows outward and rearward. A similar effect occurs during shaped charge jet penetration into sand bags.

The penetration of a ceramic armor is highly dependent upon the boundary conditions. It is known that confinement increases the penetration resistance (increasing V_{50}). This effect is not because of the strength of the confinement material. A stiffer backing also increases V_{50} to a point. There does appear to be an upper limit though. The key to good design appears to be the movement of the neutral axis out of the ceramic material and into the backing material [9].

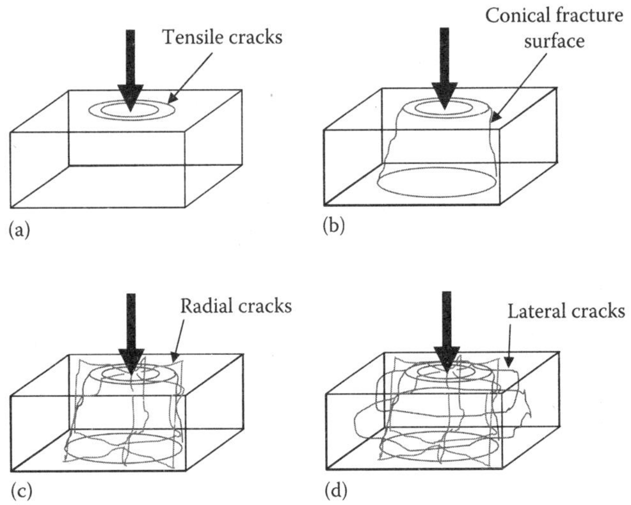

FIGURE 16.11
Ceramic fracture process illustrated: (a) tensile crack formation, (b) conical fracture surface formation, (c) radial crack formation, and (d) lateral crack formation.

One of the parameters that must be considered when designing ceramic armor is the fact that there can be large dynamic deflections during an impact. This can be more than twice the static deflection left after a penetration event. Care must be taken in mounting sensitive components in the sway space of the armor. Impact to the component may impede fighting efficiency of the vehicle. This effect is still being investigated.

One can see from the process that modeling this event (either numerically or analytically) is nontrivial. The numerical approach is the subject of intense research. A cursory look at the problem shows that we need a model for the ceramic before fracture, a crack propagation model, a microcracking model, and a model that handles the comminution, and after all that, we have to model the behavior of the backing material.

Florence [10] developed a simplified model to determine the limit velocity for an aluminum-backed ceramic armor plate. This model assumes that the projectile was a short cylindrical rod, and the conoid is idealized and the loading on the backing plate is assumed to occur across the base of this conoid. The backing material is assumed to fail when the maximum strain in it exceeds its failure strain:

$$\varepsilon_{\mathrm{r}} = 1.82 f(a) \frac{K}{S} \tag{16.160}$$

Here ε_{r} is the maximum strain in the aluminum, and the other parameters are defined later. The parameter K is the kinetic energy of the penetrator given by

$$K = m_{\mathrm{p}} \frac{V_{\mathrm{s}}^2}{2} \tag{16.161}$$

where m_{p} is the penetrator mass and V_{s} is the striking velocity.

The strength parameter S is given by

$$S = \sigma_{Y} h_{\mathrm{m}} \tag{16.162}$$

where σ_{Y} is the aluminum yield strength and h_{m} is the thickness of the aluminum plate.

The momentum parameter $f(a)$ is given by

$$f(a) = \frac{m_\mathrm{p}}{\pi a^2\left[m_\mathrm{p} + (m_\mathrm{c} + m_\mathrm{m})\pi a^2\right]} \quad (16.163)$$

Here the mass subscripts p, c, and m refer to the masses of the projectile, ceramic, and backing plate material, respectively. We can rearrange these formulas to obtain the limit velocity of the projectile–armor combination [8] as

$$V_l = \sqrt{\frac{\varepsilon_\mathrm{T} S}{(0.91)m_\mathrm{p} f(a)}} \quad (16.164)$$

This can be used exactly like the limit velocity in the Lambert model of Section 16.1.

More complicated models exist for ceramic penetrations. Walker and Anderson [11] proposed a penetration model for ball ammunition penetrating ceramic backed by a metal plate. The model assumes axisymmetric behavior, that a velocity profile in both the target and the penetrator can be analytically specified, that the rear of the projectile experiences only elastic waves (i.e., the plastic waves are arrested before reaching the rear surface), and that the shear behavior of the target can be specified as a pressure-dependent flow stress (Mohr–Coulomb) for the ceramic with a constant flow shear stress (von Mises) for the metal. The model is quite detailed, and limitations on space prevent the inclusion of the model here; however, the interested reader is directed to the paper for a full description of the model. The model still has to be solved by computer, but the nice thing is that one can program it into MathCAD or MATLAB® and quickly make many calculations. Since the model uses readily available parameters, it can be run for any materials consistent with the velocity and material behavior assumptions. The authors claim 15% accuracy, which is good.

Zaera and Sanchez-Galvez [12] proposed an interesting penetration model based on Tate's penetration equation. The model is elegant for its simplicity and seems to correlate well with medium-caliber ammunition. The model neglects mushrooming of the projectile and includes only deformation because of erosion. It assumes rigid, perfectly plastic behavior in a zone confined to be near the projectile tip.

The three basic equations are as follows. For the penetration velocity u, we have

$$\frac{1}{2}\rho_\mathrm{p}(v-u)^2 + Y_\mathrm{p} = \frac{1}{2}\rho_\mathrm{t} u^2 + R_\mathrm{t} \quad (16.165)$$

where ρ_p is the density of the projectile material; ρ_t is the density of the target material; v is the projectile velocity; u is the penetration velocity; Y_p is the dynamic yield strength in the projectile; and R_t is the ballistic resistance of the target.

The time rate of change of the projectile length owing to erosion is

$$\frac{\mathrm{d}L}{\mathrm{d}t} = -(v-u) \quad (16.166)$$

Here L is the length of the projectile. Finally, the deceleration of the projectile is given by

$$\frac{\mathrm{d}v}{\mathrm{d}t} = -\frac{Y_\mathrm{p}}{\rho_\mathrm{p} L} \quad (16.167)$$

At some point in time, the pressure on the projectile nose will be unable to erode it further; thus, Equation 16.167 will switch to

$$\frac{dv}{dt} = -\frac{R_t + \frac{1}{2}\rho_t v^2}{\rho_p L} \tag{16.168}$$

To simplify the geometry in the model, the concept of equivalent length is invoked. In this case, the length is adjusted based on the amount of material present in the projectile. The equivalent diameter is given by

$$d_{eq} = \frac{\int_0^{L_p} d^3(z)dz}{\int_0^{L_p} d^2(z)dz} \tag{16.169}$$

The equivalent length is then

$$L_{eq} = \frac{4m_p}{\pi\rho_p d_{eq}^2} \tag{16.170}$$

As described earlier, when a projectile impacts ceramic armor, a fracture conoid develops after interaction of the stress waves with the boundaries. We shall assume the time for this event to be

$$t_{conoid} = \frac{h_c}{c_L} + \frac{h_c}{v_{rad.crack}} \tag{16.171}$$

where h_c is the thickness of the ceramic (shown in Figure 16.12); c_L is the longitudinal wave speed in the material; and $v_{rad.crack}$ is the speed of radial crack growth.

We shall also assume, based on observations [10], that

$$v_{rad\cdot crack} = \frac{1}{5}c_L \tag{16.172}$$

During the penetration event, assuming the limit velocity is exceeded, the projectile tip will meet the crack front at some time. This will effectively change the mode of penetration. The equation for this is given by

$$z + s_{crack} = h_c \tag{16.173}$$

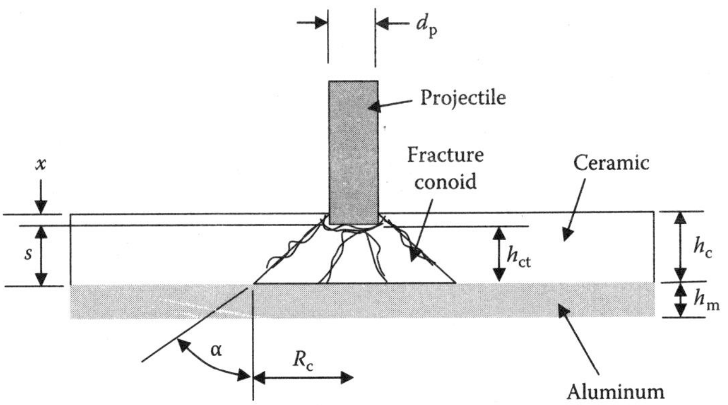

FIGURE 16.12
Model of Zaera and Sánchez-Gálvez illustrated.

The linear momentum equation assumes a constant velocity in the projectile (v), a jump discontinuity in velocity based on our flow rule at the ceramic–projectile interface (u), and a uniform velocity (w) in the metal backing plate. If we call p_c the momentum, we can write

$$\frac{dp_c}{dt} = Y_c \pi \frac{d_{eq}^2}{4} - f_m \pi R_c \tag{16.174}$$

where f_m is the force exerted by the backing plate; R_c is the base radius of the fracture conoid; and Y_c is the penetration strength of the ceramic.

If we define h_{ct} as the instantaneous thickness of the ceramic and α as the conoid semiapex angle, we can define R_c based on geometry as

$$R_c = \frac{d_{eq}}{4} + h_{ct} \tan \alpha \tag{16.175}$$

We can now integrate Equation 16.174 to yield

$$p_c = \pi \rho_c h_{ct} \left[u \left(\frac{d_{eq}^2}{16} + \frac{R_c^2}{12} + \frac{d_{eq} R_c}{12} \right) + w \left(\frac{d_{eq}^2}{48} + \frac{R_c^2}{4} + \frac{d_{eq} R_c}{12} \right) \right] \tag{16.176}$$

Since w was introduced, we have to alter Equation 16.165 to

$$\frac{1}{2} \rho_p (v-u)^2 + Y_p = \frac{1}{2} \rho_t (u-w)^2 + R_t \tag{16.177}$$

Keep in mind here that when the projectile is in the ceramic, $R_t = Y_c$, and when in the backing plate, $R_t = Y_m$.

Once the ceramic fractures and is comminuted, then its strength is significantly reduced. This is accounted for by using

$$Y_c = \begin{cases} Y_{co}, & t \le t_{conoid} \\ Y_{co} \left(\dfrac{u-w}{u_{phase1}} \right)^2 & t > t_{conoid} \end{cases} \tag{16.178}$$

In this expression, u_{phase1} is the value of u at $t = t_{conoid}$.

Zaera and Sanchez-Galvez chose an energy approach to the penetration of the metal backing plate. The work dissipated by plastic deformation is given by

$$E_p = \pi h_m Y_m \delta \left(\frac{2}{3} h_m + \frac{2}{3} \delta \right) \tag{16.179}$$

where δ is the deflection at the center of the plate and the subscript m refers to the backing plate itself.

The time rate of change of plastic work is then

$$\frac{dE_p}{dt} = \pi h_m Y_m \frac{d\delta}{dt} \left(\frac{2}{3} h_m + \delta \right) = \pi h_m Y_m w \left(\frac{2}{3} h_m + \delta \right) \tag{16.180}$$

The work to deform the interface is given by

$$T = \pi R_c^2 f_m \delta \tag{16.181}$$

Therefore, the time rate of change of this work is

$$\frac{dT}{dt} = \pi R_c^2 f_m \frac{d\delta}{dt} = \pi R_c^2 f_m w \tag{16.182}$$

The kinetic energy of the plate material is

$$E_k = \frac{1}{2}\pi R^2 h_m \rho_m w^2 \tag{16.183}$$

It then follows that the time rate of change of kinetic energy is

$$\frac{dE_k}{dt} = \pi R^2 h_m \rho_m w \frac{dw}{dt} \tag{16.184}$$

Equating Equations 16.180, 16.182, and 16.184 gives us

$$R_c^2 f_m = h_m Y_m \left(\frac{2}{3}h_m + \delta\right) + R^2 h_m \rho_m \frac{dw}{dt} \tag{16.185}$$

When the projectile reaches the backing plate, the equation for the deceleration can be written as

$$\frac{dv}{dt} = -\frac{Y_m + \frac{1}{2}\rho_m (v - w)^2}{\rho_p L} \tag{16.186}$$

We can use Equation 16.180 once more for the time rate of change of plastic energy and modify Equation 16.182 for the time rate of change of work as

$$T = \pi \frac{d_{eq}^2}{4} Y_m \delta \tag{16.187}$$

and differentiating with respect to time

$$\frac{dT}{dt} = \pi \frac{d_{eq}^2}{4} Y_m \frac{d\delta}{dt} = \pi \frac{d_{eq}^2}{4} Y_m w \tag{16.188}$$

The kinetic energy for the backing plate is

$$E_k = \frac{1}{2} m_m w^2 \tag{16.189}$$

Its time rate of change is

$$\frac{dE_k}{dt} = m_m w \frac{dw}{dt} \tag{16.190}$$

This leads us to the equation for the deceleration in the plate as

$$\frac{dw}{dt} = \frac{\pi \frac{d_{eq}^2}{4} Y_m - \pi h_m Y_m \left(\frac{2}{3}h_m + \delta\right)}{m_m} \tag{16.191}$$

We need to define m_m as the effective mass of the plate given by

$$m_m = \pi \rho_m \left[R^2 h_m - \frac{d_{eq}^2}{4}(h_m - h_{mt}) \right] \quad (16.192)$$

In this case, h_{mt} is the distance left to the free surface of the plate (i.e., distance remaining to be penetrated).

The armor is said to be perforated when

$$h_{mt} = 0 \quad (16.193)$$

This would be a piercing-/petaling-type perforation. Additionally, the armor can be defeated by plugging if at any time

$$v = w \quad (16.194)$$

Thus, the plug and projectile would be moving at the same rate.

The line-of-sight thickness can be used to handle obliquity. Thus, we would set

$$h_t' = \frac{h_t}{\cos \theta} \quad (16.195)$$

where θ is our obliquity angle measured from the plate normal. We also have to be careful that all our measurements are transformed to these lengths. Physical data show that after about a 20° obliquity, the fracture of the ceramic starts to deviate from this model. The authors show fair agreement up to 50° [12].

This model is much simpler than others and provides reasonable results. Unfortunately, it still has to be coded into a computer to solve the equations simultaneously (and as the penetrator moves into the backing plate, sequentially). It is nice because it can account for obliquity. If one generally has to get more detailed than this, direct numerical simulation is probably the best approach.

We have presented some analytic equations for the penetration of ceramic armor by projectiles. Ceramics are nearly always used with some type of backing plate. These models, although fairly complicated, allow rapid analysis of designs. They do, however, need to be coded to be used. If more detailed results are required, one must resort to direct numerical simulation.

16.5 Penetration and Perforation of Composites

Composites have arguably been used as armor materials since the Middle Ages. Advantages of using composite materials are their relatively low density, their tailorable properties, and fair to high strength. The disadvantages of composites are the inconsistency of hand layup, the dependency of strength on manufacturing process, and the somewhat expensive nature of their manufacture. Additionally, composites pose a problem to the designers because they are difficult to model.

Composites resist penetration primarily by dissipating energy. Because of the complex structure of the material, this energy dissipation manifests itself in the failure of portions of the laminate, fiber breakage, matrix cracking, and delamination. Since composite properties can vary from isotropic to a complicated anisotropic, the behavior will depend upon the configuration.

In chopped fiber composites, the material properties are usually isotropic. A notable exception to this is in injection moldings where the fibers tend to align with the flow directions near mold gate areas or areas of higher-velocity flow. An isotropic composite is usually treated as we do a metal, and those formulas should work well. We recommend that one try the Lambert model first or the Tate model.

Continuous fiber composites behave differently from metallic plates. A typical load–displacement curve is shown in Figure 16.13. In this figure, after an initial delamination point, where the load-carrying capability is degraded, we see increases and decreases in load-carrying ability based on successive delaminations of material followed by a final plug shear out. This delamination actually promotes energy dissipation by forcing the fibers to elongate. In many composites, shear failure of the fibers as well as tensile failures dominates during an impact [1].

It is extremely difficult to obtain an analytical model for the penetration of continuous fiber composites. This is due to the change of energy dissipation as the composite is damaged. Finite element methods have been utilized to determine limit velocities [1], but there are nuances to each analysis that must be explained.

The first issue that must be dealt with is how to handle the damage and its effect on the remaining strength of the composite. Some researchers have actually modeled each lamina with its correct directional properties and assumed a failure criteria based on interlaminar shear strength [1]. When the interlaminar shear strength is exceeded, the layer no longer supports shear and the overall bending stiffness is reduced. This can be accounted for explicitly having the model change internal constraints between layers or implicitly by tracking the overall smeared bending stiffness of the composite and reducing it based on the lamina that failed. Another means of handling the behavior of the composite is to average the stiffness change because of the progressive failure of lamina as shown in Figure 16.14 [1]. The issue with this approach is that test data from some sort of penetration event are required.

A second issue with analyzing fiber-reinforced composites is the actual failure of the fibers themselves. The fibers can themselves delaminate from the matrix. They can also fail in tension and are usually very sensitive to fracture. These issues of necessity complicate the analysis.

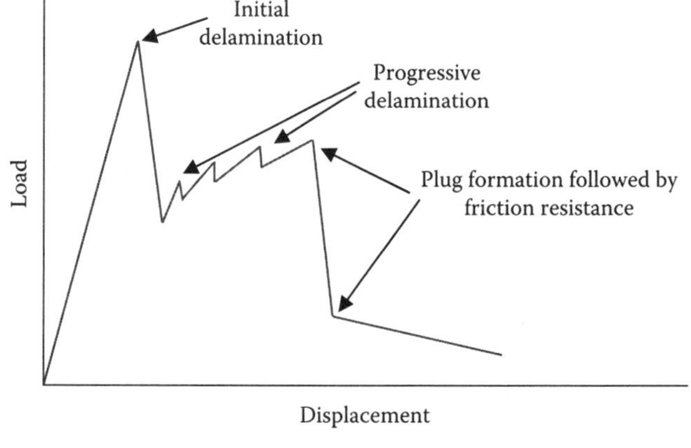

FIGURE 16.13
Load–displacement curve for a typical continuous fiber-reinforced composite.

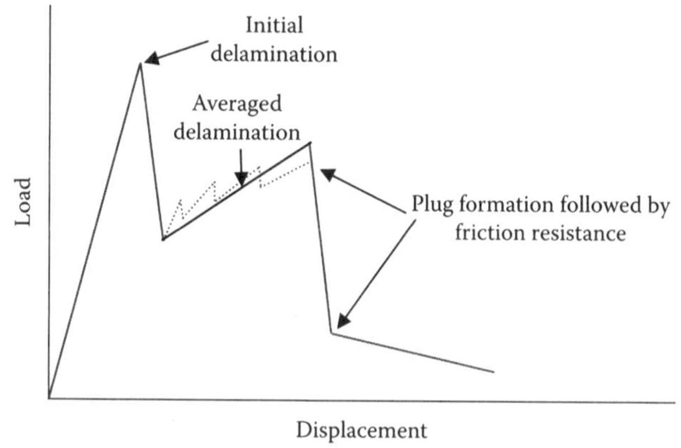

FIGURE 16.14
Load–displacement curve for a typical fiber-reinforced composite modeled with averaged properties after initial delamination.

Cheeseman [9] performed extensive work in the area of composite materials under impact loads and made the following observations regarding their behavior. First, delaminations tend to prefer moving along the fiber direction. Additionally, the compression of the composite material (e.g., at clamped locations) tends to suppress delamination, as one would expect. The extent of delamination linearly increases as the distal surface is approached if perforation occurs. However, the delamination increases then decreases if no penetration occurred. As the impact velocity increases, the delamination decreases, indicating that the bending of the target becomes less significant. This is illustrated in Figure 16.15.

With the information presented here, we have seen that the penetration of composite armor is by no means simple. We have discussed some issues with modeling these types of materials and their general behavior during a penetration event. This is an area of intense active research.

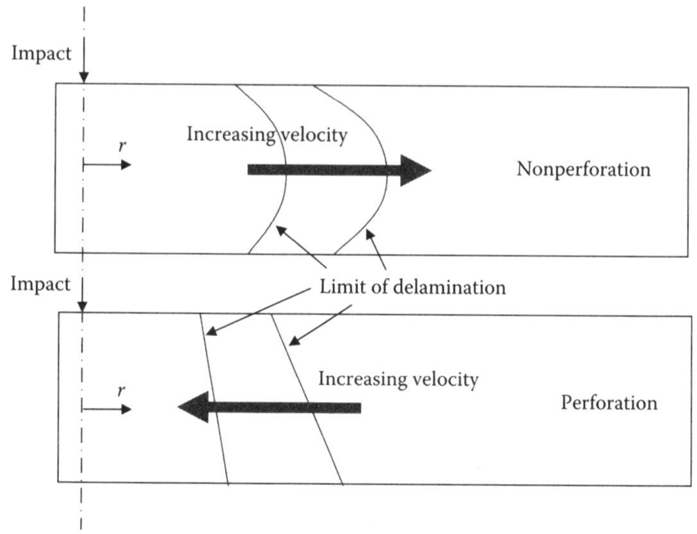

FIGURE 16.15
Extent of delamination in a composite with respect to increasing velocity during both perforating and nonperforating impacts.

References

1. Zukas, J. A., Nicholas, T., Swift, H. F., Greszczuk, L. B., and Curran, D. R., *Impact Dynamics*, Krieger, Malabar, FL, 1992.
2. Carleone, J. (Ed.), *Tactical Missile Warheads*, American Institute of Aeronautics and Astronautics, Washington, DC, 1993.
3. Forrestal, M. J., Luk, V. K., Rosenberg, Z., and Brar, N. S., Penetration of 7075-T651 aluminum targets with ogival-nose rods, *International Journal of Solids Structures*, 29(14/15), 1992, 1729–1736.
4. Forrestal, M. J., Okajima, K., and Luk, V. K., Penetration of 6061-T651 aluminum targets with rigid long rods, *Journal of Applied Mechanics*, 55, 1988, 755–760.
5. Jordan, J., and Dent, S. (Eds), *Warship 2007*, Conway, London, 2007, pp. 81–84.
6. Haykin, B., *Final Project Report on 15" Mk. I Projectile*, ME 504, 2004.
7. Forrestal, M. J., Altman, B. S., Cargile, J. D., and Hanchak, S. J., An empirical equation for penetration depth of ogive-nose projectiles into concrete targets, *International Journal of Impact Engineering*, 15(4), 1994, 395–405.
8. Forrestal, M. J., and Luk, V. K., Penetration into soil targets, *International Journal of Impact Engineering*, 12(3), 1992, 427–444.
9. Cheeseman, B., Personal correspondence with the authors and unpublished work provided to the authors, 2003.
10. Florence, A. L., *Interaction of Projectiles and Composite Armor Part II*, AMMRC CR 69-15, Stanford Research Institute, Palo Alto, CA, August 1969.
11. Walker, J. D., and Anderson, C. E., An analytical model for ceramic-faced light armors, *16th International Symposium on Ballistics*, San Francisco, CA, pp. 23–28, September, 1996.
12. Zaera, R., and Sánchez-Gálvez, V., Analytical modelling of normal and oblique ballistic impact on ceramic/metal lightweight armors, *International Journal of Impact Engineering*, 21(3), 1998, 133–148.

17

Penetration of Homogeneous, Ductile Chromium–Nickel Steel Naval Armor by Three Representative Designs of Nondeforming Hardened Steel Armor-Piercing Projectiles with Bare Noses

17.1 Introduction

The armoring of moving vehicles against enemy weapons has a long history, from Korean naval sailing ships used against an invading Japanese fleet to the placing of rows of personnel shields on the sides of Viking longships to Leonardo da Vinci's studies of armored turtle-like land vehicles with cannons (the first "tanks"). However, until the beginning of the nineteenth century, the ability to economically make anything but small handmade iron and steel objects such as swords or pieces of suits of armor did not exist.

During the early part of the nineteenth century, the Industrial Revolution, scientific research into chemistry, including that of iron and its variants, with several inventions pertaining to the smelting, alloying, and forming of large iron and steel objects in mass-production factories, finally allowed iron and steel to be applied to the large-scale manufacture of things such as railroad rails (a very large industry due to the huge numbers required), steam engines, iron and steel frames for modern high-rise buildings, motorcars and trucks (eventually including armored land vehicles, such as tanks), and ocean-going ships. It is the last we are interested in this section: iron-hulled commercial ships and most especially iron-armored "ironclad" warships (most still having wooden framing under the armor at the start, although this rapidly changed to fully iron warships, inside and out, by the late nineteenth century) that developed during the nineteenth and twentieth centuries. The development of these ironclads started in the mid-nineteenth century for the first successful warships and ended with the very large armored battleships and cruisers introduced just prior to, during, and, in a couple of cases, just after World War II (WWII).

Even though this focus on naval vessels may seem somewhat specific, it is important to the understanding of all types of armor. This is because these same materials were also extensively used in the past for armored land vehicles (tanks, armored cars, etc.), some of which are still in use and still have usefulness today in structures that have to be protected from modern projectiles. So by going through the history, uses, and formulas associated with naval armors, we shall gain a full understanding of their material behavior and how to evaluate the performance of modern projectiles against them.

The US Civil War resulted in many kinds of armored warships being constructed, from small steamships made of wood with some thin wrought iron plates on top protecting a few cannon through naval warship-sized ironclad steamships especially made for this service, such as the very radical USS. *Monitor*-type turreted ironclads and the somewhat more conventional in gun placement, but still heavily armored, ironclads such as the CSS *Virginia* (this first being made from the raised and repaired hull of the wooden frigate ex-USS *Merrimac* by adding an armored "shed" to its upper hull with rows of cannon). Foreign ironclads of the same time frame, the French wooden-hulled *Gloire* and the British iron-hulled HMS *Warrior*, were very similar to their wooden predecessors in design, being complete ocean-going warships, unlike most of the US Civil War designs, which had limited or no ocean-going capability, being for coastal or river use. The experiences with these designs, good and bad, gave the US Navy a good start as to what was required in an ironclad warship and what was to be avoided, most especially a very solid basis in the safety requirements needed when using the new weapons being developed at the time, such as high explosives (dynamite, trinitrotoluene, picric acid [British lyddite], ammonium picrate [explosive "D" in the US Army and Navy], etc.), torpedoes, and new "smokeless" gun propellants, which resulted in considerable death and destruction over the years when not manufactured, handled, and stored properly (with "properly" usually being the result of the study of a major accident causing much death and destruction ashore or on a ship).

All of this had direct effects in determining what was needed in weapons to damage and destroy ironclad warships, especially how to penetrate their protective armor and, of course, how to make the armor better to frustrate those weapons.

17.2 Properties of Iron and Steel Materials Used in Ship Construction and Armor

We are restricting ourselves here to the most widely used form of steel armor, made of metal that can deform, bend, and stretch to absorb the impact of the enemy weapon to reduce or completely negate its damage-causing effects on the portion of the ship being protected (this is also the kind of armor steel used in most tanks prior to the revolution of composite armors, explosive-reactive armors, and so forth). However, it is necessary to show why it has the properties it has compared to the entire range of what iron alloys can provide.

There are three major kinds of iron-based metals used for most purposes: wrought iron, cast iron, and steel, with this last being the material with the extremely wide range of possible properties and the one mostly used when the black art of steelmaking was finally solved for industrial mass-production purposes. There were several competing definitions of the word *steel* in the nineteenth century, which caused all sorts of confusion, both technically and legally (patents, manufacturing licenses, access to raw materials, court cases, laws, and even international politics were involved here). "Big Steel" was literally true as a dominant factor in the industrial development of the world through the nineteenth and twentieth centuries until today, with major fortunes won and lost over the years.

The most important element added to iron for most purposes is carbon, which has major effects on the properties of iron both in the final product and in the various stages of manufacture. All other elements and compounds added to iron to change its properties are secondary to the amount of carbon used and how the metal is processed to develop enhanced properties in the metal. Carbon can do this because of the following:

1. At elevated temperatures (over 723–727°C or 1333.4–1340.6°F) iron radically changes its crystal structure from the room-temperature ferrite to the elevated-temperature austenite, even though it remains solid (although softer when so heated, of course) throughout the entire manufacturing process. Note that the melting point of pure iron is 1538°C or 2800°F, which is well above what most mass-production manufacturing plant furnaces could reach during even the late nineteenth and twentieth centuries. For the most part, the amount of carbon and most alloy elements used were added during the initial smelting of the metal from its ore and remained unchanged during the later manufacturing processes (an exception is adding carbon to the outer surface of an iron object, such as a ball bearing, to harden it, which is called carburization or cementing as in the "Krupp-cemented" form of naval armor; case-hardening; or, for another particular kind of naval armor, "Harveyizing"). Austenite can absorb much higher percentages of carbon into its crystal structure (up to 2.01% by weight, although rather more by volume due to the much lighter-weight carbon atoms involved) than ferrite can usually absorb at room temperature (which is almost none). Because of this, various amounts of carbon can be introduced into the iron material sent to the manufacturing plant so that at various temperatures, the amount inside and outside the crystals can be changed and then by further changing the temperature of the material quickly or slowly, the carbon can be either squeezed out into the gaps between the forming ferrite crystals at room temperature by slow cooling (annealing or normalizing), or if cooled quickly (quenching or chilling), all or some of the carbon can be trapped in the room-temperature ferrite crystals, changing the ferrite to other crystal types, as described in the following. Mechanical working can cause internal heating and cooling inside the material as it is deformed, so this too can cause local changes to the crystal structure size and shape, as well as change the amount of carbon that stays inside each crystal so processed. Very complex and/or subtle changes can be done by both heat treatment and mechanical working, either together or in various separate stages, during manufacture of the iron-based product being produced. The discovery of these processes is what resulted in the history of iron and steel manufacture over the centuries, but most especially in the late nineteenth and twentieth centuries.
2. The true value of the carbon content of the iron comes into play when the temperature is dropped at various rates and/or mechanical working is done by different methods, such as hammering, forging, or rolling, which can greatly modify the final crystal structure of the product being made as it attempts to return into its natural, room-temperature ferrite form. The carbon, if extremely slow cooling is used and it is squeezed out almost completely into the narrow gaps between the ferrite crystals, will change the ability of the crystals to "stick" to one another, allowing the metal to remain soft and deformable, much like pure ferrite. If the temperature is lowered more rapidly, with the exact rate to get a given effect dependent on the amount of carbon in the alloy, some or all of the carbon is trapped in the forming ferrite crystals, and some or all of the ferrite never can form. This is because it would require that no carbon or other alloy element be present in the mix to interfere. Thus, a whole range of alternate crystals can be created instead of, or in addition to, ferrite, which is the

key to the extreme usefulness of carbon/iron alloys. The opposite extreme crystal to ferrite is cementite, made up of a carbon atom chemically bound to three iron atoms in an extremely hard, rigid, and brittle pyramid structure, with the larger crystals made from clusters of these interlocking pyramids. This form of iron compound is not stable at higher temperatures and dissolves back into carbon and iron to form austenite. Since the amount of carbon needed to form large amounts of cementite exceeds the 2.01% allowed in austenite crystals, cementite is virtually never the major component of any crystal structure that it is part of, usually being mixed with ferrite. This can be especially useful in a tough laminated structure of alternating cementite and ferrite crystals inside an overall crystal form. This overall mixed crystal form is called "pearlite" due to its luster under a microscope. Other crystals with greater amounts of cementite-like internal structure are various kinds of bainite and martensite, which contain carbon in cementite-like bonds interlocked in various ways with ferrite-like structures, some very brittle and others less so, although all of them are much harder and more rigid than pearlite. Martensitic steel is used in the harder forms of tank armor. In the relatively soft steel naval armors, pearlite is the primary crystal structure, displaying a good balance of toughness (crack resistance) and hardness (strength and projectile penetration resistance). The higher the carbon content, the easier it is to make the harder forms of room-temperature steel, but also it becomes more difficult to toughen the final product when a lower hardness is desired; finding the balance between the desired final properties, the difficulties in making the intermediate forms of the product during the various stages of manufacturing, and the amount of carbon and other alloying elements (which can be expensive) is the key to efficient steel manufacture.

17.3 Wrought Iron

Wrought iron was the first form of iron used in large-scale manufacture in the early to mid-nineteenth century, later being almost completely replaced by various kinds of steel, once improved steelmaking techniques were developed for mass-production of consistent, reliable products. Some uses of poor-quality steel like that used in railroad rails worked because the design was very forgiving, permitting low-quality control and variation of the properties from rail to rail since, other than compression forces, once riveted into place, rails did not need anything but the ability to withstand frictional wear and corrosion. This was definitely not the case for most load-bearing construction uses, most especially for armor that had to withstand extreme impact shock and penetration forces. Wrought iron is rather soft by iron product standards, forming more or less the bottom edge of the Brinell hardness (BHN) scale (100–105 by definition), compared to most construction steels at 140–170 BHN and the naval armor steels in the 200–230 BHN range. Cementite is roughly 700 BHN. In thin plates, loaded under strong impact forces, wrought iron deforms much like soft taffy, which causes repair of holes to be difficult, since the widely dented and bent metal is very difficult to plug for water tightness. This is not the case for thick armor plates, however, so its use in ironclad warships where the underwater portion of the ship was usually unreachable by enemy gun projectiles or protected by thick waterline armor was practical.

Wrought iron was essentially pure iron, with as low amount of impurities as possible, including a minimum amount of carbon—under 0.08% by weight, although this definition varies in places—to which 4–7% by weight (again remember, much more by volume) silicon is melted and thoroughly mixed in during manufacture. The silicon was obtained from pure quartz sand. The silicon and low carbon content has the following beneficial properties:

1. The silicon instantly on contact with the air or water forms an extremely thin surface layer silicon oxide (sapphire) and stops corrosion almost completely. Indeed, HMS *Warrior*, the very first iron-hulled ironclad completed in 1859, was rediscovered a few years ago still afloat, and although much abused over the years, its original hull was still in good shape, and she was converted to a British naval museum. This is one of its major advantages over even most modern steels.
2. Since it is so pure, wrought iron could not be readily melted in any large-scale manufacturing plant during the nineteenth century through even the end of the twentieth century. Thus, when smaller ingots of wrought iron had to be combined into single larger masses to form some large plate or forged part of, say, a ship structure (keel, for example), the smaller ingots had to be hammered or rolled into a single piece with the joint being no weaker than the rest of the object. With the large silicon content acting much like a soldering flux, two blocks of white-hot wrought iron can be hammered/rolled into a single flattened piece, and the joint becomes essentially invisible with no effect on the total structural strength of the final product. This is actually better than many modern steels that have to be specially welded at their joints to achieve equal strength there.
3. Since the material has almost no carbon, the rapidity of cooling the metal has almost no effect on its overall strength, although rapid cooling can cause unwanted, brittle "kinks" in the ferrite crystals as they rapidly form, so other effects of cooling have to be considered. Thus, the combining of the iron as in part 2 and the forming of the final object can mostly ignore mechanical working and heating techniques that would cause major problems with higher carbon-content steel, where unwanted hardening might happen and crack the object during manufacture. This is another advantage of wrought iron over virtually all steels. This is also what allowed even the rather primitive methods of manufacture during much of the nineteenth century to be able to manufacture wrought iron. It took many years to reach the same level of expertise with steel.

17.4 Cast Iron

Cast iron is the opposite of wrought iron in that it usually contains well over the 2.01% carbon that defines the maximum carbon soluble in steel, usually on the order of 4% carbon, so it always has free carbon in the form of graphite mixed in and in between with its ledeburite crystals. It also usually contains 1–3% silicon, which reduces its problems with corrosion and thus allows objects made from it to have a long-term life after being manufactured. Its main benefit in manufacturing is that it melts at a much lower temperature than wrought iron or even most steels. This melting temperature is in the range 1150–1200°C (2100–2190°F). This range can be economically reached in most manufacturing plants, so that the metal can be poured into molds for very easy forming of objects, with only surface grinding/machining needed to finish the product.

With such a large amount of carbon, it always contains a large quantity of brittle cementite-like crystal structures and although it is possible to form a somewhat ductile form by extremely slow annealing, it is usually a material used only when impact shock or large amounts of lateral shear forces are not envisioned to be supported by the design. Thus, it is used in the construction of buildings when only stable vertical compression strength is required (obviously not where earthquakes are a significant possibility).

The large amount of carbon interferes with the cohesion of the crystals inside the material, and cast iron is usually rather weak compared to well-made steels in everything but pure compressive strength. Cast iron is also used in such things as relatively cheap but very strong mounting brackets under heavy machines, although again there cannot be a large chance of unwanted shock effects that can crack this brittle material in this application.

Due to its brittleness, once a usual cast iron object has been made in a poured mold, very little or no mechanical working of the object is possible, so any variations in its properties postpouring must be completely accomplished by heat treatments alone. If the object is of a more complex shape, applying heat treatments that have no unwanted side effects such as warping can be difficult.

Due to its high carbon content, it is easy to quench after manufacture or chill by cooling a portion of the mold it is poured into, both methods resulting in a very hard cementite-like variant in the crystal structure of the material in the quenched/chilled region. While this will make the hardened "white" cast iron part even more brittle than the usual "gray" cast iron material, if the object can be made thick enough so that a deep layer of gray cast iron remains behind the white cast iron surface layer, that gray cast iron can act as a shock absorber, and the cast iron in this form can be used in armor. In this case, a form of "face-hardened" armor is created where the material consists of two distinct forms of iron with a mixed layer at their interface. In fact, huge, immensely thick (up to 31.5 in. [80 cm], although most were about half of that in thickness) dome-shaped rotating gun turrets used in some European land fortifications, starting in the year 1868 continuing through the 1890s, were produced by a German armor manufacturer named Grüson. This was accomplished by chilling the outer face of large individual wedge-shaped portions of the ring-shaped sides of the turret and soldering them together with molten steel. The assembly was then topped with a 76.2 mm (3 in.) flattened dome roof. The gun ports in the front were preshaped in the mold where the turret was thickest, since the enemy was assumed to be where the guns were pointed. These turrets were a complete success due to the fact that even steel projectiles of the time period, which had no protective caps on their noses, as later became the norm in the twentieth century, shattered to pieces on the surface of these turrets and did nothing but chip the surface and form minor cracks. What is known as Krupp cemented steel armor used in large gun-equipped warships, starting in the mid-1890s and continuing with variations through the end of WWII, was the improved, tougher steel version of this type of armor, although in the case of steel, the face-hardening was done after the plate was made by heating the face to a very high level and quickly cooling it using powerful streams of cold water to get the deep face layer.

17.5 Steel

Once the main issues with steel were solved, steel replaced both wrought iron and cast iron in all armor functions and in most construction functions as well. The first issues were the removal of impurities in the iron during smelting and development of a process to keep the steel from getting too hard during the various stages of manufacture. With this addressed, the steel was less brittle and therefore less prone to cracking. The second issue that was solved was the development of proper alloying techniques to allow the middle of a thick steel plate to properly cool without forming undesirable crystal structures. This, if not remedied, tended to create defects that were hidden until the enemy hit the plate and it broke apart.

Because there is a wide variation in the possible carbon content of steel and with the possible addition of other alloying elements to maximize the desired properties while suppressing the undesired properties in the final product, we can tailor the properties of steel to behave very similarly to wrought iron at one extreme or, like cast iron, at the other and anything in between. This is why the development of mass-produced steel essentially created our modern world.

With regard to armor, what one desires in a homogeneous, ductile steel is maximum possible strength to decelerate and, if possible, completely stop the incoming projectile if at a near-normal impact or to gradually deform while continually exerting a lateral force away from the protected region to cause the projectile to glance off if the impact is at an oblique angle. These forces should ideally continue at full amplitude throughout the entire time the armor is resisting the impact, so as not to act in a brittle manner and split apart prior to the last possible moment of resistance. Unfortunately, in steel, the higher the strength (in most cases), the closer the ultimate tensile strength (UTS) is to the yield strength (YS). Thus, in a high-strength steel, the force that progressively damages the armor and reduces its resistance ability is very close to the force that will completely defeat the armor and allow the projectile to pass through. What we desire is the armor to delay the final failure, gradually allowing penetration and, hopefully, outlasting the energy supply of the projectile so that the projectile stops its penetration prior to the armor failing completely. This desired delay requires a larger separation between the UTS and the YS to keep the resistance of the armor below, but close to the UTS at all times by yielding at just the correct rate under the impact to trade time for resistance and remove energy from the projectile. Too great a hardness, resulting in a high UTS but a small UTS/YS separation, will result in brittle failure, defeating the armor by truncating the length of time that the resistance continues, while too low hardness means that the UTS is too low and the projectile can continue to penetrate long enough to allow it to completely tear through the armor with energy to spare, despite a wide UTS/YS gap. A balance must be struck for optimum resistance to the greatest number of possible enemy projectiles that one expects the armor to be able to resist. One would not expect the armor of a cruiser to defeat a battleship-caliber weapon, but it may be possible to resist the most powerful enemy cruiser-caliber weapon under most battle conditions through proper design.

For an optimum armor steel, we desire the lowest possible amount of carbon* that will still allow the final product to have the necessary UTS to provide adequate deceleration/glancing resistance to prevent the expected threat to penetrate under the widest range of attack parameters. However, we also have a minimum carbon content that sets the melting point of the steel, since the lower the percentage of carbon, the higher the melting point and the more expensive the steel is to make. Most high-grade armor steels in the twentieth century used 0.3–0.4% carbon (again, by weight, somewhat more by the volume taken up in the steel), although some armors had as low as 0.2% and others had up to 0.5–0.55% carbon, with the latter usually implying that the homogeneous, ductile armor steel was also going

* This would present the smallest chance of overhardening during manufacture.

to be face-hardened. The best example of this was the Japanese new Vickers noncemented (NVNC) homogeneous naval armor steel. Introduced in 1931, this steel had 0.55% carbon. The Japanese developers also planned to use the identical steel for their Vickers-hardened (VH) face-hardened armor in the *Yamato* class battleships, which were only in the very earliest part of their design phase in 1931. Obviously, the Japanese Navy planned ahead.

Another desirable feature in armor is something to enhance toughness and minimize hardness by making the existing carbon more efficient as a hardening agent but without the accompanying brittleness of increased carbon. This is why most of the other alloying elements in twentieth-century armor steel were added. The most important elements found in the final armor steel product other than carbon are as follows:

1. Nickel acts as a major toughening agent by replacing the very similar iron atoms at various places in the pearlite crystals or associated ferrite crystals surrounding the pearlite. It was typically used in 2.5–4% quantities in most armor steels, although up to 7% was used in the first high-strength armor steels introduced around the year 1890. These similar but different size atoms act like pieces of cloth caught in a zipper when a crack tries to form, jamming the crack tip and greatly reducing brittle behavior, i.e., increasing toughness, even at somewhat elevated hardness levels compared to a steel without it.
2. Chromium, used in 2–3% quantities in many armor steels, is a corrosion inhibitor and increases the ductility of the steel somewhat, but its primary purpose is that when heated and quenched, it, like iron, forms carbides with the carbon, so that new high-hardness, high-strength crystal structures occur in addition to those formed by the iron with the available carbon. Thus, we obtain a higher strength and hardness with the exact same amount of carbon present had chromium not been added. An additional favorable property is that chromium slows down the transition of austenite to the various low-temperature crystals, such as pearlite, ferrite, martensite, and cementite, giving the steel more time to cool in the center of a thick plate and thus preventing the center from forming undesired crystal structures that can compromise the resistance of the plate under impact. The addition of chromium is even more important in face-hardened armors due to this last property, since the hardness change from a high value at the surface of the face to a much lower value at the point inside the armor where the face and ductile back meet critically depends on this cooling rate being precisely controlled.*
3. Silicon, when added in 0.05–0.15% quantities, does not instill the corrosion resistance of wrought iron. It is usually combined with manganese in both regular construction steels and steel armors to form a combination that has some of the effects of the combined nickel–chromium combination described earlier, although to a much lesser degree. Silicon, however, is much cheaper than nickel or chromium. Many extrahigh-strength construction steels have very successfully used the silicon–manganese combination to enhance the desired properties without excessive loss in toughness. Nickel–chromium is, however, much better at preventing cracking under impact shocks, so nickel or both nickel and chromium are added in addition to silicon in many naval construction steels when ballistic protection is also required. Many nations used compromise materials with limited toughness

* The first modern armor plate using chromium was a 15 cm (5.905 in.) thick chromium–nickel steel armor plate manufactured and tested by the German firm of Krupp in 1894, being its Test Plate #420.

against the major impact forces which provided resistance that was "good enough" for stopping projectile fragments (for example, the Japanese WWII cruisers in many cases used the extra high-strength British-developed D steel* as both construction steel and side armor). The US Navy refused to do this and never used its high-tensile steel (HTS) that was developed for construction purposes and antitorpedo bulkheads for armor. The US Navy always specified the full armor-grade Bureau of Construction and Repair/Bureau of Ships special treatment steel (STS) or Bureau of Ordnance class "B" armor (very similar materials) when any impact resistance was required.

4. Manganese, added in 0.1–0.2% quantities, was combined with silicon, as described in the previous paragraph, so as to provide moderate toughness and hardness enhancement at a much lower cost than using nickel or chromium. In addition, manganese would chemically combine with much of the remaining sulfur in the steel and render that undesired impurity ineffective, further enhancing the toughness of the armor without changing anything else.
5. Molybdenum, added usually in 0.4% quantities or sometimes less, had the effect of improving the toughness of steel at higher temperatures, allowing more rigorous mechanical working or more rapid heat treatments while preventing cracking compared to identical steels without this added alloying element. While rarely used through the end of World War I (WWI), British, German, and Japanese armor steels frequently used this element from the early 1930s through the end of WWII. US armor manufacturers used molybdenum in a more limited manner: Bethlehem used 0.35% molybdenum in its class B armor under 7 in. (17.78 cm) thickness, and Midvale used 0.25% for class B plates under 15 in. (38.1 cm) thickness, while Carnegie–Illinois never used any. It can be argued that these US manufacturers produced the best-quality steel in the world at that time, so this element appears to have been employed to ease production and not as a requirement for the optimum armor strength.
6. Vanadium was used only in the United States at 0.12% by Midvale in class B plates under 15 in. (38.1 cm) thickness. Neither Bethlehem nor Carnegie–Illinois, nor any other foreign armor manufacturer, ever used any, to the author's knowledge, through the end of WWII. Vanadium was initially used in large amounts by the US automobile industry for sheet steel plates forming the outer covering of automobile chassis and for springs. This last use, in springs, made them more resilient and less likely to lose their elasticity as the automobile was bounced up and down thousands of times each day over bumps and potholes in the road. It is still used for this purpose, including watch springs and for any design with similar requirements. The use for chassis in cars was eventually ended when better metallurgy eliminated the need for this very expensive alloying element. Why Midvale used it in its armor is somewhat unclear, but test records made at the United States Naval Proving Ground of ballistic test plates for acceptance testing shows that Midvale plates had by far the least variation in any metallurgical properties from lot to lot of any of the three aforementioned US armor manufacturers, showing a very consistent product, perhaps partially due to this alloy addition.
7. Copper, added in usually about 0.1–0.15% quantities in most post-WWI Japanese armor steels to replace an equal percentage of nickel, has some of the properties of

* D steel was introduced in the early 1920s.

nickel, but to a lesser degree. Nickel was a critical material that was in short supply in Japan, being entirely imported, so any and all steps to minimize its use in any steels were made. In thinner armor, up to 8 cm (3.15 in.) thick, special grades of copper noncemented (CNC) armor steels using up to 0.8% copper in place of an equal amount of nickel were introduced in the late 1930s that met all the specifications of the previous NVNC armor steels of that thickness or less and, in so doing, completely replaced NVNC for thinner armor, such as the antibomb armor around the funnel uptakes of the *Yamato* class battleships. While the United States and Britain were highly successful in their mid-WWII developmental efforts into low-alloy steels with acceptable properties, none of the Axis powers seem to have met with much success in this important topic, with only CNC armor steels coming anywhere near the requirements of the regular full-alloy steels substituted in WWII by those nations. To the author's knowledge, only Japan used any significant amount of copper in any armor or naval construction steel through the end of WWII.

8. Sulfur and phosphorus are major impurities in many iron ores that cause excessively brittle behavior under high-stress conditions, so they are removed to the maximum amount possible in armors. A value of 0.035% or less for each was the maximum allowed in the final ingot for WWII armor steels. This was accomplished through the addition of chemical additives mixed into the molten steel during smelting to combine with the sulfur and phosphorus or through the use of high-pressure air or, more recently, by forcing pure oxygen through the liquid metal to "burn off" these elements. When so chemically processed, the resultant compounds either float to the surface or, less likely, sink to the bottom of the liquid steel mass. During the final pouring of the liquid steel to form the ingots for selling to steel manufacturers, the compounds that floated to the top are poured off first as "slag" prior to forming the solid ingot for shipment. In addition, standard practice in many steel smelting plants is to slice a portion of the top and bottom of the ingot off and send these sliced-off portions to be remelted, and any impurities are added to those recovered from the slag and sold to other manufacturers who can use these chemicals. This produces final ingots with the minimum practical impurities. As mentioned earlier, one of the beneficial side effects of using manganese is that it chemically combines with sulfur and removes even more of that element from the final alloy. Phosphorus must be removed in the slag during smelting. The author is unaware of any equivalent to manganese to reduce its effects when still in the iron alloy.

The final description of the homogeneous, ductile armor steel we shall use as the standard because of its metallurgical properties. This steel is the average WWII US Navy grade of STS/class B armor for midthickness plates that are 3–6 in. (76.2–152.4 mm) thickness [1], derived by Dr. Allen V. Hershey, the head of the Ballistics Computation Division at the US Naval Proving Ground, Dahlgren, Virginia, during WWII through 1955, when armor research in the US Navy was cancelled and all further research sent to the US Army Materiel Command personnel who developed tank armor. He then went into the new digital computer developmental work there until his retirement in 1981. In this role, he developed much of the software that was adopted as the first text-processing portion of the Autodesk graphics engine used for the digital creation of drawings and other diagrams/pictures used in modern computer-aided manufacturing design and documentation work. Each of the three WWII US naval armor manufacturers had their own views as to how UTS should decrease with increasing plate thickness. Bethlehem Steel was able to keep the UTS virtually constant at 106,000–107,500 psi over the entire range of thicknesses. Carnegie–Illinois production practices resulted in a drop in the UTS from about 120,000 psi for the thinnest plates down to 100,000 psi at 12 in. (30.5 cm), at which point, they were able to keep it constant for all plates of greater thickness. Midvale production saw a drop in the UTS linearly from 102,500 psi for the thinnest plates down to 92,500 psi for 18 in. (45.7 cm) plates after which it remained constant for any thicker plates. As long as the plates failed by ductile bending, stretching, and tearing and not brittle fracture (cracking), the change in plate resistance as UTS went down was very gradual, so the thickest plates may have been only about 90–95% as strong as the midthickness plates, but the thinnest plates were about 105–110% as strong as these midthickness plates. This tracks with US Army hardness value drop observed in its rolled homogeneous armor (RHA) over its more limited range of thicknesses and over its higher overall hardness range. RHA has to withstand small projectiles fired at short range hitting at very high velocity which is not usually the case for naval armor. In navy applications, ogival projectiles impact at a much lower average velocity, and the projectiles are usually of a diameter as large as or larger than the armor thickness, with impacts occurring at a medium-to-high obliquity angles from the normal. As plate size decreases, given constant metallurgy, the size of each crystal in the steel goes up relative to the plate thickness. This is also true of the crystal size compared with the size of the attacking projectile, assuming it is scaled to match a given set of conditions. At the extreme, a tiny projectile would be usually hitting a single crystal. Since crystals such as pearlite have internally much higher strength than the intercrystal bonds in the ferrite layer surrounding much of the crystal surface, this means that the resistance to penetration for this tiny projectile would be much greater than with larger projectiles against proportionately larger plates. Thus, this effect, although actually rather gradual, means that thinner plates tend to have a higher strength and toughness than thicker plates and can thus be hardened somewhat more without loss of minimum required toughness, increasing the total resistance of the thin plate [2].

This default standard "Hershey" US Navy homogeneous, ductile naval armor steel, either STS or class "B," had a BHN of 225 (note that this at the low end of the similar alloy for US Army RHA); its UTS was 115,000 psi and its YS was 95,000 psi. Its percent elongation was 25% and its reduction in area was 60–70%. These properties were near the top for any such parameters for all US and foreign armors in WWII. Its Charpy impact test values were also near the top, although no direct values are provided in [2]; the properties were just discussed in comparison to foreign armors tested by the US Navy after WWII. It can be considered as the "mid-thickness" armor plate mentioned in the previous paragraph.

The massive experience in steel manufacture from building construction, ship construction, railway construction, armor manufacture, and, the largest effect, automobile manufacture seems to have had a huge positive effect on US steel manufacture for all purposes, including armor plate.

17.6 Projectiles

There are three different projectile designs used in this chapter description, giving a good example of how complex this topic is.

17.6.1 US Army WWII 15 lb (6.8 kg), 3 in. (76.2 mm) M79 Armor-Piercing Monobloc Shot

For this projectile design [2–4], we shall define the normal kinetic energy density KE_ρ as

$$KE_\rho = \frac{1}{2} w \left(\frac{(\text{NBL} \cos\theta)^2}{\pi \left(\frac{d}{2} \right)^2 t} \right) \tag{17.1}$$

Equation 17.37 describes the energy of the associated with the velocity V component at right angles to the plate face on initial impact at obliquity θ divided by the volume of a (minimum) circular hole of projectile diameter entirely through the plate of thickness t. This will be used in some of the following figures.

Figure 17.1 is a plot of normal energy density vs. plate thickness for this projectile. A side-view drawing of this projectile is included in the lower right of the figure. In this depiction of the M79 projectile, the forward bourrelet and the rotating band are shown as shiny surfaces.

This is the main standard all-steel projectile used by Hershey in most of his penetration testing and computations. It was also scaled in testing to other projectile sizes. It was a very simple design for an antitank weapon, with a bare nose, lacking any pointed windscreen to decrease air drag, nor a thin-steel hood soldered to the nose to hold the windscreen on and no thick hardened-steel armor-piercing (AP) cap to protect the nose from impacts that might cause it to shatter and fail to penetrate. The idea of a hardened cap was used in most post-1900 steel AP projectiles, termed APC by the US Army, but just AP in the US Navy. The M79 projectile contained no internal explosive filler cavity, only a tracer, and had a simple tangent ogive with a pointed tip. The ogive radius was 1.67 calibers, which extends the length of the shell by 1.19 calibers above the end of the cylindrical body. The projectile had a total length of 3.25 calibers. The M79 nose shape is about midway between the longer pointed shapes used during WWI and the rather shorter points of most foreign WWII AP projectiles and the even blunter oval-shaped noses of most WWII US Navy hard-capped AP projectiles (designed for optimum high-obliquity penetration of deck armor at long range).

Hershey had a large number of these 3 in. AP projectiles available for test purposes, since many 76.2 mm US Army antitank guns and the up-gunned M4 Sherman tanks were transitioning to the US Army M62 explosive-filled, base-fuzed APC projectile instead. This obviously also influenced his choice of the standard test projectile.*

17.6.2 Experimental 3 in. (76.2 mm) and 0.78 in. (20 mm) Flat-Nosed Projectiles

The 3 in. projectiles were either specially made 13 lb solid shot projectiles or a set of 3 in M79 AP projectiles with their noses sliced off flat to yield short projectiles of 9–12 lb weight. All were fired under Hershey's supervision against various thicknesses of roughly average STS plate at obliquities from normal to 75° from the normal. Many of the projectiles, penetrating or not, suffered severe nose damage on impact, but enough of the projectile remained more-or-less intact to give good impact results for an evaluation. Note that none of these projectiles were ever able to penetrate an STS plate thicker than one caliber at any obliquity, completely shattering in the failed attempt and not even making a hole through the

* Hershey also had access to a large number of modified M79 projectiles with a flat nose that are described in Section 17.6.2.

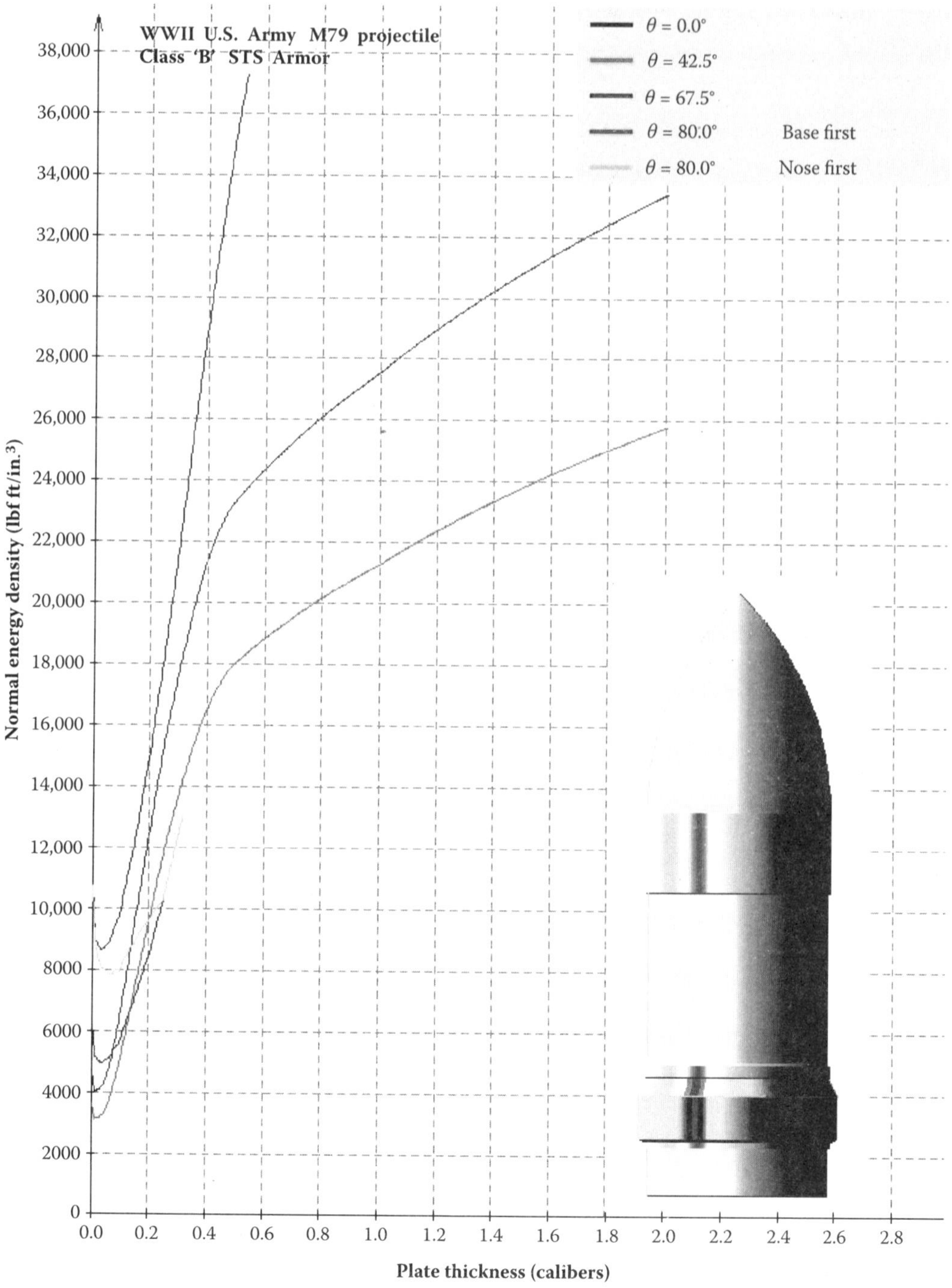

FIGURE 17.1
(See color insert.) Normal energy density vs. plate thickness at various obliquity angles (θ in degrees) for an M79 AP projectile.

plate. Normal energy density vs. plate thickness for this type of projectile is plotted as Figure 17.2.

In Figure 17.2, the sudden jump in the values in this figure is due to an estimate of the requirement to double the striking velocity to achieve a shattered projectile penetration. It is only an estimate, but it gives some idea as to why this flat-nosed projectile was not adopted after tests.

A separate set of flat-nosed projectile tests to compare the 3 in. WWII tests was found [5]. These alternate test results are very similar to the US 3 in. projectile test results, just shifted

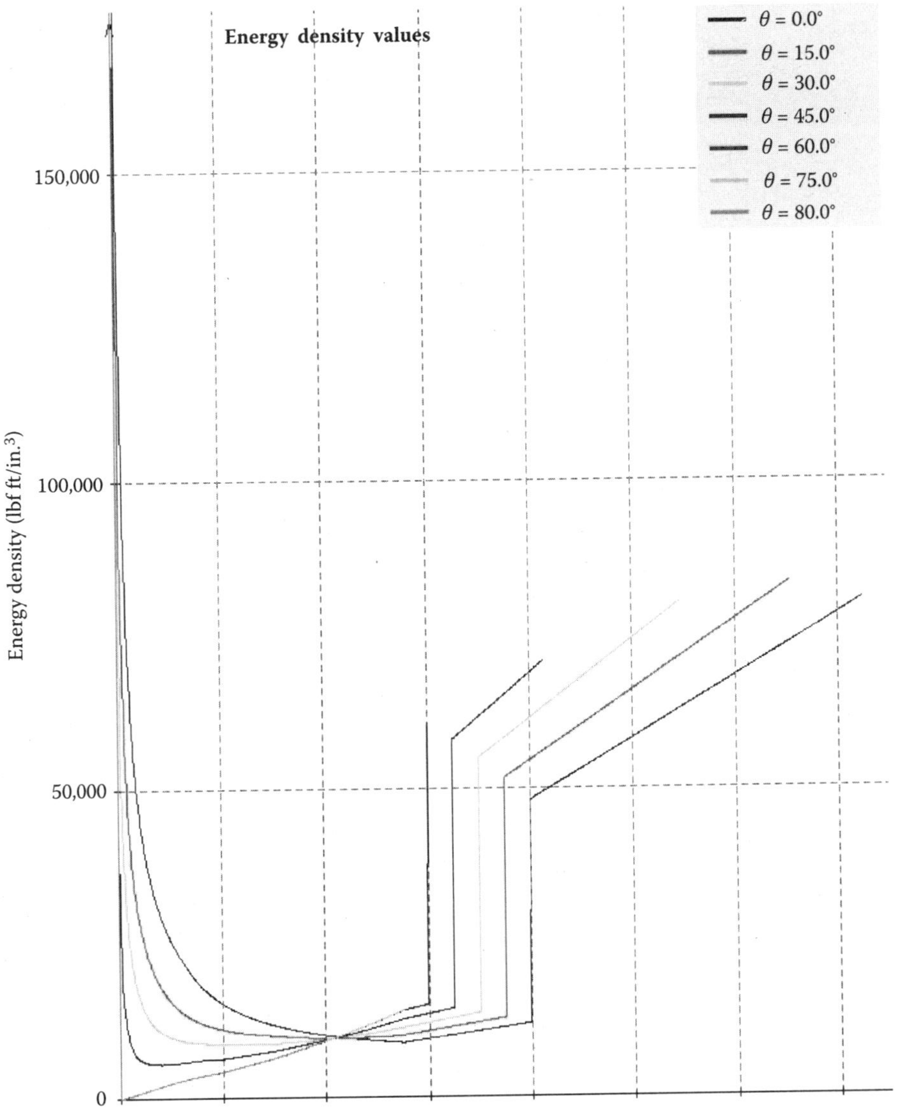

FIGURE 17.2
(See color insert.) Normal energy density vs. plate thickness at various obliquity angles (θ in degrees) for a flat-nosed projectile of 13 lb.

somewhat due to the different projectile and plates used, including the impossibility of penetrating a plate over one caliber thickness. These additional tests were performed with modern, specially made, extraheavy 20 mm (0.78 in.) flat-nosed, hardened-steel projectiles fired at modern high-tensile steel construction plates of various thicknesses at normal obliquity, which gave results that tracked with the US Navy WWII normal-impact 3 in. tests in the shape of the curve generated for penetration vs. velocity. Because a different steel was used as a target and in the projectiles and because of the significant difference in projectile size and the much heavier relative weight of these smaller, but proportionally much longer, projectiles, calibrating the effects of the scaled projectile weight difference was essentially impossible, so the M79 scaling term and weight computation was used as the default in performing the evaluations. It seems certain that this scaling and/or weight computation is not correct, but it is the best information available. A constant was determined to multiply the 20 mm US Navy Ballistic Limit (NBL) result for one plate thickness to get the identical 3 in. M79 proportional plate thickness NBL and all the other NBL test results for normal impact and intact projectiles from both sets of tests overlaid one another perfectly. The changes to the amount of energy to penetrate steel of various thicknesses with intact flat-nosed projectiles thus appear to be identical no matter what the scale, projectile weight, or plate strength. What is missing is a method to compute this constant for all possible plates and projectile tests. Thus, at the time of the present writing, the 3 in. WWII tests are enough to fully evaluate penetration results for armor up to one caliber thickness.

At the present writing, the author knows of only one test that was able to completely penetrate, in an intact condition, a steel plate over one caliber thick. This WWII test involved a Krupp experimental flat-nose 75 mm AP shot projectile fired against a 1.1-caliber Wotan Hart* (Wh) plate at normal obliquity. Based on this evidence, such a thick-plate penetration is not impossible with a full-width flat-nose projectile, but highly unlikely, with the projectile usually shattering to pieces without making a hole given the typical expected striking velocities for standard-weight steel AP projectiles fired from WWII antitank guns.†

The US NBL vs. plate thickness plots for intact projectile penetrations generated for this particular projectile were radically different from the conventional M79 projectile NBL results. The normal energy density vs. plate thickness evaluation chart for this projectile shape would be very different from those we have generated for pointed nosed M79 projectiles.

17.6.3 Experimental 13 lb, 3 in. Tapered Flat-Nose AP Projectile with Small Conical Windscreen

After several naval battles during WWII, samples of Japanese 8 in. (20.3 cm) type 91 AP projectiles were obtained, both intact and exploded. These projectiles were actually a form of uncapped semi-armor-piercing (SAP) projectile with a tapered flat nose (flat face of 0.69-caliber diameter), on top of which was placed a blunt ogival pointed "cap head" held in place only by special weakened threads, which also held the long pointed windscreen on. On impact with a ship structure or the ocean, the windscreen would be torn off. The cap head would also come off unless the impact against the target was at 45° or less. If the impact was at 45° or less, the impact forces would press the cap head onto the flat nose and

* Wotan Hart was an improved post-WWI German homogeneous naval armor in its hardened, maximum-strength form used in cruiser armor and armored battleship decks.

† Note that it is possible that this single Wh plate had a defect (lamination, internal crack, etc.) that caused it to act like a thinner plate in this test.

hold it in place to form a bluntly pointed (more-or-less) conventional nose shape until the armor was penetrated or the penetration was arrested, at which point, in either case, the cap head and the projectile would separate. This cap head could also act as an AP cap in some cases to improve the chance of penetrating face-hardened armor as the cap head shattered on the surface of the plate, but provided little protection to the internal explosive filler if this type of armor was hit, even if complete penetration was achieved. At impacts above 45°, the side forces on the cap head would cause it to be laterally flung before it could do much to the object struck. In this case, it would act as a separate subcaliber projectile with its own trajectory, while the main projectile, now with the tapered flat nose with no cap head or windscreen, would continue into the target. Normal energy density vs. plate thickness for this type of projectile is plotted in Figure 17.3.

The purpose of this unique Japanese projectile design was to allow the projectile, if it hit the ocean just short of the target at a highly oblique angle, to cut into the ocean surface without ricocheting, allowing the shell to follow an underwater trajectory of up to 200 calibers (roughly 133 ft for an 8 in. projectile) distance and hit the enemy ship well below the waterline, all while still moving at a reasonably high speed and oriented nose first. This highly oblique angle could be down to as low as a 7° angle of fall at very close range. The "diving effect" worked best in the range of 15–25° angle of fall. After impact with the water, the cap head would be torn off and the tapered flat nose shape would allow the underwater motion. This could allow the shell to completely bypass the heavy side armor of a cruiser or battleship target and punch into the lower hull before exploding inside. In response to this threat, the new WWII US battleships, the new Japanese *Yamato* Class battleships, and most rebuilt older Japanese battleships had special tapered underwater belts to stop just this kind of penetration. The 0.69-caliber méplat of the shell was exactly half of the surface area of the full projectile diameter if a full-diameter flat nose had been used. This compromise shape allowed not only the reduction of the effects of turbulence and flow separation that was generated by the rather short conical bevel going from the bourrelet to the méplat, enabling the continued nose-first trajectory, but also the reduction of the drag on the smaller flat face enough to keep the projectile speed underwater to reach the designed 200-caliber path length. The shape also generated some upward forces that prevented the projectile from curving downward and missing the target ship until it had considerably slowed at the end of its underwater trajectory. Additionally, shortening the time underwater prevented the special, very long time delay base fuze (see the following) from detonating the shell in the water before it could reach the target.

To allow this long underwater path to be realized, in addition to the tapered flat nose, an extremely long 0.4 s base fuze delay was used in all Japanese type 91 AP shells 8 in. and larger. This enabled the very long underwater path. The downside with this approach was that with such a long fuze delay, a direct target hit would pass entirely through an enemy ship without exploding unless it hit heavy armor thick enough to sufficiently slow it down after penetrating.

Thus, the type 91 AP shell was an all-or-nothing design. For full explosive effect on the target, it either hit the ocean short of the target and essentially worked like a miniature torpedo or hit and penetrated the heavy armor of the target, if present. If neither of these conditions were met, the shell simply passed through the target like a solid shot, causing minimal damage unless it impacted something substantial on the inside.

The US Navy knew nothing about this intricate Japanese design philosophy for these shells. They only had examples of the shells fired at them. So they wondered if the Japanese had figured out something that the US Navy did not know concerning flat-nosed projectiles.

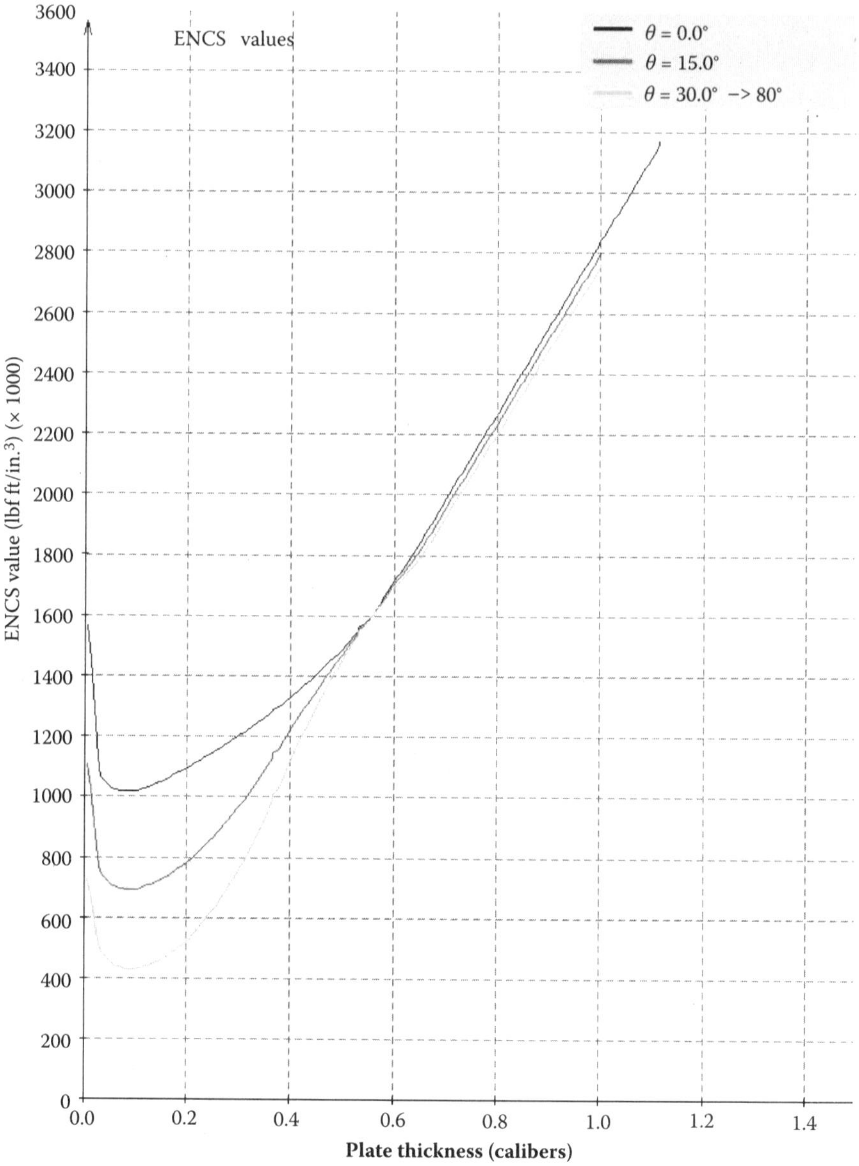

FIGURE 17.3
(See color insert.) Normal energy density vs. plate thickness at various obliquity angles (θ in degrees) for a flat-nosed projectile of 13 lb.

As a result of obtaining samples of Japanese shells, the US Navy contracted for a set of specially made 3 in., 13 lb tapered flat-nosed SAP projectiles. These special projectiles were the same weight as the regular US Navy 3 in. Mk 29/30 capped AP projectile, which was designed for use against surfaced Japanese or German submarines. These uncapped, tapered flat-nosed shells were termed *common* in the US Navy.* This set of special shells was

* Only shells with AP caps were designated as "AP" in the US Navy after about 1900.

produced by one of the major projectile manufacturers* under contract with the Navy. For the tests which followed, the explosive cavities of the projectiles were inert filled. The shells, however, were designed to allow them to be used as regular naval gun projectiles, with explosive D fillers and standard US Army M66 0.16 s delay base fuzes installed (the same fuze used in the 3 in. Mk 29/30 AP shell). The plan was that if tests proved successful, the new shell would be issued to the fleet.

These shells differed from the larger Japanese AP shells in several ways: No cap head was present, so no matter what angle of impact occurred, these tapered flat-nosed shells would have a more difficult time penetrating the thicker gauges of homogeneous, ductile armor plates at low obliquity. Also, more damage to the projectile when used against such plates would occur on impact due to the higher deceleration on the plate surface caused by the flat nose. These shells would be totally useless against face-hardened armor of any significant thickness due to the absence of an AP cap allowing them to shatter on the surface of the plate. The rather short, lightweight, conical windscreen was firmly screwed onto the edge of the flat nose, so it would not easily come off on impact. Thus, the primary goal of the Japanese shell design to prevent ricochet from the ocean surface and enable an underwater hit was eliminated. The diameter of the flat nose was expanded to 85% of the projectile diameter (2.55 in. [6.48 cm] for these 3 in. shells) to try to approximate the Japanese flat nose better. The height of the tapered portion of the nose was 0.47 caliber, roughly the same as the Japanese shell.

The test personnel likely fully expected that these common shells would act much like previously run experimental fully flat nose shells covered in B earlier. This would be a reasonable expectation since the nose taper was small, just 15% less than a fully flat nose shape.

The testing significantly differed from expectations, however. The penetration curves were almost as different from the fully flat nose test shells as they were from the M79 shell results. The small 15% taper caused an enormous difference in the way the shell penetrated the armor and how much energy was needed under various test conditions to achieve the NBL with these shells.

This test series had another rather strange result that is apparent when the normal energy density vs. plate thickness of the intact-projectile NBL values found during the testing are compared to the previous fully flat nose projectile test series [2]. Each test series included tests at a number of obliquities from normal to 75°. One would expect a cluster of data points that varied above and below an average line drawn through the center of the test data for various plate thicknesses at any given obliquity as was true for both the M79 and fully flat nose projectiles when plotted in the literature [2].

That is not what happened. Instead, other than the normal impact curve for the thinner plates, where the required energy (all of it since there was no obliquity angle here) went upward somewhat compared to the other NBL test results, *all* of the normal energy density plot data points for this special projectile for all obliquities, including normal impact against the thicker plates tested, *lay on a single straight line*. The upward curve for the thinnest plates at normal obliquities is possibly due to the plate "dishing" (denting over a wide area about the impact circle) during the impact event because the nose was not sharp enough at its edge to cut out the plate material in front of it until it had stretched as far as it could.† The data points for thicker plates at normal obliquity (where reduced dishing occurs) or at all

* Carnegie–Illinois Steel Company was the manufacturer.

† Although for an oblique impact, the concentrated asymmetric pressure on the narrow corner of the flat forward surface of the projectile pressing against the plate (the "chin" as described later) was enough to start a cut/notch into the plate prior to major dishing occurring.

oblique angles were all *exactly* on top of this straight line to the thickness of the pen used to draw the line, with no distribution above or below the line whatsoever, in other words, a *perfectly* identical function of normal energy vs. plate thickness for this projectile type under every single test condition (except for the thin plates at normal mentioned earlier)! This is difficult to comprehend, since the armor plates hit should have caused some variation due to manufacturing tolerances, if nothing else.

17.7 Details on the M79 Nose Shape Effects on Armor Penetration vs. Standard STS Plate

17.7.1 Normal Impact Results

When hit by the M79 AP projectile shape of any size at normal obliquity, the standard armor behavior can be binned into three more or less distinct regions based on the plate thickness in calibers (t/d) that have smooth transition regions between them (see the 0° curve in Figure 17.1) [3,4].

17.7.1.1 Very Thin-Plate Regime

In this regime, plate dishing dominates as the primary energy absorber. The very thin-plate regime ranges in plate thicknesses from zero through 0.02 calibers. In this regime, the plate acts much like a trampoline when hit, forming a wide conical dish or dent about the impact point. The thinner the plate, the wider the dish is. This dishing increases with decreasing plate thickness until the dish reaches the entire surface area of the plate. This dish widens at a large fraction of the speed of sound in the steel, which is far greater than the projectile velocities we are assuming here, so the dish will absorb quite a bit of impact energy as it widens and deepens under the force of the forward-moving projectile nose tip. Eventually, assuming the projectile impact velocity is above the (rather small) NBL for such a thin plate, a maximum depth is attained as the armor material is stretched to its limit, at which point, the tip of the projectile nose tears through the plate, which then curls back radially in several triangular "petals" of the entire plate thickness, to allow the projectile through. This petal-forming behavior is observed whether or not a dish actually develops in the thin plate. This final tearing open of the petals causes very little added energy loss to the projectile since the torn edges are so thin and the bending back of the thin-plate material in the petals has almost no energy associated with it. Since thinner plates have a larger diameter dish before the tear-through point is reached, the energy to penetrate, small though it is, is actually *greater* for the thinner plates in this regime than for the thicker plates in this regime, with a maximum value just prior to either the plate thickness going to zero or when the dish reaches the edge of the plate and cannot expand more due to the plate edge attachment supports stopping further expansion of the dish. We therefore observe a downward curve of required energy per plate unit thickness to penetrate. This is due to the observed shrinking of the dished area with increasing plate thickness that attains a minimum at 0.02-caliber plate thickness. The energy curve turns upward at this point because the thicker petals absorb a greater portion of the energy as they are bent back than thinner ones. The region where petaling now dominates the resistance to the projectile penetration, and dishing becomes negligible in its effects, is above 0.04-caliber plate thickness. At this point, the dished region is still rather wide, but the shrinking rate of the dish diameter is now slowing down as the width of the dish reaches a rough minimum for all thicker plates, so it essentially transitions to a rather small constant energy loss for all plate impacts against thicker armor.

17.7.1.2 Midthickness-Plate Regime

This regime encompasses plate thicknesses between 0.04 and 0.425 calibers. The behavior is dominated by petaling and cratering/wedging. We shall note here that what we have termed *wedging* is called piercing in the literature as well as in earlier sections of this book. We feel that this term is more descriptive of the physics of what happens in a penetration event where material is pushed aside by the nose of the projectile. Starting from the end point 0.04-caliber plate thickness given in Section 17.7.1.1, the NBL and required penetration energy increase almost linearly up to about 0.3 caliber thickness, at which point, the penetration energy curve changes to a very slightly downward curve. This indicates a slightly lower energy requirement to penetrate than the region up to 0.3-caliber plate thickness. This upper region ends at 0.425 caliber at normal obliquity. Above this plate thickness, the required penetration energy curve bends again and will be discussed in Section 17.7.1.3.

The 0.04- to 0.3-caliber region is completely dominated by the full plate-thickness petals that are split open at the center of the impact site by the pointed nose to form several radiating tears. These tears end up being slightly longer than the final circular hole radius in the plate if impact occurs at the NBL, with the ring of triangular petals formed by these radial tears being bent back until they are parallel with the projectile axis and, in an optimally hardened armor, are usually snapped off of the plate by the passage of the remains of the driving band through the hole.

This petal bending process requires that the projectile push against a resistance that is directly proportional to the plate thickness. The metal is homogeneous, so each unit thickness offers the same resistance force as any other unit thickness, adding up in this case for the entire plate thickness over the full penetration time acting as a beam in bending. The distance that the projectile has to move in contact with the bases of all these petals before the petals completely bend out of the path of the projectile at the NBL and above is also directly proportional to the plate thickness. Since the energy needed to accomplish this is the product of force and distance, the work W is given by,

$$E_{\text{pen}} = W = \mathbf{F} \cdot \mathbf{x} \tag{17.2}$$

We can define the force in this case as, for some constant K,

$$F = Kt \tag{17.3}$$

where here we have assumed the scalar form since we know the direction of the force and K is due to some constant metallurgical property of the plate steel. At each thickness increment in the plate of total thickness t, which is also the distance to be traversed, we can now write

$$W = Kt^2 \tag{17.4}$$

The kinetic energy at the NBL provides the mechanism to just barely do the work required to penetrate and is written as

$$\text{KE}_{\text{NBL}} = \frac{1}{2} w_{\text{p}} V_{\text{NBL}}^2 \tag{17.5}$$

At the NBL, we can equate Equations 17.4 and 17.5 to obtain

$$t = V_{\text{NBL}}\sqrt{\frac{1}{2}\frac{w_{\text{P}}}{K}} \tag{17.6}$$

Thus, the plate thickness t was barely penetrated when the impact velocity is the NBL, V_{NBL}; the required kinetic energy to penetrate and t are both increasing rapidly and linearly in this thickness interval.*

Above the 0.3 caliber plate thickness, when the nose tip of the projectile pushes into the plate, the petaling layer at the surface opposite the impact is rigid enough that the plate material in the impact face layer cannot move backward due to the resistance of the material behind it. The thickness of this rear, more rigid layer runs from zero at the 0.3-caliber plate thickness to a maximum of 0.125 caliber at the 0.425-caliber plate thickness. When this is the case, instead of immediate petaling, the nose digs into the plate face, and as it approaches the constant 0.3-caliber thick petaling layer at the rear of the plate, the face layer is bulged up slightly in a ring around the nose tip and forms a jagged, narrow, craterlike lip hugging the projectile nose. This crater widens with the widening nose as the projectile moves deeper into the plate, and only when the nose tip starts to dig into the petaling region does the plate start to bend back into petals to allow the projectile through. As a result, in this 0.3–0.425 caliber thickness range, we still get a set of petals, but they are limited to only 0.3 caliber thickness. The remaining material in the difference between 0.3 caliber and the actual plate thickness t ends up as a shallow bulge and narrow, jagged caliber-wide crater ring (also called the "coronet" in US Navy documents) tightly surrounding the final impact hole. From the 0.3 caliber plate thickness and up, the petals at the plate back always absorb the same amount of energy from the penetration energy of the projectile at and above the NBL. When the 0.425 caliber thickness is reached, the same thing occurs to the now-constant energy lost due to forming the caliber-wide, 0.125 caliber deep sideways–compressed crater zone in the plate face metal ringing the hole.

The amount of energy required to generate the crater is somewhat less than that needed for petaling since the petals have to be bent over their entire length to 90° from their original position. This requires more energy than merely bulging the crater metal radially from the impact center to a diameter only slightly larger than the projectile diameter. This is why the 0.3–0.425 caliber thickness range has a somewhat less steep upward energy requirement as plate thickness increases. The reason that the 0.3–0.425 caliber plate thickness zone is not linear as in the petaling thickness zone is that this 0.125-caliber-thick zone is so thin as to allow for a smooth transition from petaling to cratering and yields a shallow curve as the mode changes. Before the cratering zone can assert its own linear behavior, it has reached its maximum thickness and now only removes a constant amount of energy from the impacts against any plates above 0.425 caliber thickness.

In the examination of Figure 17.1, for the 3 in. M79 AP projectile at normal obliquity, the region of the curve between 0.3 and 0.425 caliber thicknesses is, as with the below 0.3-caliber petaling region, steeply increasing in a nearly straight line as plate thickness increases. As mentioned earlier, the interpretation of such an energy increase with thickness is that it is the energy that is linearly accumulating, so that the energy needed to penetrate a given thickness is the sum of the energies needed to penetrate all the thinner lamina in this thickness range, being the final energy needed to penetrate the last plate thickness in the

* Note that this approximately linear function for full thickness petaling and plugging failure was originally recognized by the famous physicist Hans Bethe when working for the US Army during WWII as described earlier in this chapter.

previous thickness range (e.g., the dishing region for a 0.04-caliber plate) plus an added constant energy increment for each additional thickness increment t. The armor plate in this 0–0.425 caliber thickness range is resisting this projectile design as a single solid body in two smoothly transitioning thickness steps and the projectile must punch through all layers of it, in effect, simultaneously to completely penetrate.

17.7.1.3 Thick-Plate Regime

In Figure 17.1, a horizontal line with a constant energy density over a continuous range of thicknesses would mean that each thickness increment is acting totally independent of any other, much like swimming in water, where it does not matter if we are in an ocean or a small swimming pool, we are only pushing against the water next to us, and the rest of the water can be ignored. For this case, the formula for penetration would be of identical form to the kinetic energy formula, with the plate thickness penetrated increasing as the work, as described earlier, but in this case, the force $F = K$ at all points (F is not a function of t). So we can write

$$t = V_{\mathrm{NBL}}^2 \sqrt{\frac{1}{2}\frac{w_{\mathrm{P}}}{K}} \tag{17.7}$$

This means that penetration will increase much faster with increasing V at the NBL than the linear case of Equation 17.5. We note that the steeper the curves in this figure, the more energy is needed to penetrate as plate thickness increases and the less penetration will increase with increasing striking velocity at the NBL. This graph is thus the inverse of a more conventional penetration vs. striking velocity plot, yet it allows greater insight into the effect of increasing impact energy and the various plate failure modes as plate thickness changes.

Above 0.425 caliber thicknesses, the penetration event is dominated by the radial compression of the hole created by the projectile. As with the crater zone, when the plate "stiff" rear layer is above 0.3 caliber, the metal nearer the impact surface of the plate cannot move backward easily. However, due to the crater material in the impact region of the plate, material in the zone that lies between the cratering front region and petaling rear region cannot move either forward to form a surface bulge nor backward to form petals. The armor here is now sandwiched between these two layers and can only move directly radially, compressing the armor surrounding the projectile nose as it widens as the projectile digs deeper into the plate. This forms a very wide, although shallow, bulge of armor both at the front and at the back of the plate due to material displaced from the center region of the plate that is made up of the steel that was in what we shall call the final wedging volume. This volume comprises the product of the cross-sectional area of the caliber-sized hole and the thickness of the plate between the 0.425 caliber value used up by cratering and petaling and the actual plate thickness. All further penetration resistance of the plate is that caused by the energy required to move material radially out of the path of the projectile as this region thickens with increasing target thickness, recalling that the cratering and petaling zones are now constant-thickness, constant- energy loss regions. As plate thickness increases, these two constant energy-draining regions at the plate face and rear become less and less important, and it is this middle zone that is thickening and absorbing all the increase in the total penetration energy. We call this centerplate zone in thicker plates the "wedging" zone since that is what the pointed nose of the projectile is doing, wedging aside the armor in front of it just like a metal wedge does in splitting open a wood log, although here the metal will smoothly deform outward from the center rather than crack apart if the plate remains in the ductile region of its hardness range, as with the standard armor we are discussing.

There is a smooth, but still rather sudden "kink" in the energy density vs. plate thickness curve over a narrow thickness range at and above the 0.425 caliber point where the rate of increase in the required penetration energy suddenly decreases. The steepness of this curve gradually decreases even further as the plate thickness goes up towards the maximum plotted in Figure 17.1 (2 calibers). What this indicates is that, as mentioned, the petaling and cratering energy drains are now constant, so they merely establish a threshold energy at the 0.425 caliber minimum thickness for the wedging zone. The curved portion above that minimum is entirely due to the added energy required to cause radial compression to penetrate the center portion of the plate.

The fact that this portion of the energy density vs. plate thickness curve is much less steep means the amount of energy needed to penetrate a still thicker plate is only partially being passed on, with much of the energy increase still being dissipated both in the front cratering and rear petaling regions and in the form of shockwaves, heat, and so forth. The rapidly changing speed of the projectile as it is slowed by the armor causes different amounts of energy loss in the wedging region near the plate face than in this region near the plate rear, so no true steady-state (horizontal line) effect is reached in a real projectile/plate impact scenario. Since the curve is slowly flattening out more and more as plate thickness increases, but not yet horizontal, the required loss in energy to penetrate the center wedging region of thinner plates is still less than thicker ones, but the difference is continually decreasing. However, if the striking velocity becomes high enough to penetrate plates much thicker than two calibers, the projectiles start to shatter and eventually even soften or begin to melt on impact, negating the penetration formulas being described here. As the plate thickness increases well above two calibers, it would seem that eventually, the extremely thick ("semi-infinite") plate would completely absorb any noticeable compression of the displaced armor in the region surrounding a hypothetical indestructible projectile, and the penetration would indeed be just like penetration into water, with a horizontal, constant energy increment needed if the plate thicknesses in Figure 17.1 were further increased.

17.7.2 Oblique Impact Results

When we look at Figures 17.1 through 17.4 for the 3 in. M79 AP projectile at various impact obliquities, we note that the basic shape of these curves is the same as with the normal incidence curve, either compressed, stretched, and/or offset from the normal incidence curve. For comparison purposes, we have also included Figures 17.5 and 17.6 with all data for all obliquities plotted [3,4]. The three oblique impact curves included in Figure 17.1 are the end points as the curves shift from one obliquity to another. The 42.5° curve is the end state of a smooth gradual compression of the normal incidence graph as obliquity increases to that point. When the obliquity exceeds 42.5°, the curves smoothly reverse the trend and stretch until they reach the 67.5-degree curve, where they again reverse and smoothly compress down to the 80° curve, after which, no more evaluations are plotted. The curves were truncated at 2 caliber plate thickness or 3500 ft/s (1067 m/s) striking velocity. The full curves for the higher obliquities above 42.5° are usually truncated too. These choices were made because the M79 projectile stops having a good chance of remaining in one piece against thicker plates and higher velocities even at normal impact and will probably have severe problems with major damage at higher obliquities.

What does this upward/downward undulation pattern with increasing obliquity mean?

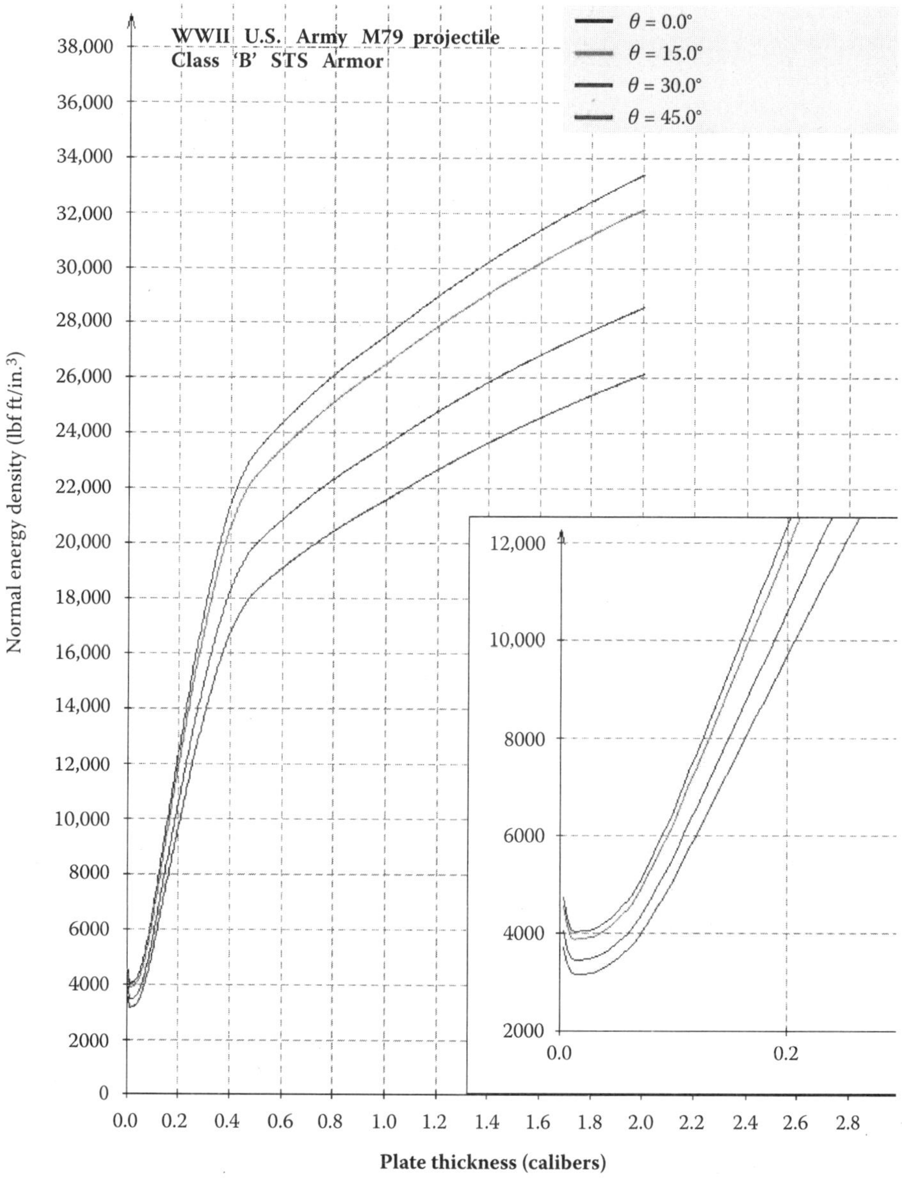

FIGURE 17.4
(See color insert.) Normal energy density vs. plate thickness from 0° to 45° (θ in degrees) for an M79 AP projectile.

Note that when computing the normal component of the impact velocity for use in determining the normal energy density value, we use the initial impact obliquity, as though the projectile passes right through the plate at the NBL, totally unaffected by the impact other than being slowed down. This is absolutely not the case in real oblique impact events, most especially when the projectile is just barely capable of penetrating. In this case, it has a relatively low average velocity as its nose traverses the plate and emerges from its back and thus the projectile spends a relatively long time in the plate during the later portion of the penetration event at the NBL.

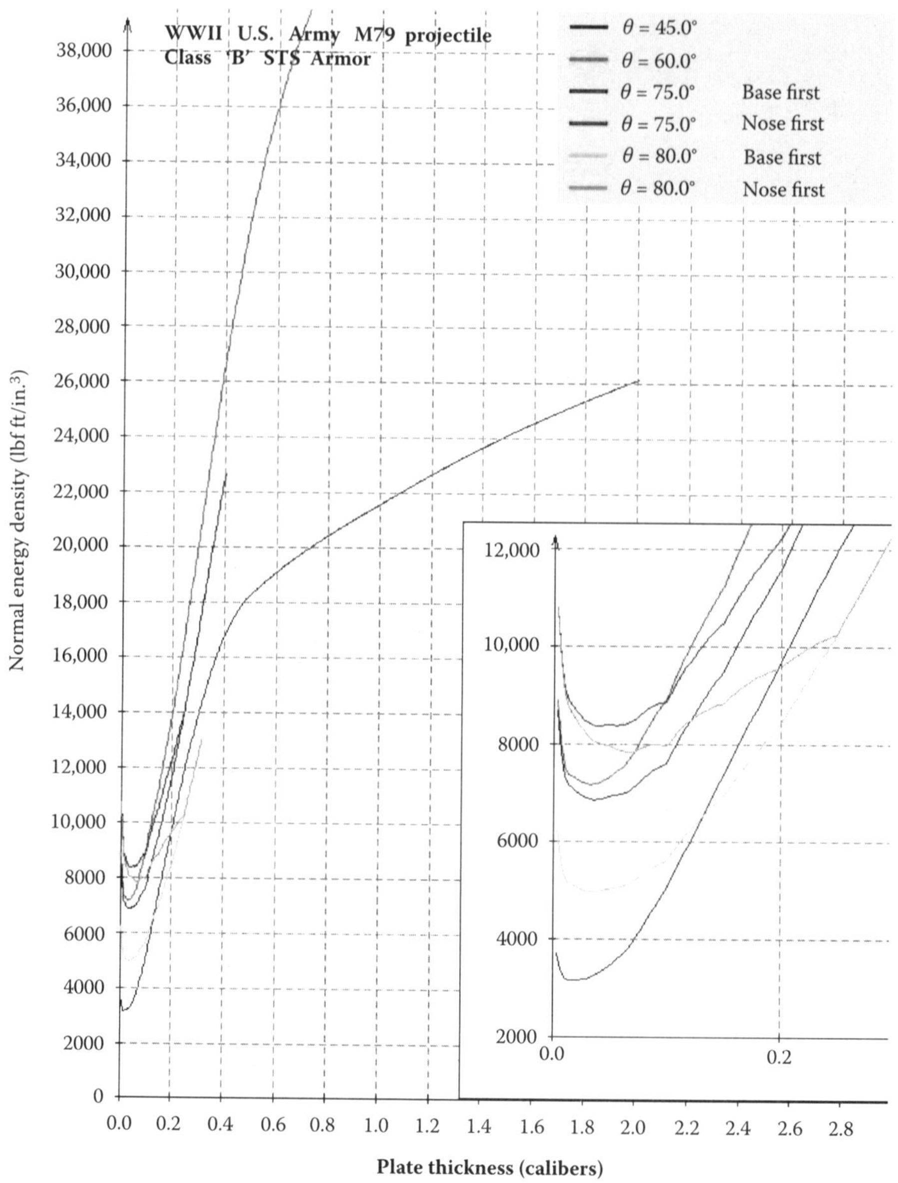

FIGURE 17.5
(See color insert.) Normal energy density vs. plate thickness from 45° to 80° (θ in degrees) for an M79 AP projectile.

At up to 42.5° obliquity, at the NBL the projectile will be turned by the target plate to point more towards the plate normal (that is, acting as if it hit at a lower obliquity if it was not so deflected). This is due to the very strong asymmetric surface forces acting on its nose as it pushes through the plate. The plate, in effect, pushes and pulls on the projectile, depending on how deep the nose is in the plate. Initially, the force is greatest on the side of the projectile nose in contact with the plate—the chin—and pushes the nose away from the plate, trying to cause it to glance off. In the under 42.5° region, however, the nose quickly digs into the plate face and the gouged-out armor on the other side of the nose bulges

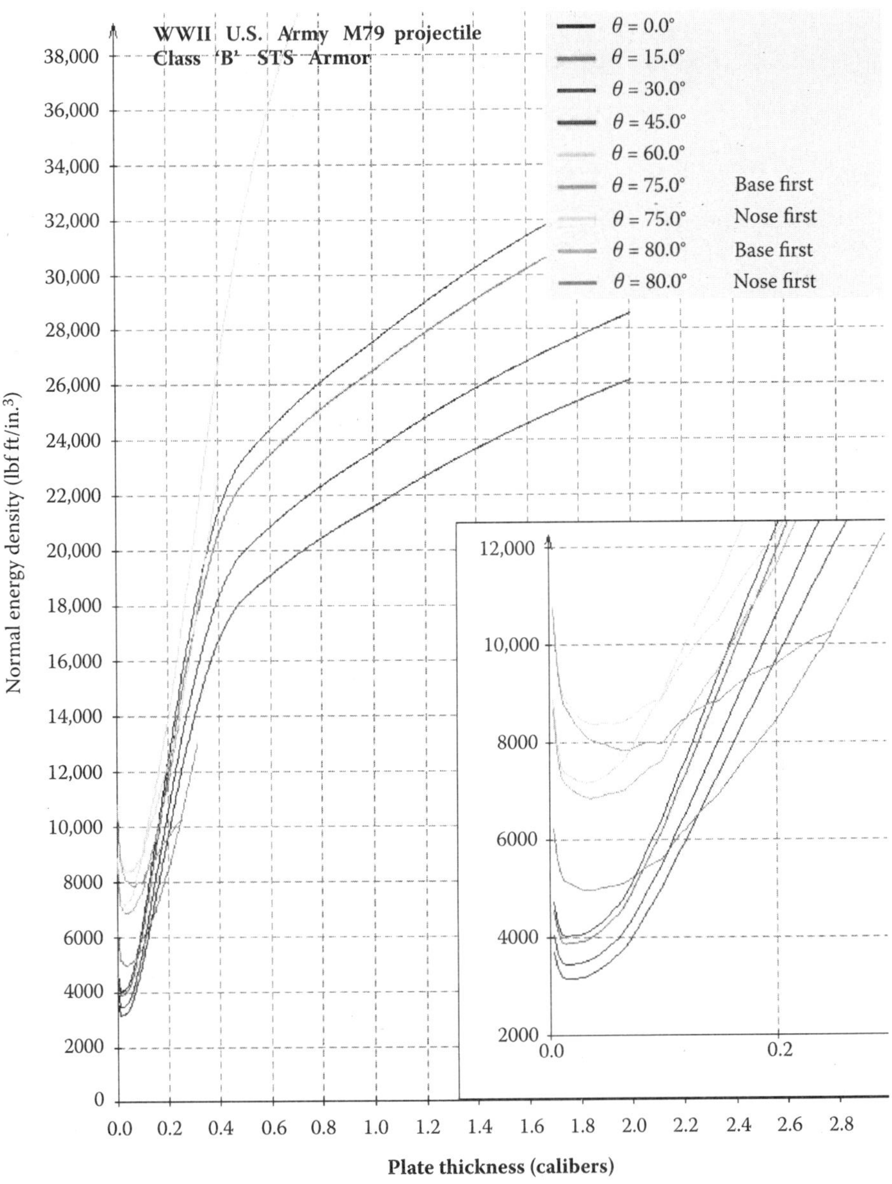

FIGURE 17.6
(See color insert.) Normal energy density vs. plate thickness from 0° to 80° (θ in degrees) for an M79 AP projectile.

upward over the nose and reverses this net force as the nose goes deeper and deeper into the plate. This causes the nose to be rotated in the direction more at right angles to the plate face, reducing the angle that it exits the plate back compared to its original obliquity angle, most strongly at and just above the NBL. Opposed to this up/down nose yawing motion is the high spin of the projectile that is trying to keep the nose going in its original direction, countering the forces generated by the armor. Since the nose deflection is causing, in effect, a lower obliquity path through the plate as the nose is pushed into the plate by the projectile inertia (sometimes termed *normalizing* of the path), the penetration is actually easier than if the projectile had remained going in its original direction unchanged. Thus, the required normal penetration energy goes down compared to a simple NBL cos θ rule, and the normal energy density curve compresses to a lower level of impact energy required to penetrate than at the NBL for normal impact ($\theta = 0$).

Note two things here: First, the initial forces on the nose are attempting to push the nose away from the plate in a glancing manner. But since we are at the NBL and expect complete penetration of the plate, the nose will rapidly dig deep enough into the plate in this obliquity range at anywhere near the NBL velocity value and reverse this behavior, as discussed earlier. Second, when the projectile strikes a plate of the same thickness, but at a velocity significantly higher than the NBL, the forces have less time to act on the projectile as it passes through the plate and the deflection will be less and less. Eventually, in the limit, at a very high velocity, assuming the projectile stays in one piece, the projectile will pass through the plate as if the plate was not even there (other than slowing it down, of course). When evaluating postimpact effects, this behavior has to be included and has been to a major extent in the HCWCALC3 and, with less confidence, in the two flat-nosed projectile evaluation programs included on the NAVWEAPS website.

When the obliquity increases above 42.5°, the initial glancing forces on the projectile nose rapidly increase due to increased leverage on the chin and increased time for these forces to continue. It therefore requires a greater impact velocity to dig into the plate deep enough to punch the nose through to the plate rear surface and trap the nose so that the armor pushes the nose down, reversing the glancing effects before the projectile ricochets off the plate. Also, note that the projectile will never slow to a stop when a nose-first penetration occurs at the NBL at an obliquity higher than 45°. The projectile is decelerated, but the plate forces trying to laterally oppose its penetration through the plate do not have as much effect on the speed of the projectile parallel to the plate surface until the last possible moment, which is not enough time to cause complete deceleration to a stop. In this case, the projectile goes from nose-up ricochet to nose-down penetration with only a small increment of additional drag applied during the complete penetration process near its end; most of the plate resistance prior to making the final hole is due to drag along the projectile side as it bends the plate material downward and sideways, forming a canoe-shaped gouge and wide dish to either side, with a centerline split-open slot entirely through the plate when near the NBL, not in slowing the projectile by direct force on its nose tip. The higher the obliquity, the longer this gouge/slot becomes before final penetration at its end. The path deflection of the projectile is also not as large at the NBL as it is at lower obliquity, for the same reason. Namely, since the projectile never stops moving when going from ricochet to penetration, it already has the reduced time in the plate that was mentioned in the previous paragraph, so the magnitude of the deflection is compromised by this high remaining speed even at the NBL.

When the obliquity reaches 67.5°, the plates that can be penetrated within a maximum striking velocity in the figure and in the code, 3500 ft/s, are thin plates. When impacted at near the NBL, the impact forms a very long, canoe-shaped gouge in the plate, bending it down and, as the velocity nears the NBL, splitting the plate open to form a long slot parallel to the projectile path down its center. Obviously, tearing open such a slot will dramatically reduce the strength of the armor at resisting the widening and deepening of the gouge, so the required penetration energy will now begin to decrease with increasing obliquity. In addition, as the obliquity increases, the maximum thickness of the plate that can be penetrated within the allowed maximum striking velocity continues to decrease, which makes the effects of the torn central slot even more compromising to the total resistance of the plate. Since dishing is still a considerable effect in such thin plates, even though petaling is the most important, anything that compromises the dishing portion of the resistance by allowing the two sides of the long gouge (which is a form of a dish) to spread open easier will reduce the required energy to penetrate. Thus, as the obliquity increases the remaining 12.5° to the maximum 80° value, the needed normal energy to penetrate nose first again drops to about the same value that was true at 42.5° for a plate of a given thickness.

To summarize:

1. In the 0-42.5° obliquity range, the forces on the nose quickly cause the nose to be refracted to less than the obliquity angle, so that at the NBL the projectile exits the plate back at an exit angle of nearly right angles to the face at very close to a zero remaining velocity. The normal energy density is less than the normal-impact value, being lowest at 42.5°.
2. In the 42.5–67.5° obliquity range, the forces on the nose cause it to dig a long gouge in the plate before finally penetrating at and above the NBL, and at above 45°, the deceleration forces never stop the projectile completely (for nose-first penetration, not the less well-defined base-first penetration [see the following]), so the final deflection above 45° is never zero and is always less than in the under 45° range even at the NBL, using a similar deflection formula that requires zero remaining velocity at the NBL to get a right angle exit angle. The normal energy density increases until it is well above the normal impact value, being highest at 67.5°.
3. In the 67.5–80° obliquity range, the plates become so thin within the velocity range allowed by the penetration formulas (3500 ft/s maximum) that they tear open a slot along their centers shortly after the initial gouge in the plate forms, reducing their ability to resist the projectile, widening this slot enough to completely penetrate, so the loss in projectile velocity as it penetrates (again, for nose-first penetration, not for the less well-defined base-first penetration) is even less than in the under 67.5° obliquity range, significantly increasing the remaining velocity at the NBL and reducing the deflection of the projectile from its original path and making the exit angle be closer to the obliquity angle at all times. The normal energy density decreases again very rapidly and, by 80° obliquity, has gone down to the point that it is not far from its lowest, 42.5° value.

17.7.3 Base-First Penetration

Starting at 67.5°, another effect occurs. When the long slot is torn at the bottom of the gouge made in plates under about 0.25 caliber thick, the slot is narrow near the tip of the projectile nose, but is rapidly widened as the pressure of the projectile side on the gouge bottom slides up the slot following the nose. By the time the slot is under the base of the projectile, this pressure has considerably widened the slot. If the nose of the projectile glances off, the base of the projectile is forced down as the projectile rotates nose up, and at around the NBL, this added force can punch the base through the plate and the projectile can be trapped and "surf," nose up, sideways through the plate, tearing a caliber-wide slot in front of it and rapidly decreasing in velocity. When the projectile slows down enough, its spin can flip it out of the slot either upward, so that it merely rolls around above the plate, or downward through the armor, where it can roll around in the region behind the plate. In neither case will the projectile be moving very fast after its long time in the slot, so its ability to penetrate any significantly thick internal bulkhead behind the armor is negligible. Note that in this case, the base fuze is highly unlikely to properly function, so the shell will usually act like a dangerous rapidly spinning heavy weight randomly moving around the area. This base-first penetration at high obliquity against thin plates has a somewhat blurry NBL, since there is a rather wide range of velocities where the projectile could flip either way near the nominal average NBL value.

This base-first NBL will occur somewhat below the minimum NBL for a nose-first penetration, causing such impacts to have two NBL values, in which case we term the base-first NBL the *true* NBL since it is the one that can penetrate at the lowest velocity. This situation is indicated by the two NBLs for the 80° obliquity in Figure 17.1.

In a deck impact on a real target that allows a base-first "surfing" action to begin, the actual armor plates are limited in size and usually have heavy reinforcement at their edges, which may include doubling plates and I-beams bracing the armor, so in reality, the projectile cannot go very far before hitting something pretty solid that will cause it to immediately have to "decide" if it will penetrate or bounce off. Some results at proving grounds are not possible with actual structural considerations incorporated into the tests.

17.8 US Army WWII M79 AP Projectile Penetration of Average-Strength US Navy WWII STS

In the following sections, we shall expand upon the discussion presented earlier and focus on the penetration of a solid, ogival AP projectile into STS and describe the empirical equations that are typical of this event. Although we introduced STS earlier in Section 17.6, we shall now go into more detail on its properties.

STS is a chromium–nickel alloy. It was used as a full-strength armor-class steel by the US Navy Bureau of Ships (BuShips) (previously the Bureau of Construction and Repair before it merged with the Bureau of Engineering in 1941). The Carnegie, later Carnegie–Illinois, Steel Company, a major part of US Steel Corporation, had the sole contract for this material in the US Navy. STS was used in most ship hull armor protection,* bullet-proof plating protecting anti-aircraft gun mounts and other lightly protected exposed hull-mounted structures that needed to be impact resistant. A very similar material made by Carnegie, Bethlehem Steel Corporation, and the Midvale Company for US Navy Bureau of Ordnance (BuOrd) called class B armor was used in larger main and secondary gun mounts, conning towers, and some other specific ship structures (mostly exclusive of the hull which was under BuShips control) requiring higher degrees of protection. This steel was used when maximum strength and/or protection from direct hits by projectiles and fragments of projectiles or bombs was required. High-strength naval construction steel, called HTS, which was about 80–90% as strong, was never allowed by the US Navy anywhere where protection from direct weapon impacts was required. HTS was allowed for protection from explosive blast and concussion, as in the deep side spaced protection systems in battleships against torpedoes. STS is a direct descendant of the original German Krupp Company 15 cm steel plate #420 tested in late 1894 and immediately introduced to replace the previous nickel-only alloy maximum-strength armor steels that were in use. The French had introduced these nickel alloy steels only four years earlier. WWII STS and class B armor was somewhat better than the original Krupp "quality 420" steel mostly because it was tougher

* STS was used because it was homogeneous and ductile and had structural properties. Thick (and brittle) hull face-hardened side belt and transverse citadel ends armor—"class A" armor—in large US Navy warships was the BuOrd responsibility, and only it could contract for its manufacture.

and cracked less and could stretch under impact more, which made it better against large-caliber projectiles than the older forms of this armor steel.

US WWII STS/class B armor was about the best armor steel in the world at the time. It was composed of a very clean ~0.2–0.3% (low) silicon–manganese carbon steel to which roughly 3–3.5% nickel and 1.75–2.5% chromium were added (small amounts of molybdenum and vanadium were added to some thicknesses of Bethlehem and Midvale class B armor, but Carnegie never used either of these additional alloy elements). The nickel added toughness. The chromium added some corrosion resistance and ductility, but mostly, it allowed the creation of chromium–carbide crystals as previously described. This allowed a lower-temperature quench for equal hardness and strength and allowed thicker plates to be made without internal problems with undesired crystals forming deep inside due to too-slow cooling.

STS/class B armor varied in its metallurgical properties with the UTS and YS decrease with increasing thickness, which varied by manufacturer, as previously described. We shall assume a "typical" plate of 225 BHN; 115,000 psi (792.9 MPa) UTS; ~95,000 psi (655.0 MPa) YS; 25% elongation (minimum); and 60–70% reduction in area (minimum). It had a rather good Charpy toughness of over 100 J/cm^2 at 0°C. It had a density of 0.283 lbm/in.^{3} (0.007843 kg/cm^3), similar to those of most high-alloy steels.

This armor had good deformation properties, being tough with little brittle cracking exhibited under projectile impacts. Its major crystal structure was a form of normalized pearlite/ferrite.

The M79 projectile is a simple design with a bare nose (no windscreen to improve streamlining or any other nose covering whatsoever), about 9.5 in. (24.1 cm) long with a simple single narrow copper rotating band. It had a tangent ogive nose shape of 1.67 calibers radius. The nose length was 1.19 calibers, which was about 38% of its total length. The projectile diameter was 3 in. (7.62 cm), and it weighed 15 lbm (6.8 kg). If we define the projectile weight density as w/d^3 for this projectile, we obtain a weight density of 0.556 lbm/in.^{3} This weight density is typical of the weight of AP projectiles of most nations in WWII, including the US Navy medium-to-large- caliber AP projectiles prior to the introduction of their unique "superheavy" or "20% overweight" AP designs used in the last groups of US heavy cruisers and battleships. These heavier AP projectiles had w/d^3 between 0.65 and 0.66. As a comparison, WWI common and AP projectiles, including smaller-caliber ammunition of 5 in. (12.7 cm) and smaller in size, had about 0.41–0.51 for this parameter, indicating lighter structure with less impact strength.

The shape of the nose of the M79 projectile is almost exactly in the halfway between the longer pointed noses used in most AP gun projectiles from the 1890s through the 1920s (usually tangent ogive noses with about 2 caliber radii) and the rather blunt points used by most later AP projectiles. These blunter projectiles significantly varied in their nose shapes. Starting in the early 1930s, newer AP projectiles used tangent ogives with radii reduced to around 1.3–1.4 calibers or even blunter, but there were more complex but still pointed noses, such as used in the Japanese battleship-sized type 91 AP shells introduced in 1931 and even round-tipped, oval-nosed AP projectiles introduced by the US Navy in the middle of WWII. These very blunt Japanese and US designs were for maximum penetration ability against medium-thickness deck armor at medium-to-high obliquity and medium-to-long range, even at the cost of some penetration ability when hitting heavy homogeneous armor at near normal obliquities at close range. Since most really heavy warship side armors used during the twentieth century were not homogeneous, but face-hardened, where nose shape was much less important, this use of such very blunt nose shapes was not a major problem. The M79 projectile thus allows a good middle-of-the-road analysis of this topic.

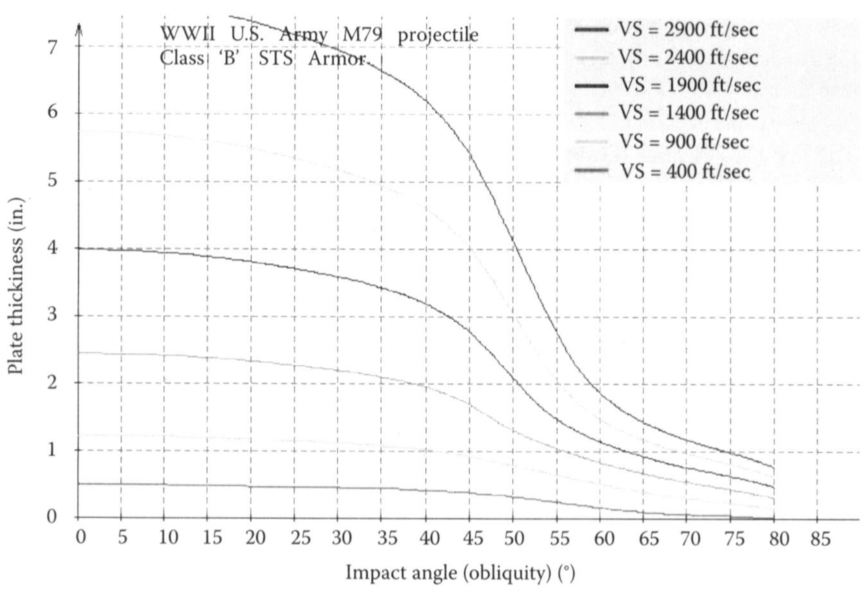

FIGURE 17.7
(See color insert.) Impact obliquity vs. NBL plate thickness from 0° to 80° (θ in degrees) for an M79 AP projectile.

The projectile contained no internal cavity for an explosive fill, just a small recess in the center of its base for a tracer. It was made from a hardened chromium–nickel steel alloy somewhat similar to STS armor steel with the nose and upper body hardened (quenched) to a very high level (around 600 BHN) to resist deformation while the center and lower body were heat treated to be softer. This hardness rapidly and smoothly decreased from the highest value at the nose to around 300 BHN near the base. The projectile was toughened through tempering to reduce breakage when hitting the armor at an oblique angle.

We are going to assume here that this projectile never breaks or deforms in this discussion to simplify the concepts involved. Figure 17.7 shows a plot of penetration of this projectile at various obliquities vs. different thicknesses of armor plate.

17.9 More Detailed Definition of Armor Penetration for Ogival Projectiles

In Section 17.8, we developed some definitions of armor penetration. Now we shall expand on this theme [3,4]. A number of definitions of what it meant for a projectile to "defeat" an armor plate have been used over the years, both for testing purposes to accept armor and projectile lots delivered from the manufacturer and for criteria as to the actual damage inflicted on the enemy by a hit. Acceptance criteria for armor have tended to be somewhat weaker than the criteria for projectiles. Much of the time, armor acceptance testing allowed that only a crack in the plate through which light can pass through and be seen from the far side needs to be met to establish success or failure as to whether the plates passed the criteria established based on the projectile used, the striking velocity, and the obliquity of impact path from the normal. Design and testing requirements only slowly changed from normal for side armor around WWI up to 30° obliquity by WWII. Thin deck armor not only continued to be tested against large-caliber shells at high obliquity (near battle conditions), but also allowed for lower-obliquity testing against smaller shells, too, just like the heavy side armor tests.

The US Navy changed its test success/failure criteria for armor to what became known as the " NBL" ca. WWI and retained it through the rest of its armor acceptance tests for normal-to-medium obliquity. By this criterion, the plate had to have a hole made in it, and the entire projectile if the shell remained intact, or about 80% of its body weight if the shell was broken up, had to pass entirely through the plate and end up somewhere on the far side. The body weight was considered the original weight of the shell minus its windscreen, hood, or AP cap, if any. Any other effects of the postimpact condition of the projectile, given it completely penetrated, by this criterion were ignored. For some face-hardened armors, added criteria were included as to how much damage the armor must do to the impacting projectile, but this was not universal (and was a mistake to enforce as it later turned out). Since this discussion is limited to STS, we shall not dwell on this other criterion.

Projectile specifications required that the projectile hit at a higher velocity and, usually, at a similar obliquity to the armor tests. To pass, the explosive cavity and base fuze insertion areas* of the projectile had to remain undamaged, whether or not a fuze and explosive charge had actually been fitted during the test. As a side note, the US Navy during WWII required obliquities up to 35–40° for the larger AP projectiles, the toughest tests ever required by anyone. Other kinds of damage to the shell were not considered of any significance. When evaluating ship armor mockups, in many cases, the term *defeat* was raised to the projectile criteria in some nations, although the US Navy decided that the regular NBL was a good all-round criterion for such testing, although sometimes, this had to revert to the through crack criteria for high-obliquity deck impact testing where complete penetrations were difficult to obtain.

We will always assume in this section that the NBL is the computed meaning of penetration, with the single modification that at very high obliquity and thin plates, there can be two NBLs, a lower one where the base of an intact projectile tears through the plate as the nose tries to glance off, called NBL(BF), and a somewhat higher one where the projectile goes through the armor nose first, just like for thicker plates or lower obliquity hits, called NBL(NF). In the former case, the projectile may surf sideways tilted nose up through the slot it is forming until it either completely punches through or flips upward and fails to penetrate, in either case, most likely with its base fuze never being set off. The lower-velocity NBL(BF) and the slightly higher NBL(NF) are noted in the following, when applicable. The lowest-velocity applicable NBL is always called the true NBL here. The NBL (BF) is somewhat more stochastic, probably indicating more like a 50% chance of going through, with the percentage going to 100% at the NBL(NF), with the chance of armor defeat averaging 75% chance in the gap between them.

17.10 Residual Velocity and Projectile Exit Angle

If a projectile impact is examined and the NBL is not achieved, no further calculations are performed [3,4,6]. The projectile is assumed to have broken up, partially penetrated or "bounced" off. If the projectile completely penetrates at or over the NBL, an estimate of how

* Usually, the base plug through which the explosive charge was inserted.

fast the projectile is moving is made and in what direction relative to the normal of the plate rear–surface in the impact obliquity plane the projectile exits. No out-of-plane deflection is considered (although, due to projectile spin, a small sideways deflection may occur).

Below 45° obliquity, the projectile will be assumed to always be stopped or rebound if it fails to completely penetrate. As a first approximation, a projectile striking at the NBL will be assumed to just fall out of the plate back at zero residual (remaining) velocity and with its centerline exit angle normal to the plate rear surface. It is this normal from the rear face from which the exit angle of the projectile is measured. When the projectile hits at an oblique angle, the force on the chin of the nose (that side pressed up against the plate face) pushes the nose as to make it glance off. As the nose digs in deeper, the armor material on the opposite side, the "forehead," quickly thickens and stops and then reverses the glancing effect before it can rotate the projectile by much (we are assuming an impact near or above the NBL here). The projectile then rotates toward the plate normal, rotating its base around to hit the far side of the hole, which, in some projectile, impacts against thick armor, most especially face-hardened armor, can break the projectile apart. We shall ignore any such damage here. As mentioned, as a first approximation, at the NBL, the result is to have the projectile nose point out the plate back, normal to the plate. This would be an exit angle of zero degrees. In reality, the nose would end up pointing somewhere in a narrow cone about that line, with a zero residual velocity. It would also be offset from the initial impact point along the vertical plane of impact by the side forces pushing the nose. In a tapering plate, we would assume the average plate thickness under the tip of the projectile nose, but add/subtract, depending on the direction of the impact, *half* of the angle difference between the face and back surfaces to/from the initial obliquity.

As the 45° obliquity is exceeded, the force on the chin increases further in the glancing direction, and the impact velocity must be greater to push the nose deep enough into the target plate to arrest the glancing forces, increasing the NBL above what a zero-residual-velocity penetration would require. Thus, the nose would be moving more rapidly than it required under 45° obliquity when glancing stops and the rotation back toward the plate normal begins. As a result, neither the residual velocity nor the exit angle would be zero at the NBL, with the higher the impact obliquity, the greater these minimum values would be. In effect, the projectile goes from ricocheting from the face to "ricocheting" off the plate back, bulging it up into a tilted "cave" topped by a forward-tilting pointed "spur" of bent armor that held the nose from glancing off of the plate at the end of the elliptical gouge. Below the NBL, the projectile will make a canoe-shaped gouge in the face until, at a higher velocity, it hits hard enough to split the gouge as a slot under the chin of the nose, after which the weakened armor that is now separated on each side of the slot can bend outward sideways like "bomb bay doors" and allow the projectile to "belly flop" downward. The behavior will become more prominent as the striking velocity is increased until, at the NBL, its nose tip can hook under the end of the slot and bulge the armor up into the spur and cave to push the nose down and allow it to completely penetrate.

The higher the velocity is above the NBL at any oblique impact, the less time the deflection has to occur, and the closer the projectile gets to having its path undeflected with the exit angle approaching the initial obliquity value, although usually offset sideways somewhat. The loss in energy to penetrate as the path straightens out would probably be lessened, too, but one usually just subtracts a constant energy loss based on the NBL and keeps the energy loss constant.

In a base-first penetration, we shall assume that the projectile, if it does go all the way through rather than flipping back upward and not penetrating, has essentially no residual velocity to speak of and will be moving completely randomly as to its exit angle direction when it tumbles through the plate due to the interaction caused by its spin rate as it was pushed around by the plate during the impact. In this case, there would be no fuze action and, therefore, no real potential to cause major damage beyond whatever it flies into in the space immediately behind the plate penetrated.

17.11 Basic Ogival Penetration Formulas and Definitions

Most of the older penetration formulas were set up to compute their ballistic limit velocity V_l as the dependent variable from the other variables as inputs [2–4]. In some cases, this limit velocity was defined slightly differently than we have done earlier, but for our purposes, we shall assume it is similar. Because of the multiplicative factors used in these equations, it is important that the inputs be entered using the correct units. The inputs normally used were as follows: projectile diameter d in inches, projectile weight w in pound-mass, obliquity angle θ in degrees, target plate thickness t in inches, and, if used, a unitless plate/projectile combined quality factor coefficient C, which adjusts the V_l value to match the armor type to the projectile type being discussed, assuming that they are not the ones used to create the original formula. The most important formula in the twentieth century was for French Schneider et Cie. 1890 nickel steel homogeneous armor impacted by bare-nosed, nondeforming, unbreakable tangent ogive-nosed projectiles with an ogive radius of 2.0, giving a nose height of 1.323 calibers. We shall use US Customary System (USCS) units throughout this discussion. This formula was the DeMarre nickel steel formula, which, interestingly enough, also used the US NBL definition of penetration well before most other people. This formula is usually seen as

$$\mathrm{NBL} = V_s = 1022C\left\{\frac{t^{1.4}}{w^{0.5}d^{1.7}}\right\} \tag{17.8}$$

Here C, the DeMarre coefficient, modifies the striking (limit) velocity V expected to give the NBL for the plate/projectile combination under study. It is about 1.21–1.22 for typical STS-type nickel–chromium armor impacted by medium-nose height pointed-nose AP projectiles at normal obliquity. There was an obliquity adjustment sometimes used, but it was not very accurate and only worked for up to about 20° obliquity. Because of this, we shall ignore oblique impact when talking about the DeMarre formula.

This formula is in a form from which it is difficult to see the underlying physics. The use of exponents 1.7, 1.4, and 0.5 clouds the issue, and no real understanding of what is going on is apparent. A better way of writing this formula and other penetration formulas that will illuminate the physics is as follows:

$$\frac{t}{d} = \left(\frac{K}{Q_A}\right)\left(\frac{S}{M_\theta}\right)\left(\frac{w}{d^3}\right)^W (V)^v \tag{17.9}$$

Let us go over each term:

1. t/d: This is the thickness of armor of the given plate type penetrated at the NBL in calibers, allowing test results of projectiles of different sizes and other test parameters to be directly compared.

2. K: Numerical constant to give the correct answer for the default projectile and armor types used to generate the formula from empirical data.
3. Q_A: Armor quality factor based on metallurgical parameters of the steel used, as demonstrated in penetration tests, replacing C and here multiplies the thickness t of the armor, not the striking (limit) velocity V. The larger the Q_A, the smaller t is that can be penetrated at a given V, since when building a warship or tank, one uses a single armor plate, so one has to be chosen to remain within a weight budget and still hopefully give adequate protection from whatever enemy threat is envisioned. Since we may not know what that threat will be, we have no idea of what the NBL will be for that specific threat until a threat assessment is performed on all weapons that might be used against whatever plate material we are studying.
4. S: The scaling factor that shifts the penetration due to absolute size of the projectile, which is one of the results from tests that shows how the armor responds when we try to create a design for a battleship, say, scaled down to a cruiser thickness against scaled-down cruiser guns. Usually, the larger the projectile, the lower the NBL is, for identical projectiles and scaled identical plates that differ only in size.
5. M_θ: Obliquity thickness multiplier, either from an empirical formula or a lookup table set or both; this term reduces the plate thickness that can be penetrated at a given NBL or increases the NBL for a fixed plate thickness, as the impact obliquity θ goes from zero (normal impact) to the maximum angle calibrated. Based on experience, it is usually cut off at 80°, but sometimes, it has to be less. Note that for some nose shapes, such as flat noses, increasing the obliquity may not always result in a higher NBL, but sometimes, an unchanged or even lower NBL can occur over a range of obliquity values. This is not typical and is not the case for the M79 nose shape.
6. w/d^3: This is the apparent density of the projectile and is the same for exact scale models of a single projectile design. For example, all German Krupp WWII naval L/4.4 AP shells were very close to a single design, with only minor differences such as the relative size of the base fuze and so forth to tell them apart if not labeled as to which was which size. This term is important because it greatly reduces the confusing effects of weight variations between projectiles, in this case, as a relative scaling effect. It also allows different designs to be compared just based on their weight differences. If there are other unexpected results during tests, the effects due to relative weight of the various shells involved can be subtracted/divided out here as part of the troubleshooting effort.
7. V: The striking velocity or, in complete penetration computations, the NBL.

Note that there is a separate exponent W for the apparent density (w/d^3) term from the exponent v for the striking velocity term. In face-hardened armor, those two values are indeed separate and must be computed that way (v is much more important than W in penetrating that kind of armor), but in homogeneous, ductile armor hit by pointed projectiles, as we are discussing here, the total kinetic energy $(½)m_PV^2$ is indeed the driving force for penetrating STS-type armor, so trading w for V^2 works reasonably well (excluding projectile damage changes), and v is always twice W. Thus, we can replace the separate w/d^3 and V terms with a single term $[(w/d^3)(V^2)]^W$.

For the DeMarre nickel steel armor penetration formula, we get

$$\frac{t}{d} = \left(\frac{5.021\times 10^{-5}}{C^{1.42857}}\right)\left(d^{0.07144}\right)\left(\frac{w}{d^3}V^2\right)^{0.71429} \tag{17.10}$$

where there is no M_θ term (it is not worth the trouble) and $Q_A = C^{(2)(0.71429)} = C^{1.42857}$, so $C = 1.21$ for typical STS plate gives a $Q_A = 1.313$. Note that the very small scaling factor $S = d^{0.07144}$ roughly matches in the magnitude that found by Dr. Allen V. Hershey as he developed most of the M79 formulas/database system that we shall use, evaluated for STS, although S in his formulation is a somewhat different format to be described in detail in the following.

The DeMarre Ni-steel formula is a one-size-fits-all equation that assumes that the method of resistance of the armor to penetration at normal obliquities is the same for all plate thicknesses against the default pointed AP projectile assumed. Between 0.2- and 1.1-caliber plate thicknesses (the upper portion of the petaling range through the low end of the wedging range), the curve is not too far off the actual resistance plot if a good choice for the Q_A value (C in Equation 17.10) is made.* It completely fails in the dishing and lower petaling range at the low end or in most of the wedging range at the top end, where the methods of energy absorption by the plate are considerably different. Also, the major changes caused by oblique impact to the energy absorption methods of the plate even in the regime that work at normal obliquities means that the DeMarre Ni steel formula is of little use in that case, too, without complicated tables of C values to match the actual curves. If we require a large library of C values, we might as well start from scratch with a formula/database that more closely matches the correct behavior.

The dependence on total kinetic energy for homogeneous, ductile armor penetration by pointed projectiles is still valid. The curves that must be matched to compute the NBL are rather more complex and will not use a simple single exponent W as in the DeMarre Formula, so we instead write

$$\frac{t}{d} = \left(\frac{1}{Q_A}\right) f\left\{K_t\left(\frac{S}{M_\theta}\right)\left(\left(\frac{w}{d^3}\right)V^2\right)\right\} \tag{17.11}$$

Here f indicates some complex function of variation of all the variables, even the constant K is now a variable K_t that changes with plate thickness t to fit the empirically obtained curves at any point with kinetic energy (KE) to match the test results over the entire range of plate thicknesses and impact obliquities that we have good data on for intact projectiles. Q_A is a constant for all plate thicknesses of a given plate type, not based on striking velocity, so it is not modified by the KE adjustments.

Although we shall not go into all the numerous penetration formulas for ogival projectiles, a particularly useful "universal" one was worked out by Dr. L. T. E. Thompson at the US Navy's Dahlgren Naval Proving Ground in the late 1920s [3,4]. It is given in two forms: One is explicit in NBL or limit velocity, while the other is explicit in t/d, which is more useful in examination of penetration physics.

---

* The optimum value of C to approximate the actual penetration curve for US Navy WWII STS with the US Army 76 mm M79 AP projectile at normal is 1.2205, slightly higher than the 1.21 value used in US Army and Navy homogeneous armor acceptance tests ca. WWI.

$$\frac{t}{d} = 1728.04\left(\frac{w}{d^3}\right)\left(\frac{V_s}{F}\right)^2 \cos^2\theta \tag{17.12}$$

$$V_1 = \left(\frac{1}{41.57}\right) F\sqrt{\frac{t}{d}}\sqrt{\frac{d^3}{w}}\frac{1}{\cos\theta} \tag{17.13}$$

In these formulas, the thickness t and diameter d are in inches, the mass w is in pounds, and the velocities are in feet/second. The factor F was used to compare test results in experiments instead of V, and a "standard form" for averaged penetration tables and specifications was calculated in 1931 from many US Navy 8 in. Mark 11 MOD 1 AP projectile tests at 0–75° obliquity as follows, with the angle θ in degrees:

$$F_{\text{STD}} = 6\left(\frac{t}{d} - 0.45\right)(\theta^2 + 2000) + 40{,}000 \tag{17.14}$$

The formulas as presented earlier were derived from averaged class B armor or STS plate tests. Sometimes, in the literature, one finds the limit velocity based on the value of F_{STD} written in terms of the "%NBL." This is defined as

$$\%\text{NBL} = \frac{V_l}{(V_l)_{\text{STD}}} \times 100 \tag{17.15}$$

where the limit velocity or NBL obtained in some actual test is compared to that obtained against an "average" pre-1931 US Navy class B or STS plate by the standard 8 in. AP projectile earlier. This gives a quick way to sort results compared to a defined standard for evaluations of whatever properties one is studying. Detailed research work requires comparing the raw F values themselves, however.

Without going into the exact details of the following sets of equations, some of them may be helpful for the reader against certain targets which will be noted. As always, be very careful in the application of these formulas. Test data to validate the models are always good. The units used in all the following formulas are those mentioned earlier, namely, the thickness t and diameter d are in inches, the mass w is in pounds, and the velocities are in feet/second. The angle θ in degrees is not required as the formulas given in the following were for low obliquity impacts.

The Fairbairn formula (UK, ca. 1865) espoused a linear increase of target thickness with limit velocity that was nearly correct for thin wrought iron plates ($t < 0.25d$); it was given by

$$\frac{t}{d} = 0.0007692\sqrt{\frac{w}{d^3}}V_s \tag{17.16}$$

$$V_l = 1300\left(\frac{t}{d}\right)\sqrt{\frac{d^3}{w}} \tag{17.17}$$

Tressider created a formula (UK, ca. 1870) which was good for medium-thickness plates ($0.25d$–$0.75d$) at low obliquity.* It was generally good for hard, brittle armor.

* It should be noted that the Tressider formula given in Equation 17.18 does not have the w/d^3 exponent, at 0.5, as half of the V_s exponent of 1.5. It is smaller than half, somewhat like, but not as extreme as, that used in the Facehard program on the CRC website. It is the only other formula outside of Facehard that does this.

$$\frac{t}{d} = 0.00003798\sqrt{\frac{w}{d^3}}V_s^{1.5} \tag{17.18}$$

$$V_l = 885\left(\frac{t}{d}\right)^{2/3}\left(\frac{d^3}{w}\right)^{1/3} \tag{17.19}$$

In all the preceding and following formulas, the d^n term implies that the scaling of the diameter has a greater effect on penetration. Larger diameter projectiles tend to pierce a similarly scaled armor plate more easily. Shearing and surface cracking also show up more as the target thickness and projectile diameter increase. At the time of the development of these equations, there may have been more manufacturing quality issues with thicker plates which has crept into the formulas.

The Gâvre formulas (France, ca. 1870) are as follows:

$$\frac{t}{d} = 0.00002887 d^{0.42857}\left(\frac{w}{d^3}\right)^{0.71429} V_s^{1.42858} \tag{17.20}$$

$$V_l = 1500\left(\frac{t}{d}\right)^{0.7} d^{-0.3}\sqrt{\frac{d^3}{w}} \tag{17.21}$$

The tests at Gâvre Naval Proving Ground may have been against poorer quality plates than the others. This shows up in the fact that the scaling term has a much more pronounced effect. Additionally, the increase in penetration with increasing kinetic energy is less marked than in the DeMarre formula.

In 1894, Krupp developed Krupp cemented (KC) armor. This cementation process was a hardening of the face of the armor plate by putting it in contact with carbon during the manufacturing process. This formed a thin, superhard layer about 1–1.5 in. (25–38 mm) thick. In addition, a thick, less-hard layer was formed under that thin face surface to reinforce it and markedly increase projectile damage. The total face-hardened layer was around 35% of the total plate thickness. The Krupp formula for uncapped projectiles fired against face-hardened armor is as follows:

$$\frac{t}{d} = 0.0006659\sqrt{\frac{w}{d^3}}V_s \tag{17.22}$$

$$V_l = 1500\left(\frac{t}{d}\right)\sqrt{\frac{d^3}{w}} \tag{17.23}$$

In the Krupp formulas earlier, penetration linearly increases with velocity. The damage to the projectile during the experiments was worse as the plate thickness increased which countered the higher impact velocity. There were large test-to-test variations potentially due to the destruction of the projectile noses by the hardened face.

In 1891, the Harvey Steel Company in the United States developed an armor that had a 1–1.5 in. thick "cemented" face "baked into" the new French nickel steel armor* that contained a large amount of nickel as an alloying element for greater impact toughness.

* This was the same face surface layer type used later in KC armor.

Around 1895, the following formulas were developed by US Navy Ensign Cleland Davis for penetration of uncapped projectiles into this material [8].

$$\frac{t}{d} = 0.00003466 d^{1/3} \left(\frac{w}{d^3}\right)^{2/3} V_s^{4/3} \tag{17.24}$$

$$V_l = 2200 \left(\frac{t}{d}\right)^{3/4} \left(\frac{1}{d}\right)^{1/4} \sqrt{\frac{d^3}{w}} \tag{17.25}$$

This was the first attempt in the United States to develop a face-hardened armor formula. The hardened face was 1–1 1/2 in. thick, regardless of total plate thickness. Plates caused less and less damage to the projectiles, as they got thicker and were attacked by progressively larger projectiles. It was noticed that poor projectiles shattered on impact. The total kinetic energy approach used in this formula may be incorrect if the hardened pieces were pushed through the plate.

Between 1895 and 1900, soft AP caps were added to almost all US Navy AP projectiles to improve penetration into face-hardened armor. Ensign Cleland Davis modified his formula to allow for capped projectiles fired into Harveyized armor as follows [8]:

$$\frac{t}{d} = 0.00008582 d^{1/4} \left(\frac{w}{d^3}\right)^{5/8} V_s^{5/4} \tag{17.26}$$

$$V_l = 1800 \left(\frac{t}{d}\right)^{4/5} \left(\frac{1}{d}\right)^{1/5} \sqrt{\frac{d^3}{w}} \tag{17.27}$$

Neither Cleland Davis formula proved to be particularly accurate for the Harvey armor, but the capped form was modified by Krupp engineers for general use for both homogeneous armor and their KC armor later, replacing the DeMarre formula, with some success.

17.12 Program Formulas, Data Sets, and Evaluation Logic

Provided on the CRC website is a version of the HCWCLCR3 program developed by the author from Dr. Hershey's database and NPG reports and the author's further analyses [2–4,6]. This is a great improvement on the empirically derived formulas of the previous section. In order to illustrate the physics and describe the computational logic, we shall walk through the code step by step. In this section, we shall describe the formulas utilized in two ways. The first (which will contain the equation number) is in the format used throughout the book. The second, immediately following the first, will use the terminology of the source code provided on the website and will not have an equation number.

We start by using the HCWCLCR3 program with the windscreen, hood, and AP cap weights zeroed to reproduce a bare-nosed M79-type AP projectile. The NBL is computed from the various plate and projectile parameter inputs and compared to a striking velocity V, also input by the user to determine if complete penetration occurred and, if so, the postimpact motion of the projectile.

The value used for f here, described in Equation 17.10, is based on Hershey's averaged curves [2] plus a failure strain adjustment formula, developed by the author, from the

Krupp G.Kdos. 100 penetration curves for projectiles over 8 in. in size [9]. The formula is given as

$$\frac{t}{d} = \left(\frac{1}{Q_{\mathrm{A}}}\right)\left\{\frac{\left(\frac{w}{d^3}\right)(\mathrm{NBL})^2}{\left(\frac{K_{\mathrm{t}}S}{(\varepsilon_{\mathrm{F}}J_{\mathrm{t}}M_{\theta})^2}\right)}\right\}^{\frac{1}{2T}} \tag{17.28}$$

$T/D = (1/Q_a)\{(W/D^3)(NBL)^2/[(K_t)(S)/(\{\%E\}\{J_t\}\{M_{Ob}\})^2]\}^{\{1/[(2)(Texp)]\}}$.

Here, ε_{F} is a function based on the strain (percent elongation) at failure of the target—which is only used (set to a value other than 1.00) for projectiles over 8 in. (20.3 cm) in diameter and when the steel has this strain under 25%—and T is an exponent term based on thickness of the target plate. Equation 17.28 has roughly the same form as the DeMarre formula (Equation 17.8), but differs in several major ways. Note the change to the t/d format versions of the numerical constant (for each thickness interval) K_{t} and the scaling factor term S; the rest of the terms remain in Hershey's original formats modifying the NBL, hence the square of their product to match the $(\mathrm{NBL})^2$.

Note the added term J_{t}, which is needed due to the rather complex shape of the curve relating the kinetic energy to the thickness of plate penetrated with plate thickness. As discussed earlier, the energy-absorbing modes in the plate vary over a large range of possibilities; these various modes overlap one another in places. The J_{t} term smoothens the transitions from one mode to another.

If we put the formula back into a format for computing the M79 AP projectile NBL from these inputs, we obtain

$$\mathrm{NBL}_{\mathrm{M79}} = J_{\mathrm{t}}M_{\theta}\varepsilon_{\mathrm{F}}\frac{\left\{Q_{\mathrm{A}}\frac{t}{d}\right\}^T}{K_{\mathrm{t}}S\sqrt{\left(\frac{w}{d}\right)^3}} \tag{17.29}$$

$\text{NAVYBLM79 (1st Estimate)} = \text{(NUM)(J)(MOBLIQUITY)(SCALEFACTOR)[(QA)(TSLASHD)]}^{\text{TEXP}}/\text{PROJDENSITY}^{0.5}$,

where the computer program parameter "SCALEFACTOR" is a function of S that already takes the exponent T into account and inverts the divisor S (see Equation 17.35).

When the parameter %E will modify the results (has a nonzero value and all use requirements are met), we modify NAVYBLM79 as follows:

```
NAVYBLM79 (Final Value) = (NAVYBLM79(1st Estimate)) (1 - %E),
```

where the computer program calculates "%E" as defined by Equation 17.36. Otherwise, %E is set to zero and the final value equals the first estimate.

Equation 17.29 is the format used in the HCWCLCR3 computer program when no windscreen, hood, or AP cap is present. It is used in this format to determine whether or not a given projectile hitting a given plate at a given striking velocity and obliquity angle will penetrate or not, so the program has to compute the NBL first and then compare the input striking velocity to the NBL to determine the penetration result. We shall now go into detail of the procedure.

First, except for the $1/Q_A$ term, all the terms in the "t/d =" format are set to the power of the exponent $1/(2T)$, which is the replacement for the W exponent in Equation 17.9. The requirement of the "2" in the denominator of $1/(2T)$ is because Equation 17.28 used $(NBL)^2$ and Hershey's various functions modify the NBL, not the plate thickness t, just like C in the DeMarre formula (Equation 17.8), so in the "t/d =" format, the equation written here requires this exponent to keep Hershey's numerical values and formula formats consistent.

Second, the equation is broken down into a number of thickness range steps for the values K_t. In the program, K_t is the inverse of the square of "NUM," with "NUM" set equal to the selected table value NC(n); J_t or "J" in the program is calculated from three selected table values (JA(n), JB(n), and JC(n)); T or "Texp" in the code is set by the selected table value TCOEF(n) in the program; and, for Ob > 45° only, part of M_θ in the calculation. When t/d ("TSLASHD" in the program) is computed from the user inputs, a lookup table counter steps through various t/d cutoff values by incrementing its index "n" ("TSLASHDMAX(n)" table entries in the program) until the correct "n" for the given thickness step is found, and this defines the correct K_t, J_t, and T values to select.

Third, after the obliquity angle θ or "Ob" in the code is input by the user, a decision is made: If θ is over 45°, then t/d and θ are both used to select values from a large two-dimensional matrix for M_θ written as "MPTBL(m,n)" in the code. These data are then interpolated both in the θ and t/d directions to calculate a current value for the M_θ (MOBLIQUITY in the code) multiplier. If θ is equal to or below 45°, the only interpolation is in the obliquity direction between the two nearest straddling values from a separate very small M_θ table called "MPLTBITE(n)." The final value of M_θ used is computed by

```
MOBLIQUITY = (Interpolated value from MPTBL(m,n) table)/COS(Ob), if Ob > 45°
```

or

```
MOBLIQUITY = (Interpolated value from MPLTBITE(n) table)/COS(Ob), otherwise.
```

This will be discussed in greater detail in the following.

We will now discuss the various parameters in this version of Hershey's procedure.

1. The values of t/d, Q_A, and w/d^3 are the same values shown in Equation 17.9.
2. The values of the upper cutoff, TSLASHDMAX(n), are obtained from Table 17.1. The table was generated to select the value of "n" used to further determine NC(n) and K_t, as well as TCOEF(n) to set equal to T (Texp in the code) and to select the three values JA(n), JB(n), and JC(n) needed to compute J_t (J in the code). Here, user-defined t/d is stepped through the many t/d threshold values shown in the following to determine the correct "n," which is the opposite of how "n" is used in all later "n"-indexed lookup tables.

Note how many steps there are at the low end of the plate thickness range, where the energy-absorbing modes in the plate rapidly shift for even a small thickness change. For example, penetrating a plate much over 2 calibers would require very low obliquity and a low value of Q_A since we are limited in the program to striking velocities only up to 3500 ft/s (1066.8 m/s). At striking velocities above this value, the thermal and shock effects

TABLE 17.1

Values of n Chosen from a t/d Ratio Initially Calculated from Equation 17.27

n	t/d	n	t/d
1	0.01156	7	1
2	0.05000	8	2
3	0.07500	9	3
4	0.10000	10	4
5	0.25490	11	6
6	0.47931		

will make the calculations using this method inaccurate because these mechanisms are not considered in the method.

3. The values of the numerical constant K_t are set from the square of the inverse of the table entries NC(n) in Table 17.2. Numerically, we have

$$K_t = \left\{\frac{1}{\mathrm{NC(n)}}\right\}^2 \tag{17.30}$$

Note that the table requires one to multiply the K_t column value by 10^{-6} to get the actual K_t.

At high t/d values, the curve flattens out to nearly a constant horizontal line, indicating that the penetration is becoming closer to t/d proportional to V^2 = KE situation, as though the projectile was passing through the armor which is behaving hydrodynamically. In this hydrodynamic situation, only the armor that is in direct contact with the projectile at any given moment is absorbing energy.

4. The values of the KE exponent T or "Texp" in the code is set from the table entries "TCOEF(n)" in Table 17.3.

As mentioned with K_t earlier, at high t/d, the curve approaches a constant horizontal line, indicating that the penetration is becoming more hydrodynamic, where T = 0.5. This theoretical completely flat horizontal penetration plot, with an exponent of 0.5 and K_t constant from then on, is never actually reached because the front and rear portions of the plate

TABLE 17.2

Values of NC(n) and K_t for n Determined by Table 17.1

n	NC(n)	$K_t \times 10^6$	n	NC(n)	$K_t \times 10^6$
1	335.253920	8.896898	7	1179.94178	0.7182553
2	516.857560	3.743331	8	1179.94178	0.7182553
3	902.414250	1.227971	9	1173.77728	0.7258195
4	1290.90181	0.6000861	10	1201.73299	0.6924430
5	1687.69560	0.3510846	11	1227.31234	0.6638803
6	1361.41055	0.5395377			

TABLE 17.3

Values of T (Texp in the code) for n Determined by Table 17.1

n	T or Texp	1/(2T) or 1/(2Texp)	n	T or Texp	1/(2T) or 1/(2Texp)
1	0.4336513	1.1530001	7	0.6196090	0.8069605
2	0.5306597	0.9422234	8	0.6387357	0.7827964
3	0.7166931	0.6976487	9	0.6462927	0.7736433
4	0.8549115	0.5848559	10	0.6248678	0.8001693
5	0.9713125	0.5147674	11	0.6095760	0.8202423
6	0.8141355	0.6141838			

crater and petal, which modify the penetration energy. As the plates get thicker, this energy loss becomes a smaller fraction of the total energy loss. As the impact energy increases with the increasing striking velocity needed to penetrate thicker plates, new significant effects due to shock waves, compression and deformation of the projectile, and increased heating of the plate and projectile at the impact site all begin to occur and make the assumptions used here less and less accurate. This problem is increased by the fact that the speed of the projectile through the plate is very high near the face and very low near the back at the NBL, so even with a homogeneous material, it reacts somewhat differently at various depths, precluding a uniform, perfectly KE-dependent relationship between the striking velocity and the thickness penetrated. This is why there is a 3500 ft/s (1066.8 m/s) maximum on the allowed striking velocity.

5. The value of the variable J_t is always 1.0, and all the JA(n), JB(n), and JC(n) are set to zero for those thickness steps except for the three thickness intervals shown in Table 17.4. This results in no influence of the J_t parameter on Equations 17.28 and 17.29.

The parameter J_t is a Green's function that interpolates between two curves calculated from different formulas, each of which is correct for its data interval. This allows us to "feather" the mixed-data interval between two curves, matching the test data over the bridging span by using a user-created, completely empirical equation. At each end of the J_t curve, the function matches the value of formula used up to that point. Physically speaking, the two intervals are where the dishing and petaling regions overlap (between n values of 2 and 3) and where the cratering/petaling and wedging regions overlap (n = 6). J_t continually changes the value of K_t for the selected interval into a variable "K_t(modified)" value that is no longer a constant over the entire interval. J_t is not defined anywhere else in the code and is neutralized by being set to 1.0 there. The formula for J_t in both intervals is the

TABLE 17.4

Values of JA(n), JB(n), and JC(n) for n Determined by Table 17.1

n	JA(n)	JB(n)	JC(n)
2	−0.02	4682.6223	54.131113
3	−0.01	7200.0000	360.00000
6	0.02	802.10329	204.45613

following (it is always 1.0 when JA(n) is zero):

$$J_\theta = \mathrm{JB(n)}\left(\frac{t}{d}\right) - \mathrm{JC(n)} \tag{17.31}$$

```
Jdeg = [JB(n)](T/D) - JC(n),
```

```
Jsin = SIN(Jdeg).
```

Note: This angle is changed to radians for trigonometry in BASIC.

If sin (J_θ) or Jsin in the code is negative, then we set Jsin = 0; otherwise, we use Jsin as is, in which case

$$J_t = 1 + \mathrm{JA(n)}\sin(J_\theta) \tag{17.32}$$

```
Jt = 1 + [JA(n)](Jsin).
```

Note that J_t is a distorted sine curve, curving up when JA(n) is negative and curving down when JA(n) is positive, bent and stretched to follow the smoothed averaged data points in the applicable interval as plotted on Hershey's graphs in his documents [2,4].

Since J_t depends on t/d, for values of "n" where J_t is not 1.0, we have to solve for t/d initially with J_t set to 1.0, then use the resulting t/d to solve the formula for J_t. We then use this new, non-1.0 value of J_t to solve Equation 17.28, which will almost certainly be different from the previous $J_t = 1.0$ solution for t/d. We can use this new t/d value to again solve for J_t and continue this iteration loop until t/d and J_t converge, which results in the final value of t/d.

6. The scale factor that allows larger projectiles of a given baseline design to have a somewhat easier time penetrating scaled-up armor plates of the same metallurgical structure used by Hershey [2] is based on actual ballistic tests. Data which fed the posttest analysis were derived from the results of STS-type steels being subjected to loading at various rates from several sources. These data were averaged and adjusted to the standard STS metal parameters used in the testing [1]. Hershey also graphed wrought iron and mild steel high-speed load tests to show that the scaling effect decreased in a predictable manner as the steel got stronger. This effectively produces a smaller difference with projectile size against scaled plates, justifying the scaling effect formula that he adopted for his analysis effort.

Since the scaling effect is based on striking velocity controlling the rates at which the armor is being forced out of the path of the projectile, and since the metal fails by several different modes, the energy absorption mechanism (dishing etc.) changes as a function of plate thickness. Therefore, the scaling effect term in this penetration description is not a simple constant as given in the DeMarre formula. Hershey's NBL modifier equation for the STS scale effect SF is

$$\mathrm{SF} = \sqrt{1 - (0.04)\ln\left(\frac{d}{3}\right)} \tag{17.33}$$

where the standard 3 in. M79 AP projectile is the default center of the scaling adjustment range. For the penetration calculation in HCWCALC, this is changed:

$$S_{\text{sq}} = \frac{1}{(\text{SF})^2} \tag{17.34}$$

The exponent "2" in Equation 17.34 cancels the "2" in the "1/[(2)(Texp)]" exponent in Equation 17.28. Since S appears in Equation 17.28, and since the Texp term varies with plate thickness, the value of S as a function of projectile size also varies with plate thickness. Thus, we write

$$S = \left(\frac{1}{(\text{SF})}\right)^{\frac{1}{T}} \tag{17.35}$$

For example, if the projectile is 6 in. in diameter and the plate is a half-caliber (3 in.) thick, we have a t/d of 0.5. Examination of Table 17.1 shows that $n = 7$. Then,

$$\text{SF} = \sqrt{1-(0.04)\ln\left(\frac{6}{3}\right)} = 0.98604 \tag{E17.1}$$

so

$$S_{0.5} = \left(\frac{1}{(0.98604)}\right)^{\frac{1}{0.6196010}} = (1.01415)^{1.61394} = 1.023 \tag{E17.2}$$

Based on this calculation, we would need a plate that is 2.3% thicker (3.069 in.) to stop that 6 in. shell when impact occurs at the NBL of the thinner test plate hit by the 3 in. shell, on average.

Similarly, for a 0.1-caliber plate, we have $n = 4$ and

$$S_{0.1} = \left(\frac{1}{(0.98604)}\right)^{\frac{1}{0.8549115}} = (1.01415)^{1.1697117} = 1.017 \tag{E17.3}$$

which is about a 25% smaller increase. Thus, scaling is variable here.

Note: This value for SF is for Ni–Cr steel of ca. 225 BHN. It has a larger change with plate thickness for significantly weaker iron alloys such as mild steel (BHN = 140–160 and coefficient 0.04 replaced by 0.06) or wrought iron (BHN = 100–105 and coefficient 0.04 replaced by 0.08), so they will require a variable Q_A as d changes. But this would be needed only if, for some reason, the NBL of these rather weak and variable ballistic-quality materials has to be determined to such a tight precision. This precision is rarely warranted.

7. The failure strain modification to Hershey's M79 AP projectile analysis methodology was developed by the author, taken from the detailed penetration curves of Krupp post-WWI Wh (or "Odin-hardened") standard STS-type homogeneous, ductile armor steel, drawn by the German firm of Krupp for the WWII German Navy in 1940 [9]. In these graphs, for up to 8 in. projectile size, the charts more or less tracked similar US Navy and British Navy computations, such as those based here on the M79 AP projectile against STS-type armor. However, in these charts, against larger shells, the Wh armor that the Germans assumed to be used by all

warships, German and enemy, had its resistance to a normal impact drop off steadily with increasing projectile size, even when using virtually identical late-model German Psgr.m.K. L/4,4 projectiles* of 8 in. (20.3 cm), 11.1 in. (28.3 cm), and 14.96 in. (38 cm) sizes. The Psgr.m.K. L/4,4 were hard-capped AP (US Army APC) projectiles with an overall of length of 4.4 calibers. The Krupp Wh metallurgical specification for 1939 (prior to changes due to wartime shortages) show this armor to be very similar to both US and British naval armors, except that it retained the WWI minimum failure strain of 18% (US WWI STS had a requirement of 19%, for comparison), while the improved WWII US and British armors now required a much better strain to failure of 25%. It is possible that this is not the reason for the Wh deficient functioning against large-caliber shells, but it is the only specification parameter that is known to be significantly different between these armors. In addition, post-WWII testing by both the Americans and British showed that this problem was indeed the case. The only Wh plate that was significantly inferior to US STS/class B armor plate was the very thickest one hit by the largest US projectile size in the test series. Detailed metallurgical analysis of the Wh armor showed that it was a good quality material similar to the US and British armors made to high standards by Krupp, so there does not seem to be any other parameter other than strain to failure related to this large-projectile weakness problem. Since the specified value is a minimum, actual plates would be somewhat more elastic than this, on average. Unless Krupp was specifically making materials far above its own design specifications in its plants and ignoring the actual metallurgical specification it created for its own Wh armor, the average failure strain value for its Wh plates would be lower than the average of both US and British WWII armors. Krupp would be the last company to accuse of such a thing because of its highly conservative mind-set, so it is unlikely that they ignored that single specification value, while meeting closely all the others, unless they purposely set the bar extremely low to assure that all their plates would pass the acceptance testing.

We are left with no alternative but to use the Wh specification value for strain to failure as the "typical" one for that armor type and do the same to any other steel whose specifications we use.

Using the G.Kdos. 100 penetration curves [10] and assuming a simple linear relationship between projectile size and resistance drop, yet being conservative and assuming that the effect was small if an even lower strain to failure was used by another armor-like steel (some steel materials had down to 16% in this parameter), the following relationship forms the basis of the effect:

If failure strain < 25% and $d > 8$ in., then the NBL multiplier (1 – %E), usually set to 1.0 for no effect, is reset to the following value:

$$\varepsilon_F = 1 - \left\{1 - \sqrt{\frac{\varepsilon_{spec}}{25}}\right\}\left[\frac{d-8}{8}\right] \qquad (17.36)$$

* In this German terminology the *P* is for "Panzer" (armor); the *s* is for "Sprengstoff" ("high explosive" is the German Navy definition); and the *gr.* is for "granate" (projectile/shell). The original German Navy AP projectiles made prior to 1902 had no explosive filler and were named *Stgr.* ("steel projectiles") to separate them from the even older chilled cast iron shot used before.

```
%E = 1 - [1 - √(Manufacturer's spec% Elongation/25)][(D - 8)/8],
```

if ($D > 8$) and (Spec% Elongation < 25). Otherwise %E calculation is skipped (%E becomes, in effect, 0), and the NAVYNBLM79 final value remains the first estimate value.

This is a second scaling effect term for the larger shells against armor that is too brittle to fully stretch (25% or more is considered fully) before it tears apart. The larger the projectile, the greater the distance the plate material at the impact center has to move to get out of the way of the projectile. If the armor tears open too early in each of a series of virtual concentric expanding rings, each ring thickness of the length of one test sample (50 mm in most metric tensile test machines or 2 in. in the usual USCS tensile test machine), as a crude first approximation, the weaker the armor will be as the number of these virtual rings (imagine a "bull's-eye" pattern) increases with projectile size. For a projectile up to 8 in. in size, the expansion radially from the center to the edge of a hole will be at most 4 in. at zero obliquity, and this means that only one full 2 in. test sample-length virtual ring, at most, is fully stretched to failure, with part of a second outer ring that is not stretched to failure. Thus, for the smaller projectiles, the metal failure results for rapid radial expansion are more or less constant, and no significant failure strain effect occurs.

For an 18% maximum strain specification requirement, as with Wh, the drop in the NBL against a 16 in. APC projectile would be

$$\mathrm{NBL}_{\mathrm{Wh}\varepsilon} = (\mathrm{NBL}_{\mathrm{Whno}\varepsilon})\left(1-\left\{1-\sqrt{\frac{\varepsilon_{\mathrm{spec}}}{25}}\right\}\left[\frac{d-8}{8}\right]\right) \tag{E17.4}$$

$$\mathrm{NBL}_{\mathrm{Wh}\varepsilon} = (\mathrm{NBL}_{\mathrm{Whno}\varepsilon})\left(1-\left\{1-\sqrt{\frac{18}{25}}\right\}\left[\frac{16-8}{8}\right]\right) = (0.8485)(\mathrm{NBL}_{\mathrm{Whno}\varepsilon}) \tag{E17.5}$$

Here $\mathrm{NBL}_{\mathrm{Wh}\varepsilon}$ is the NBL for Wh armor with the strain effect, while $\mathrm{NBL}_{\mathrm{Whno}\varepsilon}$ is the NBL for Wh armor with no strain effect. This is a very significant loss in protection against the larger AP shells, which is not evident in US or British test data. Note that very thick homogeneous, ductile armor usually has its Brinell hardness and ultimate tensile strength values somewhat reduced compared to thinner plate due to difficulties in keeping the crystal structure of the center of the plate of the desired type. However, even the modest reduction in typical heavy plate strength due to this is not a factor in the 1939 version of Krupp Wh armor, since Krupp's specifications allow no such reduction, being identical for all thicknesses.

Note that this reduction will apply to any homogeneous, ductile armor, including essentially all WWI and earlier armors that have specifications that allow an ultimate strain to be less than 25%. This shows the major improvements in metallurgy that occurred between WWI and WWII in most homogeneous, ductile, and face-hardened armors. Even a seemingly minor parameter such as average strain to failure, if improved markedly, can have major effects in improving armor resistance.

Most experimental tests done by naval proving grounds and by manufacturers are with projectiles that are economical to repeatedly use, usually at most 6 in. (15.2 cm), although sometimes up to 8 in. (20.3 cm). Sufficient funding to use larger, battleship-size projectiles in a major test series was rare. Since projectiles had to be over 8 in. in diameter to show the preceding effect and the use of homogeneous, ductile nickel–chromium armor in thick sections for protection against large projectiles was rare, these tests were simply not performed. Even though virtually all WWI-era class B armor and STS and their foreign equivalents would have shown this effect if extensive tests of thicker solid plates had been done, most battleships of the time only used rather thin laminated decks of STS-type or even only HTS-type layers. Occasionally, these layers might be overlayed/underlayed with nickel steel and/or mild steel plating. These complicated structures made detailed evaluation of the various metallurgical factors in the steels difficult to distinguish due to effects of laminating the various layers together. The moderately thick homogeneous, ductile plates that were used were of various designs where failures in many cases were due to adjacent plates pulling apart at the joints, not being penetrated in their centers when hit in AP projectile tests. These tests were usually performed at high obliquity as well. Full-scale tests with steeply falling, aircraft-dropped AP bombs against turret roofs were very rare, even in WWII, where such attacks were much more likely, as happened twice at Pearl Harbor. Nobody prior to the late 1930s, when solid single-plate heavy armor decks were introduced in the new battleships, would have discovered a high strain effect in all of the "noise" due to so many other factors. Indeed, the United States in WWII was the only nation to ever use single heavy class B armor plates for conning towers and turret faces (and even here, some laminated armor plates were employed for the thickest plates) in its new battleships. All other nations used face-hardened armor for these areas so there was virtually no testing that would have identified the high strain effect by anyone else except for the rather thorough development tests by Krupp on its own Wh armor. This thick WWII US armor had its maximum strain at or over the 25% minimum value described earlier, so tests with it would not have shown this effect even had they been performed. One can only discover something new when one has the ability to compare the new test results with known prior testing where there is a clear understanding.

8. The largest modifier of armor resistance is the effect of changing the angle at which the projectile hits the plate, the impact obliquity θ, which is measured from the normal, as used by the Americans and British in their tests. Other nations used other reference angles. German tests, for example, used 90° as the plate normal. This requires careful checking when using documents from foreign sources. We keep Hershey's methodology and tables for modifying the NBL for θ, which are based, in the code, on a modifier $M_{\text{table}}(\theta)$, multiplying the line-of-sight thickness ($1/\cos\theta$ is used in the program as it is easier to compute). Thus we can write

$$M_\theta = \frac{M_{\text{table}}(\theta)}{\cos\theta} \tag{17.37}$$

For $\theta > 45°$, the value of $M_{\text{table}}(\theta)$ resides in a large matrix "MPLT(m,n)." In the matrix "m" increments obliquity in 2.5° steps and "n" increments plate thickness in 0.05-caliber steps. Thus, this parameter depends both on θ and on plate thickness t/d, up to a plate thickness of 0.9 caliber, above which the thickness effect stops changing and only changes in θ will change the value of $M_{\text{table}}(\theta)$. Each computation linearly interpolates along the t/d axis first. The two matching θ-axis table entries for the lower of the two straddling thickness t/d-axis entries for the highest tabled $M_{\text{table}}(\theta)$ value that does not exceed the actual θ are interpolated to the actual t/d value. Then the same thing is done for the two closest θ-axis table entries immediately above the actual θ for the t/d-axis interpolation. We now have a new point below the actual θ and a new point above the actual θ, each already corrected for its t/d-axis value. Now we linearly interpolate between these two new points onto the θ-axis to obtain the final $M_{\text{table}}(\theta)$ value to use.

17.12.1 Obliquity Angles up to 45°

For $\theta < 45°$, there is only a single $M_{table}(\theta)$ value for any plate thickness, so linear interpolation between the tabled values using the actual obliquity is sufficient to compute the $M_{table}(\theta)$ to generate the final NBL multiplier M_θ via Equation 17.36 [3]. The table in the code for this low obliquity range is "MPLTBITE(n)" and is incremented in 2.5° steps.

Up to 45° obliquity, the force under the chin of the projectile nose caused by the asymmetric force on that side of the nose during initial impact is very rapidly reversed as the nose digs deeper and deeper into the plate. This effect is either physically penetrating into a thick plate or by stretching the center of the wide dish region as a funnel shape for a thin plate. Since the deeper the nose digs in, the more the plate material accumulates above the nose on the forehead (the portion of the nose on the direct opposite side from the chin). This is depicted in Figure 17.8. Also, as the nose digs in and pushes forward somewhat parallel to the surface, the armor material in front of the nose, parallel to the plate surface, to the depth of the nose tip, gets pushed straight up and folded forward like the peeling from a wood plane. This upward-bent face armor layer riding the forehead of the projectile is termed the *spur* due to its curved shape. In effect, the spur builds up the weight of armor dislocated from the surface of the plate on top of the armor further forward in the direction of penetration parallel to the plate face. This buildup of this material above the nose of the projectile greatly increases the force countering the glancing force on the chin very rapidly, and the nose almost immediately begins to tilt the other way, into the plate, toward the plate normal.

One would normally expect, in an ideal impact situation at the NBL, the nose to exit the plate back surface at exactly right angles to the face (and to the back unless the plate had a tapering cross-section) forming a more or less round hole in the plate back, even though the hole in the plate face is oval in shape due to the original obliquity of the projectile. This rotation of the projectile axis by a force surrounding the projectile nose so that the exit angle θ_e ("Ex" in the code), measured from the extension of the normal line through the plate back, causes the middle and/or lower body of the projectile to swing around using the nose as the pivot and impact the roughly vertical far side of the hole. This is depicted in Figure 17.9. This phenomenon is termed *base slap* (this slap can also occur on the plate face if the projectile completely glances off) and can cause major damage to the aft and middle portions of the projectile and possibly even break it in two. Major projectile design

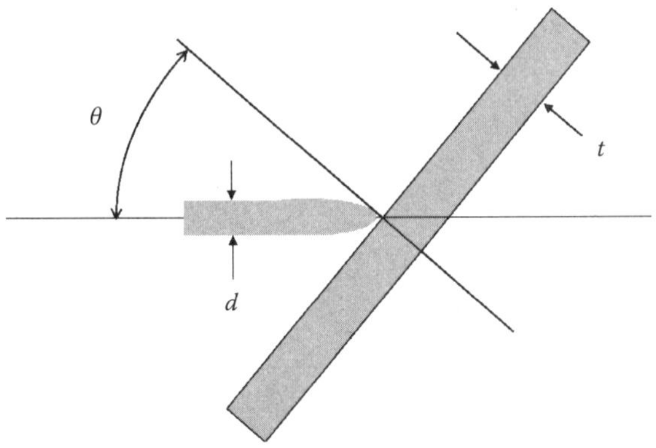

FIGURE 17.8
Impact parameters defined.

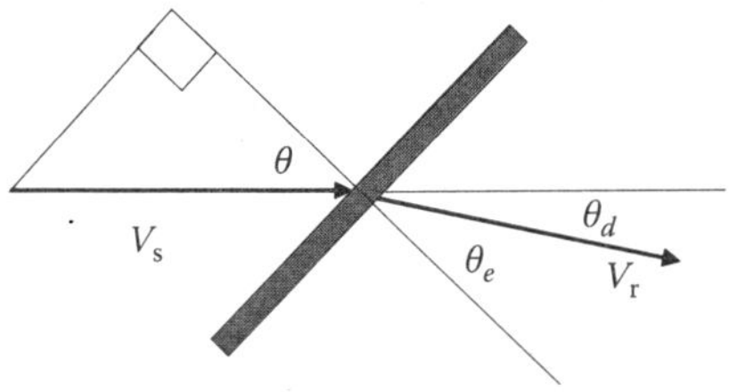

FIGURE 17.9
Change in angle of projectile velocity vector during penetration.

work to toughen the middle and lower body against this damage historically made up a considerable portion of the design effort in iron and steel AP projectiles, with significant success in some cases and bad failures in others.

Figure 17.10 is a plot of the penetration of an M79 projectile vs. various obliquities at different t/d values. This figure clearly shows the nonlinearity in the penetration of this type of projectile. Figure 17.11 is a plot of the normal kinetic energy density under the same conditions as Figure 17.10. The difference with this plot is that the NBL was held constant and the plate sloped and either thickened or thinned to provide the same NBL as the zero obliquity case. It clearly shows that below a slope value of 45–50°, sloping the armor actually results in a heavier plate to provide the same NBL than merely keeping the plate at right angles and thickening it, while above this obliquity range, the sloping of the plate can be considerably superior to thickening it. It is thus critically important to get a good idea of the direction that the threat most probably will come from.

It should be noted that in Figure 17.11, a value above the normal (0° = vertical) dashed line for that plate thickness and obliquity curve, as is true for all low obliquity tests, means that the inclined thinner plate is not as resistant as a thicker vertical plate of the same weight (i.e., sloping does not help). Only if the curve goes below this dashed normal line does the sloping and thinning of the plate to keep the weight the same gain an advantage over just making the vertical plate thicker, as seems to be true at high obliquity. Note that the curves indicate that the thicker the vertical plate is, the greater the advantage of sloping an equal-weight thinner plate at high obliquity.

As the striking velocity increases above the NBL, the projectile will pass through the plate faster, so the deflection from the value of θ to the value of θ_e steadily decreases until at some very high velocity above the NBL, the projectile will essentially pass straight through with no significant deflection.

Similarly, at under 45°, the residual velocity V_r of the projectile after it penetrates will be essentially zero at the NBL, since the plate can stop the projectile. However, as with the exit angle, the drop in the remaining energy will be that needed to penetrate at the NBL plus the decreasing amount of energy lost due to the deflection from θ to θ_e as the two approach one another with increasing striking velocity. The component of the residual velocity, measured along θ_e, is the velocity computed after subtracting the energy lost at the NBL and subtracting an additional amount due to the deflection angle θ_d. This is because V was originally down the θ-axis, and any lateral velocity component created when going from θ to θ_e is lost, too, when calculating V_r. Actually, there may be a change in the energy required to penetrate as θ_e approaches θ if V increases well above the NBL, but this is difficult to determine, so we assume no difference as a general rule. With pointed projectiles against STS armor, we assume no plugging, so no energy is lost throwing off such armor chunks (this is a major computation in face-hardened armor, though!).

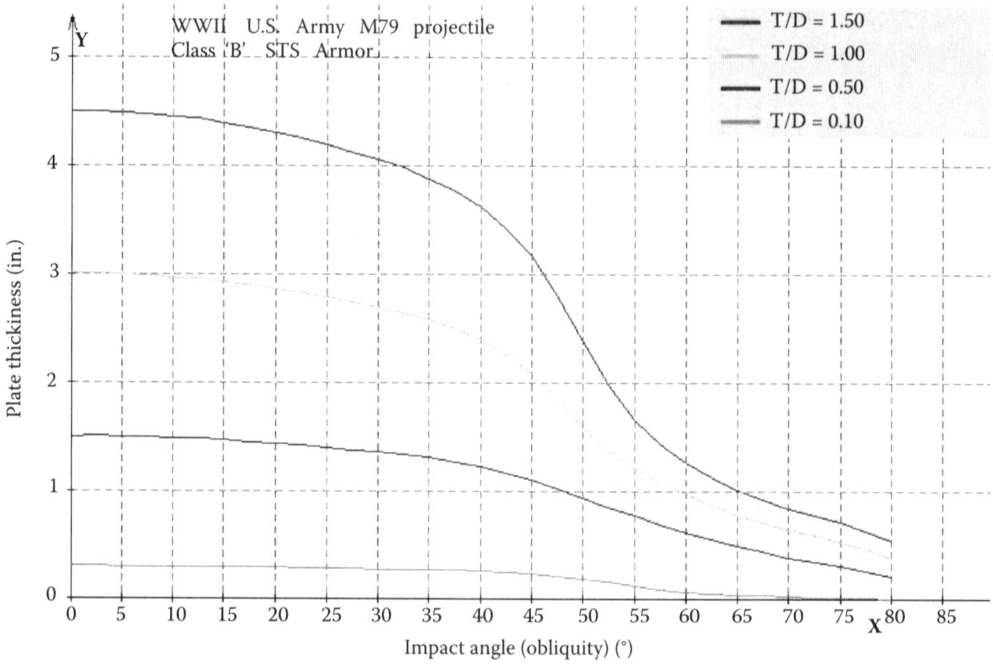

FIGURE 17.10
(See color insert.) Penetration of M79 projectile at various obliquities and t/d ratios.

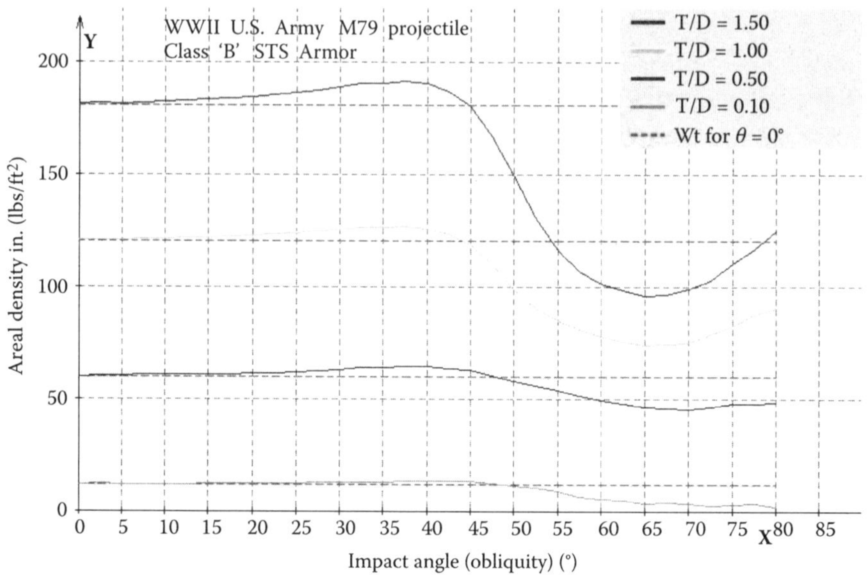

FIGURE 17.11
(See color insert.) Areal density of plate required to provide the same NBL protection given various expected impact obliquities.

The algorithm used to compute the deflection angle in the code after a complete penetration is given by

$$\theta_d = \theta - \theta_e \tag{17.38}$$

The following procedure was found in the study by Recht and Ipson [6], in which the authors fired small-caliber AP bullets at several thicknesses of high tensile steel plates at different obliquities up to 45° and developed the following rather elegant set of formulas to match the data that they obtained. It is used for all the homogeneous, ductile armor and face-hardened armor computation programs described here. The assumption that in the under-45° obliquity range, the exit angle is always at right angles to the plate back, giving a deflection angle equal to the obliquity angle, is the only difference from Recht and Ipson [6].

We define the ratio of the striking velocity to the NBL:

$$V_{\text{rat}} = \frac{V}{\text{NBL}} \tag{17.39}$$

```
VRAT = V/NBL.
```

We now define the ratio of the energy remaining with the projectile after the penetration event to the energy required to barely penetrate the plate.

$$KE_{rat} = V_{rat}^2 - 1 \tag{17.40}$$

Inserting Equation 17.39 into Equation 17.40, we obtain

$$KE_{rat} = \frac{V^2 - NBL^2}{NBL^2} \tag{17.41}$$

```
TMPVEL = (V² - NBL²)/NBL².
```

We now define a function of the total impact energy as a multiple of the penetration energy at the NBL times the sum of that total energy plus the ratio of the remaining energy after penetration to the original impact energy. This last term will approach twice the impact energy ratio for large values of V, but decrease to just the impact energy ratio itself as V is approaches the NBL.

$$KE_{tot} = V_{rat}^2 + \frac{V_{rat}}{\sqrt{KE_{rat}}} \tag{17.42}$$

If we insert Equation 17.41 into Equation 17.42 and manipulate the terms, we obtain

$$KE_{tot} = V_{rat}\left(V_{rat} + \sqrt{\frac{V^2 - NBL^2}{NBL^2}}\right) = V_{rat}\left(V_{rat} + \frac{V}{NBL}\sqrt{\frac{\left(V^2 - NBL^2\right)\left(NBL^2\right)}{\left(NBL^2\right)\left(V^2\right)}}\right) \tag{17.43}$$

which, after using Equation 17.39, yields

$$KE_{tot} = V_{rat}^2\left\{1 + \sqrt{\frac{V^2 - NBL^2}{V^2}}\right\} \tag{17.44}$$

```
TMPV = (VRAT²){1 + √[(V² - NBL²)/V²]}.
```

We now define the central function of this algorithm from Recht and Ipson [6], which is "$\sin\theta\cos\theta$" (SNCS):

```
SNCS = [SIN(Ob)][COS(Ob)].
```

This obliquity angle function goes to zero at both 0 and 90° with a maximum value of 0.5 at 45°.

We now set up values used to compute θ_d from $\tan^{-1}\theta_d$. The equations are

$$\theta_{d1} = \frac{\sin\theta\cos\theta}{KE_{tot}} \tag{17.45}$$

```
TMPDF1 = SNCS/TMPV,
```

$$\theta_{d2} = 1 - 4(\theta_{d1})^2 \tag{17.46}$$

```
TMPDF2 = 1 - (4)(TMPDF1²).
```

As V increases, KE_{tot} approaches 2 times V_{rat}^2 and θ_{d1} decreases toward zero. However, if V is near the NBL, θ_{d2} approaches the value of SNCS. For large V, θ_{d2} approaches 1 and for V near the NBL, θ_{d2} approaches $1 - (4)(\sin\theta\cos\theta)^2$.

We now set up $\tan\theta_d$ from trigonometric identities, using θ_{d1} and θ_{d2} defined earlier, with the definition

$$\tan\theta_d = \frac{\sin\theta_d}{\cos\theta_d} \tag{17.47}$$

Using Equations 17.45 and 17.46, we can write

$$\tan\theta_d = \frac{1-\sqrt{\theta_{d2}}}{2\theta_{d1}} \tag{17.48}$$

```
TAN(Df) = [1 - √(DF2)]/[(2)(DF1)]
```

and

$$\theta_d = \tan^{-1}\left(\frac{1-\sqrt{\theta_{d2}}}{2\theta_{d1}}\right) \tag{17.49}$$

```
Df = ARC[TAN(Df)]
```

Let us test the extreme values for θ_d when V is either very large or near the NBL:

If V is very large, then

$$\tan\theta_d = \frac{1-\sqrt{1-4(\sin\theta\cos\theta)^2}}{2(\sin\theta\cos\theta)} \rightarrow \frac{1-\sqrt{1-4(0)^2}}{2(0)} \tag{E17.6}$$

but the value of θ_{d2} inside the square root sign is decreasing as the square of θ_{d1} decreases, as well as the numerator approaching the value $1 - 1 = 0$ when θ_{d1} approaches zero, so the numerator is decreasing faster than the denominator, and the entire formula thus approaches zero, indicating that deflection will be close to zero, as desired for very large V.

If V approaches the NBL, then KE_{tot} rapidly decreases to V_{rat}^2, which is 1 when V equals the NBL. At this end of the V range, $\theta_{d1} = \sin\theta\cos\theta$ at the NBL and

$$\theta_{d2} = 1 - 4(\theta_{d1})^2 = 1 - 4(\sin^2\theta\cos^2\theta) = 1 - 4(\sin^2\theta)(1 - \sin^2\theta) \tag{E17.7}$$

$$\theta_{d2} = 1 - 4(\sin^2\theta) + 4(\sin^4\theta) = \{1 - 2(\sin^2\theta)\}^2 \tag{E17.8}$$

so

$$\tan\theta_d = \frac{1 - \sqrt{\{1 - 2(\sin^2\theta)\}^2}}{2\sin\theta\cos\theta} = \frac{\sin^2\theta}{\sin\theta\cos\theta} = \tan\theta \tag{E17.9}$$

Thus, the projectile exits the plate back along the normal line, since it is deflected by the value of θ, cancelling θ completely.

The residual velocity V_r is calculated rather easily now (no armor plugging and, in this case, the projectile weight is fixed).

We compute the remaining energy after penetration, assuming that the energy loss at the NBL is a constant for all impacts above it. In the code, at this point, "NBL" = "VMIN" since the program allows the projectile body to weigh less than the original projectile due to AP caps, hoods, and windscreens, which we have ignored in this analysis. We also do not need to compute the "Energy NBL" used in the above 45° computations. This will be discussed in the following as it is more complicated. So, in the code, we set

```
VMIN = NBL
```

and define

$$KE_{remtot} = V^2 - V_{\min}^2 \tag{17.50}$$

```
ELEFTTOTAL = (V² - VMIN²).
```

This is set equal to KE_{rempen} ("ELEFTPEN" in the code) since the projectile penetrator weight is the entire projectile weight w:

$$KE_{remtot} = KE_{rempen} \tag{17.51}$$

We then adjust the value of V_r due to the deflection angle θ_d since the lateral component of V_r is shifted by θ_d from θ to θ_e as energy is absorbed by the plate, leaving only the component directed down θ_e:

$$V_r = \cos\theta_d\sqrt{KE_{rempen}} \tag{17.52}$$

```
Vr = [COS(Df)] [√(ELEFTPEN)].
```

Except for using a different V_{min} and perhaps having KE_{remtot} not equal to KE_{rempen} due to changing projectile penetrator weights, this computation is identical for all V_r calculations, including for over 45° computations.

17.12.2 Obliquity Angles over 45°

For $\theta > 45°$, the process of obtaining of $M_{table}(\theta)$ was described earlier [3]. As with the under 45° calculation, Equation 17.36 still applies.

Ignoring base-first penetrations for the moment, we need θ_d, θ_e, and V_r to be computed, as with the under 45° obliquity range.

The very useful SNCS formula for computing θ_d only went up to 45° obliquity, so the author adapted it to the over 45° obliquity range. It was assumed as a first approximation that the θ_d adjustments would be symmetrical about 45°. However, there would be major differences as to the energy absorbed by the plate at the NBL (all of it when θ was below 45°). Unlike below 45°, the energy loss used to compute V_r from V would be greater than the actual value lost due to projectile deceleration if the NBL in the simple KE formula for the under 45° case was used for most of the range for the over 45° obliquity case. The NBL is much more affected by the glancing forces at higher obliquity, and they redirect the projectile, not slow it down. A method of reducing the energy loss to only those forces that did decelerate the projectile was required to get the higher actual V_r for over 45° obliquity. Physically, the minimum residual velocity must also increase as obliquity increases. This is because more and more of the penetration energy at the NBL goes into deflecting rather than slowing down the projectile. The energy loss must also decrease, increasing the residual velocity still more, just like the under 45° case, as V increases above the NBL.

The following modified version of the SNCS formula for this high-obliquity range was developed by the author. The method involves the determination of the "energy NBL," which uses a smaller V_{min} value in place of the NBL to calculate the amount of energy absorbed—that is, the projectile actually slowed down—by the plate in the KE formula when computing V_r. The rest of the needed NBL energy just caused the projectile to ricochet with a considerable remaining velocity. The over 45° θ_d and V_r computations based on the SNCS formula are identical to the under 45° computation, but using this modified V_{min} instead of the regular NBL.

We determine the ratio of the θ to 45° (bottom edge); that is, how much bigger θ is than 45°:

$$E_1 = 1 + \frac{\theta - 45^\circ}{45^\circ} \tag{17.53}$$

```
E1 = 1 + (Ob - 45)/45.
```

The value of E_1 linearly increases from 1.0 at 45° to 2.0 at 90°, although 80° is the cutoff point in these computations. This is because all projectiles are assumed to ricochet at over 80° for any iron or steel plate hit. At 80°, E_1 has a value of 1.778.

The SNCS term can be used to determine the amount of energy actually absorbed by the armor plate at various obliquities compared to the total needed to penetrate at the NBL. We normalize it to 1.0 at 45°, symmetrically dropping off to zero in both directions, with the θ range of interest here being the obliquity half above 45°:

$$E_2 = 2 \sin\theta \cos\theta \tag{17.54}$$

```
E2 = (2)(SNCS) = (2)[SIN(Ob)][COS(Ob)]
```

Note that E_2 has a value of 0.34202 at 80°.

The maximum energy absorbed at the NBL (all of it) will occur at $\theta = 45°$ (note that this must always match the result we obtain at 45° via the under 45° formulas). The absorbed energy will decrease at the NBL as θ increases, and ricochet, rather than deceleration, becomes the method used to prevent penetration. This is essentially a mirror image of the region below 45°. If we examine the ratio of the $\theta - 45°$ ratio fraction E_1 to the normalized energy loss E_2 at a given θ, we can define E_3 as

$$E_3 = \frac{E_1}{E_2} = \frac{1 + \dfrac{\theta - 45°}{45°}}{2 \sin\theta \cos\theta} \tag{17.55}$$

```
E3 = E1/E2 = [1 + (Ob - 45)/45]/{(2)[SIN(Ob)][COS(Ob)]}.
```

At $\theta = 45°$ this is equal to 1, so the energy is completely absorbed by the plate at the NBL and V_{min} equals the NBL, as required. As θ increases, the E_2 denominator gets smaller while the E_1 numerator gets larger; thus, E_3 increases more rapidly as θ goes up. The maximum value of θ is 80°. This results in

$$E_{3\,max} = \frac{1 + \dfrac{80° - 45°}{45°}}{2\sin(80°)\cos(80°)} = 5.1979 \tag{17.56}$$

We now compute the value of the energy NBL V_{min}, defining the energy actually absorbed in deforming and tearing open the armor plate at the NBL instead of merely deflecting the projectile path through ricochet, which, as specified, is less than the NBL for $\theta > 45°$ and gets smaller as θ increases. We shall use this V_{min} to compute V_r, which will be greater than it would have been in under 45° impacts. We set

$$E_4 = \frac{\sqrt{2E_3 - 1}}{E_3} \tag{17.57}$$

```
E4 = √[(2)(E3) - 1]/E3
```

and

$$V_{min} = (E_4)(\text{NBL}) \tag{17.58}$$

```
VMIN = (E4)(NBL).
```

The function for E_4 was chosen to give a good ballpark figure for this decrease in energy loss. The actual loss considerably varies with small changes in projectile yaw, plate quality, the exact obliquity angle, and the limits of these computations when the projectile hits actual plates. With real impact scenarios, we need to know many things such as exactly where the initial hit was; the direction the shell was going; how wide the plates are before a joint is encountered; how the plates are connected (welded, bolting, or riveted); and the design of the support structures of the plate behind/under them. This last consideration may mean that a shell passing through a plate will have a high chance of running into such supports, particularly at plate edges and, thus, now hitting I-beams and doubling plates and so forth. These "nuisances" greatly modify, rather randomly, what happens to the projectile as it tries to punch through the plate at a highly oblique angle. Real life is not simple!

At $\theta = 45°$, $E_4 = E_3 = 1$ and V_{min} = NBL, as required, so all impact energy is going into the plate to stop the projectile. At $\theta = 80°$, however,

$$E_{4\,\max} = \frac{\sqrt{2(5.1979) - 1}}{(5.1979)} = 0.5897 \tag{17.59}$$

so only 58.97% of the NBL is used to compute the loss of kinetic energy absorbed by the plate, with the rest remaining with the projectile when it goes through the penetration process of forcing a dish, gouge, slot, and, finally, a hole in the plate, making V_r nonzero even at the NBL and usually much larger than the V_r computed for any given value of V above the NBL for impacts below 45°.

The loss in energy to the projectile at the NBL at 80° is thus only $(0.5897)^2 = 0.3478 =$ 34.78%, meaning that other than the additional loss due to the deflection angle effect, which varies with the striking velocity above the NBL, the projectile retains 65.23% of its original energy after punching through the plate. Since such a plate has to be very thin to allow any penetration at all within the maximum striking velocity used, this is quite reasonable.

17.12.3 Base-First Penetration at θ = 65° or Greater

At high obliquity, the projectile will ricochet at impact velocities below the NBL, with only part of the energy being absorbed by the plate as the projectile skips off its surface [3]. This absorption occurs by dishing the plate, creating a canoe-shaped gouge. The entire time the projectile is in contact with the plate, it is bending the plate material downward and radially outward under the pressure of the nose of the shell as the gouge is formed by the forward motion of the shell along the surface. During this gouging, the nose is pushed from nose down to nose up. If a high enough striking velocity is achieved, the bottom of the gouge will split open into a long open slot that widens as V approaches the NBL. At the nose-first NBL (NBL(NF) in the code), the nose digs in deep enough to push through the tip of the slot and bulge the armor in front of it up. It will then bend this slot-tip armor upward and forward into a plate-thickness triangular spur at the end of the slot and ricochet downward off of the now raised plate back surface and through the spur-shaped, near-full projectile-width opening in the plate, making an oval hole at the end of the slot. The value of V_r is never zero, even at the NBL, for a nose-first penetration at over 45° obliquity, with the minimum V_r increasing rapidly with obliquity. This was discussed in part 2 earlier.

However, if the projectile hits a plate of 0.101–0.25 caliber thickness at θ = 65–67.49° or 0.1 caliber thick or less at 67.5–80°, there is a value of V, slightly under the nose-first NBL, where the slot at the bottom of the gouge will widen to the point such that when the nose finally rotates out of the gouge to ricochet away, the base slams down and finishes the widening process to 1 caliber, allowing the lower body of the projectile to tear through and the projectile to surf nose up through the gouge slot. During this motion, the projectile will be held by the pinching in of the armor on either side of the slot being forced open in front of the projectile. The steeper impact at 65° causes a shorter, deeper gouge and a more violent rotation as the nose points upward, allowing the base to slam through a somewhat thicker (up to 0.25 caliber) plate than at the slightly higher obliquity of 67.5°. At 67.5°, the maximum thickness for this to occur drops to 0.1 caliber. This is thin enough to allow a base-first penetration up to the maximum obliquity of 80°. The reason for the 0.101 caliber lower limit in the 65–67.49° range is that in this angular range, the difference between the nose-first and base-first NBLs is so small in thinner plates as to be not worth computing. Note that if the plate is thinner than mentioned at the beginning of this paragraph, the nose will tear through any plate where a base-first penetration might occur.

These base-first penetrations are rather stochastic. At the base-first NBL (NBL(BF) in the code), there is a 50% chance of the shell flipping up and away (ricochet) or down and through (penetration). Additionally, since the actual value of the NBL is not that precise, there are several percent errors possible from hit to hit either way. In the narrow velocity interval between the NBL(BF) and NBL(NF), the chance of the shell flipping through the plate rather than outward is probably at most about 75%. Also, it is highly unlikely that a base fuze would be activated by a base-first penetration as the forces are not down the length of the projectile to allow the firing pin to be thrown into the primer. In either case, up and away or down and through, the projectile would be moving at a very low residual velocity. If flipped through the plate, more or less at right angles to the plate back, the projectile might be moving sideways or even bounce backward as the bent armor straightens out when the projectile stops moving forward.

The NBL(BF) computations use the standard over 45° $M_{\text{table}}(\theta)$ values that NBL(NF) computations use when they occur for plates that do not have an NBL(BF). As previously mentioned, in the code, it is called the *true* NBL when it and the NBL(NF) can both occur; otherwise, the NBL(NF) is the true NBL since it is the only NBL. To compute the slightly higher NBL(NF) in plates, where both NBLs can occur, a separate computation using a small, six-entry modified $M_{\text{table}}(\theta)$ table called "MPNF(n)" is used to compute NBL(NF).

References

1. Hershey, A. V., Analytical Summary—Part 1: The Physical Properties of STS under Triaxial Stress, *NPG Report No. 6-46*, Naval Proving Ground, Dahlgren, VA, June 1, 1946.
2. Hershey, A. V., *Ballistic Summary—Part II: The Scale Effect and the Ogive Effect*, *NPG Report No. 4-46*, Naval Proving Ground, Dahlgren, VA, April 1, 1946.
3. Hershey, A. V., *The Theory of Projectile Ricochet*, *NPG Report No. 1041*, Naval Proving Ground, Dahlgren, VA, May 27, 1955.
4. Hershey, A. V., *Analytical Summary—Part IV: The Theory of Armor Penetration*, *NPG Report No. 9-46*, Naval Proving Ground, Dahlgren, VA, May 1, 1946
5. Barvik, T., Leinum, J. R., Solberg, J. K., Hopperstad, O. S., and Langseth, M., Observations on shear plug formation in Weldex 460 E steel plates impacted by blunt-nosed projectiles, *International Journal of Impact Engineering*, 25, 2001, 533–572.
6. Recht, R. F., and Ipson, T. W., *Ballistic perforation dynamics*, *Transactions of the ASME*, September 1963, pp. 384–389.
7. Thompson, L., Ballistics engineering problems: Empirical summaries, *Proceedings of the US Naval Institute*, 56(5), May 1930.
8. Davis, C., On the perforation of face hardened armor, *US Naval Institute Proceedings*, Annapolis, MD, Vol. 23, No. 2, 1897.
9. Geheime Kommandosache 100 (Abbr.: G.Kdos. 100) (Eng.: "Secret Command Document #100"), Unterlagen und Richtlinien zur Bestimmung der Hauptkampfentfernung und der Geschoßwahl (Eng.: "Basis and Guidance for the Determination of the Optimum Fighting Range and Projectile Type"), Heft (Eng.: "Chapter/Section") A, Introduction and General Discussion; Oberkommando der Kriegsmarine (Eng.: "German Navy High Command"), Berlin, 1940.
10. Thompson, L., Reduced scale studies in armor penetration, *Physics*, 3, September 1932, 155–158.

18

Shock Physics

18.1 Shock Hugoniots

A most lucid treatment of the Rankine–Hugoniot jump equations is found in the book *Explosives Engineering* by Cooper [1]. In the shocking of a solid, it is critical that we completely understand these equations. The purpose of this section will be to gain an understanding of the equations required to characterize the shock front in a solid (or fluid).

First, we shall describe a Hugoniot. Simply put, a Hugoniot (Hyoo' gon nee oh) is a curve that contains all possible equilibrium states at which a material can exist. It is an empirically derived curve that relates any two of the following variables to one another: pressure p; shock velocity U; particle velocity u; and specific volume v (or density ρ). It is not an equation of state, although it can be used in a similar manner. It is sometimes used as if it was an isentrope even though it is not the same. It is not the same because entropy increases across a shock. It is derived experimentally, and therefore, the experiment will have all the irreversibilities present.

A velocity Hugoniot is an empirical relationship that relates particle velocity in a material to the velocity of a shock front moving through that material. For most materials, it is a simple linear relationship expressed in the form

$$U = c_0 + su \tag{18.1}$$

where U is the speed of propagation of the shock front; c_0 is the bulk speed of sound in the medium (not really a sound speed per se but the y-intercept of the Hugoniot curve); u is the particle velocity; and s is an empirically obtained velocity coefficient.

In some materials, the curve is bilinear or trilinear, usually indicating a phase change, although some authors have fitted quadratics or cubics to the curves.

The real power of this simple relationship is seen when we use it in conjunction with our equations of mass conservation, conservation of momentum, and conservation of energy as repeated next:

$$\frac{\rho_1}{\rho_0} = \frac{V_0}{V_1} = \frac{U - u_0}{U - u_1} \tag{18.2}$$

$$p_1 - p_0 = \rho_0(u_1 - u_0)(U - u_0) \tag{18.3}$$

$$e_1 - e_0 = \frac{p_1 u_1 - p_0 u_0}{\rho_0(U - u_0)} - \frac{1}{2}\left(u_1^2 - u_0^2\right) \tag{18.4}$$

where the subscript 0 represents conditions ahead of the shock wave and 1 represents conditions after the passage of the wave. It is useful at this stage to examine an example problem.

Example Problem 1

A slab of polystyrene has the following properties:

$$\rho_0 = 1.044 \left[\frac{\text{gm}}{\text{cm}^3}\right]$$

$$c_0 = 2.746 \left[\frac{\text{km}}{\text{s}}\right]$$

$$s = 1.319$$

The particle velocity in an experiment is known to be $u_1 = 1.37$ km/s. Calculate the shock velocity and shock pressure.

The shock velocity follows from Equation 18.1, where, plugging in numbers, we have

$$U = (2.746)\left[\frac{\text{km}}{\text{s}}\right] + (1.319)(1.37)\left[\frac{\text{km}}{\text{s}}\right] = 4.553\left[\frac{\text{km}}{\text{s}}\right] \tag{18.5}$$

The pressure is obtained from conservation of momentum (with p_0 and $u_0 = 0$) by using Equation 18.3:

$$p_1 = \rho_0 u_1 U = (1.044)\left[\frac{\text{g}}{\text{cm}^3}\right](1.37)\left[\frac{\text{km}}{\text{s}}\right](4.553)\left[\frac{\text{km}}{\text{s}}\right] = 6.152\,[\text{GPa}] \tag{18.6}$$

Now wait a minute. How did those units work out? It is good to remember that with density in grams per cubic centimeter and velocities in kilometers per second, we obtain answers in gigapascals. This is done so that we do not have a lot of zeros or 10^x powers around. Here is the breakout

$$\begin{aligned}
&(1.044)\left[\frac{\text{g}}{\text{cm}^3}\right](100)^3\left[\frac{\text{cm}^3}{\text{m}^3}\right]\left(\frac{1}{1000}\right)\left[\frac{\text{kg}}{\text{g}}\right](1.37)\left[\frac{\text{km}}{\text{s}}\right](4.553)\left[\frac{\text{km}}{\text{s}}\right](1000)^2\left[\frac{\text{m}^2}{\text{km}^2}\right]\\
&= 6.512\times10^9\left[\frac{\text{kg}}{\text{m s}^2}\right]6.512\times10^9\left[\frac{\text{kg}}{\text{m s}^2}\right] = 6.512\times10^9\left[\frac{\frac{\text{kg-m}}{\text{s}^2}}{\text{m}^2}\right]\\
&= 6.512\times10^9\left[\frac{\text{N}}{\text{m}^2}\right] = 6.512\times10^9\,[\text{Pa}] = 6.512\,[\text{GPa}]
\end{aligned} \tag{18.7}$$

You can see why we will not carry the units around in these examples any longer.

If we combine Equation 18.1 with our continuity and momentum equations [1], we obtain the p–v Hugoniot in the following form:

$$p_1 = \frac{c_0^2(v_0 - v_1)}{[v_0 - s(v_0 - v_1)]^2} \tag{18.8}$$

For simplicity, we assumed p_0 and u_0 were equal to zero in Equation 18.8. We need to recall that the specific volume v is equal to $1/\rho$. This Hugoniot then tells us how pressure varies with density. Equation 18.8 is very powerful in the sense that it can tell us to what pressure a material will jump if we know the change in density or specific volume. This "jump" will occur through the formation of a shock wave. This can be seen on a p–v diagram such as in Figure 18.1. In this figure, we have noted the elastic, elastic–plastic, and plastic regions to be discussed later.

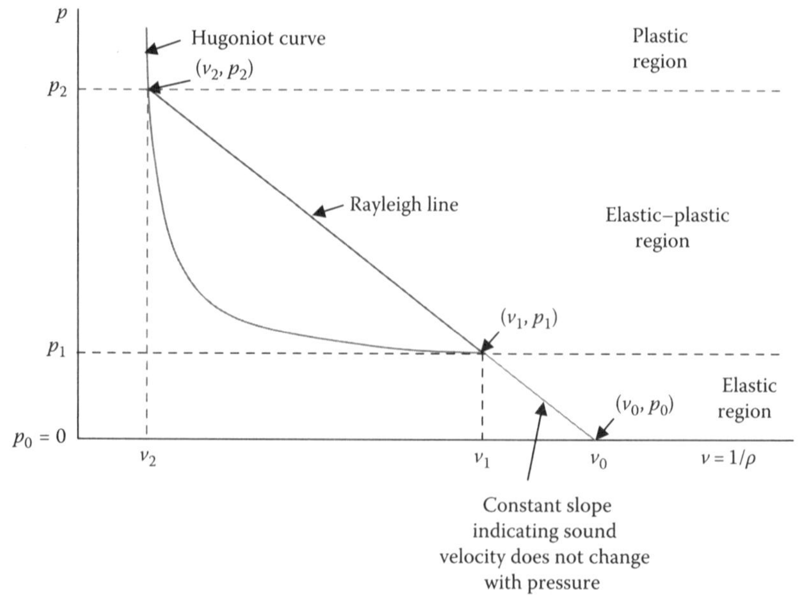

FIGURE 18.1
p–v diagram showing elastic, elastic–plastic, and plastic regions of a material.

We shall now look at another example.

Example Problem 2
A slab of aluminum has the following properties:

$$\rho_0 = 2.785 \left[\frac{\text{g}}{\text{cm}^3}\right]$$

$$c_0 = 5.328 \left[\frac{\text{km}}{\text{s}}\right]$$

$$s = 1.338$$

If we shock this material with a pressure of 40.2 GPa, what will the density of the material be behind the shock front? If the material is initially at rest, how fast will the particles move behind the shock wave and what will the velocity of the shock wave be?

To determine the density of the material behind the shock front, we need Equation 18.8. We note here, however, that this equation is in terms of the specific volume. We need to convert our initial data as follows:

$$\rho_0 = 2.785 \left[\frac{\text{g}}{\text{cm}^3}\right] = \frac{1}{v_0} \rightarrow v_0 = 1.359 \left[\frac{\text{cm}^3}{\text{g}}\right] \tag{18.9}$$

Now let us rewrite Equation 18.8. We need to rearrange our equation into a quadratic so that we can solve it easily

$$[v_0 - s(v_0 - v_1)]^2 p_1 - c_0^2(v_0 - v_1) = 0 \tag{18.10}$$

Now if we put in our values noting that kilometers per second, cubic centimeters per gram, and gigapascals are consistent units, we can write

$$v_1^2 + 0.213v_1 - 0.133 = 0 \tag{18.11}$$

If we solve this using the quadratic formula

$$x = \left[\frac{-b \pm \sqrt{b^2 - 4ac}}{2a}\right] \tag{18.12}$$

we get

$$v_1 = \frac{-0.213 \pm \sqrt{(0.213)^2 - (4)(1)(-0.133)}}{2(1)} = -0.107 \pm 0.380 \left[\frac{\text{cm}^3}{\text{g}}\right] \tag{18.13}$$

$$v_1 = 0.273 \left[\frac{\text{cm}^3}{\text{g}}\right] \tag{18.14}$$

We chose this root because it is impossible to have a negative density. Thus, our density behind the shock wave is

$$\rho_1 = \frac{1}{v_1} = \frac{1}{0.273 \left[\frac{\text{cm}^3}{\text{g}}\right]} = 3.664 \left[\frac{\text{g}}{\text{cm}^3}\right] \tag{18.15}$$

nearly double the density. To find the speed at which the shock wave will propagate, we need to do a little algebra. We know from our Hugoniot relation that

$$U = c_0 + su \tag{18.16}$$

We also know that from Equation 18.2 we can write, assuming that $u_0 = 0$,

$$\frac{v_0}{v_1} = \frac{U}{U - u_1} \tag{18.17}$$

If we put some numbers in here, we have

$$U = (5.328)\left[\frac{\text{km}}{\text{s}}\right] + (1.338)u_1 \left[\frac{\text{km}}{\text{s}}\right] = 5.328 + 1.338u_1 \left[\frac{\text{km}}{\text{s}}\right] \tag{18.18}$$

and

$$\frac{v_0}{v_1} = \frac{0.359}{0.273} = 1.315 = \frac{U}{U - u_1} \tag{18.19}$$

Substitution of Equation 18.18 into Equation 18.19 yields

$$1.315 = \frac{5.328 + 1.338u_1 \left[\frac{\text{km}}{\text{s}}\right]}{5.328 + 0.338u_1 \left[\frac{\text{km}}{\text{s}}\right]} \tag{18.20}$$

Solving for u_1 gives us

$$u_1 = 1.88 \left[\frac{\text{km}}{\text{s}}\right] \tag{18.21}$$

Our shock velocity then directly follows from Equation 18.18:

$$U = (5.328)\left[\frac{\text{km}}{\text{s}}\right] + (1.338)(1.88)\left[\frac{\text{km}}{\text{s}}\right] = 7.84\left[\frac{\text{km}}{\text{s}}\right] \tag{18.22}$$

We could also have solved this using Equation 18.25.

A jump as described in the previous paragraph will take place through the formation of a shock wave and proceed along what is called a Rayleigh line. The equation of the Rayleigh line is derived by a combination of the mass and momentum equations and, for convenience, setting $u_0 = 0$. This results in

$$p_1 - p_0 = \frac{U^2}{v_0} - \frac{U^2}{v_0^2} v_1 \tag{18.23}$$

The slope of the Rayleigh line is then

$$\text{slope} = \frac{U^2}{v_0^2} = \rho_0^2 U^2 \tag{18.24}$$

Recall from thermodynamics that the area under a p–v diagram represents the work done on or by the system. Then if we shock a system up a Rayleigh line and allow it to relax along the Hugoniot, the net work we have done on the system is determined from the area between the curves. Figure 18.2 shows how, depending on the pressure to which we shock a material, the wave speeds will vary. In fact, if we shock a material into the elastic–plastic regimes, there will be two shocks, an elastic wave (precursor) that will move at the longitudinal wave speed (speed of sound) in the solid and a plastic wave that will move at a slower speed. We shall discuss this further later.

If we assume p_0 and u_0 are equal to zero and combine the momentum equation (Equation 18.3) with our U–u Hugoniot equation (Equation 18.1), we obtain the p–u Hugoniot in the following form:

$$p_1 = \rho_0 u_1 (c_0 + s u_1) \rightarrow p_1 = \rho_0 c_0 u_1 + \rho_0 s u_1^2 \tag{18.25}$$

This relationship gives the pressure as a function of material velocity u, when the material is initially at rest. If the material was not initially at rest, our equation would be a little more complicated

$$p_1 = \rho_0 c_0 (u_1 - u_0) + \rho_0 s (u_1 - u_0)^2 \tag{18.26}$$

This equation was obtained by taking Equation 18.25 and subtracting the same equation with $u = u_0$. This would be appropriate if the wave was moving to the right ($u_1 > u_0$); thus, it is aptly called a "right-going Hugoniot" in following with the derivation set forth by Cooper [1]. If the wave was moving to the left ($u_1 < u_0$), we would have a left-going Hugoniot and the equation would be

$$p_1 = \rho_0 c_0 (u_0 - u_1) + \rho_0 s (u_0 - u_1)^2 \tag{18.27}$$

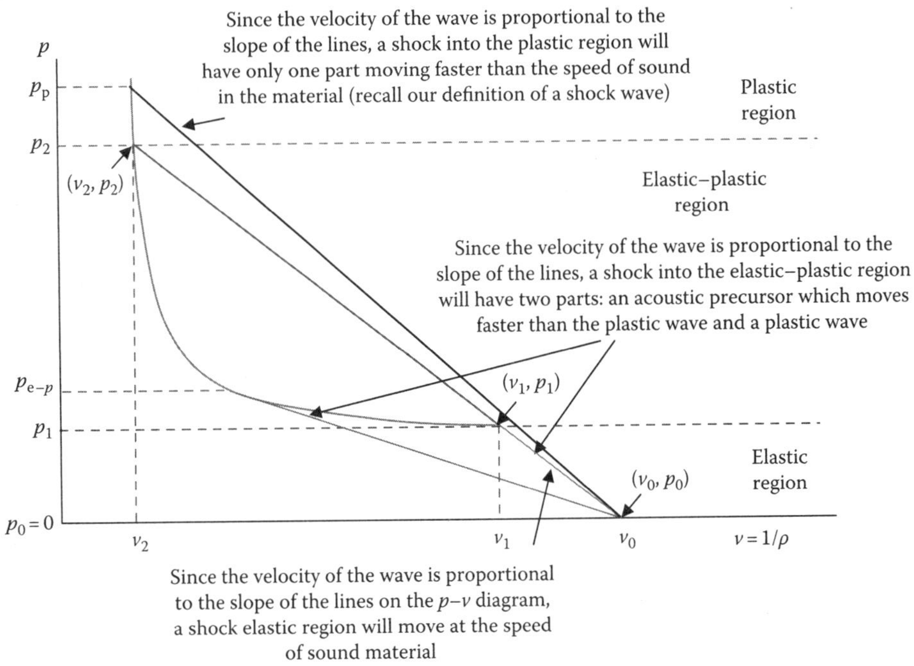

FIGURE 18.2
p–v diagram describing the wave behavior in the elastic, elastic–plastic, and plastic regimes.

The effect of having a nonzero u_0 is to shift the x-intercept of the curve as depicted in Figure 18.3.

We have these wonderful equations for left- and right-going waves (the Hugoniot) so what do we do with them? By using Equations 18.26 and 18.27, we can calculate how a wave will propagate (transmit) and reflect when two dissimilar materials impact one another or when a shock crosses an interface where they are initially in contact. First, we shall define the impedance Z. The impedance of a material is the product of its density and the velocity that a shock wave travels in that material:

$$Z = \rho U \tag{18.28}$$

The acoustic impedance of a material is the product of the material density times the speed of sound (the speed of an infinitesimally small disturbance) in that material:

$$Z_{\text{Acoustic}} = \rho c \tag{18.29}$$

When a shock wave crosses a boundary between materials of the same impedance, there will be no reflection, and all of the wave will be transmitted into the new material—the wave acts as though the interface is not there. If the materials are not in intimate contact, this will not be the case.

We shall now introduce a means of looking at shocks known as a t–x plot. A t–x (time–displacement) plot is used as a method of keeping track of material motion in a wave propagation problem. An example of this type of plot is in Figure 18.4 for two slabs, which will impact one another. Because time is the ordinate, the slopes of the lines are the reciprocal of the velocity.

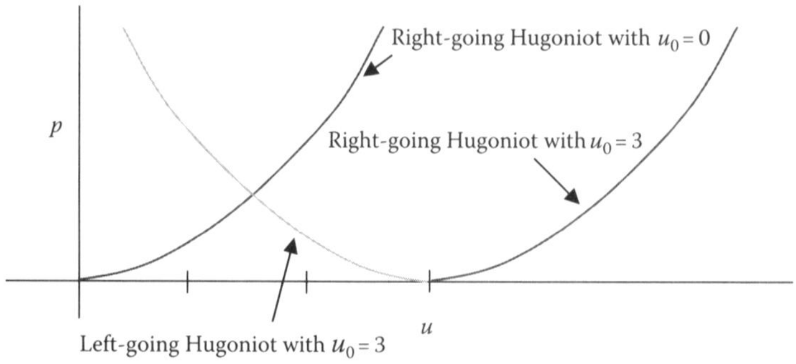

FIGURE 18.3
Effect of initial material velocity on a Hugoniot curve.

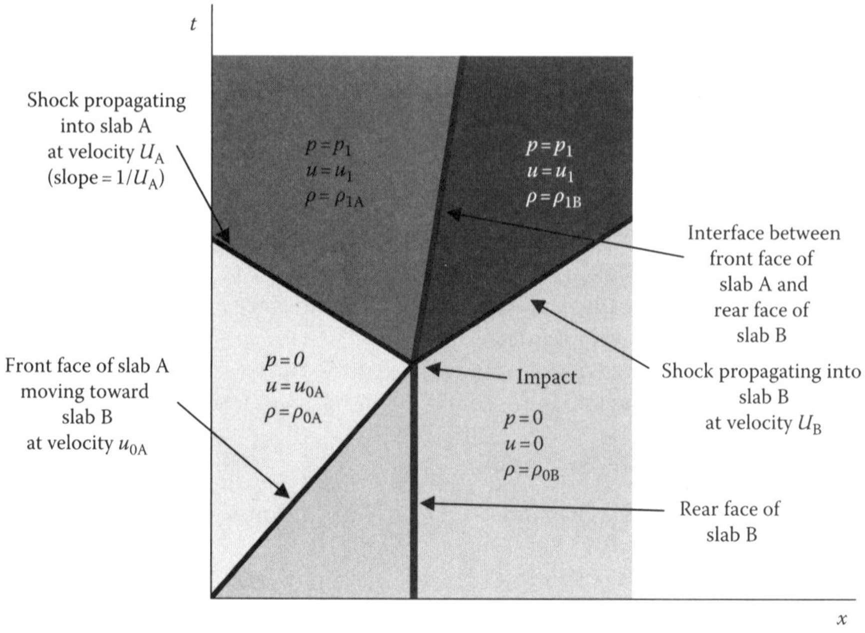

FIGURE 18.4
(See color insert.) Time–displacement plot of a slab impact problem. (Cooper, P. W.: *Explosives Engineering*. 1996. Copyright Wiley-VCH Verlag GmbH & Co. KGaA. Reproduced with permission.)

When two slabs impact one another, the following conditions must apply: The pressure at the interface must be consistent across the interface and the velocity of the particles at the interface must be the same in both materials. Consider that we have slab B sitting at rest and slab A impacts it with the initial conditions that slabs A and B are both stress free, but slab A is moving (i.e., all the particles of slab A have the same particle velocity). Once impact occurs, a shock wave of equal strength will pass into each material, a right-going wave in B and a left-going wave in A. We can see this on a p–u plot in Figure 18.5. Let us consider another example problem.

Example Problem 3

An experiment is set up in which a magnesium slab is launched at a slab of brass. The velocity at impact is measured to be 2.0 km/s. Determine

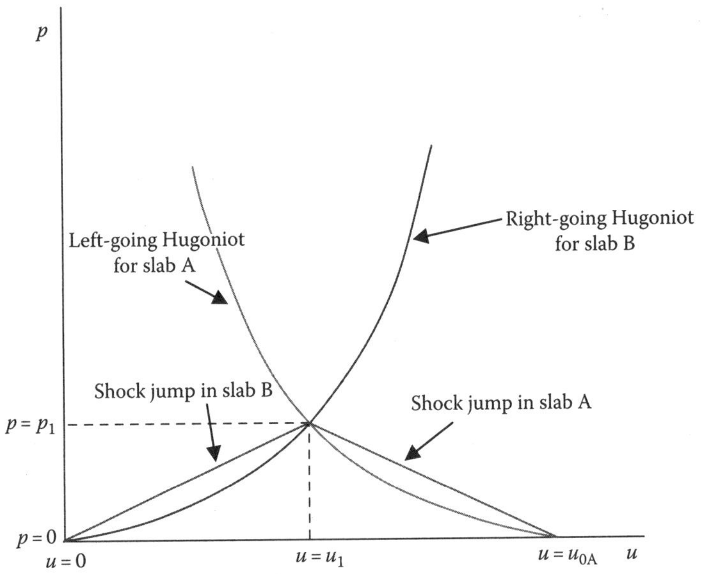

FIGURE 18.5
(See color insert.) p–u Hugoniot plot for an impact event. (Cooper, P. W.: *Explosives Engineering*. 1996. Copyright Wiley-VCH Verlag GmbH & Co. KGaA. Reproduced with permission.)

1. The particle velocity in the two materials at the interface
2. The shock pressure at the interface
3. The speed at which the shock wave travels in the brass
4. The speed at which the shock wave travels in the magnesium

The slabs have the following properties:

Magnesium	Brass
$\rho_{0_{Mg}} = 1.775 \left[\frac{g}{cm^3}\right]$	$\rho_{0_{Brass}} = 8.450 \left[\frac{g}{cm^3}\right]$
$c_{0_{Mg}} = 4.516 \left[\frac{km}{s}\right]$	$c_{0_{Brass}} = 3.726 \left[\frac{km}{s}\right]$
$s_{Mg} = 1.256$	$s_{Brass} = 1.434$

Solution: The first thing we do is write the p–u Hugoniot equations for both materials; by convention, assume that the magnesium plate is flying from left to right, then we need a right-going Hugoniot in the target (brass) and a left-going Hugoniot for the flyer (magnesium). We shall examine the brass first. A right-going Hugoniot is described by Equation 18.26, but since the brass was not initially moving, we can use Equation 18.25. Inserting values for the brass, we have

$$p_1\,[\text{GPa}] = (8.450)\left[\frac{g}{cm^3}\right](3.726)\left[\frac{km}{s}\right]u_1\left[\frac{km}{s}\right] + (8.450)\left[\frac{g}{cm^3}\right](1.434)u_1^2\left[\frac{km}{s}\right]^2 \qquad (18.30)$$

$$p_1\,[\text{GPa}] = 31.485u_1 + 12.117u_1^2 \tag{18.31}$$

Since we know that the compatibility relation requires pressure to be identical in both materials at the interface, we can write the left-going Hugoniot for the magnesium, equate the two expressions, and solve for the particle velocity (which must also be the same in both materials at the interface). The left-going Hugoniot in the magnesium is given by Equation 18.27. Inserting our values yields

$$p_1\,[\text{GPa}] = 2.229u_1^2 - 16.932u_1 + 24.948 \tag{18.32}$$

If we equate Equations 18.31 and 18.32, we obtain

$$u_1^2 + 4.897u_1 - 2.523 = 0 \tag{18.33}$$

Now if we solve this using the quadratic formula, we get

$$u_1 = 0.471\left[\frac{\text{km}}{\text{s}}\right] \tag{18.34}$$

Here we used the positive velocity since the other root is meaningless. To determine the pressure at the interface, we can put this value back into either Equation 18.25 or 18.27 to yield

$$p_1 = 17.52\,[\text{GPa}] \tag{18.35}$$

To find the speed that the shock wave moves in each material, we call upon the *U*–*u* Hugoniots for each (Equation 18.1). For the brass, we have

$$\begin{aligned} U_{\text{Brass}} &= (3.726)\left[\frac{\text{km}}{\text{s}}\right] + (1.434)\,u_1\left[\frac{\text{km}}{\text{s}}\right] \\ &= (3.726)\left[\frac{\text{km}}{\text{s}}\right] + (1.434)\,(0.471)\left[\frac{\text{km}}{\text{s}}\right] \end{aligned} \tag{18.36}$$

$$U_{\text{Brass}} = 4.401\left[\frac{\text{km}}{\text{s}}\right] \tag{18.37}$$

Note that this velocity is to the right because we used a right-going Hugoniot. For the magnesium, we have

$$\begin{aligned} U_{\text{Mg}} &= (4.516)\left[\frac{\text{km}}{\text{s}}\right] + (1.256)(u_0 - u_1)\left[\frac{\text{km}}{\text{s}}\right] \\ &= (4.516)\left[\frac{\text{km}}{\text{s}}\right] + (1.256)(2.0 - 0.471)\left[\frac{\text{km}}{\text{s}}\right], \end{aligned} \tag{18.38}$$

$$U_{\text{Mg}} = 6.436\left[\frac{\text{km}}{\text{s}}\right] \tag{18.39}$$

This velocity is to the left because we used a left-going Hugoniot. Notice that we used $u_0 - u_1$ in place of u_1 because the shock velocity is relative to the wave.

When a shock wave propagates from a lower impedance material into a higher impedance material, as always, the compatibility condition is such that the pressure must also be continuous at the interface and the particle velocities must be equal. The higher impedance material will cause the pressure to increase and this higher-pressure wave will propagate back into the lower impedance material (but at a lower velocity) and into the higher impedance material at a lower velocity than the original wave. The particle velocity will be the same (and lower) in both materials. We shall illustrate this with an example.

Example Problem 4

An experiment is set up in which a magnesium slab is shocked while in contact with a slab of brass. The particle velocity at the interface is measured to be 2.0 km/s. Determine

1. The pressure generated at the interface
2. The speed at which the transmitted shock wave travels in the brass
3. The particle velocity in the magnesium before the impact
4. The speed at which the original shock pulse traveled in the magnesium
5. The pressure of the original shock pulse in the magnesium

The slabs have the following properties:

Magnesium	Brass
$\rho_{0_{Mg}} = 1.775 \left[\frac{g}{cm^3}\right]$	$\rho_{0_{Brass}} = 8.450 \left[\frac{g}{cm^3}\right]$
$c_{0_{Mg}} = 4.516 \left[\frac{km}{s}\right]$	$c_{0_{Brass}} = 3.726 \left[\frac{km}{s}\right]$
$s_{Mg} = 1.256$	$s_{Brass} = 1.434$

Solution: If we examine Figure 18.6, we see that we should be able to determine the answer to part 1 from the right-going Hugoniot in the brass.

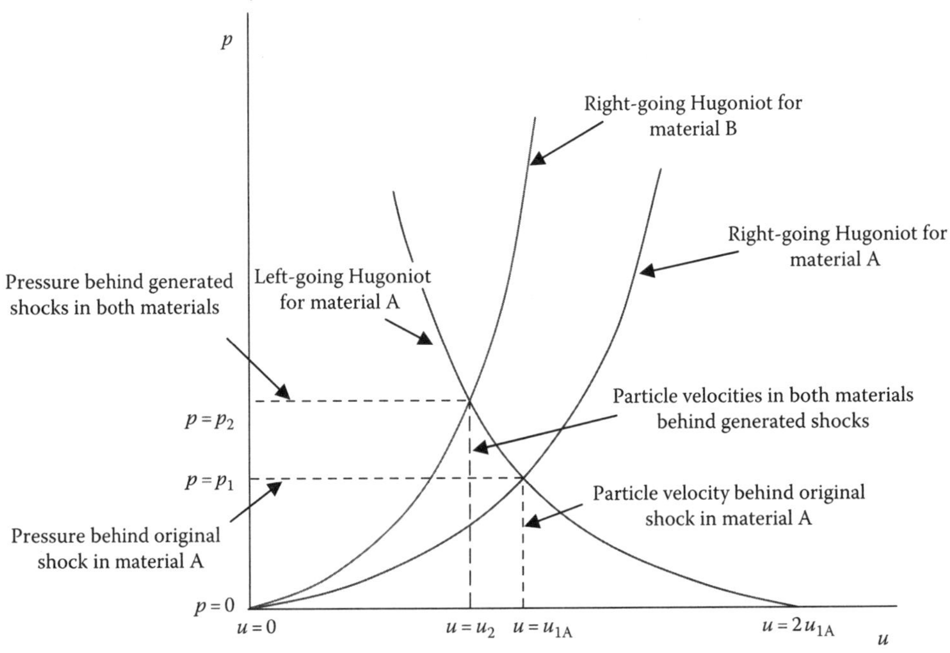

FIGURE 18.6
p–u diagram for low to high impedance shock propagation.

A right-going Hugoniot is described by Equation 18.26, but since the brass was not initially moving, we can use Equation 18.25, but to stay consistent with our diagram, we will say the particle velocity is u_2 for this case:

$$p_2 = \rho_0 c_2 u_2 + \rho_0 s u_2^2 \tag{18.40}$$

Inserting values for the brass, we have

$$p_2\,[\text{GPa}] = 31.485u_2 + 12.117u_2^2 \tag{18.41}$$

We were provided with u_2, so we can write

$$p_2\,[\text{GPa}] = 31.485(2) + 12.117(2)^2 \tag{18.42}$$

$$p_2 = 111.438\,[\text{GPa}] \tag{18.43}$$

The speed at which the transmitted shock wave travels in the brass can be directly found from our *U–u* Hugoniot Equation 18.1. For the brass, we have

$$U_{\text{Brass}} = (3.726)\left[\frac{\text{km}}{\text{s}}\right] + (1.434)u_2\left[\frac{\text{km}}{\text{s}}\right] = (3.726)\left[\frac{\text{km}}{\text{s}}\right] + (1.434)(2.0)\left[\frac{\text{km}}{\text{s}}\right] \tag{18.44}$$

$$U_{\text{Brass}} = 6.594\left[\frac{\text{km}}{\text{s}}\right] \tag{18.45}$$

The particle velocity in the magnesium before impact is found by noting that we have the point (u_2, p_2) on the left-going Hugoniot which, by definition, has to pass through point ($2u_{1A}$, 0) as well. Our equation for the left-going Hugoniot is Equation 18.27. Putting this in terms of our diagram, we can write

$$p_2 = \rho_{0A} c_{0A}(2u_{1A} - u_2) + \rho_{0A} s_A (2u_{1A} - u_2)^2 \tag{18.46}$$

Inserting our values for magnesium, we can write

$$u_{1A}^2 - 0.202u_{1A} - 13.297 = 0 \tag{18.47}$$

from which we obtain the solution

$$u_{1A} = 3.749\left[\frac{\text{km}}{\text{s}}\right] \tag{18.48}$$

The speed at which the original shock pulse travels in the magnesium falls out directly from our *U–u* Hugoniot again:

$$U_{\text{Mg}} = (4.516)\left[\frac{\text{km}}{\text{s}}\right] + (1.256)u_{1A}\left[\frac{\text{km}}{\text{s}}\right] = (4.516)\left[\frac{\text{km}}{\text{s}}\right] + (1.256)(3.749)\left[\frac{\text{km}}{\text{s}}\right] \tag{18.49}$$

$$U_{\text{Mg}} = 9.225\left[\frac{\text{km}}{\text{s}}\right] \tag{18.50}$$

The pressure of the original shock pulse in the magnesium then follows from the momentum equation:

$$p_{1A} = \rho_{0A} u_{1A} U_{\text{Mg}} \tag{18.51}$$

$$p_{1A} = (1.775)\left[\frac{\text{g}}{\text{cm}^3}\right](3.749)\left[\frac{\text{km}}{\text{s}}\right](9.225)\left[\frac{\text{km}}{\text{s}}\right] \tag{18.52}$$

$$p_{1A} = 61.388\,[\text{GPa}] \tag{18.53}$$

When a shock wave propagates from a higher impedance material into a lower impedance material, the compatibility condition still requires the pressure be continuous at the interface and the particle velocities be equal. The lower impedance material will cause the pressure to decrease, and this lower pressure (relief) wave will propagate back into the higher impedance material (at a higher velocity) and into the lower impedance material at a higher velocity than the original wave. The particle velocity will be the same (and higher) in both materials. Another example will illustrate the point.

Example Problem 5

An experiment is set up in which a brass slab is shocked while in contact with a slab of magnesium. The particle velocity at the interface is measured to be 2.0 km/s. Determine

1. The pressure generated at the interface
2. The speed at which the transmitted shock wave travels in the magnesium
3. The particle velocity in the brass before the impact
4. The speed at which the original shock pulse traveled in the brass
5. The pressure of the original shock pulse in the brass

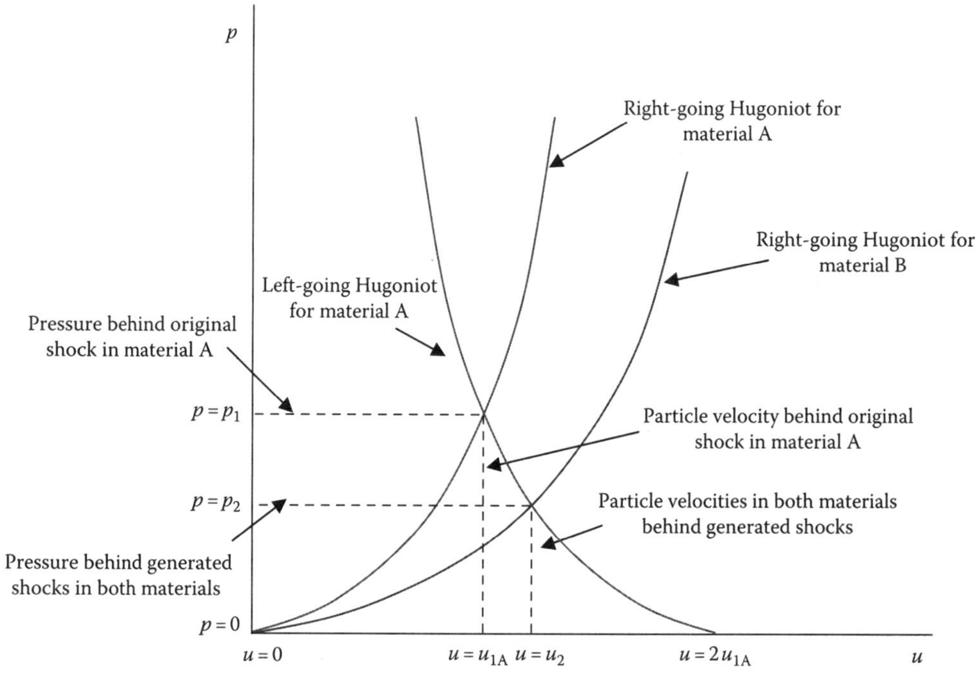

FIGURE 18.7

p–u diagram for high to low impedance shock propagation. (Cooper, P. W.: *Explosives Engineering*. 1996. Copyright Wiley-VCH Verlag GmbH & Co. KGaA. Reproduced with permission.)

The slabs have the following properties:

Magnesium	Brass
$\rho_{0_{Mg}} = 1.775\left[\frac{g}{cm^3}\right]$	$\rho_{0_{Brass}} = 8.450\left[\frac{g}{cm^3}\right]$
$c_{0_{Mg}} = 4.516\left[\frac{km}{s}\right]$	$c_{0_{Brass}} = 3.726\left[\frac{km}{s}\right]$
$s_{Mg} = 1.256$	$s_{Brass} = 1.434$

Solution: If we examine Figure 18.7, we see that we should be able to determine the answer to part 1 from the right-going Hugoniot in the magnesium.

A right-going Hugoniot is described by Equation 18.26, but since the magnesium was not initially moving, we can use Equation 18.25, and to stay consistent with our diagram, we will say the particle velocity is u_2 for this case:

$$p_2 = \rho_0 c_0 u_2 + \rho_0 s u_2^2 \tag{18.54}$$

Inserting values for the magnesium, we have

$$p_2\,[\text{GPa}] = 8.016u_2 + 2.212u_2^2 \tag{18.55}$$

$$p_2\,[\text{GPa}] = 8.016(2.0) + 2.212(2.0)^2 \tag{18.56}$$

$$p_2 = 24.879\,[\text{GPa}] \tag{18.57}$$

The speed at which the transmitted shock wave travels in the magnesium can be found directly from Equation 18.1:

$$U_{Mg} = (4.516)\left[\frac{km}{s}\right] + (1.246)u_2\left[\frac{km}{s}\right] = (4.516)\left[\frac{km}{s}\right] + (1.246)(20)\left[\frac{km}{s}\right] \tag{18.58}$$

$$U_{Mg} = 7.008\left[\frac{km}{s}\right] \tag{18.59}$$

The particle velocity in the brass before impact is found by noting that we have point (u_2, p_2) on the left-going Hugoniot which, by definition, has to pass through point $(2u_{1A}, 0)$ as well. Our equation for the left-going Hugoniot is Equation 18.27. Putting this in terms of Figure 18.7, we can write

$$p_2 = \rho_{0A} c_{0A}(2u_{1A} - u_2) + \rho_{0A} s_A (2u_{1A} - u_2)^2 \tag{18.60}$$

By inserting our values for brass, we have

$$u_{1A}^2 - 0.701u_{1A} - 0.812 = 0 \tag{18.61}$$

From this, we see that

$$u_{1A} = 1.317\left[\frac{km}{s}\right] \tag{18.62}$$

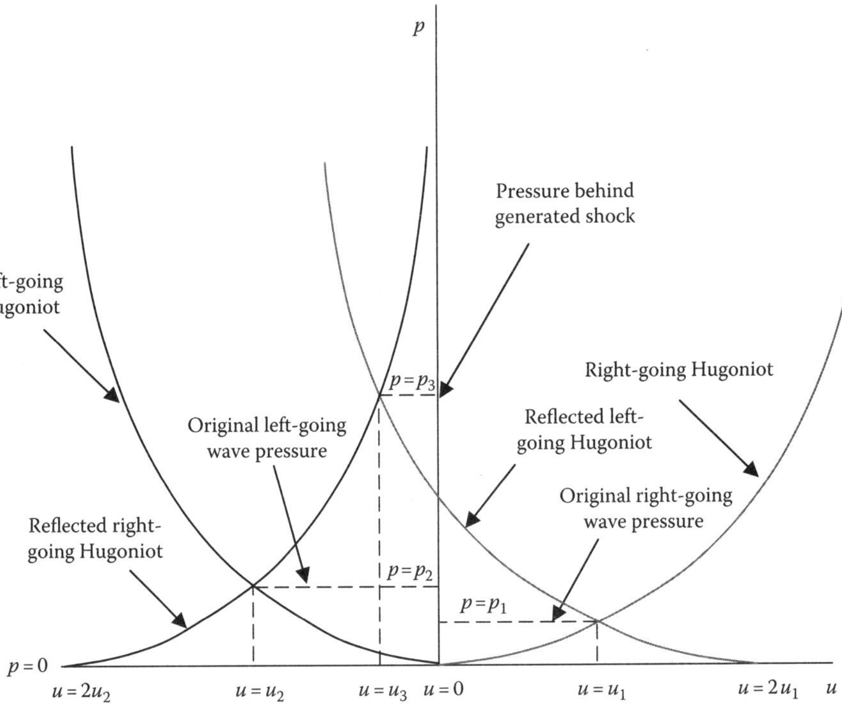

FIGURE 18.8
Collision of two shock waves. (Cooper, P. W.: *Explosives Engineering*. 1996. Copyright Wiley-VCH Verlag GmbH & Co. KGaA. Reproduced with permission.)

The speed at which the original shock pulse travels in the brass falls out directly from our U–u Hugoniot again Equation 18.1:

$$U_{\text{Brass}} = 5.615 \left[\frac{\text{km}}{\text{s}}\right] \tag{18.63}$$

The pressure of the original shock pulse in the brass then follows from the momentum equation:

$$p_{1A} = \rho_{0A} u_{1A} U_{\text{Brass}} \tag{18.64}$$

$$p_{1A} = 62.487\,[\text{GPa}] \tag{18.65}$$

When two shock waves collide in the same material, the pressure will jump to a new value that is greater than the sum of the two individual pressure pulses. Let us assume that we have a wave originally traveling to the right at pressure p_1 and a stronger wave originally traveling to the left at pressure p_2 in a material. We need to reflect the Hugoniots of these waves as shown in Figure 18.8 to solve for the resulting pressure p_3. We shall examine this again by example.

Example Problem 6

An experiment is set up in which a magnesium slab is shocked from both ends. The pressure generated in the left-going shock is 20 GPa. The pressure generated in the right-going shock is 10 GPa. Determine

1. The particle velocity in the right-going shock
2. The particle velocity in the left-going shock
3. The resultant particle velocity in the material
4. The resultant pressure generated

The slab has the following properties:

Magnesium

$$\rho_{0_{Mg}} = 1.775 \left[\frac{\text{g}}{\text{cm}^3}\right]$$

$$c_{0_{Mg}} = 4.516 \left[\frac{\text{km}}{\text{s}}\right]$$

$$s_{Mg} = 1.256$$

Solution: If we examine Figure 18.8, we see that we should be able to determine the answer to part 1 from the right-going Hugoniot in the magnesium.

A right-going Hugoniot is described by Equation 18.26, but since the magnesium was not initially moving, we can use Equation 18.25, and to stay consistent with our diagram, we will say the particle velocity is u_1 for this case:

$$p_1 = \rho_0 c_0 u_1 + \rho_0 s u_1^2 \tag{18.66}$$

Inserting values for the magnesium, we have

$$u_1^2 + 3.624u_1 - 4.521 = 0 \tag{18.67}$$

Solving this, we obtain

$$u_1 = 0.982 \left[\frac{\text{km}}{\text{s}}\right] \tag{18.68}$$

The particle velocity in the left-going shock is found again by noting that we have the left-going Hugoniot passing through the origin. Our equation for the left-going Hugoniot is

$$p_2 = \rho_0 c_0 (u_2 - 0) + \rho_0 s (u_2 - 0)^2 = \rho_0 c_0 u_2 + \rho_0 s u_2^2 \tag{18.69}$$

Inserting our values for magnesium, we have

$$u_2^2 + 3.624u_2 - 9.042 = 0$$

and, therefore,

$$u_2 = -1.699 \left[\frac{\text{km}}{\text{s}}\right] \tag{18.70}$$

The resultant particle velocity is found by taking these data, reflecting the Hugoniots around u_1 and u_2, and eliminating the pressure (since it is equal to p_3) from the equation. We shall reflect the right-going Hugoniot first. This will result in a left-going Hugoniot where we know points (u_1, p_1) and $(2u_1, 0)$:

$$p_3 = \rho_0 c_0 (u_3 - 2u_1) + \rho_0 s (u_3 - 2u_1)^2 \tag{18.71}$$

or

$$p_3 = 2.212u_3^2 - 16.705u_3 + 24.275 \tag{18.72}$$

Now we need to examine the right-going Hugoniot where we know points (u_2, p_2) and $(2u_2, 0)$:

$$p_3 = \rho_0 c_0 (u_3 - 2u_2) + \rho_0 s (u_3 - 2u_2)^2 \tag{18.73}$$

or

$$p_3 = 2.212u_3^2 + 23.049u_3 + 52.779 \tag{18.74}$$

If we now subtract Equation 18.72 from Equation 18.74, we can solve for u_3, so we have

$$39.754u_3 + 28.524 = 0 \tag{18.75}$$

Therefore,

$$u_3 = -0.717 \left[\frac{\text{km}}{\text{s}}\right] \tag{18.76}$$

The pressure then can be found from either Equation 18.72 or 18.74:

$$p_3 = 37.390 \,[\text{GPa}] \tag{18.77}$$

We have now completed our introduction of the Hugoniot curve and examined the use of Hugoniots for an impact problem. We have demonstrated the behavior of shocks across an interface and have examined infinite shock behavior in a single material (incipient shock and collision of two shocks). These shocks were considered infinite because the driving pressure was always present behind them, generating continued motion. Further reading is provided in the references. We shall now move on to discuss rarefaction waves.

Problem 1

An experiment is set up in which a steel slab is shocked from both ends. The pressure generated in the left-going shock is 20 GPa. The pressure generated in the right-going shock is 10 GPa. Draw the *p*–*u* diagram and determine

1. The particle velocity in the right-going shock

   Answer: $u_1 = 0.256 \left[\frac{\text{km}}{\text{s}}\right]$

2. The particle velocity in the left-going shock

   Answer: $u_2 = 0.479 \left[\frac{\text{km}}{\text{s}}\right]$

3. The resultant particle velocity in the material

   Answer: $u_3 = 0.223 \left[\frac{\text{km}}{\text{s}}\right]$

4. The resultant pressure generated
   Answer: $p_3 = 32.872$ [GPa]

The slab has the following properties:

Steel

$$\rho_{0_{Steel}} = 7.896 \left[\frac{g}{cm^3}\right]$$

$$c_{0_{Steel}} = 4.569 \left[\frac{km}{s}\right]$$

$$s_{Steel} = 1.490$$

Problem 2

A strange jeweler wants to make an earring by launching a quartz slab at a slab of gold. His high-tech instrumentation measures the induced velocity in the gold as 0.5 km/s. Determine

1. The impact velocity

   Answer: $u_0 = 3.420 \left[\frac{km}{s}\right]$

2. The shock pressure at the interface
   Answer: $p_1 = 36.960$ [GPa]
3. The speed at which the shock wave travels in the gold

   Answer: $U_{Au} = 3.842 \left[\frac{km}{s}\right]$

4. The speed at which the shock wave travels in the quartz

   Answer: $U_Q = 5.743 \left[\frac{km}{s}\right]$

The slabs have the following properties:

Quartz	Gold
$\rho_{0_Q} = 2.204 \left[\frac{g}{cm^3}\right]$	$\rho_{0_{Au}} = 19.24 \left[\frac{g}{cm^3}\right]$
$c_{0_Q} = 0.794 \left[\frac{km}{s}\right]$	$c_{0_{Au}} = 3.056 \left[\frac{km}{s}\right]$
$s_Q = 1.695$	$s_{Au} = 1.572$

18.2 Rarefaction Waves

We have examined infinite waves in the previous section (i.e., waves in which the pressure does not abate). In the shocking of a real material, the pressure pulse lasts for a finite time only, and then the material must expand back to a relaxed state. Nature accomplishes this expansion through a rarefaction wave.

A rarefaction wave is the manner in which nature restores a material to its unshocked state after the passage of a shock wave. Unlike a shock front (which is a nearly discontinuous jump in pressure), a rarefaction or relief wave will occur over some finite distance, which will gradually increase with time. We typically assume that rarefaction waves occur rapidly enough that the process may be considered adiabatic.

Recall that passing a shock wave through a material increases its internal energy as shown through the Rankine–Hugoniot equation:

$$e_1 - e_0 = \frac{1}{2}\left(\frac{1}{\rho_0} - \frac{1}{\rho_1}\right)(p_0 + p_1) = \frac{1}{2}(p_0 + p_1)(v_0 - v_1) \tag{18.78}$$

As a consequence of the second law of thermodynamics, we can write

$$\mathrm{d}E = T\mathrm{d}S - p\mathrm{d}V \tag{18.79}$$

Since we assumed that the rarefaction process is adiabatic, we know that

$$\mathrm{d}Q = T\mathrm{d}S = 0 \tag{18.80}$$

Since on the Rankine or Kelvin scales, T must be positive, then, except in a special nonzero case, $\mathrm{d}S$ must equal zero for this equation to be true. Thus, the rarefaction or relief process must be isentropic. This presents us with a bit of a dilemma. Except for an ideal gas, we do not have an isentropic relation to allow us to quantify the expansion process. If we had such a relationship, it would, in theory, allow us to eliminate one of the variables in our Equation 18.79, which, through Equation 18.80, can be rewritten as

$$\mathrm{d}E = -p\mathrm{d}V \rightarrow E = E(p, V) \tag{18.81}$$

We have stated before that a Hugoniot curve is neither an equation of state nor an isentrope. Here we will use it as if it was one and accept any errors that result.

We now know that we can handle a rarefaction wave through use of the Hugoniot. The simplest way to illustrate how to obtain the rarefaction wave velocity is to consider the case where the initial material velocity is equal to zero; we can then write

$$p|_{u_0=0} = \rho_0 u U_\mathrm{R} \tag{18.82}$$

Taking the first derivative, we obtain

$$\frac{\mathrm{d}p|_{u_0=0}}{\mathrm{d}u} = \rho_0 U_\mathrm{R} \tag{18.83}$$

We also saw that we can write the p–u Hugoniot as

$$p = \rho_0 c_0 u + \rho_0 s u^2 \tag{18.84}$$

Taking the derivative of Equation 18.84, we obtain

$$\frac{\mathrm{d}p}{\mathrm{d}u} = \rho_0 c_0 + 2\rho_0 s u \tag{18.85}$$

Eliminating $\mathrm{d}p/\mathrm{d}u$ between Equations 18.83 and 18.85 yields

$$\rho_0 U_\mathrm{R} = \rho_0 c_0 + 2\rho_0 s u \tag{18.86}$$

or

$$U_R = c_0 + 2su \tag{18.87}$$

which is our final relation for the speed of the head of the rarefaction wave. This is depicted in Figure 18.9. If we recall the speed of our shock wave (U–u Hugoniot), we would see

$$U = c_0 + su \tag{18.88}$$

If we were to shock a material with a certain pulse length λ_1 over a particular time t_1, the shock would have moved a distance

$$\lambda_1 = Ut_1 \tag{18.89}$$

The instant the applied load has ceased, a relief wave would begin moving into the material and at a time $t > t_1$, would be located at a distance from the point of shock initiation of

$$d_2 = U_R(t - t_1) \tag{18.90}$$

Since we saw from our examination of Equations 18.87 and 18.88 that $U_R > U$, we can determine the distance at which the relief wave will catch up to the shock wave through

$$Ut = U_R(t - t_1) \tag{18.91}$$

If we insert Equations 18.87 and 18.88 into this expression, we obtain

$$(c_0 + su)t = (c_0 + 2su)(t - t_1) \tag{18.92}$$

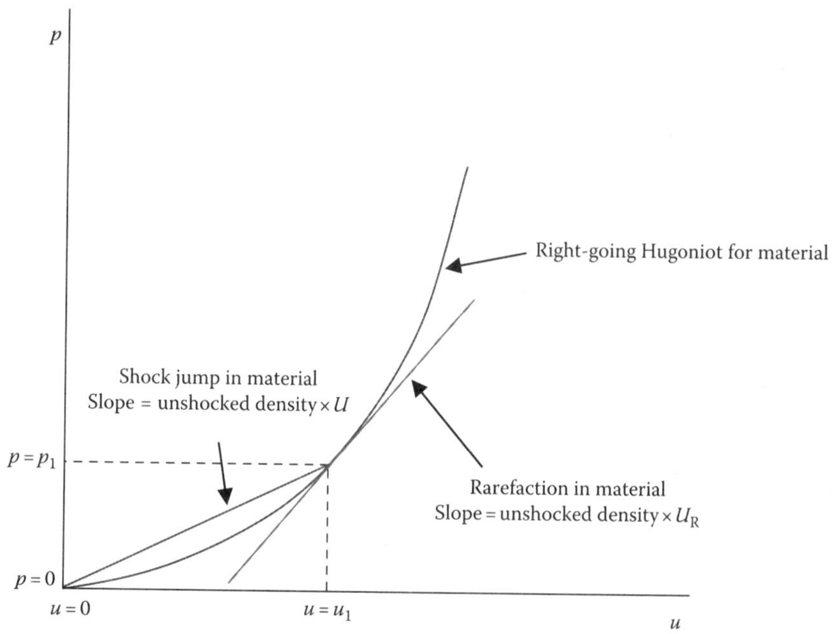

FIGURE 18.9
Speed of the rarefaction wave head.

which simplifies to

$$sut = c_0 t_1 + 2sut_1 \tag{18.93}$$

Thus, the time required for the rarefaction wave to catch up with the initial shock is determined from

$$t = \frac{c_0 t_1 + 2sut_1}{su} \tag{18.94}$$

where we should know everything on the right-hand side (RHS) from the material and the strength of the initial pulse.

We could then use Equation 18.95 to determine the catchup distance:

$$\lambda_c = Ut \tag{18.95}$$

The book by Paul Cooper [1] offers the clearest treatment of rarefaction wave physics that these authors have ever encountered. We shall endeavor to follow that method of explanation here. Consider a finite square shock pulse of wavelength λ as shown in Figure 18.10.

Recall that the shock velocity is dependent upon the pressure ratio across the disturbance. Unlike the compression shock, where the increasing pressure caused the part of the wave initially behind the leading edge of the shock to catch up and form a front, at the rear end of this disturbance, the pressure is decreasing. This causes the rearmost portions of the rarefaction wave to fall further and further behind the incident shock. Additionally, since the rarefaction wave is passing into an effectively denser material, the head of the wave will be moving faster than the compression shock.

On a p–v diagram, we would see what appears in Figure 18.11 if we considered only points 1, 2, and 3 in our square pulse shown in Figure 18.10.

From the p–v diagram in Figure 18.11, we can see that our wavelet from p_1 to p_2 will move faster than the compression shock, and our wavelet from p_2 to p_3 will move slower. Over time, the shape of the pulse will change as depicted in Figure 18.12.

Figure 18.12 is a very crude discretization to facilitate understanding. The more elements we break the wave into, the closer the rarefaction wave gets as we approach the continuous

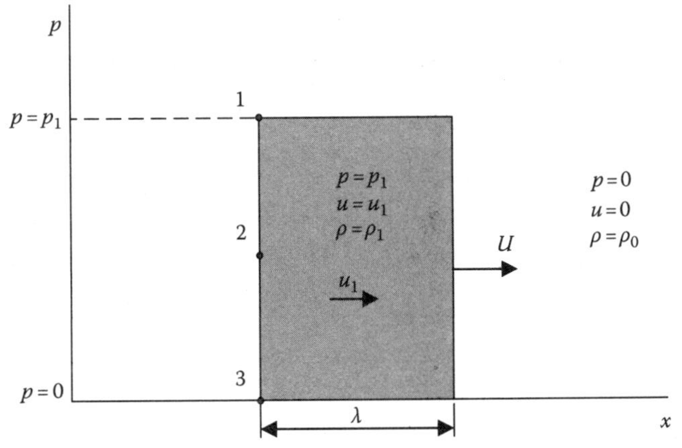

FIGURE 18.10
Simple model of a rarefaction wave. (Cooper, P. W.: *Explosives Engineering*. 1996. Copyright Wiley-VCH Verlag GmbH & Co. KGaA. Reproduced with permission.)

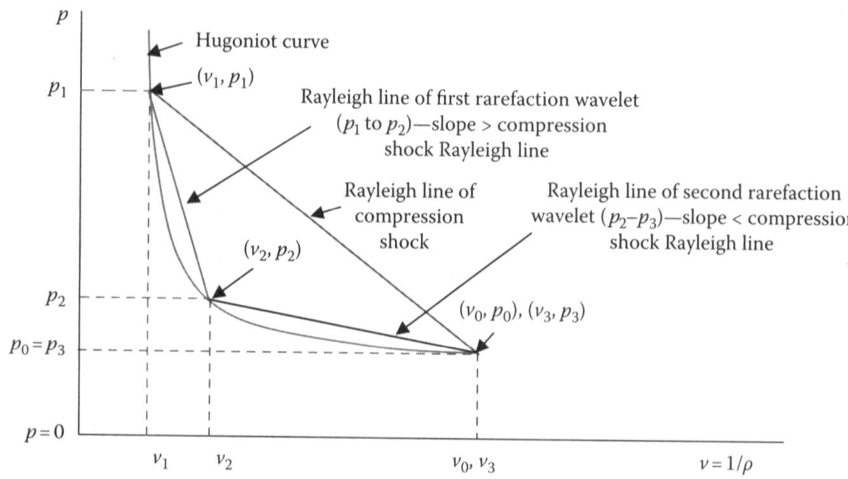

FIGURE 18.11
p–v diagram of a simple rarefaction wave. (Cooper, P. W.: *Explosives Engineering*. 1996. Copyright Wiley-VCH Verlag GmbH & Co. KGaA. Reproduced with permission.)

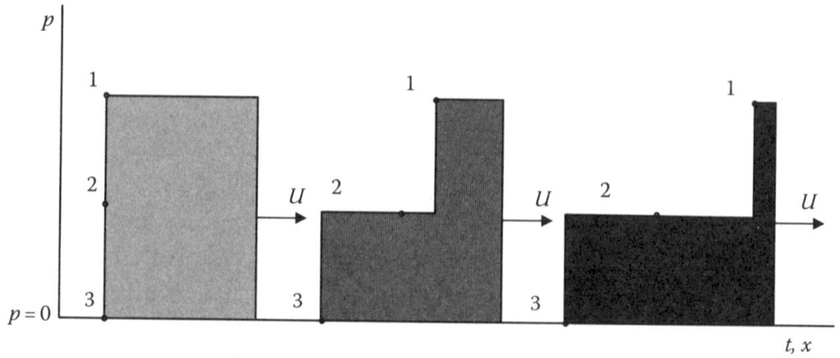

FIGURE 18.12
Rarefaction wave modeled as two wavelets catching up to incident shock. (Cooper, P. W.: *Explosives Engineering*. 1996. Copyright Wiley-VCH Verlag GmbH & Co. KGaA. Reproduced with permission.)

(actual) situation. This is illustrated in Figure 18.13. If we wanted to draw a *t–x* plot of the rarefaction wave illustrated in Figure 18.13, the result would appear as in Figure 18.14.

We shall now examine some classic rarefaction problems in detail. The first is quite important for use in terminal ballistics—the reflection of a square wave at a free surface.

When a compressive pulse reaches a free surface in a material, recall that the condition of zero stress on the surface must be maintained. Nature accomplishes this through the generation of a relief (rarefaction) wave at the surface such that the total stress is zero. The relief wave will exactly cancel the compressive wave. This has implications in stress behavior, which we shall see later, and we shall see how we can treat that scenario a little differently. The interaction with the free surface also results in a material velocity that is double the material velocity behind the original compressive pulse. Let us consider the *t–x* plot of a shock wave that encounters a free surface as depicted in Figure 18.15. The plot of this interaction on a *p–u* diagram is shown in Figure 18.16. In these figures, we see that after the compression shock encounters the free surface, a rarefaction wave propagates back into the material, dropping the pressure down to zero and doubling the material velocity.

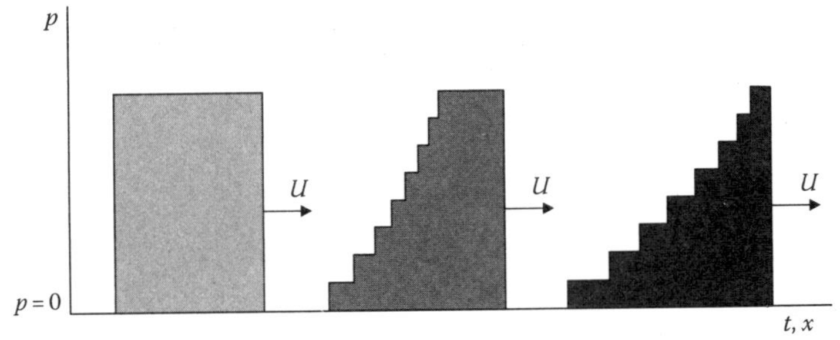

FIGURE 18.13
Rarefaction wave modeled as eight wavelets catching up to incident shock.

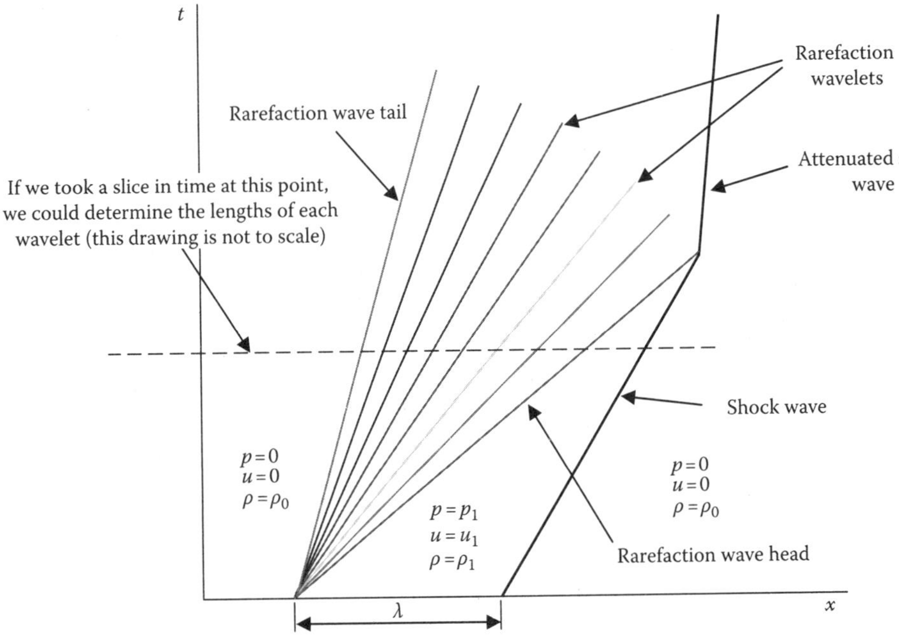

FIGURE 18.14
Rarefaction wave modeled as eight wavelets on a *t*–*x* plot. (Cooper, P. W.: *Explosives Engineering*. 1996. Copyright Wiley-VCH Verlag GmbH & Co. KGaA. Reproduced with permission.)

The rarefaction wave will have to travel back into material that is still approaching it at an induced velocity created by the incident shock. This requires us to understand the difference between Lagrangian and Eulerian coordinate systems. This is shown in Figure 18.17.

We have previously described Lagrangian coordinates as a coordinate system that is moving with the shock. Eulerian coordinates are stationary relative to the laboratory. All the velocities we examined thus far were Lagrangian (this made our equations simple). When we want velocities in Eulerian coordinates, we need to account for the motion of the material the shock is moving into. For instance, as previously mentioned, in our reflected shock, U_R is the Lagrangian velocity of the reflected wave. The Eulerian velocity of this same wave would be $U_R - u_1$ or $U_R + u_1$ if you consider U_R as negative in our lab and u_1 as positive.

The interaction with a free surface will now be illustrated with an example.

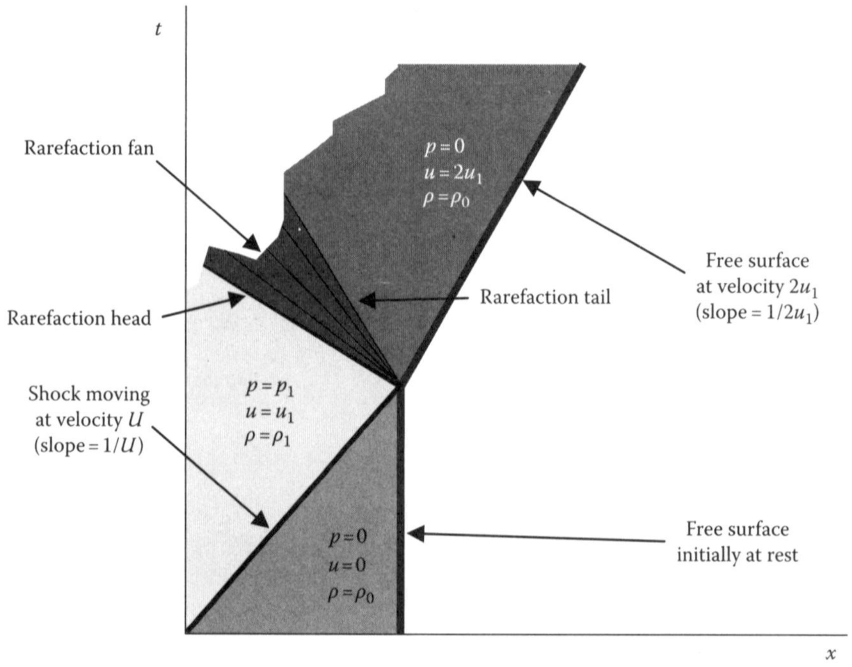

FIGURE 18.15
(See color insert.) *t*–*x* plot of a shock wave interacting with a free surface. (Cooper, P. W.: *Explosives Engineering*. 1996. Copyright Wiley-VCH Verlag GmbH & Co. KGaA. Reproduced with permission.)

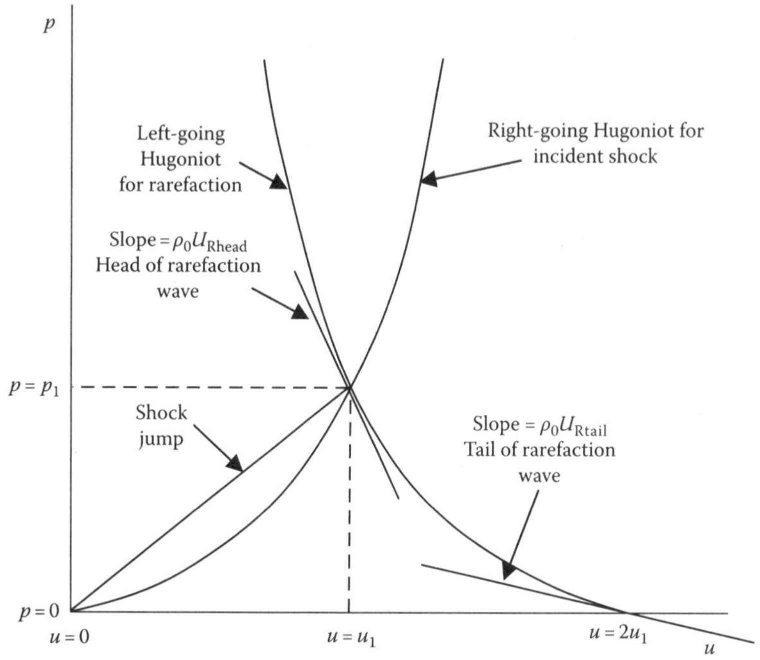

FIGURE 18.16
p–*u* Hugoniot plot of a shock wave interacting with a free surface. (Cooper, P. W.: *Explosives Engineering*. 1996. Copyright Wiley-VCH Verlag GmbH & Co. KGaA. Reproduced with permission.)

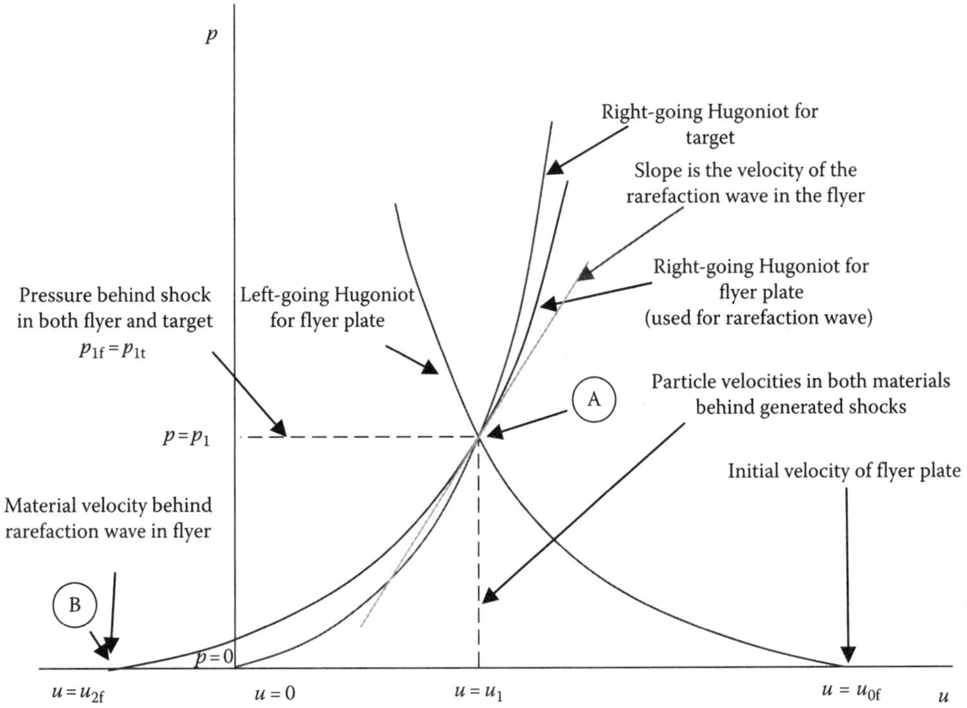

FIGURE 18.17
Rarefaction wave speed determination.

Example Problem 7

An experiment is set up in which a magnesium slab is shocked with a constant pressure of 5.0 GPa. Determine

1. The particle velocity in the magnesium behind the incident shock before an encounter with a free surface
2. The velocity of the free surface after the interaction
3. The particle velocity behind the surface after the interaction
4. The Lagrangian velocity of the leading edge of the rarefaction
5. The Eulerian velocity of the leading edge of the rarefaction

The material has the following properties:

Magnesium

$$\rho_{0_{Mg}} = 1.775 \left[\frac{\text{g}}{\text{cm}^3}\right]$$

$$c_{0_{Mg}} = 4.516 \left[\frac{\text{km}}{\text{s}}\right]$$

$$s_{Mg} = 1.256$$

Solution: We can determine the answer to part 1 from the right-going Hugoniot in the material. A right-going Hugoniot in a nonmoving material is described by Equation 18.25. Inserting values for the magnesium, we have

$$5.0\ [\text{GPa}] = (1.775)\left[\frac{\text{g}}{\text{cm}^3}\right](4.516)\left[\frac{\text{km}}{\text{s}}\right]u_1\left[\frac{\text{km}}{\text{s}}\right] + (1.775)\left[\frac{\text{g}}{\text{cm}^3}\right](1.256)u_1^2\left[\frac{\text{km}}{\text{s}}\right]^2 \tag{18.96}$$

Following through, we have

$$5.0\ [\text{GPa}] = 8.016u_1 + 2.229u_1^2 \tag{18.97}$$

or

$$u_1^2 + 3.596u_1 - 2.243 = 0, \tag{18.98}$$

which results in

$$u_1 = -1.798 \pm 2.340 \rightarrow u_1 = 0.542\left[\frac{\text{km}}{\text{s}}\right] \tag{18.99}$$

The velocity of the free surface is simply

$$u_2 = 2u_1 = (2)(0.542)\left[\frac{\text{km}}{\text{s}}\right] = (1.084)\left[\frac{\text{km}}{\text{s}}\right] \tag{18.100}$$

The particle velocity behind the reflected wave is the same as the free surface velocity. The Lagrangian velocity of the leading edge of the rarefaction is given by Equation 18.87:

$$U_R = (4.516)\left[\frac{\text{km}}{\text{s}}\right] + (2)(1.256)u_1\left[\frac{\text{km}}{\text{s}}\right] = (4.516)\left[\frac{\text{km}}{\text{s}}\right] + (2)(1.256)(0.542)\left[\frac{\text{km}}{\text{s}}\right] \tag{18.101}$$

resulting in

$$U_R = 5.877\left[\frac{\text{km}}{\text{s}}\right]. \tag{18.102}$$

The Eulerian velocity is found by noting that the reflected wave is moving in the negative direction and the material behind it is moving in the positive direction, so we can write

$$U_{R_{\text{Lab}}} = U_R + u_1 = -5.877\left[\frac{\text{km}}{\text{s}}\right] + 0.542\left[\frac{\text{km}}{\text{s}}\right] = -5.335\left[\frac{\text{km}}{\text{s}}\right]. \tag{18.103}$$

We will now examine two cases where a flyer plate (a thin plate) impacts a thick target. The flyer plate assumption allows us to ignore reflections of shocks from the free boundaries transverse to our impact direction. Case 1 is that of a flyer plate with an impedance less than or equal to that of the target (Cooper [1] treats these individually, but the case where they are equal is really the limiting case for a lower impedance flyer). Case 2 is that of a flyer plate with a greater impedance than the target. An important item to note is that Hugoniots are derived from compressive data; thus, if we have a tensile wave, we usually use a linear model on the *p*–*u* Hugoniot diagrams when negative values in pressure (tension) occur. The slope of these lines is $\rho_0 c_L$. Here c_L is the longitudinal speed of sound in the material.

If the flyer plate has an impedance less than or equal to that of the target on impact, a compressive shock will propagate into both objects. The shock in the flyer will reflect from the free surface of it and return as a rarefaction wave to the interface. When the rarefaction wave reaches the interface, two things happen: the flyer will rebound off the target and a new rarefaction wave will propagate into the flyer. Recall that waves reflect as like waves when the boundary condition stipulates a higher impedance. A rarefaction wave will also propagate into the target. This new rarefaction wave in the target will eventually catch up to the shock front in the target and reduce its strength. In the flyer, since it has free surfaces now, the waves will reflect in opposite sense until they equilibrate. The t–x plot and the p–u Hugoniots follow in Example Problem 8.

Example Problem 8

An experiment is set up in which a brass slab is shocked by impact from a magnesium flyer plate that is 1 mm in thickness. The impact velocity was measured to be 2.0 km/s. Determine

1. The material velocity behind the generated shock
2. The pressure generated at the interface
3. The time duration of the shock pulse in the target
4. The velocity with which the magnesium plate will rebound

The materials have the following properties:

Magnesium	Brass
$\rho_{0_{Mg}} = 1.775 \left[\frac{g}{cm^3}\right]$	$\rho_{0_{Brass}} = 8.450 \left[\frac{g}{cm^3}\right]$
$c_{0_{Mg}} = 4.516 \left[\frac{km}{s}\right]$	$c_{0_{Brass}} = 3.726 \left[\frac{km}{s}\right]$
$s_{Mg} = 1.256$	$s_{Brass} = 1.434$
$c_{L_{Mg}} = 5.770 \left[\frac{km}{s}\right]$	$c_{L_{Brass}} = 4.700 \left[\frac{km}{s}\right]$

Solution: Figure 18.18 tells us that to get the pressure generated at the interface, we need to calculate the left-going Hugoniot in the flyer and solve it for the pressure since we have the impact velocity and we know the target was initially at rest. The particle velocity in the magnesium before impact is given, and we have located it in our diagram on the left-going Hugoniot, which, by definition, has to pass through point (u_{0f}, 0) as well. Our equation for the left-going Hugoniot is Equation 18.27, which after insertion of the given values yields

$$p_1 = 2.229u_1^2 - 16.932u_1 + 24.949 \tag{18.104}$$

Here we assume that the units are correct, and we know the answer will be in gigapascals. Also for our right-going Hugoniot in the brass, we can use Equation 18.25 to write

$$p_1 = 12.117u_1^2 - 31.485u_1 \tag{18.105}$$

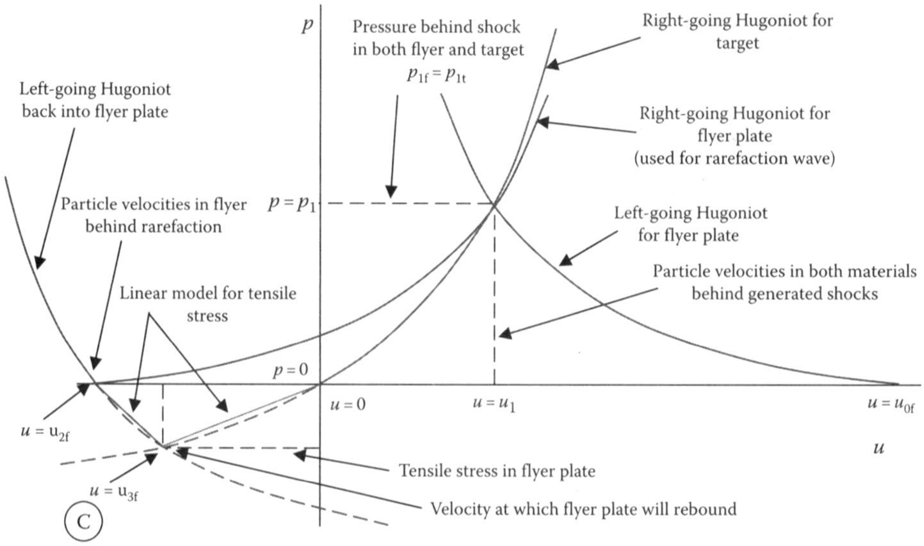

FIGURE 18.18
p–u plot of a flyer plate interaction with target of higher impedance.

Equating Equations 18.86 and 18.87 yields

$$u_1^2 + 4.897u_1 - 2.523 = 0 \tag{18.106}$$

Then

$$u_1 = 0.470 \left[\frac{\text{km}}{\text{s}}\right] \tag{18.107}$$

With Figure 18.18, it had to be positive. The pressure now comes from inserting this value in either Equation 18.104 or 18.105:

$$p_1 = (2.229)(0.470)^2 - (16.932)(0.470) + 24.949 = 17.483\ [\text{GPa}] \tag{18.108}$$

We have previously stated that the flyer plate will remain in contact with the target until the shock wave propagates to the rear face of the flyer, reflects as a rarefaction wave, and then reaches the front face. After this occurs, waves will continue moving back and forth in the flyer until the material velocity equilibrates. To determine the time of impact, we break the problem into two parts: the time it takes for the shock to reach the rear face and the time it takes for the first rarefaction to reach the impact surface.

The time it takes the shock to reach the rear face is determined by noting that the speed of wave propagation is the slope of the jump on the p–u Hugoniot divided by the initial density. Thus, we can write

$$U = \frac{p_1 - p_0}{\rho_0(u_1 - u_0)} \tag{18.109}$$

By inserting our values, we obtain

$$U = -6.438\left[\frac{\text{km}}{\text{s}}\right] = -6.438\left[\frac{\text{mm}}{\mu s}\right] \tag{18.110}$$

Why did not we use Equation 18.80 or 18.78? The reason is that if we used Equation 18.80, we would actually obtain the Eulerian velocity, which would be

$$p_1 - p_0 = \rho_0(u_1 - u_0)(U - u_0) \tag{18.111}$$

$$U = -4.438\left[\frac{\text{km}}{\text{s}}\right] = -4.438\left[\frac{\text{mm}}{\mu s}\right] \tag{18.112}$$

If we were interested in the velocity alone, relative to the lab, this would be the correct answer. However, the material in the flyer is moving toward the interface during the shock event so it would appear to an observer on the shock that the face will move to meet the wave.

We shall return to the problem. If the shock was moving toward the rear surface of our flyer plate at 2.450 mm/μs, then it would reach the rear of the plate in

$$\Delta t = \frac{l}{U} = \frac{1\,[\text{mm}]}{6.438\left[\frac{\text{mm}}{\mu s}\right]} = 0.155\,[\mu s] \tag{18.113}$$

To determine the speed of the leading edge of the rarefaction wave, we need to examine Figure 18.19. Here we see that the speed of the head of the rarefaction wave is the slope of

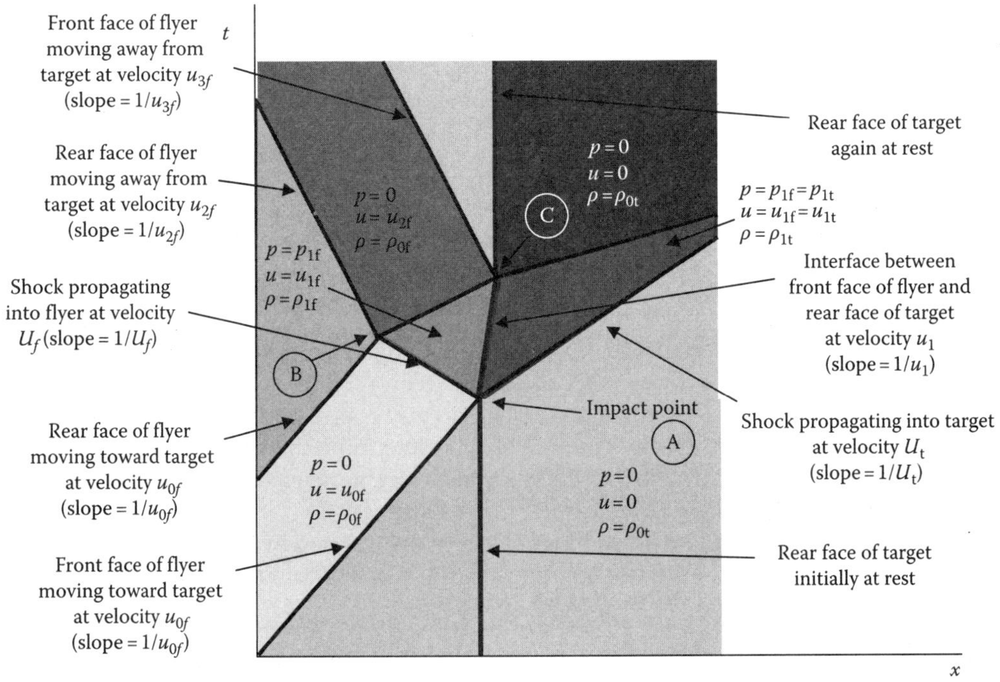

FIGURE 18.19
(See color insert.) t–x plot of flyer plate interaction with a target of higher impedance. (Cooper, P. W.: *Explosives Engineering*. 1996. Copyright Wiley-VCH Verlag GmbH & Co. KGaA. Reproduced with permission.)

the p–u Hugoniot curve at the material pressure. Equation 18.83 was written for a left-going rarefaction wave. In our case, the slope is the negative of this value which we know. Here we need to use Equation 18.104 since this is the Hugoniot for the flyer. Taking the derivative, we have

$$\left.\frac{dp}{du}\right|_{u=u_1} = 4.458u_1 - 16.932 = (4.458)(0.470) - 16.932 = -14.837 \tag{18.114}$$

The rarefaction velocity is the negative of this value divided by the density of the material, so we have

$$U_{\text{Rhead}} = \frac{14.837}{1.775}\left[\frac{\text{km}}{\text{s}}\right] = 8.359\left[\frac{\text{mm}}{\mu\text{s}}\right] \tag{18.115}$$

So the time it takes the rarefaction to reach the front face is

$$\Delta t = \frac{1}{U} = \frac{1\,[\text{mm}]}{8.359\left[\frac{\text{mm}}{\mu\text{s}}\right]} = 0.120\,[\mu\,\text{s}] \tag{18.116}$$

Then, the total time for the shock pulse is the time between impact and the rarefaction wave reaching the interface or

$$t_{\text{shock}} = 0.155\,[\mu\,\text{s}] + 0.120\,[\mu\,\text{s}] = 0.275\,[\mu\,\text{s}] \tag{18.117}$$

To determine the velocity at which the magnesium plate will rebound, let us look at Figure 18.18.

We have the densities of both materials, we have the longitudinal sound speeds, and we have u_{2f}, so we can find u_{3f} by solving the following equations simultaneously:

$$\rho_{0_{\text{Mg}}} c_{\text{L}_{\text{Mg}}} = \frac{p_3 - 0}{u_{3f} - u_{2f}} \rightarrow p_3 = \rho_{0_{\text{Mg}}} c_{\text{L}_{\text{Mg}}} (u_{3f} - u_{2f}) \tag{18.118}$$

$$\rho_{0_{\text{Mg}}} c_{\text{L}_{\text{Brass}}} = \frac{p_3 - 0}{u_{3f} - 0} \rightarrow p_3 = \rho_{0_{\text{Brass}}} c_{\text{L}_{\text{Brass}}} u_{3f} \tag{18.119}$$

Combining Equations 18.118 and 18.119 gives us

$$\rho_{0_{\text{Brass}}} c_{\text{L}_{\text{Brass}}} u_{3f} = \rho_{0_{\text{Mg}}} c_{\text{L}_{\text{Mg}}} (u_{3f} - u_{2f}) \tag{18.120}$$

$$(8.450)(4.700)u_{3f} = (1.775)(5.770)(u_{3f} - u_{2f}) \tag{18.121}$$

A neat way to find u_{2f} is to note that the two Hugoniots for the magnesium are reflected about the velocity u_1. So we can write

$$u_{2f} - u_1 = u_1 - u_{0f} \tag{18.122}$$

$$u_{2f} = (2)(0.470)\left[\frac{\text{km}}{\text{s}}\right] - 2.0\left[\frac{\text{km}}{\text{s}}\right] = -1.060\left[\frac{\text{km}}{\text{s}}\right] \tag{18.123}$$

Then, we can rewrite Equation 18.120 as

$$(39.715)u_{3f} = (10.242)(u_{3f} - 1.060) \tag{18.124}$$

$$u_{3f} = -0.368 \left[\frac{\text{km}}{\text{s}}\right] \tag{18.125}$$

A t–x plot of this event is shown in Figure 18.19.

If the flyer plate has an impedance greater than that of the target on impact, a compressive shock will again propagate into both objects. This shock will again reflect from the free surface of the flyer and return as a rarefaction wave to the interface. When the rarefaction wave reaches the interface, several things will happen: The rarefaction will again reflect in the opposite sense (as a shock) because the material into which it is propagating is of lower impedance, a new shock wave will propagate into the flyer as it digs into the target, and the rarefaction wave will propagate into the target. This new rarefaction wave in the target will again eventually catch up to the shock front in the target and reduce its strength. In the flyer, the process will repeat until equilibrium is reached. The physics of this event is again best described by an example problem.

Example Problem 9

An experiment is set up in which a magnesium slab is shocked by impact from a brass flyer plate that is 1 mm in thickness. The impact velocity was measured to be 2.0 km/s. Determine

1. The material velocity behind the generated shock
2. The pressure generated at the interface
3. The time duration of the initial shock pulse in the target
4. The material velocity behind the first rarefaction
5. The pressure behind the first rarefaction
6. The speed of the head of the first rarefaction wave in the target

The materials have the following properties:

Magnesium	Brass
$\rho_{0_{Mg}} = 1.775 \left[\frac{\text{g}}{\text{cm}^3}\right]$	$\rho_{0_{Brass}} = 8.450 \left[\frac{\text{g}}{\text{cm}^3}\right]$
$c_{0_{Mg}} = 4.516 \left[\frac{\text{km}}{\text{s}}\right]$	$c_{0_{Brass}} = 3.726 \left[\frac{\text{km}}{\text{s}}\right]$
$s_{Mg} = 1.256$	$s_{Brass} = 1.434$
$c_{L_{Mg}} = 5.770 \left[\frac{\text{km}}{\text{s}}\right]$	$c_{L_{Brass}} = 4.700 \left[\frac{\text{km}}{\text{s}}\right]$

Solution: Figure 18.20 tells us that to obtain the pressure generated at the interface, we need to calculate the left-going Hugoniot in the flyer and solve it for the pressure since we have the impact velocity and we know that the target was initially at rest. We again do this by simultaneously solving the left-going Hugoniot in the flyer and the right-going Hugoniot in the target.

The particle velocity in the brass after impact is located at point A in Figure 18.20 on the left-going Hugoniot, which, by definition, has to pass through point (u_{0f}, 0) as well. Our equation for the left-going Hugoniot (Equation 18.87) with the appropriate numbers

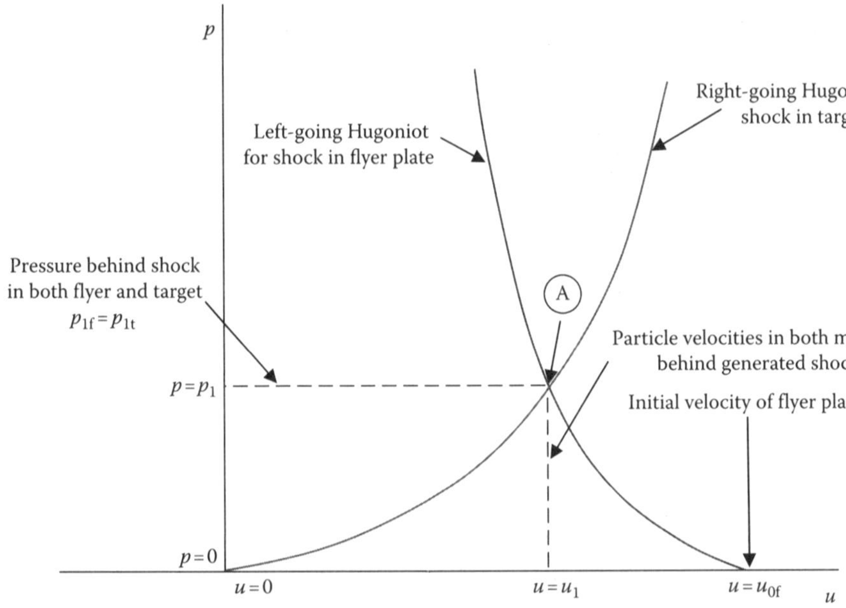

FIGURE 18.20
p–u plot of a flyer plate's initial interaction with target of lower impedance. (Cooper, P. W.: *Explosives Engineering*. 1996. Copyright Wiley-VCH Verlag GmbH & Co. KGaA. Reproduced with permission.)

inserted is

$$p_1 = (31.485)(2.0 - u_1) + (12.117)(2.0 - u_1)^2 \tag{18.126}$$

or

$$p_1 = 12.117u_1^2 - 80.165u_1 + 111.650 \tag{18.127}$$

Here we again know the answer will be in gigapascals. Also for our right-going Hugoniot in the magnesium, we can write using Equation 18.85

$$p_1 = (1.775)(4.516)u_1 + (1.775)(1.256)u_1^2 \tag{18.128}$$

$$p_1 = 2.229u_1^2 + 8.016u_1 \tag{18.129}$$

Equating Equations 18.127 and 18.129 yields

$$u_1^2 - 8.918u_1 + 11.291 = 0 \tag{18.130}$$

Then,

$$u_1 = 1.528 \left[\frac{\text{km}}{\text{s}}\right] \tag{18.131}$$

Here we used the least positive value because the velocity u_1 has to be less than our initial velocity. The pressure now comes from inserting this value in either Equation 18.127 or 18.129:

$$p_1 = (2.229)(1.528)^2 + (8.016)(1.528) = 17.453\,[\text{GPa}] \tag{18.132}$$

We have previously stated that the flyer plate will remain in contact and dig into the target in this case. Even though this is the case, the shock wave will still propagate to the rear face of the flyer, reflect as a rarefaction wave, and reach the front face. It is at this time that the initial pulse into the target will end. To determine the time of this event, we again break the problem into two parts: the time it takes for the shock to reach the rear face and the time it takes for the first rarefaction to reach the impact surface.

The time it takes the shock to reach the rear face is determined by noting that the speed of wave propagation is the slope of the jump on the p–u Hugoniot divided by the initial density. Thus, we can write

$$U = \frac{p_1 - p_0}{\rho_0(u_1 - u_0)} \tag{18.133}$$

$$U = \frac{(17.453 - 0)}{(8.450)(1.528 - 2.0)} \tag{18.134}$$

$$U = -4.376\left[\frac{\text{km}}{\text{s}}\right] = -4.376\left[\frac{\text{mm}}{\mu\text{s}}\right] \tag{18.135}$$

The shock will thus reach the rear of the plate in

$$\Delta t = \frac{l}{U} = \frac{1\,[\text{mm}]}{4.376\left[\dfrac{\text{mm}}{\mu\text{s}}\right]} = 0.228\,[\mu\text{s}] \tag{18.136}$$

To determine the speed of the leading edge of the rarefaction wave, we need to examine Figure 18.21. Here we recall that the speed of the head of the rarefaction wave times the initial density is the slope of the p–u Hugoniot curve at the material pressure. The slope is the negative of this value which we know. We need to use Equation 18.127 since this is the Hugoniot for the flyer. Then,

$$\left.\frac{dp}{du}\right|_{u=u_1} = 24.234u_1 - 80.165 = (23.234)(1.528) - 80.165 = -43.135 \tag{18.137}$$

The rarefaction velocity is the negative of this value, so we have

$$U_{\text{Rhead}} = \frac{(43.135)}{(8.450)\left[\dfrac{\text{g}}{\text{cm}^3}\right]} = 5.105\left[\frac{\text{mm}}{\mu\text{s}}\right] \tag{18.138}$$

So the time it takes the rarefaction to reach the front face is

$$\Delta t = \frac{l}{U} = \frac{1\,[\text{mm}]}{5.105\left[\dfrac{\text{mm}}{\mu\text{s}}\right]} = 0.196\,[\mu\text{s}] \tag{18.139}$$

Then, the total time for the shock pulse is the time between impact and the rarefaction wave reaching the interface or

$$t_{\text{shock}} = 0.228\,[\mu\text{s}] + 0.196\,[\mu\text{s}] = 0.424\,[\mu\text{s}] \tag{18.140}$$

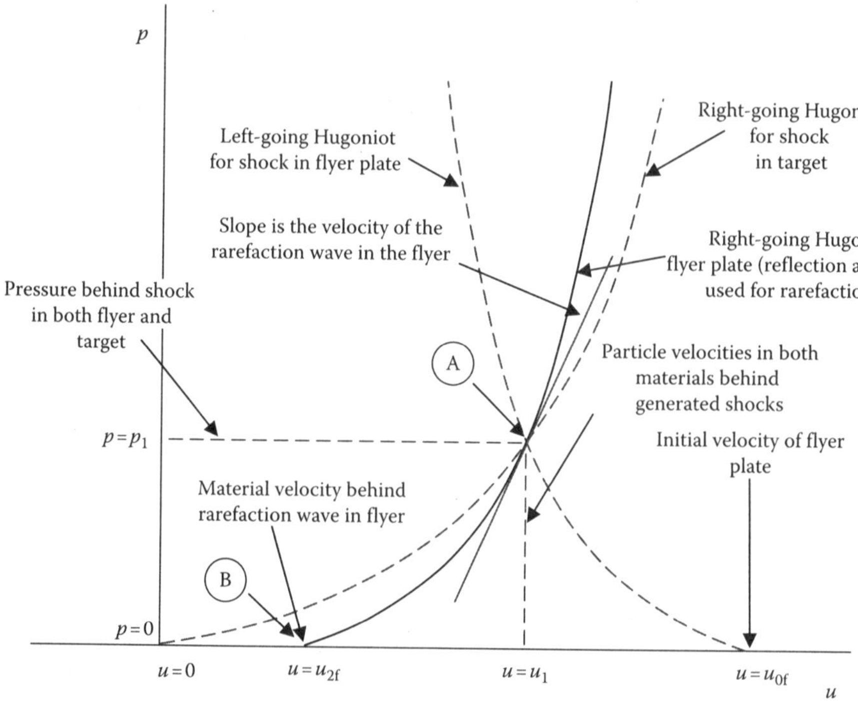

FIGURE 18.21
p–u plot of a flyer plate's rarefaction behavior during an interaction with target of lower impedance.

The material velocity behind the first rarefaction in the flyer is found by solving the right-going Hugoniot in the flyer plate for $p_2 = 17.453$ GPa. So we have

$$p_2 = 17.453 = \rho_0 c_0 (u_1 - u_{2f}) + \rho_0 s (u_1 - u_{2f})^2 \tag{18.141}$$

Inserting some numbers in here, we have

$$u_{2f}^2 - 5.654 u_{2f} + 4.865 = 0 \tag{18.142}$$

Then,

$$u_{2f} = 1.059 \left[\frac{\text{km}}{\text{s}}\right] \tag{18.143}$$

Again u_{2f} had to be less than u_1.

Our next task is to find the pressure behind the first rarefaction wave in the flyer plate. We shall refer to Figure 18.22 throughout this part of the discussion.

The rarefaction will drop our pressure along the Hugoniot from point A to point C as shown in Figure 18.22. We need to reflect our right-going Hugoniot in the flyer plate about material velocity u_{2f} and simultaneously solve with our right-going Hugoniot in the target. To reflect our flyer plate Hugoniot, we shall write the equation for a left-going Hugoniot centered at u_{2f}:

$$p_3 = \rho_0 c_0 (u_{2f} - u_{3f}) + \rho_0 s (u_{2f} - u_{3f})^2 \tag{18.144}$$

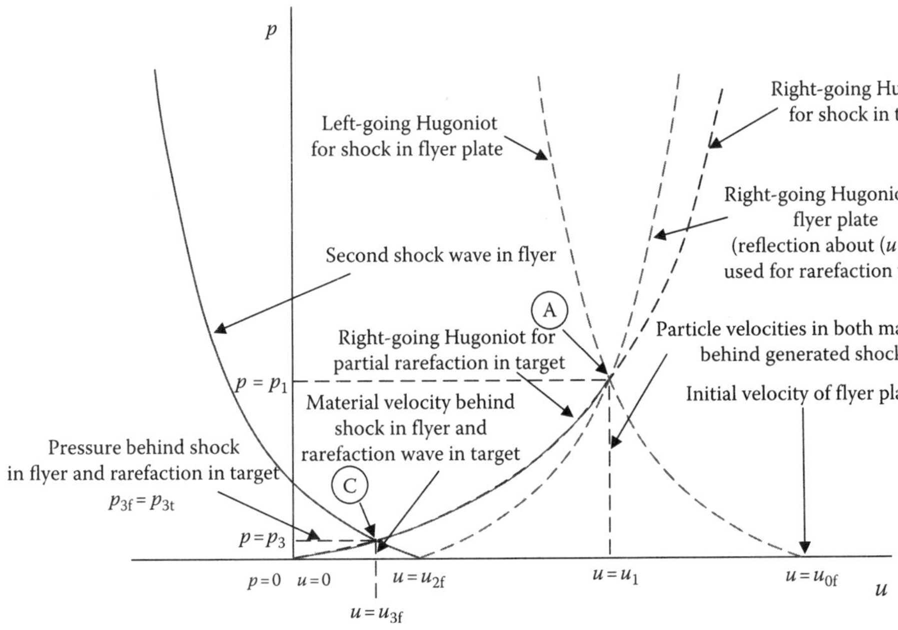

FIGURE 18.22
p–u plot of a flyer plate's behavior during the second shock interaction with target of lower impedance. (Cooper, P. W.: *Explosives Engineering*. 1996. Copyright Wiley-VCH Verlag GmbH & Co. KGaA. Reproduced with permission.)

or

$$p_3 = 12.117u_{3f}^2 - 57.149u_{3f} + 46.932 \tag{18.145}$$

You know the drill by now. We have to simultaneously solve this equation with Equation 18.129 from before since we are looking for the intersection of the two Hugoniots:

$$p_3 = 2.229u_{3f}^2 + 8.016u_{3f} \tag{18.146}$$

This leaves us with

$$u_{3f}^2 - 6.590u_{3f} + 4.746 = 0 \tag{18.147}$$

Then,

$$u_{3f} = 0.822 \left[\frac{\text{km}}{\text{s}}\right] \tag{18.148}$$

The speed of the head of the rarefaction wave in the target will be different from the speed of the rarefaction wave in the flyer. Recall that the speed of the head of the rarefaction wave is the slope of the Hugoniot at the shock pressure. An examination of Figure 18.23 shows this clearly.

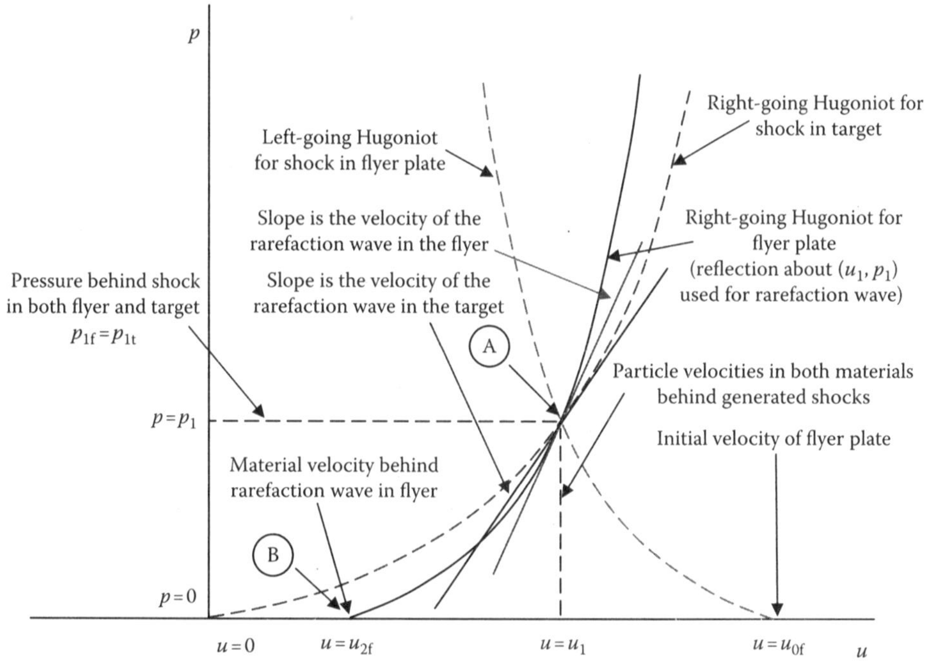

FIGURE 18.23
p–u plot of the rarefaction behavior into the target during a flyer plate impact into a lower impedance target.

We can find this slope by differentiating the Hugoniot for the target, Equation 18.129, at $u = u_1$:

$$\left.\frac{dp}{du}\right|_{u=u_1} = 4.458u_1 + 8.016 = (4.458)(1.528) + 8.016 = 14.828 \quad (18.149)$$

$$U_{\text{R head target}} = \frac{(14.828)}{(1.775)}\left[\frac{\text{km}}{\text{s}}\right] = 8.353\left[\frac{\text{mm}}{\mu\text{s}}\right] \quad (18.150)$$

A *t–x* plot of this event is shown in Figure 18.24.

To close out the subject of rarefaction waves, we will discuss how to use our previous techniques to determine if spalling or scabbing of a material will occur. We shall discuss a different method in the following section, but this is a good way to introduce the physics involved.

Recall that in earlier discussions, we stated that in a compressive wave the material velocity follows the wave, and in a rarefaction, the opposite is true. This behavior implies that if two rarefaction waves collide, the tension of the material will result (tensile waves will propagate away from the plane of collision). If this tensile stress exceeds the (dynamic) ultimate tensile stress of the material, the material will scab or spall. Also recall that we generally assume linear behavior of the material in tension (so the Hugoniots of the generated tensile waves will be straight lines). Once more, we shall illustrate the theory through an example problem.

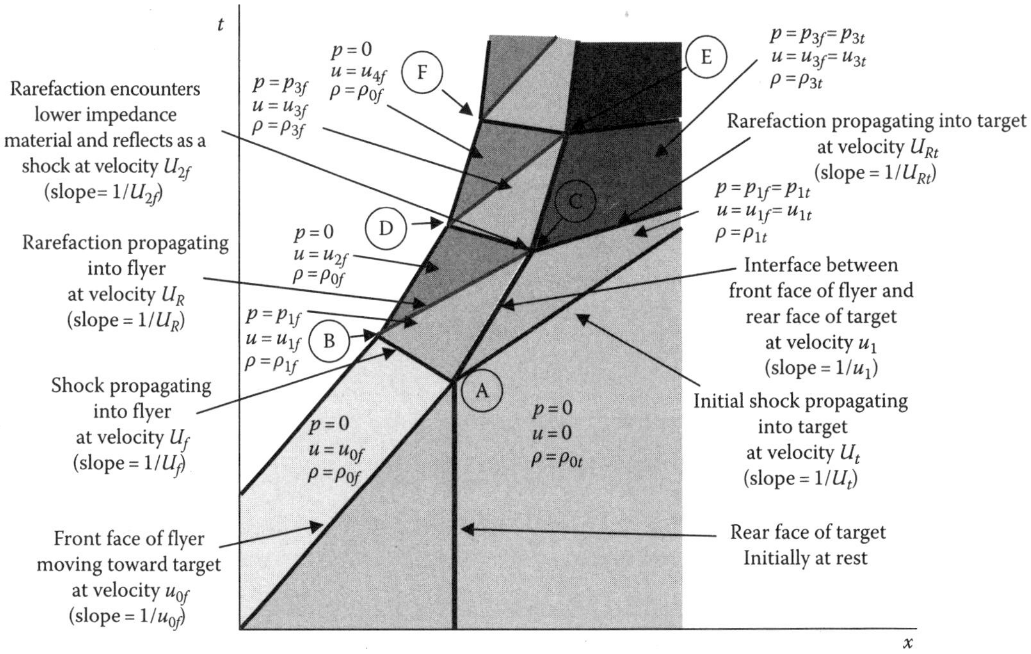

FIGURE 18.24
(See color insert.) t–x plot of flyer plate interaction with a target of lower impedance. (Cooper, P. W.: *Explosives Engineering*. 1996. Copyright Wiley-VCH Verlag GmbH & Co. KGaA. Reproduced with permission.)

Example Problem 10

An experiment is set up in which a brass plate is shocked by an explosive from both sides. The shock pressure was measured to be 4.0 GPa. Determine if the brass will spall.

The material has the following properties:

Brass

$$\rho_{0_{\text{Brass}}} = 8.450 \left[\frac{\text{g}}{\text{cm}^3}\right]$$

$$c_{0_{\text{Brass}}} = 3.726 \left[\frac{\text{km}}{\text{s}}\right]$$

$$s_{\text{Brass}} = 1.434$$

$$c_{L_{\text{Brass}}} = 4.700 \left[\frac{\text{km}}{\text{s}}\right]$$

$$\sigma_{\text{UTS}_{\text{Dynamic}}} = 2.1\ [\text{GPa}]$$

Solution: The only piece of information we have is the shock pressure (p_1), but we do know the equations for the two Hugoniot curves and the approximate tensile isentropes. The situation is illustrated in Figure 18.25.

We can locate u_1 on the left-going Hugoniot, which, by definition, has to pass through point (u_1, 0) as well. Our equation for the left-going Hugoniot is

$$p_1 = \rho_0 c_0 (u_1 - 0) + \rho_0 s (u_1 - 0)^2 \tag{18.151}$$

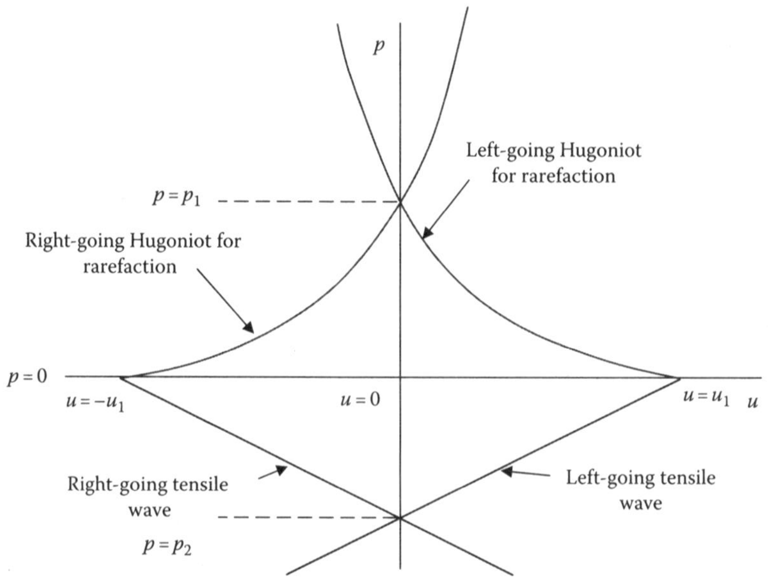

FIGURE 18.25
p–u plot of the collision of two rarefaction waves. (Cooper, P. W.: *Explosives Engineering*. 1996. Copyright Wiley-VCH Verlag GmbH & Co. KGaA. Reproduced with permission.)

By inserting values,

$$4.0 = (8.450)(3.726)(u_1 - 0) + (8.450)(1.434)(u_1 - 0)^2 \tag{18.152}$$

$$u_1^2 + 2.598u_1 - 0.330 = 0 \tag{18.153}$$

$$u_1 = 0.121 \left[\frac{\text{km}}{\text{s}}\right] \tag{18.154}$$

Now we have located our *x*-axis intercept on the aforementioned diagram. All that is left to do is determine the equation for the tensile isentrope and solve for the pressure. Recall that the slope of this isentrope is defined as

$$\rho_{0_{\text{Brass}}} c_{L_{\text{Brass}}} = \frac{0 - p_2}{0 - u_1} \rightarrow p_2 = \rho_{0_{\text{Brass}}} c_{L_{\text{Brass}}} u_1 \tag{18.155}$$

$$p_2 = (8.450)(4.700)(0.121) = 4.805\,[\text{GPa}] \tag{18.156}$$

Since this value is greater than the dynamic tensile strength of the material, the part will spall.

Problem 3

An experiment is set up in which a tungsten penetrator is fired against a rigid target. The impact velocity is 500 m/s. Determine the shock pressure and tensile stress and if the penetrator will break up.

Answer: p_1 = 44.672 [GPa], p_2 = 53.26 [GPa], and it will spall.

The material has the following properties:

Tungsten

$$\rho_{0_W} = 19.224 \left[\frac{g}{cm^3}\right]$$

$$c_{0_W} = 4.029 \left[\frac{km}{s}\right]$$

$$s_W = 1.237$$

$$c_{L_W} = 5.541 \left[\frac{km}{s}\right]$$

$$\sigma_{UTS_{Dynamic}} = 2.0\ [GPa]$$

Problem 4

A 4 in. long steel bar impacts a 12 in. thick slab of 4340 steel at 1000 m/s and bounces off. Assuming the impact is normal and using one-dimensional equations, determine

1. The duration of the impact event
   Answer: $\Delta t = 33.00$ [μs]
2. The pressure developed at the interface
   Answer: $p_1 = 20.980$ [GPa]

4340 steel

$$\rho_{0_{Steel}} = 7.896 \left[\frac{g}{cm^3}\right]$$

$$c_{0_{Steel}} = 4.569 \left[\frac{km}{s}\right]$$

$$s_{Steel} = 1.490$$

$$c_{L_{Steel}} = 5.941 \left[\frac{km}{s}\right]$$

Problem 5

We would like to determine how fast a large fragment will be propelled by an explosion using Hugoniots. Assume we have a 6 in. cube of steel that has a pressure of 5.0 GPa applied to one face for 3 μs.

1. Determine the velocity of the piece of steel by using the Hugoniot curve.
2. Determine the velocity of the piece of steel by using impulse and momentum (it might help to reference a statics and dynamics book).
3. Compare the two methods and comment on the differences.
4. What can you do to the problem parameters to make the answers the same?

Assume that the material does not spall at all. The material has the following properties:

Steel

$$\rho_{0_{\text{Steel}}} = 7.896 \left[\frac{\text{g}}{\text{cm}^3}\right]$$

$$c_{0_{\text{Steel}}} = 4.569 \left[\frac{\text{km}}{\text{s}}\right]$$

$$s_{\text{Steel}} = 1.490$$

$$c_{L_{\text{Steel}}} = 5.900 \left[\frac{\text{km}}{\text{s}}\right]$$

Problem 6

One of the ways that British troops tried to defeat snipers hiding behind a sniper plate was to pull their bullets from the cartridge cases and replace them backward so that they could be fired base first. This caused the plate to spall into the face of the sniper even though the bullet had no prayer of penetrating. In theory, this might put more of an abrupt shock pulse into the target. Assume that the bullet in Problem 8 of Chapter 18 was fired backward and was able to impact at 1512 ft/s. Using Hugoniot curves for lead and steel, determine if the 1 in. thick plate will spall.

18.3 Stress Waves in Solids

A stress wave is generated in a solid whenever an impact occurs—it is the way nature reacts to this violent event. The stress wave affects both the penetrator and the target. It is a major consideration in the breakup of the penetrator and is the primary cause of scabbing and spalling of the target.

Stress waves in solids are either elastic or elastic–plastic in nature. By this we mean that in the elastic regime, the material returns to its original shape, while in the plastic regime, the material is permanently distorted. How we treat the materials involved depends on the rate and intensity of loading. If these loads and rates are high enough, we can treat the materials as fluids. We will often refer to a target as being semi-infinite with the effect that geometrically, only the impact surface is present and there is no reflection of the stress wave once it enters the target. This further implies that the material can only compress or move backward from the free surface.

We also classify materials for the purpose of modeling as follows: isotropic (material properties are independent of direction), anisotropic (material properties are dependent upon direction), or orthotropic (material properties vary in three-orthogonal directions). Inertial effects are said to be important when the motion of the mass of the material is a major consideration in the behavior. We further stipulate that a dilatational wave is one that only involves normal stresses, and a distortional wave is one where shear stresses are involved [2].

When an impact occurs in a material, several things happen simultaneously [3]: longitudinal (dilatational) waves propagate into the material; transverse (distortional) waves propagate at right angles to the longitudinal waves; Rayleigh surface waves propagate along the surface and a small distance into the material; in a material that has layers with different properties (such as a laminate or a composite), a Love shear wave may occur; and depending on the geometry of loading, torsional or flexural waves may be generated. We shall examine only the first two in detail, and we will call the velocity of a longitudinal and a shear waves as c_L and c_S, respectively.

The acoustic velocity (velocity of sound) in a solid medium is greatly influenced by the boundary conditions. Using a cylindrical steel bar as an example, the material is considered "bounded" if the wave encounters a boundary in the radial direction. Otherwise, the material is "unbounded" [2].

We say the following about the acoustic velocities:

Extended (unbounded) Bounded

$$c_L^2 = \frac{\lambda + 2\mu}{\rho} = \frac{E(1-v)}{\rho(1+v)(1-2v)} \qquad \frac{E}{\rho} \tag{18.157}$$

$$c_S^2 = \frac{\mu}{\rho} = \frac{G}{\rho} = \frac{E}{2\rho(1+v)} \qquad \frac{G}{\rho} \tag{18.158}$$

where E is the modulus of elasticity; λ and μ are the Lamé parameters; v is the Poisson's ratio; G is the shear modulus; and ρ is the density.

We must note that since shear waves are, by definition, perpendicular to the main wave front, the form of the equation does not change between the bounded and the unbounded conditions. In a real wave, some mechanical energy is converted to heat. This is not considered in the models that we have just introduced.

In our discussions of compressible fluids, a wave simply rebounded off a solid boundary. However, in a solid medium, a compression wave will reflect off a free surface as a tensile wave. If the intensity of this tensile wave is greater than the ultimate tensile strength of the material, the material will fracture. If the intensity of the loading is such that the yield strength is exceeded, there will be two waves: an elastic wave (precursor in a rate-independent [RI] material) and a plastic wave (very intense but rapidly attenuated in most materials). At high loading rates, with a material that has a concave-up strain rate dependency, a shock can form with the plastic wave overtaking the elastic wave. We have seen this in our earlier work.

The stress–strain behavior of a material is characterized as either rate independent or rate dependent. An RI material has stress–strain curves that are unaffected by a change in loading rate. Examples of RI materials are aluminum and some steels. Examples of rate-dependent materials are titanium and most steels. If the intensity of the load is about two orders of magnitude above the strength of the material, we can consider both target and penetrator as viscous fluids. In computer solutions, to impact phenomena, this is where the term *hydro-code* comes from.

Proceeding into the analysis, we need to introduce indicial notation because this is a compact way of writing the equations. For any vector **F**, in an x, y, and z space, we can write it based on its components as

$$\mathbf{F} = F_x + F_y + F_z \tag{18.159}$$

In indicial notation, this vector is written as F_i, where $i = 1, 2, 3$, which is equivalent to our x, y, and z space. We then have

$$F_i = F_1 + F_2 + F_3 \tag{18.160}$$

In this notation, a pair of distinct indices indicate a tensor:

$$\sigma_{ij} = \begin{bmatrix} \sigma_{11} & \sigma_{12} & \sigma_{13} \\ \sigma_{21} & \sigma_{22} & \sigma_{23} \\ \sigma_{31} & \sigma_{32} & \sigma_{33} \end{bmatrix} \tag{18.161}$$

Repeated indices indicate a sum; for instance, the trace of our previous tensor is

$$\sigma_{ii} = \sigma_{11} + \sigma_{22} + \sigma_{33} \tag{18.162}$$

A derivative with respect to a coordinate is indicated by a comma; thus,

$$\sigma_{ij,j} = \frac{\partial \sigma_{ij}}{\partial x_j} \begin{bmatrix} \dfrac{\partial \sigma_{xx}}{\partial x} & \dfrac{\partial \sigma_{xy}}{\partial y} & \dfrac{\partial \sigma_{xz}}{\partial z} \\ \dfrac{\partial \sigma_{yx}}{\partial x} & \dfrac{\partial \sigma_{yy}}{\partial y} & \dfrac{\partial \sigma_{yz}}{\partial z} \\ \dfrac{\partial \sigma_{zx}}{\partial x} & \dfrac{\partial \sigma_{zy}}{\partial y} & \dfrac{\partial \sigma_{zz}}{\partial z} \end{bmatrix} \tag{18.163}$$

Two repeated subscripts after the comma indicate a second derivative as follows:

$$u_{i,jj} = \frac{\partial^2 u_i}{\partial x_j \partial x_j} = \begin{bmatrix} \dfrac{\partial^2 u_x}{\partial x^2} + \dfrac{\partial^2 u_x}{\partial y^2} + \dfrac{\partial^2 u_x}{\partial z^2} \\ \dfrac{\partial^2 u_y}{\partial x^2} + \dfrac{\partial^2 u_y}{\partial y^2} + \dfrac{\partial^2 u_y}{\partial z^2} \\ \dfrac{\partial^2 u_z}{\partial x^2} + \dfrac{\partial^2 u_z}{\partial y^2} + \dfrac{\partial^2 u_z}{\partial z^2} \end{bmatrix} \tag{18.164}$$

Two other terms are frequently seen: the tensor called the Kronecker delta δ_{ij} and the alternating tensor ϵ_{ijk}. The Kronecker delta takes on the values as

$$\delta_{ij} = 1 \text{ if } i = j \quad \text{or} \quad \delta_{ij} = 0 \text{ otherwise} \tag{18.165}$$

The alternating tensor takes on the values as

$$\epsilon_{ijk} = \begin{cases} 1 & \text{if } ijk = 123,\ 231, \text{ or } 312 \\ 0 & \text{if any two indices are alike} \\ -1 & \text{if } ijk = 321,\ 213, \text{ or } 132 \end{cases} \tag{18.166}$$

Let us return to the physics of stress waves in a solid. In an elastic solid, we require three relations to describe the material behavior: an equation of motion that requires the force to be converted into stress (force/unit area), an equation relating stress to strain for which we will use Hooke's law, and an equation relating strain to displacement.

If we begin with an equation of motion (Newton's second law), we have, using indicial notation to change from the vector form,

$$\mathbf{F} = m\mathbf{a} = m\ddot{\mathbf{u}} \rightarrow F_i = ma_i = m\ddot{u}_i \tag{18.167}$$

Note that u here is the material displacement/position. If we divide Equation 18.167 by a unit volume, we get

$$\frac{F_i}{V} = \frac{m}{V}\ddot{u}_i \tag{18.168}$$

We know that the mass per unit volume is defined as the density, and if we call the body force per unit mass f_i, we get

$$\frac{F_i}{V} = \frac{mf_i}{V}\left(\rho\frac{V}{m}\right) = \left(\frac{m}{V}\right)\ddot{u}_i \rightarrow \rho f_i = \rho\ddot{u}_i \tag{18.169}$$

We need to consider the internal forces in terms of stresses in our equation, so we shall add another term to the left-hand side (LHS) to account for this with a derivation to follow. Thus, we have

$$\frac{\partial\sigma_{ij}}{\partial x_j} + \rho f_i = \rho\ddot{u}_i \tag{18.170}$$

This is the equation of motion for a differential element of a continuum.

Although Equation 18.170 is a three-dimensional equation, we shall illustrate its derivation in two dimensions. Assume we have a cube of material with volume dxd$y$dz. The mass of the cube is the density times this volume, and the body forces on the cube are f_x and f_y for simplicity. We can then draw the situation (with dz into the paper) in two dimensions as shown in Figure 18.26.

If we write the force balance in the x-direction, we obtain

$$\begin{aligned}&\rho f_x(\mathrm{d}x\mathrm{d}y\mathrm{d}z) + \sigma_{xx}(\mathrm{d}y\mathrm{d}z) + \frac{\partial\sigma_{xx}}{\partial x}\mathrm{d}x(\mathrm{d}y\mathrm{d}z) - \sigma_{xx}(\mathrm{d}y\mathrm{d}z)\\&+\tau_{yx}(\mathrm{d}x\mathrm{d}z) + \frac{\partial\tau_{yx}}{\partial y}\mathrm{d}y(\mathrm{d}x\mathrm{d}z) - \tau_{yx}(\mathrm{d}x\mathrm{d}z) = \rho(\mathrm{d}x\mathrm{d}y\mathrm{d}z)\ddot{u}_x\end{aligned} \tag{18.171}$$

After we cancel terms and divide by the volume dxd$y$dz, we obtain

$$\rho f_x + \frac{\partial\sigma_{xx}}{\partial x} + \frac{\partial\tau_{yx}}{\partial y} = \rho\ddot{u}_x \tag{18.172}$$

Examined in three dimensions, the equation would be

$$\rho f_x + \frac{\partial\sigma_{xx}}{\partial x} + \frac{\partial\tau_{yx}}{\partial y} + \frac{\partial\tau_{zx}}{\partial z} = \rho\ddot{u}_x \rightarrow \rho f_i + \frac{\partial\sigma_{ij}}{\partial x_j} = \rho\ddot{u}_i \tag{18.173}$$

If we recall Hooke's law in its one-dimensional form, we get

$$\sigma = E\varepsilon \tag{18.174}$$

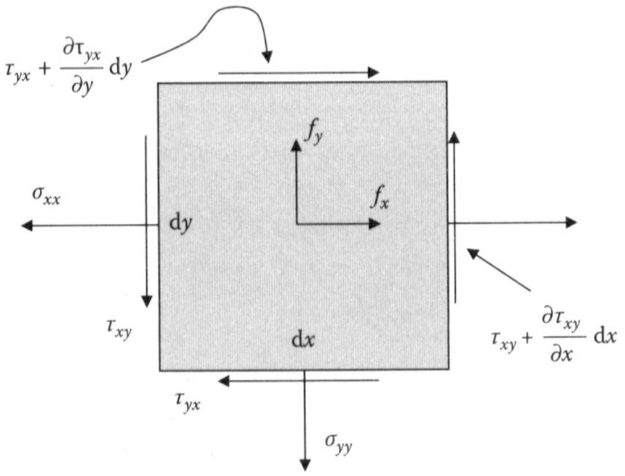

FIGURE 18.26
Differential element for calculation of stresses.

In three dimensions, it is written for a homogeneous material using two material constants, called the Lamé constants, as

$$\sigma_{ij} = \lambda \varepsilon_{kk} \partial_{ij} + 2\mu \varepsilon_{ij} \tag{18.175}$$

Here we define the constants as

$$\lambda = \frac{vE}{(1+v)(1-2v)} \tag{18.176}$$

$$\mu = G = \frac{E}{2(1+v)} \tag{18.177}$$

where G is the shear modulus and v is Poisson's ratio.

A strain–displacement relationship is the final equation necessary for our description of wave motion. For a homogeneous continuum, it is usually written as

$$\varepsilon_{ij} = \frac{1}{2}\left(u_{i,j} + u_{j,i}\right) + \frac{1}{2}\left(\frac{\partial u_i}{\partial x_j} + \frac{\partial u_j}{\partial x_i}\right) \tag{18.178}$$

To obtain the material displacement as a function of forces and accelerations, we shall first combine Equations 18.175 and 18.178

$$\sigma_{ij} = \lambda \frac{\partial u_k}{\partial x_k} \delta_{ij} + \mu\left(\frac{\partial u_i}{\partial x_j} + \frac{\partial u_j}{\partial x_i}\right) = \lambda \frac{\partial u_j}{\partial x_j} \delta_{ij} + \mu\left(\frac{\partial u_i}{\partial x_j} + \frac{\partial u_j}{\partial x_i}\right) \tag{18.179}$$

If we take the derivative of Equation 18.179 with respect to x_j, we get

$$\frac{\partial \sigma_{ij}}{\partial x_j} = \lambda \frac{\partial^2 u_j}{\partial x_j \partial x_j} \delta_{ij} + \mu\left(\frac{\partial^2 u_i}{\partial x_j \partial x_j} + \frac{\partial^2 u_j}{\partial x_i \partial x_j}\right) \tag{18.180}$$

Since δ_{ij} is not equal to zero only when $i = j$, we can freely interchange i and j in the first term on the RHS of Equation 18.180 to yield

$$\frac{\partial \sigma_{ij}}{\partial x_j} = \lambda \frac{\partial^2 u_j}{\partial x_j \partial x_j} + \mu \left(\frac{\partial^2 u_i}{\partial x_j \partial x_j} + \frac{\partial^2 u_j}{\partial x_j \partial x_i} \right) = \mu \frac{\partial^2 u_i}{\partial x_j \partial x_j} + (\lambda + \mu) \frac{\partial^2 u_j}{\partial x_j \partial x_i} \tag{18.181}$$

If we insert Equation 18.181 into Equation 18.173, we get

$$\mu u_{i,jj} + (\lambda + \mu) u_{j,ji} + \rho f_i = \rho \ddot{u}_i \tag{18.182}$$

or

$$\mu \frac{\partial^2 u_i}{\partial x_j \partial x_j} + (\lambda + \mu) \frac{\partial^2 u_i}{\partial x_j \partial x_i} + \rho f_i = \rho \ddot{u}_i \tag{18.183}$$

Equations 18.172, 18.174, 18.178, and 18.183 are the equations necessary to describe wave motion in a material.

We want to simplify these equations to look like the wave equation. To do so, first we define

$$\Delta = \varepsilon_{jj} = \frac{\partial u_i}{\partial x_j} \tag{18.184}$$

If we ignore the body forces, we can rewrite Equation 18.183 as

$$\mu \frac{\partial^2 u_j}{\partial x_j \partial x_j} + (\lambda + \mu) \frac{\partial \Delta}{\partial x_i} = \rho \ddot{u}_i \tag{18.185}$$

If we differentiate the aforementioned equation, we get

$$\rho \frac{\partial \ddot{u}_i}{\partial x_j} = \mu \frac{\partial^3 u_i}{\partial x_j \partial x_j \partial x_i} + (\lambda + \mu) \frac{\partial^2 \Delta}{\partial x_i \partial x_i} \tag{18.186}$$

From our earlier definition, we can see that

$$\frac{\partial \ddot{u}_i}{\partial x_i} = \frac{\partial^2 \Delta}{\partial t^2} \tag{18.187}$$

We can also see that

$$\frac{\partial^3 u_i}{\partial x_j \partial x_j \partial x_i} = \frac{\partial^2 \Delta}{\partial x_j \partial x_j} = \frac{\partial^2 \Delta}{\partial x_i \partial x_i} \tag{18.188}$$

So we now rewrite Equation 18.186 as

$$\rho \frac{\partial^2 \Delta}{\partial t^2} = (\lambda + 2\mu) \frac{\partial^2 \Delta}{\partial x_i \partial x_i} \rightarrow \frac{\partial^2 \Delta}{\partial t^2} = \left(\frac{\lambda + 2\mu}{\rho} \right) \frac{\partial^2 \Delta}{\partial x_i \partial x_i} \tag{18.189}$$

which is the classical wave equation of the form

$$\frac{\partial^2 \psi}{\partial t^2} = c^2 \frac{\partial^2 \psi}{\partial x_i \partial x_i} \tag{18.190}$$

The solution to this equation is

$$\psi = f(x - ct) + g(x + ct) \tag{18.191}$$

We previously stated that boundaries have a significant effect on wave propagation. If the medium were infinite, waves would spherically propagate at the speed of sound (wave velocity) in the material. The wave velocity in a material is defined for one-dimensional wave motion as

$$c = \sqrt{\frac{E}{\rho}} \tag{18.192}$$

For a bar impact, if the ratio of the radius of the bar to the wavelength is much less than 1, we can use these simplified equations. If we limit our study to longitudinal waves, our wave equation reduces to

$$\frac{\partial^2 u}{\partial t^2} = c^2 \frac{\partial^2 u}{\partial x^2} \tag{18.193}$$

Much like the discussion we had about the fluid in a shock tube after the bursting of a diaphragm, when a bar is stressed by a suddenly applied load, not all parts of the bar immediately feel the impact. The waves created traverse the material and distribute the stresses and strains accordingly. We will examine first the longitudinal wave (also called dilatational, irrotational, or primary [*P*] wave). This wave moves in the same direction as the pulse was applied. Next we will examine a transverse wave (also called a distortional, rotational, shear, or secondary [*S*] wave). This wave moves normal to the applied pulse.

As in compressible flow, there are several ways we can describe the motion of the material: stress vs. time; particle velocity vs. time; stress vs. distance; or particle velocity vs. distance. The two velocities we will use quite frequently are the speed of sound in the material c and the particle velocity at a point v. The symbol u represents axial displacement. We shall make some simplifying assumptions in this treatment. We assume that the bar has a length to diameter ratio of at least 10:1. We shall neglect transverse strain. We shall neglect lateral inertia. We shall neglect body forces and internal dissipation (i.e., friction and damping).

If we look at Newton's second law for a longitudinal impact of force F_L and bar mass m, we have

$$F_L \mathrm{d}t = \mathrm{d}(m v_L) \tag{18.194}$$

If we note that the stress $\sigma = F_L/A$ and the mass $m = \rho A \mathrm{d}l$, we can rewrite the aforementioned equation as

$$\sigma A \mathrm{d}t = \rho A \mathrm{d}l \mathrm{d}v_L \tag{18.195}$$

where $\mathrm{d}l$ is the distance the pulse has moved in time $\mathrm{d}t$. We can simplify the aforementioned equation to

$$\sigma = \rho \frac{\mathrm{d}l}{\mathrm{d}t} \mathrm{d}v_L \tag{18.196}$$

But the speed of the pulse is dl/dt, so we can write for either a longitudinal or a shear wave (changing the differential to a finite difference) as

$$\sigma = \rho c_L \Delta v_L \tag{18.197}$$

$$\tau = \rho c_S \Delta v_S \tag{18.198}$$

As in the case of a wave in a fluid, when a wave in a solid reaches a boundary, it is reflected. The normal stress on a free surface must be equal to zero so a compression wave reflects as a tensile wave and vice versa. It can be shown that the shape of the reflected pulse is the same as that of the incident pulse but opposite in sign. The position (displacement) of the incident and reflected pulses (right and left running characteristics) is

$$u_I = f(x - ct) \tag{18.199}$$

$$u_R = g(x + ct) \tag{18.200}$$

In these and all subsequent equations, displacements, velocities, stresses, and strains with the subscript I denote those occurring due to the incident pulse, whereas the subscript R denotes the reflected pulse effects. At the boundary ($x = l$), we have

$$u_I\big|_{x=l} = f(l - ct) \tag{18.201}$$

$$u_R\big|_{x=l} = g(l + ct) \tag{18.202}$$

Also we need to note that the strain at $x = l$ is

$$\varepsilon_I\big|_{x=l} = \frac{\partial u_I}{\partial x}\bigg|_{x=l} = \frac{\partial}{\partial(x - ct)} f(x - ct) \frac{\partial(x - ct)}{\partial x}\bigg|_{x=l} = f'(l - ct) \tag{18.203}$$

$$\varepsilon_R\big|_{x=l} = \frac{\partial u_R}{\partial x}\bigg|_{x=l} = \frac{\partial}{\partial(x + ct)} g(x + ct) \frac{\partial(x + ct)}{\partial x}\bigg|_{x=l} = g'(l + ct) \tag{18.204}$$

At the free boundary, the stress must be zero, so we have

$$\sigma_{net}\big|_{x=l} = \sigma_I + \sigma_R = 0 \tag{18.205}$$

But since $\sigma = E\varepsilon$, we can write

$$\sigma_{net}\big|_{x=l} = 0 = E\left[f'(l - ct) + g'(l + ct)\right] \tag{18.206}$$

$$f'(l - ct) = -g'(l + ct) \tag{18.207}$$

We can define the net velocity at a point as

$$v_{net} = v_I + v_R = \frac{\partial u_I}{\partial t} + \frac{\partial u_R}{\partial t} \tag{18.208}$$

The terms on the RHS are

$$v_I\big|_{x=l} = \frac{\partial u_I}{\partial x}\bigg|_{x=l} = \frac{\partial}{\partial(x - ct)} f(x - ct) \frac{\partial(x - ct)}{\partial x}\bigg|_{x=l} = -cf'(l - ct) \tag{18.209}$$

$$v_R|_{x=l} = \frac{\partial u_R}{\partial x}\bigg|_{x=l} = \frac{\partial}{\partial(x+ct)} g(x+ct)\frac{\partial(x+ct)}{\partial x}\bigg|_{x=l} = -cg'(l+ct) \quad (18.210)$$

But at $x = l$, we can insert Equation 18.207 giving us

$$v_{net} = 2cg(l + ct) \quad (18.211)$$

Thus, with a free boundary, the particle velocity and displacement are both double the incident value when the waves overlap.

If the boundary was rigid, Equations 18.205 through 18.207 are no longer true, but we know that the velocity must be zero, so we can write

$$v_{net} = 0 = -cf'(l - ct) + cg'(l + ct) \quad (18.212)$$

$$cf'(l - ct) = cg'(l + ct) \quad (18.213)$$

We can then write Equation 18.206 as

$$\sigma_{net}|_{x=l} = E\left[f'(l - ct) + g'(l + ct)\right] = 2Ef'(1 - ct). \quad (18.214)$$

Thus at a rigid boundary, the stress is doubled while the displacement and particle velocities are zero.

These equations allow us to visualize wave interactions with fixed or free ends as follows. When a tensile wave encounters a free boundary, it is reflected as a compressive wave. If we have a free surface, we can imagine a phantom pulse coming in from outside the bar as depicted in Figure 18.27. With a fixed boundary, the imagined pulse is in the same sense as the incident pulse as depicted in Figure 18.28.

When a bar elastically impacts a surface, a stress wave of strength $\rho v_0 c_L$ moves into the bar, stopping the motion behind it. At time $t = l/c_L$, the bar is stationary and in compression and all of the kinetic energy has been converted to strain energy, which can be written as

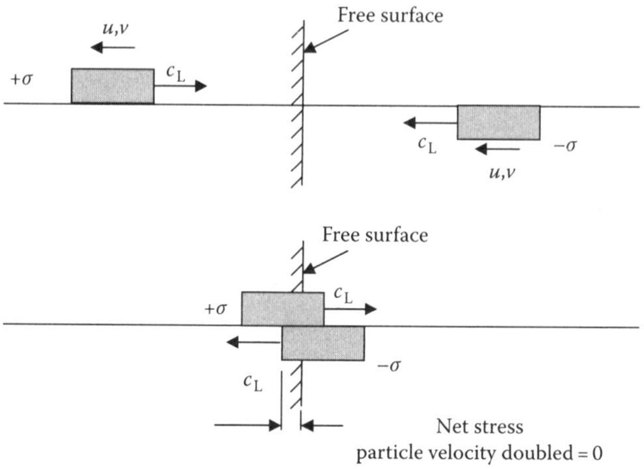

FIGURE 18.27
Wave interaction at a free boundary. (From Zukas, J. A. et al., *Impact Dynamics*, Krieger, Malabar, FL, 1992. With permission.)

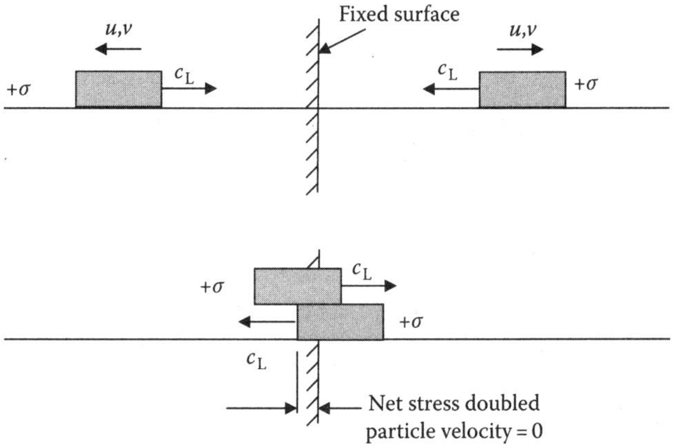

FIGURE 18.28
Wave interaction at a fixed boundary. (From Zukas, J. A. et al., *Impact Dynamics*, Krieger, Malabar, FL, 1992. With permission.)

$$\frac{1}{2}A_0 l\rho v_0^2 = \frac{A_0 l}{2E}\left(\rho c_L v_0^2\right) \tag{18.215}$$

When this wave encounters the free end, it reflects as a tensile wave with all the particles behind it moving at velocity v_0 away from the impact surface. This is depicted in Figure 18.29.

When a wave encounters a change in cross section (as illustrated in Figure 18.30) or in a new material, part of it is transmitted and part is reflected. The conditions that must be satisfied at the interface are that the forces must be equal and the velocities must be equal. The general equations for this interaction are

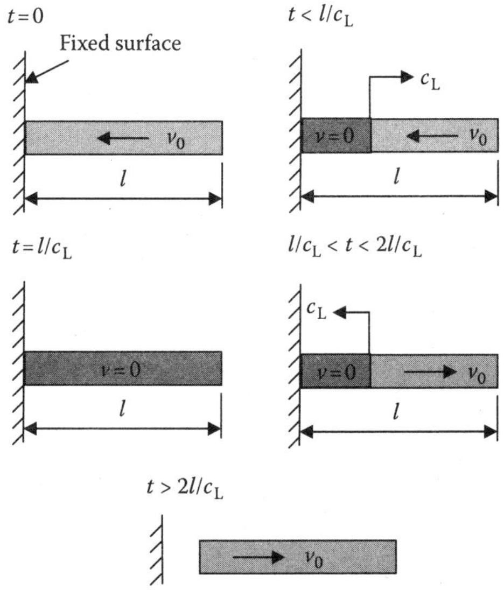

FIGURE 18.29
Elastic bar impact. (From Zukas, J. A. et al., *Impact Dynamics*, Krieger, Malabar, FL, 1992. With permission.)

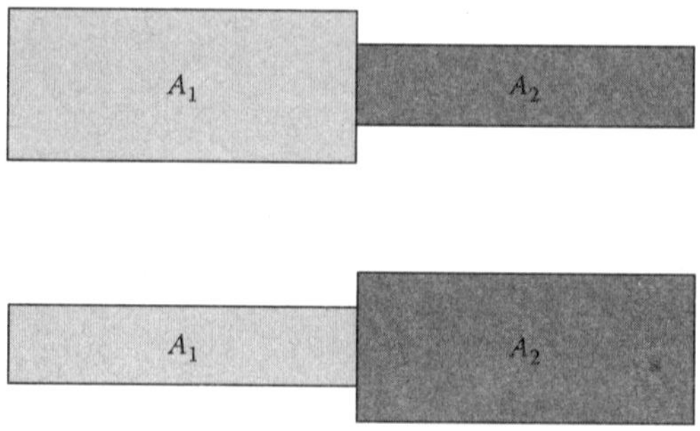

FIGURE 18.30
Bars of varying cross section.

$$\sigma_T = \frac{2A_1\rho_2 c_2}{A_1\rho_1 c_1 + A_2\rho_2 c_2}\sigma_I \tag{18.216}$$

$$\sigma_R = \frac{A_2\rho_2 c_2 - A_1\rho_1 c_1}{A_1\rho_1 c_1 + A_2\rho_2 c_2}\sigma_I \tag{18.217}$$

where σ_T is the transmitted stress; σ_R is the reflected stress; and σ_I is the incident stress.

The implications of these equations are that if $A_2/A_1 \to 0$, the bar is effectively free and σ_R approaches $-\sigma_I$. If $A_2/A_1 \to \infty$, the bar is effectively fixed and σ_R approaches σ_I. Also, σ_R equals 0 if $A_2\rho_2c_2 = A_1\rho_1c_1$, and if $\rho_2c_2 \gg \rho_1c_1$, the stress in the transmitted pulse is approximately twice the incident stress.

When we look at shock waves in solids, we usually use plates to simplify the problem. In plates, we assume uniaxial strain (three-dimensional stress). In bars, we assume uniaxial stress (three-dimensional strain). The stress–strain diagrams of these two behaviors are illustrated in Figure 18.31. The following analysis was originally developed by Zukas et al. [2] and neglects thermomechanical coupling as well as assuming one-dimensional deformation (i.e., the constraints are set up such that lateral strains are zero).

If we break the strain up into an elastic part (superscript *e*) and a plastic part (superscript *p*), we can write the strain in three orthogonal directions as

$$\varepsilon_1 = \varepsilon_1^e + \varepsilon_1^p, \quad \varepsilon_2 = \varepsilon_2^e + \varepsilon_2^p, \quad \varepsilon_3 = \varepsilon_3^e + \varepsilon_3^p \tag{18.218}$$

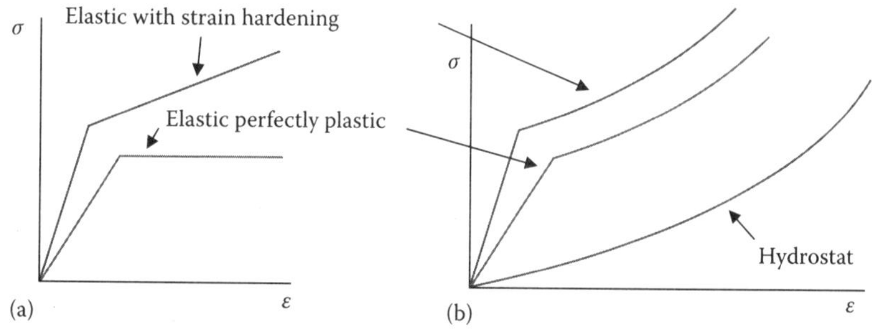

FIGURE 18.31
Comparison of (a) uniaxial stress and (b) uniaxial strain models in stress–strain diagrams.

In uniaxial strain, we have

$$\varepsilon_2 = \varepsilon_3 = 0 \rightarrow \varepsilon_2^e = -\varepsilon_3^p \quad \text{and} \quad \varepsilon_3^e = -\varepsilon_3^p \tag{18.219}$$

Because of symmetry, we can write

$$\varepsilon_2^p = \varepsilon_3^p \tag{18.220}$$

The material is incompressible so

$$\varepsilon_1^p + \varepsilon_2^p + \varepsilon_3^e = 0 \rightarrow \varepsilon_1^p = 2\varepsilon_2^e \tag{18.221}$$

This behavior is illustrated in Figure 18.32.

Thus, the total strain is

$$\varepsilon_1 = \varepsilon_1^e + \varepsilon_1^p = \varepsilon_1^e + 2\varepsilon_2^e \tag{18.222}$$

If we note that $\sigma_3 = \sigma_2$, we can write

$$\varepsilon_1 = \frac{\sigma_1(1-2\nu)}{E} + \frac{2\sigma_2(1-2\nu)}{E} \tag{18.223}$$

If we use a yield criterion such as von Mises, we can write

$$\sigma_1 - \sigma_2 = Y_0 \tag{18.224}$$

$$\sigma_1 = \frac{E}{3(1-2\nu)}\varepsilon_1 + \frac{2}{3}Y_0 = K\varepsilon_1 + \frac{2}{3}Y_0 \tag{18.225}$$

The bulk compressibility term K causes the stress to increase regardless of yield strength or strain hardening. This is important as we shall later see and is depicted in Figure 18.31. The reason that uniaxial strain is applicable in our work is that in the initial phases of

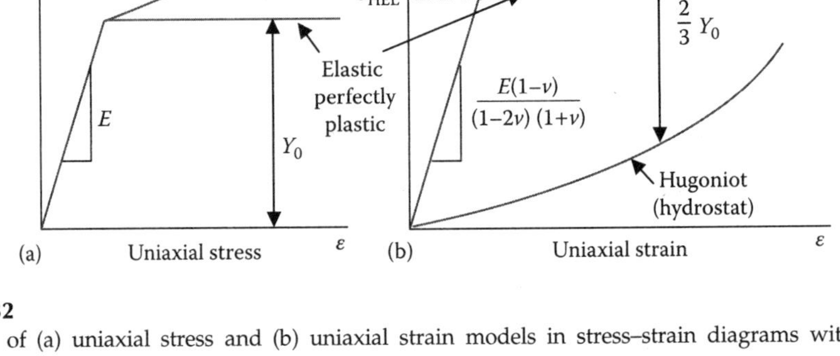

FIGURE 18.32
Comparison of (a) uniaxial stress and (b) uniaxial strain models in stress–strain diagrams with parameters established.

impact, the material does not have time to laterally expand. Later on in the impact, a condition closer to uniaxial stress may occur as the lateral deformation progresses. At extremely high pressures (~100 GPa, ~14.5 × 10^6 psi), the material will behave like a compressible fluid and will follow the Hugoniot curve (hydrostat). At lower pressures, deviation from the Hugoniot curve will occur.

If the applied stress is above the Hugoniot elastic limit (HEL), two stress waves will propagate through the material as was discussed in the previous sections. The first is an elastic wave with speed

$$c_E^2 = \frac{E(1-\nu)}{\rho_0(1-2\nu)(1+\nu)} \tag{18.226}$$

The second is a plastic wave with speed

$$c_p^2 = \frac{\sigma_B - \sigma_{HEL}}{\rho_{HEL}(\varepsilon_B - \varepsilon_A)} \tag{18.227}$$

In the aforementioned expression, σ_B and ε_B are the stress and strain caused by the pulse, respectively; ε_A is the strain at the HEL; and ρ_{HEL} is the material density at the HEL. After the applied pulse is over, an elastic unloading wave is generated. This unloading wave usually travels faster than the compressive wave, and, if the material region is long enough, we will eventually catch up and unload the initial pulse. The point at which this occurs is called the catchup distance. This behavior is illustrated in Figure 18.33.

The spalling of armor from a nonpenetrating or partially penetrating hit can be significant. Some projectiles are even designed so that they simply create spall.

When a finite thickness material is impacted on one side by an object that either does or does not penetrate, a stress wave will be generated, which can cause spalling or scabbing. This is to be expected in materials that are strong in compression but weak in tension. We are going to examine the impact event as a saw-tooth pulse in one dimension and assume that the pulse propagates without change in stress or intensity. We define the failure

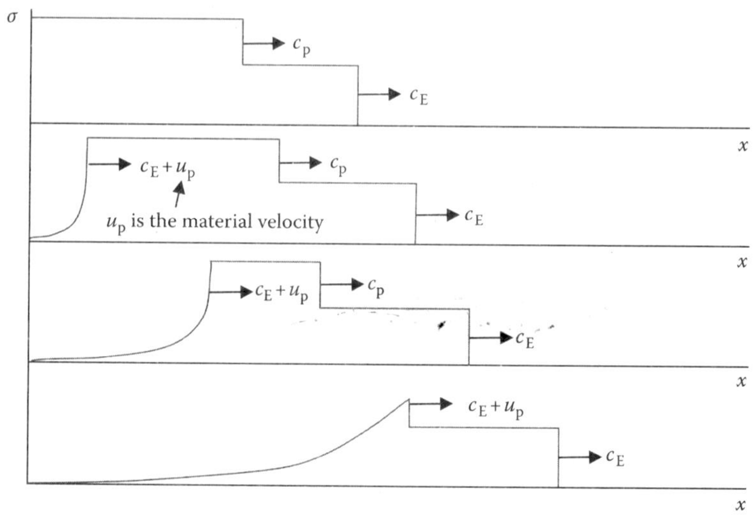

FIGURE 18.33
Diagram depicting plastic wave attenuation. (From Zukas, J. A. et al., *Impact Dynamics*, Krieger, Malabar, FL, 1992. With permission.)

strength of a material as the point where the tensile stress reaches some critical value σ_F. The length of the incident compressive pulse is defined as λ and its magnitude is specified as σ_m. The wave is reflected from the free surface with a net maximum tensile stress σ_T, which will always occur at the leading edge of the wave (Figure 18.34).

At any time, we can write

$$\sigma_T = \sigma_m - \sigma_I \tag{18.228}$$

where σ_I is the part of the compression wave remaining at an instant in time. If σ_T ever exceeds σ_F, a fracture will occur. Thus at fracture, we can write

$$\sigma_F = \sigma_m - \sigma_I \tag{18.229}$$

If we assume that this occurs at some instant, we will generate a spall thickness t_1, and we can write this spall thickness as

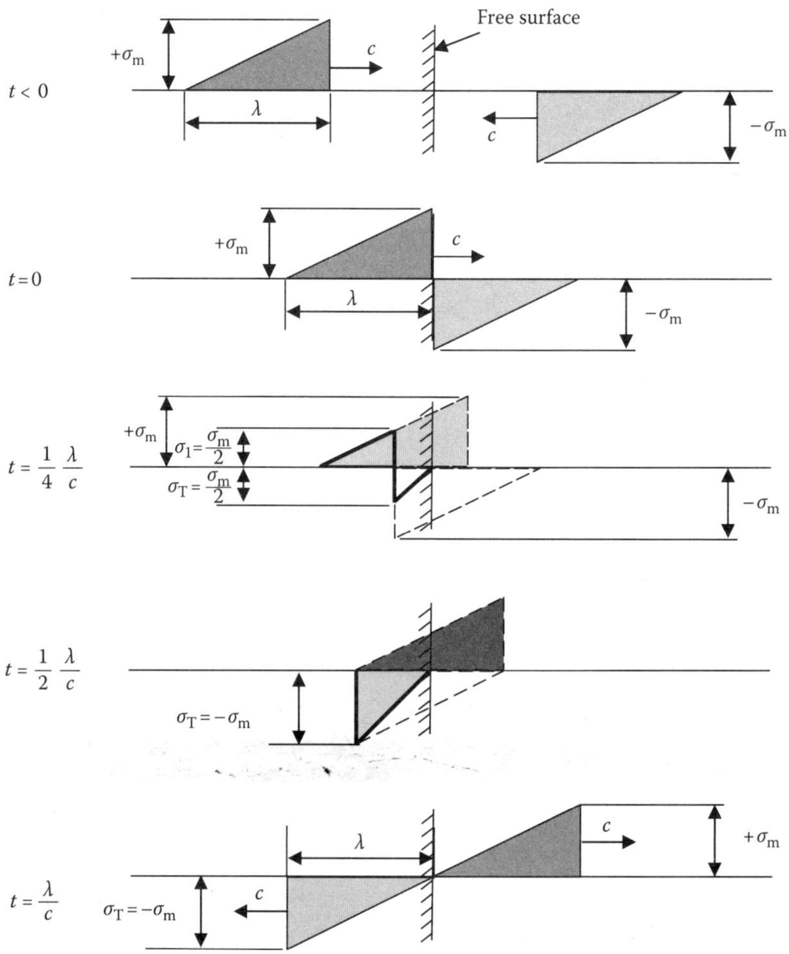

FIGURE 18.34
Triangular pulse encounter with a free surface.

$$\frac{\sigma_I}{\lambda - 2t_1} = \frac{\sigma_m}{\lambda} \tag{18.230}$$

It can be shown that by eliminating σ_I between Equations 18.228 and 18.229, and using Equation 18.230, we can write the spall thickness as

$$t_1 = \frac{\sigma_F}{\sigma_m}\frac{\lambda}{2} \tag{18.231}$$

Thus, if the initial pulse amplitude into the material is equal to its tensile strength, the material will fail at a distance one-half of the pulse wavelength from the rear face. We also need to note that if $\sigma_m < \sigma_F$, there will be no fracture, and if $\sigma_m \gg \sigma_F$, there will be multiple fractures.

If multiple fractures occur, the portion of the pulse trapped in a fractured piece will leave with that piece (actually forcing it away), and the part of the pulse that remains in the original target plate is defined through

$$\lambda_2 = \lambda - 2t_1 \tag{18.232}$$

$$\sigma_{m_2} = \sigma_I \tag{18.233}$$

If this occurs, we would enter these values back into our original equations to obtain

$$t_2 = \frac{\sigma_F}{\sigma_{m_2}}\frac{\lambda_2}{2} \tag{18.234}$$

This process is repeated until conditions no longer permit spalling (i.e., $\sigma_{m_n} < \sigma_F$).

We shall use the principle of impulse and momentum to determine the velocity of the spalled piece. The momentum of the spall is

$$mv_{t_1} = (\rho t_1 A)v_{t_1} \tag{18.235}$$

The impulse imparted to the spall is

$$\int F\mathrm{d}t = \left[\frac{(\sigma_m + \sigma_I)}{2}A\right]\frac{2t_1}{c} \tag{18.236}$$

Here the average stress acting over the time the wave is trapped in the spalled piece has been used. If we make the substitution for σ_I and combine Equations 18.235 and 18.236, we get

$$vt_1 = \frac{2\sigma_m - 3\sigma_F}{\rho c} \tag{18.237}$$

If there is a second spall layer, the velocity of that will be

$$vt_2 = \frac{2\sigma_m - 3\sigma_F}{\rho c} \tag{18.238}$$

If there are more spall layers, their velocities will be

$$vt_n = \frac{2\sigma_m - (2n-1)\sigma_F}{\rho c} \tag{18.239}$$

The number of spall layers a wave will produce is given by

$$n = \frac{\sigma_m}{\sigma_F} \tag{18.240}$$

Unlike a triangular pulse, a theoretically square pulse (shown in Figure 18.35) can spall only one piece of material because of its discontinuous nature. The thickness will be either zero (if $\sigma_m < \sigma_F$) or $\lambda/2$ (if $\sigma_m \geq \sigma_F$). The velocity imparted to the spalled piece will be given by

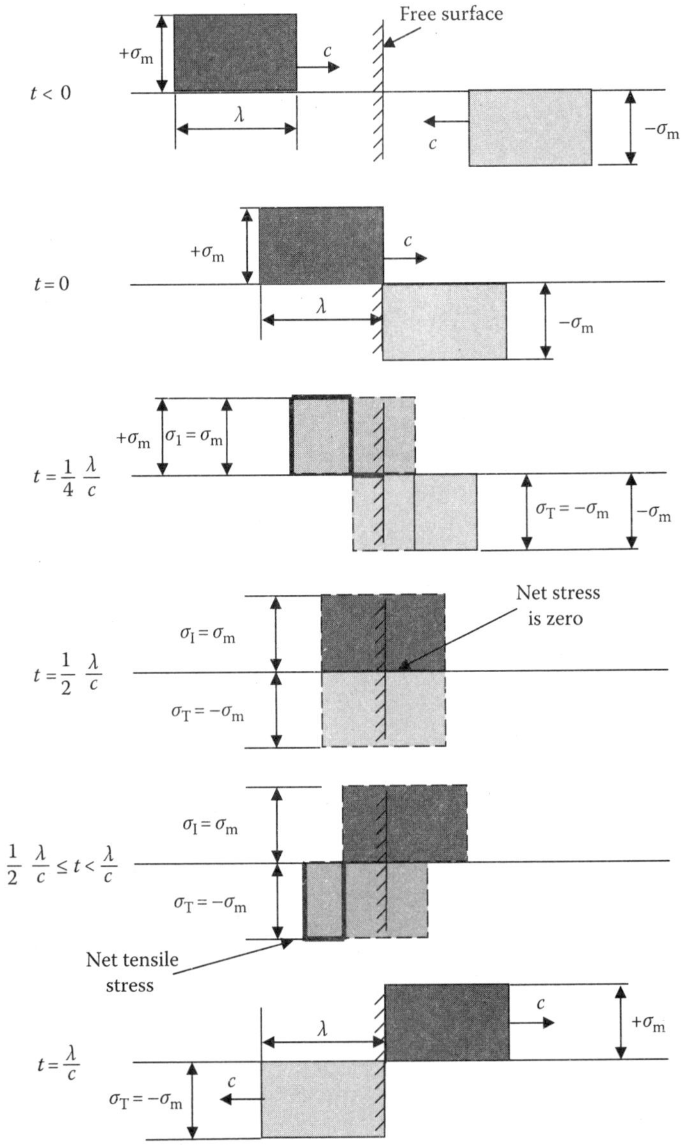

FIGURE 18.35
Square pulse encounter with a free surface.

$$v_t = \frac{\sigma_m}{\rho c} \tag{18.241}$$

The previous formulas yield qualitative results only. Dynamic fracture can be divided into four phases: nucleation of microcracks at many locations in the material, symmetric growth of the fracture nuclei, coalescence of the fractures, and spallation owing to formation of a large fracture surface.

Spallation is such a common occurrence in armor that some terms have been established to describe it. The incipient spall threshold is that combination of stress amplitude and pulse duration below which no damage is detected in a specimen at ×100 magnification. The complete spall threshold is the combination of stress amplitude and pulse duration at which a large piece of material will spall. Because of the complicated nature of the phenomena, it is difficult to predict exactly when and how a material will spall. There are various models all of which attempt to describe the spallation process by some physical means, one of which was introduced in Section 18.2.

Problem 7

A 4 in. long steel bar impacts a 12 in. thick slab of 4340 steel at 1000 m/s and bounces off. Assuming the impact is normal and using one-dimensional equations, determine

1. The duration of the impact event (use Hugoniots)
   Answer: $t_{shock} = 35.87$ [μs]
2. The pressure developed at the interface (use Hugoniots)
   Answer: $p_1 = 20.980$ [GPa]
3. The thickness of the first spalled piece (if any) assuming the input pulse is a constant square wave pulse throughout the impact event
   Answer: $t_1 = 3.75$ [in.]

Illustrate your answer to part 2.
Illustrate your answer to part 3.

4340 steel

$$\text{Modulus of elasticity} = 30.0\ (\times 10^6\text{psi})$$

$$\text{Modulus of rigidity (shear)} = 11.5\ (\times 10^6\text{psi})$$

$$\text{Poisson's ratio} = 0.29$$

$$\text{Ultimate tensile stress} = 250,000\ (\text{lbf/in.}^2)$$

$$\rho_{0_{\text{Steel}}} = 7.896 \left[\frac{\text{g}}{\text{cm}^3}\right]$$

$$c_{0_{\text{Steel}}} = 4.569 \left[\frac{\text{km}}{\text{s}}\right]$$

$$s_{\text{Steel}} = 1.490$$

$$c_{L_{\text{steel}}} = 5.941 \left[\frac{\text{km}}{\text{s}}\right]$$

Problem 8

A Japanese 20 mm projectile with the following properties impacts a 7 in. thick concrete wall at 0° obliquity. The concrete has a 1500 psi unconfined compressive strength and density of 0.080 lbm/in.^{3} The concrete dynamic tensile strength is 1000 psi. If the projectile has the properties given later,

1. Determine the duration of the impact event by using the assumption of nonpenetration (use Hugoniots).
   Answer: $t_{shock} = 24.85\ [\mu s]$
2. Determine whether the concrete will spall, and if so, determine the extent (in inches of thickness) of the total spallation—list all assumptions.
   Answer: $t_1 = 2.32$ [in.]
3. Determine if the projectile perforates the concrete accounting for the spallation.
   Answer: The projectile will perforate.
4. Using your ability to determine the timing of the penetration events, explain why or why not the aforementioned model is valid; i.e., prove it using the numbers.

Estimated penetrator information:

$$s = 40\ [\text{mm}] \quad m = 128\ [\text{g}]$$

$$d = 20\ [\text{mm}] \quad V_s = 550 \left[\frac{\text{m}}{\text{s}}\right] \quad \rho_p = 0.283 \left[\frac{\text{lbm}}{\text{in}^3}\right]$$

$$L = 60\ [\text{mm}]$$

Steel	Concrete
$\rho_{0_{Steel}} = 7.896 \left[\frac{\text{g}}{\text{cm}^3}\right]$	$\rho_{0_{Concrete}} = 2.232 \left[\frac{\text{g}}{\text{cm}^3}\right]$
$c_{0_{Steel}} = 4.569 \left[\frac{\text{km}}{\text{s}}\right]$	$c_{0_{Concrete}} = 4.0 \left[\frac{\text{km}}{\text{s}}\right]$ (estimate)
$s_{Steel} = 1.490$	$s_{Concrete} = 1.4$ (estimate)
$c_{L_{steel}} = 5.941 \left[\frac{\text{km}}{\text{s}}\right]$	$c_{L_{Concrete}} = 4.0 \left[\frac{\text{km}}{\text{s}}\right]$ (estimate)

Problem 9

If we include the effects of spallation in a concrete penetration problem, the resultant concrete thickness will have to increase. Let us look at the penetration of 8 in. of reinforced concrete wall at 0° obliquity with the projectile of Problem 11 in Chapter 16. Let us assume the concrete (assuming reinforcement is included) has a 1500 psi unconfined compressive strength and density of 0.084 lbm/in.^{3}. The concrete dynamic tensile strength is 1000 psi. If the projectile has the properties,

1. Determine the duration of the impact event using the assumption of nonpenetration—use Hugoniots.

2. Determine whether the concrete will spall, and if so, determine the extent (in inches of thickness) of the total spallation—use any method but list all assumptions.
3. Using your ability to determine the timing of the penetration events, explain why or why not the aforementioned model is valid.

Steel	Concrete
$\rho_{0_{Steel}} = 7.896\left[\frac{g}{cm^3}\right]$	$\rho_{0_{Concrete}} = 2.232\left[\frac{g}{cm^3}\right]$
$c_{0_{Steel}} = 4.569\left[\frac{km}{s}\right]$	$c_{0_{Concrete}} = 4.0\left[\frac{km}{s}\right]$ (estimate)
$s_{Steel} = 1.490$	$s_{Concrete} = 1.4$ (estimate)
$c_{L_{steel}} = 5.941\left[\frac{km}{s}\right]$	$c_{L_{Concrete}} = 4.0\left[\frac{km}{s}\right]$ (estimate)

Problem 10

A 2 in. long steel bar impacts a 6 in. thick slab of 4340 steel at 1000 m/s and bounces off. Assuming the impact is normal and using one-dimensional Hugoniot equations, determine a material type and thickness of a backing material that is required to keep the interface pressure below 2.5 GPa for the duration of the impact event. For full credit state, all your assumptions.

Use these properties for the steel

4340 steel

Modulus of elasticity $(\times 10^6 \text{psi}) = 30.0$

Modulus of rigidity (shear) $(\times 10^6 \text{psi}) = 11.5$

Poisson's ratio $= 0.29$

Ultimate tensile stress $(\text{lbf/in.}^2) = 250{,}000$

$$\rho_{0_{Steel}} = 7.896\left[\frac{g}{cm^3}\right]$$

$$c_{0_{Steel}} = 4.569\left[\frac{km}{s}\right]$$

$$s_{Steel} = 1.490$$

$$c_{L_{Steel}} = 5.941\left[\frac{km}{s}\right]$$

18.4 Detonation Physics

Now that we have talked about shock in nonreacting solids, it is appropriate to discuss how these shocks behave in detonating materials. While the interested reader is again referred to the references to find more detailed treatment, we shall endeavor to introduce the concept of detonation by building on what we have previously discussed.

In 1950, Zel'dovich, von Neumann, and Doering developed the so-called ZND model for detonation [4]. This model is sometimes known as "the simple model" for a reaction. The model is a one-dimensional model that neglects transport properties. In this model, the leading part of the detonation wave is a nonreacting shock, a jump discontinuity called the von Neumann spike. In the model, shocks of sufficient strength raise the density (and the temperature) above the ignition point beginning the reaction. In the gas behind the final state of reaction zone is the following flow that was denoted as moving with velocity u_p in our previous work.

In the ZND model, there are essentially two conditions that can exist: the unsupported case and the overdriven case. In the unsupported case, an initial shock starts the reaction, and it can continue if the conditions are right or it can die out. In the overdriven case, there is a force that continually drives the wave forward similar to the infinite shock pulses that we have examined earlier. Our approach here will be to physically describe the types of waves on a p–x diagram and then to relate these descriptions to the Hugoniot curves.

The unsupported case is depicted in Figure 18.36 as a p–x diagram. In this figure, there is an initial shock that begins the reaction. The detonation wave velocity is D. This is commonly known as the von Neumann spike. This spike begins the chemical reaction that takes place in the reaction zone immediately behind the shock. The reacted products are said to be in their final state when they leave the reaction zone. Once the reaction is completed, there is a rarefaction wave that follows the reaction zone. This is followed by the constant state where the chemically altered gases follow the rarefaction. Sometimes, we like to imagine that there is a piston that causes the induced velocity u_p, and this is depicted in the figure. Later on, we shall introduce restrictions on this piston velocity that is consistent with our unsupported definition.

The overdriven case is depicted in a p–x diagram as in Figure 18.37. Again, there is an initial shock that begins the reaction. This spike begins the chemical reaction that takes place in the reaction zone immediately behind the shock. The reacted products are said to be in

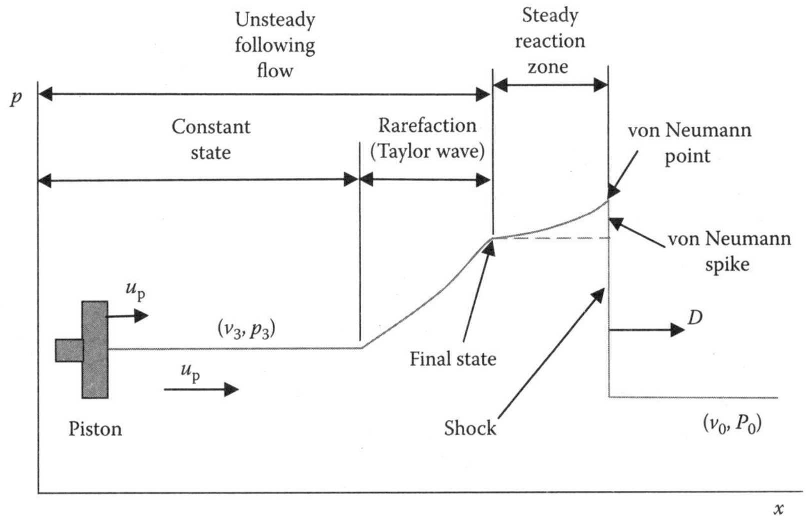

FIGURE 18.36
Unsupported detonation wave.

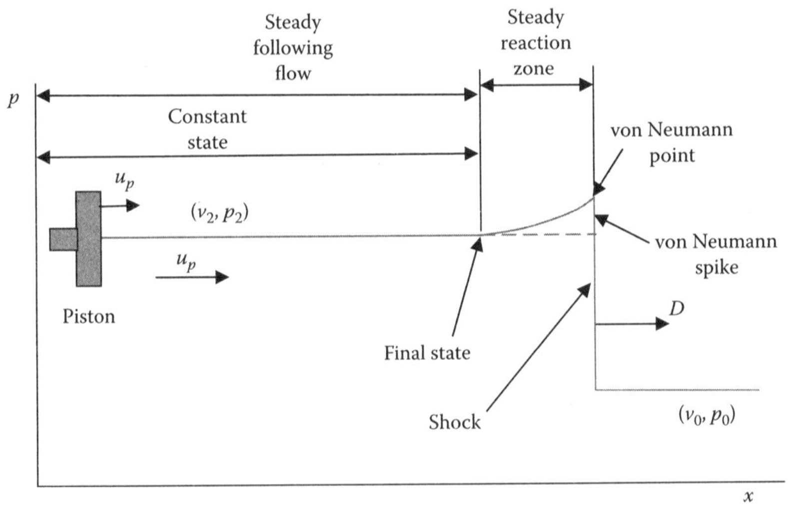

FIGURE 18.37
Overdriven detonation wave.

their final state when they leave the reaction zone. In this case, however, there is no rarefaction wave. Our imaginary piston is pushing the reacted gas at such a velocity that the rarefaction cannot form. We shall soon see that this piston velocity, in either the unsupported or overdriven case, completely determines the geometry and the velocity of the detonation wave.

The ZND model has two main parts. First, we must determine all possible steady solutions for the detonation wave velocity D. This will determine what the final state is. Then we must find a following flow (piston velocity u_p) that is a function of the detonation velocity. If this is greater than the minimum value of D, the wave is overdriven. If it is less than the minimum D, the wave is unsupported. If it is equal to the minimum D, the wave is a steady detonation wave. For now, we shall assume that the reaction takes place instantaneously. Thus, the steady reaction zone is a jump discontinuity.

With a reactive flow, there are some nuances associated with the Hugoniot curves. The first we must recognize is that once the reaction has taken place, we have a different material than the solid unreacted material we started with. Because of this material change, we have a different Hugoniot. It will be shifted toward the concave side as depicted in Figure 18.38. Thus, any further shocks or rarefactions take place using this new curve.

If we assume that the products of the reaction are instantaneously produced by the shock (i.e., the reaction zone is infinitesimally small in thickness), we obtain the simplest theory. If we rewrite the conservation of mass equation using the detonation velocity, we obtain

$$\rho_0 D = \rho_1 \left(D - u_p\right) \tag{18.242}$$

Similarly, we can write the conservation of momentum as

$$p_1 - p_0 = \rho_0 D u_p \tag{18.243}$$

If we eliminate u_p from these two equations, we obtain the equation for the Rayleigh line:

$$\rho_0^2 D^2 - \frac{(p_1 - p_0)}{(v_0 - v_1)} = 0 \tag{18.244}$$

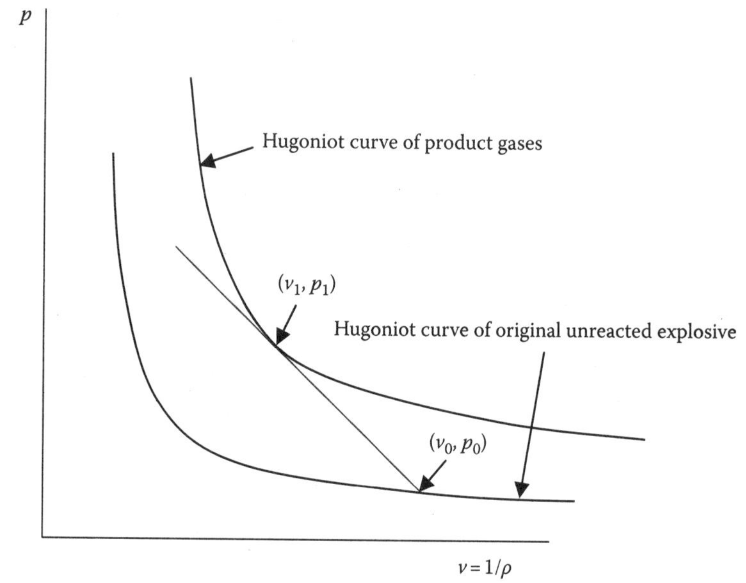

FIGURE 18.38
Hugoniot curve for reacted and unreacted material—overdriven detonation wave.

Here we have used the specific volume because we like to deal with p–v diagrams.

From our Rayleigh line Equation 18.244, we can see that it passes through point (v_0, p_0) and has a slope of $-\rho_0 D^2$. Some interesting things can be gleaned from this. First, we know that ρ_0 is positive and finite. If the Rayleigh line was horizontal, it would represent a detonation velocity of zero; hence, the detonation would not go anywhere. This is known as a constant pressure detonation. If the line was vertical, this would represent an infinite detonation velocity, so the detonation would happen everywhere at once. This is known as a constant volume detonation. This is illustrated in Figure 18.39.

If we eliminate D between Equations 18.242 and 18.244, we obtain the equation for the Hugoniot curve:

$$u_p^2 = (p_1 - p_0)(v_0 - v_1) \tag{18.245}$$

Thus, if we are given u_p and D, everything else is known because we can find the intersection of the Rayleigh line and the Hugoniot curve. If we write the energy equation using specific volume, we obtain

$$e_1 - e_0 - \frac{1}{2}(p_1 + p_0)(v_0 - v_1) = 0 \tag{18.246}$$

In this case, remember that the reaction is complete at state 1, and we have the energy of the unreacted explosive at state 0. We can then intersect this with the Rayleigh line (Equation 18.244) to determine the state of the explosive products. This is illustrated in Figure 18.40.

If we assume a polytropic gas (an ideal gas with constant specific heats), we can write the equation of state as

$$pv = Rt \tag{18.247}$$

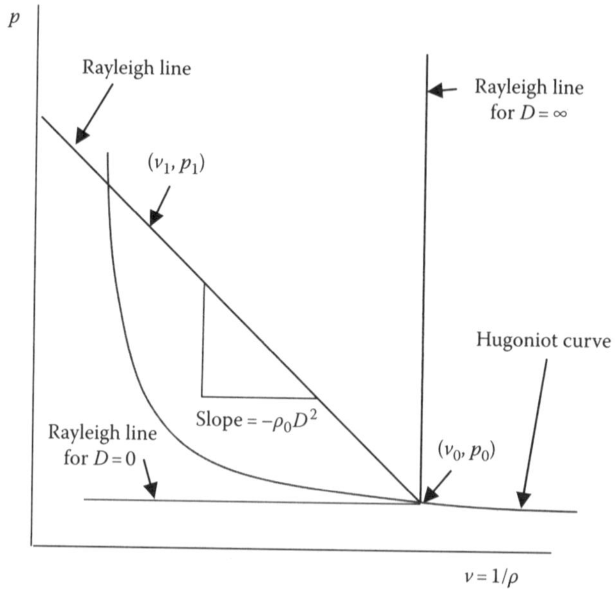

FIGURE 18.39
Constant-pressure and constant-volume detonation.

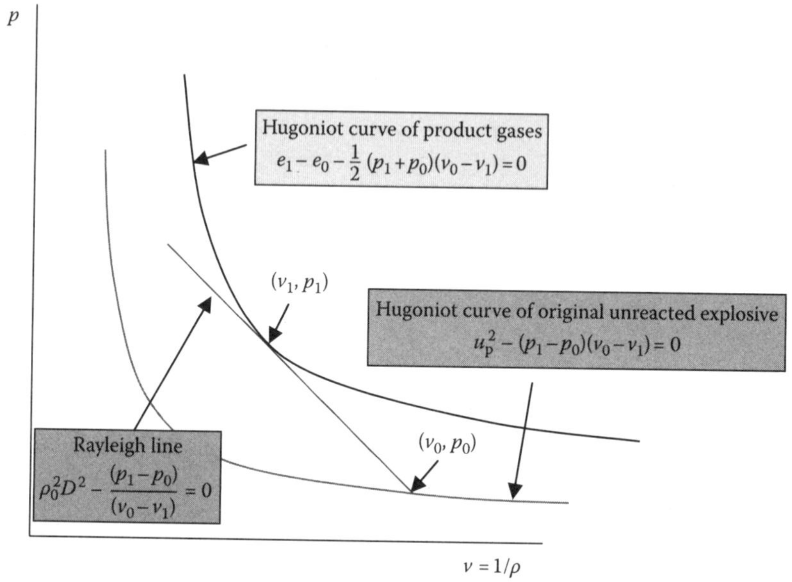

FIGURE 18.40
Hugoniots of unreacted and reacted explosive.

The energy equation then would be

$$e = C_v T - \lambda q \tag{18.248}$$

with

$$q = \Delta h_r^0 \tag{18.249}$$

where C_v is the (constant) specific heat at constant volume; T is the absolute temperature; q is the heat released from the reaction; and Δh_r^0 is the heat of reaction of the complete reaction.

In this equation, λ is a parameter that varies from 0 to 1 indicating the degree of reaction:

- $\lambda = 0$ means the reaction has not even begun.
- $\lambda = 1$ means the reaction is complete.

In this simplest model, there are only two states, 0 and 1.

We can rearrange Equation 18.247 as follows:

$$T = \frac{pv}{R} \tag{18.250}$$

If we recall the relationship between specific heat at constant volume and the gas constant as

$$R = C_v(\gamma - 1) \tag{18.251}$$

where γ is the ratio of specific heats, we can then say that

$$T = \frac{pv}{C_v(\gamma - 1)} \tag{18.252}$$

Inserting Equation 18.252 into Equation 18.248 yields

$$e = \frac{pv}{(\gamma - 1)} - \lambda q \tag{18.253}$$

Putting this result directly into our Hugoniot equation gives us

$$\frac{p_1 v_1}{(\gamma - 1)} - \frac{p_0 v_0}{(\gamma - 1)} - \lambda q - \frac{1}{2}(p_1 + p_0)(v_0 - v_1) = 0 \tag{18.254}$$

By defining

$$\mu^2 = \frac{(\gamma - 1)}{(\gamma + 1)} \tag{18.255}$$

we can express Equation 18.254 as

$$\left(\frac{p_1}{p_0} + \mu^2\right)\left(\frac{v_1}{v_0} - \mu^2\right) - 1 + \mu^4 - \mu^2 \frac{2\lambda q}{p_0 v_0} = 0 \tag{18.256}$$

This is the equation of a hyperbola in the $(v/v_0, p/p_0)$ plane centered at $v/v_0 = \mu^2$ and $p/p_0 = -\mu^2$. This is a Hugoniot curve that defines all possible end states of the detonation reaction. If this is simultaneously solved with a Rayleigh line (Equation 18.244), their intersection defines the state of the gas emerging from the reaction. The issue now is that the slope of the Rayleigh line is dependent upon the detonation velocity so one of three families of solutions exists:

- Two intersections of the Hugoniot by the Rayleigh line
- One intersection of the Hugoniot by the Rayleigh line
- No intersections of the Hugoniot by the Rayleigh line

This is depicted in Figure 18.41.

If the detonation wave speed D is sufficiently high, say, $D = D_1$, then there will be two intersections of the Rayleigh line with the Hugoniot. If the detonation wave speed D is sufficiently high, say, $D = D_{CJ}$, then there will be one intersection of the Rayleigh line with the Hugoniot. If the detonation wave speed D is sufficiently low, say, $D = D_2$, then there will be no intersection of the Rayleigh line with the Hugoniot. If there are no solutions, then the detonation will (under the assumptions of the model) die out. If there are two solutions, we generally call the upper solution the strong solution, and the lower one, the weak solution (S and W in Figure 18.41). If there is only one solution, we call this the Chapman–Jouguet (C–J) solution.

For the strong solution, any disturbance created behind the wave will overtake the wave. Examine the Hugoniot in Figure 18.41. The slope of a line tangent at S is greater than the detonation wave Rayleigh line; therefore, any disturbance will move faster than the detonation wave and will eventually catch up with it. Induced flow is subsonic relative to the wave (i.e., $c > D_1 - u$). In the weak solution, the induced velocity is supersonic with respect to the detonation wave. The slope of a line tangent at W is smaller than the detonation wave Rayleigh line; therefore, any disturbance will move slower than the detonation wave and will fall farther and farther behind. Induced flow is supersonic relative to the wave (i.e., $c < D_1 - u_p$).

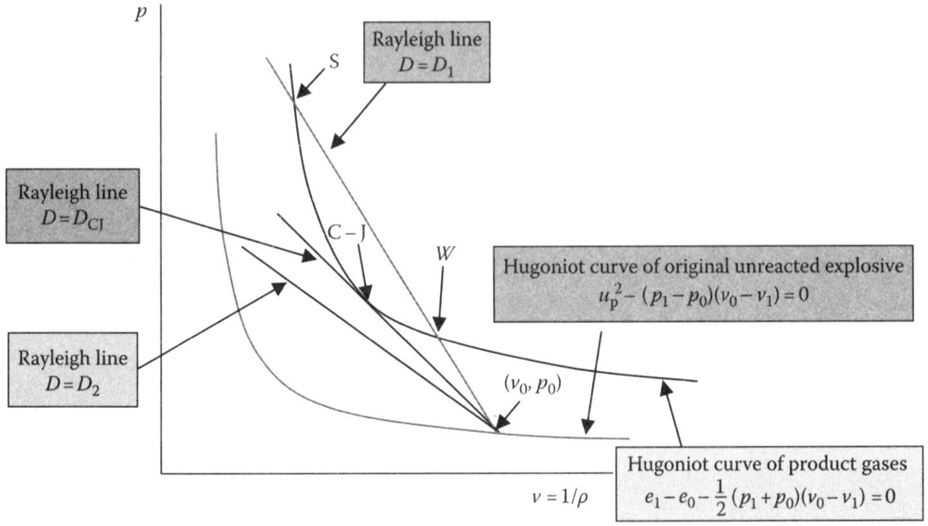

FIGURE 18.41
Possible intersections of Rayleigh line and Hugoniot curves.

For the C–J solution, any disturbance created behind the wave will maintain its distance from the wave. Once more, look at the Hugoniot of Figure 18.41. Since the line tangent at the C–J point is the Rayleigh line, any disturbance will propagate at the same speed as the detonation wave and will keep pace with it. Induced flow is sonic relative to the wave (i.e., $c = D_1 - u_p$). If we recall the slope of the Rayleigh line as

$$\left(\frac{dp}{dv}\right)_{Rayleigh} = -\frac{p_1 - p_0}{v_0 - v_1} \tag{18.257}$$

we shall divide our Hugoniot Equation 18.246 by $(v_0 - v_1)^2$ to obtain

$$\frac{(e_1 - e_0)}{(v_0 - v_1)^2} - \frac{1}{2}\frac{(p_1 + p_0)}{(v_0 - v_1)} = 0 \tag{18.258}$$

Now we multiply by 2 and separate the first term into

$$\frac{2\left(\frac{de}{dv}\right)_{Hugoniot}}{(v_0 - v_1)} + \frac{(p_1 + p_0)}{(v_0 - v_1)} = 0 \tag{18.259}$$

Let us distribute the negative sign on the second term to write

$$\frac{2\left(\frac{de}{dv}\right)_{Hugoniot}}{(v_0 - v_1)} + \frac{(-p_1 - p_0)}{(v_0 - v_1)} = 0 \tag{18.260}$$

We can add and subtract $p_1/(v_0 - v_1)$ to obtain

$$\frac{2\left(\frac{de}{dv}\right)_{Hugoniot}}{(v_0 - v_1)} + \frac{(-p_1 - p_0)}{(v_0 - v_1)} + \frac{2p_1}{(v_0 - v_1)} = \left(\frac{dp}{dv}\right)_{Hugoniot} \tag{18.261}$$

The only way for Equation 18.261 to equal Equation 18.257 is for

$$\left(\frac{de}{dv}\right)_{Hugoniot\text{-}CJ} = -p_1 \tag{18.262}$$

If we recall from thermodynamics that on an isentrope,

$$\left(\frac{de}{dv}\right)_s = -p \tag{18.263}$$

therefore, the Rayleigh line and Hugoniot curve lie on the isentrope at the C–J point. The implications of this are

$$\gamma = \frac{C_p}{C_v} = \frac{\left(1 - \frac{p_0}{p_1}\right)}{\left(\frac{v_o}{v_1} - 1\right)} \tag{18.264}$$

We can use this fact and assuming $p_0 \approx 0$ by substituting back into our Rayleigh and Hugoniot equations to state that at the C–J point, the following are true:

$$p_{CJ} = \frac{\rho_0 D^2}{(\gamma + 1)} \tag{18.265}$$

$$v_{CJ} = \frac{v_0 \gamma}{(\gamma + 1)} = \frac{1}{\rho_{CJ}} \tag{18.266}$$

$$u_{pCJ} = \frac{D}{(\gamma + 1)} \tag{18.267}$$

$$c_{CJ} = \frac{D\gamma}{(\gamma + 1)} \tag{18.268}$$

We have stated that in this simplest theory, the reaction occurs instantaneously. Thus, as soon as unreacted material passes through the detonation wave, it is instantaneously converted to a new material. We can determine this final state by the intersection of the Rayleigh line with the reacted material Hugoniot curve. In this theory, there are three cases we must consider: $D < D_{CJ}$, $D = D_{CJ}$, and $D > D_{CJ}$.

If $D < D_{CJ}$, the Rayleigh line does not intersect the Hugoniot curve of the reaction products and we will not have a steady reaction—the reaction will die out. If $D = D_{CJ}$, the Rayleigh line intersects the Hugoniot curve of the reaction products at one point; the detonation wave will continue to move into the unreacted material; and the detonation products will move away from the wave, relative to the wave, at the sonic velocity. There is only one solution—the reaction will be steady. If $D > D_{CJ}$, the Rayleigh line intersects the Hugoniot curve of the reaction products at two points (strong and weak). We will ignore the weak solution as inadmissible because the pressure will have to drop. For the strong solution, the detonation wave will continue to move into the unreacted material. In this case, the detonation products will move away from the wave, relative to the wave, at a subsonic velocity.

The speed of the reaction products u_p is also a parameter we must consider. Sometimes, this problem is known as the piston problem since we can imagine a piston pushing the reaction products at a speed u_p. Once we have determined the detonation velocity, we can then find u_p.

First, we shall examine a strong solution where

$$u_p > u_{p_{CJ}} \tag{18.269}$$

In this case, any decrease in piston velocity will generate a rarefaction wave that will catch up to the detonation wave and the flow will equilibrate to the new velocity. If we have a situation where

$$u_p = u_{p_{CJ}} \tag{18.270}$$

and there is a rarefaction generated, it cannot catch up to the front because it will move at the sonic velocity. If we have a situation where

$$u_p < u_{p_{CJ}} \tag{18.271}$$

then we need a rarefaction wave to reduce the flow velocity from the detonation wave speed at the front (which, recall, must move at a speed of at least D_{CJ}) to the speed of the piston. This rarefaction wave will be time dependent. If the piston was moving at zero velocity, then the tail of the rarefaction would stay attached to the detonation wave while the head of the rarefaction would remain about halfway between the detonation wave and the piston. This would be exactly halfway for a polytropic gas with $p_0 = 0$. In common problems, it will be typical to have the piston velocity less than or equal to zero. All these conditions are illustrated in Figure 18.42.

If we initiate a detonation at a point $x = 0$ and $t = 0$ and we have $u_p < u_{p_{CJ}}$, then a t–x plot of this situation would look like Figure 18.43. The detonation front would move at velocity D_{CJ}, and after a time $t = t_1$, it would be at position $x = D_{CJ}t_1$. There would also be a centered rarefaction wave that, at the same time, would move to position $x = u_p t_1$. This centered rarefaction wave is sometimes called a Taylor wave. A particle path is also depicted in the figure.

An equation of state is required to close the set of equations and solve a reacting flow problem. There are some equations of state that do not explicitly treat the chemical reaction. When we have such a case, empirical values are obtained for the relationships. Thus, each new reaction must be calibrated through an experiment. We shall look at an equation of state that does treat the reaction. In this case, all that is needed is the composition of the reactants, the initial density, and the heats of formation.

The Kistiakowsky–Wilson equation of state is given by

$$\frac{pv}{RT} = 1 + xe^{bx} \tag{18.272}$$

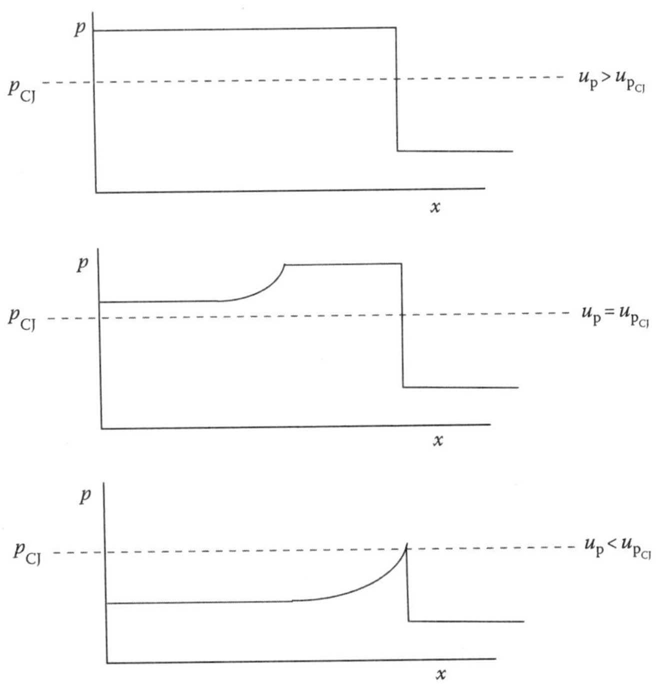

FIGURE 18.42
Varying behavior of explosive reaction products—overdriven detonation wave. (From Fickett, W., and Davis, W. C., *Detonation: Theory and Experiment*, Dover, New York, 1979. With permission.)

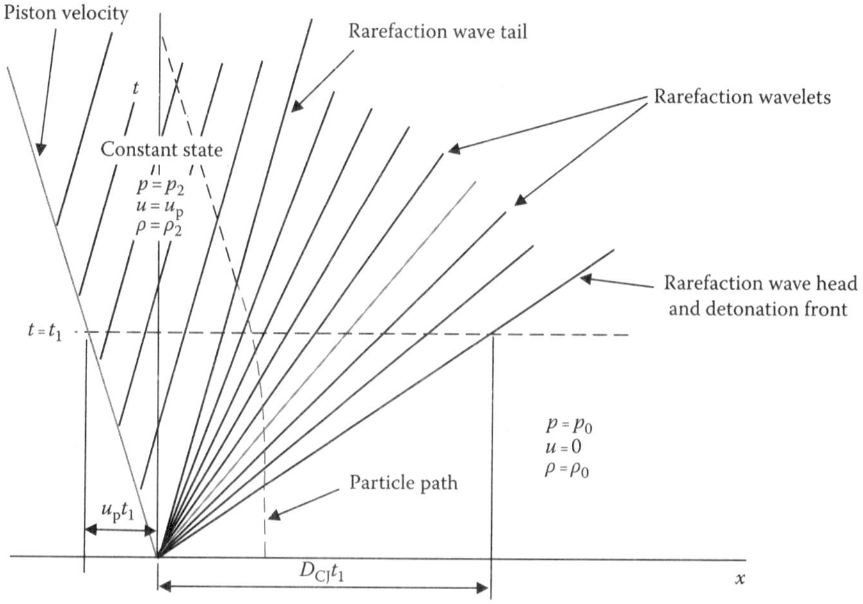

FIGURE 18.43
t–x diagram of reaction products—overdriven detonation wave. (From Fickett, W., and Davis, W. C., *Detonation: Theory and Experiment*, Dover, New York, 1979. With permission.)

Here

$$x = \frac{k}{v(T+\theta)^a} \tag{18.273}$$

where k is the effective mixture covolume determined through

$$k = \kappa \sum_{i=1}^{m} \chi_i k_i \tag{18.274}$$

In these equations, a, b, κ, θ, and k_i are empirical constants; k_i is the covolume of each species i; *and* χ_i is the mole fraction of each species i.

Unless better data are available, it is common to use $a = 0.25$, $b = 0.30$, $\kappa = 1$, and $\theta = 0$.

Kamlet and Jacobs empirically fit data to come up with the following definitions at the C–J state:

$$p_{\mathrm{CJ}} = \zeta \rho_0^2 \phi \tag{18.275}$$

$$D_{\mathrm{CJ}} = A\sqrt{\phi}(1 + B\rho_0) \tag{18.276}$$

$$\phi = N\sqrt{MW_{\mathrm{avg}} \Delta h_{\mathrm{r}}^0} \tag{18.277}$$

In these equations, ζ, A, and B are empirical constants in SI units (m, kg, s); $\zeta = 0.762$; $A = 22.3$; $B = 0.0013$; N is the number of moles per unit mass in kilogram-moles per kilogram; MW_{avg} is the average molecular weight of the gaseous products in kilograms per kilogram-mole; Δh_r^0 is the specific heat of reaction of the gaseous products in joules per kilogram; D will be in meters per second; and p will be in pascals if ρ_0 is in kilograms per cubic meter.

If there are no solids in the reaction products, then [4]

$$N = \frac{1}{MW_{avg}} \tag{18.278}$$

Equation 18.277 then reduces to

$$\phi = \sqrt{N\Delta h_r^0} \tag{18.279}$$

Equation 18.276 would then be

$$D_{CJ} = A(1 + B\rho_0)\left(\sqrt[4]{N\Delta h_r^0}\right) \tag{18.280}$$

Equation 18.275 would correspondingly be

$$p_{CJ} = \zeta\rho_0^2\sqrt{N\Delta h_r^0} \tag{18.281}$$

With this in mind, we shall now discuss a procedure for the simplest theory that allows us to calculate the behavior of the reaction.

To estimate reaction product behavior, we must first develop the balanced chemical reaction. With this, we need to estimate the heat of detonation. Usually, we know the heat of formation of the unreacted explosive. We then calculate the heat of formation of the gas mixture through

$$\Delta\bar{h}^0_{\text{product gas}} = \sum_i N_i \Delta\bar{h}_f^0 \tag{18.282}$$

At this point, we must guess at the ideal temperature of the explosive products. This guess is T_2^*. We next calculate the ideal ratio of specific heats through

$$\gamma = 1 + \frac{R}{C_v} \tag{18.283}$$

We also know that

$$R = \frac{R_u}{MW} \tag{18.284}$$

The universal gas constant is

$$R_u = 1.99\left[\frac{\text{cal}}{\text{g-mol}\cdot\text{K}}\right] \tag{18.285}$$

We can obtain the specific heat at constant volume through

$$C_v = A + BT \tag{18.286}$$

where the constants A and B are provided in Table 18.1. We can calculate the average specific heat of the products at our assumed temperature, and then use this value in Equation 18.283.

TABLE 18.1
Coefficients for Specific Heat at Constant-Volume Calculation

Molecule	Heat of Formation Δh_f^0 (cal/mol)	A	B	Covolume k (cm^3/g mol)
H_2	0	5.02	0.28	153
CO_2	94,450	10.30	0.42	687
CO	26,840	5.82	0.33	386
H_2O (g)	57,801	7.13	0.67	108
N_2	0	5.68	0.37	353
OH	5,930	5.20	0.26	108
O_2	0	5.86	0.28	333
NO	−21,600	6.00	0.15	233
C (s)	0	4.52	0.20	0

Note: C_v (cal/g mol K) = $A + B\,[T\,(K)]$

If we use our notation for averages and estimated values, Equation 18.283 becomes

$$\gamma_2^* = 1 + \frac{NR_u}{C_v^*} \tag{18.287}$$

If we recall the energy equation that we will rewrite as

$$\Delta e = \overline{C_v}(T_2 - T_1) - q \tag{18.288}$$

where

$$q = \Delta h_r^0 \tag{18.289}$$

we can rearrange this to

$$T_2 = \frac{\Delta e}{\overline{C_v}} + \frac{q}{\overline{C_v}} + T_1 \tag{18.290}$$

The first term on the RHS is the kinetic energy; the second is heat released. We know from the energy equation that

$$\Delta e = \frac{1}{2}(p_2 + p_1)(v_1 - v_2). \tag{18.291}$$

If we factor p_2 and v_2 out of Equation 18.291, we get

$$\Delta e = \frac{1}{2}p_2 v_2\left(1 + \frac{p_1}{p_2}\right)\left(\frac{v_1}{v_2} - 1\right) \tag{18.292}$$

If we state here that $p_1 \ll p_2$, we can write

$$\Delta e = \frac{1}{2}p_2 v_2\left(\frac{v_1}{v_2} - 1\right) \tag{18.293}$$

Recall our definition of the specific heat ratio

$$\gamma \equiv \frac{C_p}{C_v} = \frac{\left(1 - \dfrac{p_1}{p_2}\right)}{\left(\dfrac{v_1}{v_2} - 1\right)} \tag{18.294}$$

Again, if $p_1 \ll p_2$, we can write

$$\gamma \equiv \frac{C_p}{C_v} = \frac{1}{\left(\dfrac{v_1}{v_2} - 2\right)} \tag{18.295}$$

Substitution of Equation 18.295 into 16.293 yields

$$\Delta e = \frac{1}{2}\frac{p_2 v_2}{\gamma} \tag{18.296}$$

If we now use the ideal gas relation, we obtain

$$\Delta e = \frac{1}{2}\frac{NR_u T_2}{\gamma} \tag{18.297}$$

We can now write Equation 18.290 as

$$T_2 = \frac{\dfrac{1}{2}NR_u T_2}{\gamma \overline{C_v}} + \frac{q}{\overline{C_v}} + T_1 \tag{18.298}$$

Now we can use Equation 18.297 to estimate T_2^*:

$$T_2^* = \frac{\dfrac{1}{2}NR_u T_2^*}{\gamma_2^* \overline{C_v}^*} + \frac{q}{\overline{C_v}^*} + T_1 \tag{18.299}$$

To use this equation, we substitute our guessed temperature into the RHS with our calculated γ_2^* and C_v^*. If the LHS comes out reasonably close to the RHS, we are done and our guess was correct. If it does not agree, we use the new value to calculate a new γ_2^* and C_v^* and repeat the process until the solution converges.

To determine the detonation velocity, recall Equation 18.244 that we can rearrange as

$$D = v_0 \sqrt{\frac{(p_1 - p_0)}{(v_0 - v_1)}} \tag{18.300}$$

We can factor this equation and use our definition of γ to make it look as follows:

$$D = v_0 \sqrt{\frac{p_1\left(1 - \dfrac{p_0}{p_1}\right)}{v_1\left(\dfrac{v_0}{v_1} - 1\right)}} = v_0 \sqrt{\frac{p_1 \gamma}{v_1}} \tag{18.301}$$

If we multiply and divide the inside by v_1^2, we obtain

$$D = \frac{v_0}{v_1}\sqrt{p_1 v_1 \gamma} \tag{18.302}$$

We can use Equation 18.266 to alter v_0/v_1 to yield

$$D = \frac{(\gamma + 1)}{\gamma}\sqrt{p_1 v_1 \gamma} \tag{18.303}$$

This can be rearranged as

$$D = (\gamma + 1)\sqrt{\frac{p_1 v_1}{\gamma}} \tag{18.304}$$

And inserting the ideal gas equation of state, we obtain

$$D = (\gamma + 1)\sqrt{\frac{NRT}{\gamma}} \tag{18.305}$$

We can now calculate the ideal detonation velocity D^* through

$$D^* = (\gamma_2^* + 1)\sqrt{\frac{NR_u T_2^*}{\gamma_2^* \left(MW_{\text{explosive}}\right)}} \tag{18.306}$$

Once we have these ideal values T_2^*, γ_2^*, and D^*, we need to calculate the real values based upon the covolume correction of Equation 18.272. Using Table 18.1, we determine a covolume for the product gas mixture through

$$k = \sum_i N_i k_i \tag{18.307}$$

Now we find our correction factor x_1 from Equation 18.273 modified as

$$x_1 = \frac{k}{v_2 (T_2^*)^a} \tag{18.308}$$

We can now use Tables 18.2 through 18.5 with interpolation to obtain $\frac{D}{D^*}$, $\frac{T_2}{T_2^*}$, x_2, and $\frac{\gamma_2}{\gamma_2^*}$. These are the actual (nonideal) detonation wave velocity, temperature, and specific heat ratio. To determine the pressure, we now can use

$$p_2 = \rho_0 D^2 \left(1 - \frac{x_1}{x_2}\right) \tag{18.309}$$

and to find the induced or material velocity, we use

$$u_P = D\left(1 - \frac{x_1}{x_2}\right) \tag{18.310}$$

TABLE 18.2

Specific Heat Ratio Table for Simple Formula Calculation

	$\gamma_2^* = 1.15$		$\gamma_2^* = 1.19$		$\gamma_2^* = 1.23$		$\gamma_2^* = 1.27$		$\gamma_2^* = 1.31$		$\gamma_2^* = 1.35$	
x_1	γ_2/γ_2^*	Ø	γ_2/γ_2^*	Ø	γ_2/γ_2^*	Ø	γ_2/γ_2^*	Ø	γ_2/γ_2^*	Ø	γ_2/γ_2^*	Ø
0.1	0.991	−0.004	0.989	−0.006	0.987	−0.007	0.985	−0.008	0.984	−0.010	0.982	−0.011
0.2	0.987	−0.003	0.983	−0.004	0.980	−0.004	0.977	−0.005	0.974	−0.006	0.971	−0.007
0.3	0.984	−0.001	0.979	−0.001	0.976	−0.002	0.972	−0.003	0.968	−0.004	0.964	−0.004
0.4	0.983	−0.001	0.978	−0.001	0.974	−0.002	0.969	−0.002	0.964	−0.002	0.960	−0.003
0.5	0.982	0.000	0.977	0.000	0.972	−0.001	0.967	−0.001	0.962	−0.001	0.957	−0.001
0.6	0.982	0.000	0.977	0.000	0.971	0.000	0.966	0.000	0.961	−0.001	0.956	−0.001
0.7	0.982	0.000	0.977	0.000	0.971	0.000	0.966	−0.001	0.960	−0.001	0.955	−0.001
0.8	0.982	0.001	0.977	0.000	0.971	0.000	0.965	0.000	0.959	0.000	0.954	−0.001
0.9	0.983	0.000	0.977	0.001	0.971	0.001	0.965	0.001	0.959	0.000	0.953	0.000
1.0	0.983	0.001	0.978	0.000	0.972	0.000	0.966	0.000	0.959	0.000	0.953	−0.001
1.1	0.984	0.001	0.978	0.001	0.972	0.001	0.966	0.000	0.959	0.000	0.952	0.000
1.2	0.985	0.001	0.979	0.001	0.973	0.000	0.966	0.000	0.959	0.000	0.952	0.000
1.3	0.986	0.000	0.980	0.000	0.973	0.001	0.966	0.001	0.959	0.001	0.952	0.000
1.4	0.986	0.001	0.980	0.001	0.974	0.000	0.967	0.000	0.960	0.000	0.952	0.000
1.5	0.987	0.001	0.981	0.000	0.974	0.001	0.967	0.000	0.960	0.000	0.952	−0.001
1.6	0.988	0.000	0.981	0.001	0.975	0.000	0.967	0.000	0.960	−0.001	0.951	0.000
1.7	0.988	0.001	0.982	0.000	0.975	0.000	0.967	0.000	0.959	0.000	0.951	−0.001
1.8	0.989	0.000	0.982	0.001	0.975	0.000	0.967	0.000	0.959	0.000	0.950	0.000
1.9	0.989	0.001	0.983	0.000	0.975	0.000	0.967	0.000	0.959	−0.001	0.950	−0.001
2.0	0.990	0.000	0.983	0.000	0.975	0.000	0.967	0.000	0.958	0.000	0.949	0.000
2.1	0.990	0.000	0.983	0.000	0.975	0.000	0.967	0.000	0.958	−0.001	0.949	−0.001
2.2	0.990	0.000	0.983	0.000	0.975	0.000	0.967	−0.001	0.957	−0.001	0.948	−0.001
2.3	0.990	0.000	0.983	0.000	0.975	0.000	0.966	0.000	0.956	0.000	0.947	−0.001
2.4	0.990	0.000	0.983	0.000	0.975	−0.001	0.966	−0.001	0.956	−0.001	0.946	−0.002
2.5	0.990	0.000	0.983	0.000	0.974	0.000	0.965	−0.001	0.955	−0.002	0.944	−0.002
2.6	0.990	0.000	0.983	−0.001	0.974	−0.001	0.964	−0.001	0.953	−0.001	0.942	−0.001
2.7	0.990	0.000	0.982	0.000	0.973	−0.001	0.963	−0.001	0.952	−0.001	0.941	−0.002
2.8	0.990	0.000	0.982	−0.001	0.972	−0.001	0.962	−0.001	0.951	−0.002	0.939	−0.002
2.9	0.990	−0.001	0.981	0.000	0.971	−0.001	0.961	−0.002	0.949	−0.002	0.937	−0.002
3.0	0.989	0.000	0.981	−0.001	0.970	−0.001	0.959	−0.001	0.947	−0.001	0.935	−0.002
3.1	0.989	−0.001	0.980	−0.001	0.969	−0.001	0.958	−0.002	0.946	−0.002	0.933	−0.003
3.2	0.988	0.000	0.979	−0.001	0.968	−0.002	0.956	−0.002	0.944	−0.002	0.930	−0.002
3.3	0.988	−0.001	0.978	−0.001	0.966	−0.001	0.954	−0.002	0.942	−0.003	0.928	−0.002
3.4	0.987	−0.001	0.977	−0.002	0.965	−0.002	0.952	−0.002	0.939	−0.002	0.926	−0.003

TABLE 18.3

Temperature Ratio Table for Simple Formula Calculation

	$\gamma_2^* = 1.15$		$\gamma_2^* = 1.19$		$\gamma_2^* = 1.23$		$\gamma_2^* = 1.27$		$\gamma_2^* = 1.31$		$\gamma_2^* = 1.35$	
x_1	T_2/T_2^*	Ø	T_2/T_2^*	Ø	T_2/T_2^*	Ø	T_2/T_2^*	Ø	T_2/T_2^*	Ø	T_2/T_2^*	Ø
0.1	0.994	−0.005	0.992	−0.006	0.991	−0.008	0.989	−0.009	0.988	−0.011	0.986	−0.012
0.2	0.989	−0.005	0.986	−0.006	0.983	−0.007	0.980	−0.008	0.977	−0.010	0.974	−0.011
0.3	0.984	−0.004	0.980	−0.006	0.976	−0.007	0.972	−0.009	0.967	−0.009	0.963	−0.011
0.4	0.980	−0.004	0.974	−0.005	0.969	−0.007	0.963	−0.008	0.958	−0.010	0.952	−0.011
0.5	0.976	−0.005	0.969	−0.006	0.962	−0.007	0.955	−0.008	0.948	−0.009	0.941	−0.011
0.6	0.971	−0.004	0.963	−0.006	0.955	−0.007	0.947	−0.009	0.939	−0.010	0.930	−0.011
0.7	0.967	−0.005	0.957	−0.006	0.948	−0.008	0.938	−0.009	0.929	−0.011	0.919	−0.012
0.8	0.962	−0.005	0.951	−0.006	0.940	−0.008	0.929	−0.009	0.918	−0.010	0.907	−0.012
0.9	0.957	−0.005	0.945	−0.007	0.932	−0.008	0.920	−0.010	0.908	−0.011	0.895	−0.012
1.0	0.952	−0.006	0.938	−0.007	0.924	−0.008	0.910	−0.010	0.897	−0.012	0.883	−0.013
1.1	0.946	−0.006	0.931	−0.007	0.916	−0.009	0.900	−0.010	0.885	−0.012	0.870	−0.013
1.2	0.940	−0.006	0.924	−0.008	0.907	−0.010	0.890	−0.011	0.873	−0.012	0.857	−0.014
1.3	0.934	−0.006	0.916	−0.008	0.897	−0.010	0.879	−0.011	0.861	−0.013	0.843	−0.014
1.4	0.928	−0.007	0.908	−0.009	0.887	−0.010	0.868	−0.012	0.848	−0.013	0.829	−0.015
1.5	0.921	−0.007	0.899	−0.009	0.877	−0.010	0.856	−0.012	0.835	−0.014	0.814	−0.015
1.6	0.914	−0.008	0.890	−0.009	0.867	−0.011	0.844	−0.013	0.821	−0.014	0.799	−0.015
1.7	0.906	−0.008	0.881	−0.010	0.856	−0.012	0.831	−0.013	0.807	−0.014	0.784	−0.015
1.8	0.898	−0.008	0.871	−0.010	0.844	−0.012	0.818	−0.013	0.793	−0.015	0.769	−0.016
1.9	0.890	−0.009	0.861	−0.011	0.832	−0.012	0.805	−0.014	0.778	−0.015	0.753	−0.016
2.0	0.881	−0.009	0.850	−0.011	0.820	−0.013	0.791	−0.014	0.763	−0.015	0.737	−0.017
2.1	0.872	−0.009	0.839	−0.011	0.807	−0.013	0.777	−0.014	0.748	−0.016	0.720	−0.016
2.2	0.863	−0.010	0.828	−0.012	0.794	−0.013	0.763	−0.015	0.732	−0.015	0.704	−0.017
2.3	0.853	−0.010	0.816	−0.012	0.781	−0.014	0.748	−0.015	0.717	−0.016	0.687	−0.017
2.4	0.843	−0.011	0.804	−0.012	0.767	−0.014	0.733	−0.015	0.701	−0.017	0.670	−0.017
2.5	0.832	−0.011	0.792	−0.013	0.753	−0.014	0.718	−0.016	0.684	−0.016	0.653	−0.017
2.6	0.821	−0.011	0.779	−0.013	0.739	−0.014	0.702	−0.015	0.668	−0.016	0.636	−0.017
2.7	0.810	−0.011	0.766	−0.013	0.725	−0.015	0.687	−0.016	0.652	−0.017	0.619	−0.017
2.8	0.799	−0.012	0.753	−0.014	0.710	−0.015	0.671	−0.016	0.635	−0.016	0.602	−0.017
2.9	0.787	−0.012	0.739	−0.014	0.695	−0.015	0.655	−0.016	0.619	−0.017	0.585	−0.017
3.0	0.775	−0.013	0.725	−0.014	0.680	−0.015	0.639	−0.016	0.602	−0.017	0.568	−0.017
3.1	0.762	−0.012	0.711	−0.014	0.665	−0.015	0.623	−0.016	0.585	−0.016	0.551	−0.017
3.2	0.750	−0.013	0.697	−0.015	0.650	−0.016	0.607	−0.016	0.569	−0.017	0.534	−0.016
3.3	0.737	−0.014	0.682	−0.015	0.634	−0.016	0.591	−0.016	0.552	−0.016	0.518	−0.017
3.4	0.723	−0.013	0.667	−0.015	0.618	−0.016	0.575	−0.016	0.536	−0.016	0.501	−0.016

TABLE 18.4

Detonation Velocity Ratio Table for Simple Formula Calculation

	$\gamma_2^* = 1.15$		$\gamma_2^* = 1.19$		$\gamma_2^* = 1.23$		$\gamma_2^* = 1.27$		$\gamma_2^* = 1.31$		$\gamma_2^* = 1.35$	
x_1	D_2/D_2^*	Ø	D_2/D_2^*	Ø	D_2/D_2^*	Ø	D_2/D_2^*	Ø	D_2/D_2^*	Ø	D_2/D_2^*	Ø
0.1	1.094	0.090	1.093	0.089	1.090	0.089	1.090	0.087	1.089	0.086	1.088	0.085
0.2	1.184	0.088	1.182	0.087	1.179	0.087	1.177	0.086	1.175	0.085	1.173	0.083
0.3	1.272	0.088	1.269	0.087	1.266	0.086	1.263	0.084	1.260	0.083	1.256	0.082
0.4	1.360	0.088	1.356	0.087	1.352	0.085	1.347	0.084	1.343	0.083	1.338	0.082
0.5	1.448	0.088	1.443	0.087	1.437	0.085	1.431	0.084	1.426	0.082	1.420	0.081
0.6	1.536	0.089	1.530	0.087	1.522	0.086	1.515	0.084	1.508	0.082	1.501	0.080
0.7	1.625	0.089	1.617	0.087	1.608	0.086	1.599	0.084	1.590	0.082	1.581	0.081
0.8	1.714	0.090	1.704	0.088	1.694	0.086	1.683	0.084	1.672	0.082	1.662	0.080
0.9	1.804	0.090	1.792	0.088	1.780	0.086	1.767	0.084	1.754	0.082	1.742	0.079
1.0	1.894	0.091	1.880	0.089	1.866	0.086	1.851	0.084	1.836	0.082	1.821	0.080
1.1	1.985	0.092	1.969	0.089	1.952	0.087	1.935	0.085	1.918	0.082	1.901	0.079
1.2	2.077	0.093	2.058	0.090	2.039	0.087	2.020	0.084	2.000	0.081	1.980	0.078
1.3	2.170	0.094	2.148	0.091	2.126	0.088	2.104	0.084	2.081	0.081	2.058	0.079
1.4	2.264	0.095	2.239	0.091	2.214	0.088	2.188	0.085	2.162	0.081	2.137	0.078
1.5	2.359	0.095	2.330	0.092	2.302	0.087	2.273	0.084	2.243	0.081	2.215	0.077
1.6	2.454	0.096	2.422	0.092	2.389	0.088	2.357	0.084	2.324	0.081	2.292	0.077
1.7	2.550	0.097	2.514	0.092	2.477	0.088	2.441	0.084	2.405	0.080	2.369	0.076
1.8	2.647	0.097	2.606	0.093	2.565	0.088	2.525	0.084	2.485	0.079	2.445	0.075
1.9	2.744	0.098	2.699	0.093	2.653	0.088	2.609	0.083	2.564	0.079	2.520	0.074
2.0	2.842	0.099	2.792	0.093	2.741	0.088	2.692	0.082	2.643	0.078	2.594	0.074
2.1	2.941	0.099	2.885	0.093	2.829	0.088	2.774	0.082	2.721	0.077	2.668	0.073
2.2	3.040	0.100	2.978	0.093	2.917	0.087	2.856	0.082	2.798	0.076	2.741	0.071
2.3	3.140	0.100	3.071	0.093	3.004	0.087	2.938	0.081	2.874	0.076	2.812	0.071
2.4	3.240	0.100	3.164	0.093	3.091	0.086	3.019	0.080	2.950	0.074	2.883	0.069
2.5	3.340	0.100	3.257	0.093	3.177	0.086	3.099	0.080	3.024	0.074	2.952	0.069
2.6	3.440	0.101	3.350	0.092	3.263	0.085	3.179	0.078	3.098	0.072	3.021	0.067
2.7	3.541	0.100	3.442	0.092	3.348	0.084	3.257	0.077	3.170	0.072	3.088	0.066
2.8	3.641	0.100	3.534	0.092	3.432	0.083	3.334	0.076	3.242	0.070	3.154	0.064
2.9	3.741	0.100	3.626	0.091	3.515	0.082	3.410	0.076	3.312	0.069	3.218	0.063
3.0	3.841	0.100	3.717	0.090	3.597	0.082	3.486	0.075	3.381	0.067	3.281	0.061
3.1	3.941	0.100	3.807	0.089	3.679	0.081	3.561	0.071	3.448	0.066	3.342	0.060
3.2	4.041	0.099	3.896	0.089	3.760	0.080	3.632	0.072	3.514	0.065	3.402	0.059
3.3	4.140	0.099	3.985	0.088	3.840	0.078	3.704	0.071	3.579	0.064	3.461	0.057
3.4	4.239	0.098	4.073	0.086	3.918	0.077	3.775	0.069	3.643	0.062	3.518	0.056

TABLE 18.5

Correction Factor Table for Simple Formula Calculation

	$\gamma_2^* = 1.15$		$\gamma_2^* = 1.19$		$\gamma_2^* = 1.23$		$\gamma_2^* = 1.27$		$\gamma_2^* = 1.31$		$\gamma_2^* = 1.35$	
x_1	x_2	Ø	x_2	Ø	x_2	Ø	x_2	Ø	x_2	Ø	x_2	Ø
0.1	0.177	0.159	0.173	0.159	0.171	0.157	0.169	0.156	0.167	0.155	0.165	0.154
0.2	0.336	0.152	0.332	0.150	0.328	0.150	0.325	0.148	0.322	0.147	0.319	0.146
0.3	0.488	0.145	0.482	0.145	0.478	0.144	0.473	0.143	0.469	0.142	0.465	0.141
0.4	0.633	0.142	0.627	0.141	0.622	0.139	0.616	0.139	0.611	0.138	0.606	0.138
0.5	0.775	0.139	0.768	0.137	0.761	0.137	0.755	0.136	0.749	0.135	0.744	0.134
0.6	0.914	0.135	0.905	0.136	0.898	0.134	0.891	0.133	0.884	0.133	0.878	0.132
0.7	1.049	0.133	1.041	0.132	1.032	0.132	1.024	0.131	1.017	0.131	1.010	0.130
0.8	1.182	0.132	1.173	0.131	1.164	0.130	1.155	0.130	1.148	0.129	1.140	0.129
0.9	1.314	0.130	1.304	0.129	1.294	0.129	1.285	0.128	1.277	0.127	1.269	0.127
1.0	1.444	0.129	1.433	0.128	1.423	0.127	1.413	0.127	1.404	0.127	1.396	0.126
1.1	1.573	0.128	1.561	0.127	1.550	0.126	1.540	0.126	1.531	0.125	1.522	0.125
1.2	1.701	0.126	1.688	0.126	1.676	0.126	1.666	0.124	1.656	0.124	1.647	0.124
1.3	1.827	0.126	1.814	0.125	1.802	0.124	1.790	0.124	1.780	0.124	1.771	0.123
1.4	1.953	0.124	1.939	0.124	1.926	0.124	1.914	0.123	1.904	0.122	1.894	0.122
1.5	2.077	0.124	2.063	0.123	2.050	0.122	2.037	0.123	2.026	0.122	2.016	0.122
1.6	2.201	0.123	2.186	0.122	2.172	0.122	2.160	0.122	2.148	0.122	2.138	0.121
1.7	2.324	0.122	2.308	0.122	2.294	0.122	2.282	0.121	2.270	0.121	2.259	0.121
1.8	2.446	0.122	2.430	0.122	2.416	0.121	2.403	0.120	2.391	0.120	2.380	0.120
1.9	2.568	0.121	2.552	0.121	2.537	0.120	2.523	0.120	2.511	0.120	2.500	0.119
2.0	2.689	0.121	2.673	0.120	2.657	0.120	2.643	0.120	2.631	0.119	2.619	0.119
2.1	2.810	0.120	2.793	0.120	2.777	0.120	2.763	0.119	2.750	0.119	2.738	0.119
2.2	2.930	0.120	2.913	0.119	2.897	0.119	2.882	0.119	2.869	0.119	2.857	0.119
2.3	3.050	0.119	3.032	0.119	3.016	0.119	3.001	0.119	2.988	0.118	2.976	0.118
2.4	3.169	0.119	3.151	0.119	3.135	0.118	3.120	0.118	3.106	0.118	3.094	0.118
2.5	3.288	0.119	3.270	0.118	3.253	0.118	3.238	0.118	3.224	0.118	3.212	0.118
2.6	3.407	0.118	3.388	0.118	3.371	0.118	3.356	0.118	3.342	0.118	3.330	0.118
2.7	3.525	0.118	3.506	0.118	3.489	0.118	3.474	0.117	3.460	0.118	3.448	0.117
2.8	3.643	0.118	3.624	0.117	3.607	0.117	3.591	0.117	3.578	0.117	3.565	0.117
2.9	3.761	0.117	3.741	0.117	3.724	0.117	3.708	0.117	3.695	0.117	3.682	0.117
3.0	3.878	0.117	3.858	0.117	3.841	0.117	3.825	0.117	3.812	0.117	3.799	0.117
3.1	3.995	0.117	3.975	0.117	3.958	0.117	3.942	0.117	3.929	0.117	3.916	0.117
3.2	4.112	0.117	4.092	0.117	4.075	0.117	4.059	0.117	4.046	0.117	4.033	0.117
3.3	4.229	0.117	4.209	0.117	4.192	0.116	4.176	0.116	4.163	0.116	4.150	0.116
3.4	4.346	0.116	4.326	0.116	4.308	0.116	4.292	0.116	4.279	0.116	4.266	0.116

While more realistic models exist for examining detonation, we will refer the interested reader to the references for further study.

Problem 11

Tetryl ($C_7H_5N_5O_8$) is detonated in standard sea level air. Assuming nonideal behavior and

$$\rho = 0.86 \left[\frac{\text{g}}{\text{cm}^3}\right], \quad MW_{\text{mix}} = 213 \left[\frac{\text{g}}{\text{g-mol}}\right], \quad \text{and} \quad \Delta\bar{h}_f^0 = +4.67 \left[\frac{\text{kcal}}{\text{g-mol}}\right]$$

1. Determine the reaction equation assuming no dissociation
2. Determine the temperature of the products behind the detonation wave T_2
   Answer: $T_2 = 3308$ [K]
3. Determine the speed of the detonation wave D
   Answer: $D = 4742 \left[\frac{\text{m}}{\text{s}}\right]$
4. Determine the pressure behind the detonation wave p_2
   Answer: $p_2 = 5.29$ [GPa]
5. Determine the induced velocity of the gas behind the wave u_2
   Answer: $u_2 = 1296 \left[\frac{\text{m}}{\text{s}}\right]$

18.5 Explosive's Equations of State

Modeling the detonation and postdetonation dynamics produced by high explosives relies on accurate descriptions of the high-explosive behavior. Often such modeling is done using high-rate continuum models. In this modeling, the equation of state for the detonation products is the primary modeling description of the work output from the explosive that causes the subsequent effects. The Jones–Wilkins–Lee (JWL) equation of state for detonation products is probably the currently most used equation of state for detonation and blast modeling. The Jones–Wilkins–Lee–Baker (JWLB) equation of state is an extension of the JWL equation of state that is also commonly used. This section provides a thermodynamic and mathematical background of the JWL and JWLB equations of state, as well as parameterization methodology.

18.5.1 JWL Equation of State

The JWL thermodynamic equation of state [5] was developed to provide an accurate description of highly explosive product expansion work output and detonation C–J state. For blast applications, it is vital that the total work output from the detonation state to high expansion of the detonation products be accurate for the production of appropriate blast energy. The JWL mathematical form is

$$p = A\left(1 - \frac{\Gamma}{R_1 V^*}\right)\exp\left[-R_1 V^*\right] + B\left(1 - \frac{\Gamma}{R_2 V^*}\right)\exp\left[-R_2 V^*\right] + \frac{\Gamma E}{V^*} \tag{18.311}$$

where V^* is the relative volume; E is the product of the initial density and specific internal energy; and Γ is the Grüneisen parameter.

The equation of state is based upon a first-order expansion in energy of the principal isentrope. The JWL principle isentrope form is

$$p_s \equiv A\exp[-R_1V^*] + B\exp[-R_2V^*] + V^{*-(\Gamma+1)} \tag{18.312}$$

For JWL, the Grüneisen parameter is defined to be a constant:

$$\Gamma \equiv \left.\frac{V^*dp}{dE}\right|_{V^*} \tag{18.313}$$

Energy along the principal isentrope is calculated through the isentropic identity:

$$dE_s = -p_s\,dV^* \tag{18.314}$$

Inserting Equation 18.312 and integrating yields

$$E_s = \frac{A}{R_1}\exp[-R_1V^*] + \frac{B}{R_2}\exp[-R_2V^*] + \frac{C}{\Gamma V^{*\Gamma}} \tag{18.315}$$

This relationship defines the internal energy referencing for consistency, so that the initial internal energy release is

$$E_0 = E_{CJ} - \frac{1}{2}p_{CJ}(V_0^* - V_{CJ}^*) \tag{18.316}$$

The general equation of state is derived from the first-order expansion in energy of the principal isentrope:

$$p = p_s + \left.\frac{dp}{dE}\right|_{V^*}(E - E_S) = p_S + \frac{\Gamma}{V^*}(E - E_S) \tag{18.317}$$

Combining Equations 18.312, 18.314, and 18.317 results in Equation 18.311, repeated here for reference:

$$p = A\left(1 - \frac{\Gamma}{R_1V^*}\right)\exp\left[-R_1V^*\right] + B\left(1 - \frac{\Gamma}{R_2V^*}\right)\exp\left[-R_2V^*\right] + \frac{\Gamma E}{V^*}$$

From Equations 18.314 through 18.316, it can be seen that E_0 represents the total work output along the principal isentrope. For blast, this would represent the total available blast energy from the explosive.

18.5.2 JWLB Equation of State

The JWLB thermodynamic equation of state [6] is an extension of the JWL equation of state. JWLB was developed to more accurately describe overdriven detonation, while maintaining an accurate description of highly explosive products expansion work output and detonation C–J state. The equation of state is more mathematically complex than the JWL equation of state, as it includes an increased number of parameters to describe the principle isentrope, as well as a Grüneisen parameter formulation that is a function of specific volume. The increased mathematical complexity of the JWLB highly explosive equations of state provides increased accuracy for practical problems of interest. The JWLB mathematical form is

$$p = \sum_n A_i\left(1 - \frac{\Gamma}{R_iV^*}\right)\exp[-R_iV^*] + \frac{\lambda E}{V^*} \tag{18.318}$$

$$\lambda = \sum_i (A_{\lambda i} V^* + B_{\lambda i}) e^{-R_{\lambda i} V^*} + \Gamma \quad (18.319)$$

where V^* is the relative volume; E is the product of the initial density and specific internal energy; and λ is the Grüneisen parameter.

The JWL equation of state may be viewed as a subset of the JWLB equation of state where two inverse exponentials are used to describe the principal isentrope ($n = 2$), and the Grüneisen parameter is taken to be a constant ($\lambda = \Gamma$).

18.5.3 Analytic Cylinder Model

An analytic cylinder test model that uses JWL or JWLB equations of state has been developed, which provides excellent agreement with high-rate continuum modeling. Gurney formulation has often been used for highly explosive material acceleration modeling [7], particularly for liner acceleration applications. The work of Taylor [8] provides a more fundamental methodology for modeling exploding cylinders, including axial flow effects by Reynolds hydraulic formulation. A modification of this method includes radial detonation product flow effects and cylinder thinning. The modifications were found to give better agreement with cylinder expansion finite element modeling [9]. One method of including radial flow effects is to assume spherical surfaces of constant thermodynamic properties and mass flow in the detonation products. The detonation products mass flow is assumed to be in a perpendicular direction to the spherical surfaces. A diagram of a products constant spherical surfaces cylinder expansion due to highly explosive detonation is presented in Figure 18.44. It should be noted that flow velocities are relative to the detonation velocity D. If constant detonation product properties are assumed across spherical surfaces, the following model results using the JWLB thermodynamic equation of state.

The equation for mass conservation can be written as

$$\rho_{CJ} U_{CJ} A_0 = \rho U A \quad (18.320)$$

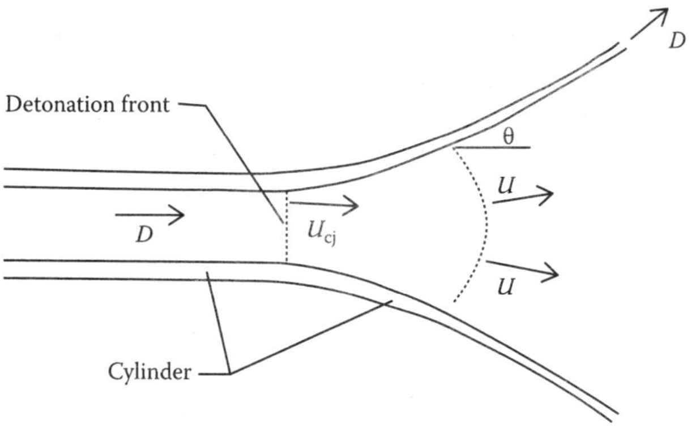

FIGURE 18.44
Diagram of the analytic cylinder expansion model.

The conservation of momentum equation can be expressed as

$$p_{CJ}r_0^2 - pr^2 = \frac{m}{\pi}D^2\cos\theta - \frac{m}{\pi}D^2 + \rho U^2 r^2 - \rho_{CJ}U_{CJ}^2 r_0^2 \tag{18.321}$$

The conservation of energy can be written, in intensive form, as

$$\rho_{CJ}U_{CJ}A_0\left(\frac{U_{CJ}^2}{2} + e_{CJ}\right) + p_{CJ}U_{CJ}A_0 = \rho UA\left(\frac{U^2}{2} + e\right) + pUA \tag{18.322}$$

Assuming an isentropic process, we can express the pressure in terms of the principal isentrope as

$$p = \sum_i A_i \exp\left[\frac{-R_i\rho_0}{\rho}\right] + C\left(\frac{\rho_0}{\rho}\right)^{-(\Gamma+1)} \tag{18.323}$$

where

$$de = -p dv = -p d\left(\frac{1}{\rho}\right) \tag{18.324}$$

We can express the Taylor angle (to be discussed in detail in the next chapter)

$$\alpha_T = \frac{V}{2D} = \sin\frac{\theta}{2} \tag{18.325}$$

We can write an expression for the area of a sphere as

$$A = \pi r^2 \frac{2(1 - \cos\theta)}{\sin^2\theta} \tag{18.326}$$

Inserting Equation 18.323 into Equation 18.324 and integrating yields

$$e_{CJ} - e = \sum_i \frac{A_i}{\rho_0 R_i}\left\{\exp\left[\frac{-R_i\rho_0}{\rho_{CJ}}\right] - \exp\left[\frac{-R_i\rho_0}{\rho}\right]\right\} + \frac{C}{\Gamma\rho_0}\left\{\left(\frac{\rho_0}{\rho_{CJ}}\right)^{-\Gamma} - \left(\frac{\rho_0}{\rho}\right)^{-\Gamma}\right\} \tag{18.327}$$

Using Equation 18.323 in Equation 18.322 yields

$$\frac{U^2}{2} = \frac{U_{CJ}^2}{2} + \frac{p_{CJ}}{\rho_{CJ}} - \frac{p}{\rho} + e_{CJ} - e \tag{18.328}$$

Using Equation 18.321, we obtain

$$\frac{\alpha_T^2}{2} = \left\{p\left(\frac{r}{r_0}\right)^2 - p_{CJ} + \rho\left(\frac{r}{r_0}\right)^2 U^2 - \rho_{CJ}U_{CJ}^2\right\}\frac{C}{m\rho_0} \tag{18.329}$$

Finally, combining Equations 18.320, 18.324, and 18.325, we obtain

$$\rho = \frac{\rho_{\mathrm{CJ}} U_{\mathrm{CJ}}}{U\left(\dfrac{r}{r_0}\right)^2} \left\{1 - \alpha_{\mathrm{T}}^2\right\} \tag{18.330}$$

This set of equations is solved for a given area expansion $(r/r_0)^2$ using Brent's method [10]. The spherical surface approach has been shown to be more accurate for smaller charge-to-mass ratios without any loss of agreement at larger charge-to-mass ratios. It should be recognized that this analytic modeling approach neglects initial acceleration due to shock processes [11] and is therefore anticipated to be more accurate as the initial shock process damps out. The model as expressed does not consider the fact that the cylinders thin during radial expansion. One simple way to account for this wall thinning is to assume that the wall cross-sectional area remains constant and r_{in} and V represents the inside radius and inside surface wall velocity:

$$V_{\mathrm{out}} = V \frac{r_{\mathrm{in}}}{r_{\mathrm{out}}} \tag{18.331}$$

$$r_{\mathrm{out}}^2 = r_{\mathrm{in}}^2 + r_{\mathrm{out0}}^2 - r_{\mathrm{in0}}^2 \tag{18.332}$$

Figure 18.45 presents Arbitrary Lagrangian Eulerian 3D Code (ALE3D) high-rate continuum modeling of an explosively filled copper cylinder that is explosively expanded. Figure 18.46 presents copper cylinder velocity histories from ALE3D compared to analytic

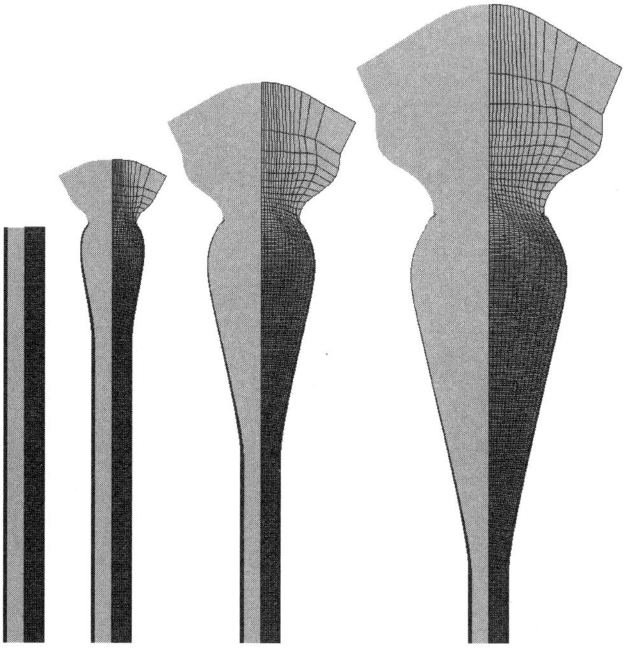

FIGURE 18.45
Modeling at 10 μs intervals for 0.1 in. thick copper cylinder.

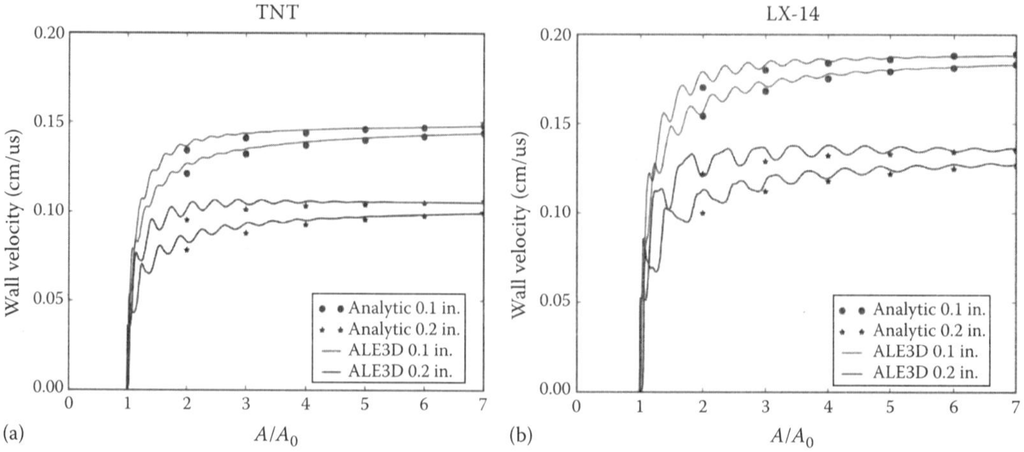

FIGURE 18.46
Copper cylinder velocity histories from ALE3D compared to analytic cylinder test modeling using identical JWLB equations of state. (a) Copper cylinder expansion using trinitrotoluene as the explosive fill and (b) copper cylinder expansion using LX-14 as the explosive fill.

cylinder test modeling using identical JWLB equations of state for TNT and LX-14 using 1 in. diameter charges and 0.1 in. and 0.2 in. thick copper cylinders.

References

1. Cooper, P. W., *Explosives Engineering*, Wiley-VCH, New York, 1996.
2. Zukas, J. A., Nicholas, T., Swift, H. F., Greszczuk, L. B., and Curran, D. R., *Impact Dynamics*, Krieger, Malabar, FL, 1992.
3. Rinehart, J. S., *Stress Transients in Solids*, Hyperdynamics, Santa Fe, NM, 1975.
4. Fickett, W., and Davis, W. C., *Detonation: Theory and Experiment*, Dover, New York, 1979.
5. Lee, E. L., Hornig, C., and Kury, J. W., *Adiabatic Expansion of High Explosive Detonation Products*, Rept. UCRL-50422, Lawrence Livermore Laboratory, Livermore, CA, 1968.
6. Baker, E. L. An application of variable metric nonlinear optimization to the parameterization of an extended thermodynamic equation of state, *Proceedings of the Tenth International Detonation Symposium*, Edited by J. M. Short and D. G. Tasker, Boston, MA, pp. 394–400, July 1993.
7. Gurney, R. W., *The Initial Velocities of Fragments from Bombs, Shells, and Grenades*, BRL Report 405, US Army Ballistic Research Lab, Aberdeen Proving Ground, MD, 1943.
8. Taylor, G. I., Analysis of the explosion of a long cylindrical bomb detonated at one end, *Scientific Papers of Sir G. I. Taylor*, Vol. 111:2770286, Cambridge University Press, Cambridge, UK (1963), 1941.
9. Baker, E. L., *Modeling and Optimization of Shaped Charge Liner Collapse and Jet Formation*, Picatinny Arsenal Technical Report ARAED-TR-92017, Picatinny Arsenal, Morris County, NJ, January 1993.
10. Brent, R., *Algorithms for Minimization without Derivatives*, Prentice-Hall, Englewood Cliffs, NJ, 1973.
11. Backofen, J. E., Modeling a material's instantaneous velocity during acceleration driven by a detonation's gas-push, *Proceedings of the Conference of the American Physical Society Topical Group on Shock Compression of Condensed Matter*, American Institute of Physics, New York, July 28, 2006, Vol. 845, pp. 936–939, 2005.

Further Reading

Achenbach, J. D., *Wave Propagation in Elastic Solids*, North Holland/Elsevier, Amsterdam, Netherlands, 1975.

Anderson, J. D., *Modern Compressible Flow with Historical Perspective*, 3rd ed., McGraw-Hill, New York, 2003.

Billingham, J., and King, A. C., *Wave Motion*, Cambridge University Press, Cambridge, UK, 2000.

Dremin, A. N., *Toward Detonation Theory*, Springer, New York, 1999.

Drumheller, D. S., *Introduction to Wave Propagation in Non-linear Fluids and Solids*, Cambridge University Press, Cambridge, UK, 1998.

Kolsky, H., *Stress Waves in Solids*, Dover, New York, 1963.

Lieber, C.-O., *Assessment of Safety and Risk with a Microscopic Model of Detonation*, Elsevier, Amsterdam, Netherlands, 2003.

Lopanov, A. M., ed., *Theory of Combustion of Powder and Explosives*, Nova Science Publishers, New York, 1996.

Zel'dovich, Y. B., and Raizer, Y. P., *Physics of Shock Waves and High Temperature Hydrodynamic Phenomena*, Dover, New York, 2002.

Zukas, J. A., Nicholas, T., Swift, H. F., Greszczuk, L. B., and Curran, D. R., *Impact Dynamics*, Krieger, Malabar, FL, 1992.

Zukas, J. A., and Walters, W. P. (Eds), *Explosive Effects and Applications*, Springer, New York, 1997.

19

Introduction to Explosive Effects

Explosive effects are an important consideration when dealing with projectiles that are designed to deliver blast, fragments, or even deep penetrating effects such as a shaped charge jet. Chapter 16 focuses on the penetration events that occurred when a relatively solid projectile impacted the target. This impact resulted in either a nonpenetration/partial penetration or a perforation. The latter effect was the sole cause of damage considered. Before the advent of the kinetic energy (KE) long rod, even armor-piercing projectiles carried some explosive that would burst the projectile (hopefully) after passage through the armor of the target. This further damage mechanism would use fragmentation to destroy the soft targets protected by the armor.

Some projectiles are designed as strictly high-explosive (HE) carriers. While these projectiles may have some armor-penetration capability, their primary job is to kill soft targets. A soft target is one that does not require a large amount of KE to kill or one that requires a large number of small perforations to destroy. Classically, soft targets are personnel, trucks, aircraft, radars, etc. While a single, well-placed KE projectile would kill these targets, their vulnerable areas are small; so to increase the probability of kill, a large number of slower-moving or lower-mass fragments are required.

A further adaptation of focused explosive energy is the shaped charge, which is the subject of Chapter 18. These devices can penetrate deep into armor and do not require any delivery KE to be effective. The explosive effects we shall discuss here are used in Chapter 18 but further adapted for shaped charge jet analysis.

In this chapter, we will first discuss how an explosive wave propagates to generate velocity in the metal casing that it is adjacent to. This will allow us to calculate the velocity and direction of fragment flight. After this, we will discuss the penetration mechanisms (very similar to ogival-nosed projectiles and KE long rods) of fragments.

19.1 Gurney Method

The objective of the Gurney method is to obtain algebraic relationships for metal velocity when an explosive in contact with it is detonated. R. W. Gurney was a researcher who worked at the US Army Ballistic Research Laboratory in the 1940s and studied explosively driven metal plates during that time. The method is valid for both shaped charge analysis and fragmentation problems. The Gurney method assumes that all explosive chemical energy is converted into the KE of the fragments and expansion of the explosive products. We call the Gurney energy E the energy that is converted from chemical energy to KE and thus propels the metal and explosive products. This is in actuality only a portion of the energy generated during an explosion. We further shall assume that the gaseous detonation products expand uniformly with constant density.

The method is based on the conservation of both momentum and energy and results in answers that are usually within 10% of experimental results. The governing parameter in the Gurney method is the mass-to-charge (m/c) ratio. This ratio is actually mass per unit area (or length in some configurations) divided by charge per unit area (or length). The method works in its basic form for $0.1 \le m/c \le 10.0$. It is believed that the accuracy of this method comes about through offsetting errors [1]. The method ignores rarefaction waves in the explosive that would cause the calculated velocity to be too high, while at the same time the method assumes that density is constant rather than being greatest at the surface of the charge. This latter assumption causes the calculated velocity to be too low. With these offsetting errors, the method is surprisingly accurate.

A slapper detonator or open-faced sandwich consists of explosive on one side and a metal plate on the other side. This configuration is depicted in Figure 19.1. This configuration is used extensively in explosive characterization tests but has been used in ordnance as well. When the explosive is detonated, a velocity gradient is assumed to be set up as depicted in the figure. In Figure 19.1, the y-coordinate is associated with a layer of particles (a Lagrangian system) and thus can move. The velocities are interpreted as velocities after all the detonation product gases have expanded to several times their initial volume.

If we assume a constant density throughout the gas products, we can show that

$$\rho_{\text{gas}} y_0 = c \tag{19.1}$$

where y_0 is typically taken as the initial thickness of the explosive since; based on our assumptions, Equation 19.1 holds true for all time.

The velocity distribution for this configuration is given as

$$V_{\text{gas}} = (V_0 + V)\frac{y}{y_0} - V \tag{19.2}$$

Without going into the detailed derivation (the derivation can be found in the book by Walters and Zukas [1], pp. 47–49), we can write the final expression for an open-faced sandwich as depicted in Figure 19.2 as

$$V = \sqrt{2E}\left\{\frac{1}{3}\left[\left(\frac{2m}{c}\right)^2 + \frac{5m}{c} + 1\right]\right\}^{-1/2} \tag{19.3}$$

The velocities for the metal fragments in the flat sandwich, cylinder, and tamper configurations can also be derived [1] as follows. For the flat sandwich as depicted in Figure 19.3, we have

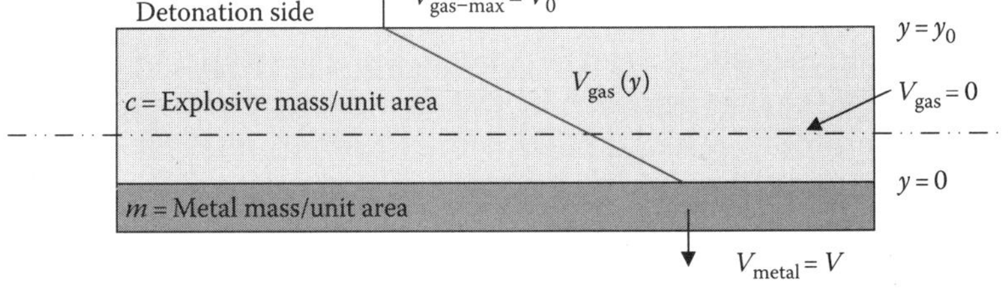

FIGURE 19.1
Open-faced sandwich configuration with velocity gradient. (From Walters, W. P., and Zukas, J. A., *Fundamentals of Shaped Charges*, CMC Press, Baltimore, MD, 1989. With permission.)

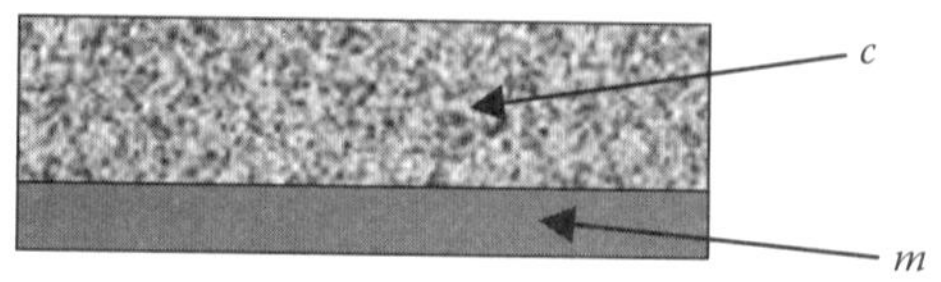

FIGURE 19.2
Open-faced sandwich.

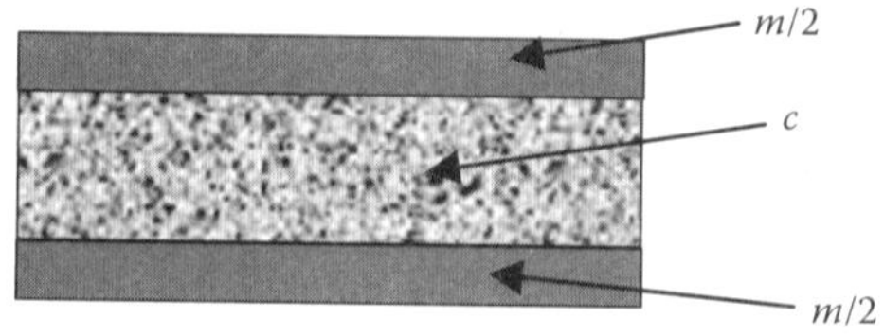

FIGURE 19.3
Flat sandwich.

$$V = \sqrt{2E}\left(\frac{m}{c} + \frac{1}{3}\right)^{-1/2} \tag{19.4}$$

Many configurations in common use for military applications require a cylindrical configuration where a tube of metal is filled with explosive material. This is also a common configuration for use in shaped charge jet analysis. For a cylindrical geometry as depicted in Figure 19.4, we can write

$$V = \sqrt{2E}\left(\frac{m}{c} + \frac{1}{2}\right)^{-1/2} \tag{19.5}$$

In some instances, it is necessary that the metallic plates are not of the same mass. This is commonly referred to as the tamper configuration. The formula that expresses the metal velocities for this configuration is

$$V_m = \sqrt{2E}\left[\frac{1 + A^3}{3(1 + A)} + \frac{n}{c}A^2 + \frac{m}{c}\right]^{-1/2} \tag{19.6}$$

where

$$V_n = AV_m \tag{19.7}$$

$$A = \frac{1 + 2(m/c)}{1 + 2(n/c)} \tag{19.8}$$

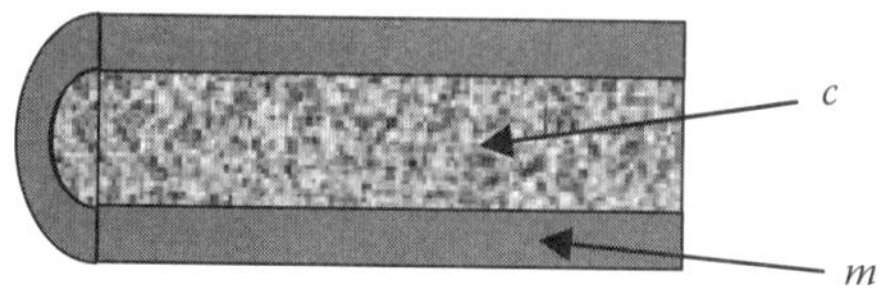

FIGURE 19.4
Cylindrical geometry.

In these cases, the subscript "n" refers to the thicker tamper plate and the subscript "m" refers to the thinner driven plate. This is illustrated in Figure 19.5.

In some instances, it is informative to examine the behavior of a spherical geometry. The equation that describes the metal velocity for the configuration illustrated in Figure 19.6 is given as Equation 19.9. The derivation for this expression is found in the books by Cooper [2] and Carleone [3]:

$$V = \sqrt{2E}\left(\frac{m}{c} + \frac{3}{5}\right)^{-1/2} \tag{19.9}$$

Since m/c, n/c, and therefore A are dimensionless, the term $\sqrt{2E}$ has units of velocity, and it is sometimes called the Gurney characteristic velocity, Gurney velocity, or the Gurney constant. If analyzing an explosive for which there is no Gurney velocity, an approach recommended by Kennedy (1970) as mentioned by Marshall and Sanow [2] is to use $E \sim 0.7H_D$, where H_D is the heat of detonation. For most explosives, $0.61 < E/H_D < 0.76$.

As the m/c ratio approaches zero, the velocity of the fragments approaches a constant value. For a flat sandwich, an open-faced sandwich, and an asymmetric sandwich (tamper), this value is $\sqrt{6E}$. For a cylinder, this value is $\sqrt{4E}$. And for a sphere, this value is $\sqrt{(10/3)E}$.

The Gurney method is fairly accurate, but of all the configurations, it is the least accurate for the open-faced sandwich configuration. In this case, the metal velocity would be predicted too high. Unfortunately, more complex methods are not always worth the increased accuracy.

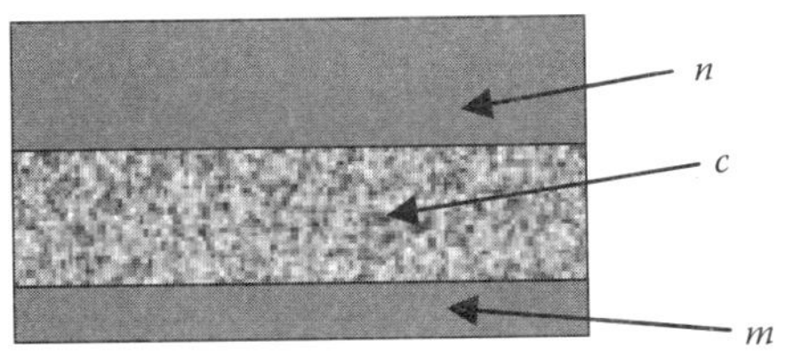

FIGURE 19.5
Tamper configuration.

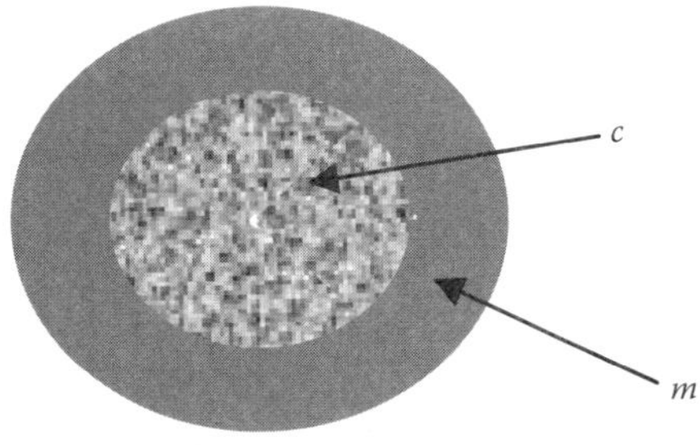

FIGURE 19.6
Spherical geometry.

19.2 Taylor Angles

The previous section explains a means of determining the velocity to which a metal, initially in contact with the explosive, will be projected. This section focuses on the Taylor method that predicts the angle at which the metal will be thrown given a detonation event.

In the Gurney method, the equations assume that the metal moves normal to its surface. If an explosive wave strikes the metal at some angle, this assumption is no longer valid and the metal will be projected at some angle. It is in these instances that we need to invoke the Taylor angle approximation. In this method, we assume that the metal is accelerated to its final velocity instantaneously. We also assume this is a pure rotation, so no thickness change or change in length of the metal occurs.

Consider a detonation wave that is propagating from right to left at velocity D as depicted in Figure 19.7. During this time, the explosive wave moves from the initial position to point O, the point initially at P moves to P'. If the detonation wave passes point P at time $t = 0$, then we can show that

$$\overline{OP} = Dt \tag{19.10}$$

and

$$\overline{PP'} = Vt \tag{19.11}$$

Then, it follows from geometric arguments noting that V is perpendicular to OP'

$$\sin \alpha_T \approx \alpha_T = \sin\frac{\theta}{2} = \frac{\overline{PP'}}{2\overline{OP}} = \frac{Vt}{2Dt} = \frac{V}{2D} \tag{19.12}$$

If we know D from the explosive properties and we can estimate V from the Gurney method, we can get an idea of what θ will be. Experiments usually use smear cameras and measure V_A, which relates to V through

$$V_A = D\tan\theta = \frac{V_N}{\cos\theta} \tag{19.13}$$

Usually V, V_N, and V_A are within a few percent of one another. This allows us to use them somewhat interchangeably. Also, for most explosives, $V/2D$ is approximately constant [1].

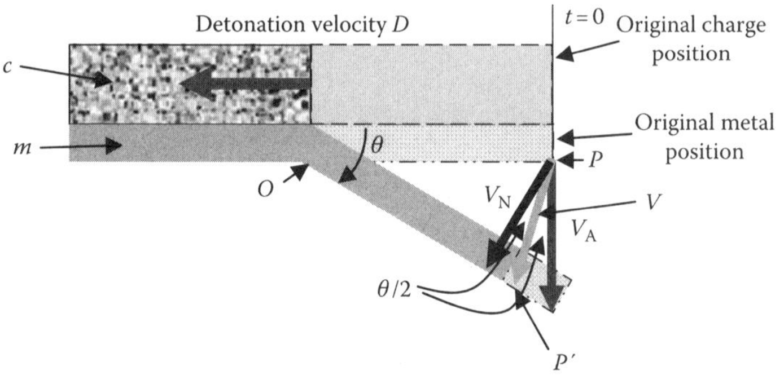

FIGURE 19.7
Taylor angle geometry. (From Walters, W. P.m and Zukas, J. A., *Fundamentals of Shaped Charges*, CMC Press, Baltimore, MD, 1989. With permission.)

If we examine a typical HE shell and assume a detonation velocity D from the fuze and given that we know the geometry, we can generate a reasonable estimate for the spray pattern of the fragments. We do this by dividing the shell into segments and solving for the Gurney velocities and Taylor angles in each segment. We can curve-fit the data. Spreadsheet programs are great for this task. However, there are specialized codes that perform this task for us as well.

We shall illustrate the procedure with an example.

Example Problem 1

A projectile is to be fabricated from steel and filled with trinitrotoluene (TNT) as depicted in Figure 19.8. For a detonation of the fill, graph the fragment velocities in meters per second and Taylor angles in degrees vs. distance from the nose of the projectile. The required properties for this calculation are given as follows:

TNT Gurney velocity $(2E)^{1/2} = 2.039$ km/s
TNT detonation velocity $(D) = 6730$ m/s
TNT density = 1.63 g/cm^3
Steel density = 0.283 lbm/in.^{3}

Solution: Let us get everything in consistent units, the density of TNT first:

$$\rho_{TNT} = (1.63)\left[\frac{g}{cm^3}\right](2.45)^3\left[\frac{cm^3}{in.^3}\right]\frac{(2.046)}{(1000)}\left[\frac{lbm}{g}\right] = 0.059\left[\frac{lbm}{in.^3}\right] \tag{19.14}$$

The next step is to get the sectional densities calculated for the fill and the case. We need to use only four stations as depicted in Figure 19.9 because in the areas of constant cross section, we need only one data point but the data will be slightly different at the transition from the cone. We shall list only the calculations for the first location and depict the results in a table by using the same procedure.

For cross section 1, we have

$$m_1 = \rho_{steel}\left(A_{1_{case}}\right) = (0.283)\left[\frac{lbm}{in.^3}\right]\pi\left(1.003^2 - 0.750^2\right)\left[in.^2\right] = 0.394\left[\frac{lbm}{in.}\right] \tag{19.15}$$

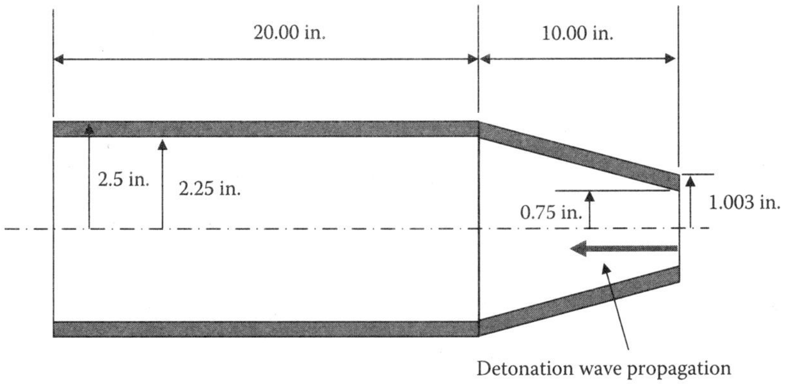

FIGURE 19.8
Projectile with an HE fill.

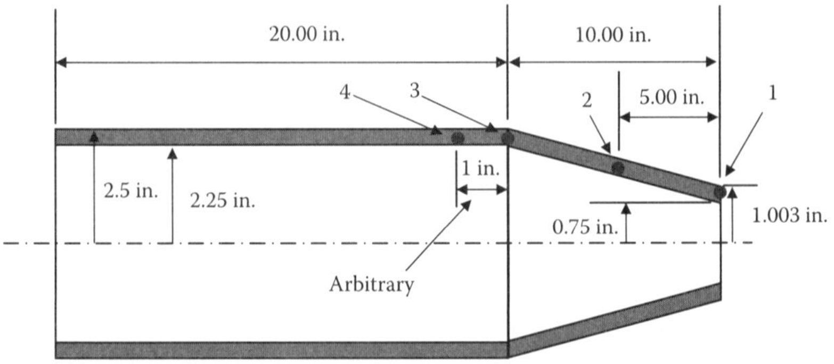

FIGURE 19.9
Projectile with an HE fill discretized.

For the fill, we want the dimension normal to the surface, so we need to determine the angle of the surface as

$$\alpha = \tan^{-1}\left(\frac{2.25-0.75}{10}\right) \rightarrow \alpha = 8.531° \tag{19.16}$$

$$c_1 = \rho_{TNT}\left(A_{1_{fill}}\right) = (0.059)\left[\frac{\text{lbm}}{\text{in.}^3}\right]\pi\left[\frac{(0.750^2)}{\cos^2(8.531°)}\right]\left[\text{in.}^2\right] = 0.107\left[\frac{\text{lbm}}{\text{in}}\right] \tag{19.17}$$

Now the fragment velocity follows directly from

$$V = \sqrt{2E}\left(\frac{m}{c}+\frac{1}{2}\right)^{-1/2} \tag{19.18}$$

$$V_1 = (2.039)\left[\frac{\text{km}}{\text{s}}\right](1000)\left[\frac{\text{m}}{\text{km}}\right]\left(\frac{0.394}{0.107}+\frac{1}{2}\right)^{-1/2} = 997\left[\frac{\text{m}}{\text{s}}\right] \tag{19.19}$$

For the Taylor angle, we first need to find the angle $\theta/2$ from our formula,

$$\sin\frac{\theta}{2} = \frac{V}{2D} = \frac{(997)[\text{m/s}]}{2(6730)[\text{m/s}]} = 0.074 \rightarrow \frac{\theta}{2} = 4.25° \tag{19.20}$$

This Taylor angle would tend to tilt the fragment at 4.25° in the direction of the detonation wave (toward the base), but at this point, our nose is canted 8.531° toward the projectile axis; so the actual angle is 4.25° − 8.531° or −4.281° (see Figure 19.10).

If we take all of our data and put these in a table, we get Table 19.1. Figure 19.11 shows the graph of these data.

A similar plot could be drawn using the Taylor angles listed in Table 19.1. It must be noted that the slight velocity increase at the ogive/bourrelet transition (10 in. from the nose) is an artifact of the way the projectile was discretized. We would normally assume that there is a smooth tangency point at that location.

Problem 1

A Bangalore torpedo was a device built by the United States during the Second World War to clear beach (or any other) obstacles. It consisted of a long tube filled with explosive

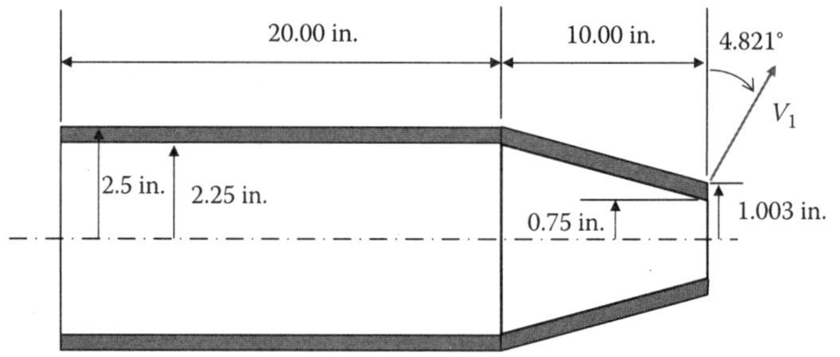

FIGURE 19.10
Taylor angle at the projectile nose tilted to account for ogive angle.

TABLE 19.1
Gurney Velocities and Taylor Angles for Projectile Fragments

Position	Axial Location (in.)	$m = rV/L$	$c = rV/L$	Fragment Velocity (m/s)	Taylor Angle α_T (°)	Projected Angle α (°)
1	0.000	0.394	0.107	995	4.240	−4.291
2	5.000	0.732	0.426	1370	5.841	−2.690
3	10.000	1.056	0.959	1612	6.877	−1.653
4	11.000	1.056	0.938	1599	6.825	6.825
5	20.000	1.056	0.938	1599	6.825	6.825

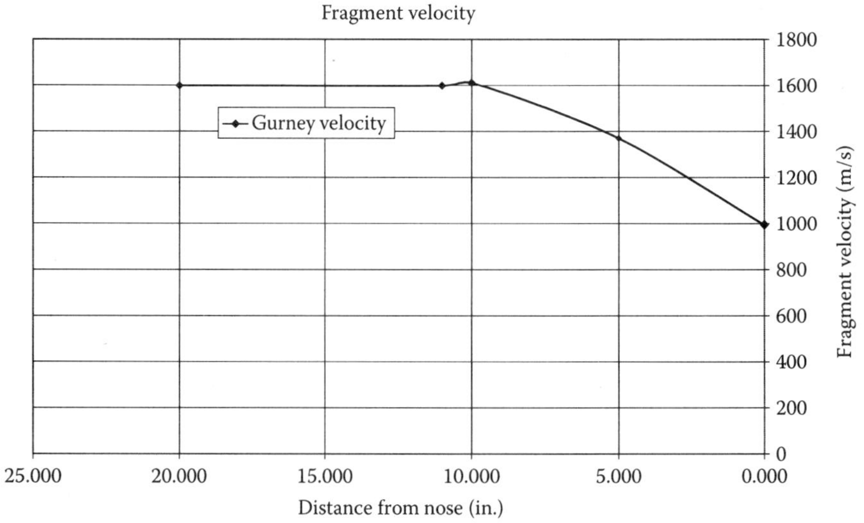

FIGURE 19.11
Gurney velocity vs. distance from projectile nose.

that was detonated on the end. Assume that we have a similar device made of steel and filled with Composition B. The device is 3 ft long. The inside diameter (ID) is constant at 2 in. The outside diameter (OD) varies with length. The first foot of length is 2 1/4 in. in diameter, the next foot of length is 2 3/4 in. in diameter, and the last foot of length is 3 in. in diameter. Assuming that we detonate the device at the 2 1/4 in. end, perform the following:

1. Draw a graph of the fragment velocities vs. length in feet and feet per second.
2. Draw a graph of the Taylor angles in feet and degrees from the device axis.

Assume that the tube is steel with a density of 0.283 $lbm/in.^3$. Assume that the filler density is 1.70 g/cm^3. Assume that the detonation velocity is 7.89 mm/μs and the Gurney constant is 2.7 mm/μs.

Problem 2
Assume that we used the Paris gun so often that it finally blew up. We want to determine the velocity of the fragments and their Taylor angles. Assume the section where the explosion took place is centered over a jacket transition. Therefore, the analysis consists of two sections, each 4 ft long. The ID of the weapon is 210 mm. The OD of the forward section is constant at 350 mm. The OD of the jacketed section is also constant at 420 mm. Assume the explosion begins at the projectile and propagates rearward. Assume that the Gurney constant for the filler/propellant combination is 1.8 km/s.

1. Draw a graph of the fragment velocities vs. length in feet and feet per second.
2. Draw a graph of the Taylor angles in feet and degrees from the bore axis.

Assume that the tube is steel with a density of 0.283 $lbm/in.^3$. Assume that the filler/propellant density averages to about 0.6 g/cm^3. Assume that the detonation velocity is 16,500 ft/s.

Problem 3
A projectile is to be fabricated from steel and filled with TNT as depicted in Figure 19.12. For a detonation of the fill, graph the fragment velocities in meters per second and the Taylor angles in degrees vs. distance from the nose of the projectile. The required properties for this calculation are given as follows:

TNT Gurney velocity $(2E)^{1/2}$ = 2.039 km/s
TNT detonation velocity (D) = 6730 m/s
TNT density = 1.63 g/cm^3
Steel density = 0.283 $lbm/in.^3$

19.3 Mott Formula

The preceding sections outline the procedure for determining the velocity and directions that the fragments of an exploded projectile will fly when the fuze is initiated. In this

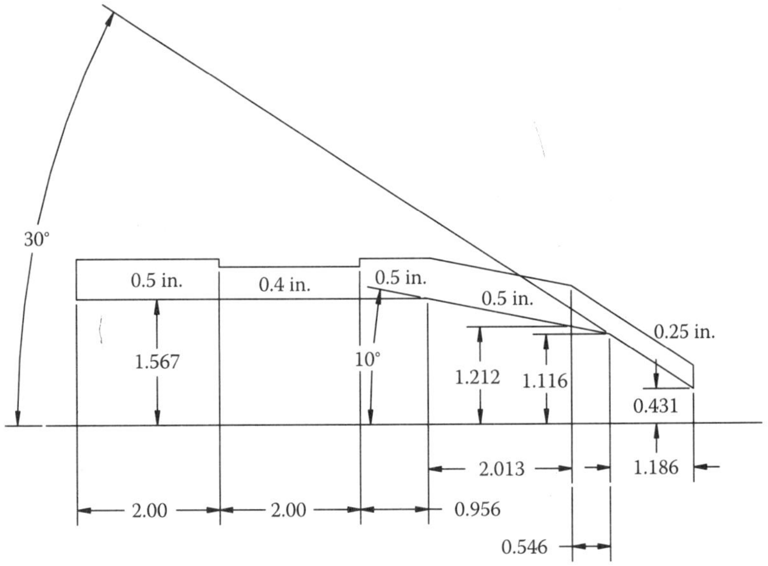

FIGURE 19.12
Projectile geometry for Problem 3.

section, we examine the Mott formula, a method by which we can estimate the mass of the fragments. We begin by describing the fragmentation process itself.

When we detonate an HE fill in a metallic cylinder (projectile), several things occur. First, a detonation wave propagates along the axis of detonation. This results in pressure being generated with the attendant stress being transferred to the metallic casing. At this point, the case expands and ruptures by shear or brittle failure. If the case expands significantly and removes significant energy from the detonation products, we have a condition known as a terminal detonation. If the case expands very little before fragmenting, the result is known as a prompt detonation. Once the case ruptures, fragments fly in directions dependent upon the Taylor angle and their individual geometries. At some point, the fragments may impact a target. The processes of detonation, acceleration, and flight have been dealt with in detail in our prior work. Here, we shall concentrate on the fragmentation process and penetration of the fragments themselves.

There are several factors that affect the fragmentation process: explosive brisance (see the glossary in Appendix A), charge-to-mass ratio, casing diameter, casing wall thickness, and mechanical properties of the casing. The fragmentation of the casing usually begins at the OD through the formation of sharp radial cracks. These cracks then join with shear cracks from the inside of the material (or not, if the material is extremely brittle). The cracks then coalesce into long, longitudinal cracks. If the casing material is ductile enough, as the case expands radially and during this process, the wall will thin out somewhat. Finally, the casing will fragment completely. This is depicted in Figure 19.13.

Some general rules for case fragmentation based on material properties are presented here. In general, a more brittle material such as gray cast iron will produce a very large number of small fragments. This is desirable when lethal effects are to be localized to the projectile area. A precision delivery would be required to use this property most effectively. A more ductile material will generally produce a smaller number of large fragments.

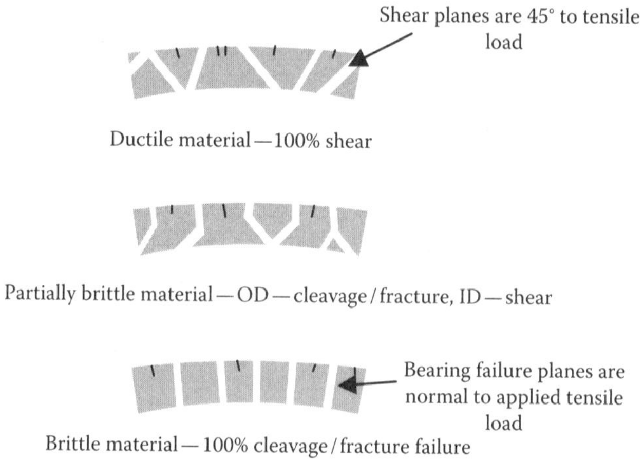

FIGURE 19.13
Fragmentation process.

These fragments will be more lethal at longer ranges. This has the advantage of being able to account for some inaccuracy in projectile delivery. It is generally accepted that changes in the material microstructure affect this phenomenon.

The fragmentation process directly relates to the effectiveness of the weapon system. More fragments means a greater probability of a fragment hit P_h. A larger fragment size translates to a greater probability of a kill, given a hit, P_{hk}. This trade-off must be made through an effectiveness analysis. In other words, if the target we are looking to kill is susceptible to even small fragment impacts, then we are better off with smaller fragment sizes as that will maximize our probability of killing more targets. If, however, we can kill the target of interest with a large fragment only, we must take the degradation in the hit probability. Mathematically, we want to maximize the effectiveness through

$$E_{hk} = E_h P_{hk} \tag{19.21}$$

where E_{hk} is our expected number of impacts that kill a given target and E_h is the expected number of fragments that impact the target.

So what we have learned here is that more, small fragments means greater E_h and lower P_{hk}, while fewer, larger fragments means smaller E_h and larger P_{hk}. If we would like to quantify the total probability of a kill P_k, on a given target, we can write

$$P_k = 1 - e^{-E_{hk}} \tag{19.22}$$

There are several ways the fragmentation process can be controlled: explosive selection, case material selection, heat treatment of the casing, prestressing, preforming, or explosive wave shaping. One of the important things to remember is that the projectile body design has to survive rough handling and gun launch. Sometimes, this is at odds with the desired fragmentation effect and trades must be made. For a given target as well as any collateral damage effects, control of the fragmentation process translates to control of the following: fragment velocity, number of fragments, mass of the fragments, shape of the fragments, and distribution of the fragments (i.e., the fragmentation pattern). We have already mentioned how some of these contradict one another.

We have discussed some simple analytical approaches to determining fragment velocities and patterns in previous sections. However, experimentally, an arena test is the best verification. An arena test is one in which we detonate the projectile of interest and surround it with evaluation panels. A typical arena test setup is depicted in Figure 19.14. Two types of panels are commonly used: velocity panels and fragment recovery panels. Velocity panels are thin aluminum sheets between which there are sometimes placed light sources. High-speed films taken during the fragmentation event reveal bright spots caused by perforation. Since the distance is well known, the average velocity can be calculated from the speed of the camera and time of arrival (appearance of the bright spot).

The recovery panels allow the velocity to be estimated from depths of penetration into the panels. In mild steel panels, the depth of penetration can be estimated through

$$P = cm_{\mathrm{P}}^{1/3}\left(\frac{V_{\mathrm{s}}}{1000}\right)^{4/3} \tag{19.23}$$

Here for mild steel, P is the depth of penetration (in.), $c = 0.112$, m_{P} is the fragment weight (oz), and V_{s} is the striking velocity (ft/s).

For composition board (Celotex) panels, we can write

$$V_{\mathrm{s}} = 1865\frac{P^{1/3}}{m_{\mathrm{P}}^{0.1}} \tag{19.24}$$

where P is the depth of penetration (in.); m_{P} is the fragment weight (g); and V_{s} is the striking velocity (ft/s).

In all cases, if the projectile that creates the fragment is moving at a high velocity, this must be vectorially added to the fragment velocity in the effectiveness analysis. Mathematically, this is given by

$$V_0^2 = V_{\mathrm{projectile}}^2 + V_{\mathrm{frag}}^2 \tag{19.25}$$

where V_0 is the resultant initial fragment velocity; V_{frag} is the fragment velocity resulting from the detonation; and $V_{\mathrm{projectile}}$ is the projectile velocity at the time of detonation.

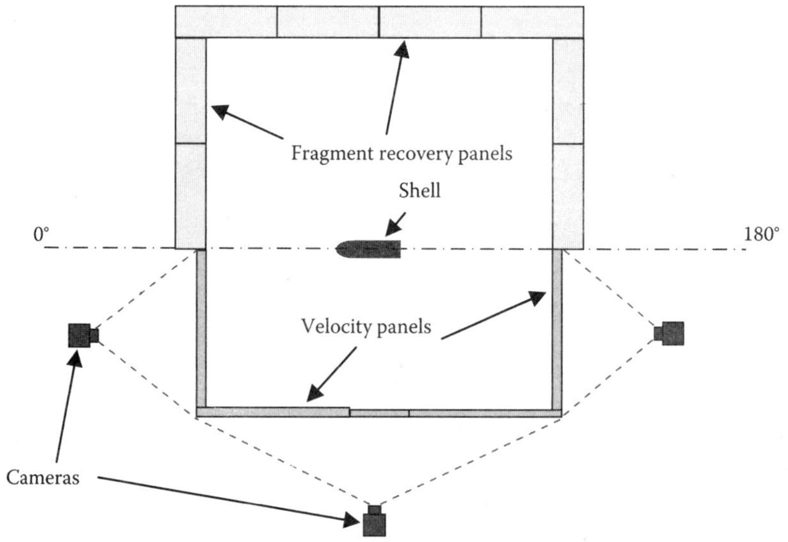

FIGURE 19.14
Typical arena test setup.

As we have discussed in Part II, an object that moves through air will lose velocity because of the mechanisms of drag. This effect is usually more pronounced on fragments because of their irregular and sometimes inconsistent shapes that present varying frontal areas to the air stream. To simplify matters somewhat, it is typical to use a drag model that assumes a constant drag coefficient for fragments. This model is given by

$$V_s = V_0 e^{-k_1 x} \tag{19.26}$$

Here we define the constant k_1 as we have in the exterior ballistics section by using

$$k_1 = \frac{\rho S}{2m} C_D \tag{19.27}$$

In these equations, V_s is velocity of the fragment at impact; V_0 is the initial fragment velocity caused by the explosion (Gurney velocity); x is the distance from the point of detonation to the point of impact; S is the presented area of the fragment; C_D is the fragment drag coefficient; ρ is the density of the ambient air in the vicinity of the detonation; and m is the mass of the fragment.

Typical drag curves for fragments can be found in the book by Hoerner [4].

The mass of fragments is a critical piece of data in any effectiveness analysis. It is a daunting task to determine how a naturally fragmenting warhead breaks up. If a warhead contains preformed fragments, we can assume that the fragment size will be based on the preformed geometry. Mott [5] proposed the following semiempirical equation for predicting the number of fragments in a naturally fragmenting warhead:

$$N(m) = \frac{M_0}{2M_K^2} \exp\left(-\frac{\sqrt{m}}{M_K}\right) \tag{19.28}$$

where $N(m)$ is the number of fragments greater than mass m; m is the mass of the fragment (lbm); M_0 is the mass of the projectile (lbm); and M_K is a distribution factor defined in Equation 19.29 ($\text{lbm}^{1/2}$):

$$M_K = Bt^{5/6} d^{1/3} \left(1 + \frac{t}{d}\right) \tag{19.29}$$

where B is a constant specific for the particular explosive/metal combination; t is the wall thickness (in.); and d is the ID of the projectile (in.).

The Mott coefficients B for mild steel cylinders combined with particular explosives is given in Table 19.2 [2]. We also know that the charge-to-mass ratio has an effect; this is implicit in the combination of B, t, and d.

TABLE 19.2

Mott Formula Coefficients for Typical Projectile Fills

Explosive	B ($\text{Ib}^{1/2}$ $\text{in.}^{-7/6}$)
Composition B	0.0554
Cyclotol (75/25)	0.0493
Pentolite (50/50)	0.0620
TNT	0.0779
Composition A-3	0.0549
RDX/wax (95/5)	0.0531
Tetryl	0.0681

When an HE warhead explodes, fragments fly in all directions. As previously mentioned, these fragments seldom penetrate heavily armored targets—they are effective only against light armor or soft targets. Because of this, we usually examine fragment impacts against thin targets. Usually, this means that the target is thinner than any characteristic dimension of the fragment. Simple shapes are usually considered for ease of analysis; the shapes are usually cubes and spheres. The penetration behavior of a fragment is typically characterized by its residual mass and velocity once it has perforated the target material.

The fragment momentum equation is given by [3]

$$m_0 V_s = m_{rp} V_{rp} + m_p V_{rm} + I \tag{19.30}$$

where m_0 and V_s are the mass and impact velocity of the fragment relative to the target, respectively; m_{rp} and V_{rp} are the residual mass and velocity of the mass center of the fragment pieces that perforate the target, respectively; m_p and V_{rm} are the residual mass and velocity of the mass center of the target pieces that have broken free of the target, respectively; and I is the impulse transmitted to the target owing to both the target stopping pieces of the penetrator and the absorption of the shear energy by the target that is required to set the mass m_p free.

The energy equation for a fragment impact is given by [3]

$$\frac{1}{2} m_0 V_s^2 = \frac{1}{2} m_{rp} V_{rp}^2 + \frac{1}{2} m_p V_{rm}^2 + \frac{1}{2}\left(m_0 - m_{rp}\right) V_0^2 + E_f + W_s \tag{19.31}$$

where E_f is the energy associated with the plastic deformation of masses m_0 and m_p. It is calculated as if mass m_p were not attached to the target. W_s is the work associated with the shearing mass m_p while it is attached to the target. The third term on the right-hand side represents the KE of the initial impact that remains with the target.

The residual velocity of a fragment after it perforates a soft target is important in estimating its lethality. Recht [3] showed that an equation can be written for residual velocity of a fragment as

$$V_r = \frac{\sqrt{V_s^2 - V_x^2}}{1 + \left(\dfrac{m_p}{m_{rp}}\right)} \tag{19.32}$$

where V_x is a characteristic velocity that is normally replaced by V_{50}. After one calculates V_r, the impulse transmitted to the target can be calculated as a function of V_x through

$$\frac{I}{m_0 V_s} = 1 - \left(\frac{m_{rp}}{m_0}\right)\sqrt{1 - \left(\frac{V_x}{V_s}\right)^2} \tag{19.33}$$

This impulse can be normalized to V_{50} to determine the optimum velocity of a fragment. For a thin plate, if the penetration velocity is close to V_{50}, the impulse transmitted to the plate is maximized. In most damage theories, more damage occurs to a component with more impulse applied. This means that if one would like to damage a component behind thin armor, for maximum effect, one would like a fragment that gets through the outer armor without a problem yet impacts the component near its V_{50}.

Much like long-rod penetrators, fragments tend to lose mass as the penetration event progresses. When a blunt fragment impacts a plate, material is eroded from the contact surface. This process occurs continually until the relative velocity between what remains of the fragment and the contact surface drops below the plastic wave velocity in the fragment material. Recht [3] developed the following equation for determination of fragment residual mass:

$$\frac{m_{re}}{m_0} = 1 + \frac{m_p}{2m_0} \ln \left[\frac{1 + \left(\frac{1}{Q} \right)}{1 + \left(\frac{\left(\frac{V_s}{U_c} \right)^2}{Q} \right)} \right] \tag{19.34}$$

In this expression, m_p is the plate plug mass (same as earlier); $Q = \sigma_e / \rho_p U_c^2$ (dimensionless parameter); σ_e is the dynamic yield strength of fragment material; ρ_p is the density of fragment; and U_c is the plastic wave speed in the fragment material.

With this material, we have completed the treatment of fragmentation. These formulas can be used with fair accuracy to predict fragment behavior from HE devices.

References

1. Walters, W. P., and Zukas, J. A., *Fundamentals of Shaped Charges*, CMC Press, Baltimore, MD, 1989.
2. Cooper, P. W., *Explosives Engineering*, Wiley-VCH, New York, 1996.
3. Carleone, J. (Ed.), *Tactical Missile Warheads*, American Institute of Aeronautics and Astronautics, Washington, DC, 1993.
4. Hoerner, S. F., *Fluid Dynamic Drag*, Hoerner Fluid Dynamics, Vancouver, WA, 1965.
5. Mott, N. F., Fragmentation of shell cases, *Proceedings of the Royal Society*, London, Vol. A189, pp. 300–308, 1947.

Further Reading

Zukas, J. A., and Walters, W. P. (Eds.), *Explosive Effects and Applications*, Springer, New York, 1997.

20

Shaped Charges

Although shaped charges can trace their origin to the early 1900s (and some authors suggest even further back), it was not until the Second World War that their use proliferated. Monroe in the United States and von Foerster and von Neumann in Europe discovered that a hollow charge, i.e., a block of explosive with a cavity on the target side, caused a deeper penetration than a similar charge that had no cavity. About the time of the Second World War, the combatants determined that if they lined this cavity with a metal and pulled the charge back from the surface, they achieved an even deeper penetration. The penetration depths achieved were on the order of several warhead diameters. These warheads were and still are so effective that they continue to be developed by nearly every nation. It is the goal of this chapter to describe their behavior and analysis.

Shaped charge warheads fall under the category of chemical energy warheads because they do not require any kinetic energy (KE) from the delivery system to be effective. This property makes them ideal for use in items such as shoulder fired weapons, grenades, mines, and even static cutting charges. The oil industry as well as the steel industry use them in large numbers to clear plugs or open up pores in rock to allow oil to flow into well shafts. These devices are also used to cut large masses of steel plate and bars.

The process through which a shaped charge works is as follows:

1. An explosion is generated which passes a detonation wave over the liner.
2. The liner collapses from the rear forward and is squeezed by the pressure of the expanding gases.
3. A jet of material forms, the tip of which moves at high velocity toward the target.
4. The remaining liner material is formed into a slug which follows the jet at a much lower velocity (approximately 1/10 the tip velocity).
5. The tip then penetrates the target material and the overall length of the jet is decreased until either the target is perforated or the entire jet is consumed.

This process generates high temperatures and pressures. As we have previously discussed, pressure much higher than the ultimate stress in the material allows us to model the material as an inviscid fluid. This has led to several common misconceptions. Shaped charges do not burn through the armor plate. This is believed to have been the misconstrual of the acronym HEAT. which actually stands for high explosive anti-tank. As we have stated earlier, high temperatures are generated during a penetration event, but it is the KE of the jet that does the work. Shaped charges do not turn the liner into a liquid. When pressures are orders of magnitude above the yield strength of the material (and they are during a jet formation), we can treat the problem as a fluid dynamics problem even though the liner material really is not a fluid. If we could somehow magically stop the detonation process, we would have a solid rod of material. The formation of a typical shaped charge jet is shown as Figure 20.1.

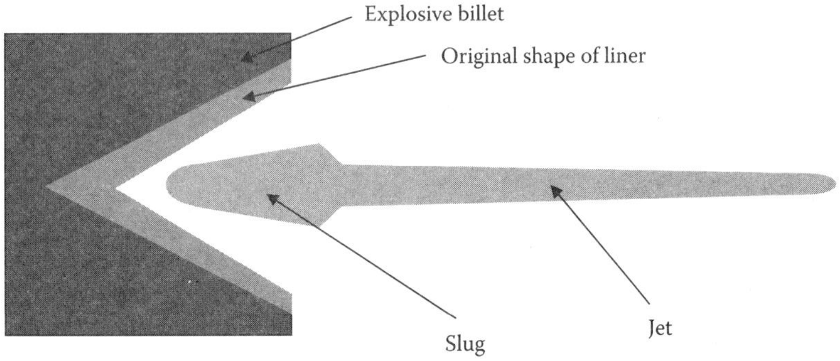

FIGURE 20.1
Shaped charge jet formation.

The standoff *s* of a shaped charge is the distance from the base of the liner or cavity to the target. This is illustrated in Figure 20.2. It is known that the standoff distance in shaped charges has an optimum value for armor penetration. This is depicted in Figure 20.3. The penetration performance is very sensitive to the standoff and performance rapidly decays if it is too large or too small. Explosive reactive armor is an effective way to defeat a shaped charge by breaking the jet up on impact, feeding additional material to erode the jet, and altering the standoff. Standoff plates (you can see these in many Second World War photographs of German vehicles) and sandbags defeat shaped charges by respectively affecting the standoff or forcing the jet to be consumed.

In addition to standoff, detonation symmetry is also very important. A slight asymmetric geometry of the liner or charge ignition will result in inefficient or improper formation. This is why most liners designed for military use are machined to precise tolerances. Charge-to-liner mass (c/m) ratio greatly affects the velocity of the jet. If this ratio is too high, the liner can fragment and fail to penetrate. If this ratio is too low, the jet velocity will not be high enough for efficient penetration. Many authors use the inverse of this parameter as the (m/c) ratio. The liner geometry has a pronounced effect on the jet formation because it affects how the explosive wave collapses the liner and forms the jet.

Liner material also has an effect on penetration. This is illustrated in Figure 20.3 for several different materials.

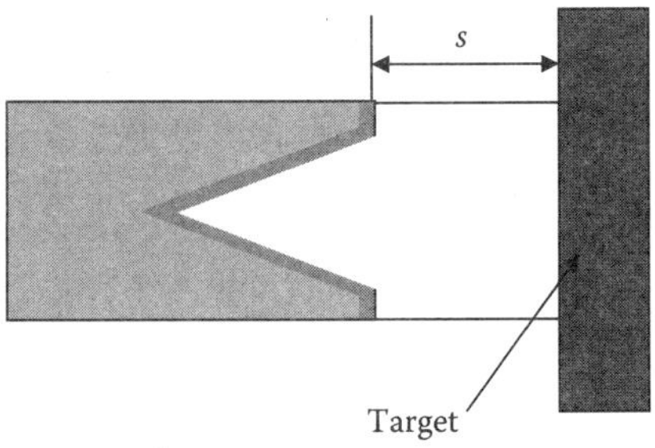

FIGURE 20.2
Standoff *s* of a shaped charge.

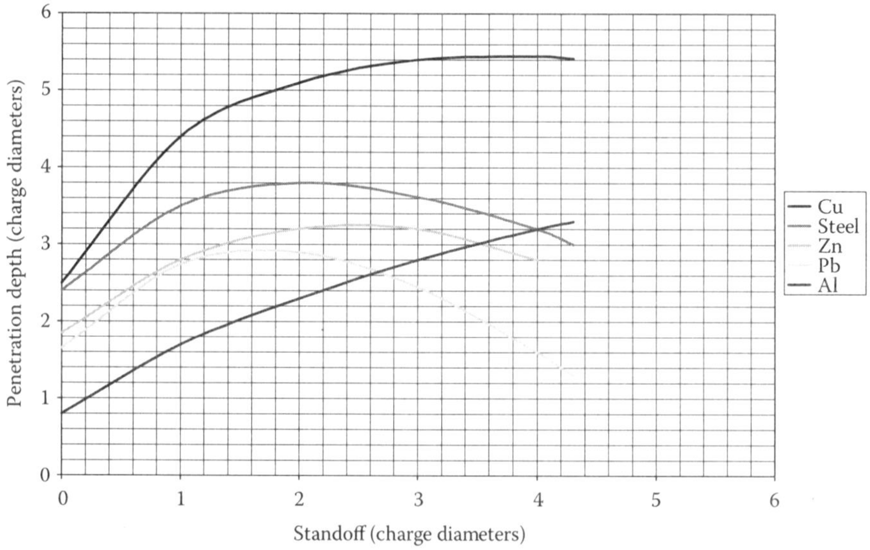

FIGURE 20.3
Effect of standoff on jet penetration using 45° conical liners.

20.1 Shaped Charge Jet Formation

The previous section introduced some general terms commonly used in discussing shaped charges. In this section, we shall examine methods of predicting jet formation. Shaped charge jet penetration is critically dependent upon proper formation of the jet. The ability to predict this formation allows the designer to predict performance and even to optimize the design. Although computational techniques now allow great accuracy in predicting jet formation and penetration, it is always good practice to use a simplified analytical technique as a check of the computer models. While the analytic solution, with its associated idealizations, is not as accurate as the computational solution, it will be close enough to gain an appreciation of whether the code is outputting erroneous answers or not.

Birkhoff and others developed a theory in 1948 [1] that assumed that the pressures generated by the explosive products are so great that the liner material strength could be neglected. Because of this, liners are typically modeled as inviscid, incompressible fluids. This was important because the modeling was greatly simplified. Birkhoff assumed that the liner particles were instantly accelerated to their final collapse velocity. It was further assumed that this velocity was constant throughout the formation. We know from experience that this is incorrect, as the tip of the jet moves faster than the tail or slug. This analysis method was later modified by Pugh in 1952 to include the velocity gradient. The model became only slightly more complicated, but the accuracy improved.

The theory that was developed is now known as the Birkhoff–MacDougal–Pugh–Taylor theory. It is a fairly accurate, simple-to-use theory that allows for rapid estimates of jet and slug velocities. The theory assumes no velocity gradient in the jet and that the particles of the liner are instantly accelerated to their final velocity.

The theory models the liner collapse as follows. We shall use the nomenclature introduced by Walters and Zukas [2] to describe this process, which is illustrated in Figure 20.4.

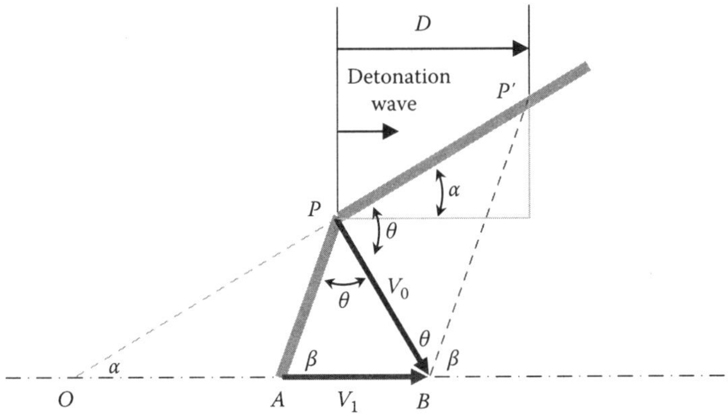

FIGURE 20.4
Illustration of liner collapse. (From Walters, W. P., and Zukas, J. A., *Fundamentals of Shaped Charges*, CMC Press, Baltimore, MD, 1989. With permission.)

When we initiate an explosive behind a liner, after a time, the detonation wave will pass any point of interest as depicted in Figure 20.4. The liner is assumed to collapse inward at a velocity V_0. We assume an instantaneous angle (2β) between the moving walls of the liner, which is greater than the initial angle (2α). We assume that the detonation wave moves at a constant velocity D. If we imagine ourselves in a Lagrangian reference frame attached to point P in Figure 20.4, the liner material can be assumed to move inward along $P'P$ and out along PA with the pressure forces perpendicular to this motion. From the geometry in Figure 20.4, we can show that [1]

$$V_1 = \frac{V_0 \cos((\beta - \alpha)/2)}{\sin \beta} \tag{20.1}$$

The trigonometry for this is fairly detailed and well developed in the book by Carleone [1]. If an observer was moving with point A as depicted in Figure 20.5, he/she would see point P approaching at a velocity

$$V_2 = V_1 \cos\beta + V_0 \sin\left(\frac{\beta - \alpha}{2}\right) \tag{20.2}$$

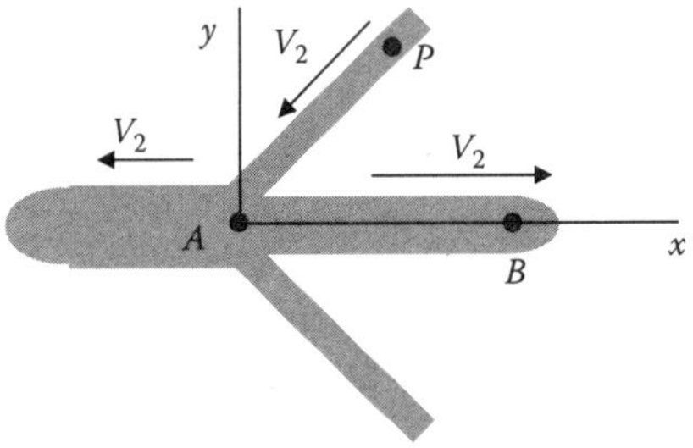

FIGURE 20.5
Jet formation in the Lagrangian frame. (From Walters, W. P., and Zukas, J. A., *Fundamentals of Shaped Charges*, CMC Press, Baltimore, MD, 1989. With permission.)

We can solve for the detonation velocity D through

$$\frac{D}{\cos\alpha} = \frac{V_0 \cos((\beta - \alpha)/2)}{\sin(\beta - \alpha)} \tag{20.3}$$

If we were riding along in our coordinate system at point A, we would see both the slug and the jet moving away from us at velocity V_2 and the liner moving toward us at the same velocity. As a reminder, we are assuming inviscid, incompressible flow in this case. If our coordinate system was stationary (Eulerian), however, we would see the jet velocity as

$$V_j = V_1 + V_2 \tag{20.4}$$

and the slug velocity as

$$V_s = V_1 - V_2 \tag{20.5}$$

The mass of the system must be conserved, therefore at any time t, we can write

$$m = m_j + m_s \tag{20.6}$$

where m_j is the jet mass per unit length into the paper; m_s is the slug mass per unit length into the paper; *and* m is the liner mass per unit length into the paper.
If we now write the conservation of axial momentum, we obtain

$$mV_2 \cos\beta = m_s V_2 - m_j V_2 \tag{20.7}$$

We can solve Equations 20.6 and 20.7 simultaneously to write

$$m_j = \frac{1}{2} m(1 - \cos\beta) \tag{20.8}$$

$$m_s = \frac{1}{2} m(1 + \cos\beta) \tag{20.9}$$

It must be noted that this model assumes that the jet and slug velocities as well as their cross-sectional areas are constant. With all these assumptions, we can write the velocities of the jet and slug, respectively, in terms of our known detonation velocity as

$$V_j = \frac{D}{\cos\alpha} \sin(\beta - \alpha)\left[\operatorname{cosec}\beta + \cot\beta + \tan\left(\frac{\beta - \alpha}{2}\right)\right] \tag{20.10}$$

$$V_s = \frac{D}{\cos\alpha} \sin(\beta - \alpha)\left[\operatorname{cosec}\beta - \cot\beta - \tan\left(\frac{\beta - \alpha}{2}\right)\right] \tag{20.11}$$

We can see from these equations that as $\alpha \to 0$, the jet velocity approaches a theoretical maximum.

$$V_{\max} = D\left[1 + \cos\beta - \sin\beta \tan\left(\frac{\beta}{2}\right)\right] \tag{20.12}$$

But $\beta \to 0$ as $\alpha \to 0$, so

$$V_{\max} = 2D \tag{20.13}$$

Thus, the maximum jet velocity can never exceed twice the detonation velocity of the explosive.

Another noteworthy observation is that as $\alpha \to 0$ and $\beta \to 0$, $V_s \to 0$. Also, as $\alpha \to 0$, we approach a cylindrical geometry of the liner. Cylindrical liners are well known for their high velocity and low mass jets. If we could somehow generate an explosive wave that moved perpendicular to a conical liner, we would see that $\beta = \alpha$ and the velocities of the jet and slug, respectively, could be expressed as

$$V_j = \frac{V_0}{\sin\alpha}(1 + \cos\alpha) \tag{20.14}$$

$$V_s = \frac{V_0}{\sin\alpha}(1 - \cos\alpha) \tag{20.15}$$

With this type of detonation wave, the jet velocity could be increased without bound by decreasing α. However, we must note that as $\alpha \to 0$, $V_0 \to 0$ and $m_j \to 0$. Therefore, the momentum would also approach zero as shown in Equation 20.16:

$$m_j V_j = \frac{mV_0}{2}\sin\alpha \to 0 \tag{20.16}$$

To perform calculations either by hand or with the help of a spreadsheet, the following steps are provided:

- Determine the steady-state jet and slug velocities from Equations 20.10 and 20.11.
- Calculate the masses from Equations 20.8 and 20.9.
- Determine the momentum or energy or other parameters of interest from the results.

This procedure tends to overpredict jet velocities somewhat. Also, since no velocity gradient is present, jet stretching will not be predicted. Let us now look at an example of the procedure.

Example Problem 1

A conical-shaped charge liner is to be fabricated from steel and filled with trinitrotoluene (TNT) as the explosive. The thickness of the liner is to be 0.1 in., and the half angle α is to be 45°. The length of the liner is 5 in., and the charge outside diameter (OD) is 12 in. Determine the following using the Birkhoff et al. theory:

1. Mass of the jet
2. Mass of the slug
3. Velocity of the jet
4. Velocity of the slug

The required properties for this calculation are given as follows:

- TNT Gurney velocity $(2E)^{1/2} = 2.039$ km/s
- TNT detonation velocity $(D) = 6730$ m/s
- TNT density = 1.63 g/cc
- Steel density = 0.283 lbm/in.^{3}

Solution:

The first thing we need to do is get everything in consistent units, the density of TNT first.

$$\rho_{TNT} = (1.63)\left(\frac{\text{g}}{\text{cm}^3}\right)(2.54)^3\left[\frac{\text{cm}^3}{\text{in.}^3}\right]\frac{(2.046)}{(1000)}\left[\frac{\text{lbm}}{\text{g}}\right] = 0.059\left[\frac{\text{lbm}}{\text{in.}^3}\right]$$

Next, we need to break the problem into sections and determine the Gurney velocity for each section. (For this case, we shall use five 1 in. long sections as shown in Figure 20.6.)

We need to determine, for each section, the liner mass-to-charge mass ratio to determine our velocity V_0 for our later calculations. With our truncated cones, we will simply assume that each section is a cylinder at the average radius of the section. Bill Walters [3] suggested that to determine this ratio, we use dimensions of the charge perpendicular to the liner. Then we can write the masses of the liner and charge as follows:

For cross section 1, we have

$$m_1 = \rho_{\text{steel}} 2\pi\left(\frac{r_1 + r_0}{2}\right)t = (0.283)\left[\frac{\text{lbm}}{\text{in.}^3}\right](2)\pi(0.1)[\text{in.}]\left(\frac{1+0}{2}\right)[\text{in.}] = 0.088\left[\frac{\text{lbm}}{\text{in.}}\right] \tag{20.17}$$

$$\begin{aligned} c_1 &= \rho_{\text{TNT}}\pi\left[\frac{r_c^2}{\cos^2\alpha} - \left(\frac{r_1 + r_0}{2}\right)^2\right](0.059)\left[\frac{\text{lbm}}{\text{in.}^3}\right]\pi\left[\frac{(6)^2}{\cos^2(45)} - \left(\frac{1+0}{2}\right)^2\right]\left[\text{in.}^2\right] \\ &= 13.299\left[\frac{\text{lbm}}{\text{in.}}\right] \end{aligned} \tag{20.18}$$

Now the liner segment velocity follows directly from

$$V = \sqrt{2E}\left(\frac{m}{c} + \frac{1}{2}\right)^{-1/2} \tag{20.19}$$

$$V_{0_1} = (2.039)\left[\frac{\text{km}}{\text{s}}\right](1000)\left[\frac{\text{m}}{\text{km}}\right]\left(\frac{0.088}{13.299} + \frac{1}{2}\right)^{-1/2} = 2864\left[\frac{\text{m}}{\text{s}}\right] \tag{20.20}$$

We can now use Equation 20.3 to find the angle β:

$$\frac{D}{\cos\alpha} = \frac{V_0 \cos[(\beta - \alpha)/2]}{\sin(\beta - \alpha)} \tag{20.21}$$

It is convenient to solve this using iteration. Once we have these results, we can determine the jet mass and the slug mass using an average of the angles β. As you can see from

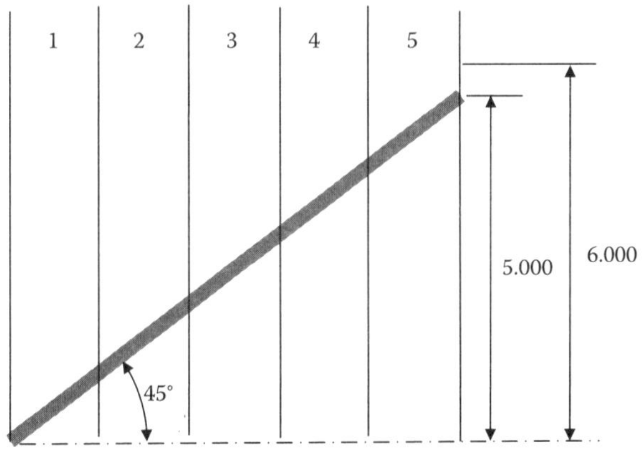

FIGURE 20.6
Discretization of a shaped charge liner.

our overall results contained in Table 20.1, when using this method, this angle does not vary too much. Our average β is 61.768°, so we have
Answer:

$$1.\quad m_j = \frac{1}{2}(2.223)\,[\text{lbm}]\,[1 - \cos(61.768°)] = 0.586\,[\text{lbm}] \tag{20.22}$$

$$2.\quad m_s = \frac{1}{2}(2.223)\,[\text{lbm}]\,[1 + \cos(61.768°)] = 1.637\,[\text{lbm}] \tag{20.23}$$

The overall liner mass is the sum of all our individual masses tabulated in Table 20.1 (or it could be calculated directly from the geometry). It is 2.223 lbm.
The jet and slug velocities are obtained for a conical liner from Equations 20.10 and 20.11:

$$V_j = \frac{(6730)[\text{m/s}]}{\cos(45^\circ)}\sin\left(\beta - 45^\circ\right)\left[\operatorname{cosec}\beta + \cos\beta + \tan\left(\frac{\beta - 45^\circ}{2}\right)\right] \tag{20.24}$$

The answers are shown in Table 20.1. We could also have taken an average as well. For the slug velocity, we have

$$V_s = \frac{(6730)[\text{m/s}]}{\cos(45^\circ)}\sin\left(\beta - 45^\circ\right)\left[\operatorname{cosec}\beta - \cot\beta - \tan\left(\frac{\beta - 45^\circ}{2}\right)\right] \tag{20.25}$$

All our data for this problem are summarized in Table 20.1.

We shall just briefly discuss the PER theory, details of which can be found in the book by Carleone [1]. The PER theory was developed by Pugh, Eichelberger, and Rostoker at the US Army Ballistic Research Laboratory. The theory assumes a variable velocity during liner collapse, which improves the correlation with experiment. Typically, as a liner collapses, the collapse velocity decreases as the detonation wave progresses from the apex of the cone to its base. This makes sense based on what we have learned so far since there is usually a smaller explosive mass compared to the liner mass. The end result is that the tip of the formed jet moves faster than the tail or slug, stretching the jet.

When the velocity of collapse decreases with time, the collapse angle β actually increases as does the amount of material entering the jet. This is illustrated in Figure 20.7. If we examine this figure, we see that as the detonation wave travels from point P to Q, the element originally at P collapses to J. From the figure, we also see that the element at P' arrives at M at the same time that P reaches J. If the collapse velocity were constant, point P' would arrive at N instead, and the collapsed shape would be conical as we have seen in Figure 20.4.

TABLE 20.1

Results of Computations for Jet and Slug Velocities

Position	$m = rV/L$	$c = rV/L$	Segment Velocity V_0 (m/s)	β (°)	V_j (m/s)	V_s (m/s)
1	0.089	13.299	2864	62.310	5115.384	1280.944
2	0.267	12.928	2826	62.074	5063.423	1262.104
3	0.445	12.187	2784	61.818	5006.654	1241.659
4	0.622	11.075	2734	61.516	4939.138	1217.534
5	0.800	9.592	2669	61.122	4850.150	1186.366
Total	2.223	59.082	Average	61.768	4994.950	1237.721

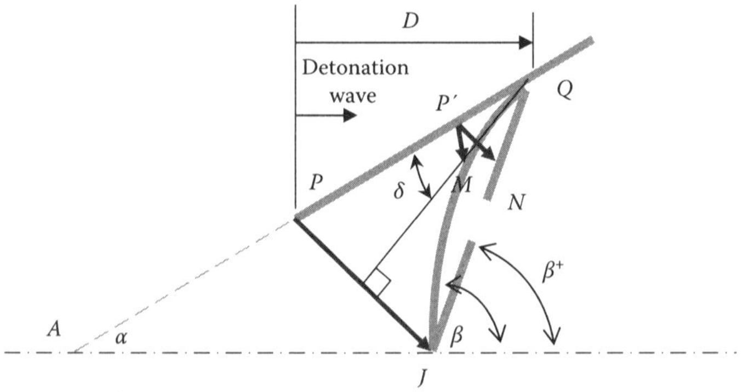

FIGURE 20.7
Geometry of the PER theory. (From Walters, W. P., and Zukas, J. A., *Fundamentals of Shaped Charges*, CMC Press, Baltimore, MD, 1989. With permission.)

Since the derivation of this theory is adequately addressed in the book by Carleone [1], we will not derive the detailed mathematics behind it. The interested reader is referred to that work for the details.

The results based on Figure 20.7 yield an instantaneous velocity for the tip of the jet and the tail of the slug as given in the following:

$$V_j = V_0 \operatorname{cosec}\frac{\beta}{2}\cos\left[\alpha - \frac{\beta}{2} + \sin^{-1}\left(\frac{V_0}{2u}\right)\right] \tag{20.26}$$

$$V_s = V_0 \sec\frac{\beta}{2}\sin\left[\alpha - \frac{\beta}{2} + \sin^{-1}\left(\frac{V_0}{2u}\right)\right] \tag{20.27}$$

Here u is defined as

$$u = \frac{D}{\cos\alpha} \tag{20.28}$$

At any time, the mass must be in either the liner, the slug, or the jet, so we can write

$$dm = dm_j + dm_s \tag{20.29}$$

where

$$\frac{dm_j}{dm} = \sin^2\frac{\beta}{2} \tag{20.30}$$

$$\frac{dm_s}{dm} = \cos^2\frac{\beta}{2} \tag{20.31}$$

We can now see that Equations 20.17 through 20.22 depend upon the cone angle 2α, the detonation velocity D, the collapse angle β, and V_0.

Now we shall let t be the elapsed time between the instant the detonation wave passes the apex of the cone and define

$$T = \frac{x}{D} = \frac{x}{u\cos\alpha} \tag{20.32}$$

We can then express the position of any particle of the liner, initially at a distance x from the apex in cylindrical coordinates as

$$Z = x + V_0(t - T)\sin A \tag{20.33}$$

$$r = x \tan\alpha - V_0(t - T)\cos A \tag{20.34}$$

where we define

$$A = \alpha + \delta \tag{20.35}$$

From this, the angle β can be shown to be

$$\tan\beta = \frac{\sin\alpha + 2\sin\delta\cos\alpha - x\sin\alpha(1 - \tan A\tan\delta)(V'_0/V_0)}{\cos\alpha - 2\sin\delta\sin A + x\sin\alpha(\tan A + \tan\delta)(V'_0/V_0)} \tag{20.36}$$

where

$$V'_0 = \frac{dV_0}{dx} \tag{20.37}$$

Equations 20.17 through 20.27 are typically solved by computer to determine the formation parameters and describe the jet formation. It is beyond our scope to discuss the coding of the equations. Results of this model are shown in the book by Carleone [1].

Problem 1

A conical-shaped charge liner is to be fabricated from copper and filled with composition B as the explosive. The thickness of the liner is to be 0.1 in. and the half angle α is to be 45°. The length of the liner is 3 in., and the charge OD is 7 in. Determine the following using the Birkhoff et al. theory:

1. Mass of the jet
   Answer: $m_j = 0.302$ [lbm]
2. Mass of the slug
   Answer: $m_s = 0.990$ [lbm]
3. Velocity of the jet
   Answer: $V_j = 4752$ [m/s]
4. Velocity of the slug
   Answer: $V_s = 1086$ [m/s]

Note that depending on how you discretize the problem, you may get a somewhat (but not too) different answer.
The required properties for this calculation are given as follows:

- Composition B Gurney velocity $(2E)^{1/2} = 2.35$ km/s
- Composition B detonation velocity (D) = 7890 m/s
- Composition B density = 1.717 g/cm^3
- Copper density = 0.323 lbm/in.^{3}

Problem 2

A conical-shaped charge liner is to be fabricated from copper and filled with composition B as the explosive. The thickness of the liner is to be 0.15 in., and the half angle α is to be 30°. The length of the liner is 5 in., and the charge OD is 8 in.

1. Determine the following using the Birkhoff et al. theory:

   a. Mass of the jet
      Answer: $m_j = 0.410$ [lbm]
   b. Mass of the slug
      Answer: $m_s = 1.787$ [lbm]
   c. Velocity of the jet
      Answer: $V_j = 7500$ [m/s]
   d. Velocity of the slug
      Answer: $V_s = 960$ [m/s]

2. Estimate the jet length assuming constant velocity of the tip and slug if the standoff is 1 m (use the fastest tip velocity and the average slug velocity).
   Answer: $L \approx 0.875$ [m]

The required properties for this calculation are given as follows:

- Composition B Gurney velocity $(2E)^{1/2} = 2.79$ km/s
- Composition B detonation velocity $(D) = 7910$ m/s
- Composition B density = 1.717 g/cm^3
- Copper density = 0.323 lbm/in.
- Steel density = 0.283 lbm/in.^{3}

20.2 Shaped Charge Jet Penetration

Now that we have discussed how shaped charge jets are formed, we will move to how they penetrate their targets. As previously mentioned, shaped charge jets are formed at relatively close standoffs. The jet stretches from the instant it is formed with velocities ranging from 10 (tip) to 2 km/s (tail). Because of this stretching, the jet will eventually break up, thereby reducing penetration because of drift/tumbling of the jet segments. This is known as particulation.

The penetration performance of shaped charge jets is dependent upon whether or not they are continuous. The further away from a target that the jet is formed, the more the jet will stretch. If this standoff distance is large enough, the jet will particulate. This particulation complicates the penetration calculation.

The simplest penetration formula is attributed to Birkhoff [2], who assumed a constant velocity of the jet and thus described jet penetration through a momentum balance

$$\frac{1}{2}\rho_j(V-U)^2 = \frac{1}{2}\rho_t U^2 \tag{20.38}$$

where ρ_j is the jet density; ρ_t is the target density; U is the velocity of the bottom of the hole in the target; and V is the (constant) velocity of the jet.

By solving for U in the aforementioned equation and noting that the total penetration can be described as follows, we can obtain an expression for the depth of penetration:

$$P(t) = \int_0^t U(t)\mathrm{d}t \tag{20.39}$$

Here $P(t)$ is the total penetration of the jet at time t.

From the aforementioned integral, we obtain the formula from the penetration of a continuous velocity jet (called the density law):

$$P = l_j \left(\frac{\rho_j}{\rho_t} \right)^{1/2} \tag{20.40}$$

Here l_j is the length of the jet. Equation 20.40 states that for a constant velocity jet, the penetration is dependent upon only the jet length and the density ratio. If the jet is segmented, Pack and Evans [2] proposed the following relation:

$$2\rho_j (V - U)^2 = \rho_t U^2 \tag{20.41}$$

which implies

$$P = l_j \left(\frac{2\rho_j}{\rho_t} \right)^{1/2} \tag{20.42}$$

where l_j is the length of the jet including the gaps between segments and ρ_j is the jet density calculated based on the length (including gaps) so that the overall density will be lower than a continuous jet.

In this case, P ends up usually being lower. We must note that there are cases in which a particulated jet can actually penetrate deeper into the target material than a nonparticulated jet [2].

As the jet velocity decreases, there is a point where the constitutive strength of the target material becomes important. There are formulas by Pack and Evans as well as by Eichelberger that account for this [2].

The expressions developed so far assume that the jet velocity is constant. If this assumption does not provide an accurate enough answer, we can use the formulas derived by DiPersio and Simon [2] to account for jet stretching. This technique uses three different formulas dependent upon where particulation occurs.

If the jet is continuous throughout the penetration event, we can write

$$P = s\left[\left(\frac{V_0}{V_{min}} \right)^{1/\gamma} - 1 \right] \tag{20.43}$$

If particulation occurs sometime during the penetration event,

$$P = \left\{ \frac{(1+\gamma)(V_0 t_1)^{(1/(1+\gamma))(\gamma/(S^{1+\gamma}))} - V_{min} t_1}{\gamma} \right\} - s \tag{20.44}$$

If particulation occurs before penetration,

$$P = \frac{(V_0 - V_{min})t_1}{\gamma} \tag{20.45}$$

In Equations 20.43 through 20.45, V_0 is the jet tip velocity; s is the distance from the target surface to the virtual origin of the jet (this is a theoretical origin derived from examination of a velocity–distance curve—to be explained later); t_1 is the time from jet formation to particulation; V_{min} is the minimum jet velocity capable of penetrating the target material; and γ is defined as

$$\gamma = \left(\frac{\rho_t}{\rho_j}\right)^{1/2} \tag{20.46}$$

V_{min} is a value that is usually between 2 and 8 km/s. There are various methods to calculate V_{min}, but we usually assume 2 km/s for the purposes of rough analysis. Some authors use U_{min}—as we shall see later.

V_{min} (or U_{min}) for a metallic target can be calculated through [4]

$$V_{min}\left[\frac{cm}{\mu s}\right] = U_{min}\left[\frac{cm}{\mu s}\right] = 0.044 + (0.000206)(BHN) \tag{20.47}$$

Note that this expression uses the Brinnell hardness number (BHN) of the target as a parameter affecting penetration.

Another form of the nonuniform velocity jet equations similar to the DiPersio and Simon equations is as follows [4]. For very short standoffs, defined as

$$0 \le s \le (1+\gamma)V_{min}t_1\left[\frac{(1+\gamma)V_{min}}{V_0}\right]^{1/\gamma} \tag{20.48}$$

the depth of penetration can be found through

$$P = s\left\{\left[\frac{V_0}{(1+\gamma)V_{min}}\right]^{1/\gamma} - 1\right\} \tag{20.49}$$

or

$$P = \left\{\frac{(1+\gamma)(V_0t_1)^{(1/(1+\gamma))(\gamma/(S^{1+\gamma}))} - V_{min}t_1}{\gamma}\right\} - s \tag{20.50}$$

For moderate standoffs, defined as

$$(1+\gamma)V_{min}t_1\left[\frac{(1+\gamma)V_{min}}{V_0}\right]^{1/\gamma} \le s \le V_0t_1 \tag{20.51}$$

the depth of penetration is given by

$$P = \frac{(1+\gamma)}{\gamma}(V_0t_1)^{(1/(1+\gamma))(\gamma/(S^{1+\gamma}))} - \frac{1}{\gamma}\sqrt{(1+\gamma)(V_{min}t_1)(V_0t_1)^{(1/(1+\gamma))(\gamma/(S^{1+\gamma}))}} - s \tag{20.52}$$

and for long standoffs where

$$V_0t_1 \le s \le \frac{V_0t_1}{\gamma}\left(\frac{V_0}{V_{min}} - 1\right) \tag{20.53}$$

the penetration can be found through

$$P = \frac{V_0t_1}{\gamma^2}\sqrt{(V_{min}t_1)(V_0t_1 + \gamma s)} \tag{20.54}$$

Mott, Pack, and Hill [4] developed a theory that accounts for the material behavior of the jet. This theory is known as the MPH theory. In this theory, the density of a shaped charge jet is given by

$$\rho_j = \frac{m_j}{V_j} = \frac{m_j}{A_j l_j} \tag{20.55}$$

In Equation 20.55, we have used the density ρ_j; mass m_j, cross-sectional area A_j, length l_j, and volume V_j, of the jet. The MPH theory states that the penetration depth is given by

$$P = l_j \sqrt{\frac{\lambda \rho_j}{\rho_t}} \tag{20.56}$$

Here the parameter λ is a factor which accounts for how the material behaves:

- For pure hydrodynamic behavior, $\lambda = 1$.
- For a particulated jet, $\lambda = 2$.
- For a jet which is hydrodynamic but particulates, $1 < \lambda < 2$.

If we insert Equation 20.46 into Equation 20.55, we obtain

$$P = \sqrt{\frac{\lambda m_j l_j}{\rho_t A_j}} \tag{20.57}$$

We can modify this formula for the effect of standoff distance by assuming the distribution of the jet mass is linear with its length or, mathematically,

$$l_j = l_0(1 + \alpha s) \tag{20.58}$$

Here l_0 and α are constants and s is the standoff distance.

If we assume pure hydrodynamic behavior, then the volume of the jet is a constant and $\lambda = 1$. We know that

$$V_j = A_j l_j \tag{20.59}$$

and

$$P = \sqrt{\frac{m_j l_j}{\rho_t A_j}} = \sqrt{\frac{m_j l_j^2}{\rho_t V_j}} = l_j \sqrt{\frac{m_j}{\rho_t V_j}} \tag{20.60}$$

If we include the effects of standoff, we can write

$$P = l_0(1 + \alpha s)\sqrt{\frac{m_j}{\rho_t V_j}} \tag{20.61}$$

This states that P linearly varies with s or, mathematically,

$$P \propto (1 + \alpha s) \tag{20.62}$$

This behavior is graphically shown in Figure 20.8.

If we assume the jet particulates, then the cross-sectional area of the jet is a constant and $\lambda = 2$. Then Equation 20.57 can be written as

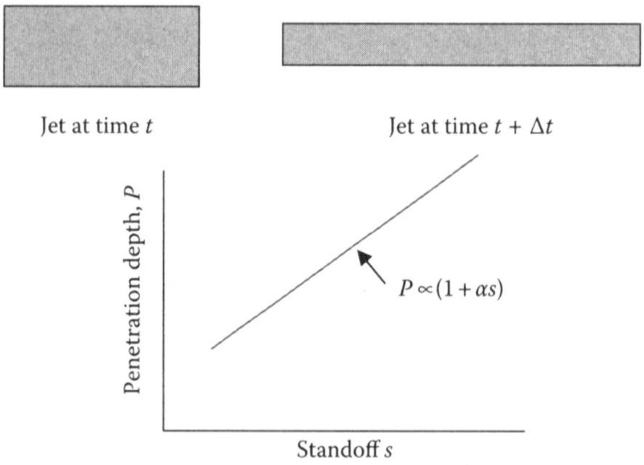

FIGURE 20.8
Hydrodynamic jet behavior.

$$P = \sqrt{\frac{2m_j l_j}{\rho_t A_j}} \tag{20.63}$$

If we include the effects of standoff, we can write

$$P = \sqrt{\frac{2m_j l_0(1 + \alpha s)}{\rho_t A_j}} \tag{20.64}$$

This states that P varies with the square root of $(1 + \alpha s)$ or, mathematically,

$$P \propto (1 + \alpha s) \tag{20.65}$$

This behavior is illustrated in Figure 20.9.

If we assume that the jet is somewhat hydrodynamic and particulates, then both the volume and area of the jet are variables and $1 < \lambda < 2$. In this case, Equation 20.57 applies directly. If we modify this expression to include the effects of standoff, we can write

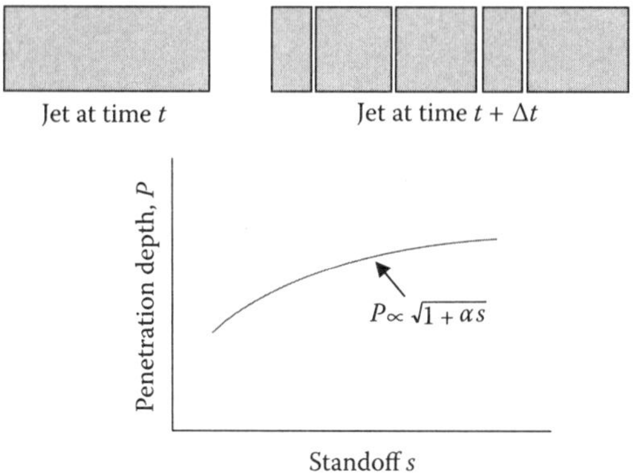

FIGURE 20.9
Particulating jet behavior.

$$P = \sqrt{\frac{\lambda m_j l_0 (1 + \alpha s)}{\rho_t A_j}} \tag{20.66}$$

This states that P varies with the square root of $\lambda(1 + \alpha s)$ or, mathematically,

$$P \propto \sqrt{\lambda(1 + \alpha s)} \tag{20.67}$$

This behavior is illustrated in Figure 20.10.

The MPH theory can be modified to account for jet waver. Jet waver is the phenomenon whereby the particles in the jet move off the flight axis as illustrated in Figure 20.11. This is caused by imperfections in the formation, strain hardening of the jet material, and subsequent breakup that provides for asymmetric particles. These particles begin to rotate with the end result being a jet that does not completely exert its energy in deepening the hole in the target but widens the hole as the particles impact the sides. We can account for this by adjusting the area term through

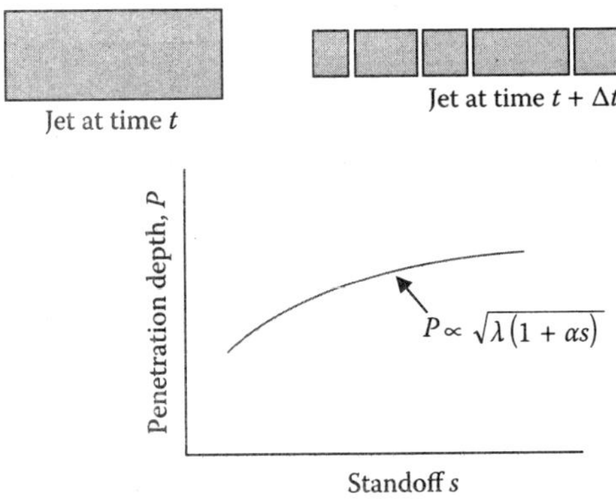

FIGURE 20.10
Mixed mode jet behavior.

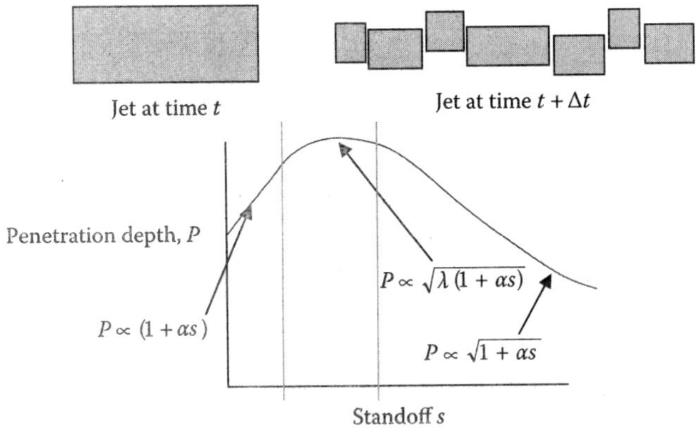

FIGURE 20.11
Wavering jet behavior.

$$A_j = A_1\left(1 + Bs^2\right) \tag{20.68}$$

Here A_1 and B are empirically obtained constants. This expression can be substituted directly into our three penetration equations to yield

$$P = \sqrt{\frac{m_j l_0 (1 + \alpha s)}{\rho_t A_1 (1 + Bs^2)}} \tag{20.69}$$

$$P = \sqrt{\frac{2 m_j l_0 (1 + \alpha s)}{\rho_t A_1 (1 + Bs^2)}} \tag{20.70}$$

$$P = \sqrt{\frac{\lambda m_j l_0 (1 + \alpha s)}{\rho_t A_1 (1 + Bs^2)}} \tag{20.71}$$

Here Equations 20.69 through 20.71 replace Equations 20.61, 20.64, and 20.66, respectively. As we can see from Figure 20.11, we can use each of these equations to determine the depth of penetration dependent upon the standoff distance.

Once we have decided upon the proper particulation model to use for the penetration event, we determine the depth of penetration. To do this, we need to examine the penetration event from a Lagrangian viewpoint. If we were watching the stationary target as shown in Figure 20.12, we would see a hole that is deepening while the jet was shortening. In this figure, the rear of the jet would have a faster velocity V than the speed at which the hole was advancing U. It is convenient to analyze this problem from a Lagrangian viewpoint. If we invoke the principle of superposition, we will obtain a situation as depicted in Figure 20.13.

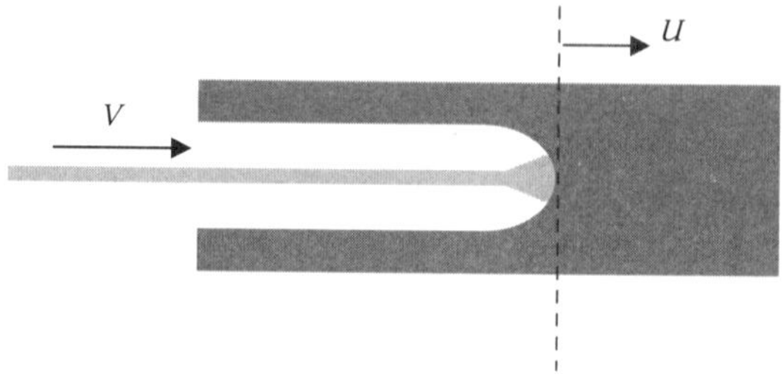

FIGURE 20.12
Eulerian view of a jet penetration.

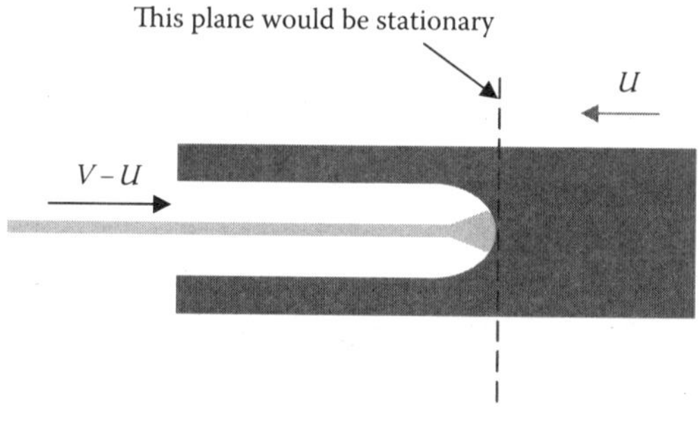

FIGURE 20.13
Lagrangian view of a jet penetration.

In this case, the jet velocity, relative to the hole velocity, would be $V - U$, and an observer moving with the hole would see target material approaching them at velocity U.

If we use an analysis technique that lets us imagine a jet of constant length passing through the target material and somehow relate this to a hole depth, we would have the visualization depicted in Figure 20.14.

With this model, we can write the conservation of momentum equations in the variables that we have previously defined as

$$\rho_j (V - U)^2 = \frac{\rho_t U^2}{\lambda} \tag{20.72}$$

With the exception of the coefficient λ, this equation is identical to the Birkhoff equation (Equation 20.38). If we collect the velocity terms and take the square root of both sides, we get

$$\frac{U}{(V - U)} = \sqrt{\frac{\lambda \rho_j}{\rho_t}} \tag{20.73}$$

The depth of penetration is still given by Equation 20.39 and essentially results in the penetration velocity times the penetration time, so we can write

$$P = U t_P \tag{20.74}$$

But we can state t_P as

$$t_P = \frac{l_j}{(V - U)} \tag{20.75}$$

Substitution of Equation 20.75 into Equation 20.74 yields

$$P = l_j \sqrt{\frac{\lambda \rho_j}{\rho_t}} \tag{20.76}$$

Again similar to the Birkhoff et al. theory with the exception of λ, for different configurations, we would assign a different value of λ:

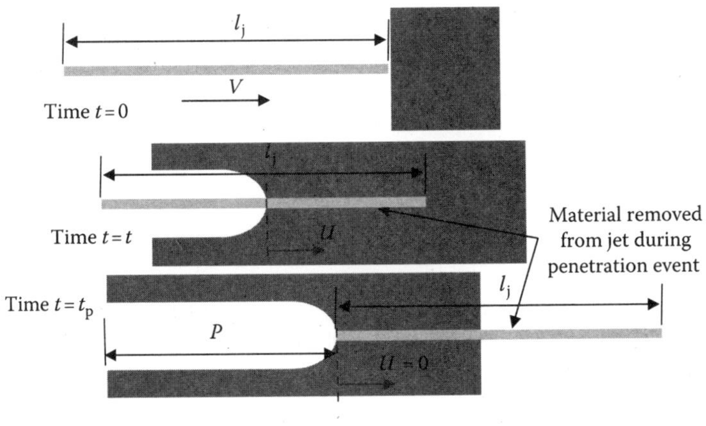

FIGURE 20.14
Model that assumes constant jet length during penetration event.

- For fluid (hydrodynamic) jets, $\lambda = 1$.
- For particulating jets, $\lambda = 2$.
- For mixed mode jets, $1 < \lambda < 2$.

We could also account for standoff by adjusting l_j accordingly.

We shall now discuss the virtual origin concept. Many researchers have determined relationships that use the virtual origin to describe shaped charge jet behavior [1,2,5]. The virtual origin is an empirically derived distance that is obtained from multiple jet tests. We have stated (repeatedly) that a real-shaped charge jet has a gradient in velocity from the tip of the jet to the tail. The velocity is highest at the tip. If we assume that this velocity gradient is linear, then when the jet impacts the target, we can say that the distance the tip (or any part) has traveled can be written as

$$x = Vt + s \tag{20.77}$$

Let us consider a situation where we fire three identical jets at different stand offs, say, 1, 1.5, and 2.5 m. If the tip velocity in each case is (say) 8 km/s and the tail velocity in each case is variable, further assume that we get a time–distance curve for each jet as shown in Figure 20.15. With this, the virtual origin will be where all parts of each jet with the same velocity line up and intercept the x-axis. This is depicted in the figure.

We have examined several models for the penetration behavior of shaped charge jets. This is greatly dependent on their formation and particulation. These models are by no means the end of all jet penetration analytical tools. There is still a great deal of work that is ongoing to describe this important behavior.

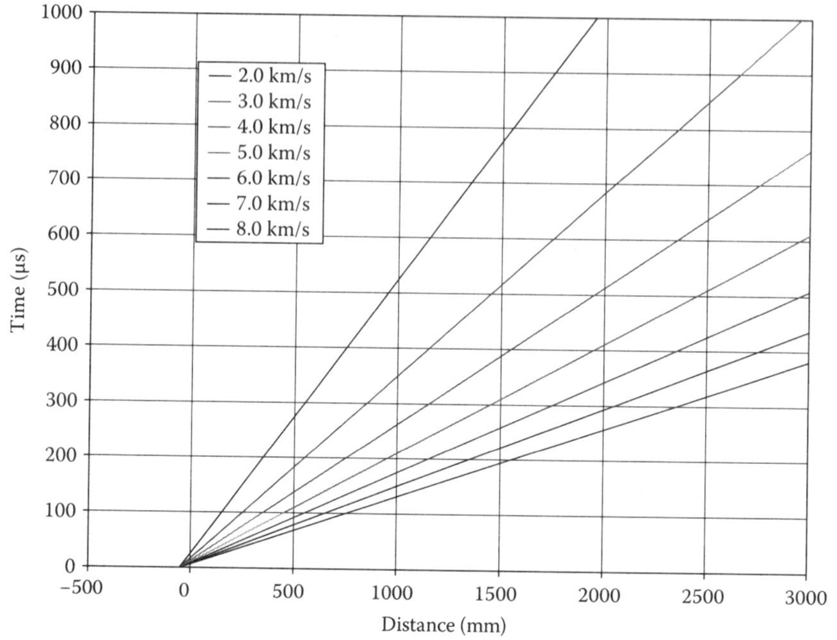

FIGURE 20.15
Virtual origin concept.

Problem 3

A conical liner as shown in Figure 20.16 is to be fabricated from copper and filled with composition A-3 as the explosive. The thickness of the liner is to be 0.10 in. The length of the 20° conical liner is 4 in., and the charge OD is 4 in. The case is fabricated from steel and is an 8 in. long cylinder and 0.12 in. thick.

1. Determine the following by using the Birkhoff et al. theory and ignoring effects of confinement (if the region over the liner is discretized into four segments, that should be sufficient):

   a. Masses of the jet and slug
   Answer: $m_j = 0.067$ [lbm] and $m_s = 0.488$ [lbm]
   b. Velocities of the jet and slug
   Answer: $V_j = 8800$ [ft/s] and $V_s = 570$ [ft/s]
   c. The velocities of the case material (plot this as fragment velocity vs. case length)
   d. The direction in which the case material fragments will be projected (plot this as departure angle vs. case length)

2. If the standoff is 8 in., determine the maximum penetration into a rolled homogeneous armor (RHA) plate at a 15° angle of obliquity by using the formula of DiPersio and Simon and assuming no particulation.
   Answer: $P = 11.89$ [in.]

The required properties for this calculation are given as follows:

- Composition A-3 Gurney velocity $(2E)^{1/2} = 2.63$ km/s
- Composition A-3 detonation velocity $(D) = 8.14$ km/s
- Composition A-3 density = 1.59 g/cm^3
- Copper density = 0.323 lbm/in.^{3}
- Steel density = 0.283 lbm/in.^{3}

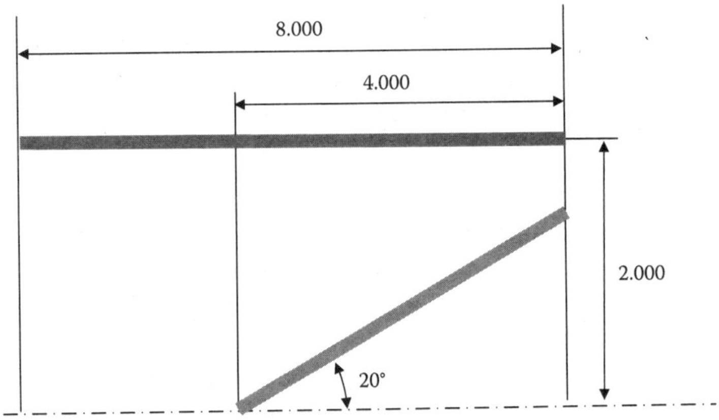

FIGURE 20.16
Shaped charge of Problem 3.

Problem 4

A conical-shaped charge liner as shown in Figure 20.17 is to be fabricated from copper and filled with composition B as the explosive. The thickness of the liner is to be 0.15 in., and the half angle α is to be 30°. The length of the liner is 5 in., and the charge OD is 8 in.

1. Determine the following using the Birkhoff et al. theory:
   a. Mass of the jet
      Answer: $m_j = 0.410$ [lbm]
   b. Mass of the slug
      Answer: $m_s = 1.787$ [lbm]
   c. Velocity of the jet
      Answer: $V_{jAvg} = 7491$ [m/s]
   d. Velocity of the slug
      Answer: $V_{sAvg} = 965$ [m/s]

2. Estimate the jet length assuming constant velocity of the tip and slug if the standoff is 1 m (use the fastest tip velocity and the average slug velocity).
   Answer: $L \approx 0.875$ [m]
3. Using the aforementioned data and assuming that the virtual origin is 10 cm behind the standoff measurement, estimate the penetration ability into mild steel assuming the jet does not particulate using the formula of DiPersio and Simon.
   Answer: $P = 2.98$ [m]
4. Compare the aforementioned answer in part 3 to that for the density law.
   Answer: $P = 0.935$ [m]

The required properties for this calculation are given as follows:

- Composition B Gurney velocity $(2E)^{1/2} = 2.79$ km/s
- Composition B detonation velocity $(D) = 7910$ m/s

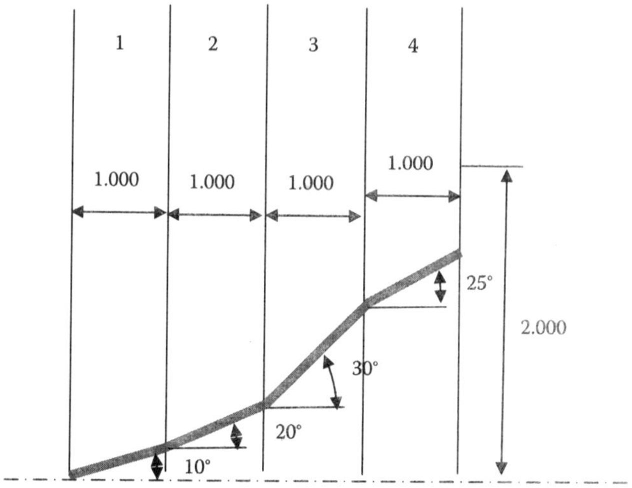

FIGURE 20.17
Shaped charge of Problem 4.

- Composition B density = 1.717 g/cm^3
- Copper density = 0.323 lbm/in.^{3}
- Steel density = 0.283 lbm/in.^{3}

Problem 5

A trumpet liner is to be fabricated from copper and filled with composition A-3 as the explosive. The thickness of the liner is to be 0.12 in. We shall approximate the trumpet liner as indicated in the following where the half angle α is to be variable. The length of the liner is 4 in., and the charge OD is 4 in.

1. Determine the following using the Birkhoff et al. theory:

   a. Masses of the jet and slug
   Answer: $m_s = 0.545$ [lbm] and $m_j = 0.045$ [lbm]
   b. Velocities of the jet and slug
   Answer: $V_j = 5200$ [ft/s] and $V_s = 200$ [ft/s]

2. If the standoff is 8 in., determine if the jet will perforate 5 in. of an RHA plate at a 70° angle of obliquity by using the formula of DiPersio and Simon and assuming no particulation.
   Answer: No, $P = 3.3$ [in.]

The required properties for this calculation are given as follows:

- Composition A-3 Gurney velocity $(2E)^{1/2} = 2.63$ km/s
- Composition A-3 detonation velocity (U_D) = 8.14 km/s
- Composition A-3 density = 1.59 g/cm^3
- Copper density = 0.323 lbm/in.^{3}
- Steel density = 0.283 lbm/in.^{3}

References

1. Carleone, J. (Ed.), *Tactical Missile Warheads*, American Institute of Aeronautics and Astronautics, Washington, DC, 1993.
2. Walters, W. P., and Zukas, J. A., *Fundamentals of Shaped Charges*, CMC Press, Baltimore, MD, 1989.
3. Walters, B., Personal communications with author June 20, 2002.
4. Cooper, P. W., *Explosives Engineering*, Wiley-VCH, New York, 1996.
5. Zukas, J. A., and Walters, W. P. (Eds), *Explosive Effects and Applications*, Springer, New York, 1997.

Further Reading

Lindner, V., *Theory of Propellants and Explosives Course Notes*, Randolph, NJ, Set 20, p. 30, 1990.

21

Wound Ballistics

Until this point, we have dealt with the penetration of projectiles into inanimate objects. One of the more distressing aspects of ballistics is the fact that it is used against living creatures. This is not meant to imply that hunting is good or bad, but simply that there are instances when people, intentionally or not, fire weapons at other people or animals, and the effects of the bullet impact must be understood.

When a projectile is fired at a living creature, some amount of incapacitation is desired. If a projectile is of the nonlethal type, trauma to the target must be minimized, and either a fluid must be injected, the victim must be rendered physically immobile, or some other effect must be obtained. If the projectile is of the lethal type, ideally one hit should subdue the victim (through any protection), rendering them incapable of harm.

Because the subject of wound ballistics is as complicated as the anatomy of the target, we shall conduct only a cursory review here, pointing the interested reader to some excellent references for further detail. We shall only treat subjects that affect humans, although these may affect animals in a similar manner.

An interesting statistic is that over 58% of combat casualties in the British army during the First World War were caused by fragments rather than bullets [1]. This is interesting since we all have seen movies (accurate or not) of wild charges into machine gun fire. This is probably the case with most conflicts.

In the sections on aeroballistics, we have learned to treat projectile flight through a fluid medium (air). While these equations still hold in a human body, the simplifications we made do not always hold, and we must take steps to include properties such as the elasticity of tissue. The initial conditions such as entrance angle become important when dealing with a wound. Additionally, a bullet is usually unstable in a human body, causing it to greatly yaw or even tumble. Thus, bullet geometry, mass properties, and material strength matter a great deal as far as the extent of damage is concerned.

Before we discuss the details any further, it must be understood that there are many people who have diligently studied the field of wound ballistics during their entire careers. These researchers have drawn on their wide experience, some from the engineering viewpoint and some from the medical viewpoint, to reach conclusions and develop theories about wound physics. They are probably all correct even though their viewpoints may be vastly different. The reality is that "anything" can happen when a bullet interacts with a human. It has been our experience that the experts can be categorized into two broad camps: the medical camp and the engineering camp. The medical camp sees the wounds (even wounds that were caused by an identical bullet at an identical entrance angle into an identical location) as individually different, and each must be treated through a medical procedure based on the caregivers' experience, observations, and understanding. The medical camp believes that the psychological and physiological effects of a wound will always be different and that no conclusions can be drawn based on weapon type, etc. The engineering camp believes that wounding can be quantified through physics. They believe that relationships (potentially very complex) can be drawn based on energy, momentum, material properties, etc., which can be used to quantify the effects of projectiles against persons. The truth is probably a combination of the ideas of both groups, but to date, no one has found the Holy Grail that would bring it all together.

Based on the aforementioned discussion, it is very difficult to absolutely define incapacitation to cover all possible aspects of wounding. When defining incapacitation, we must set some criteria. We like three simple (although general) ones:

1. Whether the targets survive the wounding
2. What the targets are able to do after they are hit
3. How long the targets are able to do it after they are hit

Other authors define things differently and consider things such as whether the targets are ambulatory after being hit and how far they can move; can they operate a vehicle; and if armed, whether they can fire back or not. Criteria that we like are given by Marshall and Sanow [2].

The study by Peters [3] states that there are several misconceptions about wounding that must be addressed. One misconception is that the temporary cavity is the major cause of tissue damage. This has probably grown out of extremely interesting videos that have been published showing massive temporary cavities in projectile firings into gelatin blocks. It is difficult to imagine, as a human, that these cavities would not cause huge amounts of damage. In fact, this topic is rather hotly debated by experts. We shall pass no judgment here, but simply state that the work of Peters suggests that less than 20% of all tissue damage is caused by the temporary cavity.

Another misconception pointed out by Peters is that tissue damage is proportional to kinetic energy of the projectile. Peters suggested that there is a relationship, but it is nonlinear. It was thought (and possibly still is) that the sizes of the maximum temporary cavity and the permanent cavity were somehow proportional to energy deposited in the target by the projectile. Peters suggests that there is a nonlinear relationship, but additionally, over some ranges of the data, it can be linearized, which is possibly why the conclusion was drawn.

Engineers who look at a person as an engineering structure at some point assume that the volume of the permanent cavity must, in some way, result from material ejected from the wound. That is, that the permanent cavity volume must equal the volume of material ejected. This is not the case since a permanent cavity remains even when the bullet stops in the target. The cause of this permanent cavity is primarily due to the inelastic deformation of the tissue.

Peters and other researchers have shown that temporary cavities in humans or animals will be of different sizes than those developed in gelatin blocks. Currently, this is a very active area of research. There are even differences in cavity formation between animals and humans to the extent that no scaling law has been universally established.

One of the most interesting aspects of wound ballistics is the inertial effect on a human body. In many Hollywood action films, we routinely see people being picked up and thrown several feet backward by impacts of small arms projectiles. When the numbers are worked out with a 7.62 mm projectile at point blank range, the energy exchange (assuming the bullet remains lodged in the target) is such that the rearward velocity is less than 0.2 mph. In fact, most human targets usually fall toward the shooter (unless they were running away when hit).

We will next discuss some bullet types that are illustrated in Figure 21.1. A solid slug is nothing more than a soft metal (usually a lead alloy) projectile that is engraved along its

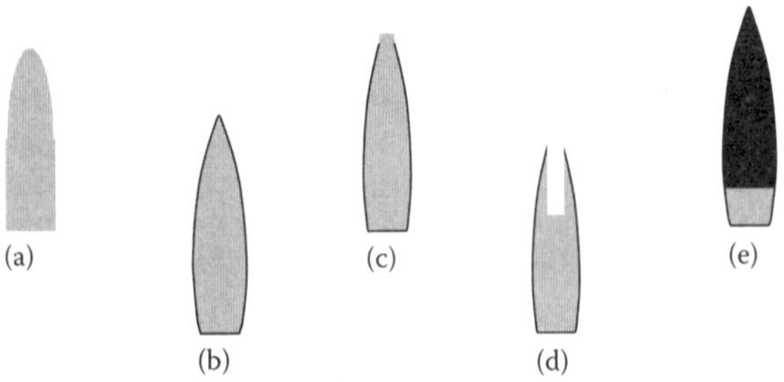

FIGURE 21.1
Geometry of several small arm bullet types: (a) solid slug, (b) full metal jacket, (c) semijacketed, (d) hollow point, and (e) steel core.

body length by the rifling to impart spin. A full metal jacket (FMJ) projectile is a solid slug that is coated with a material such as copper to better withstand firing stresses and whose residue can easily be removed from the inside of the gun tube. A semijacketed projectile or open-tipped projectile is jacketed up to a small region of the nose. This region, being softer than the jacketed region and unable to withstand the radial stresses upon impact, expands as it enters the target, theoretically causing a more extensive wound. A hollow point projectile is similar to a semijacketed projectile except that the tip is actually concave. It uses fluid mechanics coupled with the lower radial strength upon penetration to open larger. Finally, the steel-core projectile has a hard core for penetration of metallic structures or textile armor. One common type that is not shown is the wad cutter type, which can be fully jacketed or not but have a cuplike shape to the nose so that they punch a nice, clean hole through paper targets. There are many other projectile types such as slit jackets and dumdums, but usually, they fall into one of the aforementioned categories.

In the earlier paragraphs, we mentioned some terms such as temporary cavity and permanent cavity. We will now define some of these terms.

A laceration is a cut through tissue. The primary means of incapacitation of a projectile is through laceration. Because of the complicated nature of the human body, a projectile that penetrates can do anything from causing minor bleeding if no major organ or artery is damaged to rapid death if a vital organ is hit. If a projectile impacts bone tissue or even meets a severe gradient in density, it can be considerably deflected.

We learned a great deal about stress waves previously. When a projectile enters a human being, it sends stress waves through the body. These waves and associated rarefactions can cause damage, but it is generally agreed that these waves will primarily damage nerves and can possibly collapse organs.

The temporary cavity is created through the process of cavitation introduced earlier in the fluid mechanics section (Figure 21.2). It results from the adherence of the fluid molecules to the surface of the projectile, and when the shear stress drops to zero on the surface, the flow separates. This separation bubble can grow to 40 times the projectile diameter as the projectile passes through the body. Once the projectile has passed by, however, the radial energy that it imparted to the tissue is removed, and the elasticity of the tissue causes it to immediately collapse to a much smaller size. The largest extent that this bubble reaches is known as the maximum temporary cavity while the small, equilibrium cavity is known as the permanent cavity.

Projectile yaw has a dramatic effect on cavitation. As stated earlier, a projectile is usually unstable in a human body. This causes it to considerably yaw and possibly tumble. As one

FIGURE 21.2
Temporary cavity.

can imagine, because of the relatively immense presented area of a projectile flying with a large yaw, the separation and associated cavitation can be huge. In fact, if a projectile rotates 180°, it will usually exit the target base first. This is depicted in Figure 21.3.

The analysis of this flight behavior is extremely difficult because projectiles perform differently depending upon what tissue they happen to be passing through. The following is a short list of just a few of the different types of tissues that affect bullet passage through a living creature:

- Bone
- Skull and brain
- Thorax/ribs
- Lung
- Intestine/stomach/bladder
- Muscle

Each of these tissue types will have a different effect on the projectile. It is even important if an organ is flaccid (empty) or not or whether the target is living or dead. For simplicity, the most general research is carried out on muscle tissue, and that is where a great deal of work has been expended to come up with a suitable surrogate material for testing.

FIGURE 21.3
Cavitation due to projectile yaw.

Assuming we are discussing muscle tissue penetration, the first thing we must recognize is that tissue has a nonnegligible tearing stress that must be overcome. This additional stress must be incorporated into our drag model. We cannot emphasize the complexity of the problem enough. Even though, in the discussion that follows, we shall assume a penetration into homogeneous muscle tissue, we must always keep in mind that a penetration event is much more complicated. We know that as a projectile enters muscle tissue, what was once relatively simple aeroballistics becomes a more complicated problem of continuum mechanics: in air, there was no yield stress to overcome (this is the major difference); the viscosity and density of muscle are different from those of air. If the impact angle is low enough, the nose of the projectile will enter first. The usual decrease in shear stress as we progress along the projectile will occur, and at some point, the shear stress will reach zero and the tissue will separate from the projectile forming a cavitation bubble. Throughout this event, the projectile will slow down due to drag. There will also be a larger overturning moment than in air because of the large force on the small area of the nose (higher density in the dynamic pressure term), and in addition, the separation will take place ahead of the center of gravity, increasing the moment arm. The drag force will also include the force required to overcome the cohesive stresses in the tissue (tearing stress), which is not usually included in aerodynamic models. What was once a transonic/supersonic flow field becomes a transonic (at best) or subsonic flow field. This is because the speed of sound in muscle tissue is around 1500 m/s (4920 ft/s).

In comparison to the aerodynamic models we have presented earlier, Peters et al. [4] developed a drag model that accounts for the tearing of the tissue. The equation of motion is given by

$$-m\frac{\mathrm{d}V}{\mathrm{d}t} = \frac{1}{2}\rho V^2 AC_\mathrm{D} + \frac{1}{2}\rho(aU)^2 AC_\mathrm{D} \tag{21.1}$$

or it can be written in terms of distance traveled as

$$-mV\frac{\mathrm{d}V}{\mathrm{d}x} = \frac{1}{2}\rho AC_\mathrm{D}\left[V^2 + (aU)^2\right] \tag{21.2}$$

In these equations, m is the mass of the projectile; V is its velocity; ρ is the density of the tissue; A is the presented area* of the projectile; C_D is the projectile drag coefficient; x is the distance the projectile has progressed into the tissue; a is a modification to C_D; and U is a characteristic velocity of the tissue (more on these last two terms will follow).

If we examine Equation 21.2, we see that if we exclude the second term on the right-hand side, we get our classic equation for aerodynamic drag (assuming, of course, that the area is a cross-sectional area S of the projectile). The second term accounts for the energy loss associated with the tearing of the tissue and its movement away from the projectile.

The characteristic velocity is defined as

$$U = U_6\left(\frac{d}{d_6}\right)^{-1/3} \tag{21.3}$$

These are empirically derived values. In Equation 21.3, d is the diameter of the actual projectile, d_6 is the diameter of a 6 mm projectile (in case you want the units of d in a different

* Here the presented area is different from the cross-sectional area we used in the sections on exterior ballistics, which is why A is used and not S.

system), and U_6 is a characteristic velocity for different materials determined through experiments with a 6 mm diameter projectile. The parameter a in Equations 21.1 and 21.2 is a function of the projectile type and the angle of attack of the projectile.

The stability criteria developed in Part II of this book work fairly well for behavior in the human body. As stated earlier, the density terms must be increased as well as the effect of Mach number. It is also recommended to add a force term as was included in Equation 21.2; however, that would require a rederivation of the stability equations, which is beyond our scope.

If a projectile has features that would cause it to expand upon impact with the more dense human tissue, it will cause greater trauma. These were mentioned earlier as hollow point and slit-jacketed bullets. The opening of the hollow point or jacket allows more of the projectiles energy to be transferred to the body and the flatter surface directs the flow of the tissue in a more radial direction. If a bullet is unstable in the body and it tumbles, there is more surface area presented for the body to slow the projectile down, and thus, more energy would be expended on the body. A greater amount of cavitation will occur as well due to greater radial flow of the tissue. Depending on whether this expansion happens at the entrance to the body, the exit, or somewhere in between, the wound would be affected as depicted in Figure 21.4.

Some cursory remarks on the physical observations of wounds based on forensic science are worth mentioning at this point. This should allow the reader to tie some of the physics examined in the earlier sections of the book to the forensics literature.

Tattooing or stippling is a phenomenon where the target exhibits small marks on its body near the entry wound. It is caused by the impact of burnt, partially burnt, and unburnt propellant on the skin. It is less prevalent (but still may occur) if the wound was through clothing. From the color of the marks, the forensic expert can discern whether the victim was dead or alive prior to being shot. If the victim was alive at the time of the wounding, the marks will be reddish brown or orange in color. If the victim was deceased at the time of the shooting, the marks will be grey or yellow. The marks themselves are not actually burns; they are due to the impact of unburnt propellant grains and other ejecta from the firing. Nitrocellulose-based propellants often quench quickly when subjected to the rapid pressure decay as they leave the muzzle of the weapon. If the propellant was black powder, however, they could actually be burns because black powder does not quench as readily.

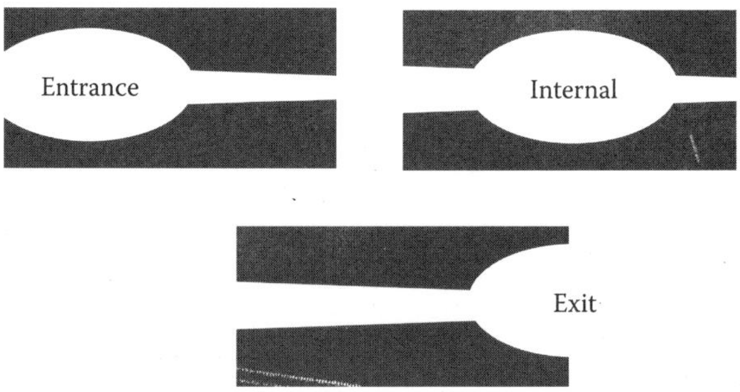

FIGURE 21.4
Wounds that are affected by the time at which tumbling or bullet head expansion occurs.

The range at which a target is fired upon affects the nature of the wound. Following Dodd [5] we shall define the following basic range bands:

1. Contact
   a. Hard contact—muzzle pressed into the skin
   b. Contact—muzzle touching the skin
2. Near contact—muzzle very close to the skin
3. Intermediate
4. Distant

The distinctions between these will be elaborated upon. There are many texts specifically devoted to the forensic evaluation of gunshot wounds [5,6], and these should be referenced for more detailed descriptions of what is described next.

Close-range wounds are evidenced by the following: parts of shot shells, wads, pistons in the body (if weapon was a shotgun); unburnt propellant (sometimes multicolored) in wound; stippling/tattooing due to impact of unburnt propellant on skin which cannot be wiped away in a postmortem examination (as mentioned previously, this is because most of the "stuff" that was involved in the interior ballistic cycle is ejected at firing). In many cases, the presence and type of clothing worn by the target will have some effect on the wound.

Contact wounds result when the muzzle of the weapon is touching or very near to the skin when fired. In the case of hard contact, one normally sees a bruising or imprint from the muzzle of the weapon and possibly the front sight. Because in the hard contact case the bore is sealed against the skin, the propellant gases are forced deep into the wound, blackening the cavity. The edges of the entry hole may be abraded. The abraded region is called the abrasion rim. Grease and dirt from the bullet and bore may get wiped into the wound. When this is present, it is called the grease rim or seared zone. The wound may show radial indications of gas wash from an imperfect seal. In hard contact, there is generally no tattooing or stippling. Since the propellant gases are at a high pressure and temperature, shredded muscle from gas wash, which is bright pink/red, is usually observed. If the weapon was pressed against a bony area, the wound may be ragged, and the propellant gas often strips skin from the bone.

In the contact case, a black ring will usually be present where muzzle gases exited and a soot ring or sear ring will be elongated in the direction of fire.

The near contact case occurs when the muzzle of the weapon is roughly 1–3 calibers away from the target. Usually there will be a wider soot ring than contact case about 4–6 calibers in diameter. One may see both abrasion and grease rings under some soot. The soot buildup will be opposite the direction of fire.

The intermediate range case is characterized as follows. Tattooing or stippling are usually present and of red-brown or orange color. The victims may have propellant particles embedded in their skin. These particles are most prevalent when the cartridge used ball, cylindrical, or single perforated grains. There may be a grease and/or an abrasion rim. For pistols, intermediate range is about two to three barrel lengths. For rifles, intermediate range is about 3 ft.

The distant range case is probably the one most encountered in military operations. In most distant range cases, there will only be a bullet hole present. It is possible to observe an abrasion or grease rim. A low-velocity (subsonic) impact may exhibit a neat hole and a well-defined abrasion rim, while a high-velocity (supersonic) impact may exhibit a hole with radiating splits. The presence of the abrasion rim is largely dependent on the aerodynamic shape of the bullet: A pointier projectile yields less abrasion rim while a blunter bullet yields more abrasion rim.

Since in military operations one is more likely to encounter distant range and perhaps intermediate range wounding, we shall discuss the behavior of different weapon-type wounding in these situations. At these ranges, round nose bullets will typically exhibit less well-defined entry holes and will bruise the target more. With handgun wounds at intermediate ranges, bullets sometimes (but not always) mushroom due to their construction of softer core material. Handgun wounds at distant range usually exhibit circular holes with small tears.

With 0.22 rimfire cartridge wounds at intermediate range, the following is generally observed. If the victim is shot in the head with a 0.22 short cartridge, the bullet rarely exits the cerebral cavity, while the same location impacted with a 0.22 long or magnum cartridges may exhibit secondary skull fractures. At distant range, holes caused by 0.22 rimfire cartridges are ice pick like. Hollow points in this caliber do not generally mushroom unless they hit bone, and even then, they often just penetrate into it. If mushrooming does occur, 0.22 long rifle and magnum are more likely to do so.

Musket balls and canister shot rarely exit the body, and when they do, the exit wound is large. In most cases, the entry wound is about the size of the ball, but the wound track is usually larger than the ball. Minié balls are far more damaging than round balls and even more damaging than FMJ bullets.

Center-fire rifle wounds at intermediate range can injure organs not in direct path due to shock effects (although this is debated). These cartridges create a very large temporary cavity. Occasionally clothing imprints are found on the entry area. At distant range, these cartridges exhibit behavior that is very similar to that at intermediate range. The entrance wound will usually be about two to three times the bullet diameter with the exit wounds usually being larger than entry wound.

The location in which a victim is hit will have a great deal of influence on the incapacitating effect of the projectile [2]. Nearly immediate incapacitation will occur when the impact is in the region between the victims' eyebrows and sternum. With a powerful cartridge, an impact in the pelvic area will cause a target to collapse. This is because the pelvis is the main load-bearing region of the body. An impact in the stomach area usually takes the longest time for incapacitation. With a handgun, unless the brain stem or upper spine is hit, it usually takes about 5 s to incapacitate a victim. Less time is possible, but this is usually due to the psychological state or medical particulars of the victim. Even when hit in the heart, a handgun wound can take 10–15 s to incapacitate a victim. A hit in the liver, spleen, or kidneys takes about 30–90 s to incapacitate, and a hit in the lungs takes over 90 s to incapacitate the victim.

As mentioned earlier, the behavior of the projectile will be affected by the different tissue types through which it passes. The different tissue types are similarly affected by the projectile differently based on their properties.

Bones are the hardest materials in the body. They behave somewhat like multilayer composites. The skull has some unique behavior because of its makeup and geometry. The calvarium has a softer layer called the diploe, which is sandwiched between two hard bony layers. Bullet impact normal to the skull causes a condition called beveling. This is chipping or spalling on the surface opposite entry. Radial fracture lines will typically emanate from the hole. A trained pathologist can sometimes tell which hole occurred first in multiple hits based on how these cracks arrest themselves on previous cracks. Keyholing is a phenomenon observed in head wounds and occurs when the projectile enters the skull at an oblique angle. Keyholing usually occurs on the entry surface of the entry wound with some beveling on the inner surface of the entry wound. If the victims' skull were empty, there would only be a small entrance and exit hole. Beveling can also occur at the pelvis or on other bones. The concept of spallation applies here equally well. Bullets have been observed to ricochet off bones and even travel along them.

Internal organs have very different properties from one another and, therefore, respond differently to bullet impacts. The effects will vary depending on the organ hit and the velocity of the projectile. With a subsonic bullet, there will normally be a small, neat hole with minimal cavitation. In the case of a supersonic bullet, we would generally see large holes with large temporary cavities. Bullet breakup is also a major factor in wounding. When, where, and how the projectile breaks up make a difference in wound severity.

Brain penetrations usually result in immediate collapse of the victim. The bullet path can result in catastrophic damage. A large temporary cavity may cause the skull to explode or the brain to be ejected from the skull. Impact in the frontal lobe may not cause immediate incapacitation.

Impacts to the spinal cord usually result in impairment of the victims' motor skills. Possible complete immobilization of the victim can occur depending on the impact site and whether the spinal cord is severed.

Penetrations of the heart or major arteries are generally not immediately fatal. There will be major internal or external loss of blood from the victim.

Impacts to the liver vary from victim to victim. This is because liver tissue can vary in elasticity and hardness due to the victims' health. If the victims' liver is hardened, it may actually spall. Death of the victim usually results from bleeding and is therefore not immediate. Higher velocity round impacts to the liver are far more catastrophic. A human liver generally has a specific gravity between 1.02 and 1.04, so it is close to water in density. The liver is not elastic, and this results in large regions of bullet destruction. Because of this inelasticity, the size of the permanent cavity is on the order of the temporary cavity.

Impacts to the stomach, gall bladder, and intestines are similar. These organs are known as hollow viscera. Death from bullet penetrations of these organs usually results from bleeding and is therefore not immediate. More damage results if the body is in the process of digestion.

The kidneys are similar in structure to the liver. Impacts to the kidneys are more survivable than impacts to the liver because the fatty tissue present in the kidneys may contain the resultant bleeding.

The bladder responds to impacts in a manner similar to the stomach. More damage results from the penetration if the bladder is distended by urine. The bladder is protected to some degree by the pelvis, although spall from the pelvis may contribute to bladder damage.

The spleen is a solid organ. Death from a spleen penetration usually comes about through bleeding and is therefore not immediate. High-velocity projectiles may shatter the spleen completely.

Muscle tissue was previously mentioned. It has a specific gravity between 1.01 and 1.02. As mentioned earlier, it is a cohesive material that resists bullet motion. There will usually only be a small region of bullet destruction. The elasticity of muscle tissue results in large temporary cavities but small permanent cavities.

Lungs are extremely different from all of the other organs in the body. Lung penetrations are fairly survivable dependent upon the extent of damage. If major blood vessels are ruptured, then death comes through suffocation. The specific gravity of lung tissue is between 0.4 and 0.5. Lung penetrations exhibit very small temporary cavity formations and a correspondingly small region of bullet destruction.

Incapacitation is probably the most controversial topic in all the ballistics disciplines. One can safely say that there is no one right answer. As stated earlier, living beings are very complicated targets influenced by their psychological as well as their physical state. Some current incapacitation theories (as interpreted by us) are the following:

1. Kinetic energy deposition theory
2. Hit location theory
3. Judicious mixture theory

In all these cases, common sense must be your guide.

In the first two cases, the theory is usually couched as "all else being equal." In reality, all things are never equal. No one theory can explain anything, although people struggle to define that one thing that can compare their favorite bullet to all others—the elusive "silver bullet."

The kinetic energy deposition theory states that projectile impact velocity has the greatest influence on wound severity—all other things being equal. Damage to the target results from laceration followed by bleeding in addition to tissue disruption and crushing. Shot location is important—so all things are not really equal. This theory goes on to state that a projectile that deposits more of its kinetic energy in the target will cause more damage. Thus, if a projectile stays in the body, it has deposited *all* of its kinetic energy. Fackler [7] disagreed with this. The shot location statement was incorporated due to the ability of different organs to absorb kinetic energy. This comes about because of the density of the organs and cohesiveness of the tissue. Organ density affects the ability to absorb kinetic energy—more dense means more likely to be damaged. The cohesiveness of tissue is the ability to dissipate kinetic energy by stretching. DiMaio [6] attributed kinetic energy loss to four factors: the initial kinetic energy of the projectile, the yaw of projectile at impact, the caliber and structure of projectile, and the tissue properties along the projectile path.

The greater the initial kinetic energy of the projectile at impact, the more there is available to deposit into the target. The greater the kinetic energy, the greater the propensity for projectile deformation and/or breakup. If the projectile does not deform, too much kinetic energy may be viewed as undesirable as the projectile may pass through the target; however, the target may still be incapacitated and may bleed to death, but this takes time. A pass-through certainly deposits less energy in the target (however, there was more to begin with, so this may be a circular argument).

More yaw at impact means that it is more difficult to penetrate the target (remember the effect of yaw on penetration?). If the projectile does penetrate the body, then the hole will be bigger and the projectile will drag down faster—that means more kinetic energy deposited in a target. It is an interesting side note that a projectile that does not penetrate and falls to the ground actually deposits all of its kinetic energy into the target—Is this a hole in the theory?

Larger caliber and/or blunter projectiles deposit more kinetic energy in a target because they have greater drag. Projectiles that expand deposit more kinetic energy into the target. This is why there are projectiles that are designed to expand by using a hollow point, slit jackets, lower hardness core, etc. Higher drag projectiles deposit more kinetic energy into the target. Projectiles that fragment deposit more kinetic energy into the target as well.

Tissue properties along the projectile path are another factor in this theory. Greater density tissue would allow more kinetic energy to be deposited by the projectile along the path.

Greater tissue strength would allow more kinetic energy to be absorbed by the body. Greater elasticity in the tissue would allow more kinetic energy to be dissipated. Fackler [7] stated that projectiles with the same kinetic energy that impact different locations will have completely different effects. This seems to be supported by at least parts of this theory.

DiMaio [6] had a theory which states that the same amount of damage occurs once some critical projectile impact velocity is achieved. According to the theory, different projectile types have different critical velocities. For FMJ projectiles and steel balls, the critical velocity is between 800 and 900 m/s. For soft or hollow point projectiles, the critical velocity is between 457 and 610 m/s.

To defend against projectile impacts, body armor has been developed since projectiles were first fired. Metallic armors were good against ball ammunition, but armor-piercing rounds can go through them easily. Textile/composite armor has met with better success at stopping penetration, but it can still happen. Even with textile armor, some depth of penetration or organ damage is still possible. In addition, the same mechanisms that we discussed about nonpenetrating damage are applicable here as well such as shock waves and momentum transfer (which is even greater for a nonpenetrating hit than a pass-through).

In summary, we have touched upon several aspects of wound ballistics. A more comprehensive treatment is provided by Sellier and Kneubuehl [8]. It is a complicated and hotly debated subject, yet one that is extremely fascinating.

References

1. Winter, J. M., *The Experience of World War I*, Oxford University Press, New York, 1989.
2. Marshall, E. P., and Sanow, E. J., *Stopping Power, a Practical Analysis of the Latest Handgun Ammunition*, Paladin Press, Boulder, CO, 2001.
3. Peters, C. E., Common misconceptions about the physical mechanisms in wound ballistics, *Journal of Trauma (China)*, 6(2), 1990, Supplement, 319–326.
4. Peters, C. E., Sebourn, C. L., and Crowder, H. L., Wound ballistics of unstable projectiles—Part 1: Projectile yaw growth and retardation, *Journal of Trauma, Injury, Infection and Critical Care*, 40(3), 1996, S10–S15.
5. Dodd, M. J., *Terminal Ballistics: A Text and Atlas of Gunshot Wounds*, Taylor & Francis, Boca Raton, FL, 2006.
6. DiMaio, V. M. J., *Gunshot Wounds: Practical Aspects of Firearms, Ballistics and Forensic Techniques*, 2nd ed., CRC Press, Boca Raton, FL, 1999.
7. Fackler, M. L., Wound ballistics, a review of common misconceptions, *Journal of the American Medical Association*, 259, 1988, 2730–2736.
8. Sellier, K. G., and Kneubuehl, B. P., *Wound Ballistics and the Scientific Background*, Elsevier, Amsterdam, 1994.

Further Reading

Peters, C. E., *Defensive Handgun Effectiveness*, Self-published, 1997.

Appendix A

Glossary

Active homing: A method of guidance whereby the device is guided by electronics that contain both a transmitter and a receiver, so that the munition can be adjusted onto the target.

Autofrettage: A process by which the inner layer of material in a gun tube is plastically yielded and held in compression by the outer layer. This increases the fatigue life of the weapon by limiting the cyclic stress amplitude during repeated firings.

Autonomous munition: A munition that needs no input from an outside source once fired.

Azimuth: The rotation of a weapon about the pintle or turret ring (side to side) as opposed to elevation (q.v.) (which is up and down).

Bag charge: A propelling charge that is not contained in a cartridge case. It usually takes the form of either a number of silk (or alternate material) bags or nitrocellulose based hard casings that are filled with granular propellant.

Ballistic cap: See *Windshield*.

Balloting: The lateral motion of the projectile in the gun tube. This can be one of three modes: the whole projectile moving side to side with its centerline remaining parallel to the bore axis, the projectile nose and base rotating about the center of gravity (centerline of projectile at an angle to bore axis), and projectile remaining pushed to one side of the bore and rotating in a cyclic motion at the rifling twist rate.

Band seat: The annular groove in a projectile into which the rotating band is swaged or welded.

Base: The rear end of a projectile.

Base gap: A gap between the explosive fill and metal base or wall of a projectile that can be very dangerous. If the weapon is fired, setback forces compress the air in the gap with a resultant heating. This process occurs over milliseconds so that the heat cannot be transferred away. The resultant heat can detonate the explosive fill in the bore of the weapon usually resulting in a loss of the weapon and the crew.

Battery: A group of three to six field artillery pieces.

Battery (second definition): The position of a weapon in its carriage when it is or is not ready to fire. For instance, a weapon "out of battery" is not ready to fire, while a weapon "in battery" is ready to fire.

Bayonet: A knife or spike that attaches to the muzzle of a rifle used in hand-to-hand combat.

Bayonet lug: A boss or protrusion located near the muzzle of a rifle that is the attachment point for the bayonet.

Bent: A latch which engages the sear, preventing the firing pin from moving forward until released by the sear.

Berdan primer: A primer whose anvil is an integral part of the cartridge case.

Bipod: A pair of supports that are used to steady a mortar or a gun, so that it can be aimed or to increase accuracy by limiting muzzle movement.

Boat tail: The angled rear end of a projectile.

Bolt: The device in a small arm that houses the firing pin. A bolt can be manually operated or automatically operated. The bolt usually obturates the breech of the weapon as well.

Booster: A section of explosive charge, usually attached to the fuze whose purpose is to accept the initiation from the primary detonator and amplify the detonation to more reliably and completely initiate the main charge of a projectile. It can be made from the same material as the main charge or different material. Its key characteristic is proximity to the primary initiation train so that reliable and timely ignition is assured.

Bore evacuator: A device connected to the bore of a gun by ports that fills with high-pressure gas upon firing. After the projectile exits the muzzle of the weapon, this high-pressure gas pushes any remaining smoke and burning embers out of the tube before the breech is opened. It is used with vehicles that have closed firing compartments so that the crew is not affected by smoke or any burning debris entering the compartment. On warships, there is an external system that blows the hot gases out (a bore scavenger).

Bourrelet (pronounced Boor′ rel lay): Regions of the projectile where the diameter is full caliber (usually divided into a forward and an aft bourrelet and separated by the undercut (q.v.)).

Boxer primer: A primer whose anvil is enclosed as part of the primer itself.

Breechblock: Device which allows access to the chamber for the loading of ammunition into the weapon and closes to maintain pressure in the chamber during the ballistic cycle. Normally, a breechblock is designed so that gravity drops it into place. Used almost exclusively with cartridge cased ammunition.

Breech plug: Device which allows access to the chamber for the loading of ammunition into the weapon and closes to maintain pressure in the chamber during the ballistic cycle. It generally screws into the breech of the weapon with an interrupted thread and can obturate the propellant gases if a cartridge case is not used.

Brilliant munition: A precision munition that can classify potential targets and potentially select the one with the highest value.

Brisance: A property of an explosive that relates its shattering effect. This is related to the rate of energy release in the explosive. A "brisant" explosive will shatter its container rather than expand it to burst like a balloon.

Burster: A charge of energetic material in a projectile or munition that is intended to burst the outer casing of the device and spread the contents over some defined area.

Butt: The end of a rifle that rests on the shoulder of the firer.

Caliber: The smallest internal diameter of a gun tube. Also, a unit of measure for a tube length. A 155 mm 39-caliber gun tube is 39 × 155 mm or 6045 mm in length.

Canards: Control surfaces mounted to an airframe or projectile ahead of the center of gravity or center of pressure.

Candle: The device carried in an illumination projectile that burns upon expulsion from the projectile after parachute deployment. The purpose of the candle is to illuminate the battlefield to allow combat to take place at night. Some candles illuminate in the infrared spectrum so that they aid only soldiers equipped with infrared optical equipment.

Canister: A projectile resembling a shotgun shell containing a burster charge and a large number of metal balls or flechettes. The purpose of a canister round is to incapacitate personnel in relatively close proximity to the weapon.

Cannelure: Circumferential groove cut in a rotating band or projectile jacket to reduce engraving pressure and allow a place for material to flow during the engraving process.

Cap: A relatively thin metal device attached to the nose of an armor-piercing projectile to grease the main penetrator by biting into the armor. It can also help penetration if the projectile strikes at an oblique angle by rotating the projectile on impact normal to the armor plate.

Carriage: The component on a weapon platform that connects the gun assembly to the trails and wheels. See also *Upper carriage* and *Lower carriage*.

Cartridge: The assembly that contains the case, propellant, and projectile.

Cartridge case: The metal or energetic material that is attached to the base of some projectiles. The purposes of the cartridge case are to contain the proper amount of propellant, keep the propellant protected against the environment, help obturate the breech, and, in the case of a combustible cartridge case, provide additional propelling energy to the projectile. The breech plug must obturate when a combustible cartridge case is used.

Cartridge rim: The flange on the cartridge case that has several functions. It retains the cartridge when the bullet is loaded into the chamber. It allows the extractor a surface to interact with to remove the spent cartridge case from the weapon.

Center of gravity (CG): The location on a body where, analytically, all the mass can be concentrated and the resultant force vector directed toward the center of the earth. The resultant force vector is equivalent to the distributed load.

Center of pressure (CP): The location on a body in motion through a fluid where, analytically, all the pressure force (integrated over the surface of the body) can be concentrated. The resultant force vector is equivalent to the distributed load.

Centering band: A band made of soft material attached ahead of the threaded region of a projectile for the purpose of maintaining concentricity of the parts. Centering bands have also been used on the exterior of projectiles to limit balloting or maintain a central position in the bore.

Click: A military term for 1 km.

Clip: A device which contains several cartridges that is fed into the magazine of a weapon.

Closing plug: A threaded plug which seals the base end of a projectile (if base fuzed) or a hi–low cartridge case (q.v.).

Commencement of rifling: The point in a gun tube at which the lands have attained full size.

Conical ogive: An ogive that is conical in shape.

Coppering: The deposition of copper from either rotating bands or jacketed projectiles along the bore of the weapon.

Cradle: Device on a weapon platform that connects the sleigh to the trunnions and allows the sleigh to rotate about the trunnions (i.e., rotate in elevation).

De-coppering agent: A material added to the propelling charge to react with the copper deposited by the projectile during firing to eliminate the buildup of copper or fouling of the gun tube.

Down bore: The direction from the breech toward the muzzle (in the direction of projectile travel).

Elevation: The rotation of a gun about the trunnions (up and down).

Equilibrators: Devices which overcome the effect of gravity when a weapon is elevated because the center of gravity of the weapon is usually ahead of the center of rotation (the trunnions).

ET fuze (electronic time fuze): A fuze that utilizes electrical energy and timing circuits to count time to initiation. This type of fuze is much more accurate than an MT (mechanical time) fuze.

Eutectic alloy: An alloy that has a physical state, under normal environmental conditions, at the eutectic point on a phase diagram (near its melting point). These alloys are used in designs where the high temperature of a fire will melt them and allow some mechanism to drop or just open a vent hole.

Expulsion charge: A charge placed in a projectile whose purpose is to expel cargo.

Fin shroud: A ringlike structure used to tie fins on mortar rounds or rockets for structural support. These devices usually have an adverse effect on drag.

Fins: Control surfaces mounted to an airframe or projectile aft of the center of gravity or center of pressure.

Flash hole: A hole through which hot gases may pass to ignite energetic material in a separate chamber or area.

Flash reducer: A mixture of material whose purpose is to reduce muzzle flash by either lowering the temperature of the combustion or inhibiting the reaction of the propellant combustion products with the air. Reduction of flash usually results in increased smoke.

Flechette: Small dart used for antipersonnel rounds.

Forcing cone: The region immediately down bore of the chamber where the internal diameter of the tube tapers to the correct caliber.

Fusable lifting plug: A lifting plug containing a eutectic alloy that melts out if the projectile is exposed to high temperature during storage allowing a pressure vent for the expulsion charge material.

Fuze: A device that is attached to the nose, base or, in some instances, buried within a projectile that contains the initiation mechanism for initiating end effects. Note that in the US Army, it is correctly spelled with a *z*. This is not the case in the Air Force or Navy documentation.

Gain twist: A scheme of rifling where the twist increases with down bore distance. The intention is to minimize wear and angular acceleration of the projectile.

Grommet: A device used with copper rotating bands to protect the soft copper from damage during rough handling; removed before ramming the projectile.

Grooves: The part of the rifling which is cut into the tube material. The internal diameter of the grooves is larger than the internal diameter of the lands.

Guided munition: A munition or projectile that has onboard guidance to steer it to the target.

Head: The portion of the bolt which presses up against the rear face of the cartridge case through which the firing pin passes. The head obturates the breech with the assistance of the cartridge case.

Headspace: The space between the head of the bolt and the forward lip of the chamber that accepts the rim of the cartridge. It is important that the headspace not be too large or small so that operation of the weapon can proceed smoothly.

HEAT (high explosive anti-tank): A projectile which uses a shaped charge for terminal effects.

HEP (high explosive plastic): A projectile with a thin, soft shell that will mash upon impact with a target.

HESH (high explosive squash head): Another name for a HEP projectile.

High explosive (HE): An energetic material that detonates, given a proper stimulus, regardless of confinement.

Hi-low: A propelling charge configuration in which there are two chambers: a high-pressure chamber and a low-pressure chamber. The propellant burns in the high-pressure chamber and exits through vent holes to pressurize the low-pressure chamber. The gases in the low-pressure chamber actually push on the projectile to impart the proper velocity.

Igniter core: A cylinder of pyrotechnic material whose purpose is to ignite the propelling charge as uniformly as possible. The igniter core is usually initiated by an igniter pad or a primer.

Igniter pad: A cloth pouch containing a sensitive pyrotechnic mixture sewn to the rear of a bag charge. The function of the igniter pad is to accept the input flame from the primer, amplify it, and either ignite the propellant or begin the burning of the igniter core.

Jacket: A hoop of metal assembled around a gun tube to increase its strength.

Jet: See *Shaped charge jet*.

Laced jacket: The outer casing of some bag charges.

Lands: The part of the rifling with an internal diameter, i.e., the caliber of the weapon.

Laser designator: A device carried or mounted on a vehicle that can illuminate (sometimes called "paint") a target using laser energy, so that a semiactive laser-guided projectile can ride the beam to the target.

Lifting plug: A device threaded into the fuze well of a nose-fuzed projectile to lift the projectile. It is removed before firing.

Lifting plug (energy absorbing): A device threaded into the fuze well of a nose-fuzed projectile to lift the projectile. It is removed before firing. This device differs from a standard lifting plug in that it is designed to shear off if the ogive of the projectile is impacted, thereby preventing fuzing of the round. Its design came about because HE projectiles would crack when dropped on the nose, the crack going unnoticed, and the projectile would detonate in-bore when fired due to structural failure.

Liner: A conical, hemispherical, or other shape manufactured out of metal or glass that, when exposed to the properly conditioned detonation of an explosive, will form a jet which will penetrate armor plate.

Loading density: The ratio of the weight of the powder charge to the volume of the empty cartridge case or chamber. Also the density to which an explosive is consolidated.

Lock time: The amount of time between when a trigger of a weapon is pulled and the weapon discharges.

Low explosive: An energetic material that requires the proper stimulus and confinement to detonate. Gun propellants are low explosives.

Lower carriage: A platform-like structure on a field piece that contains the pintle and connects the trails to the wheels or axle.

Lunette: Ring welded to the trails (or muzzle brake on some newer weapons) of a field piece that allows the weapon to be towed.

Magazine: The device in a weapon that contains the cartridges.

Man-in-the-loop: A technique (frowned upon at one time by the US Army) whereby a soldier is required to designate a target until the impact of the projectile.

Méplat: The blunt forward end of a projectile.

Mercy mission (MRSI—multiple rounds, simultaneous impact): A fire mission where the weapons fire multiple projectiles, varying the elevation and charge, so that all the projectiles impact the target area simultaneously.

Mil: Angular unit of elevation or deflection, 1/6400 of a circle approximately 1/1000 of the range. When used in a statement such as "The projectile had 5 mils of right deflection," means that the projectile fell 5 m to the right of the line of fire for every kilometer of range.

MT fuze (mechanical time fuze): A fuze that utilizes stored mechanical energy in the form of springs and gearing to count time from firing until initiation.

MTSQ fuze (mechanical time, super-quick fuze): A fuze that utilizes stored mechanical energy in the form of springs and gearing to count time from firing until initiation and has a point detonating mode that will initiate on contact with a surface. This allows a backup if the time setting is in error and will detonate before the projectile buries itself into the ground (which limits its effectiveness).

Mushroom: Device mounted in the breech plug of weapons that use bag charges to seal (obturate) the breech upon pressurization of the chamber.

Obturation: The sealing of propelling gases behind the projectile and in the chamber of a weapon.

Obturator: Plastic band which seals propelling gases behind the projectile during gun launch (in spin-stabilized projectiles this device is used in conjunction with a rotating band).

Ogive (pronounced Oh' jive): Nose region of the projectile where the shape changes from cylindrical to curved or conical.

Origin of rifling: The point in a gun tube at which the lands begin to rise from the forcing cone.

PD fuze (point-detonating fuze): A fuze which must impact an object to detonate.

PIBD fuze (point-initiating, base-detonating fuze): This type of fuze is used in HEAT, HEP, or HESH ammunition to ignite the rear of the explosive column, thereby setting up the proper conditions for jet formation or target spall. It initiates upon impact of the projectile.

PIMP (permissible individual maximum pressure): Also called PMP (permissible maximum pressure). The three sigma upper limit on the pressure produced from a propelling charge conditioned to its maximum operating temperature. This is the charge used to proof a weapon.

PIMP + 5%: The PIMP charge conditioned so as to produce 5% higher pressure.

Pintle: Pin on a field piece that connects the lower carriage to the upper carriage and allows the weapon to traverse in azimuth.

Precision munition: A munition dispensed from a projectile or other device which uses a type of on-board or off-board electronics to improve its accuracy over standard munitions.

Pressure plate: A device used in a rifled mortar projectile to press on and expand a rotating disk when the propellant burns and applies pressure to its face.

Primer: A device containing small amounts of sensitive energetic material that is ignited first in a firing train. It may either be attached to the cartridge case or provided separately with bag loading ammunition. There are several types of primers: percussion primers rely on impact to begin the chemical reaction; stab primers rely on friction; and electric primers rely on the proper supply of electrical energy.

Propellant increment: A bag or C-shaped container of propellant that allows the range of a projectile to be altered by increasing or decreasing the amount of propellant.

Proximity fuze: See VT *fuze*.

Pusher plate: A device used to transmit the pressure generated by an expulsion charge to a cargo stack. The pusher plate protects the cargo stack from damage during the expulsion event.

Receiver: The portion of a small arm that comprises the interface between the barrel, the magazine, and the bolt.

Recoil cylinders: Cylinders filled with hydraulic fluid on a weapon platform that slow down and stop the rearward motion of the weapon during and immediately after firing.

Recuperators: Devices which push a weapon back into battery after recoil.

Rifling: Grooves cut into the bore of a weapon to impart spin to a projectile for stability.

Rotating band: A swaged, shrink-fit, or welded metallic or plastic band attached to the projectile which is designed to engage the rifling of the bore and impart spin to the projectile.

Rotating disk: A disk of soft material used in rifled mortar projectiles that is subcaliber to allow a mortar round to drop down the tube initially, but when reacted upon by the pressure plate, it expands into the rifling of the mortar tube and imparts spin to the projectile.

Sabot (pronounced Sa bo'): A device used to increase the diameter of a subcaliber projectile to stabilize it in the bore of a weapon. These devices are usually discarded upon muzzle exit.

Sear: The protrusion mechanically interfaced to the trigger which locks the bent. When the trigger is pulled, the sear moves off the bent allowing the firing pin to impact the primer in the cartridge.

Secant ogive: An ogive in which the radius is centered at a point behind the end of the cylindrical section of the projectile.

Semiactive homing: A method of guidance whereby the device is guided by electronics that contain only a receiver (the transmitter being located off the munition), so that the munition trajectory can be corrected onto the target.

Set forward: The rapid unloading of the projectile as it leaves the muzzle, i.e., the unspringing of the compressed projectile structure when the base pressure drops off.

Setback: The compressive reaction of the projectile mass to forward acceleration.

Shaped charge jet: A stream of metal in a high state of strain resulting from proper detonation of an explosive encasing a liner. The jet has tremendous penetrating power, and this form of terminal effect is utilized where kinetic energy of the projectile is limited.

Sheathed core: The central penetrator in some projectiles. It is usually a solid slug of material whose purpose is to penetrate a target by kinetic energy.

Shell splinters: Another name for fragments produced when a shell explodes.

Shot exit: Sometimes called muzzle exit. This is the moment at which the base of the projectile clears the muzzle or muzzle device attached to a weapon.

Shot start: The moment at which the rotating band of a projectile shears and the projectile begins moving into the rifled section of the weapon (separate loaded ammunition) or the moment at which the projectile moves from the cartridge case (fixed ammunition).

Shrapnel: A projectile, invented by Lt. Henry Shrapnel in 1784 that contained 1 in. diameter steel balls for fragmentation effects. The name became synonymous to shell body fragments when a projectile detonates.

Shroud lines: The lines on a parachute connecting the body being supported to the canopy of the parachute.

Sleigh: Device on a weapon platform that allows the gun tube to axially move during firing and recoil and during transportation.

Sling: The fabric or leather strap that allows the weapon to be carried on the back of a soldier.

Smart munition: A precision munition that can distinguish between targets and nontargets or countermeasures.

Soft recoil: Recoil system where the recoiling parts are accelerated forward to reduce the rearward momentum as the projectile leaves the weapon.

Spades: Part of the trails on a field piece that dig in to the ground upon firing to arrest the rearward motion of the weapon during recoil.

Split rotating band: A rotating band made up of multiple segments, either located on both the shell wall and base or simply separated by shell wall material.

Stacking swivel: A swivel located near the muzzle of a rifle which allows several weapons to be stacked in a pyramid shape, limiting the exposure of the weapons to dirt and corrosion.

Standoff: The distance between the base of a liner and the intended target. There is an optimum value of the standoff where penetration of a particular shaped charge is optimum.

Standoff spike: A cylindrical protrusion at the nose of a HEAT projectile that impacts the target, thus setting the proper standoff for the formation of a shaped charge jet.

Stock: The portion of a rifle which supports the barrel and by which the weapon is held.

Subcaliber: A term which describes anything with a diameter smaller than the bourrelet diameter of a projectile or sabot.

Supercaliber: A term which describes anything with a diameter larger than the bourrelet diameter of a projectile or sabot.

Supplementary charge: A charge added to HE rounds to further amplify the shock from a booster for added assurance that the main fill will properly and completely detonate.

Swivel: A loop which can either be fixed or pivoted, through which the sling passes and allows the weapon to be carried on the back of a soldier.

Tangent ogive: An ogive whose radius begins exactly at the end of the cylindrical section.

Torsional impulse: The sudden rotation of a projectile as it engages the rifling after it has acquired some forward velocity (a common condition in worn gun tubes).

Tracer: A device containing a pyrotechnic mixture which is inserted into the base of training projectiles and some tactical projectiles. The purpose of the tracer is to allow the firer to see where the projectiles are flying. The pyrotechnic composition in a tracer is usually initiated by the propelling charge.

Trails: Part of a field piece that supports the weapon during firing and allows it to be towed.

Treeburst: A technique where PD fuzes are fired into trees over the enemies' head to maximize fragment lethality.

Trigger: The device pulled by the finger of an operator to rotate the sear and fire the weapon.

Tripod: A trio of supports to maintain a weapon such as a recoilless rifle or a machine gun steady, used primarily when portability is essential.

Trunnion: Pins on a weapon platform that connect the cradle to the upper carriage and allow the weapon to elevate.

Undercut: Region of the projectile which separates the bourrelets and is subcaliber to reduce friction and tube wear.

Upper carriage: A fork-like component on a field piece that contains the trunnions and connects to the lower carriage through the pintle.

Volley fire: When multiple guns fire simultaneously at the same target. It is called a *Salvo* in Navy parlance.

VT fuze (variable time fuze): An ET fuze that initiates in the vicinity of an object through the use of a signal or other means, sometimes called a proximity fuze.

Wear additive: A material added to the propelling charge to reduce the wear on the gun tube through either protectively coating the tube, flame temperature reduction, lubrication, or reduction of corrosive reactions.

Wheel base: Distance between forward and aft bourrelets; the size of the wheel base affects stability in the tube.

White phosphorus (WP): Smoke-producing compound used in smoke rounds which produces a very dense obscuring smoke. White phosphorus reacts with air when exposed and tends to burn very hot, creating an updraft which tends to lift the smoke skyward which is not very desirable. Despite this, it is used frequently.

Windshield: A device used to make a projectile more aerodynamically efficient by reducing drag, sometimes called a ballistic cap.

Wings: Lifting surfaces mounted near the center of gravity or center of pressure.

Wooden round: A projectile that does not require maintenance over its lifetime.

Appendix B

Tabulated Properties of Materials

The properties given in Tables B.1 through B.14 have been assembled from the references at the end of this appendix. Although not complete, these represent sufficient values to do the problems included in the book. Since this is not thermodynamics or combustion text, the tables are coarse. None of the problems in this text requires interpolation between values in these tables. In fact, *never* interpolate with these tables. If the reader is performing an analysis that requires more refined tables, the authors suggest any of the texts in the references.

We have used the SI system for the tables since that was common among the references. The reader will also note that the specific internal energies and enthalpies contain an overbar—indicating that they are on a molar basis. This is reinforced in the units.

TABLE B.1

Enthalpies of Formation for Select Materials

Material	Enthalpy of Formation $(\bar{h}_f^0)_{298}$ (kJ/kg mol)	Molecular Weight (MW) (kg/kg mol)
Carbon monoxide (CO)	−110,541	28.010
Carbon dioxide (CO_2)	−393,546	44.011
Hydrogen (H_2)	0	2.016
Hydrogen, atomic (H)	217,997	1.008
Hydroxyl (OH)	38,985	17.007
Water (H_2O)	−241,845	18.016
Nitrogen (N_2)	0	28.013
Nitrogen, atomic (N)	472,629	14.007
Nitric oxide (NO)	90,297	30.006
Nitrogen dioxide (NO_2)	33,098	46.006
Oxygen (O_2)	0	31.999
Oxygen, atomic (O)	249,197	16.000
Carbon, solid (C)	0	12.010
Air	n/a	28.97

TABLE B.2

Ideal Gas Properties of Carbon Monoxide (CO)

Temperature (K)	$\bar{h}(T)$ (kJ/kg mol)	$\bar{h}(T) - (\bar{h}_{298}^0)$ (kJ/kg mol)
298	110,541	0
500	116,484	5,943
1000	132,238	21,697
1500	149,388	38,847
2000	167,278	56,737
2500	185,577	75,036
3000	204,103	93,562
3500	222,776	112,235
4000	241,573	131,032
4500	260,489	149,948

TABLE B.3

Ideal Gas Properties of Carbon Dioxide (CO_2)

Temperature (K)	$\bar{h}(T)$ (kJ/kg mol)	$\bar{h}(T) - (\bar{h}_{298}^0)$ (kJ/kg mol)
298	393,546	0
500	401,847	8,301
1000	426,971	33,425
1500	455,227	61,681
2000	484,966	91,420
2500	515,490	121,944
3000	546,437	152,891
3500	577,666	184,120
4000	609,159	215,613
4500	640,919	247,373

TABLE B.4

Ideal Gas Properties of Hydrogen (H_2)

Temperature (K)	$\bar{h}(T)$ (kJ/kg mol)	$\bar{h}(T) - (\bar{h}^0_{298})$ (kJ/kg mol)
298	0	0
500	5,874	5,874
1000	20,664	20,664
1500	36,307	36,307
2000	52,968	52,968
2500	70,492	70,492
3000	88,733	88,733
3500	107,566	107,566
4000	126,897	126,897
4500	146,672	146,672

TABLE B.5

Ideal Gas Properties of Atomic Hydrogen (H)

Temperature (K)	$\bar{h}(T)$ (kJ/kg mol)	$\bar{h}(T) - (\bar{h}^0_{298})$ (kJ/kg mol)
298	−217,997	0
500	−213,801	4,196
1000	−203,408	14,589
1500	−193,015	24,982
2000	−182,622	35,375
2500	−172,229	45,768
3000	−161,836	56,161
3500	−151,443	66,554
4000	−141,050	76,947
4500	−130,657	87,340

TABLE B.6

Ideal Gas Properties of Hydroxyl (OH)

Temperature (K)	$\bar{h}(T)$ (kJ/kg mol)	$\bar{h}(T) - (\bar{h}^0_{298})$ (kJ/kg mol)
298	−38,985	0
500	−32,984	6,001
1000	−18,057	20,928
1500	−2,125	36,860
2000	14,791	53,776
2500	32,435	71,420
3000	50,605	89,590
3500	69,152	108,137
4000	87,977	126,962
4500	107,023	146,008

TABLE B.7

Ideal Gas Properties of Water (H_2O)

Temperature (K)	$\bar{h}(T)$ (kJ/kg mol)	$\bar{h}(T) - (\bar{h}^0_{298})$ (kJ/kg mol)
298	241,845	0
500	248,792	6,947
1000	267,838	25,993
1500	290,026	48,181
2000	314,650	72,805
2500	340,957	99,112
3000	368,408	126,563
3500	396,640	154,795
4000	425,427	183,582
4500	454,635	212,790

TABLE B.8

Ideal Gas Properties of Nitrogen (N_2)

Temperature (K)	$\bar{h}(T)$ (kJ/kg mol)	$\bar{h}(T) - (\bar{h}^0_{298})$ (kJ/kg mol)
298	0	0
500	5,920	5,920
1000	21,468	21,468
1500	38,404	38,404
2000	56,130	56,130
2500	74,305	74,305
3000	92,730	92,730
3500	111,315	111,315
4000	130,028	130,028
4500	148,860	148,860

TABLE B.9

Ideal Gas Properties of Atomic Nitrogen (N)

Temperature (K)	$\bar{h}(T)$ (kJ/kg mol)	$\bar{h}(T) - (\bar{h}^0_{298})$ (kJ/kg mol)
298	−472,629	0
500	−468,433	4,196
1000	−458,040	14,589
1500	−447,644	24,985
2000	−437,253	35,376
2500	−426,858	45,771
3000	−416,416	56,213
3500	−405,857	66,772
4000	−395,092	77,537
4500	−384,016	88,613

TABLE B.10

Ideal Gas Properties of Nitric Oxide (NO)

Temperature (K)	$\bar{h}(T)$ (kJ/kg mol)	$\bar{h}(T) - (\bar{h}^0_{298})$ (kJ/kg mol)
298	−90,297	0
500	−84,218	6,079
1000	−68,056	22,241
1500	−50,565	39,732
2000	−32,440	57,857
2500	−13,966	76,331
3000	4,698	94,995
3500	23,487	113,784
4000	42,383	132,680
4500	61,384	151,681

TABLE B.11

Ideal Gas Properties of Nitrogen Dioxide (NO_2)

Temperature (K)	$\bar{h}(T)$ (kJ/kg mol)	$\bar{h}(T) - (\bar{h}^0_{298})$ (kJ/kg mol)
298	−33,098	0
500	−24,980	8,118
1000	−728	32,375
1500	26,197	59,295
2000	54,149	87,247
2500	82,581	115,679
3000	111,211	144,309
3500	139,940	173,038
4000	168,763	201,861
4500	197,685	230,783

TABLE B.12

Ideal Gas Properties of Oxygen (O_2)

Temperature (K)	$\bar{h}(T)$ (kJ/kg mol)	$\bar{h}(T) - (\bar{h}^0_{298})$ (kJ/kg mol)
298	0	0
500	6,097	6,097
1000	22,721	22,721
1500	40,590	40,590
2000	59,169	59,169
2500	78,346	78,346
3000	98,036	98,036
3500	118,173	118,173
4000	138,705	138,705
4500	159,586	159,586

TABLE B.13

Ideal Gas Properties of Atomic Oxygen (O)

Temperature (K)	$\bar{h}(T)$ (kJ/kg mol)	$\bar{h}(T) - (\bar{h}^0_{298})$ (kJ/kg mol)
298	−249,197	0
500	−244,852	4,345
1000	−234,336	14,861
1500	−223,898	25,299
2000	−213,485	35,712
2500	−203,070	46,127
3000	−192,623	56,574
3500	−182,116	67,081
4000	−171,519	77,678
4500	−160,811	88,386

TABLE B.14

Ideal Gas Properties of Carbon (Graphite) (C)—in Solid Form

Temperature (K)	$\bar{h}(T)$ (kJ/kg mol)	$\bar{h}(T) - (\bar{h}^0_{298})$ (kJ/kg mol)
298	0	0
500	2,365	2,365
1000	11,795	11,795
1500	23,253	23,253
2000	35,525	35,525
2500	48,289	48,289
3000	61,427	61,427
3500	74,889	74,889
4000	88,646	88,646
4500	102,685	102,685

Further Reading

Borman, G. L., and Ragland, K. W., *Combustion Engineering*, McGraw-Hill, New York, 1998.

Chase, M. W., *NIST-JANAF Thermochemical Tables*, 4th ed., American Chemical Society and the American Institute for Physics, Woodbury, NY, 1988.

Turns, S. R., *An Introduction to Combustion*, 2nd ed., McGraw-Hill, New York, 2000.

Van Wylen, G. J., and Sonntag, R. E., *Fundamentals of Classical Thermodynamics*, 3rd ed., John Wiley & Sons, New York, 1986.

Wark, K., *Thermodynamics*, 5th ed., McGraw-Hill, New York, 1988.

Index

Page numbers followed by f and t indicate figures and tables, respectively.

B

C

T